Introduction to Nanoscale Science and Technology

NANOSTRUCTURE SCIENCE AND TECHNOLOGY

Series Editor: David J. Lockwood, FRSC
National Research Council of Canada
Ottawa, Ontario, Canada

Current volumes in this series:

ALTERNATIVE LITHOGRAPHY: UNLEASHING THE POTENTIALS OF NANOTECHNOLOGY
Edited by Clivia M. Sotomayor Torres

INTRODUCTION TO NANOSCALE SCIENCE AND TECHNOLOGY
Edited by Massimiliano Di Ventra, Stephane Evoy, and James R. Heflin

NANOPARTICLES: BUILDING BLOCKS FOR NANOTECHNOLOGY
Edited by Vincent Rotello

NANOSCALE STRUCTURE AND ASSEMBLY AT SOLID-FLUID INTERFACES (TWO-VOLUME SET)
Vol I: Interfacial Structures versus Dynamics
Vol II: Assembly in Hybrid and Biological Systems
Edited by Xiang Yang Liu and James J. De Yoreo

NANOSTRUCTURED CATALYSTS
Edited by Susannah L. Scott, Cathleen M. Crudden, and Christopher W. Jones

NANOTECHNOLOGY IN CATALYSIS (TWO-VOLUME SET)
Edited by Bing Zhou, Sophie Hermans, and Gabor A. Somorjai

POLYOXOMETALATE CHEMISTRY FOR NANO-COMPOSITE DESIGN
Edited by Toshihiro Yamase and Michael T. Pope

SELF-ASSEMBLED NANOSTRUCTURES
Jin Z. Zhang, Zhong-lin Wang, Jun Liu, Shaowei Chen, and Gang-yu Liu

SEMICONDUCTOR NANOCRYSTALS: FROM BASIC PRINCIPLES TO APPLICATIONS
Edited by Alexander L. Efros, David J. Lockwood, and Leonid Tsybeskov

---

Introduction to Nanoscale Science and Technology

Edited by

Massimiliano Di Ventra
Department of Physics
University of California, San Diego
La Jolla, CA

Stephane Evoy
Department of Electrical and Systems Engineering
The University of Pennsylvania
Philadelphia, PA

James R. Heflin, Jr.
Department of Physics
Virginia Polytechnic Institute and State University
Blacksburg, VA

Library of Congress Cataloging-in-Publication Data

Introduction to Nanoscale Science and Technology /
Edited by: Massimiliano Di Ventra; Stephane Evoy; James R. Heflin, Jr.
p. cm. – (Nanostructure science and technology)
Includes index.
ISBN: (acid-free paper) – 1-4020-7720-3
ISBN: (electronic) – 1-4020-7757-2
ISBN: (electronic plus paper) – 1-4020-7758-0
1. Nanostructures. 2. Nanostructured materials. 3. Nanotechnology.
I. Di Ventra, Massimiliano. II. Evoy, Stephane. III. Heflin, James R., Jr. IV. Series

QC176.8N35N3556 2004
620'.5—dc22 2003070222

Printed in the United States of America.

9 8 7 6 5 4 3 2 SPIN 11426301

springeronline.com

Preface

Over the past few years, numerous universities and colleges have been developing courses in nanotechnology, both to respond to the growing interest by their student population and to provide a well-educated next generation of scientists and engineers to this emerging field. This textbook originated from such goals, through a course entitled "Nanotechnology" developed by two of us (SE and JRH). First taught at Virginia Tech in spring 2001 and at the University of Pennsylvania in fall 2002, this course was designed to serve a broad clientele of seniors and graduate students from chemistry, physics, electrical and computer engineering, mechanical engineering, chemical engineering, and materials science. Given the unavailability of a textbook written with suitable breadth of coverage, sixty articles from journals and magazines were selected to provide the necessary reading material. While the course has been very well received, many students expressed difficulty at absorbing material from such a wide collection of journal articles. We therefore realized the need for a textbook that covers the same broad spectrum of topics, but which is *designed, written, and organized* for a student audience. "*Introduction to Nanoscale Science and Technology*" has been conceived to provide such a broad and thorough introduction aimed at undergraduate seniors and early graduate students in all of the disciplines mentioned above. It may also serve as a valuable reference desk resource for academic, industrial, and government researchers interested in a primer in this field.

The textbook consists of twenty-three chapters arranged in seven sections. All chapters have been written by experts from each respective field. Each chapter is intended to provide an overview, not a review, of a given field with examples chosen primarily for their educational purpose. The student is encouraged to expand on the topics discussed in the book by reading the references provided at the end of each chapter. The chapters have also been written in a manner that fits the background of different science and engineering disciplines. Therefore, most technical terms and jargon have been avoided giving the subjects a primarily qualitative structure rather than providing detailed mathematical analysis. Based on our own experience, the complete set of topics contained in this book can be covered in a single semester, provided one does not delve too deeply into any single subject. If time constraints require a choice to be made between sections, we strongly recommend that the first two be covered completely, as they provide the foundation upon which the remainder of the material is developed. Beyond that, individual chapters, or even entire sections, may be omitted without having an adverse effect on the comprehension of the remaining material.

Each chapter contains several problems/questions that can be used as homework assignments. The solutions of these problems can be found on a password-protected web site.

In addition, the CD-ROM accompanying the book contains copies of the figures appearing in each chapter, including color versions of many of them. Faculty can incorporate these, e.g., into slide presentations to support their lectures.

Finally, we cannot conclude this preface without thanking the outstanding researchers who have written each chapter. They are the real craftsmen of this book. Their enthusiasm for this project has been heartening and inspiring; especially since writing for a student audience can be much more difficult than writing for specialists. We would also like to thank Danette Knopp and Gregory Franklin of Kluwer for encouraging us to begin this project and shepherding it to its completion, respectively, as well as Carol Day and Jill Strathdee for their excellent attention to detail. In addition, several students have also made important contributions. We would like to thank Saifuddin Rayyan, Martin Duemling, and Bill Barnhart for their assistance in developing the original course at Virginia Tech, and Mike Zwolak for assistance in editing some of the chapters.

Massimiliano Di Ventra
San Diego, California

Stephane Evoy
Philadelphia, Pennsylvania

Randy Heflin
Blacksburg, Virginia

Contents

Introduction

Nanoscale science and technology is a young and burgeoning field that encompasses nearly every discipline of science and engineering. With rapid advances in areas such as molecular electronics, synthetic biomolecular motors, DNA-based self-assembly, and manipulation of individual atoms via a scanning tunneling microscope, nanotechnology has become the principal focus of a growing cadre of scientists and engineers and has captured the attention and imagination of the general public. This field is defined primarily by a unit of length, the nanometer ($1 \text{ nm} = 10^{-9}$ m), at which lies the ultimate control over the form and function of matter. Indeed, since the types of atoms and their fundamental properties are limited by the laws of quantum physics, the smallest scale at which we have the freedom to exercise our creativity is in the combination of different numbers and types of atoms used to fabricate new forms of matter. This is the arena of nanotechnology: to build materials and devices with control down to the level of individual atoms and molecules. Such capabilities result in properties and performance far superior to conventional technologies and, in some cases, allow access to entirely new phenomena only available at such scales.

The rapid growth of the field in the past two decades has been enabled by the sustained advances in the *fabrication* and *characterization* of increasingly smaller structures. Section I provides an overview of such enabling technologies. The fabrication side has seen the emergence of two paradigms, respectively referred as "top-down" and "bottom-up". The *top-down* method begins with large homogeneous objects and removes material as needed to create smaller-scale structures. Similar to the work of a sculptor in carving a face from a block of marble, this approach is epitomized by lithography techniques, the cornerstone of microelectronics fabrication. On the other hand, a *bottom-up* approach involves putting together smaller components (such as individual atoms and molecules) to form a larger and more complex system by leveraging naturally occurring chemical, physical, and biological processes. Advances on the fabrication side have been supported by equally important abilities in the imaging and characterization of nanometer scale features. The advent in the late 1980s of scanned-probe technologies such as scanning tunneling and atomic force microscopy has indeed provided the atomic scale resolution critical for such developments. In addition, these microscopy techniques have enabled the nanoscale modification of surfaces, and thus supported their own set of approaches for the fabrication of small objects.

In addition to the above enabling technologies, nanotechnology has also seen the emergence of entirely new classes of materials such as fullerenes, nanocomposites, and quantum dots. These materials represent the building blocks of completely new structures and devices. Section II provides an overview of such materials and their unique properties, such as the ability of carbon to form different fullerenes, including high aspect ratio cylinders known as nanotubes; nanocomposites, namely structures made of different materials that display improved properties compared to the individual components; and quantum dots, structures that confine charge carriers in all spatial dimensions such that their electronic and optical properties are uniquely altered.

While nanoscience focuses on the fundamental aspects of nanoscale fabrication, characterization, and assembly, the ultimate goal of nanotechnology is to develop materials and devices that will outperform existing technologies, and even create new ones with completely novel functionalities. Sections III to VII focus on applications of nanoscale materials and tools in five general areas: electronics, magnetism, mechanics, optoelectronics, and biology.

Section III focuses on nanoscale electronics. Following an overview of current microelectronic technologies, it outlines the ongoing challenges towards delivering electronic devices of increasingly smaller dimensions. The use of individual molecules (such as carbon nanotubes or other organic compounds) as electronic components offers promising alternatives to current microelectronic devices. Physical properties of such molecular devices and possible integration strategies are therefore outlined in this section. In addition, transistors that operate with a single electron at a time are described, together with their possible applications in metrology, magneto-electronics, and information technology.

An overview of recent advances in nanoscale magnetic systems is provided in Section IV. The section opens with the fundamentals of spintronics, and its application to quantum computing. It continues with an overview of the science and technology of spin-related electronic transport, and the related development of materials and devices showing enhanced magnetoresistive effects. It concludes with a technological overview of the applications of these new technologies in the specific market of magnetic storage.

Section V moves on to applications of nanotechnology in micro- and nanoelectromechanical systems. Also referred to as *integrative systems*, the development and integration of micro- and nanomachined mechanical structures has opened new areas of development in microelectronics by enabling complete single-chip systems that can *sense, compute, and communicate*. Following a review of integrative systems, the section provides an overview of the science and technology of nanoscale mechanical structures, including the prospects of their integration in complete single-chip systems. The section concludes with an outline of hybrid devices that leverage nanometer scale synthesis technologies for the development of sensors of tunable and controllable responses to external stimuli.

The development of novel photonic materials and devices designed at the nanometer scale is described in Section VI. Here, nanometer scale engineering allows access to size-dependent phenomena that can enhance overall device performance and enable the tunability of its behavior. The section opens with a review of quantum confined inorganic semiconductor systems and their applications in lasers and photodetectors. For organic optoelectronic devices, self-assembly has an important role in increasing the efficiency of organic light-emitting diodes and the electro-optic response and stability of nonlinear optical materials. Advances in organic solar cells have made use of composites of semiconducting polymers with nanostructures such as fullerenes, nanotubes, and nanorods. Furthermore, the advent of photonic crystals has enabled new materials with forbidden ranges of photon energies, and has supported new approaches for the design and development of novel waveguiding and coupling devices.

Overlapping the physical, chemical, and biological sciences, Section VII concludes with an overview of the engineering of nanoscale biological systems. The section opens with a review of biomimetic systems, an area that aims to replicate the function of natural structures and membranes in synthetic vesicles and devices. Still within a biomimetic framework, the section follows with a review of biomolecular motors, with the aim to comprehend and replicate naturally occurring mechanical phenomena in artificial constructs. The section concludes with a review of the technology of nanofluidic devices, which aspires to understand fluidic motion in natural nanoscale systems, as well as to develop novel structures and devices for the sieving, analysis, and separation of biologically relevant molecular systems.

The field of nanotechnology is rapidly growing with novel ideas appearing at a swift pace. Some of the technologies outlined in this book have already made their appearance into commercial products, and more are sure to follow. Still in its infancy, nanometer scale science and technology is an area that is only bounded by the interest, creativity, and imagination of a new generation of multidisciplinary scientists eager to take it to its limit.

I

Nanoscale Fabrication and Characterization

1

Nanolithography

L. R. Harriott[1] and R. Hull[1,2]

[1]*Department of Electrical and Computer Engineering*
University of Virginia
Charlottesville, VA

[2]*Department of Materials Science and Engineering*
University of Virginia
Charlottesville, VA

1.1. INTRODUCTION

Lithographic patterning at ever-decreasing dimensions has been a major technological driver over the past few decades, particularly in the microelectronics industry, where realization of "Moores Law" (see Section 9.2) has driven extraordinary increases in processing power and electronic data storage. However, realization of the full potential of nanotechnology requires development of a broader range of patterning techniques than the ultra-violet optical projection methods currently used in microelectronics manufacturing, to enable higher resolution, lower cost, and application to a broader range of materials and surface geometries. Nanolithography refers to the ability to define patterns on surfaces at ever-decreasing length scales. Such capabilities are central to an enormous range of research fields and emerging technologies, ranging from microelectronics[1] to nanomechanical systems to biomedical applications. As such, nanolithography is a keystone of the nanotechnology revolution.

A handful of approaches dominate existing and emerging nanolithography technologies. The most prevalent approach employs exposure of *resist* materials, which are most usually polymeric, by energetic photons or particles (electrons or ions). Under such exposure, irradiated areas of the resist undergo structural/chemical modification such that they have differential solubility in a developing solution with respect to unexposed areas. This structural modification may reduce or enhance solubility (referred to as *negative* and *positive* resists respectively), e.g., by cross-linking or scission of polymeric chains.

Resist exposure may be accomplished in *serial* or *parallel* modes. In the serial or *direct write* mode, the resist is exposed point-by-point, e.g., by electron or ion beams. In the parallel mode, an image of a suitable *mask* pattern is transferred onto the resist surface, either by *projection* through an optical system, or by *contact* or *proximity* alignment of the mask to the resist surface. In serial mode, the ultimate lithographic definition is generally defined by the diameter of a focused beam of the exposing radiation. In parallel mode, the wavelength of the exposing radiation is a primary limitation. In either case, a suitable high intensity, monochromatic source is required, and resist and mask (in parallel mode) technologies are also critical.

Lithographic definition by energetic ions can also be achieved through direct sputtering of the surface—a direct write technique requiring neither mask nor resist, as will be described in Section 1.5. Lithographic patterning may also be realized through local deposition from the vapor phase via interaction with focused optical, electron and ion beams.

Other emerging nanolithographic techniques are diverging from the source–mask-resist paradigm. These include nanoimprinting by mechanical impression of a topographical master into a soft resist, micro-contact printing by transfer of self-assembled monolayers from an elastomer mold, delivery of fluidic tracks from a scanning probe microscope (SPM) tip ("dip pen" lithography), and selective oxidation from a SMP tip. These and other emerging techniques are reviewed in Section 1.6.

Generic challenges associated with lithography include:

Throughput: the number of features that can be patterned per second. Generally, parallel techniques have substantially higher throughput than serial techniques.

Field: the area of target surface that is patterned in a single parallel exposure or serial pattern, e.g., between mechanical motions of the target surface with respect to the lithographic source.

Alignment/Registration: the ability to align features within a pattern to each other, either field by field within a particular *level* of a pattern, or between levels.

Source Technology: In general, sources need to be of high intensity (for throughput), of high monochromacity (for high resolution of a focused probe or projected pattern), and high stability (to ensure linearity of dose on exposure time). In addition, photon sources need to be of ever-decreasing wavelength to enable demands for higher and higher resolution.

Mask Technology: In projection lithography, a mask is required for fabrication of each unique pattern. Such masks must accurately transmit radiation in desired regions, and absorb or deflect it otherwise. Energy absorption in opaque regions of the mask can cause heating, stress generation, and distortions from the desired pattern.

Resist materials: should have high *sensitivity* (i.e., low dose required for exposure, to enable high throughput), and high *contrast* (i.e., abrupt transition between sub- and supercritical doses for exposure, to enable high resolution of the transferred pattern), among other attributes.

Cost: State-of-the-art optical projection lithography systems cost several tens of millions of dollars. In addition, a mask has to be generated for each required level of a final process. While modern microelectronic fabrication plants cost several billion dollars, lithography costs a fraction of that total cost that is of order one half and growing. Such lithographic facilities and economics are best suited to mass production of single products.

TABLE 1.1. Comparison of primary nanolithography techniques. Note that all numbers quoted are approximate and highly configuration/application dependent

Technique	Optical	E-Beam (Direct Write)	E-Beam (Projection)	Focused Ion Beam	EUV	Nano-Imprint
Resolution (nm)	100	20	50	30	30	10
Alignment (nm)[a]	30	10	20	10	10	100
Throughput (Feature/s)[b]	10^{10}	10^4	10^{10}	10^1–10^2	10^{11}	10^{12}
Instrument Cost ($)[c]	10^7	10^6	10^7	10^6	5×10^7?	10^5

[a] Inter-level alignment.
[b] Highly dependent on feature size for many techniques. Here we quote approximate numbers for features at the technique resolution.
[c] Approximate cost for basic state-of-the-art instrument, where available. This does not include the usage costs such as masks which can be significant and even dominant in a manufacturing mode.

Serial lithographic techniques are more adaptable to relatively low cost production of single or low volume runs (e.g., mask generation/repair or many research applications), but are not economically scaleable to large areas, high throughput, or large volume production.

Table 1.1 compares many of the above attributes for the primary nano lithographic techniques.

To summarize the structure of the rest of this chapter, generic issues associated with resists and masks will be described in Section 1.2. Implementations of "optical" (i.e., ultra-violet, extreme ultra-violet, and X-Ray) lithography will be described in Section 1.3. Electron beam lithographic techniques will be described in Section 1.4, and ion beam lithography in Section 1.5. Finally, in Section 1.6, we will briefly summarize the concepts and status of an emerging set of nanolithography techniques.

1.2. CROSS-CUTTING TECHNOLOGIES: RESISTS AND MASKS

Two technologies are of relevance to the majority of nanolithography technologies: resists and masks. The essential concepts relevant to each technology are reviewed in this section.

1.2.a. Resist Materials

In the majority of nanolithographic processes, a polymer resist material is first applied to the substrate. The patterned radiation is then used to change the structure of the polymer. This can work in two basically different ways: the action of the radiation can serve to lengthen the polymer chains or it can serve to shorten them. In either case, the dissolution rate or solubility of the radiated material in a suitable solvent will be different than that of the portions of the resist that have not been irradiated. In the case of a positive tone process, the energy from the radiation serves to break bonds along the polymer backbone, lowering the molecular weight and increasing the solubility in the developer solution. Therefore, the irradiated areas of polymer will be removed in the developer after exposure. In a negative tone process, the energy of the radiation serves to crosslink the polymer to increase

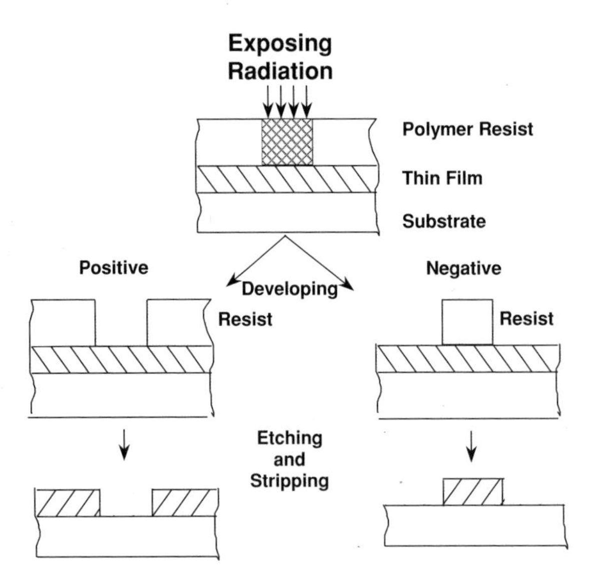

FIGURE 1.1. Schematic of positive and negative resists.

the molecular weight and thus reduce the solubility in the developer. Therefore the irradiated areas will remain on the substrate and the non-irradiated areas will be removed by the developer. These processes are illustrated in Figure 1.1.

Resist materials for nanolithography must combine a number of optical, chemical, mechanical, and process properties in order to be useful. First of all, they must be capable of forming high resolution images. This generally requires that the chemical response to the radiation input be highly nonlinear. These resists are said to have a high contrast. The effect is to act as a thresholding function or binary spatial filter on the patterning information. That is, a high contrast resist can "sharpen" a blurred exposure pattern. Figure 1.2 shows the response curve of a hypothetical positive tone resist. The thickness of the resist remaining after exposure and development is plotted against exposure dose on a semi-log scale. It can be seen that for low doses, very little happens and then, above a threshold dose, the thickness remaining decreases rapidly with increasing exposure until all resist is removed at the clearing dose. The log slope of this portion of the curve is the resist contrast, γ. All exposure methods whether using photons, electrons, or ions have some amount of blur or edge slope in the projected image. A high contrast resist will, in general, lead to the highest resolution final patterns with the sharpest edge slopes. Typical resist contrast values range from about $\gamma = 2$ for older conventional resists (such as PMMA) to as high as 15 for modern chemically amplified resists (such as the Shipley UV series, e.g., UV-6) used in integrated circuit manufacturing.

Resist materials must also have the necessary mechanical and chemical properties to be useful in subsequent processes. Resists can be used to mask etch pattern transfer process, ion implants, or other processes. The maximum thickness for a given feature size (aspect ratio) is a very important property and is usually limited by residual stresses in the resist film.

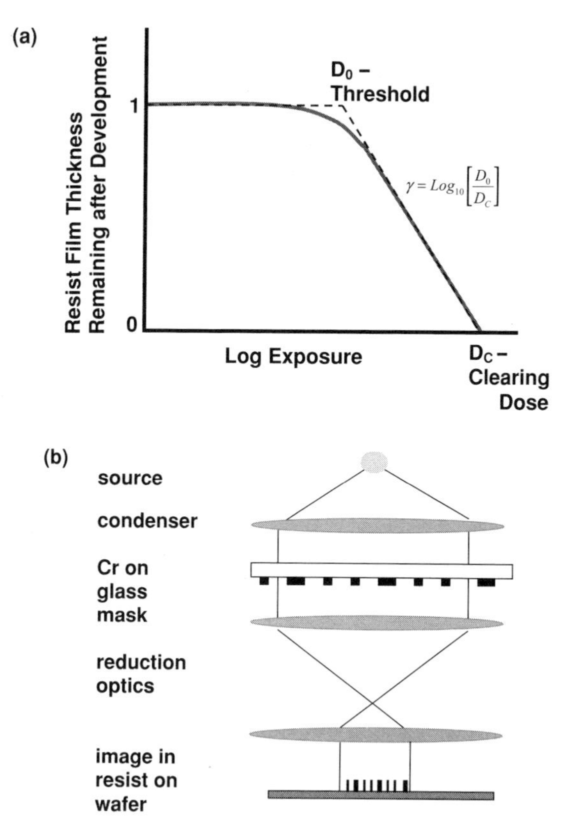

FIGURE 1.2. (a) Characteristic curve of a hypothetical positive tone resist (b) optical projection lithography schematic.

Resists can be classified by their chemistry in two broad categories: conventional and chemically amplified. In conventional resists, energy from the radiation is converted directly to a chemical reaction, while in chemically amplified resists, an intermediary catalytic reaction (usually driven by heat) happens prior to development.

Polymethylmethacralate (PMMA) resist has been used for nanolithography for more than thirty years and is still the work-horse for many laboratories. It is a positive tone conventional material that can be applied by spin casting and developed using a mixture of isopropyl alchohol (IPA) and methyl-isobutyl ketone (MIBK). PMMA is available in many different starting molecular weights (usually from 500 k to 1000 k) and its properties vary with average molecular weight. Generally it is a low contrast ($\gamma < 3$ depending on the exact process and development conditions) material that responds to UV light, X-rays, electrons, or ions.

The most commonly used resists in integrated circuit fabrication are the class of diazo-napthoquinones (DNQ). These materials use a Novalac resin (glue) that is normally soluble in an aqueous base developer (KOH, NaOH, tetramethyl ammonium hydroxide (TMAH)) mixed with a photoactive compound (PAC). Initally, the PAC acts to inhibit the dissolution of the Novalac resin. Upon exposure, the PAC is converted to a soluble acid allowing the exposed areas to be removed during development. Typical exposure doses of 100 mJ/cm^2 are required for UV light. Sometimes, a post-exposure bake (thermal treatment) is used to cause some short-range diffusion (10% of the minimum feature size) of the PAC before development to minimize standing wave effects from the UV exposure (i.e., reflections from the substrate interfering with incoming light). These resists are readily available, relatively inexpensive, simple, and exhibit moderate contrast ($\gamma \sim 6$). Spatial resolution using UV exposure at I-line (365 nm) wavelengths is usually limited to about 350 nm. These resists and processes are commonly used for the larger features in nano-device fabrication schemes while the fine features are made some other way.

Chemically amplified resists (CAR) also use a polymer backbone but in this case, a dissolution inhibitor is added to render it insoluble in the aqueous base (TMAH typically). A photo acid generator compound (PAG) is added that will change to an acid upon radiation. The PAG itself does not directly change the development properties of the polymer but instead reacts with the dissolution inhibitor groups in a catalytic fashion such that that the polymer is rendered soluble and the acid molecule is free to render other sites on the polymer backbone soluble. In this way, chemical amplification (up to about 100) is achieved since one PAG molecule can react with many sites on the backbone. CAR materials were originally developed for their high sensitivity to the exposure radiation (~5 mJ/cm^2) for deep ultra violet (DUV, $\lambda = 248$ nm) light. In addition to high sensitivity, these materials also tend to exhibit very high contrast ($\gamma > 10$ with 14 or 15 possible) making them suitable for nanolithography with enhanced optical lithography as well as electron beam lithography.

1.2.b. Mask Technologies

In photolithograpy, the pattern is contained on a mask that is then replicated onto the substrate for further processing. In some cases, the pattern on the mask is the same size as that of the final pattern on the substrate while, in other cases, the mask pattern is four or five times larger than the final pattern and is printed using a reduction or de-magnified imaging method. Masks are also used in X-ray, extreme ultra-violet (EUV), and projection electron beam (EPL), and ion beam lithographies, although the materials vary considerably. Proximity X-ray lithography necessarily uses 1:1 masks while the other forms generally use reduction printing. The primary reason for doing reduction printing is to make the job of making masks easier. In a replication process, all features, including defects, are reproduced in the final image. Defects tend to be more numerous and more difficult to remove as their size gets smaller. Masks for reduction printing have the advantage that all the errors and defects are also reduced in size during printing.

Photomasks typically consist of a glass or synthetic quartz (fused silica) substrate coated with 80 to 100 nm of Cr as a light absorber. Other absorbers such as $MoSi_2$ are sometimes used, but Cr is still the most common. The mask substrate with Cr coating (mask blank) is coated with a resist material and patterned by a primary pattern generator. This is usually an electron beam writing system, but laser-based writing systems are becoming more commonly used. The electron beam machines offer the advantage of high resolution patterning at the expense of slow speed and high machine cost. Electron beams are also subject to non-systematic beam placement errors caused by electron column charging and other effects. Laser-based writing systems are faster and less expensive by comparison but cannot usually match the resolution of the electron beam systems, although this has improved dramatically in recent years. Masks made by laser writers tend to have smoother feature edges and fewer non-systematic placement errors (the fact that light cannot be deflected by electromagnetic fields is both an advantage and a disadvantage).

Once the pattern is defined in the resist on the mask blank, pattern transfer is accomplished by wet chemical etching through the Cr layer. Wet chemical etching has the advantage of being a simple and clean process since it tends to help remove particulate defects during processing. The main disadvantage of wet etching is lack of anisotropy leading to undercutting of the Cr layer. Fortunately, the aspect ratio of the features is low enough that this is not usually a serious problem (e.g., for 200 nm features on the wafer, the features are 800 nm on the mask (4:1 reduction printing) and the Cr is usually about 80 nm thick resulting in an aspect ratio of 0.1 and a maximum undercut of 10% which can usually be tolerated). Dry etch processes have been developed but tend to have defect densities which are too high for IC manufacturing.

The final steps of mask making are defect inspection and repair. Photomasks are inspected optically by comparing images of like regions of masks and flagging differences or by comparing mask images with database models. In a good mask making process, a mask (150 × 150 mm) will have on the order of 10 or fewer defects. These defects can then be repaired using either laser-based or focused ion beam methods. Defects in the form of excess light absorber (opaque) can be removed by laser ablation (for relatively large defects) or by Ga focused ion beam sputtering (milling). Missing absorber (clear) defects are usually less common and harder to repair. They can be repaired by chemical vapor deposition (laser or ion induced) or by a multi-step local lift-off lithography process. After repairs have been made to the mask, it is then re-inspected to ensure that it is defect-free. For photomasks, the mask is then cleaned and sealed in a transparent covering called a pellicle. This pellicle is a transparent film membrane that stands off a few millimeters from the mask surface (one on each side of the mask). During use, if a particle falls on the mask/pellicle composite structure, it will be kept out of the imaging focal plane (depth of field) of the printing tool and thus rendered non-printing as long as the defects are not too large. Many of the newer lithography technologies have no equivalent of the optical pellicle and thus may be problematic in practice.

Masks are also required for EPL, EUV, and X-ray lithography. In each case, the materials are different than those used for photomasks as described in the following sections. These masks do have some common characteristics with photomasks, however. In each case, a blank substrate is coated with resist and patterned with a primary pattern generator, inspected, and repaired. The details vary due to the materials differences, but the overall processes are similar to those employed in making photomasks, as will be described further in later sections of the chapter.

1.3. PHOTON–BASED NANOLITHOGRAPHY TECHNIQUES

1.3.a. Contact and Proximity Optical Printing

Photons (whether ultra-violet, extreme ultra-violet, or X-ray) remain the favored choice for lithographic patterning of decreasing feature sizes in the microelectronics industry, which is by far the largest technological application of nanolithography. In conventional photolithography, there are three main modes of printing: contact, proximity, and projection. In contact printing, the photomask is placed in direct mechanical contact with the resist-coated substrate. This configuration is then illuminated uniformly with ultraviolet light since photoresists are most sensitive in that wavelength regime. This process has the advantage of being very simple and can yield very good high spatial resolution. The main disadvantage of contact printing is defectivity. After exposure, the mask and substrate must be separated sometimes leaving photoresist residue on the mask. Further, dust particles and other particulates can prevent the mask from contacting the sample across its entire area which can effect feature size and resolution and even cause wafer breakage. Masks must be cleaned after each exposure which will eventually degrade the quality of the mask features. Mainly for reasons of defect density, contact printing is no longer used in the IC industry but remains a very common practice in research labs and universities.

A close relative to contact printing is proximity printing. In this case, there is a small gap between the mask and wafer. The gap eliminates or reduces the defect problems associated with contact printing. In practice, most attempts at contact printing end up being proximity printing due to particulates, non-flat substrates, or other problems. The spatial resolution of proximity printing is limited by diffraction of the light by the features on the mask. In this case the diffraction patterns are usually modeled as near-field or Fresnel diffraction. A common approximation for the minimum printable feature size is:

$$W_{\min} \approx \sqrt{(\lambda g)} \tag{1.1}$$

Here $W_{\min}$ is the minimum feature size, λ is the illumination wavelength, and g is the gap between the mask and wafer. One of the disadvantages of this method is control of the gap. The physical gap size is usually very small for printing of small features and it can be difficult to place the mask such that the small gap is maintained across the sample to achieve reproducible results. The other major disadvantage of both contact and proximity printing is that the printed features are the same size on the mask and final image. Resolution and other performance parameters are often limited more by the mask making process than diffraction or other physical limitations of the printing process itself.

1.3.b. Projection Optical Printing: Fundamental Concepts and Challenges

The third, and most generally used in technological applications, method of photolithography is projection printing (see Figure 1.2b). Projection printing can be accomplished with or without image demagnification from mask to substrate. In the IC industry, it is most often practiced with a 4 or 5 times image reduction. As mentioned above, the primary reason for this is to reduce the requirements (cost) of the masks rather than anything inherent to the printing process itself. In projection printing, the mask is illuminated uniformly. Light passing through the mask is used to form an image of the mask in the plane of the resist-coated substrate. This is accomplished with an optical system composed of lenses, mirrors, or a combination. The image resolution is again limited by diffraction but, in this case, the mask to wafer distance is much much larger than the wavelength of the illuminating light (0.5 meters typically) so that far-field or Fraunhoffer diffraction models are used to describe the image intensity pattern. The simple analysis that is usually used to estimate resolution is the Rayleigh Criterion which is derived from Lord Rayleigh's analysis of diffraction in the imaging of stars with a telescope. The far field diffraction of a point object (a star or in this case any point on the photomask) produces an image intensity pattern that is represented by an Airy disk. The function has a central maximum, a finite width and decreasing intensity oscillations away from the central maximum. Rayleigh's criterion for resolving two stars within the field of the telescope was that the maximum intensity point of one star should be at least at the first minimum point of the Airy disk of the second star. The diameter of the Airy disk of a point object is given by:

$$D = 1.22\,\lambda f/d \tag{1.2}$$

where f is the focal length of the lens, d is the lens diameter, and λ is the wavelength of the light. This is usually represented as the resolution of the system. It can also be written as:

$$R = 1.22\,\lambda f/d = 1.22\,\lambda f/2f\,\sin\alpha = 0.61\,\lambda/\sin\alpha = 0.61\lambda/\mathrm{NA} \tag{1.3}$$

where α is the convergence angle of the lens $(d/2f)$, and NA is the numerical aperture of the lens ($\sin\alpha$). This formula is taken to approximate resolution in projection lithography systems without detailed justification. It is more commonly written in terms of the constant, k_1 as:

$$R = k_1\lambda/\mathrm{NA} \tag{1.4}$$

Although only roughly representative of the performance of a lithography system, it does provide some insight into the phenomenon. The nonlinear response of the eye (or photographic negative) is not expected to be the same as that of photoresist. As we have seen, the response (contrast) of photoresists varies considerably from one material to the next. The Rayleigh Criterion above cannot be taken as a universal formula but only as a guide.

In practice, the empirical value of k_1 using a moderately high contrast DNQ type photoresist is about 0.8. In IC manufacturing, k_1/NA was, until recently, estimated to be near unity so that the rule-of-thumb that developed was that features could not be printed that were smaller than the wavelength of the light used to print them.

Equation 1.4 above has three factors that can be manipulated to improve resolution. In fact, this is precisely what has happened in the IC industry over the last several decades and is the technological driving force behind "Moore's Law". The equation says that resolution can be improved by using shorter illumination wavelength and higher numerical aperture (larger) lenses. Historically, wavelengths have been shortened through the ultraviolet spectrum to improve resolution. The illumination wavelengths used have been picked based on the availability of a suitable light source. Initially, mercury arc lamps were used and filtered to obtain a narrow bandwidth output. This is necessary in order to design the projection optics for diffraction limited operation. The lenses can be compensated at a particular wavelength so that they are essentially perfect. That is, the effects of the principle aberrations (chromatic, spherical) are corrected for and the image resolution is limited by diffraction alone. Mercury arc lamps produce strong emission at wavelengths of 435 nm (G-line) and 365 nm (I-line). Each time a new wavelength is used, the photoresist must be reformulated for optimum performance. G-line and I-line sources were and still are used with DNQ-type resists.

In order to achieve shorter wavelengths and still maintain high intensity, laser-based sources are now used. The first to be used widely was KrF excimer lasers that emit strongly at 248 nm. The gas in the laser can be changed (with some other modifications) to ArF to produce light at 193 nm and F_2 at 157 nm. Optical materials for lenses and masks begin to be problematic at these wavelengths. Ordinary synthetic quartz begins to absorb light below 248 nm and other materials such as CaF_2 have been introduced. At 193 nm, a mixture of fused silica and CaF_2 is used in optical designs. At 157 nm (still in the research and development phase at this writing), quartz is too absorbing to be used and CaF_2 or MgF materials must be used. One principle difficulty with materials such as CaF_2 is optical birefringence (different focusing properties for different polarizations). This adds considerable complexity to the optical designs. Again, at each of these new wavelengths, a new photoresist chemistry is needed. Each of these wavelengths uses the chemically amplified resists approach to achieve high sensitivity and good contrast. But the polymer backbones, dissolution inhibitors, and photo-acid generator materials are different in each case due to differences in optical absorption. At 157 nm, virtually all hydrocarbon materials are too strongly absorbing to be used in a single layer photoresist. Flourocarbon based materials are being investigated for this wavelength since they are still somewhat transparent. Beyond (i.e., of lower wavelength than) 157 nm there are some possible light sources, but none has emerged strongly. At these wavelengths, the optical systems must employ mirrors rather than lenses in order to avoid excessive absorption of the light. Mask and photoresist materials are again a serious issue.

The general idea of decreasing wavelength and increasing numerical aperture has proved successful but at the cost of decreased depth of focus (DOF) and therefore more stringent requirements on planarity and wafer flatness. Again, borrowing from Lord Rayleigh, the depth of focus is modeled as the deviation in focal plane position that would cause a phase shift of $\lambda/4$ between the central ray and one at the edge of the lens (numerical aperture). Simple geometry leads to the formula:

$$\text{DOF} = \pm 0.5\lambda/(\text{NA})^2 = \pm k_2\lambda/(\text{NA})^2 \tag{1.5}$$

In practice, the factor k_2 is determined empirically and can differ from 0.5. However, the scaling of DOF with wavelength and numerical aperture do tend to follow the simple formula above. As an example, for 250 nm CMOS technology (i.e., around 1997—the Nintendo 64 era), lithography is accomplished with DUV illumination at 248 nm and numerical aperture of about 0.6, yielding a theoretical depth of focus of about 300 nm. This requires that the substrate be planar and flat to this extent across the exposure field (usually about 25 mm × 25 mm), a very difficult requirement. So, although shorter wavelengths and higher numerical apertures offer improved resolution, flatness requirements become quite severe.

Lithographic process parameters such as exposure dose and focus directly affect the quality of the lithographic image and printed linewidth and may be coupled in a complex way. In practice, exposure-focus experiments are conducted to determine the best focus and dose for the desired feature size. In addition, the tolerance in focus and dose may be determined by measuring linewidths for each combination of focus and dose and determining which combinations result in feature sizes that are within the allowed variation for the device being fabricated (usually ±10% for integrated circuits). The portion of the matrix yielding features within the specification is sometimes referred to as the process window. In this way, a usable depth of focus can be determined. It will depend on many parameters including resist contrast, linewidth required, numerical aperture, coherence of the illumination source and other factors.

1.3.c. Resolution Enhancement Technologies in Optical Projection Lithography

The resolution limit of photolithography is often loosely described as being equal to the wavelength due to diffraction. However, the performance of photolithography has been actually extended to feature sizes well below the wavelength of the exposure light through resolution enhancement technologies (RET). There are many such approaches, but each is essentially aimed at reducing the effective value of k_1 in equation 1.4 and can also affect the value of k_2 in equation 1.5 (sometimes to improve it, sometimes to make it worse). RET can be classified as to which part of the lithography system is modified: resist, mask, or exposure tool.

One of the most direct methods of RET is resist performance improvements. In diffraction limited images, the modulation of the image intensity impinging on the resist is generally less than unity. That is, if the bright portions of the image are assigned 100% brightness, the dark portions of the image are not completely dark but may have a significant fraction of the brightness of the bright areas. The result of this sort of intensity pattern on a high contrast resist (consider Figure 1.2a) can be favorable if the minimum intensity in the image does not exceed the threshold dose and the maximum intensity is equal to the clearing dose. In this way, an image with only modest intensity modulation can result in a photoresist pattern with 100% modulation. That is, the resist is completely cleared in the exposed areas and remains at the original thickness in the dark areas. Therefore, for higher resist contrast, less image modulation is required to form a completely modulated developed resist structure. This can significantly reduce the effective value of k_1 (it has little effect on k_2). This idea has been exploited in integrated circuit manufacturing to allow fabrication of features approximately half the wavelength but tends to work better for isolated features (such as transistor gates) than for closely packed features, since the diffraction limited modulation transfer function (MTF) is larger for isolated rather than dense features.

Another approach to RET through photoresist is surface imaging resists. In these materials (or material systems), the image formation process takes place only at or very near to the surface of the resist rather than throughout its thickness. In this way, less of the error budget for depth of focus is used up within the resist leaving more available for other parts of the process. This can lead to apparent improvements in printing performance and resolution. However, the general disadvantage of this approach is the need for subsequent pattern transfer. The requirements for the thickness of the resist are generally set by the next step in the device processing such as etching or ion implantation. Most processes require resist features that are a significant fraction of a micron and therefore the image imparted to the surface must be transferred through the thickness of the material below prior to the next step in processing. This pattern transfer step adds additional complexity, defects, and linewidth errors. For those reasons, surface imaging resists have not been widely adopted in manufacturing.

Many versions of RET are implemented in the mask. These generally fall into two categories: modifications to the feature geometries and modifications to the phase, as well as amplitude, of the light passing through the mask (or combinations). The most common form of shape enhancement is referred to as optical proximity effect correction (OPC).[2] It can be used to address some of the issues of diffraction limited resolution (and/or reduced process window) in some cases. The simplest form (1-D) accounts for the fact that features will have different MTF and will print at a different size depending on whether they are isolated (e.g., transistor gates) or in dense patterns (e.g., memory cells). The compensation is simply to change the size of the features on the mask to "pre-distort" the pattern such that the final printed image has all features come out the correct size regardless of gross local pattern density (optical proximity). The next level of sophistication (2D) is adding serifs or extra features to compensate for rounding of square corners by diffraction. The added intensity in the vicinity of the feature corners tends to sharpen the corners and make the features in the final image resemble more closely those on the mask. The disadvantage of this method is the added complexity and cost of the masks. The additional serif features can add significantly to the size of the data required to describe the mask, the writing time of the mask making electron beam system, and to the defect inspection time. The ultimate (3D) OPC techniques use models based on complete solutions of Maxwell's equations in the image plane to compensate for all feature distortions caused by diffraction. This method is very computationally intensive and adds considerable cost to the masks.

In addition to image amplitude information, masks can also be used to impart phase information to the pattern to improve image performance referred to as phase-shift masks (PSM). The original version of this is sometimes referred to as alternate aperture phase shifting, strong phase shifting, or Levinson phase shifting (after its inventor).[3] This particular approach was developed for printing of dense lines and spaces. Normally, the mask pattern would be alternating clear and opaque areas. In this approach, every other clear aperture is modified such that the phase of the light is shifted by 180°. When the mask is illuminated, the electric field vector representing the amplitude of the light coming out of the mask changes sign from one region to the next due to the phase shift in the mask. This sign change forces the intensity pattern (square of the amplitude) to cross through zero intensity at some point. For images at (or beyond) the conventional diffraction limit, this zero crossing will increase the intensity modulation in the image and thus improve the ultimate resolution. In theory, a factor of two improvement can be made this way but only for an infinite array of equal lines and spaces. Unfortunately, real circuits are not made up of infinite arrays of equal lines and spaces, but the idea is nonetheless useful with some modifications and reduction in performance for real patterns. It has been used for DRAM circuits in limited applications. The practical implementation aspects include difficulties with how to terminate the shifted and non-shifted areas at the ends of the features (sometimes called the three-color map problem). As a result, most implementations have been done manually in a cell based or repetitive pattern and are not amenable to automated CAD-based approaches with the exception of the two-mask approach for gate levels described below.

A much more practical approach to PSM uses two masks.[4] In this approach, a binary mask is used to print the gate level of a circuit with conventional design rules. Then a second mask is used to re-expose the same resist pattern with a phase shift pattern. This "trim" mask acts to reduce the size of the gates and can, in principle achieve linewidths as small as half that achieved using conventional masks. The "trim" mask is created using software which compares the gate level pattern with that of the underlying gate oxide (thinox) layer to identify the gate features in the pattern data. The PSM mask pattern is then created automatically. This method has been recently applied to the gate level of a 3-million transistor digital signal processor chip yielding gate lengths of 120 nm using 248 KrF lithography. A micrograph of this is shown in Figure 1.3. Currently this method is limited to gate lithography because the automated software must have a logical basis upon which to decide which areas to phase-shift.

There are many other variations of PSM including so-called rim-shifters where 180 degree phase shift areas are placed at feature edges to improve the resolution and chromeless shifters where the pattern uses only phase modulation to produce line patterns. In general, these methods are difficult to apply to real circuit applications, but illustrate the potential of this approach.

An important aspect of the two-mask PSM approach and many other of the resolution enhancement technologies is that while they improve the minimum feature size that can be printed, they do not improve the density of features over conventional binary mask lithography.[5] The phase shift approaches improve the printing of isolated features such as gates but do not improve the minimum distance (half-pitch) for the patterns. In terms of IC performance, smaller gates will improve speed and power consumption of circuits, but they do not increase the packing density or number of circuit elements per chip. This has lead to a bifurcation of the roadmap for the IC industry over the last few years. The progress in the minimum half-pitch or density of circuit elements has been driven mainly by memory applications and has continued to follow the Moore's Law trend with some amount of acceleration due to improvements in resist and numerical aperture. Meanwhile, minimum feature sizes are driven by the speed requirements of microprocessors and have been following a more aggressive trend. In general, gate sizes are about one generation ahead of circuit density or minimum half-pitch as a result of RET.

Improvements in illumination methods have also been used to improve lithographic performance. The resolution limit expressed as the Rayleigh Critereon reflects the fact that the first diffracted order of the light from the mask must be captured in the lens for the image information to be transferred by the optical system. Off-axis illumination (OAI)[6] uses a tilted illumination to capture one of the first diffracted orders while allowing the zero order also to pass. In other words, resolution performance can be doubled in principle since the aperture of the lens now must cover the zero and one of the first diffracted orders whereas in a conventional system, the aperture covers the span of the -1 to $+1$ orders. In practice, however, this method is limited by the fact that the illumination must be tailored to the mask pattern since the diffraction pattern will be different for different mask patterns. Specific illumination patterns (annular or quadrupole symmetry) are chosen for specific types of circuit patterns to emphasize the performance of specific features such as line-space gratings.

RET can also be applied in conjunction with resist post-processing or trimming to achieve sub 100 nm isolated features. In this case, resist features, usually isolated lines or dots, are fabricated using optical lithography with any combination of RET to achieve minimum linewidths. For example, high contrast (chemically amplified) resist combined with PSM can print lines at about 100 nm using 248 nm (DUV) light with some effort. The resist features generally have a fairly high aspect ratio, that is, they are thicker than they are wide. If some loss of resist thickness can be tolerated, the resist features can be subjected to isotropic wet or dry etching that will simultaneously reduce the height and width of the features. If one begins with 100 nm wide lines in 400 nm thick resist as a starting point,

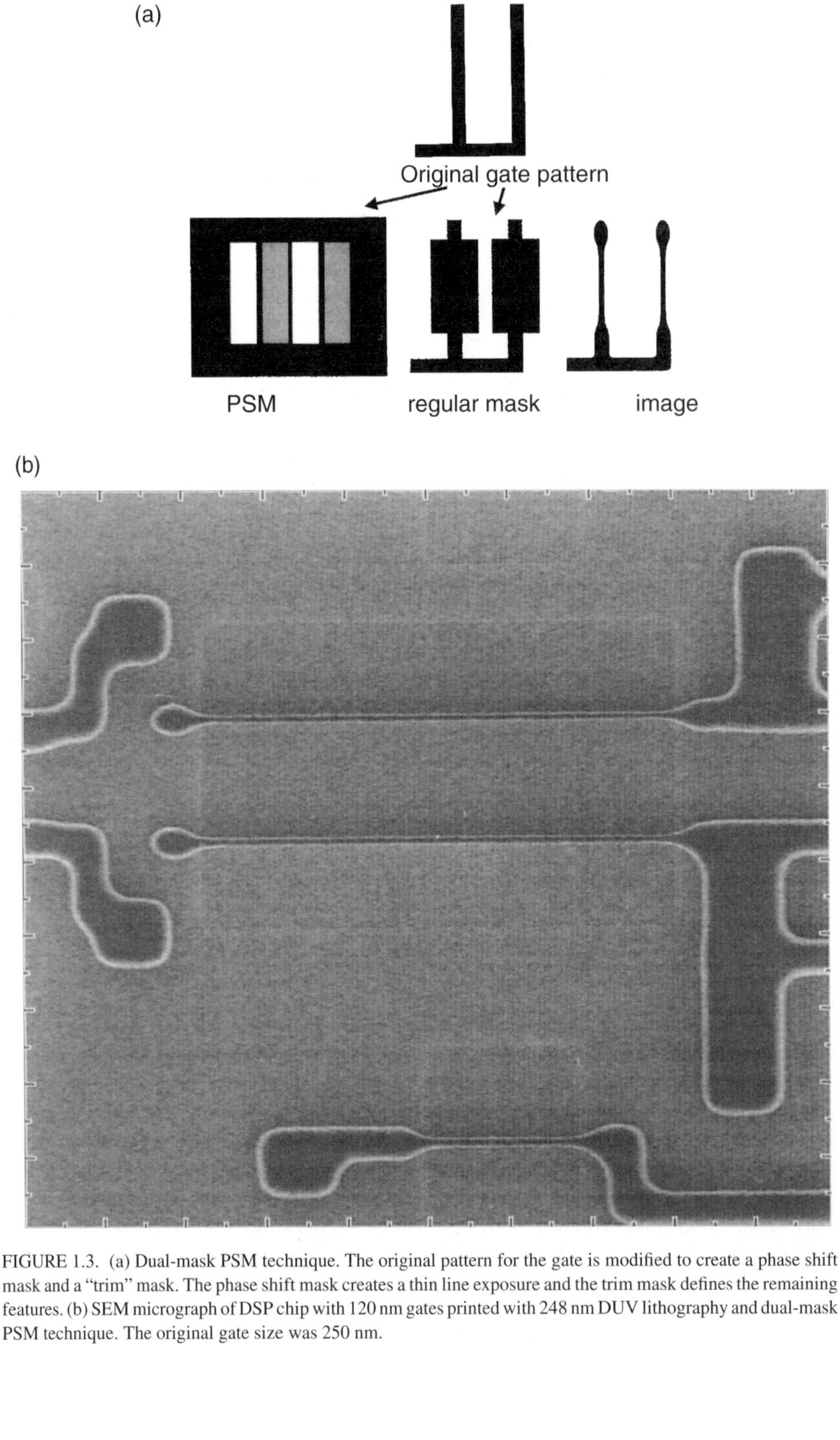

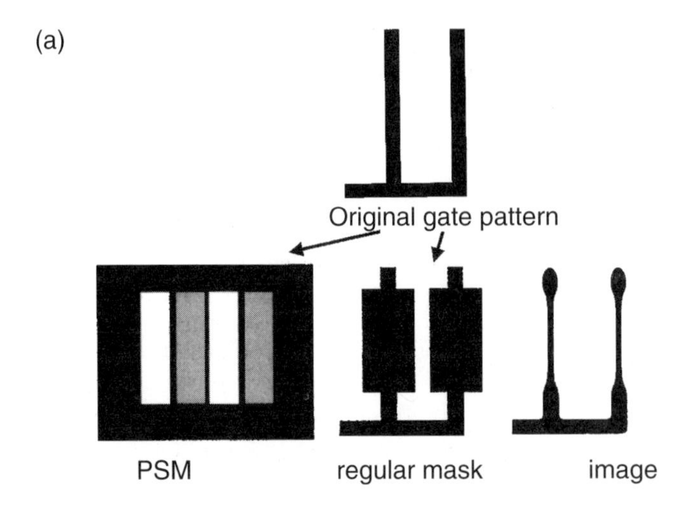

(b)

FIGURE 1.3. (a) Dual-mask PSM technique. The original pattern for the gate is modified to create a phase shift mask and a "trim" mask. The phase shift mask creates a thin line exposure and the trim mask defines the remaining features. (b) SEM micrograph of DSP chip with 120 nm gates printed with 248 nm DUV lithography and dual-mask PSM technique. The original gate size was 250 nm.

then this etching or trimming can be used to reduce the linewidth to 60 nm, for example, by etching away 20 nm from each side of the line and from the top. The resulting lines would then be 60 nm wide and 380 nm thick. Sometimes, residual stresses in the photoresist limit the extent to which trimming is effective. If the lines are trimmed too much, the stress can cause them to de-laminate or fall over. Nonetheless, this is a very effective and simple way to extend optical lithography well beyond the perceived conventional limits. The main difficulty in this method for manufacturing is the uniformity and reproducibility of the etch process associated with trimming, but it has been widely used for research applications for fabricating relatively small numbers of nanodevices.

1.3.d. Extreme Ultraviolet Lithography

In principle, Extreme Ultraviolet (EUV) lithography is a logical extension of optical lithography to very short wavelengths (10–14 nm) in the soft X-ray spectrum.[7] Extension of conventional refractive optical lithography to wavelengths below the F_2 excimer (157 nm) is problematic due to absorption in the refractive elements. It is also very difficult to design and construct reflective optical systems with sufficiently large numerical apertures to allow printing at or below the wavelength of the illumination source. The idea of EUV lithography is to use small numerical aperture reflective optical systems at wavelengths much shorter than the circuit dimensions. This combination of small numerical aperture and small (compared to the feature sizes being printed) wavelength allows simultaneous achievement of high resolution and large depth of focus. For example, an EUV system with wavelength of 14 nm and numerical aperture of 0.1 can yield 100 nm resolution and 1 micron depth of focus (assuming a k_1 factor of about 0.7, which is conservative). The reflective elements for EUV mirror optics use multilayer (Bragg) mirrors to produce reflectivities up to nearly 70% at 14 nm. The mirrors are composed of many (~80) alternating layers of dissimilar materials (e.g., Mo and Si). The mismatch in refractive index at each layer interface causes a weak reflection of the EUV radiation. If the layers are fabricated with the right thicknesses, constructive interference will occur for the EUV illumination wavelength causing a net strong reflection (only at one particular wavelength). The mask in an EUV system is also reflective and uses the same type of multilayers. A plasma or synchrotron based source is used to illuminate the mask which is imaged by a system of mirrors onto the resist-coated wafer with a reduction factor of four. The optical systems require mirrors with unprecedented tolerances with respect to figure and finish. That is, the shape of the mirror must be correct in addition to the surface being smooth. The specifications are in the angstrom and, in some cases, sub-angstrom range, posing serious challenges for mirror fabrication, coating, and mounting. The masks are made by depositing a multilayer coating on a bare silicon wafer or other flat substrate. An absorber is then deposited and patterned to complete the mask. The most serious issues for EUV masks are in creating multilayer coatings across a mask blank with no defects. Even very small (30 Angstrom) defects in the multilayers can print unwanted features on the wafers.

EUV radiation is absorbed strongly by most materials over a very short range, typically 20–50 nanometers. Resist layers required for IC processing must be nearly one micron in thickness therefore requiring some sort of surface imaging method for EUV. This differs significantly from current practice of single thick resist layers and poses many challenges as described previously.

1.3.e. X-Ray Proximity Lithography

X-ray proximity printing very simply uses high energy X-ray photons to overcome resolution limits imposed by diffraction in proximity printing.[8] According to equation 1.1, the resolution of lithographic features is proportional to the square root of the product of the exposure wavelength and the gap between the mask and wafer. Resolution in the 100 nm range is relatively easy to obtain with keV energy X-rays and micron sized gaps. This technique is appealing in its simplicity. X-ray sources can be simple Cu target systems or electron synchrotrons for higher fluxes (and throughputs).

X-ray lithography has not seen widespread adoption for technological applications despite numerous examples of impressive lithographic performance. The main issues for X-ray lithography are related to the mask.

The masks for X-ray lithography are generally constructed of a relatively low atomic number membrane (under some tension to keep it flat) such as silicon and a high atomic number X-ray absorbing material such as gold or tantalum nitride. Mask blanks are made by depositing the membrane and absorber layers on a silicon wafer substrate and then removing the substrate by backside etching in the area to be patterned. Patterning proceeds in a manner similar to that for photomasks except that the patterning is almost always done by electron beam and pattern transfer is by dry etching. Defect inspection can be optical or electron beam based. Defect repairs are accomplished using a focused ion beam for sputter removal or ion-induced deposition (in this case metals such as Pt are deposited).

X-ray lithography is a 1× printing technique and thus does not enjoy the advantage of feature size and defect size reduction during printing. The most technically difficult aspect of X-ray lithography, or any 1× method, is the mask making. All of the specifications on feature size, linewidth variations, defect size, etc. are roughly 4 times smaller than the corresponding specifications for reduction printing methods. Mask making has been the main reason for limited use of X-ray lithography in manufacturing, although its inherent simplicity has enabled its use in research labs.

1.4. ELECTRON BEAM LITHOGRAPHY

Writing with electron beams to produce small features has been used for quite a long time and is fundamentally based on technology developed for scanning electron microscopes. One of the earliest references to this idea was presented in a talk entitled "There's Plenty of Room at the Bottom", given by Richard P. Feynman at an American Physical Society meeting in December of 1959 (often cited as being the first articulation of nanotechnology).[9] In this talk, he issued his famous challenge to fit a page of written text on the head of a pin (this translates to roughly 100 nm feature sizes). He speculated that electron microscopes could be modified for this task, although he warned that the technique would probably be very slow due to space charge (mutual repulsion) effects.

1.4.a. Direct Write Electron Beam Lithography

The earliest form of Direct Write Electron Beam Lithography (DWEB) directly utilized scanning electron microscopes (SEMs) for writing, either under the control of a computer or with a flying spot scanner for imparting pattern information.[10] Today, many laboratory

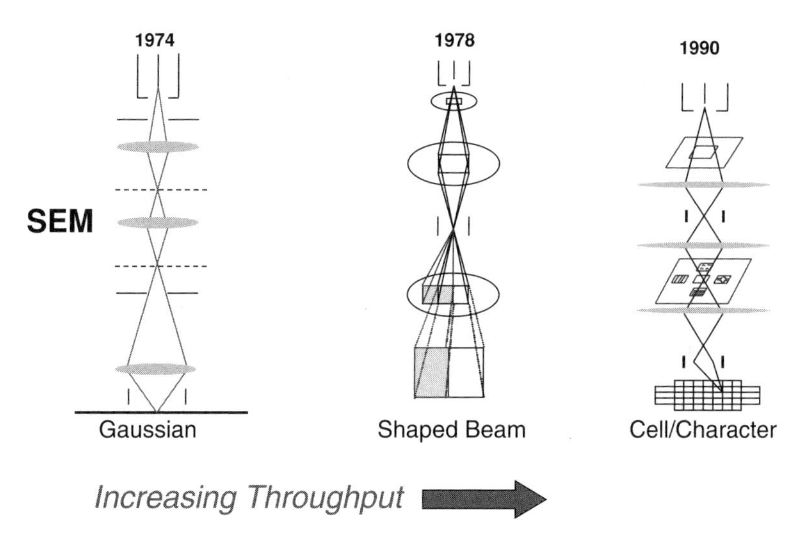

FIGURE 1.4. Gaussian beam, shaped beam, and cell projection DWEB schematics.

SEMs have been modified and are widely used for this task. These systems use the small (10 nm or so in diameter) beam produced by the microscope to expose an electron sensitive resist material. These systems are sometimes referred to as Gaussian beam systems because of the intensity of profile of the beam.

A schematic diagram of such a system is shown in Figure 1.4. In this case, the source of electrons can be a thermionic emission filament or a field emission tip. Electrons emitted from the source are focused and shaped into a spot with a roughly Gaussian profile by a series of magnetic lenses and apertures (which limit the current and define the numerical apertures of the lenses). The systems also use electrostatic plates and/or magnetic deflection coils to deflect the beam (they can also be used to correct for astigmatism). Generally, one deflector within the electron column is used to blank the beam on and off at the appropriate times while a second deflector or deflectors is used to scan the beam across the sample. In the simplest implementation, the beam is scanned in a raster pattern while the beam blanker turns the beam on and off according to the data input (bitmap) to reproduce the image in the resist (much like a television set scans an electron beam over phosphor pixels to render an image). Electron beam scanning at the sample is limited to small distances (100 to 1000 microns) due to aberrations and distortions introduced by the deflection fields. Therefore, to cover the entire sample with patterns, a combination of mechanical sample motion and electronic beam deflection is used. There are many schemes that have been employed to accomplish this. Some systems use feedback to the beam deflectors based on the sample position error (measured by optical interferometers) to produce large area coherent patterns that are "stitched" together from the individual beam deflection fields.

The electron optics for lithography systems uses magnetic lenses to focus and shape the electron beam. The principle contributors to the final diameter of the beam are usually the image of the effective source (demagnification is usually used to minimize this), chromatic aberration (energy spread of the beam leads to different focal points for each part of the energy distribution from the source), geometric (primarily spherical) aberration, and space charge effects (mutual electrostatic repulsion of the electrons). The contributions to the final spot size are usually assumed to add in quadrature. That is, the net beam diameter is estimated as the square root of the sum of the squares of the individual contributions. (This is not quite correct but usually gives reasonable estimates). The size of the aberration coefficients tends to be quite large compared to those in light optics since it is very difficult to do corrections as you can with glass (quartz) lenses. Axial chromatic aberration increases linearly with the beam aperture (NA of the lens system) size, and geometric aberrations increase with the cube of the aperture. The total beam current onto the sample increases with the square of the aperture size (area). Space charge effects tend to scale with a power law in beam current. The value of the exponent depends on the geometry of the electron column. The exponents are always positive and result in larger space charge effects (and spot sizes) for larger beam currents. The electron-electron interactions and their effect on the beam essentially depend on how many electrons there are (current), how close together they are (current density, beam cross-over points), and how much time they spend in the column (beam energy).

Ultimately, the time necessary to expose a resist film scales with the sensitivity of the resist and the total beam current on the sample. This combination of factors leads to a well-known trade off of beam size and spatial resolution. One obtains finer spot sizes and better resolution only at low beam currents (small apertures) and must live with the associated long exposure times. Details of beam scanning methods, spot shaping, overheads, etc. can and do effect the exact writing time, but an upper limit on writing speed can be obtained by considering the total beam current (expressed in Amperes or Coloumbs per second) divided by the resist sensitivity (expressed in Coulombs per square centimeter). Typically, beam currents for Gaussian beam systems are in the nanoamp (or less) range and resist sensitivities are a few microCoulombs/cm^2. Therefore, it can take many hours to write a pattern at high resolution.

Shaped beam systems were developed in the 1970's to improve the throughput of DWEB.[11] As illustrated in Figure 1.4, the current from the source impinges on a shaping aperture (usually a square) and then passes through a beam deflector and is imaged onto a second shaping aperture. With no deflection, a square beam is produced on the wafer with a relatively large beam current. With some beam deflection between the apertures, smaller squares, lines, and dots can be produced with correspondingly lower total beam currents. The current density on the sample is roughly constant (exposure times are adjusted if it varies from that). In this way, complex patterns can be written faster by filling the large features with large squares (or other shapes). The total maximum beam current for this type of system is perhaps 1 μA, a significant improvement over Gaussian beam systems, in some cases.

Unfortunately, compromises in the electron optics must be made and this usually results in poorer ultimate resolution (smallest usable feature) with shaped beam systems compared to Gaussian beam systems. For this reason, most nanolithography work is done using Gaussian systems rather than shaped beam, again, trading resolution for speed. Shaped beam systems have been used for custom circuits (where mask costs can be prohibitive), and for mask making where feature sizes are larger (at least for 4:1 reduction masks).

An extension of shaped beam lithography systems is cell or character projection beam systems.[12] In this case, the apertures are made in the shape of repeating circuit patterns (cells). In this way, for example, in a memory chip, repeating areas can be replicated rapidly. This has been used for advanced development and prototyping work but at relative large (>100 nm) feature size and modest throughput.

1.4.b. Electron Beam Projection Lithography

The next logical extension of electron beam lithography is projection printing. In this case, a mask is used to project an image of all or at least a large portion of the pattern. The imaging systems used for these systems use relatively low current density (compared to Gaussian or shaped beam systems) and beams with larger extent. This results in a reduced space charge effect which can allow larger total beam currents than other e-beam approaches. Currents of tens of microamps at resolution of 100 nm are possible.

Electron projection lithography (EPL) is not an entirely new idea[13] and has been practiced in the past using stencil mask technology. In this case, the mask is a solid membrane with holes in it representing the pattern (stencil). The electron beam is absorbed in the solid parts and passes through the holes thus imparting the pattern onto the beam. One of the principle difficulties with this approach is that the electrons absorbed in the stencil will deposit a significant amount of energy causing it to heat up and distort. Accurate image placement is critical to any lithography technology since as many as 25 or more lithographic layers must be overlaid to make a chip. Typically each layer has an overlay tolerance of one third of the minimum feature size or about 23 nm for 70 nm lithography. Mask heating and the subsequent overlay errors have prevented this type of technology from being widely used.

Modern EPL uses masks with scattering contrast to overcome both the mask heating problem and the requirement for multiple exposures in stencil mask technology (in the case of the continuous membrane approach).[14] There are two approaches to the mask construction for EPL: scattering stencil and continuous membrane. Scattering stencil masks use a membrane which is thin enough to absorb only a small fraction of the beam energy as it passes through (2 micron thick Si has been used for this). Electron beam energies of about 100 keV are typically used so that thin membranes are “transparent”. Electrons passing through the membrane are scattered through a small angle as they pass through but not absorbed. An aperture further on in the electron optical system stops the scattered electrons from reaching the substrate (for the most part). The membrane is patterned by etching holes in the shape of the circuit patterns.

For continuous membrane EPL masks (Figure 1.5), the mask is a thin membrane of low atomic number material (100 nm of SiN for example) which is transparent to the 100 keV electron beam. The pattern is formed in a thin layer of high atomic number material (e.g., 250 Angstroms of Tungsten) which is also transparent to the electron beam but will scatter the electrons more strongly than the membrane. The thickness of the scatterer layer is usually three mean free path lengths or more. After passing through the mask, contrast is produced in the image with an aperture at the back focal plane of the imaging system which stops the scattered beam and allows the unscattered beam to pass. In fact, the same imaging optical system can be generally used with either scattering stencil or membrane type EPL masks.

The electron beam is formed from a conventional thermionic emission filament source and focused with magnetic lenses. The image of the mask is projected onto the resist-coated wafer with a 4× demagnification.

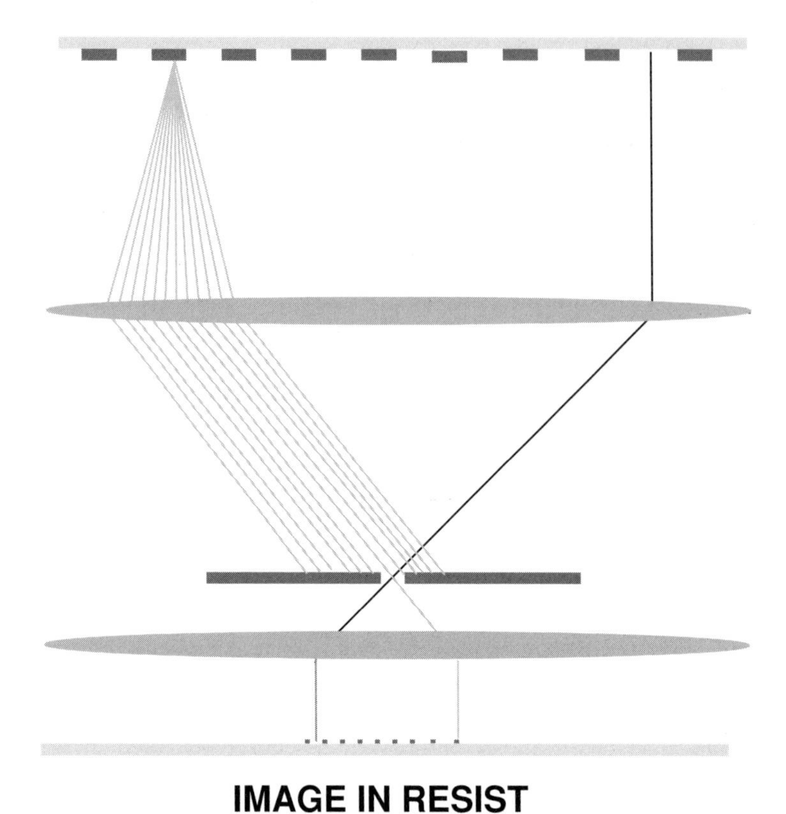

FIGURE 1.5. Schematic of Electron Projection Lithography employing scattering contrast.

The resists used for EPL are based on the same chemical platform as those used in today's DUV lithography. In some cases, the same material performs well for both electron and DUV exposures. The critical issue for EPL technology is throughput. As the beam current is increased to increase the exposure rate, electron-electron scattering tends to degrade the resolution of the image (stochastic space charge effect). Thus, charged particle lithography has an inherent trade-off between resolution and throughput.[15] However, EPL is a true 4× imaging technology which does not require any form of RET which has a significant effect on mask costs and complexity.

1.5. FOCUSED ION BEAM LITHOGRAPHY

1.5.a. Overview

In focused ion beam (FIB) lithography, a focused energetic ion probe is scanned across a surface to enable either subtractive lithography (through sputtering of atoms from the target surface) or additive lithography (through ion beam-induced decomposition of an organic vapor). Simultaneous generation of secondary electrons and ions allows real-time imaging and monitoring of the sputtering/deposition processes.

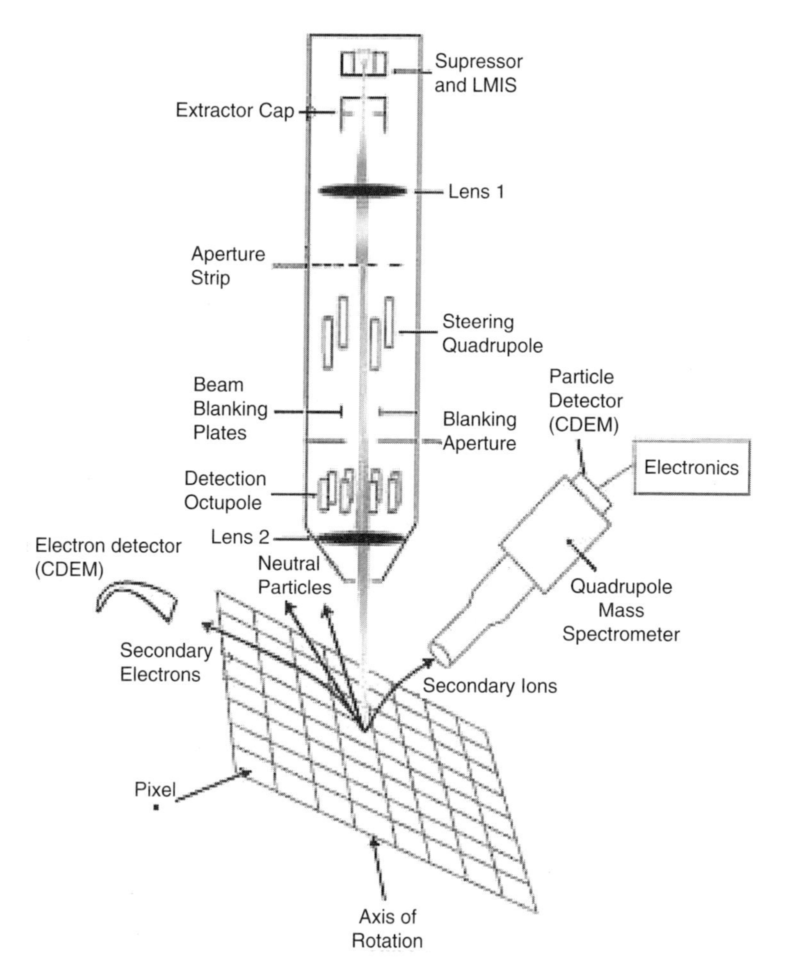

FIGURE 1.6. Schematic of a focused ion beam system. (Courtesy: A. Kubis, U. Von)

The essential elements of a focused ion beam system are shown in Figure 1.6.[16] A liquid metal ion source (LMIS) is used to generate a stream of ions that are accelerated to energies of order tens of keV and focused onto the target surface using electrostatic lenses. The LMIS material is typically Ga^+, because it is liquid at close to room temperature, alleviating the need for constant source heating and increasing source lifetime to typical values >1000 hours, although many other LMIS materials have been demonstrated. Final beam diameters at the target surface are limited by the total ion current, by the virtual source size (typically of order tens of nanometers for an LMIS), by lens aberrating and by the Boersch effect (lateral ion repulsion) at the cross-overs in the ion optics. The latter effect is much larger than at comparable charged particle densities in electron optics because of the much lower velocities of energetic ions than electrons (typically $\sim 10^6$ m/s vs. 10^8 m/s at relevant acceleration energies). Typical current densities in the focused ion probe at the target surface are of order 10 A/cm^2 with minimum ion probe diameters (defined by the full width at half maximum intensity (FWHM) of order 5–10 nm) for state-of-the-art instruments.

Conventional FIB lithography is a serial technique, where the sample surface is modified point-by-point (either through sputtering, deposition, or resist exposure) by pixel-by-pixel control of the ion beam location. This makes its throughput inherently limited, as discussed previously for direct write electron beam lithography (DWEB) in Section 1.4. In comparison to DWEB, the throughput of FIB lithography is reduced by lower blanking rates (because of the much lower velocities of charged ions with respect to electrons when accelerated to comparable energies), but enhanced by greater resist sensitivities where ion beam exposure of resist is employed. In direct modification (i.e., sputtering and deposition) modes, throughput is several orders of magnitude lower than EBL. Practical resolution is comparable to EBL at the few tens of nanometers level—whereas the limits of focused ion probe diameters are typically somewhat larger than for focused electron probes, the relative lack of a proximity effect (whereby backscattered electrons from the primary particles cause broadening of the exposed resist region) makes attainable feature resolutions broadly comparable. This reduced proximity effect in ion lithography arises because of the much lower ion range and lateral straggle in the target compared to electrons–values for range and straggle are typically of order a few tens of nanometers for 30 keV Ga^+ at normal incidence on targets of intermediate atomic weight.

The great advantage of FIB lithography is that it is eminently suitable for rapid prototyping of limited size arrays, requiring neither mask nor resist in direct modification sputtering or deposition modes. The primary limitations are throughput and the invasive nature of the LMIS species (most usually Ga^+) which cause substantial damage in the target surface, can dope semiconductor materials, or can produce new eutectic compounds (often of low melting temperature) in many target materials.

1.5.b. Fundamental Mechanisms and Throughput

The three fundamental mechanisms that can provide patterning in focused ion beam lithography are:

i) Physical sputtering of the target surface. Transfer of energy from the incident ions to atoms in the target causes physical ejection of target atoms, with an efficiency depending upon the mass, energy, and incident angle of the primary ions and the atomic structure, orientation, and chemical species in the target.[17] FIB systems designed for nanofabrication generally have incident ion energies (usually Ga^+) in the range 25–50 keV, where the sputtering yield is typically maximized for most target materials. The typical minimum dimensions of sputtered features are of order 30 nm using a 10 nm diameter ion probe (the reduction in resolution being due to the tails on the ion current distribution in the probe and lateral spreading of incident ions in the target).
ii) Focused ion beam–induced deposition. Here an organic vapor containing the deposition species is introduced at a partial pressure of order 1–10 mT to the sample surface (corresponding to a chamber partial pressure typically of order 10^{-6}–10^{-5} T) using a thin capillary tube. The precise mechanisms of FIB-induced deposition are still somewhat unclear, but it is believed that a thin film of the organic chemisorbs onto the target surface. Where the focused ion beam is incident upon the substrate, energy is locally imparted initiating a dissociation reaction of the organic, local deposition of the metallic

or insulating species, and desorption of the corresponding organic fragments.[18, 19, 20] A very broad range of deposition sources and species have been described in the literature, but the most common metallic deposition species are Pt, W, and Au from corresponding metalorganics and SiO_2 from tetraethylorthosiliciate (TEOS). The major limitations upon the purity (and hence resistivity) of the deposited species are the effects of incorporated ions from the primary beam and residual organic material, as will be discussed later. The typical minimum dimensions of deposited features are of order 50–70 nm using a 10 nm diameter ion probe (the reduction in resolution being due to the tails on the ion current distribution in the probe and the lateral distribution of incident energy in the target, allowing decomposition of the organic over a broader area).

iii) Focused ion beam modification of the internal molecular structure of resist materials, with direct analogy to electron and optical lithography techniques. Given the much lower range of energetic ions within a target, compared to electrons of equivalent energy, the rate of energy loss per unit length of target traveled is much higher, and each ion modifies the resist structure far more substantially than an electron. Thus, the doses for successful ion exposure of a resist are typically lower than for electron exposure. Line widths below 20 nm for ion exposure of PMMA have been reported.[21]

Throughput rates in direct modification mode (number of sputtered or deposited features per second) are defined by the available ion current at the required resolution and the relevant sputtering or deposition yields. These yields quantify the number of atoms removed from the target surface, or deposited on it, per incident Ga^+ ion. These parameters are highly dependent upon the species, energy, and incident angle of the incident ions, and upon the geometry and atomic species of the target. However, typical sputtering and deposition yields are of order 1–10 for incident Ga^+ ions at energies of tens of keV. This allows material removal or deposition at rates of order a few tenths of cubic microns per nano-Coloumb of incident ion current. With typical maximum ion current ranges of tens of nA in state-of-the-art instruments, this then permits maximum material removal/deposition rates of order $10\ \mu m^3\ s^{-1}$. Thus structures spanning dimensions from (tens of nm)^3 to (tens of μm)^3 may be fabricated in reasonable time spans (i.e., a few hours or less).

Another issue in defining fabrication rates is the alignment accuracy required between features, which is intimately linked to the mechanical and thermal stabilities of the FIB system. Thus, if each feature in an FIB-fabricated pattern needs to be registered at an accuracy of D with respect to other features, then the maximum fabrication time, t_m, between re-alignments to other features or to a global origin is given by: $t_m = D/d$ where d is the systematic stage drift rate per second. Of course, equivalent metrics apply to optical and electron beam lithography systems, but because FIB lithography is typically not used for large volume technological applications (an emerging example, however, is FIB trimming of magnetic heads in high performance hard drives),[22] such mechanical and thermal stability issues have typically not been refined to a comparable degree in FIB systems. For the FEI FIB 200 instrument at the University of Virginia, we have shown that d can be reduced to values of order a few Angstroms per minute,[23] allowing fabrication runs up to about 10 minutes for alignment precision of 30 nm (i.e., a standard alignment of one third of the

TABLE 1.2. FIB fabrication times in Si. Calculated approximate fabrication times for arrays of deposited or sputtered features assuming a deposition/sputtering rate of 0.5 μm³ per nA-sec of ion current. Those times shown in bold face are in excess of the maximum fabrication time for total drift distances of $D/3$ assuming the best measured drift rate of 1.5 nm/sec.

Feature Size (nm)	Beam Current (pA)	Feature Spacing (nm)	Number of Features	Field of View L × L (μm)²	Fabrication Time (mins)
100⁽¹⁾	70	300	10,000	30 × 30	5
			250,000	150 × 150	**120**
		1,000	10,000	100 × 100	5
			250,000	500 × 500	**120**
1,000⁽²⁾	6,000	3,000	10,000	300 × 300	7
			100,000⁽³⁾	1,000 × 1,000	70
		10,000	10,000	1,000 × 1,000	7

Key to notations: ⁽¹⁾Assumes feature height is 100 nm. ⁽²⁾Assumes feature height is 500 nm. ⁽³⁾Limited by field of view.

feature size for 100 nm elements). Such alignment can of course be done visually in the FIB, using secondary electron images to locate with respect to a suitable origin. This requirement can be relaxed if *local* rather than *global* requirement is required of a pattern (i.e., if accurate alignment is only required of features with respect to their near neighbors, rather than with respect to every other feature in the pattern). In this case, the dwell time of the ion beam on each feature may be adjusted such it may be completed in a single pass of the ion beam, if the feature is sufficiently small. This is in practice more effective in sputtering mode than in deposition mode, as in the latter case too long a dwell time per feature tends to deplete the organic vapor at the sample surface, causing substantial drops in deposition yield.[18] Thus, efficient FIB-induced deposition tends to require multiple beam passes over each feature. This can then lead to considerable distortions in each feature, as the shape of each feature is then the summation of multiple deposition pulses, each pulse separated by relatively large times during which the sample may have mechanically moved. In this case, a maximum fabrication time of $t_m = D/d$ will need to be employed, where D may now also represent the maximum allowable distortion of a given feature. Note that another potential issue is the accuracy of re-positioning of the beam between successive sweeps of the array (i.e., does it return to the exact same location in successive passes), but we have found that to be an insignificant factor compared to stage drift on our system.[23] A table of fabrication times for different FIB-fabricated arrays in Si, together with requirements for realignment, is given in Table 1.2.

In a recent discovery, we have found that the sputtering yield of polymethylmethacrylate (PMMA) can be orders of magnitude higher (100–1000) [24] than conventional sputtering yields under a range of sputtering conditions with Ga^+ ions. The mechanism associated with this anomalously high yield appears to be an ion beam-induced "unzipping" or depolymerization reaction. This reduces the resolution of the sputtered features (to c. 100 nm), but allows enormously higher throughput of sputtered features (we have successfully fabricated arrays with diameters of about 100 nm and depths of tens of nanometers at rates greater than 10^4 s^{-1})[24] as illustrated in Figure 1.7. This has application to the formation of master molds for microcontact printing and nanoimprinting methods to be discussed in Section 1.6.

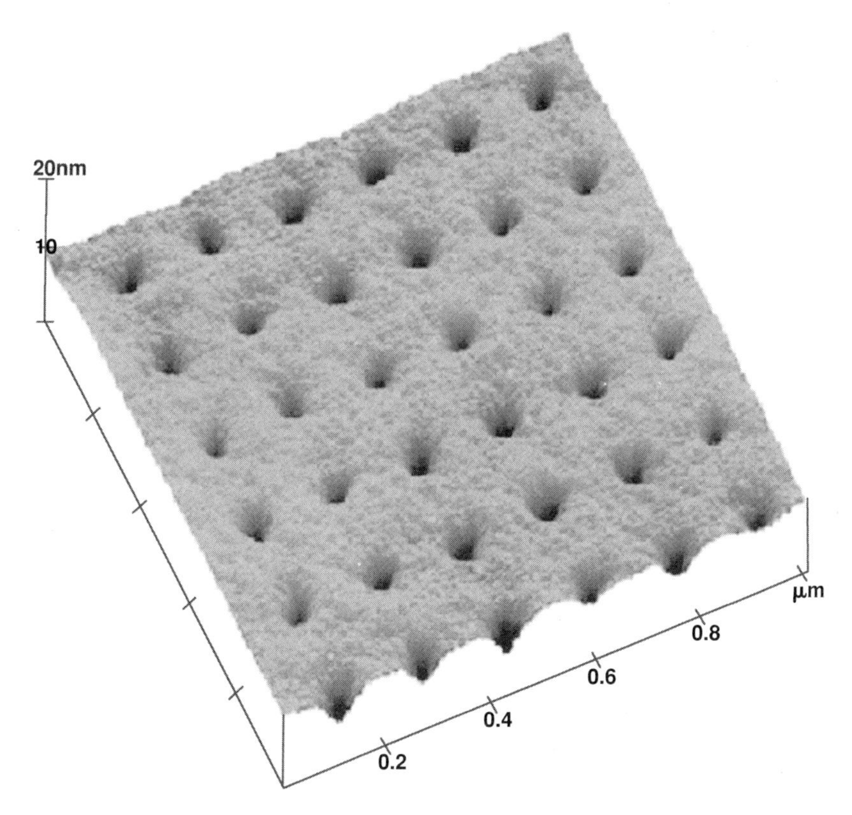

FIGURE 1.7. Atomic force microscope image of topography in PMMA following FIB exposure at 1 pA beam current and a total irradiation time of 20 µs per feature. (From Ref. 24 by permission of American Institute of Physics.)

With respect to focused ion beam writing of resist, ion exposure doses ($\sim 10^{12}$ ions cm^{-2} for 30 keV Ga^+)[25] are typically much lower than for electron exposure. Thus, in principle, ion beam patterning of resist might be expected to be faster than electron beam writing. However, the much lower velocities of energetic ions than electrons of equivalent energy (under the limit where the accelerating energy for the charged particle is fully transformed to kinetic energy, 0.5 mv^2, the c. 10^5 times higher mass of a Ga^+ ion with respect to an electron translates into a velocity more than two orders of magnitude slower), means that the necessary time to blank or reposition an ion beam is much longer. Thus, in practice, serial or direct write electron beam lithography is typically faster than serial ion beam resist lithography.

1.5.c. Damage Mechanisms

A major fundamental drawback of FIB lithography is the damage created in the target. Whereas the photons and electrons used in lithographic systems typically do not have sufficient energy to directly displace atoms in the target, each incident 30 keV Ga^+ ion will create multiple point defects in the structure. If features are created through sputtering, a high ion dose penetrates beyond the final depth of the sputtered feature to create a highly damaged (typically amorphous in semiconductors) zone of very high Ga^+ concentration. The depth of this damage zone will correspond approximately to the range of the ions in the target (approximately 30 nm for normally incident 30 keV Ga^+ ions in Si), although very high point and extended defect concentrations will extend to substantially greater depth, particularly from channeling in crystalline materials. Within the primary damage zone, the atomic Ga fraction will be approximately $1/(1 + Y)$ where Y is the sputter yield of the target material. This high implanted concentration may affect electrical (for example in Si, where Ga is an acceptor), optical, magnetic, mechanical, thermal (for example if Ga forms a low melting point eutectic alloy with the target), and electrochemical properties. Another issue is re-deposition of sputtered species. Other standard liquid metal ion sources have comparable limitations with respect to damage in the substrate. Some promising development of focused gas ion sources has been reported in the literature,[26,27] allowing the use of far less invasive noble ion species, but integration into a commercial system has not yet been realized.

With respect to lithographic patterning through FIB-induced deposition, some limited sputtering/damage/implantation into the target is generally observed before deposition commences and substantial incorporation of the incident ion species (again at an atomic fraction $\sim 1/(1 + Y)$, where Y is now the deposition yield) into the deposited material occurs. This can have substantial ramifications for the maximum resistivity of deposited insulators, and to a lesser extent for the minimum resistivity of deposited metals (typically of order 10–1000 $\mu\Omega$cm for deposited metals and 1–100 MΩcm for deposited insulators). Matching of Ga^+ ion ranges to resist thicknesses can minimize Ga contamination/damage effects during resist exposure, but some penetration beyond the resist and subsequent contamination is still possible, depending upon the resist thickness and incident ion parameters.

1.5.d. Other Factors

A highly advantageous property of focused ion beam fabrication is the high depth of focus due to the limited beam convergence angles. Thus a spot size of 100 nm is maintained over a depth of focus of 100 μm or better, as illustrated in Figure 1.8(a). This allows high resolution patterns to be fabricated over surfaces with high curvature or non planarity, without the need to refocus the beam or adjust the sample height, as illustrated in Figure 1.8(b). This property compares very favorably with optical or serial electron lithography, where depth of foci are much more severely limited.

The field of view in FIB lithography is, however, substantially limited. To maintain ion optical resolution of 100 nm in our system, the largest field area (i.e., area patterned in a single ion scan, without moving the sample) is a few hundreds of μm on a side.[23] This is limited both by pixel dimensions in the ion beam raster and by ion optics. However, coupled with the high depth of focus described in the previous paragraph, this does mean that fields of very high curvature (approaching aspect ratios of 1) may be patterned at high resolution in a single scan.

A major advantage of direct modification FIB lithography is the ability to align a pattern visually in both sputtering and deposition modes, using imaging of secondary electrons. This is of great benefit both in aligning to existing features on the target and in correcting for sample drift during longer fabrication runs. In single column (ion beam only) systems,

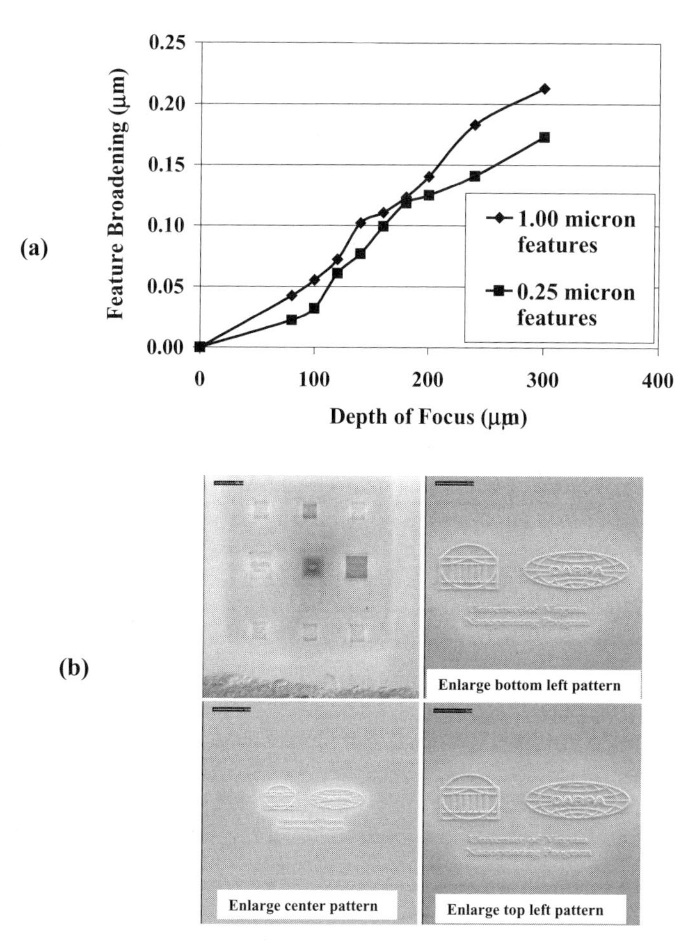

FIGURE 1.8. (a) Variation of feature size with distance of sample from focus position in FIB. (From Ref. 23 by permission of American Institute of Physics.) (b) FIB-induced Pt deposition onto the periphery of a 5 cm radius of curvature gold-coated glass lens, corresponding to height differences of order 30 µm. All images and patterns are recorded without refocusing of the ion beam. Sub 100 nm resolution is maintained over the entire field.

the secondary electron image is generated from the primary ion beam. This leads to the possibility of broad area contamination of the target, but ion doses can be very low in image acquisition mode (each primary ion typically generates several secondary electrons depending upon ion parameters and the target materials. Thus, imaging at 10 nm pixel resolution will require ion doses of order 10^{13} cm^{-2} for acceptable signal-to-noise). In double column (ion and field emission source electron beam) systems, imaging with the primary electron beam is highly advantageous, both in terms of avoiding unwanted contamination/sputtering and because of higher spatial resolution. Of course, in combination with other resist exposure techniques, visual realignment with electron or ion beams is generally impractical during FIB resist exposure lithography, both because of the danger of broad area exposure of the resist, and because alignment features may not be visible below the resist.

Integration of secondary ion mass spectroscopy (SIMS) systems allows end point detection to terminate sputtering at a chemically distinct layer, among other applications. In practice, the highly material dependent nature of secondary electron emission efficiencies means that changes in secondary electron yield from different materials at different depths in the target can often be used for the same purpose.

Another major advantage of FIB lithography in sputtering/deposition modes is the ability to correct individual errors in the fabricated pattern. Thus, if a pattern is created through sputtering and on final inspection it is found that one feature is missing, it is a simple matter (given the visual alignment capability in the FIB) to sputter the additional feature. If an additional feature has been created in error, it can be filled with FIB-induced deposition of metal or insulator. Similar arguments apply to correction of errors in deposited patterns. This ability for point-by-point error correction in a fabricated pattern is a salient advantage of FIB lithography, for example in microelectronic circuit modification and in photolithography mask repair.

1.5.e. Ion Projection Lithography

In ion projection lithography (IPL), light ions (H^+ or He^+) are used to illuminate a stencil mask and demagnified to form an image in resist on the target surface.[28] Potential advantages with respect to electron beam projection lithography (EPL) include lack of a proximity effect (from backscattered electrons generated by the primary beam) and higher resist sensitivities. However, at this stage, EPL is a more developed technology (see Section 1.4.b). A major IPL development project in Europe, MEDEA, has a target of 50 nm resolution over a 12.5 mm exposure field, with $4\times$ demagnification.[29] Of course, the incident ions can also in principle be used for resistless surface modification, as in the focused ion beam, although intensities are necessarily much lower than in the FIB.

1.6. EMERGING NANOLITHOGRAPHIES

A number of new nanolithography techniques have emerged in the past decade and offer great promise for a variety of applications, ranging from three dimensional nanofluidics to high throughput, large area patterning. These techniques most generally rely upon mechanical contact, chemical transfer, or scanning probe–induced surface modification. Many of these techniques have already matured to the stage where they are viable laboratory research tools, and some are already in early commercialization stages.

1.6.a. Microcontact Printing Techniques

Microcontact printing was invented by the Whitesides group.[30] The basic stages of this process are summarized in Figure 1.9. A "standard" lithography technique (photon, electron, or ion based) is used to pattern a topographical master. While this master may take many hours, even days, to fabricate, it can be used multiple times for subsequent mold creation, and each mold can itself be used multiple times. In a sense, then, this master is analogous to the mask used in projection lithography techniques.

An elastomer mold is then cast from the surface (generally by pouring over the master in liquid form, then curing), replicating the topography of the master. The most commonly used

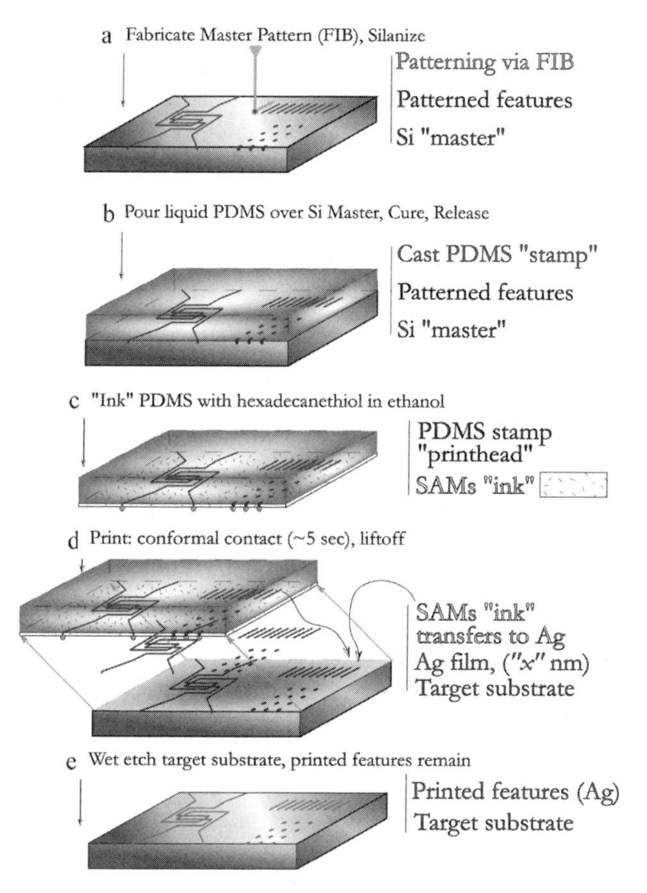

FIGURE 1.9. Schematic illustration of the microcontact printing process. (From Ref. 31 by permission of Elsevier).

elastomer is polydimethylsiloxane (PDMS). This is a very low elastic modulus polymer. It has the advantage of high conformability over complex surface morphologies (or over contaminant particulates on the master surface). Its low rigidity, however, means that very fine features in the cast mold lack the structural integrity to accurately transfer on contact to the target surface. The minimum transferable feature size using PDMS is generally reported to be in the 100–200 nm range.[23] Higher resolution features, below 100 nm, can be transferred using polymer blends with higher elastic moduli,[31,32,33] but this higher rigidity comes at the expense of lower ductility, which means less durability during peeling the mold from the master and less conformability over surface asperities.

Following release of the mold from the master, it is coated with an organic self assembling monolayer (SAM), such as hexadecanethiol. Such thiol molecules terminate coating after a film one monolayer thick is formed. The chemistry of the thiol is such that the free end of the chain has a high affinity to a metal surface. Thus, if the thiol coated elastomer mold is put into contact with a noble metal (e.g., Au, Ag) surface, thiol transfers from the contact points (the raised features in the topographic mold, corresponding to depressed features in the master) to the metal surface. In this fashion, the original set of features etched or sputtered into the master surface is reproduced as a pattern of adsorbed thiol on the target metal surface. The thiol coating can then act as an effective wet etch barrier for appropriate etches (e.g., 0.001 molar potassium ferrocyanide(II)trihydrate, 0.001 molar potassium ferricyanide, 0.1 molar sodium thiosulfate pentadydrate for hexadecanethiol on Ag), transferring the features into the target surface. This technique can thus be used to pattern virtually any surface, if it is coated with a sacrificial Au or Ag layer. Using appropriate SAM layers and etch chemistries, a wide range of other surfaces have also been directly patterned.

Salient advantages of the microcontact printing technique are:

1) The ability to print features over a wide range of length scales, from sub 100 nm to tens of μm.
2) The ability to print over a wide range of surface geometries, and for the mold to adapt to surface asperities and surface geometries.
3) Relatively high throughput is possible through step and repeat or roller applications.
4) Three dimensional structures can be fabricated, for example for nano- and micro-fluidic applications.[34]
5) Cost is relatively low, once an appropriate master has been fabricated.

The primary challenges are alignment of subsequent levels of high resolution printing, the cost of fabricating high resolution masters, and potential transfer defect densities associated with adaption of the mold around surface particulates etc. Notwithstanding these challenges, it is clear that microcontact printing offers a relatively inexpensive, rapid, and adaptable technique, particularly at the micron scale and in the research laboratory.

1.6.b. Nanoimprinting Techniques

The most advanced of the non-traditional lithographies, at least in terms of commercial development, is *nanoimprinting*. Here, a topographic master is impressed into a polymer that is heated above its glass transition temperature, allowing the pattern in the master to be transferred in relief to the target surface.[35] This is essentially the technique (albeit at a substantially lower resolution) by which compact discs are created from a master. The target polymer may be softened either thermally or by absorption of photons (the so-called "step and flash" technique).[36]

The fundamental steps in nano-imprinting are illustrated in Figure 1.10. A master mold is first fabricated by electron beam, optical, ion beam, or other lithographic technique. This mold is then used to deform the softened polymer resist film. Following imprinting, the polymer is cured, either by cooling below the glass transition temperature (for a thermoplastic resist), or by UV irradiation (for a UV-curable resist). Anisotropic etching, most generally reactive ion etching (RIE), is then used to remove the compressed resist arising from mold contact.

In this fashion, structures less than 10 nm in dimension have been fabricated.[37] Semiconductor wafer-scale masters have been fabricated,[35] and sub-micron inter-level alignment demonstrated over a 4″ silicon wafer.[38] Commercial imprinting machines are also starting to appear.[39] Continuous improvement of imprint materials and processes[40] is reducing transfer defect densities, and improving reproducibility and fidelity of the transferred patterns.

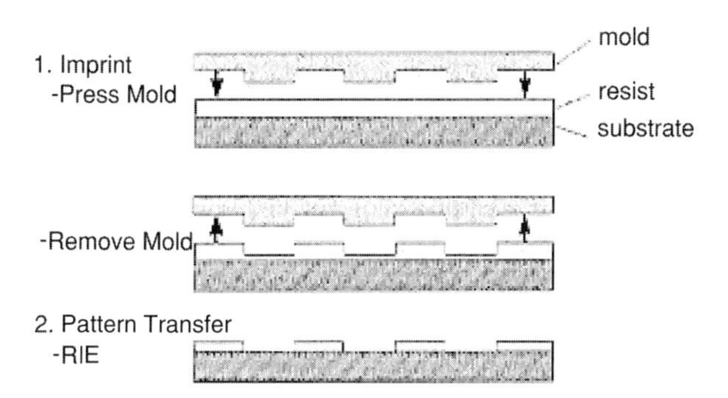

FIGURE 1.10. Schematic of nanoimprinting lithography process. (From Ref. 37 by permission of American Vacuum Society.)

Overall, nanoimprinting is undergoing extensive research and development world wide, and appears to be poised to become a plausible nano-manufacturing technology.

1.6.c. Scanning Probe Based Techniques

At one extreme, scanning probe techniques offer the ability to manipulate structure atom-by-atom, as in the extraordinary work of Eigler's group in fabricating "quantum corrals" of Fe atoms on Cu surfaces.[41] Scanning probe microscopes (see Chapter 3) generally operate on the principle of observing the interaction of a sharp probe tip with a sample across which it is scanned, generally providing resolution of 1 nm or less. This then offers the ultimate in spatial resolution, affording the ability to fabricate exquisitely atomically-engineered structures in which new quantum phenomena can be demonstrated and explored. Several other scanning probe approaches to nanolithography have been demonstrated, e.g., through positive or negative resist writing, or surface oxidation of Si.[42,43] Feature resolution in such systems can be very high (nanometers or tens of nanometers), but all suffer from inherently low throughput, defined by the probe "write rate" as it moves across the surface, with typical tip speeds of order 1–100 μm/s. Arrays of cantilever tips can be used to enhance throughput, and 50 tip arrays have been used to pattern areas up to 1 cm^2 with micron linewidths.[44]

A relatively new version of scanning probe lithography, dip-pen nanolithography (DPN) has been developed by the Mirkin group.[45] In DPN, the water meniscus that forms naturally between tip and sample enables controlled molecular transport from a liquid covered tip to the target surface. If the transfer liquid (or "ink") is reactive with the surface to be patterned, this provides a driving force for molecular transport from tip to substrate, where chemisorption can lead to stable nanostructures.[46] The inherent advantages of DPN are high spatial resolution (linewidths of order 10 nm have been demonstrated), high registration capability (as the probe can be used to both read and write, as for the focused ion beam described earlier, visual alignment is straight-forward), and the ability to directly pattern complex molecular species, including biomolecular species (initial applications have included patterning of DNA arrays via transfer of oligonucleotides).[47] The primary disadvantage is the inherently slow speed of single tip writing. Multiple arrays have been demonstrated to enhance throughput, including the ability for individual tip addressing.[46]

1.7. SUMMARY

- A set of lithographic techniques exist for patterning of surfaces with sub 100 nm resolution, generally employing photons, electrons, ions, scanning probe microscope tips, or mechanical contact.
- Sub–100 nm lithographic techniques employ projection (photon, electron, ion), direct write (ion, electron, scanning probe tips), proximity (X-ray), or mechanical contacting (nanoimprinting, microcontact printing).
- Key components of lithographic techniques are source (photon, electron, ion) technologies, mask technologies (for projection or proximity methods), and resist technologies.
- Key figures of merit include: resolution, alignment (inter- and intra-level), throughput, and cost.
- Currently, the most pervasive, sophisticated, and developed lithographic technology is optical projection lithography employing deep uv photons. With shorter wavelength sources, new forms of projection optics, and resolution enhancement technologies, this can be employed for patterns substantially smaller than 100 nm. This technology is ubiquitous in the microelectronics industry.
- Evolution of optical lithography to much smaller features employing either projection methods for extreme uv photons or X-ray proximity methods has been demonstrated in the laboratory and is under extensive commercial development.
- Direct write e-beam exposure of resists is a highly developed method for laboratory-scale patterning of limited areas with resolution down to the tens of nanometers level.
- Projection electron lithography methods have been developed with higher throughput than direct write electron beam methods, and higher resolution than current optical projection methods.
- Focused ion beam techniques (sputtering, deposition, or resist exposure) of surfaces offers a very adaptive method for high resolution (tens of nanometers) patterning of a wide range of materials and surface geometries. However, throughput and specimen damage are significant issues. Projection ion lithography methods are also under development.
- Emerging nanolithographic techniques include nanoimprinting, microcontact printing and scanning probe–based methods.

QUESTIONS

1. Why are chemically amplified resists better for printing features just resolved at the Rayleigh resolution limit?
2. If a given photoresist process was entirely linear, then the resist film thickness remaining after development (negative tone process) would be linearly proportional to the light intensity delivered to the resist. In order to resolve features, the resist thickness would be required to go to zero somewhere between the features. Using the assumption of a linear resist, derive a new resolution criterion analogous to that developed by Rayleigh for observing stars (the eye is very non-linear).

3. With the advent of DUV lithography and chemically amplified resists, the cost of resist has risen to about $2,000 per gallon. In order to hold down costs for resist, processes have been developed which use only about 2 cm^3 of resist to coat a 200 mm wafer. If the final resist film is 0.6 μm thick, how much of the resist actually ends up on the wafer (in percent)?
4. Explain why it is difficult if not impossible to make a circuit where the transistor gates in the logic part of the circuit are as small as the leading logic manufacturers while also having memory as dense as the leading memory manufacturers on the same chip at the same time.
5. Describe a process to fabricate gates for MOS transistors using electron beam lithography. Is it better to use positive or negative tone resist? Would it matter if the substrate was GaAs rather than Si?
6. Given an electron beam lithography system and a focused ion beam system, with identical beam size (e.g., 30 nm defined as the full width at half maximum of the Gaussian current profile), which one will produce smaller features and why?
7. Show that the deflection sensitivity of a parallel plate electrostatic deflector does not depend on the mass of the particle. Deflection sensitivity is defined as the magnitude of the deflection on the target per unit voltage applied to the deflector. That is, a given deflector would work the same way for electrons or any variety of singly-charged ions.

REFERENCES

1. Semiconductor Industry Association (SIA), "International roadmap for Semiconductors 2002 update edition", International SEMATECH, Austin, TX, 2003.
2. M. D. Levinson, *Jpn. J. Appl. Phys.*, **33,** 6765 (1994).
3. M. D. Levinson, N. S.Viswanathan, and R. A. Simpson, *IEEE Trans. Electron Devices*, Vol. **ED-29**, 1828 (1982).
4. I. C. Kizilyalli, G. P. Watson, R. A. Kohler, O. Nalamasu, and L. R. Harriott, *Electron Devices Meeting*, 2000. IEDM Technical Digest. International, 2000 Page(s): 829–832; H.-Y. Liu, L. Karklin, Y.-T. Wang, and Y. C. Pati. *Proceedings of SPIE*, **3334,** p. 2 (1998); M. E. King, N. Cave, B. J. Falch, C.-C. Fu, K. Green, K. D. Lucas, B. J. Roman, A. Reich, J. L. Sturtevant, R. Tian, D. Russell, L. Karklin, and Y.-T. Wang, *Proceedings of SPIE*, **3679**, p. 10 (1999).
5. M. D. Levenson, *Proceedings of SPIE*, **3051**, p. 2 (1997).
6. K. Kamon, T. Miyamoto, Y. Myoi, H. Nagata, M. Tanaka, and K. Horie, *Jpn. J. Appl. Phys.* **30**, 3012 (1991).
7. W. T. Silvast and O. R. Wood II, *Microelectronic Engineering*, **8**, 3 (1988); A. M. Hawryluk and L. G. Seppala, *J. Vac. Sci. Technol.* B**6**, 2162 (1988); C. W. Gwyn, R. Stulen, D. Sweeney, and D. Attwood, *J. Vac. Sci. Tehnol.* B**16**, 3142 (1998).
8. J. P. Silverman, *J. Vac. Sci. Technol.* B**16**, 3137 (1998).
9. R. Feynman, "There's pleanty of room at the Bottom: an invitation to enter a new field of physics", Talk at the annual meeting of the American Physical Society, 29 Dec. 1959. Reprinted in Engineering and Science **23**, 22 (1960).
10. A. N. Broers and M. Hatzakis, Scientific American, **227**, 33 (1972); D. R. Herriott, *et al., IEEE Trans. Electron Devices*, **ED-29**, 385 (1975).
11. E. Goto, T. Soma, and M. Idesawa, *J. Vac. Sci. Technol.* **15**, 883 (1978); H. C. Pfeiffer, *J. Vac. Sci. Technol.* **15,** 887 (1978); M. G. R. Thomson, R. J. Collier, and D. R. Herriott, *J. Vac. Sci. Technol.* **15**, 891 (1978).
12. Y. Nakayama, *et al., J. Vac. Sci. Technol.* **B8**, 1836 (1990).
13. M. B. Heritage, *J. Vac. Sci. Technol.* **12**, 1135 (1975).
14. J. M. Gibson and S. D. Berger, *Appl. Phys. Lett.* **57**, 153 (1990); L. R. Harriott, *J. Vac. Sci. Technol.* B**15**, 2130 (1997); H. C. Pfeiffer and W. Stickel, *Microelectronic Engineering*, **27**, 143 (1995); H. C. Pfeiffer and W. Stickel, *Proceedings of SPIE*, **2522**, 23 (1995).

15. L. R. Harriott, S. D. Berger, J. A. Liddle, G. P. Watson, and M. M. Mkrtchyan, *J. Vac. Sci. Technol.* **B13**, 2404 (1995).
16. For an excellent review of focused ion beam sources, systems, and principles, see *J. Orloff, Rev. Sci. Instrum.* **64**, 1105–30 (1993).
17. For a review of relevant sputtering mechanisms, see A. Benninghoven, F. G. Rudenauer, and H. W. Werner, Chapter 2 (John Wiley, New York, USA, 1987).
18. J. Melngailis, in SPIE Proceedings Vol. **1465**, "Electron-Beam, X-Ray and Ion-Beam Submicrometer Lithographies for Manufacturing", ed. M. C. Peckerar 36-49 (SPIE, Bellington, WA, 1991).
19. A. D. Dubner and A. Wagner, *J. Appl. Phys.* **66**, 870–4 (1989).
20. M. H. F. Overwijk and F. C. van den Heuvel, *J. Appl. Phys.* **74**, 1762–9 (1993).
21. R. L. Kubena, F. P. Stratton, J. W. Ward, G. M. Atkinson, and R. J. Joyce, *J. Vac. Sci. Technol.* B**7**, 1798–801 (1989).
22. See, for example, A. Moser, C. T. Rettner, M. E. Best, E. E. Fullerton, D. Weller, M. Parker, and M. F. Doerner, *IEEE Trans. Magnetics*, **36**, 2137–9 (2000).
23. D. M. Longo, W. E. Benson, T. Chraska, and R. Hull, *Appl. Phys. Lett.* **78**, 981–3 (2001).
24. Y. Liu, D. M. Longo, and R. Hull, *Appl. Phys. Lett.* 82, 346–8 (2003).
25. J. Melngailis, *Nuc. Inst. Meth.* B**80/81**, 1271–80 (1993).
26. Ch. Wilbertz, Th. Maisch, D. Huttner, K. Bohringer, K. Jousten, and S. Kalbitzer, *Nuc. Inst. Meth.* B**63**, 120–4 (1992).
27. L. Scipioni, D. Stewart, D. Ferranti, and A. Saxonis, *J. Vac. Sci. Technol.* B**18**, 3194–7 (2000).
28. J. Melngailis, A. A. Mondeli, I. L. Berry, and R. Mohondro, *J. Vac. Sci. Technol.* **B16**, 927–57 (1998).
29. W. H. Bruenger, R. Kaesmaier, H. Loeschner, and R. Springer, *Mat. Res. Soc. Symp. Proc.* **636**, D5.5.1–12 (2001).
30. G. M. Whitesides and Y. Xia, *Ann. Rev. Mater. Sci.* **28**, 153–84 (1998).
31. R. Hull, T. Chraska, Y. Liu, and D. Longo, *Mat. Sci. Eng.* C**19**, 383–92 (2002).
32. H. Schmid and B. Michel, *Macromolecules*, **33**, 3042–9 (2000).
33. T. W. Odom, V. R. Thalladi, J. C. Love, and G. M. Whitesides, *J. Amer. Chem. Soc.* **124** 12112–3 (2002).
34. See, for example, J. C. Love, J. R. Anderson, and G. M. Whitesides, *Mat. Res. Soc. Bull.* **26**, 523–29, and references therein.
35. See, for example, S. Y. Chou, *Mat. Res. Soc. Bull.* **26**, 512 (2001), and references therein.
36. M. Colburn, T. Bailey, B. J. Choi, J. G. Ekerdt, S. V. Sreenivasan, and C. G. Willson, Solid State Technology, **44**, 67–78 (2001).
37. S. Y. Chou, P. R. Krauss, W. Zhang, L. Guo, and L. Zhuang, *J. Vac. Sci. Technol.* B**15**, 2897–904 (1997).
38. W. Zhang, S. Y. Chou, *Appl. Phys. Lett.* **79**, 845–7 (2001).
39. Molecular Imprints, Inc. 1807-C West Braker Lane, Suite 100 Austin, TX 78758; Nanonex Corp., P.O. Box 334, Princeton, NJ, 08543.
40. See, for example, H. Schulz, H.-C. Scheer, T. Hoffman, C. M. Sotomayor Torres, K. Pfeiffer, G. Bleidiessel, G. Grutzner, Ch. Cardinaud, F. Gaboriau, M.-C. Peognon, J. Ahopelto, and B. Hediari, *J. Vac. Sci. Technol.* **B18**, 1861 (2000).
41. M. F. Crommie, C. P. Lutz, and D. M. Eigler, *Science*, **262**, 218–20 (1993).
42. S. W. Park, H. T. Soh, C. F. Quate, and S.-I. Park, *Appl. Phys. Lett.* **67**, 2415–7 (1995).
43. E. S. Snow and P. M. Campbell, *Science*, **270**, 1639 (1995).
44. S. C. Minne, J. D. Adams, G. Yaralioglu, S. R. Manalis, A. Atalar, and C. F. Quate, *Appl. Phys. Lett.* **73**, 1742–4 (1998).
45. R. D. Piner, J. Zhu, F. Xu, S. Hong, and C. A. Mirkin, *Science* **283**, 661–3 (1999).
46. C. A. Mirkin, *Mat. Res. Soc. Bull.* **26**, 535–8 (2001).
47. L. M. Demers, D. S. Ginger, S.-J. Park, Z. Li, S.-W. Chung, and C. A. Mirkin, *Science* **296**, 1836–8 (2002).

2

Self-Assembly and Self-Organization

Roy Shenhar, Tyler B. Norsten, and Vincent M. Rotello
Department of Chemistry, University of Massachusetts, Amherst, MA

2.1. THE ADVANTAGES OF SELF-ASSEMBLY

Early in the history of science it became evident that to a large extent, material properties are determined at the molecular level. As technology progressed, allowing the observation and detailed analysis of increasingly smaller objects, novel material properties that are unique to the nanometer length scale also emerged. Combining the promise provided by the unique behavior at small scales with the obvious utility of miniaturization (perhaps the most prominent example being the computer chip) affords the main drive for this field.

There are two basic methods for material fabrication. One is to start with rough, large scale material and carve a shape into it, often referred to as the "top-down" approach. The other method is to create a construct from scratch by assembling simple building blocks according to a pre-designed scheme. The latter is termed the "bottom-up" approach; as it is obviously a much more versatile methodology, most macroscopic man-made constructions are fabricated this way.

The last two decades have seen increasing efforts directed at miniaturization of electrical components—a fundamental demand for high-speed operation—with technological progress closely following the famous Moore's law (exponential growth with time of the number of transistors on a computer chip). Industrial fabrication of microelectronic components exclusively employs the "top-down" approach described in Chapter 1. Nevertheless, technological difficulties and inherent limitations are predicted to inhibit the capability of this approach to continue feeding the miniaturization race. Obtaining decreasing feature size by photolithography demands the use of shorter wavelengths, reaching into the deep and far UV, which poses inherent difficulties on the creation of masks and resists. The potent electron-beam lithography technique (see Section 1.4), which allows the creation of nanometer-scale features, is serial by nature and therefore relatively inefficient for mass production. Both lithographic methods are limited when the formation of three-dimensional structures is desired. Additionally, to continue device miniaturization, new materials are required to replace silicon oxide as the insulating layer in the transistor construction, since it has already been scaled to near its limit.[1]

Molecular self-assembly as a "bottom-up" methodology provides a direct access to the nanometer regime, having potential for a much greater versatility than methods based on the "top-down" approach. Inspired by Nature, where millions of years of evolution have resulted in working molecular machines (such as enzymes), scientists are looking for ways to design molecular building blocks that spontaneously assemble into defined, desired structures. The strength of the self-assembly approach lies in the plethora of synthetic methods available for the creation of desirable molecular building blocks. The main challenge, however, is the assembly process. As the building blocks involved here are molecules, which are by no means a structural continuum, the assembly scheme must be embedded within the inherent properties of the building blocks, making their design a crucial step in the process. Self-assembly is currently more of a science than it is a technology and is still in its infancy, but in contrast to "top-down" technologies—the investigation of self-assembly has just begun.

2.2. INTERMOLECULAR INTERACTIONS AND MOLECULAR RECOGNITION

Self-assembly,[2] based on the non-covalent assembly of smaller subunits to generate higher ordered aggregates, offers an efficient alternative to the classic covalent approach, requires fewer synthetic steps, and results in higher yields. The kinetically labile interactions that are used as the supramolecular "glue" allow for self-correcting or self-healing, resulting in the formation of stable, defect-free aggregates. The feasibility of this approach hinges on the rational design of simple building blocks that are capable of selective and spontaneous assembly. Design of these components demands the consideration of two key issues: what functionality will act as effective "glue", and how will the recognition features be incorporated and placed into the components to facilitate a discriminating self-assembly process? These two points are frequently referred to as *molecular recognition elements* and *molecular programming*, respectively.[3]

The different types of "glue" that are used to hold supramolecular architectures together have been typically defined as "non-covalent" and include: ionic interactions (ion-ion, ion-dipole, and dipole-dipole), hydrogen bonds, π–π stacking, dispersion forces, coordination or dative bonds, and the hydrophobic effect. A summary of each type of interaction is given in Table 2.1. The supramolecular synthons (discrete building blocks) can incorporate one or more of the interactions listed in Table 2.1. The combination of two, or more, either similar or different interactions increases selectivity and adaptability of the building blocks and also increases the stability of the complexes resulting from the self-assembly process.

Self-assembly processes employing molecular sized building blocks have been used to create a wealth of complex and ordered structures extending into the nanoscale. The following sections will describe how scientists are beginning to incorporate molecular recognition elements and molecular programming into macromolecular and nanoscale

TABLE 2.1. Intermolecular interactions

Interaction	Interaction Strength[a] (kJ mol^{-1})	Description	Example
Electrostatics	>190 (ion-ion) 40–120 (ion-dipole) 5–40 (dipole-dipole)	coulombic interactions between opposite charges	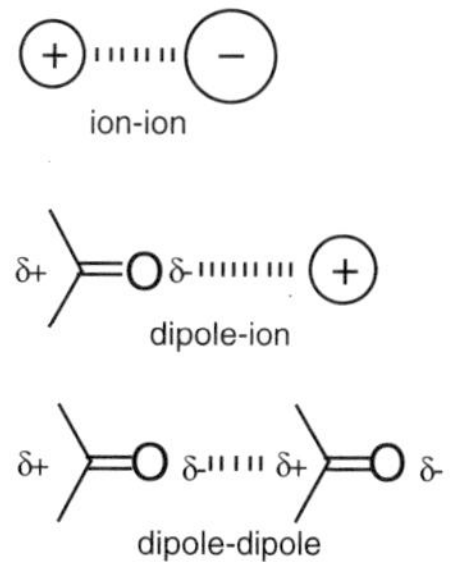 ion-ion dipole-ion dipole-dipole
Hydrogen bonding	15–40 (strong) 5–15 (moderate) <5(weak)	donor-acceptor interactions specifically involving hydrogen as the proton donor and a base as the proton acceptor	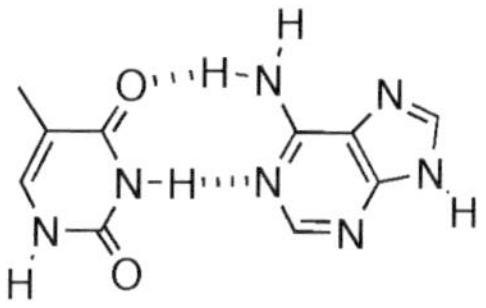
π-π Interactions	10–15 (face to face) 15–20 (edge to face)	attractive forces between electron-rich interior of an aromatic ring with the electron-poor exterior of an aromatic ring	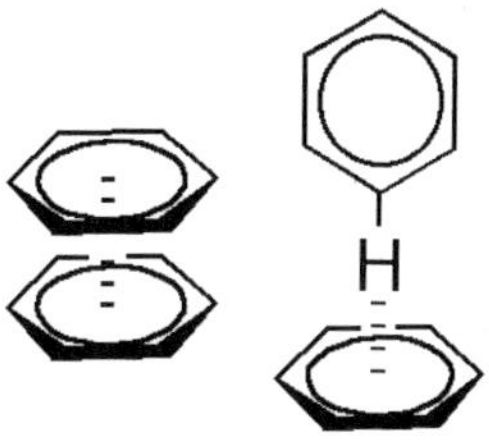
Dispersion forces	<5	momentary induced dipole-dipole interactions (also called London forces)	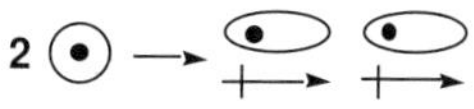
Hydrophobic effects	varied 5–40	association of non-polar binding partners in an aqueous medium or vice versa	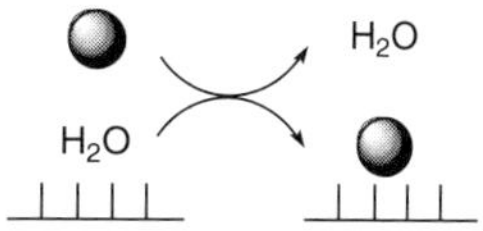 H_2O H_2O
Dative bonding	varied 20–380	coordination of a metal by a ligand donating two electrons	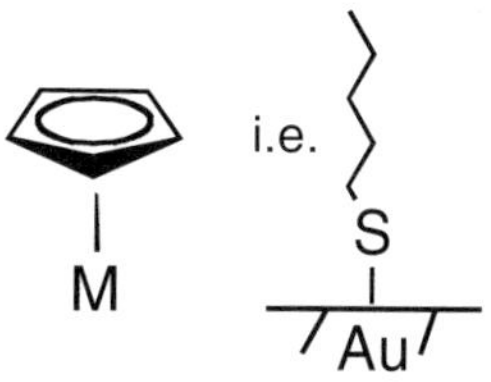 M i.e. S Au

[a] Association constants are for systems in chloroform.

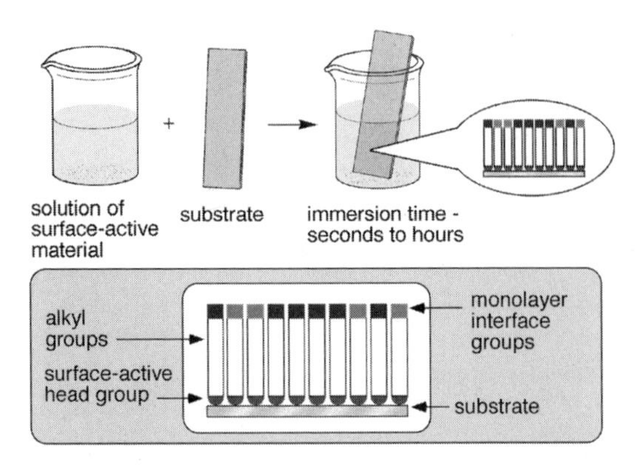

FIGURE 2.1. The process of forming a self-assembled monolayer. A substrate is immersed into a dilute solution of a surface-active material that adsorbs onto the surface and organizes via a self-assembly process. The result is a highly ordered and well-packed molecular monolayer. (Adapted from Ref. 9 by permission of American Chemical Society.)

systems in an effort to generate nanosized functional materials exhibiting order extending into the macroscale regime.[4,5,6]

2.3. SELF-ASSEMBLED MONOLAYERS (SAMs)

The formation of monolayers (layers that are a single molecule thick) by spontaneous chemisorption of long-chain amphiphilic molecules (that have both hydrophilic and hydrophobic functionalities) at surfaces is an excellent example of how self-assembly can create long-range order. Many different types of self-assembled monolayer (SAM) systems have been explored over the years.[7] Although different SAMs contain a variety of different molecules and substrate combinations, common to all SAM systems is a surface-active head group that attaches to its corresponding substrate via a chemisorption process (Figure 2.1). The result of the adsorption process is an ultrathin monolayer with a thickness that is dictated by the length of the stabilizing alkyl (series of CH_2 groups) chain. Many other factors contribute to the usefulness of various types of SAMs for application purposes such as their stability, functional group versatility, and substrate/monolayer composition.

2.3.1. Organothiol Monolayers on Flat Gold Surfaces

The most frequently studied SAM system is that of long-chain alkanethiolates (SH group at the end of an alkane chain) on gold surfaces,[8] owing to its simple preparation and relative stability once formed. The assembly process is driven primarily by the attachment of the sulfur atom to the gold surface. Once tethered, the alkyl chains of the molecules organize themselves laterally through van der Waals interactions to form a densely packed monolayer. The formation of gold SAMs is mild enough to permit a variety of functionality to be incorporated into the monolayer, providing a versatile route to the assembly of complex surfaces. As such, SAMs of alkanethiolates on gold have increased the basic understanding of such interfacial phenomena as adhesion, lubrication, wetting, and corrosion. More

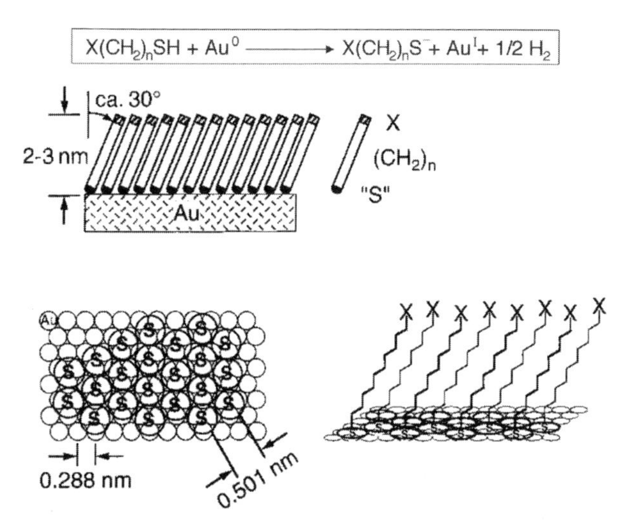

FIGURE 2.2. Schematic representations of an alkanethiolate monolayer on gold. The alkyl chains tilt at a 30° angle from the surface normal, maximizing the inter-chain van der Waals interactions. The sulfur atoms are arranged in a hexagonally close packed formation on the Au(111) surface. The X groups terminating the alkyl chains provide chemical versatility at the monolayer interface. (From Ref. 12 by permission of WILEY-VCH Verlag GmbH & Co.)

recently, techniques have become available that allow for SAM surfaces to be decorated with complex patterns. These emerging patterning technologies combined with SAM chemistry demonstrate the usefulness of self-assembly in micro- and nanofabrication techniques.

2.3.1.a. Formation, Structure, and Characterization of SAMs on Gold The chemistry of SAM formation as well as characterization of SAM structure has been extensively reviewed.[9] We will provide a brief overview of these topics here, as they are the foundation of SAM systems. SAMs of alkanethiolates on gold can be prepared either through a solution or vapor phase adsorption process. The reaction thought to take place during the adsorption of the sulfur atom to the gold is described in the equation in Figure 2.2, although the exact mechanistic details have not yet been fully worked out. Kinetic studies on SAM formation show that the adsorption process is consistent with a first-order Langmuir isotherm where the growth rate is proportional to the number of unoccupied gold sites.

It has been demonstrated that the sulfur atoms for long-chain alkanethiolates (X-$(CH_2)_n$-SH) form a hexagonally packed arrangement on the Au(111) surface. The methylene groups tilt at an angle of ca. 30° from the surface normal to maximize the favorable van der Waals interactions between adjacent chains. When the alkyl chain is adequately long ($n > 11$), densely-packed, highly-ordered pseudocrystalline monolayers can be achieved (Figure 2.2). If functionalities other than methyl groups terminate the monolayer (e.g., $X =$ bulky groups or polar groups), the packing density and overall order of the SAM may be compromised.

It should be mentioned that although SAMs in general tend to reject errors through a self-correcting mechanism during the equilibrium process leading to their formation, defects still arise due to factors such as imperfections on the surface and SAM preparation

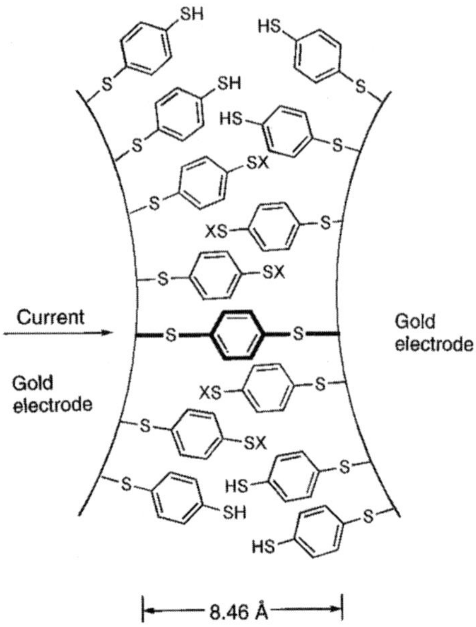

FIGURE 2.3. A schematic representation of Reed and Tour's molecular junction containing a benzene-1,4-dithiolate SAM that bridges two proximal gold electrodes. (From Ref. 10 by permission of American Association for the Advancement of Science.)

conditions. This has been described as one of the major obstacles that currently limits SAMs as a pragmatic fabrication tool for microelectronics.

2.3.1.b. Gold SAMs: Tools for Studying Molecular-Based Electronics and Creating Nanostructured Materials and Patterned Surfaces The versatility of gold SAM chemistry has allowed it to become a platform on which to study and advance nanoscale science and engineering. Aside from the purely chemical advantages that the thiol/gold combination offers, the conductive nature of the underlying gold substrate permits a variety of microscopy characterization techniques to be performed and has allowed for fundamental studies to be conducted towards the realization of molecular scale devices.

The incorporation of molecules as electronic components is of great interest as current technology progresses towards the nanoscale. Using SAM chemistry, Reed, Tour and coworkers demonstrated that it was possible to measure the conductance of a junction containing a single molecule of benzene-1,4-dithiol self-assembled between two gold electrodes (Figure 2.3).[10] This study served as a proof-of-principle that molecular scale electronic systems could be fabricated and set the stage for more complex systems.

McEuen, Ralph, and coworkers devised a molecular scale electronic device that functions as a single atom transistor.[11] The fabrication of the transistor employed a combination of assembly techniques including dative bonding interactions and typical SAM formation. Two terpyridine ligands were used to coordinate a single cobalt atom to form a SAM that is capable of bridging a pair of gold electrodes (Figure 2.4). By fine-tuning the length of alkyl chain on the terpyridine, they were able to control the electronic coupling between

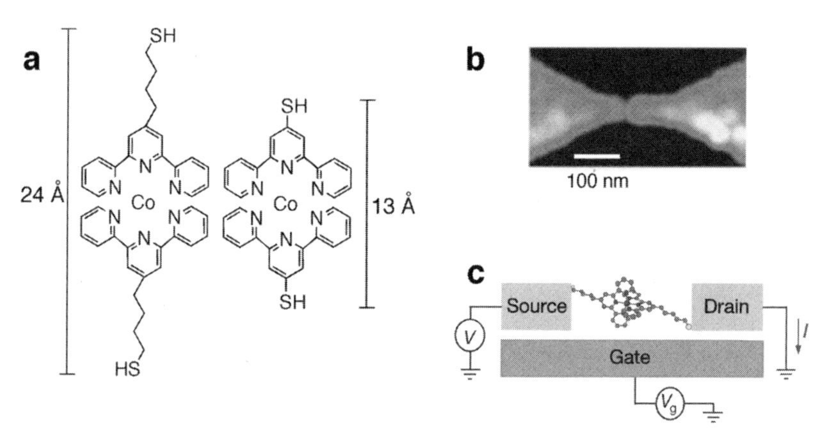

FIGURE 2.4. (a) Structures of the long and short linked cobalt coordinated terpyridine thiols used as gate molecules. (b) A topographic AFM image of the gold electrodes with a gap. (c) A schematic representation of the assembled single atom transistor. (From Ref. 11 by permission of Macmillan Magazines Ltd.)

the central cobalt ion and the flanking electrodes. For example, longer linker molecules (weaker coupling) permitted the device to behave as a single-electron transistor on which the electronic properties of the device could be effectively tuned by controlling the electrochemical properties (gate voltage) of the cobalt atom. Experiments employing short linkers (strong coupling) exhibited greatly enhanced conductance measurements that suggest a Kondo-assisted tunneling mechanism. The field of molecular electronics is described in more detail in Chapter 10.

The patterning of SAMs on gold is another area that is actively being pursued as a fabrication technique to produce well-defined micro- and nano-structures. The major impetus behind developing new fabrication techniques employing self-assembly is to provide a rapid and inexpensive route to structured nanoscale materials. Self-assembly provides a "bottom-up" approach to structure fabrication that is complementary to current photolithographic techniques and may, in the future, provide access to size regimes that are physically unattainable to photolithographic patterning. The subject has been extensively reviewed;[12] here, we give several examples demonstrating different approaches that have been used to pattern SAMs.

Scanning tunneling microscopy (STM, described in Chapter 3) has traditionally been used as a visualization tool capable of achieving molecular scale resolution. Recently it was shown that STM could also be used as a lithographic SAM patterning tool. Patterning of dodecanethiol SAMs was demonstrated by applying voltage pulses through the STM at different locations, which resulted in complete removal of the insulating monolayer at these positions (Figure 2.5).[13] Conjugated molecules could then be reintroduced at the subsequent bare positions in the monolayer. The resulting STM images clearly demonstrate insertion of the more conducting molecules into the specified locations in the monolayer.

The same STM patterning technique has been used to incorporate molecular recognition functionality (diacyl 2,6-diaminopyridine) into holes produced in the insulating monolayer.[14] Electroactive functionalization of the monolayer was then achieved through noncovalent hydrogen bonding of a complementary ferrocene-terminated uracil

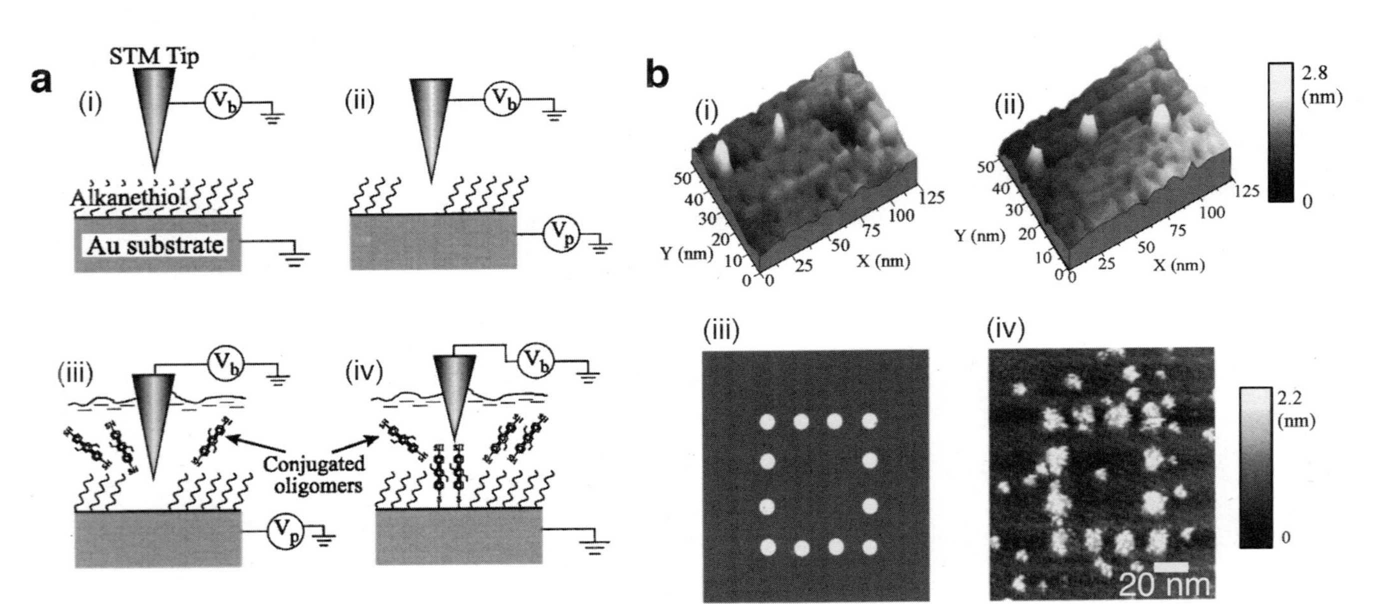

FIGURE 2.5. (a) A schematic representation of the STM patterning of SAMs. (i) Normal STM imaging of the SAM with tip bias V_b; (ii) Removal of SAM by applying a pulse V_p to the gold substrate; (iii) The same as (ii) in solution of conjugated oligomers; (iv) insertion of conjugated oligomers in the patterened sites. (b) STM image of dodecanthiol and conjugated oligomeric patterned SAMs. (i) The STM image after consecutive pulsing at three different locations indicating insertion of molecules (two peaks) and one pit without insertion. (ii) The same region imaged a few minutes later showing adsorption into the remaining pit. (iii) A programmed pattern consisting of circles tracing out a rectangle. (iv) The resulting image of the patterned dodecanthiol SAM after chemisorption of the conjugated oligomers showing the produced rectangular frame. (From Ref. 13 by permission of American Institute of Physics.)

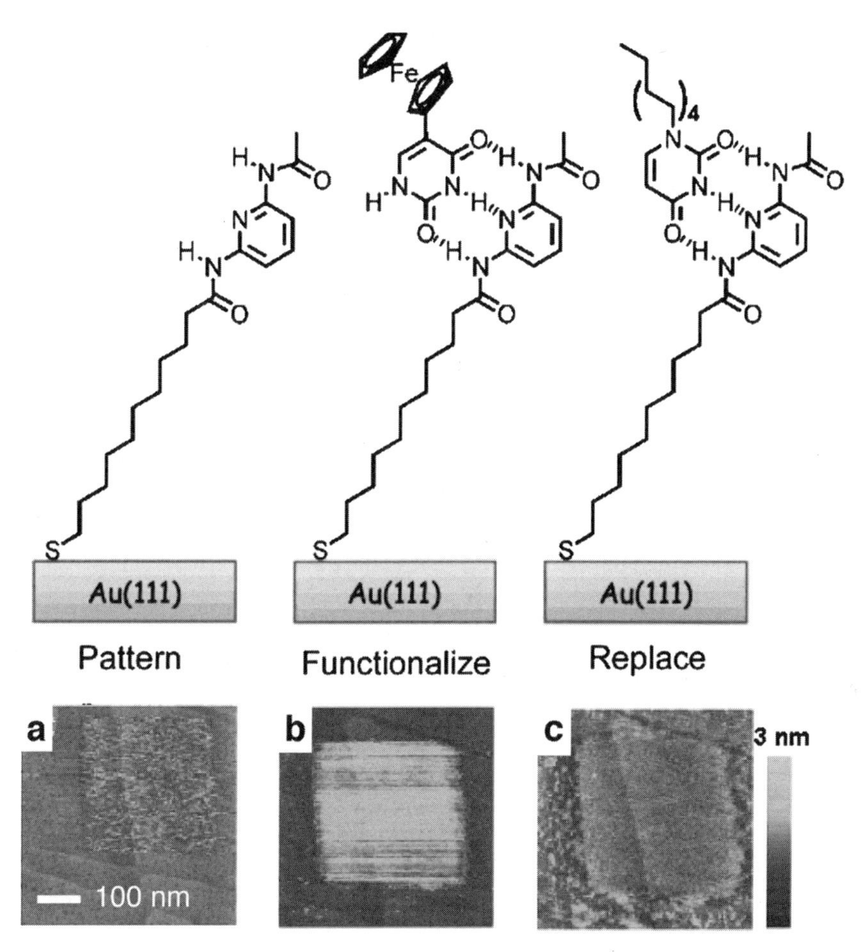

FIGURE 2.6. Formation of molecular assemblies and corresponding STM images. (a) STM image of diacyl 2,6-diaminopyridine (DAP) decanethiol binder inserted into a surrounding decanthiol monolayer. (b) Image after binding the complementary electroactive Fc-uracil showing an increase in the current-dependent apparent height contrast. (c) "Erased" pattern after replacing the electroactive guest with a more insulating dodecyl functionalized uracil. (From Ref. 14 by permission of American Chemical Society.)

(Figure 2.6a,b). The electroactive ferrocene-uracil could subsequently be "erased" by replacing it with a more insulating molecule containing a dodecyl chain (Figure 2.6c). These studies demonstrate the usefulness of STM as a nanoscale patterning technique and show how noncovalent interactions can be used to reversibly control the electronic properties of SAM-based molecular assemblies. It should be noted, however, that STM lithography, being a serial process, is a relatively slow and time-consuming approach to patterning SAMs and therefore it is currently not an economically feasible microfabrication technique.

Microcontact printing (μCP, see Section 1.6), developed by Whitesides and coworkers, is a powerful approach for creating patterns of SAMs on gold in a parallel process.[12,15] The technique is a non-photolithographic method that forms patterned SAMs with submicron lateral dimensions. In this method, an elastomeric polydimethylsiloxane (PDMS) stamp is loaded with "ink", typically a solution of alkanethiol molecules that are then simply transferred to the "paper", typically a gold substrate (Figure 2.7). A different SAM can then be introduced into the underivatized regions after the initial pattern is created or the gold underneath the underivatized regions of the SAM can be etched employing a chemical etching procedure.

The pattern on the PDMS stamp is complementary to its template, which is typically created using standard lithographic microfabrication techniques (i.e., photolithography, *e*-beam lithography, etc.). This technique is a simple, rapid, and inexpensive way to create multiple copies of patterned SAMs because the time consuming and expensive lithographic techniques are used only to create the master, and the PDMS stamp can be used multiple times before pattern degradation occurs. Another advantage to μCP over photolithography is the ability to pattern curved surfaces.[16] The limitations of the technique, on the other hand, are poor edge resolution and the number of defect sites in the created SAMs.

Dip pen nanolithography (DPN, see Section 1.6), developed by Mirkin and coworkers, represents a similar "ink" and "paper" approach to patterning SAMs on gold.[17] This approach involves using the tip of an atomic force microscope (AFM, see Section 3.4) as a "pen" to transfer the alkanethiol "ink" through capillary action to the underlying gold substrate (Figure 2.8). Currently, this technique is not amenable to rapid pattern fabrication as is the stamp methodology in μCP, however, DPN can deliver minuscule amounts of molecules from the AFM tip to the substrate with resolutions comparable to more expensive high resolution patterning techniques such as e-beam lithography (~10 nm).

2.3.1.c. Organosilicon Monolayers Organosilicon SAMs created from chemisorption of alkylsilanes ($RSiX_3$, R_2SiX_2 or R_3SiX, where R is an alkyl chain and X is chloride or hydroxyl-terminated alkyl chain) to hydroxylated silica surfaces are another important and frequently encountered class of SAM.[9] Similar to alkanethiolate SAMs, monolayer self-assembly can be carried out in solution or the vapor phase, and is driven primarily by attachment of the surface-active head group (a silane) to the substrate (a silanol). As compared to gold SAMs, it is difficult to achieve high-quality SAMs from solution based alkylsilane chemistry as many controlling factors contribute to their formation such as temperature and the amount of water present during the monolayer assembly process. The monolayer molecules are anchored to the substrate through individual silanol linkages as well as being interconnected to neighboring molecules through an extensive polysiloxane network at the surface. Once formed, however, organosilicon monolayers are extremely robust due to the network of strong Si–O–Si bonds covalently connecting the monolayer to the surface and itself.

A technique has also been developed for creating complex patterns of gold on silicon substrates employing interfacial interactions based on organothiol and organosilicon monolayer chemistries.[18] The technique, referred to as nanotransfer printing (nTP), uses SAMs as covalent "glue" and "release" layers for transferring material from patterned stamps to a substrate. A variety of complex patterns of single or multilayers can be generated on SAMs with nanometer resolution on both flexible and rigid substrates using either hard or soft polymer stamps.

In the nTP procedure (Figure 2.9), a monolayer of 3-mercaptopropyltrimethoxysilane (MPTMS) is created on a silicon wafer, leading to a reactive thiol-containing interface. A PDMS stamp coated with thermally evaporated gold is then used to print on this highly

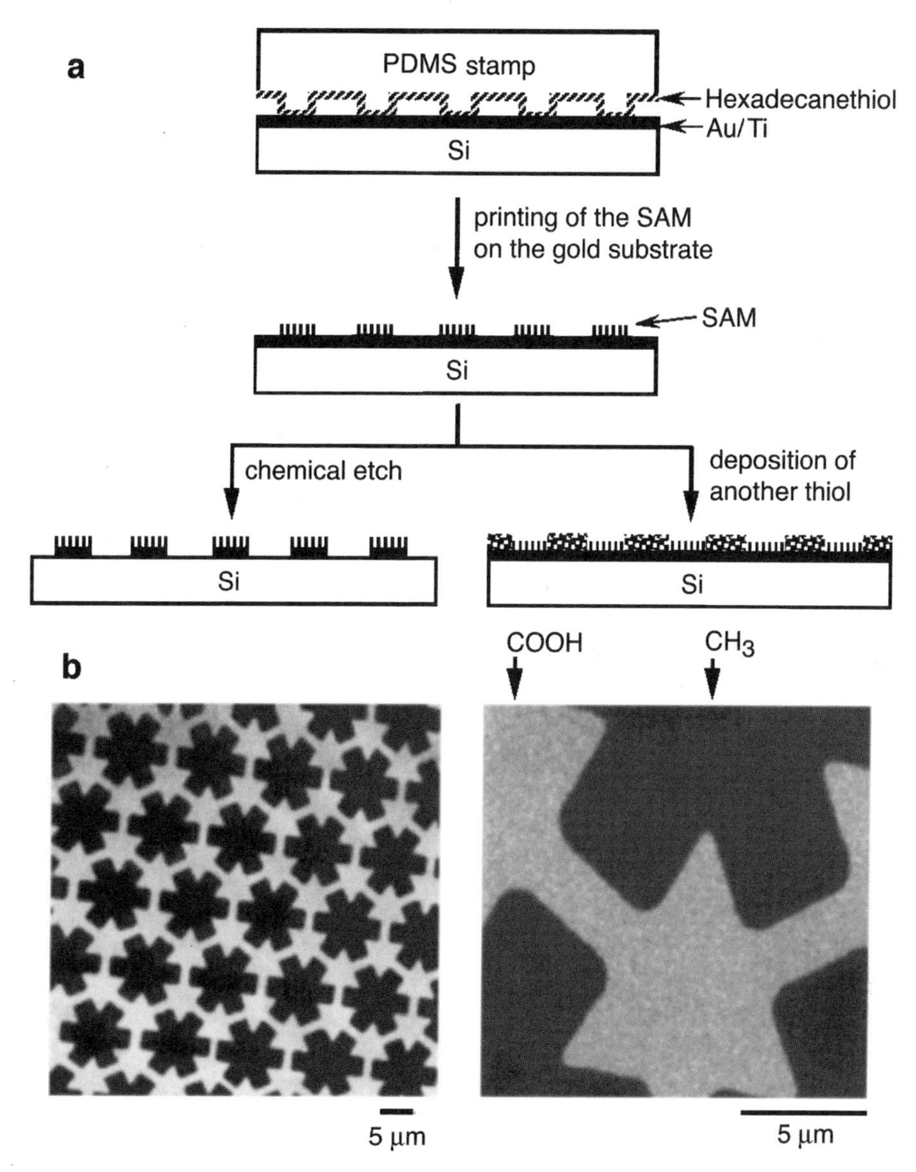

FIGURE 2.7. (a) Schematic illustration of the μCP procedure for patterning an alkanethiol (hexadecanethiol-HDT) on a flat gold substrate. (b) Lateral force microscope (LFM) images (at two different magnifications) of a patterned gold substrate with SAMs terminated in chemically different head groups (HDT-CH_3 and 16-mercaptohexadecanoic acid-COOH). The image contrast results from differences in frictional forces between the surface and the probe tip. Carboxylic acid terminated SAM show high measured frictional forces (light regions) and methyl terminated SAM show low measured frictional forces (dark regions). (From Ref. 12 by permission of WILEY-VCH Verlag GmbH & Co.)

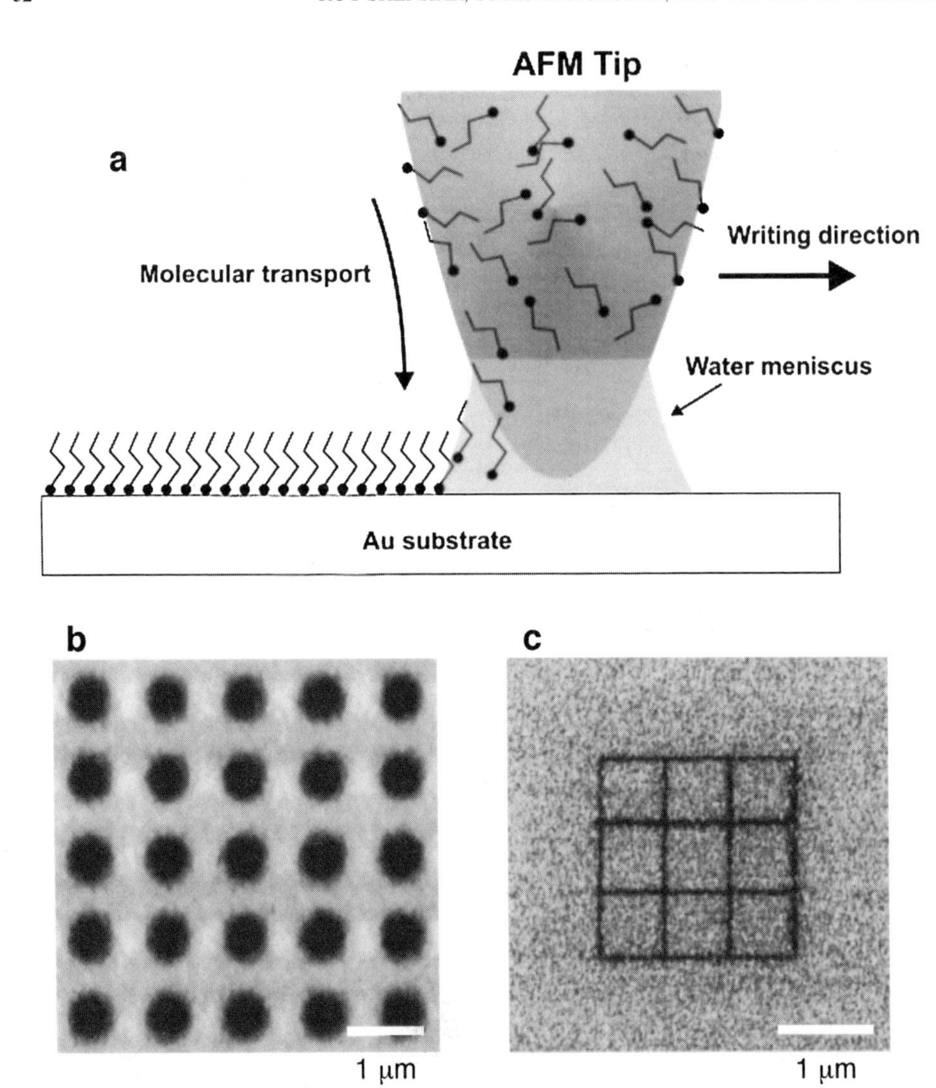

FIGURE 2.8. (a) Schematic representation of dip-pen nanolithography (DPN). A water meniscus formed between AFM tip and the gold substrate directs the thiol molecules onto the substrate. The size of the water meniscus is controlled by the relative humidity that in turn affects the overall resolution of DPN. (b) LFM image of an array of octadecanethiol dots on a gold surface generated by holding an ODT-coated AFM tip in contact with the surface for ca. 20 s. (c) LFM image of a molecule-based grid consisting of eight lines 100 nm in width and 2 μm in length. (From Ref. 17 by permission of American Association for the Advancement of Science.)

reactive surface. Because the gold does not adhere strongly to the PDMS, it is transferred efficiently to the thiol groups on the SAM when brought into contact with the substrate. Similar to μCP, only the regions of the stamp that contact the SAM are transferred to the substrate creating the pattern on the monolayer. The edge resolution of the created patterns is excellent and is comparable to the resolution of the stamp itself.

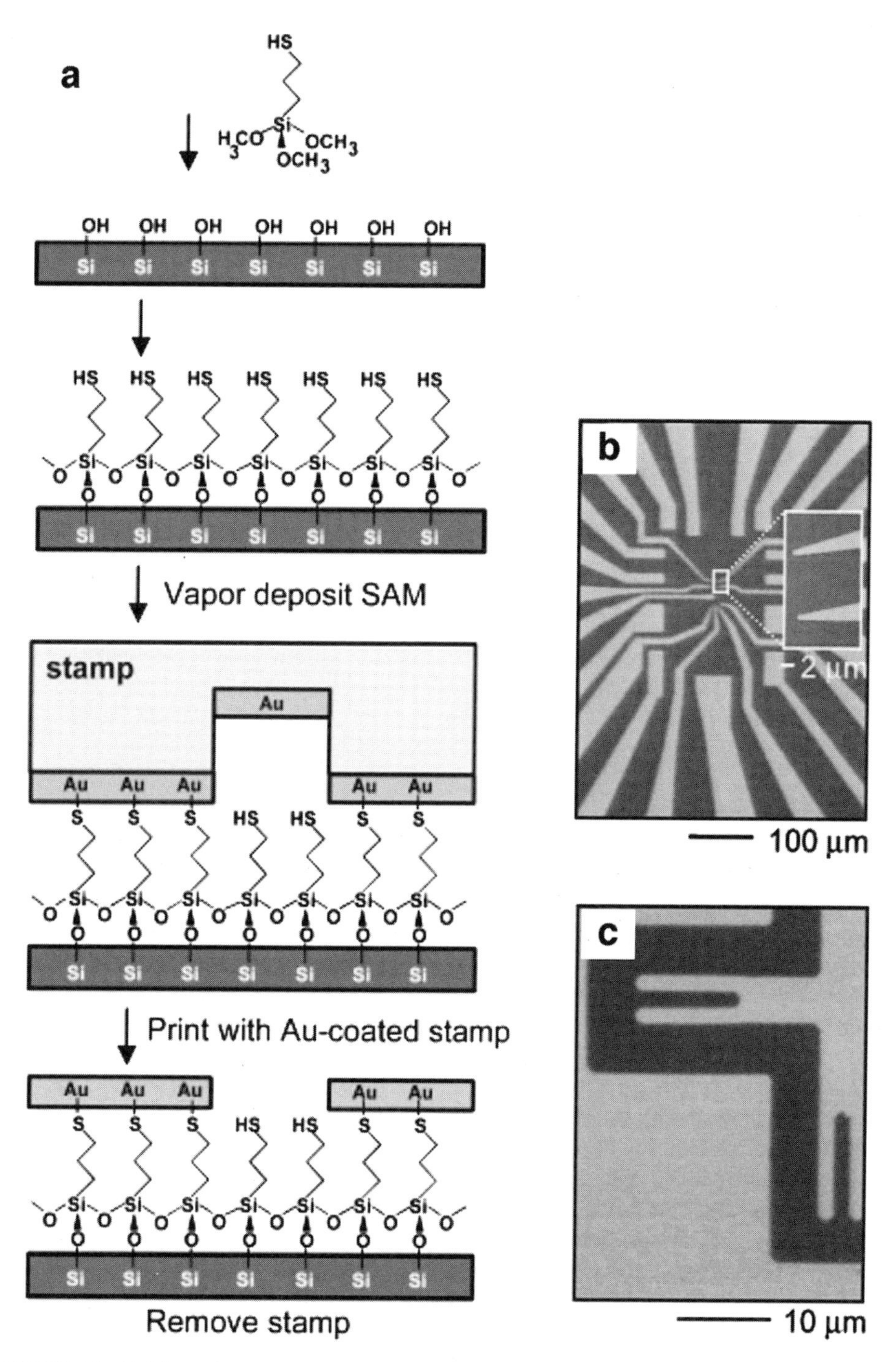

FIGURE 2.9. (a) Schematic representation of the nanotransfer printing (nTP) procedure to create gold patterns on Si substrates. Optical micrographs of a gold pattern formed by nTP on (b) a silica wafer, and (c) a plastic sheet [organosilsesquioxane modified poly(ethylene terephthalate)], demonstrating the wide applicability of the technique. (From Ref. 18 by permission of American Chemical Society.)

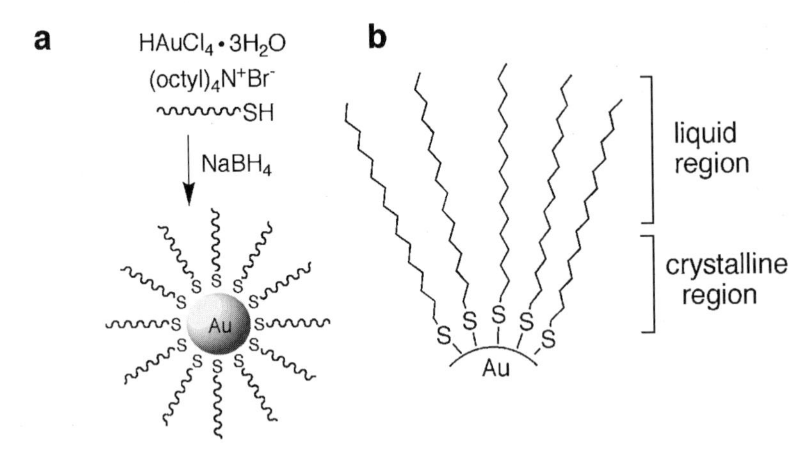

FIGURE 2.10. (a) The solution-based procedure developed by Brust and Schiffrin for synthesizing alkanethiolate stabilized gold nanoparticles. (b) Schematic illustration showing the curved surface and different regions of a nanopartice SAM.

The technique is quite versatile as other substrates containing surface hydroxyl groups can be used to create micro- and nano-scale patterns of gold on SAMs. The nTP procedure is one potential approach for fabricating molecular scale electronic devices employing SAMs as semiconductor and dielectric layers.

2.3.2. *Organothiol Monolayers on Faceted Metal Clusters*

Besides providing a versatile route to the functionalization of flat gold surfaces, SAM chemistry has played an integral role in the development and advancement of Monolayer Protected Clusters (MPCs), commonly referred to simply as nanoparticles. Metallic nanoparticles comprise a unique and relatively unexplored niche of materials, whose physical, chemical, and electronic properties can be tuned according to their size.

One of the most frequently encountered MPCs is that of the alkanethiolate-stabilized gold cluster synthesized by solution-phase methods.[19] In this procedure, chemical reduction of a gold salt by a hydride reducing agent in the presence of thiol capping ligands furnishes the desired monolayer-protected nanoparticles (Figure 2.10a). The monolayer assembled during nanoparticle formation shields the gold core from agglomeration. Careful control of the reaction conditions can yield a variety of different core sizes (1.5–8 nm).

SAMs assembled on MPCs are quite different than SAMs on flat gold surfaces. First, the nanoparticle surface is highly faceted and hence contains many defect sites such as edges and vertices. Second, due to the curved surface and unlike the monolayers on flat gold surfaces, nanoparticle SAMs radiate outward from the core. Third, the monolayer structure exhibits both pseudo-crystalline internal regions (providing long-term stability against agglomeration both in air and in solution) and fluid-like packing at the periphery (Figure 2.10b). These attributes make nanoparticles amenable to studies of interfacial and chemical phenomena, and allow for easy characterization using standard solution-phase techniques such as nuclear magnetic resonance (NMR), infrared (IR), and UV-Vis spectroscopy.

An important advantage of nanoparticles is the ability to modify their surface through place exchange reactions, where new incoming thiol ligands displace monolayer thiols on the parent MPC to afford Mixed Monolayer Protected Clusters (MMPCs).[20] Additionally, it has been shown that the alkanethiolate-supporting monolayer is robust enough to allow a variety of covalent modification reactions to be achieved at the monolayer interface. This provides a route to synthetically modify the surface of the particle after synthesis, which greatly enhances the diversity of functionalities that can be attached to the nanoparticle. Moreover, the nanoparticle SAM is dynamic, providing some degree of mobility to the alkanethiolates on the surface. To this end, Rotello and coworkers have demonstrated the radial control of intramolecular interactions within the SAM,[21] as well as templation and molecular recognition processes of small molecules through intermolecular interactions at the interface.[22]

2.4. ELECTROSTATIC SELF-ASSEMBLY

While van der Waals interactions prove very powerful for the fabrication of self-assembled monolayers and for other purposes such as ordering spherical particles,[23,24] this type of interaction offers only limited control and diversity for the creation of assembled structures. This arises from the fact that van der Waals interactions are relatively weak and cannot be easily tailored on the molecular level.

Electrostatic interactions, including hydrogen-bonding interactions, offer an important alternative. From a chemical point of view, the ability to put charged groups on a polymer skeleton or on nanoparticles greatly enhances the modularity of the building blocks employed in the assembly process. Nanoscopic building blocks of desired properties can be engineered to carry assembly-inducing components such as charged groups and hydrogen-bonding functionalities, and thus can be assembled into macroscopic structures such that their properties can be studied and exploited. Moreover, novel materials and nanocomposites may be constructed possessing unique assembly properties that are different from the properties of the individual corresponding building blocks.

The following subsections will cover different approaches for creating and assembling three-dimensional structures using electrostatic interactions, going from the simple and versatile layer-by-layer deposition technique to methods aimed at the creation of three dimensional composites with defined shapes.

2.4.1. Layer-By-Layer (LBL) Deposition

A straightforward way to bridge the gap between the molecular world and the macroscopic world and at the same time to induce some level of ordering is to deposit matter on surfaces. Solid substrates can be easily handled and manipulated. As supports for thin films, they define spatial boundary conditions for the deposited matter, hence they induce ordering at the molecular level with respect to a macroscopic coordinate (the substrate surface). Besides acting as supports, different substrates can also provide functional properties. For example, employment of transparent substrates enables the expression of photo-physical properties of the supported matter, and conducting substrates carrying a pre-designed film can be used as novel electrodes for sensing and other applications.

Since the early 1990s, the so-called "layer-by-layer" (LBL) deposition technique has evolved to be a prominent method for the fabrication of tailor-designed surface coatings.[25] LBL was initially applied to polymers but later was demonstrated with nanoparticles and numerous other multivalent species.[26] The method utilizes the electrostatic attraction between oppositely charged groups on different polymers/nanoparticles as the driving force for their adhesion to the surface. The power of LBL lies in part on its remarkable simplicity. First, the surface of a solid support is modified to bear charged groups. This is usually achieved by modification of common surfaces either by a chemical reaction or by adsorption of molecules bearing charged groups. The charged surface is then dipped into a solution of an oppositely charged polyion for a short time (ranging from minutes for polyelectrolytes to hours in the case of gold nanoparticles). Under the proper conditions, the amount of polymer adsorbed contains more than the stoichiometric number of charges on the substrate; as a result the sign of the charge of the exposed surface is reversed. This has two important consequences: (i) repulsion of surplus material, which is washed away by solvent rinsing before the next layer is deposited (thus regulation of the deposition process is achieved), and (ii) enabling the deposition of a second layer on top of the first one. The coating film is thus grown by consecutive deposition cycles of polymers of alternating charge (Figure 2.11) and is often monitored with ellipsometry and UV-Vis spectrometry.

Films prepared by LBL can be highly uniform and exhibit good thermal stability and resistance against dissolution in most organic solvents. The fact that the electrostatic interaction imposes the least steric demand (compared to all other interactions) proves beneficial as it renders LBL somewhat insensitive to local defects in internal layers. Another advantage of LBL is its applicability to a wide variety of charged polymers and nanoparticles, with reduced sensitivity to the specific nature of the counterions.

The key to the wide applicability of the LBL method as well as to its feasibility in general is, apparently, the entropy gain arising from the massive liberation of the counterions and the solvent shell upon adhesion of the charged polymer to the surface.[26] To a first approximation, the adsorption merely trades the electrostatic interactions between the polymer's charged groups and their counterions with similar interactions with the substrate's charged surface, hence resulting in a small enthalpy change; therefore it is *entropy* that governs the process, and not enthalpy.

The outcome of the entropy dominance is that the deposition process is kinetically controlled. While this feature has advantages as indicated above, it lies also in the basis of the method's main limitation: the structures formed represent kinetically-trapped arrangements of the adsorbed polymers. Although the overall film thickness usually grows linearly with the deposition of each bilayer, the internal stratified structure is nevertheless fuzzy, with polymers interdigitating and penetrating each others' domains. Additionally, the method is very sensitive to working conditions such as humidity, drying or not between cycles, and even the relative deposition order of the polymer layers.

Nevertheless, within the limitations arising from the kinetic control of the LBL deposition process, results are generally fairly reproducible, and the benefits of the LBL method exceed its limitations in many applications. Further elaboration of the polymers by other functional units, incorporation of inorganic materials (e.g., clays) and nanoparticles make LBL amenable to the fabrication of coatings with unique combination of mechanical and electrical properties. Films made using LBL can function as biosensors, separation membranes, insulating/semiconducting layers, etc. Prominent examples of such applications are described in the following paragraphs.

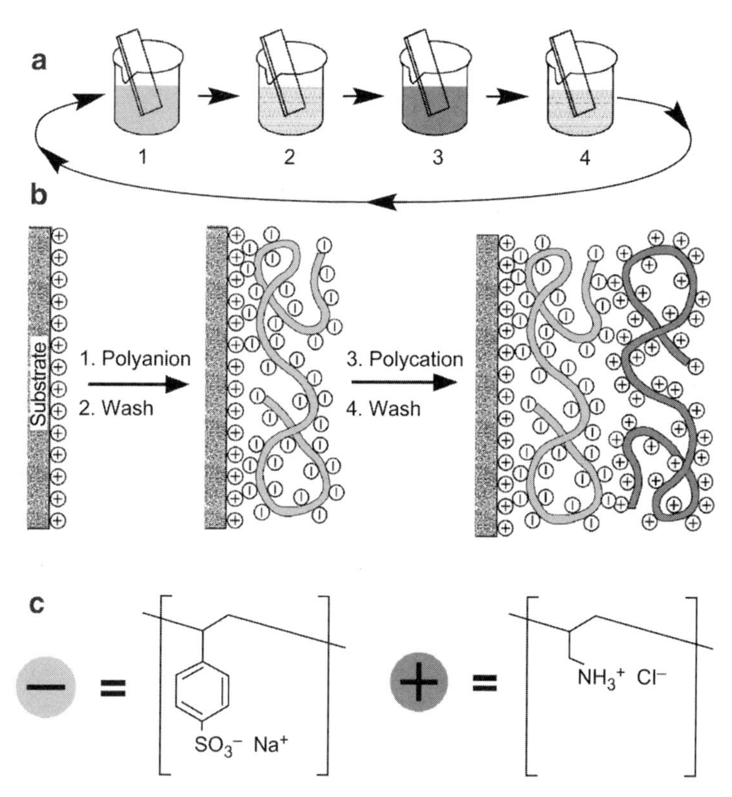

FIGURE 2.11. Film growth by the LBL technique: (a) Repeated dipping of a charged-surface slide into polyanion and polycation solutions (steps 1 and 3, respectively) with intermediate washing steps (2 and 4) results in alternate deposition of the corresponding polyions (b). (c) Chemical structures of two typical polyions: sulfonated polystyrene (SPS) and poly(allylamine hydrochloride) (PAH). (From Ref. 25 by permission of American Association for the Advancement of Science.)

2.4.1.a. Conducting LBL Films and Light-Emitting Diodes (LEDs) Ordered deposition of multiple layers of polymers of interest allows combination, enhancement, and fine-tuning of their properties in the resulting film. Rubner and coworkers exploited the built-in charge in *p*-type-doped electrically conductive polymers to fabricate conductive multi-layer films.[27] Their LBL-based approach avoids derivatizing the polymers with charged groups, which usually results in reduced environmental stability and poor conductivity. Known conductive polymers were employed (e.g., doped polypyrrole or polyaniline); sulfonated polystyrene (SPS) was used as the assembly counterpart. The fuzzy layer structure (overlapping of the layers along the substrate normal), which is normally taken as a disadvantage, proved useful in this case, as it allowed electrical continuity between nonadjacent conducting layers, thus enhancing the film's conductivity. Films produced with this technique were uniform and transparent, and showed conductivity as high as 40 S/cm, in some cases requiring only a few cycles of bilayer deposition to achieve satisfactory conductivity. A possible application for such films would be anti-static coatings of plastics.

Success in assembling conjugated polymers by the LBL method[28] paved the way to the manufacturing of films that function as light-emitting diodes (LEDs).[29,30] The first studies, by Rubner's group, were based on multilayer films of the conjugated poly(*p*-phenylene

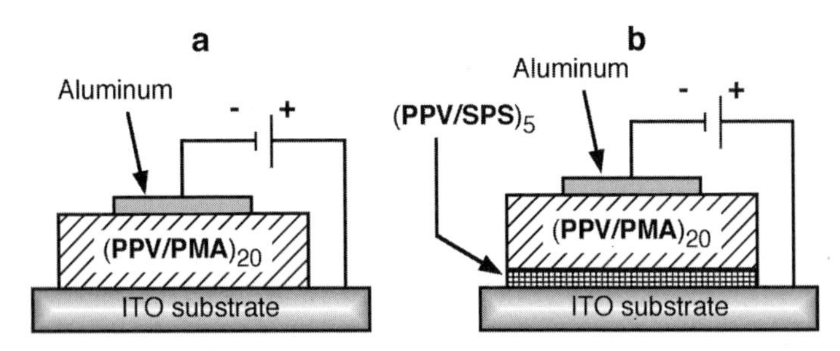

FIGURE 2.12. LBL engineering of a light-emitting device: (a) Multilayer of PPV/PMA performs similarly to PPV alone. (b) Combined architecture of PPV/PMA multilayer on top of PPV/SPS multilayer allows better hole-transport between the PPV/PMA and the ITO anode, and results in enhanced electroluminescence. (From Ref. 32 by permission of American Institute of Physics.)

vinylene) (PPV) with polyanion counterparts on indium tin oxide (ITO, a transparent conductor) anode, with evaporated aluminum on top of the film serving as the cathode (Figure 2.12).[31] The cationic precursor of PPV was assembled with a suitable anionic polymer into multilayer thin films and polymerized *in situ* to its fully-conjugated, light-emitting PPV form (another demonstration of the versatility of LBL).

PPV films produced by this technique are inexpensive, emit light uniformly across large areas with similar luminance level exhibited by films of PPV alone, and have the added advantage of molecular level control of film thickness and homogeneity as well as overall device performance.[31] For example, a remarkable effect was shown in the comparison between using SPS and poly(methacrylic acid) (PMA) as the assembling anionic polyelectrolyte counterpart. Films based on PPV/SPS showed significantly *higher* current densities while delivering much *lower* luminance levels compared to PPV/PMA devices. This was attributed to some *p*-type doping of the PPV by the relatively acidic SPS, which creates polarons and bipolarons. These sites, while enhancing conductivity, are also known as effective luminescence quenching sites. Evidently, this feature could be harnessed to enhance device performance by facilitating hole transport between the anode and the PPV/PMA light-emitting film.[32] Indeed, devices created by placing a few bilayers of PPV/SPS between the ITO and PPV/PMA film (Figure 2.12b) exhibited luminance levels nearly an order of magnitude higher than a device made of PPV alone and were typically two to four times more efficient.

2.4.1.b. LBL with Metallic Nanoparticles for Sensing Applications Modification of conductive surfaces by films of deposited material leads to functional electrodes. As shown before, this is of great importance to the development of electronic components, but it also has promising prospects with respect to electrochemical sensing applications. Negatively charged, citrate-stabilized gold nanoparticles were used to assemble positively charged host molecules on an ITO electrode in an LBL fashion (Figure 2.13).[33,34] The main idea is to create a selective membrane for electroactive analyte molecules in the shape of a nanoparticle network inlaid with host molecules. This membrane would attract only suitable analytes from the bulk solution and concentrate them at the surface of the electrode while acting as a barrier to unfitting analyte molecules. Integration of metallic nanoparticles into the LBL assembly process has some clear advantages. First, a three-dimensional stratified

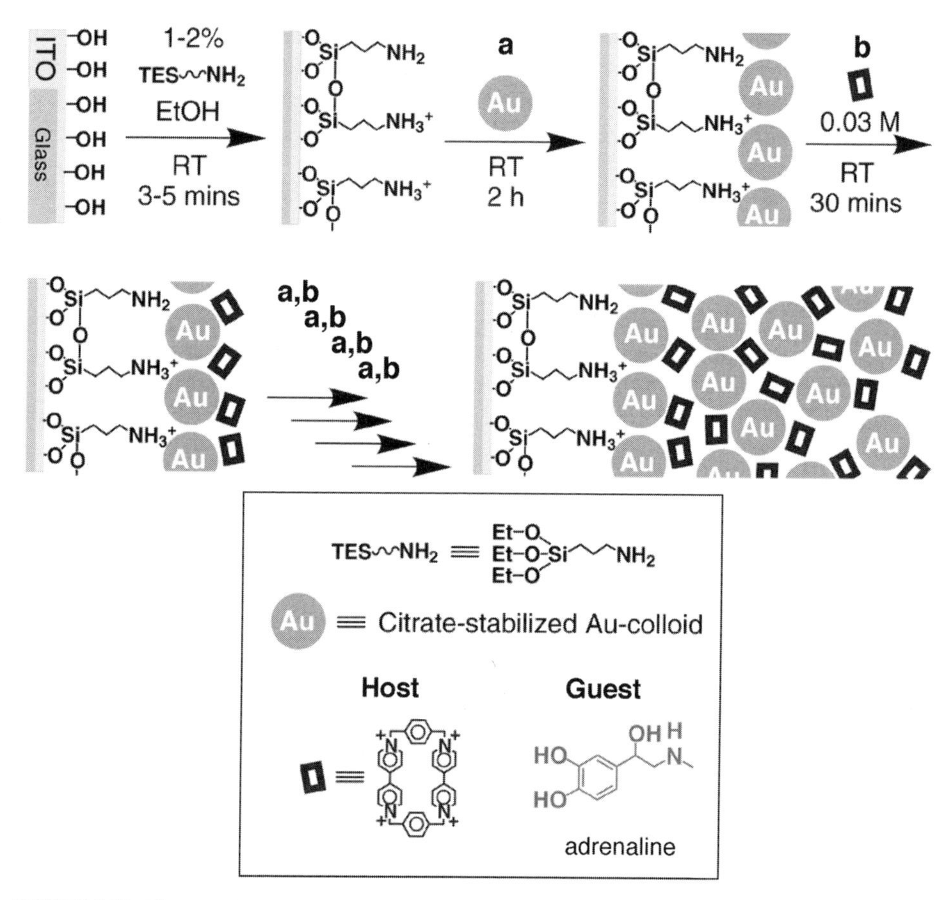

FIGURE 2.13. Electrostatic assembly method for creation of functionalized sensing electrodes: alternating deposition steps of (a) negatively charged gold nanoparticles, and (b) positively charged host molecules. Electrodes modified in this fashion with the host shown proved highly sensitive for detection of adrenaline (as the guest molecule). (From Ref. 33 by permission of WILEY-VCH Verlag GmbH & Co.)

array of nanoparticles provides a porous scaffold for the assembly of the host molecules, resulting in a high surface area electrode and thus leading to enhanced sensing efficiency. Second, the conductivity of the metallic nanoparticles permits sensing of host-complexed electroactive analytes by electrochemical means. Third, the LBL method allows tuning of the device sensitivity, where increasing the number of nanoparticles/hosts layers enhances the electrochemical sensing signal and allows detection of analytes at very low concentrations (10^{-5}–10^{-6} M). Electrodes constructed of gold nanoparticles and bipyridinium cyclophanes (multiply charged compounds, as required for LBL, that contain size-specific, electron poor receptor cavities) successfully detected electroactive aromatic molecules such as neurotransmitters adrenaline and dopamine.

The selectivity of the electrodes fabricated by this technique originates from the structure of the host molecules. It was shown that a perfect match between the size of the analyte molecules and the hosts' cavity dimensions is required for analyte detection. In addition, construction of multifunctional devices that combine layers consisting of different hosts was demonstrated. Applying the correct order of deposition of the different host layers (which proved to be a critical factor[33]) enabled creation of bifunctional devices that can detect two types of analytes.

2.4.1.c. Hollow LBL Spheres The generality of the LBL method allows its application to different types of charged surfaces, including surfaces that have non-planar topologies. For example, an LBL-based technique has been developed to create polymer capsules (hollow spherical polymer shells) from charged particle templates.[35,36] In this approach, the particles (e.g., positively charged, weakly cross-linked melamine formaldehyde particles) are coated with alternating layers of negatively and positively charged polyelectrolytes [e.g., sulfonated polystyrene (SPS) and poly(allylamine hydrochloride) (PAH)]. This is done by dispersing the particles in the respective polymer solution and subsequent centrifugation to settle and separate the coated particles. Each adsorption step is followed by repeated centrifugation and washing to remove non- or weakly adsorbed polymer. Subjecting the polyelectrolyte-coated particles to either strongly acidic conditions or oxidative degradation results in decomposition of the core; the resulting oligomeric particle fragments are expelled from the core and permeate through the polymer shell, leaving hollow polymer microcapsules behind (Figure 2.14a-d).

Hollow spheres produced in this technique possess novel features. As with other LBL techniques, the composition and wall thickness of the shell can be easily fine-tuned in the nanometer range by altering the number of layers adsorbed or the deposition conditions. Particle templates of a wide variety of compositions, shapes, and sizes (from 70 nm latex spheres to biocolloids larger than 10 μm) can be used, defining the resulting capsules' dimensions. Shells formed with as little as nine polymer layers (ca. 20 nm wall thickness) are remarkably chemically and physically stable against rupture or hole formation (see Figure 2.14e). Most importantly, the fabricated shells are permeable by small (ca. 1–2 nm in diameter) polar molecules, suggesting possible applications such as drug delivery and micro-reactors for chemical reactions in restricted volumes. For example, it was shown that controlling the charges of the interior and exterior of the capsules could be used to create charge-selective nucleation centers for crystal growth of compatible molecules and hence to direct it to occur exclusively inside or outside the capsules (Figure 2.14e-f). Crystal growth in the capsules' interior was restricted by the polymer walls and resulted in size-controlled crystallization (Figure 2.14h).[37]

2.4.2. *Three-Dimensional Nanocomposites*

As with the LBL method, oppositely charged groups immobilized on nanoparticle surfaces also induce creation of large three-dimensional aggregates.[38] Nevertheless, as pointed out in the previous section, using pure electrostatic interaction for assembly leads to kinetically trapped structures with reduced control over their shape. Additionally, pure electrostatic interactions do not distinguish between different functional groups of the same charge; this severely hinders the ability to use selectivity in the assembly process and thus results in limited diversity of assembly possibilities.

Hydrogen bonding, which is a subtler variant of the electrostatic interaction, provides a suitable alternative. Since a hydrogen bond takes place between *three* nuclei that are only

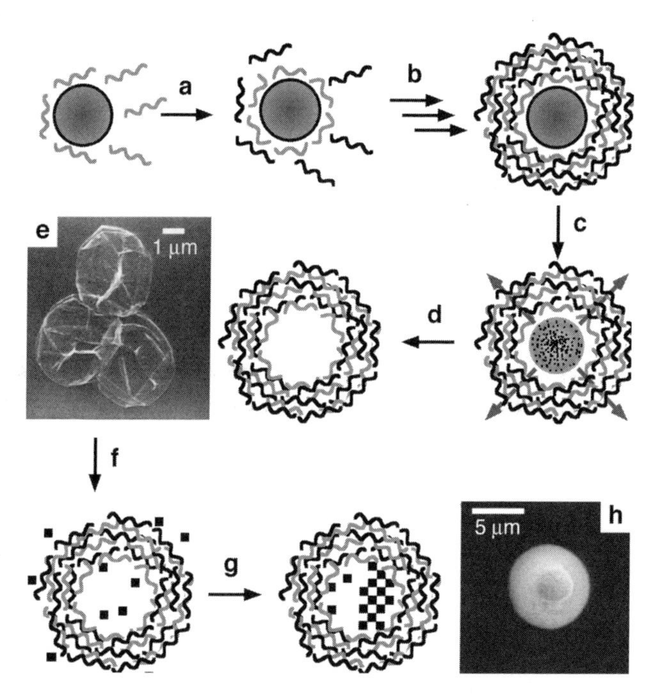

FIGURE 2.14. Polymer capsules: Consecutive adsorption of positively (grey) and negatively (black) charged polyelectrolytes onto negatively charged particles (a,b) and subsequent decomposition of the colloidal core (c,d) yields a suspension of hollow polyelectrolyte capsules (e, TEM image). Addition of crystalline molecules bearing charged groups (f) and changing the ionic or solvent composition drives the molecules into the capsules, where nucleation centers form and crystals are grown. (h) TEM image of a volume-confined crystal of 6-carboxyfluorescein grown inside a capsule that was based on discocyte biocolloid. (From Refs. 35 and 37 by permission of WILEY-VCH Verlag GmbH & Co.)

partially charged, hydrogen bonding allows assembly to take place at near-equilibrium conditions, facilitating defect corrections and the creation of ordered constructs. Moreover, as mentioned in Section 2.2, specific recognition between molecular functionalities can be achieved through the use of multiple hydrogen bonding moieties in each group and complementarity between the interacting groups. This option greatly elaborates the collection of tools available for the creation of three-dimensional nanocomposites.

2.4.2.a. DNA-Induced Nanoparticle Assembly Mirkin and coworkers and Alivisatos and were among the first to employ the hydrogen-bonding strategy for self-assembly of inorganic nanoparticles.[39,40] These researchers took advantage of the natural tendency of complementary DNA sequences to hybridize spontaneously into double stranded helices. In one approach, two batches of gold nanoparticles were each modified with different, non-complementary oligonucleotides (single stranded DNA sequences), which were functionalized with thiol groups at their 3′ termini to enable displacement into the nanoparticles' protective monolayers. In the next step, a linker was added to the mixture of the two types of modified nanoparticles to form the assembly (Figure 2.15a). This linker is a DNA fragment (single or double stranded) with floppy ends that are complementary to the

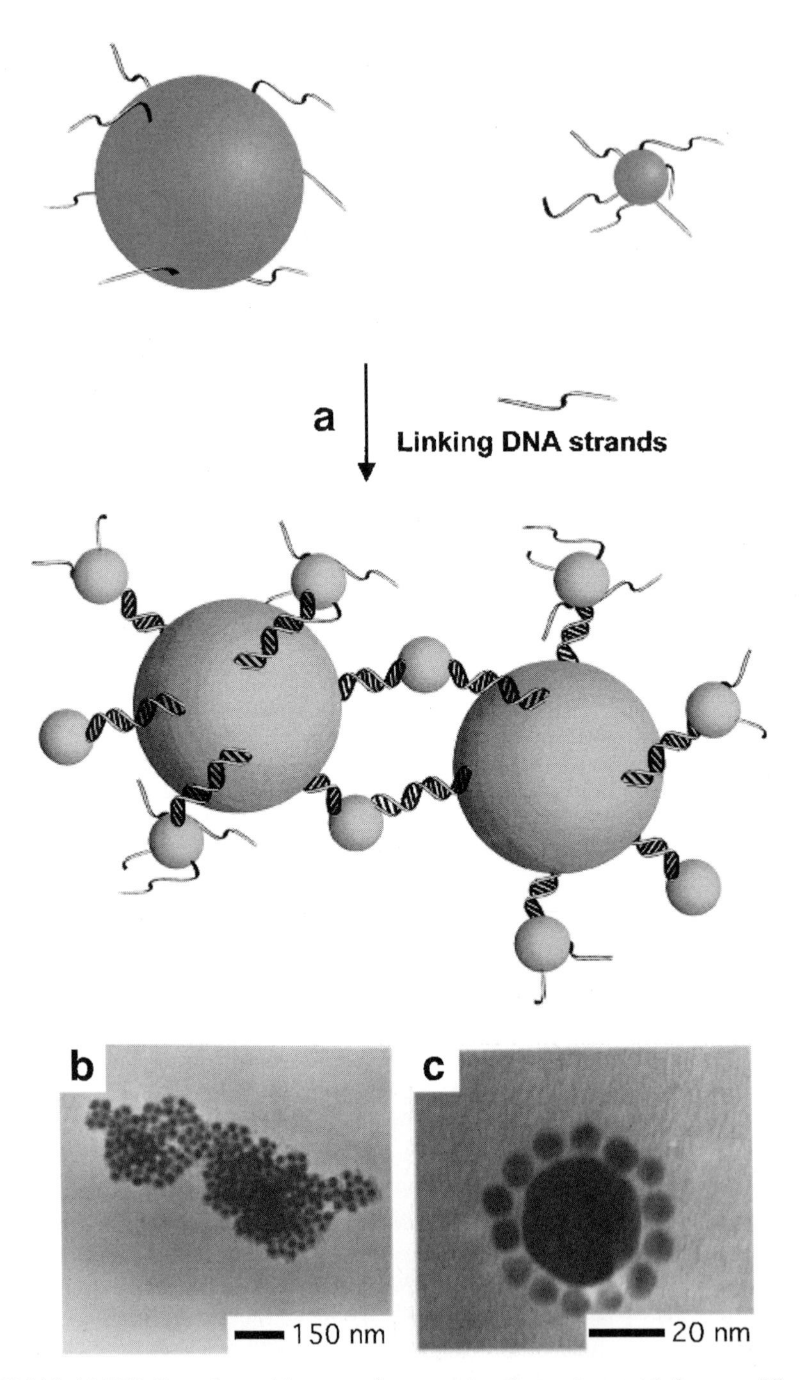

FIGURE 2.15. (a) DNA-directed assembly strategy for preparation of network materials from two different-sized oligonucleotide-functionalized nanoparticles. TEM images of: (b) Aggregate formed by linking 8-nm and 31-nm gold nanoparticles, (c) "Satellite" structure obtained in 120-fold excess of the 8-nm colloids. (From *Inorg. Chem.* **39**(11), 2258–2272 (2000) by permission of American Chemical Society and Ref. 41 by permission of American Chemical Society.)

oligonucleotides immobilized on the nanoparticles (each end to a different oligonucleotide). The required recognition between the linker ends and the corresponding oligonucleotides assembled the nanoparticles in an AB-AB periodic structure, which was easily detected when two different sizes of nanoparticles were used (Figure 2.15b,c).[41] Especially interesting is the type of "satellite" structures shown in Figure 2.15c, which under certain conditions (overwhelming excess of small nanoparticles) is the only aggregated structural mode observed. Since the small nanoparticles at the rim have unhybridized oligonucleotide chains extending to the outward of the satellite structure, this structure can be further implemented as a new type of building block in the fabrication of multi-component materials.

The DNA-induced aggregation mechanism can be thermally reversed (since at high temperatures the DNA duplex unravels) and the clusters could be broken and rebuilt many times by changing the temperature with no apparent degradation. Importantly, it was shown that having nanoparticles tethered to the hybridized DNA substantially narrows its "melting" profile. Combining these features with the strong visible light absorption of gold nanoparticles and its wavelength dependence on aggregation mode, it was shown that these clusters could serve as detection probes for target DNA sequences (e.g., known sequences of viruses or those attributed to genetic diseases). This approach proved highly sensitive even to single-point mutations (base-pair mismatches), far exceeding the performance of conventional DNA detection methods based on fluorescence probes.[42]

2.4.2.b. Spherical Polymer-Nanoparticle Composites As shown above, nanoparticles bearing functional groups can be assembled into structures of macroscopic dimensions. Nevertheless, the assembly itself is only one facet of creating novel materials, and the ability to control all other parameters related to the aggregation process is highly desired. One such goal is the control over the final size and shape of the aggregates, which might have implication on later processing of the material. Although under certain conditions the DNA assembly method yields small clusters of controlled shape (the satellite structures shown above), in general the larger aggregates formed with this approach are of variable shapes and sizes, and when supplied with enough material they grow indefinitely.

Aggregate shape and size regulation was demonstrated by Rotello and coworkers. These researchers employed a polymer-mediated approach for the assembly of nanoparticles using hydrogen bonding recognition units. In their approach, flexible polystyrene-based polymers that were randomly functionalized with recognition units were used to assemble gold nanoparticles bearing complementary recognition units (Figure 2.16a). Interestingly, it was found that certain types of recognition units induce the formation of large spherical nanoparticle clusters of moderate size distribution (Figure 2.16b).[43] The size and morphology of these aggregates were thermally controlled, as at –20°C, for example, spherical clusters were formed that were an order of magnitude larger than those created at room temperature (Figure 2.16c). A better control over aggregate size distribution was further achieved using diblock copolymers, where the growth of the aggregates was restricted by the length of the functionalized block and the presence of an "inert" polystyrene block.[44]

2.5. SELF-ORGANIZATION IN BLOCK COPOLYMERS

Polymers play an important role as materials suited for nanotechnological applications. This originates from the fact that these macromolecules span the size range from single to

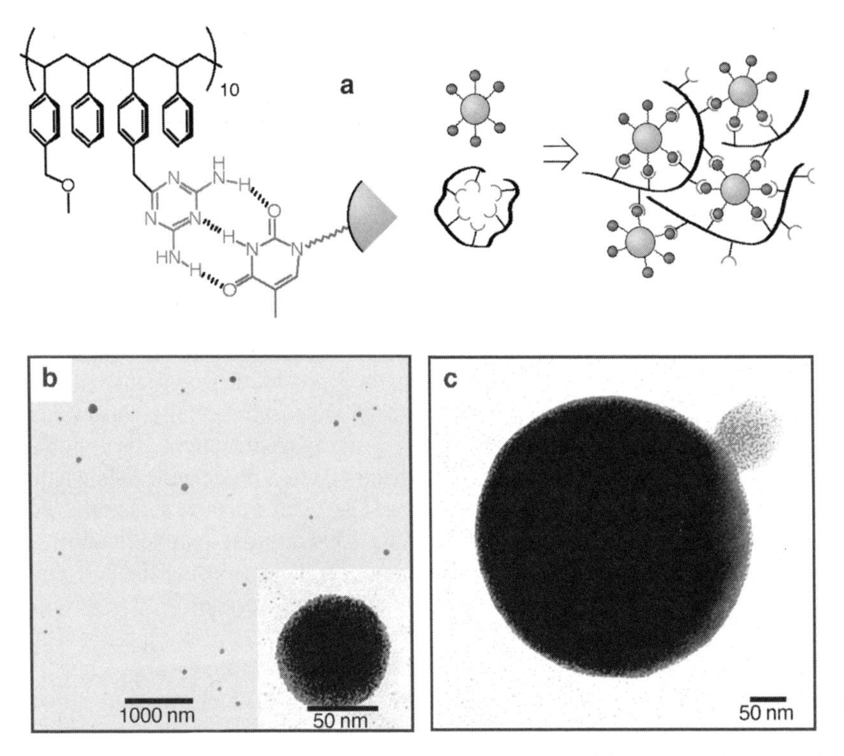

FIGURE 2.16. (a) Multiple recognition events are translated into nanoparticle aggregate formation. (b) TEM images of spherical aggregates formed at ambient temperature and (c) at −20°C, showing thermal control over aggregate size. (From Ref. 43 by permission of Macmillan Magazines Ltd.)

hundreds of nanometers, while their properties can be tailored at the molecular level (in terms of the monomer synthesis). Most importantly, the variety of polymer architectures gives rise to different types of behavior. When considering technological implications on the nanometer scale, the unique physics exhibited by the *block copolymer* architecture requires special attention.

Copolymers (in contrast to *homopolymers*) are polymers composed of at least two different types of monomers. *Block copolymers* consist of two (or more) segments of chemically distinct polymer chains, which are interconnected by a covalent bond (Figure 2.17), the simplest example being two incompatible homopolymers (denoted as polyA-*b*-polyB or PA-*b*-PB). Block copolymers are readily synthesized using living polymerization methods, in which the polymer end remains reactive after the polymerization of the first monomer, so that polymerization can be continued with the second monomer to create the second block (Figure 2.17).

Block copolymers are employed in a wide variety of applications, utilizing the inherent combination of properties arising from the essence of the different polymer blocks. In the context of nanotechnology, however, an important attribute of block copolymers is the physics of microphase separation, which offers a totally different approach to ordering on the nanometer scale than those outlined in the previous sections of this chapter.

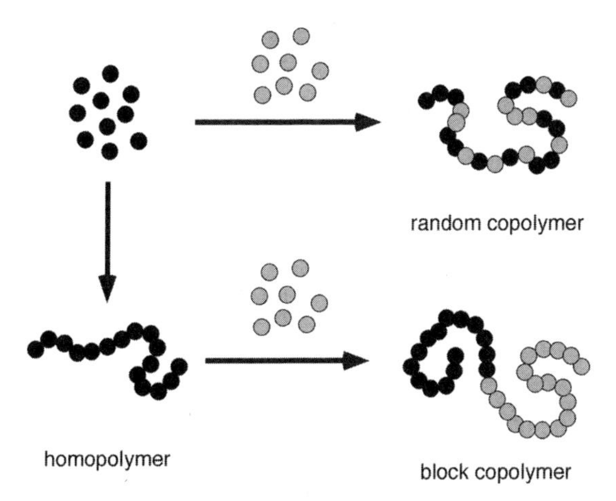

FIGURE 2.17. Schematic illustration of the synthetic routes to homopolymers, random copolymers, and block copolymers.

2.5.1. The Physics of Microphase Separation

A blend of incompatible polymers would phase separate on a macroscopic scale into different domains (phases), each one rich with one polymer and depleted with the others.[45,46] In block copolymers, however, the constituting incompatible blocks cannot phase separate on a macroscopic level since they are interconnected by a covalent bond. As a result, block copolymers form periodic arrays of domains that are restricted in size by the length of the corresponding blocks in the polymer, typically in the range of 10–100 nm (Figure 2.18a).[47,48] The degree of order associated with the phase separation phenomenon can therefore be harnessed to create structural periodicity on the nanometer scale.

The volume fraction of each component in the copolymer is of major importance: whereas volume-symmetric block copolymers (in which the volume fraction of each component is similar, e.g., is equal to 0.5 in diblock copolymers) form lamellar structure, other morphologies are obtained by volume-asymmetric block copolymers. Driven by the thermodynamic tendency to lower the surface area at the interfaces between the domains, curved interfaces are preferred in highly volume-asymmetric block copolymers. This gives rise to cylindrical, spherical, and other more complex domain morphologies of the less abundant component surrounded by a matrix of the other component (Figure 2.18b). Characterization of the morphology and ordering behavior of block copolymers is performed mainly using transmission electron microscopy (TEM), small-angle neutron/X-ray scattering methods (SANS/SAXS), optical microscopy, and atomic force microscopy (AFM).

Although studies in the bulk contributed significantly to the understanding of the phenomenon of microphase separation in block copolymers,[47,48] bulk systems are of limited value for practical nanotechnological applications since they lack long range ordering of the domains with respect to each other. In the last few years, efforts have been directed toward understanding microphase separation in thin films,[49] where the presence of an impenetrable substrate (a boundary condition) is used to induce ordering of the polymer domains. Homogeneous thin films of block copolymers with controlled film thickness are obtained by spin casting the polymer from a non-preferential solvent (i.e., a solvent with similar solvation

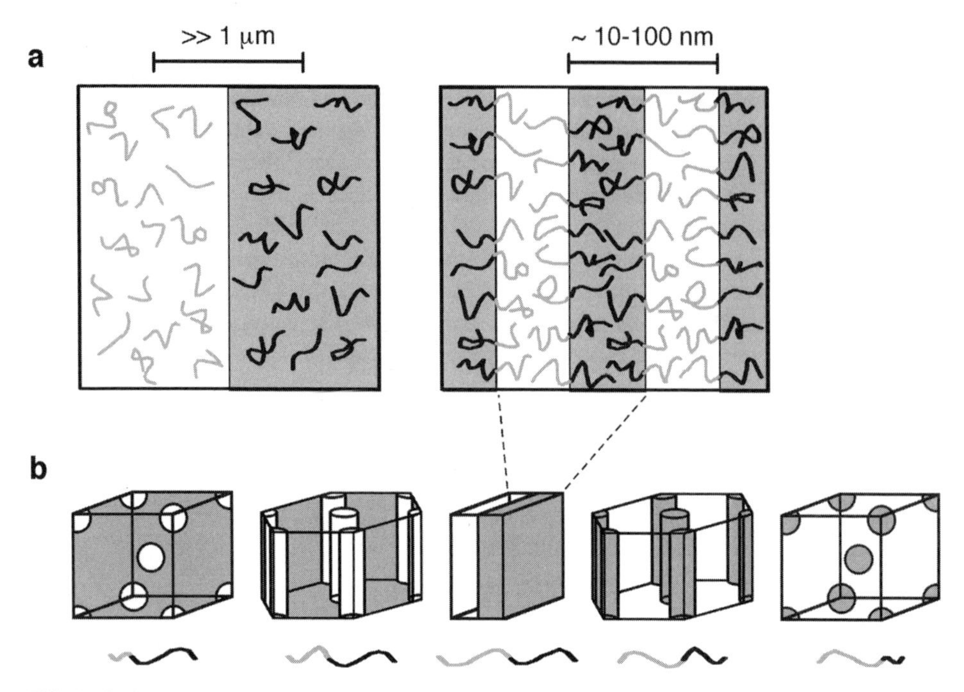

FIGURE 2.18. (a) Fundamental differences in phase behavior of different polymers: A blend of two incompatible homopolymer separates into distinct phases on a large scale (left), whereas block copolymers microphase separate into periodic domains on the scale of a single polymer strand (right). (b) Basic morphologies obtained by different block copolymer compositions.

properties for all blocks) on flat substrates. Solution concentration and spin rate enable control of film thickness within one nanometer. Subsequent annealing leads to self-organization of the block copolymers in the film into different domains.

For technological applications, full control of the ordering of the block copolymer domains with respect to the substrate is of crucial importance. Apart from block-block incompatibility, two more factors that strongly influence the ordering behavior enter into the thermodynamic considerations: interactions between the blocks and the substrate (interfacial energy) and between the blocks and the air (surface tension). Strongly interacting substrates will induce orientation of the features (domains) in layers parallel to the substrate. Where vertical alignment of domains is desired, substrate passivation (i.e., neutrality to either block) is often the key. One way to achieve that is to anchor to the substrate a polymer brush that is neutral to the block copolymer used (usually a random copolymer that consists of a specific ratio of the same monomers used in the block copolymer).[50] The simplest way to induce perpendicular orientation of the domains then is to cast a film of thickness that is slightly less than the polymer bulk periodicity. This situation results in frustration due to incommensurability between the film thickness and the block copolymer bulk periodicity, which can only be relieved by perpendicular domain orientation that allows the accommodation of the polymer period in the lateral dimensions.

Other methods exist for the creation of perpendicularly orientated domains, such as application of electric fields,[51] substrate patterning,[52] and film confinement between two walls.[53] The control over both morphology and ordering of microphase segregated block copolymer domains has direct technological implications. A few representative examples of existing applications, which hint for the potential of the field in nanotechnology, are outlined in the following paragraphs.

2.5.2. *Applications Based on Microphase Separation of Block Copolymers*

2.5.2.a. Directed Metal Assembly Microphase separation of diblock copolymers in a thin film creates a pattern of nanometric features. If the features are exposed half cylinders lying parallel to the surface or vertically arranged lamellae, the pattern formed can be employed as a template for the preparation of conductive metal wires. Recently, the successful production of continuous metal wires and chains of separate metal nanoparticles of different types of metals has been achieved by selective segregation of evaporated metal atoms to specific domains.[54] The polymer used was volume-asymmetric polystyrene-*b*-poly(methylmethacrylate) (PS-*b*-PMMA), which forms cylindrical domains of PMMA of 50 nm periodicity surrounded by PS matrix. The general procedure involves a few steps. First, a solution of PS-*b*-PMMA is spin cast onto a Si_3N_4 substrate to create a thin film of thickness corresponding to one repeat-distance (50 nm). Subsequent annealing results in laterally alternating domains. Second, a small amount of metal (of nominal thickness about 0.5 nm) is thermally evaporated on top of the polymer template. At this stage, some selectivity is already observed for most metals (Figure 2.19a): gold and silver preferentially wet the PS domain whereas indium, lead, tin, and bismuth segregate into the PMMA domain. Nearly 100% selectivity (Figure 2.19b) is obtained by a second, short-term annealing above the glass transition of the polymer that allows the composite system to reorganize.

High metal loading is required in order to obtain continuous conductive paths. Metals, however, exhibit high surface energies, which eventually lead to aggregation of the metal while ignoring the polymer template. Hence the selective deposition process should be performed under conditions that are far from equilibrium, i.e., small metal amounts and very

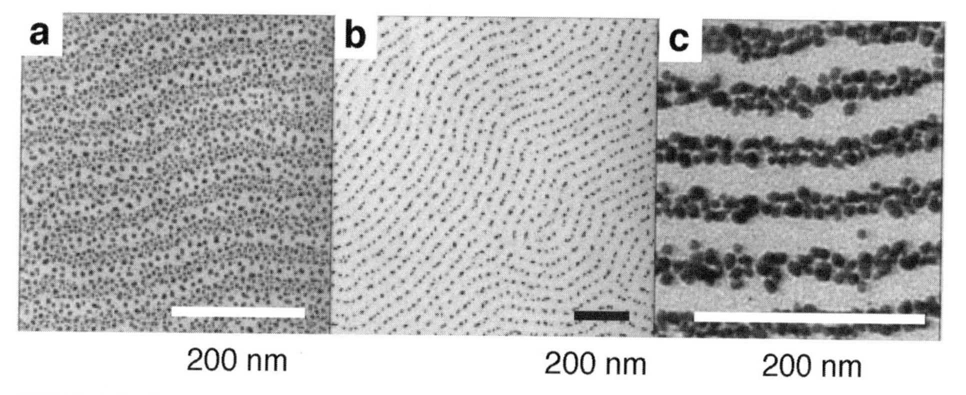

FIGURE 2.19. Metal nanowire formation using a block copolymer template: Shown are TEM images of: (a) Gold metal vapor-deposited onto a preformed PS-*b*-PMMA template. (b) After annealing at 180°C for 1 min., gold nanoparticles segregate selectively to the PS domains and form chains. (c) Repeated deposition and short-time annealing increases the metal loading, forming continuous conductive nanowires. (From Ref. 54 by permission of Macmillan Magazines Ltd.)

short annealing times (minutes). Increasing the amount of metal loading in the preferred domain is therefore achieved by repeating the deposition and annealing treatment, and results in densely packed nanoparticle chains and eventually in nanoparticle fusion into continuous metallic wires (Figure 2.19c). The electrical properties of the patterned surface can thus be varied from insulating (no metal) and semi-conductive (spaced nanoparticles, in which coulomb blockades allow electron tunneling only at high applied voltages) to fully conductive paths [nano-wires, which exhibit ohmic response (linear I–V curves)].

2.5.2.b. Porous Membranes Apart from inducing microphase separation, the chemical distinction between the blocks can be used to create porous arrays. Particularly advantageous are films made of volume-asymmetric block copolymers, in which the minority component can be etched out leaving behind a porous matrix. This principle has been used to create porous and relief ceramic nanostructures.[55] The researchers used triblock copolymers of the type A_1BA_2, in which A is polyisoprene (PI) and B is poly(pentamethyldisilylstyrene) [P(PMDSS)]. A single-step, bifunctional oxidation (using ozone and ultraviolet light) selectively removes the hydrocarbon blocks and simultaneously converts the silicon containing block to silicon oxycarbide ceramic. By careful selection of the volume fractions, either three-dimensional (3D) connected pores (of a 24/100/26 kg mole^{-1} composition, forming a double gyroid PI domains that are converted into pores, Figure 2.20a,b) or a 3D connected strut network [of a 44/168/112 kg mole^{-1} composition, forming an inverse double gyroid morphology of P(PMDSS), Figure 2.20c] were obtained. It should be noted that such complex 3D structures are not obtainable using conventional lithographic techniques.

Calculated interfacial area of ca. 40 m^2 per gram and pore/strut sizes of ca. 20 nm evidence a highly exposed and densely packed array of features. Feature size and spacing may be tailored through control of molecular weight or through blending of the block copolymer with corresponding homopolymers of comparable molecular weight (that swell the corresponding domain). An added advantage comes from the heat and solvent stability of

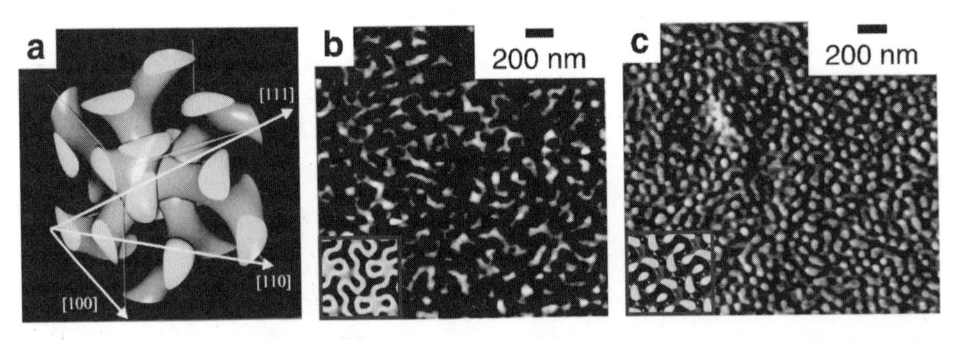

FIGURE 2.20. (a) Schematic of the unit cell of the double gyroid morphology (space group *Ia3d*) found in certain triblock copolymers of specific block compositions, where the minority component forms three-dimensionally continuous networks of cylinder-like connectors. Selective removal of the PI component of PI-*b*-P(PMDSS)-*b*-PI yields: (b) A porous ceramic matrix (for a 24/100/26 kg mole^{-1} polymer composition, where the PI forms networks), or (c) Silicon oxicarbide strut networks (for a 44/168/112 kg mole^{-1} composition, where the PI forms the matrix). AFM height images (maximum height 10 nm, high regions appear bright); insets: computer simulations of corresponding volume-rendered surfaces in the double gyroid unit cell. (From Ref. 55 by permission of American Association for the Advancement of Science.)

the resulting ceramics. Immediate possible applications are high-temperature nano-porous membranes for fine particle separation, featuring reduced likelihood of being clogged by the filtrate (due to the redundancy of interconnected pathways characteristic by the double gyroid morphology). The nano-relief structure may serve as a support for the production of novel high-temperature catalysts, since efficient catalysis involves high exposure of the catalytic material (regularly obtained by high surface area of the support).

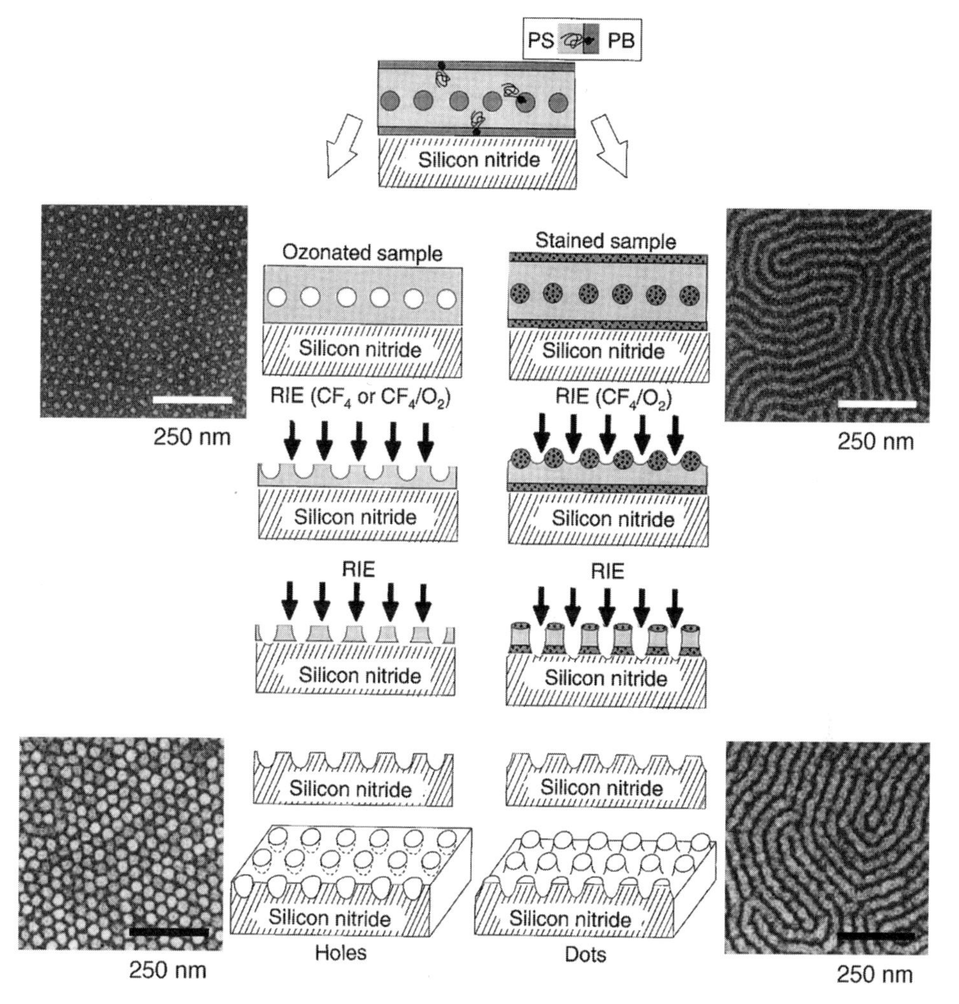

FIGURE 2.21. Lithographic mask fabrication from PS-*b*-PB templates with spherical PB domains. Spherical voids are created by ozonolysis (which degrades the PB into water dispersible fragments), while OsO_4 staining increases the PB domain resistance toward etching, and thus negative or positive etching masks are produced. Shown are corresponding TEM images: top: Negative/positive masks that were used to create holes/lines (respectively), where light spots in the left image are voids and dark lines in the right image are stained PB domains; bottom: Transferred patterns on the silicon nitride substrates, where light regions in the left image are ~15-nm-deep etched-out holes in the substrate, and dark regions in the right image are ~15-nm-thick ridges in the substrate, which were protected from etching. (From Ref. 56 by permission of American Association for the Advancement of Science.)

2.5.2.c. Nanolithographic Masks Feature size and spacing of asymmetric block copolymer domains hints to their potential to serve as novel lithographic masks with improved performance both in resolution (compared to photolithography) and in time of production (compared to the serial electron-beam lithography). Block copolymer thin films that accommodate only a single layer of features (cylinders or spheres) can serve as lithographic masks by using reactive ion etching (RIE) as the feature transfer mechanism.[56] Through manipulation of the resistance of the different domains to RIE (using ozonolysis or staining), the domain features could either be positively or negatively transferred to the underlying substrate (Figure 2.21). The array formed displays uniform holes 20 nm across and 40 nm apart with exceptional density of ca. 7×10^{10} holes/cm^2.

The aforementioned nanolithography is a general and useful technique. For example, it was shown that by combining it with other microfabrication techniques, an equally dense array of semiconducting GaAs nanocrystals could be obtained.[57] For comparison, if each

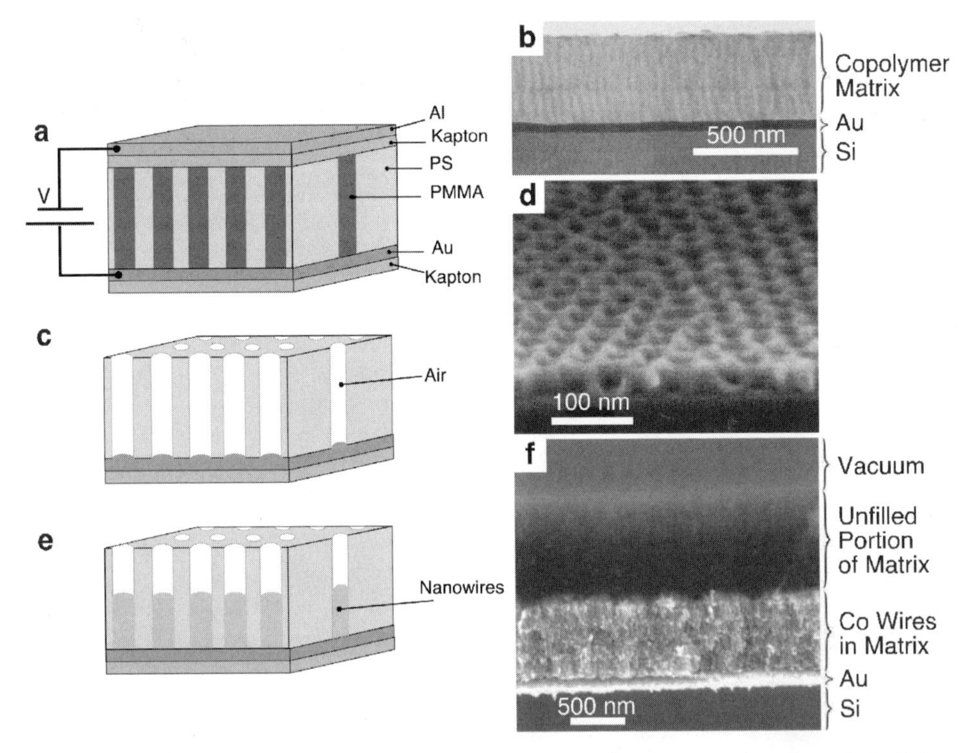

FIGURE 2.22. High-density nanowire fabrication in a polymer matrix. (a) Array of hexagonally-packed cylinders aligned normal to the substrate is obtained by annealing an asymmetric diblock copolymer above its glass transition under an applied electric field. (b) Cross-sectional TEM image of an 800-nm thick film of PS-*b*-PMMA on gold after annealing in an electric field of 25 V/μm reveals that the PMMA cylinders penetrate all the way through the film. (c) The minor component is degraded and removed, leaving a porous film. (d) Field emission scanning electron microgram (FE-SEM) of PS-*b*-PMMA thin film after removal of the PMMA cylindrical domains (cross-sectional view). (e) Metal nanowires are grown inside the pores by electrodeposition. (f) SEM image of a fracture surface showing cobalt nanowires partially filling the pores in the block copolymer template. (From Ref. 58 by permission of WILEY-VCH Verlag GmbH & Co and Ref. 59 by permission of American Association for the Advancement of Science.)

GaAs crystal functioned as a memory bit, a regular-sized compact disc manufactured by this technique would contain 750 gigabytes of memory, more than a 1000-fold improvement over current technology.

2.5.2.d. Long-Range Ordered Arrays of Magnetic Elements Addressability is a key issue in memory applications. If block copolymer templating is to be finally employed in production of ultra-high density data storage, long-range ordering of the domains much better than that offered by thermodynamics is essential, since local ordering defects can result in accidental switching of bits. Additionally, in many such applications a high aspect ratio of the features is required (e.g., in magnetic data storage). Creation of substrate-perpendicular cylinders in thick polymer films provides an answer.

Application of an electric field during the annealing process of asymmetric PS-*b*-PMMA produces aligned PMMA cylinders that penetrate all the way through the film to the substrate (Figure 2.22a,b).[51,58] After creation of the array, exposure to UV irradiation simultaneously degrades the PMMA and causes predominantly cross-linking of the PS matrix, rendering it insoluble. Subsequent rinsing with acetic acid leaves a porous film with deep cylindrical holes (Figure 2.22c,d) that are densely packed and hexagonally ordered.

The fact that the cylindrical holes expose the substrate is advantageous in many ways. For example, a 1 μm-thick PS template on top of a gold substrate was created, which allowed filling the large aspect-ratio holes with direct current electrodeposition of different metals (Figure 2.22e,f).[59] Especially interesting was the creation of 500-nm-long cobalt magnetic nanowire array, which has promise as a magnetic data storage medium.

2.6. SUMMARY

- Self-assembly is a "bottom-up" methodology.
- Self-assembly is achieved through non-covalent interactions.
- Self-assembled monolayers (SAMs) are formed spontaneously between molecules with surface-active head groups and reactive substrates.
- Layer-by-layer (LBL) deposition technique involves oppositely charged polymers that are deposited in an alternant fashion onto a substrate bearing charged groups.
- Hydrogen bonding and molecular recognition allow the creation of ordered nanoparticle assemblies.
- The physics of microphase separation in block copolymer thin films allows their self-organization into distinct domains of controlled morphology, periodicity, and alignment.

QUESTIONS

1. Name the main interaction types used in molecular self-assembly. Which interactions induce the self-assembly in SAMs, LBL, block copolymers, and nanoparticle aggregates?
2. What is the advantage of using non-covalent interactions in the construction of nanometer-sized constructs?

3. What are the advantages of self-assembly over "top-down" methodologies? What are the difficulties?
4. What dictates the phase morphology of a given block copolymer?
5. How can the ordering of block copolymer domains be directed and enhanced?
6. Calculate the density of the memory bits of a 650 MB compact disc (1 byte = 8 bits). Given that there are no voids between the bits, what is the area occupied by a single bit? If each bit occupies a circle—what is its diameter? Compare to the example given in this chapter.

REFERENCES

1. M. Lundstron, Moore's law forever? *Science* **299**(5604), 210–211 (2003).
2. G. M. Whitesides, J. P. Mathias, and C. T. Seto, Molecular self-assembly and nanochemistry: A chemical strategy for the synthesis of nanostructures, *Science* **254**(5036), 1312–1319 (1991).
3. J.-M. Lehn, *Supramolecular chemistry* (VCH, Weinheim, 1995).
4. G. M Whitesides and B. Grzybowski, Self-assembly at all scales, *Science* **295**(5564), 2418–2421 (2002).
5. G. U. Kulkarni, P. J. Thomas, and C. N. R. Rao, in: *Supramolecular organization and materials design*, edited by W. Jones and C. N. R. Rao (University Press, Cambridge, 2002), pp. 265–294.
6. J.-M. Lehn, in: *Supramolecular polymer chemistry—scope and perspectives*, edited by A. Ciferri (Marcel Deker, New York, 2000), pp. 615–641.
7. A. Ulman, *An introduction to ultrathin organic films: From Langmuir-Blodgett to self-assembly* (Academic Press, San Diego, 1991).
8. L. Isaacs, D. N. Chin, N. Bowden, Y. Xia, and G. M. Whitesides, in: *Supramolecular materials and technologies*, edited by D. N. Reinhoudt (John Wiley & Sons, New York, 1999), pp. 14–24.
9. A. Ulman, Formation and structure of self-assembled monolayers, *Chem. Rev.* **96**(4), 1533–1554 (1996).
10. M. A. Reed, C. Zhou, C. J. Muller, T. P. Burgin, and J. M. Tour, Conductance of a molecular junction, *Science* **278**(5336), 252–254 (1997).
11. J. Park, A. N. Pasupathy, J. I. Goldsmith, C. Chang, Y. Yaish, J. R. Petta, M. Rinkoski, J. P. Sethna, H. D. Abruña, P. L. McEuen, and D. C. Ralph, Coulomb blockade and the Kondo effect in single-atom transistors, *Nature* **417**(6890), 722–725 (2002).
12. Y. Xia and G. M. Whitesides, Soft Lithography, *Angew. Chem. Int. Ed.* **37**(5), 551–575 (1998).
13. J. Chen, M. A. Reed, C. L. Asplund, A. M. Cassell, M. L. Myrick, A. M. Rawlett, J. M. Tour, and P. G. Van Patten, Placement of conjugated oligomers in an alkanethiol matrix by scanned probe microscope lithography, *Appl. Phys. Lett.* **75**(5), 624–626 (1999).
14. G. M. Credo, A. K. Boal, K. Das, T. H. Galow, V. M. Rotello, D. L. Feldheim, and C. B. Gorman, Supramolecular assembly on surfaces: Manipulating conductance in noncovalently modified mesoscale structures, *J. Am. Chem. Soc.* **124**(31), 9036–9037 (2002).
15. A. Kumar and G. M. Whitesides, Features on gold having micrometer to centimeter dimensions can be formed through a combination of stamping with an elastomeric stamp and an alkanethiol "ink" followed by chemical etching, *Appl. Phys. Lett.* **63**(14), 2002–2004 (1993).
16. R. J. Jackman, S. T. Brittain, A. Adams, M. G. Prentiss, and G. M. Whitesides, Design and fabrication of topologically complex, three-dimensional microstructures, *Science* **280**(5372), 2089–2091 (1998).
17. R. D. Piner, J. Zhu, F. Xu, S. Hong, and C. A. Mirkin, "Dip-Pen" nanolithography, *Science* **283**(5402), 661–663 (1999).
18. Y.-L. Loo, R. L. Willett, K. W. Baldwin, and J. A. Rogers, Interfacial chemistries for nanoscale transfer printing, *J. Am. Chem. Soc.* **124**(26), 7654–7655 (2002).
19. M. Brust, M. Walker, D. Bethell, D. J. Schiffrin, and R. Whyman, Synthesis of thiol-derivatized gold nanoparticles in a 2-phase liquid-liquid system, *J. Chem. Soc., Chem. Commun.* (7), 801–802 (1994).
20. A. C. Templeton, W. P. Wuelfing, and R. W. Murray, Monolayer protected cluster molecules, *Accounts Chem. Res.* **33**(1), 27–36 (2000).

21. R. Shenhar and V. M. Rotello, Nanoparticles: Scaffolds and building blocks, *Accounts Chem. Res.* **36**(7), 549–561 (2003).
22. A. K. Boal and V. M. Rotello, Fabrication and self-optimization of multivalent receptors on nanoparticle scaffolds, *J. Am. Chem. Soc.* **122**(4), 734–735 (2000).
23. K. P. Velikov, C. G. Christova, R. P. A. Dullens, and A. van Blaaderen, Layer-by-layer growth of binary colloidal crystals, *Science* **296**(5565), 106–109 (2002).
24. C. J. Kiely, J. Fink, M. Brust, D. Bethell, and D. J. Schiffrin, Spontaneous ordering of bimodal ensembles of nanoscopic gold clusters, *Nature* **396**(6710), 444–446 (1998).
25. G. Decher, Fuzzy nanoassemblies: Toward layered polymeric multicomponents, *Science* **277**(5330), 1232–1237 (1997).
26. P. Bertrand, A. Jonas, A. Laschewsky, and R. Legras, Ultrathin polymer coatings by complexation of polyelectrolytes at interfaces: Suitable materials, structure and properties, *Macromol. Rapid Commun.* **21**(7), 319–348 (2000).
27. J. H. Cheung, A. F. Fou, and M. F. Rubner, Molecular self-assmebly of conducting polymers, *Thin Solid Films* **244**(1–2), 985–989 (1994).
28. M. Ferreira, J. H. Cheung, and M. F. Rubner, Molecular self-assembly of conjugated polyions—a new process for fabricating multilayer thin-film heterostructures, *Thin Solid Films* **244**(1–2), 806–809 (1994).
29. H. Hong, D. Davidov, Y. Avny, H. Chayet, E. Z. Faraggi, and R. Neumann, Electroluminescence, photoluminescence, and x-ray reflectivity studies of self-assembled ultra-thin films, *Adv. Mater.* **7**(10), 846–849 (1995).
30. J. Tian, C.-C. Wu, M. E. Thompson, J. C. Sturm, R. A. Register, M. J. Marsella, and T. M. Swager, Electroluminescent properties of self-assembled polymer thin-films, *Adv. Mater.* **7**(4), 395–398 (1995).
31. A. C. Fou, O. Onitsuka, M. Ferreira, M. F. Rubner, and B. R. Hsieh, Fabrication and properties of light-emitting diodes based on self-assmbled multilayers of poly(phenylene vinylene), *J. Appl. Phys.* **79**(10), 7501–7509 (1996).
32. O. Onitsuka, A. C. Fou, M. Ferreira, B. R. Hsieh, and M. F. Rubner, Enhancement of light emitting diodes based on self-assmebled heterostructures of poly(*p*-phenylene vinylene), *J. Appl. Phys.* **80**(7), 4067–4071 (1996).
33. A. N. Shipway, M. Lahav, and I. Willner, Nanostructured gold colloid electrodes, *Adv. Mater.* **12**(13), 993–998 (2000).
34. A. N. Shipway, E. Katz, and I. Willner, Nanoparticle arrays on surfaces for electronic, optical, and sensor applications, *ChemPhysChem* **1**(1), 18–52 (2000).
35. E. Donath, G. B. Sukhorukov, F. Caruso, S. A. Davis, and H. Möhwald, Novel hollow polymer shells by colloid-templated assembly of polyelectrolytes, *Angew. Chem. Int. Ed.* **37**(16), 2202–2205 (1998).
36. F. Caruso, Nanoengineering of particle surfaces, *Adv. Mater.* **13**(1), 11–22 (2001).
37. G. Sukhorukov, L. Dähne, J. Hartmann, E. Donath, and H. Möhwald, Controlled precipitation of dyes into hollow polyelectrolyte capsules based on colloids and biocolloids, *Adv. Mater.* **12**(2), 112–115 (2000).
38. T. H. Galow, A. K. Boal, and V. M. Rotello, A "building block" approach to mixed-colloid systems through electrostatic self-organization, *Adv. Mater.* **12**(8), 576–579 (2000).
39. C. A. Mirkin, R. L. Letsinger, R. C. Mucic, and J. J Storhoff, A DNA-based method for rationally assembling nanoparticles into macroscopic materials, *Nature* **382**(6592), 607–609 (1996).
40. A. P. Alivisatos, K. P. Johnsson, X. G. Peng, T. E. Wilson, C. J. Loweth, M. P. Bruchez, and P. G. Scultz, Organization of "nanocrystal molecules" using DNA, *Nature* **382**(6592), 609–611 (1996).
41. R. C. Mucic, J. J. Storhoff, C. A. Mirkin, and R. L. Letsinger, DNA-directed synthesis of binary nanoparticle network materials, *J. Am. Chem. Soc.* **120**(48), 12674–12675 (1998).
42. T. A. Taton, C. A. Mirkin, and R. L. Letsinger, Scanometric DNA array detection with nanoparticle probes, *Science* **289**(5485), 1757–1760 (2000).
43. A. K. Boal, F. Ilhan, J. E. DeRouchey, T. Thurn-Albrecht, T. P. Russell, and V. M. Rotello, Self-assembly of nanoparticles into structured spherical and network aggregates, *Nature* **404**(6779), 746–748 (2000).
44. B. L. Frankamp, O. Uzun, F. Ilhan, A. K. Boal, and V. M. Rotello, Recognition-mediated assembly of nanoparticles into micellar structures with diblock copolymers, *J. Am. Chem. Soc.* **124**(6), 892–893 (2002).
45. M. Doi, *Introduction to Polymer Physics* (Clarendon Press, Oxford, 1996), p. 38.
46. F. S. Bates, Polymer-polymer phase behavior, *Science* **251**(4996), 898–905 (1991).

47. F. S. Bates and G. H. Fredrickson, Block copolymer thermodynamics: Theory and experiment, *Annu. Rev. Phys. Chem.* **41,** 525–557 (1990).
48. I. W. Hamley, *The Physics of Block Copolymers* (Oxford University Press, Oxford, 1998).
49. M. J. Fasolka and A. M. Mayes, Block copolymer thin films: Physics and applications, *Annu. Rev. Mater. Res.* **31,** 323–355 (2001).
50. P. Mansky, Y. Liu, E. Huang, T. P. Russell, and C. Hawker, Controlling polymer-surface interactions with random copolymer brushes, *Science* **275**(5305), 1458–1460 (1997).
51. T. L. Morkved, M. Lu, A. M. Urbas, E. E. Enrichs, H. M. Jaeger, P. Mansky, and T. P. Russell, Local control of microdomain orientation in diblock copolymer thin films with electric fields, *Science* **273**(5277), 931–933 (1996).
52. L. Rockford, Y. Liu, P. Mansky, T. P. Russell, M. Yoon, and S. G. J. Mochrie, Polymers on nanoperiodic, heterogeneous surfaces, *Phys. Rev. Lett.* **82**(12), 2602–2605 (1999).
53. G. J. Kellogg, D. G. Walton, A. M. Mayes, P. Lambooy, T. P. Russell, P. D. Gallagher, and S. K. Satija, Observed surface energy effects in confined diblock copolymers, *Phys. Rev. Lett.* **76**(14), 2503–2506 (1996).
54. W. A. Lopes and H. M. Jaeger, Hierarchical self-assembly of metal nanostructures on diblock copolymer scaffolds, *Nature* **414**(6865), 735–738 (2001).
55. V. Z.-H. Chan, J. Hoffman, V. Y. Lee, H. Iatrou, A. Avgeropoulos, N. Hadjichristidis, R. D. Miller, and E. L. Thomas, Ordered bicontinuous nanoporous and nanorelief ceramic films from self assembling polymer precursors, *Science* **286**(5445), 1716–1719 (1999).
56. M. Park, C. Harrison, P. M. Chaikin, R. A. Register, and D. H. Adamson, Block copolymer lithography: Periodic arrays of $\sim 10^{11}$ holes in 1 square centimeter, *Science* **276**(5317), 1401–1404 (1997).
57. R. R. Li, P. D. Dapkus, M. E. Thompson, W. G. Jeong, C. Harrison, P. M. Chaikin, R. A. Register, and D. H. Adamson, Dense arrays of ordered GaAs nanostructures by selective area growth on substrates patterned by block copolymer lithography, *Appl. Phys. Lett.* **76**(13), 1689–1691 (2000).
58. T. Thurn-Albrecht, R. Steiner, J. DeRouchey, C. M. Stafford, E. Huang, M. Bal, M. Tuominen, C. J. Hawker, and T. P. Russell, Nanoscopic templates from oriented block copolymer films, *Adv. Mater.* **12**(11), 787–791 (2000).
59. T. Thurn-Albrecht, J. Schotter, G. A. Kästle, N. Emley, T. Shibauchi, L. Krusin-Elbaum, K. Guarini, C. T. Black, M. Tuominen, and T. P. Russell, Ultrahigh-density nanowire arrays grown in self-assembled diblock copolymer templates, *Science* **290**(5499), 2126–2129 (2000).

3

Scanning Probe Microscopes

K.-W. Ng
Department of Physics and Astronomy
University of Kentucky
Lexington, KY

3.1. INTRODUCTION

Scanning probe microscopes are a class of tools that scan a sharp probe tip across a sample and, by observation of the interaction between the two, provide nanometer-scale information concerning the sample. The first scanning probe microscope, the scanning tunneling microscope (STM), was invented by G. Binnig and H. Rohrer in the early 1980s. They used it to directly observe the positions of the individual atoms in the reconstruction of the silicon surface, for example.[1] Since then, STM has been widely used as a surface characterization tool. In an STM, a sharp tip (the probe) is positioned and scanned very close to the sample surface. The topographical image is formed by plotting the tip to sample distance as a function of position. The distance is measured by the electrical tunneling current between the tip and the sample that results when a voltage is applied between them. Besides the distance, the tunneling current also depends on the local electronic properties of the sample. For this reason, STM can also be used to measure the local electronic density of states and work function. In addition to serving as a local probe of the sample, the tip of an STM has even been used to manipulate individual atoms on the sample surface.

One major limitation of STM is that it can only image conducting surfaces, since electric current is involved in the measurement. Not long after the invention of STM, G. Binnig et al. suggested that if the probe was allowed to make direct contact with the surface, it would follow the topography by deforming itself.[2] The probe deflection can be used to generate an image and can be measured either by another STM or by optical methods. In this case, the interaction force between the tip and the sample surface is being measured. For this reason, this type of microscope is called an atomic force microscope (AFM) or scanning force

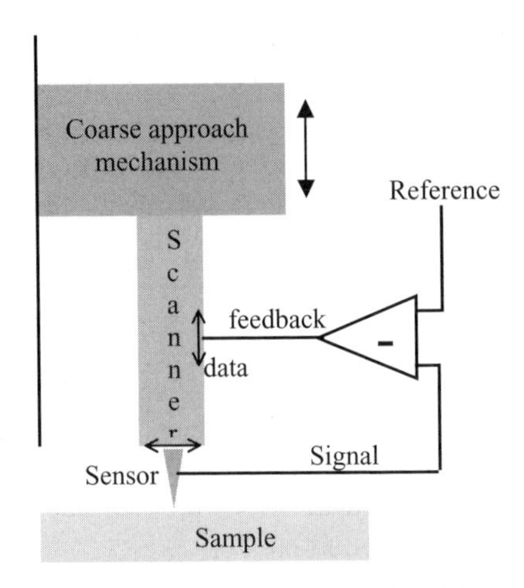

FIGURE 3.1. Schematic showing all major components of an SPM. In this example, feedback is used to move the sensor vertically to maintain a constant signal. Vertical displacement of the sensor is taken as topographical data.

microscope (SFM). AFM has been broadly used in research laboratories as an extremely high power microscope. It has a significant advantage over STM and the scanning electron microscope (SEM) in that it can image insulating or semiconducting surfaces directly in ambient atmosphere or even in liquids.

AFM provides information on the mechanical properties of the surface, but not the electronic properties as the STM does. Other sensing methods have been developed, in accordance to the type of sample properties desired to be measured. For example, the near field scanning optical microscope (NSOM or SNOM) uses visible light for topographical imaging.[3] It can provide information concerning the optical properties (e.g., transmittance or fluorescence) of the sample. We can map out the magnetic properties of a sample by using Hall's device[4] or a superconducting quantum interference device (SQUID)[5] as the scanning probe. If we are interested in the dielectric properties of the sample, we can measure the capacitance between the probe and the sample. Although different sensing techniques are used in these microscopes, their scanning mechanism is the same. For small scanning area, they employ a small device called piezoelectric tube to provide the actual scanning motion. All these different types of microscopes are categorized as scanning probe microscopes (SPMs). Schematically, an SPM is composed of the following parts for different functions (Figure 3.1):

1. *The sensor*. Different types of sensors or sensing methods can be used to probe a particular surface property. Additional electronics specific to the sensing technique may be required. If the measured property can be converted to a voltage, it is possible to feed the data to commercial SPM electronics for control and data processing purposes.
2. *The scanner*. The scanner physically holds the sensor and provides the scanning motion along the sample surface. In most cases, the scanner can move the sensor in all three (x-, y-, and z-) directions. The scanner is constructed with piezoelectric components. These piezoelectric components will expand or contract to produce

the motion when a voltage is applied to them. Most commercial SPM electronics provide high voltage outputs to drive the scanner. Typically, the scanner can move the probe over a range of several thousand angstroms, with a fraction of an angstrom in resolution. A motor has to be used for longer scan range.

3. *The feedback control.* If the sample surface is rough, there is a good chance for the probe to crash into the surface. In this case, a feedback control is needed to maintain the distance between the tip and the sample. The circuit controls the scanner to ensure the measured value stays close to a preset value. Most commercial SPM electronics provide a standard feedback loop for this purpose.
4. *The coarse approach.* The piezoelectric scanner can provide only a few thousand angstroms of motion. All SPMs have a coarse approach system to provide long-range motion so that the probe can be placed close to the sample safely. There are many different coarse approach designs. The most common ones are mechanical machines with combinations of differential springs, fine thread screws, and levers. Piezoelectric motors are also commonly used in modern SPM.

SPM is a very important and versatile tool for nanotechnology. Although some SPMs (like STM and AFM) are commercially available, a basic understanding of its operation is essential for obtaining quality results. This is especially true if we want to modify the instrument for our special research needs. In this chapter, we will describe how some of these SPMs work. We will discuss the above components of an SPM in more detail in the next section.

3.2. BASICS OF SPM

3.2.a. Piezoelectric Scanner

All SPMs use piezoelectric components to produce the scanning motions of the scanner. Many materials display piezoelectric effect in which electric charges are produced at the surfaces when the material is under stress. In reverse, stress or strain can be generated if the material is placed in an electric field. The effect can be measured by the piezoelectric coefficients (Figure 3.2). Let S_1 and S_3 be the strain components along the x- and z-axis respectively when an electric field E_3 is applied along the z-axis ($S_1 = \delta x/x$ and $S_3 = \delta z/z$, where x and z are the initial lengths of the material and δx and δz are the changes when the

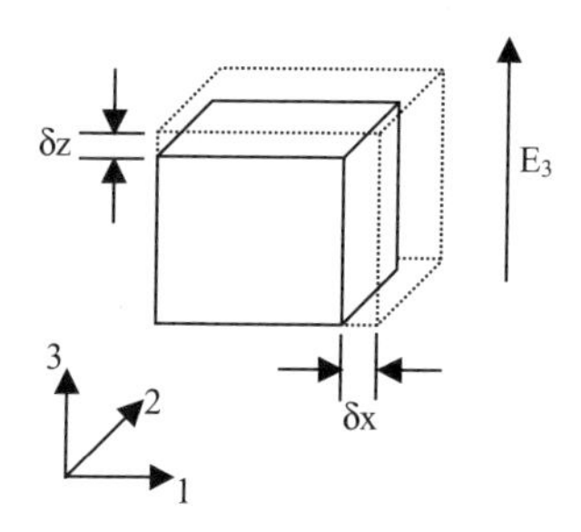

FIGURE 3.2. Deformation of piezoelectric material in an electric field and the defining coefficients.

TABLE 3.1. d_{31} and d_{33} values for several common piezoelectric materials

	PZT-4D	PZT-5H	PZT-7D	PZT-8
D_{31} (10^{-10} m/V)	−1.35	−2.74	−1.00	−0.97
D_{33} (10^{-10} m/V)	3.15	5.93	2.25	2.25

field is applied). The two common piezoelectric coefficients are given as

$$d_{31} = \frac{S_1}{E_3} \quad \text{and} \quad d_{33} = \frac{S_3}{E_3} \tag{3.1}$$

Values for some typical piezoelectric materials are given in Table 3.1. The two opposite ends of the device are coated with metal (e.g., nickel) so that applying a voltage between these electrodes can produce the needed electric field. Most SPM scanners use a piezoelectric tube. The inner and outer surfaces of the tube are coated with metal to form the inner and outer electrodes. The tube will expand or contract when a voltage V is applied across these two electrodes and the change in length can be estimated as

$$\Delta L = \frac{2d_{31}VL}{b-a}. \tag{3.2}$$

where b and a are the outer and inner diameter of the tube, respectively.

The capacitance between the electrodes is given by

$$C = \frac{2\varepsilon_r \varepsilon_0 \pi L}{\ell n \left(\frac{b}{a}\right)}. \tag{3.3}$$

With a typical dielectric constant (ε_r) of 1000, the capacitance per unit length (C/L) is about 10 nF/cm. Measuring this capacitance provides a convenient way to ensure the high voltage leads are attached to the electrodes properly.

The most common piezoelectric material used is lead zirconate titanate (a mixture of $PbTiO_3$ and $PbZrO_3$) ceramic, known as PZT. It is actually ferroelectric and possesses a permanent electric dipole even when there is no electric field. A poling process induces the permanent dipole when the piezoelectric component is made. For this reason, the piezoelectric component has polarity and that is why it can either expand or contract depending on the sign of the applied voltage. The polarity can be determined by measuring the polarity of the voltage generated as the temperature is raised slightly. Care has to be taken to avoid depolarization of the piezoelectric tube. It will depolarize if (i) the temperature is too high, exceeding the Curie temperature (this may happen for ultra high vacuum SPM when the chamber is baked at high temperature); or (ii) too high a reverse bias is applied to the electrodes. Once a piezoelectric tube is depolarized, it has to be re-poled or replaced with a new one.

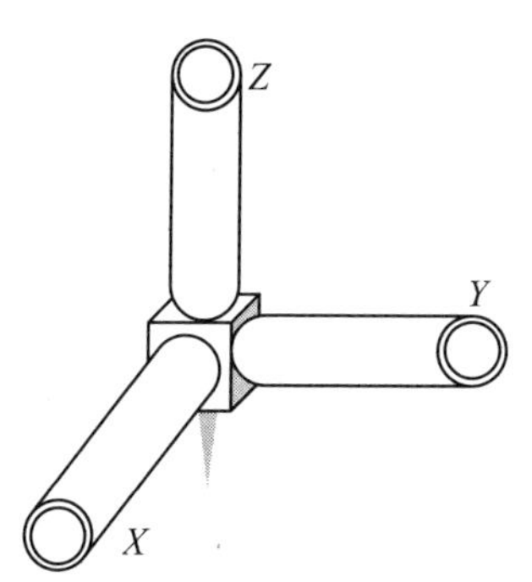

FIGURE 3.3. Tripod design of SPM scanner with the sensor tip located at the bottom.

There are two major designs in assembling the scanner with these piezoelectric tubes. The first one is the simple tripod in which three piezoelectric tubes (or rods) are aligned like the x-, y-, and z-axes and are glued together with the sensor holder at the origin, as shown in Figure 3.3. As these three rods expand or contract, they can move the sensor in all three directions. At least four leads are needed to apply the controlling voltages to these piezoelectric tubes, one to each of the three tubes and the fourth wire serves as the common. One important consideration in the design of SPM is the resonance frequency of the structure. In general, we want the resonance frequency to be as high as possible, so that the microscope is more tolerant against low frequency ambient mechanical noise. The tripod design involves the fixation of several components, and its resonance frequency cannot be too high. The resonance frequency of most tripod designs is about 1 kHz.

A more compact scanner is to use one single piezoelectric tube. In this design the outer electrode of the piezoelectric tube is sliced into four equal sections as shown in Figure 3.4. The sensor holder is attached to one end of the tube. If a voltage is applied to one of the sectional electrodes, only that portion of the piezoelectric tube will expand or contract. The whole tube will bend accordingly and hence produce scanning motion in the x- and y-directions. The deflection in the x- or y-direction is given by:

$$\Delta x \text{ (or } \Delta y) = \frac{\sqrt{2} d_{31} V L^2}{\pi D h} \tag{3.4}$$

where L is the length of the tube, D is the diameter of the tube, and h is the thickness of the tube. V is the voltage applied to one of the quadrants. If we vary the voltage applied to the inner electrode, the whole tube will expand or contract in the z-direction. With

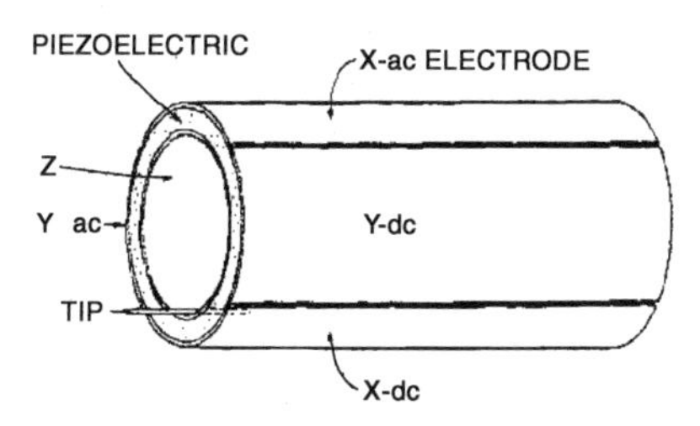

FIGURE 3.4. Tube scanner. (From Ref. 6 by permission of American Institute of Physics.)

such compactness, the resonance frequency can easily be improved by a few to ten times in comparison with the tripod design. To improve the linearity and orthogonality of the scanning motion, it is advisable to apply voltages of equal magnitude but opposite sign to the opposite quadrants of the electrodes. The deflection will then be double the one given by equation 3.4. In a more sophisticated case, we can fine tune the voltages applied to each individual quadrant either by software or external circuitry for perfect orthogonality.

3.2.b. Coarse Approach Mechanism

While it can provide sub-angstrom resolution, a piezoelectric scanner can only move over a very limited range of distance (several thousand angstroms). This scanning range is much smaller than the distance we can see even under an optical microscope. Many sensors (like STM or AFM) require a close working distance on the order of nanometers to tens of nanometers between the tip and sample. A coarse approach mechanism is needed to place the scanner within range of the sample. The resolution of the coarse motion does not need to be very high. It will work as long as it can move a distance shorter than the scan range of the scanner, with a reasonable safety margin. However, the coarse approach system should be able to move the scanner over a long distance by multiple steps, depending on how close we can place the probe next to the surface with the naked eye. With the coarse approach mechanism holding the scanner, we can approach the probe towards the surface as follows. With all electronics and the feedback loop on, the scanner will push the sensor towards the surface with a ramping voltage. If a signal comparable to the preset value is detected, the scanner will stop at the right place and the approach is successful. If no signal is detected even after extending to its full length, this means the surface is at least one scan range away. The piezoelectric tube will then retract to its original length, and the coarse approach mechanism will push everything (the scanner and the sensor) forward by the distance that has been tested by the scanner to be safe. This process will be cycled until a signal is detected.

There are many different designs in coarse approach system. This will ultimately determine the performance of the SPM. A good coarse approach mechanism should be small and rigid for the highest possible resonance frequency. Earlier SPMs often use mechanical devices like differential springs, differential screws, and levers. They can be combined together in the design to give the proper motion reduction needed. Sometimes commercially available parts like an optical translator, gearbox, speed reducer, and step motor can also be used. These mechanical devices can provide reliable motion and a great mechanical advantage. However, a forceful approach may not be an advantage for SPM, because this will cause serious damage to the sample and scanner if there is an unrealized crash. The major disadvantage for these mechanical devices is bulkiness and loose mechanical coupling. This will make the SPM vulnerable to external mechanical noise. Furthermore, if the SPM is installed inside a high vacuum chamber or low temperature cryogenics, mechanical feed-through is needed. Transmission of mechanical motion over a long distance will make it unreliable.

For these reasons, many newer SPMs use a "piezoelectric motor" to provide the coarse approach motion. A piezoelectric motor can produce the small motion directly and no further reduction is needed. It can be installed right next to the scanner. It is interesting to point out that Binnig and Rohrer used a coarse approach mechanism called "louse" in their STM. Louse actually belonged to this category because it used piezoelectric actuators to produce the motion. Some piezoelectric motors are commercially available, but many

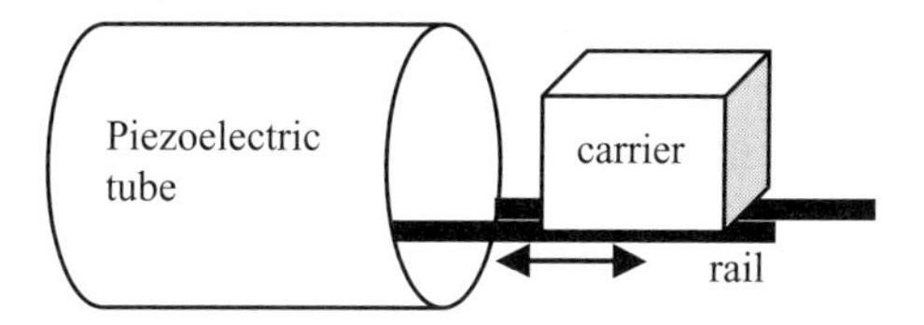

FIGURE 3.5. An example of an inertia motor used for coarse approach in SPM.

of them are too bulky to be effective in SPM. Here, we will discuss one coarse approach method known as an inertia motor, commonly used in commercial or home-built SPMs.

The inertia motor uses a piezoelectric actuator, often a tube, to produce the needed motion.[7] In our example set-up, two parallel rails are glued at the end of a piezoelectric tube as shown in Figure 3.5. The carrier (mass M) holding the sample is placed on the rail and is free to slide on it. If the coefficient of static friction between the carrier and the rail is μ, then the rail can hold the carrier by friction up to an acceleration of μg. If the piezoelectric tube now pushes the rail forward with an acceleration a_f larger than μg through a distance d, the carrier will slide and move through a distance of $d' = (\mu g/a_f)d$ during this time. If the piezoelectric tube now retracts slowly with an acceleration much less than μg, the carrier will follow the rail and move backwards for the whole distance d. In one cycle like this, the carrier will move a distance $d - d' = (1 - \mu g/a_f)d$ closer to the piezoelectric tube. Obviously, this simple setup will not work vertically. However, with simple modification like clamping the carrier to the rail with spring load clips, the setup can actually stand up and the carrier moves in the vertical direction. Note that the scanner can be installed concentrically inside the coarse motion tube. This will reduce the size of the SPM significantly.

There are many other good designs for using piezoelectric components for coarse motion. The inertia motor discussed above is quite commonly used. Some commercial SPM electronics provide special outputs for the driving voltage of the motor. If the SPM head is home-built, it is a good idea to construct the driving circuit also. Figure 3.6 shows the proper waveform for the inertia motor.

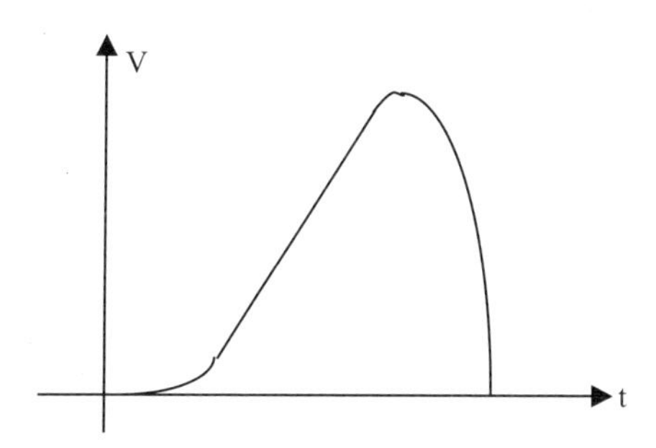

FIGURE 3.6. One possible waveform for the inertia motor. The carrier is first pushed with a near constant speed and then the rail withdraws with great acceleration to cause slipping. The carrier will move in opposite direction by reversing the voltage polarity.

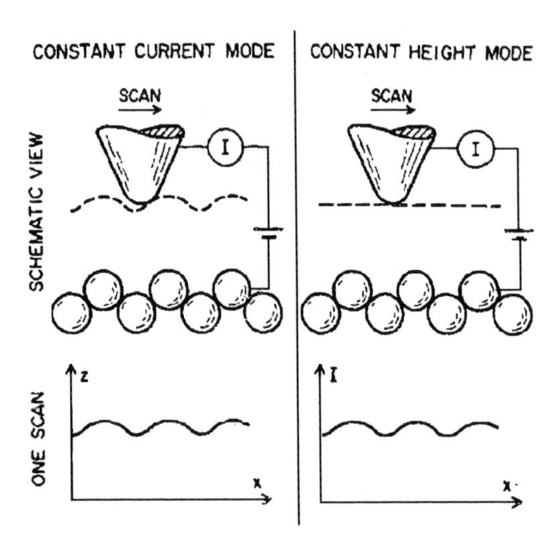

FIGURE 3.7. Left: Constant current mode, with feedback turned on to maintain a constant tunneling current. Right: Constant height mode, feedback is turned off. (From Ref. 8 by permission of American Institute of Physics.)

3.2.c. Feedback Loop

For rough surfaces and a sensing technique (like tunneling) that requires to scan the probe very close to the surface (within a few angstroms), a feedback loop is needed to maintain a safe distance between the probe and the sample. In the case of STM, the microscope can be operated in two possible modes (Figure 3.7): constant height mode (without feedback loop) or constant current mode (with feedback loop). In the constant current mode, the feedback loop varies the height of the scanner to maintain the constant current value. This variation in height is then the output variable of the STM. Although feedback is necessary to prevent crashing between probe and sample, it also unavoidably reduces the measurement speed and resolution. For high resolution and small area imaging, after confirming the flatness of the surface by first imaging with the feedback loop on, it is always advisable to image again with the feedback loop off.

Many newer commercial SPM electronics use digital signal processing in the feedback loop. From the user point of view, the operation is similar to the analog ones. A feedback loop is schematically represented in Figure 3.8. The purpose of the feedback loop is to bring the measured quantity O as close to the reference value R as possible. For simplicity, we

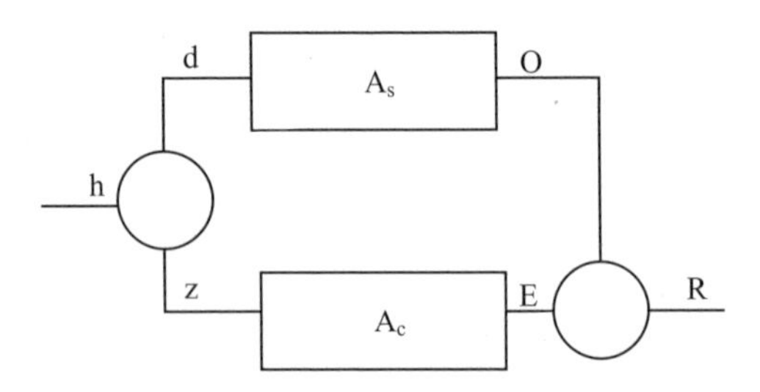

FIGURE 3.8. Schematic for the feedback loop used in constant current mode of STM.

can assume both O and R are voltages. The difference between O and R is $E = R - O$. Within all possibility, we want to make E (the error) as small as possible. This error signal is amplified (which often involves a high voltage operational amplifier) and applied to the z-electrode of the scanner. Let the gain be A_c (in Angstrom/Volt) as indicated in the figure. This will adjust the probe position z. As the probe scans along the sample surface, the surface height h will change in accordance to the topography. The sample probe distance d is given by $d = z - h$. This distance will be measured by the probe and magnified by the following electronics. If A_s (in Volt/Angstrom) is the gain, then $\delta O = A_s\ \delta d = A_s(\delta z - \delta h)$. Assume the reference R is kept constant during operation, so that $\delta R = 0$. We have $\delta E = -\delta O = -A_s(\delta z - \delta h)$. On the other hand, we have $\delta z = A_c\ \delta O$. From these we have $\delta z = A_c A_s(\delta z - \delta h)$ and hence $\delta z = A_c A_s \delta h/(A_c A_s - 1)$. $A_c A_s$ is called the loop gain. If we want the probe to follow the surface as close as possible, this will require a large loop gain ($A_c A_s \gg 1$) so that $\delta z = \delta h$.

The above argument is over-simplified, because is assumes a static condition. In reality, it takes time for each stage in the loop to respond to a signal change. In other words, if the probe has scanned over a surface step δh, the adjustment δz will not come instantly. Instead, it will take some time τ for it to come to the fully compensated value of δh. This time delay can also cause instability and oscillation in the loop. For a sinusoidal signal of a particular frequency f, the loop causes a one cycle $(1/f)$ delay and then adds to the signal. The oscillation will sustain itself even if the signal is not there. The frequency f_0 when this occurs is the natural frequency of the loop. Note that $\delta E = -\delta O$, so the negative sign introduces an automatic half cycle (180° phase shift) in the loop for all frequencies already.

The feedback is known as proportional (P) if A_c is a constant. In general, the loop gain decreases with increasing frequency because A_s behaves this way. To avoid oscillation, the loop gain $A_c A_s$ at f_0 has to be small. The frequency when the loop gain is 0 dB (or 1) is called the cross over frequency f_c. f_c can be varied by adjusting A_c. To determine the maximum A_c we can set, it is customary to require the phase shift at f_c to be around 120°. The difference between this phase shift and the troublesome 180° is known as the phase margin (60° in the present case). If the phase margin is too small, there is a good chance of oscillation. However, if it is too large, the response time for the loop will be slow, and A_c is unnecessarily too small.

One common way to further increase the gain of A_c is to add an integrator in parallel to the proportional amplifier. In this case, we have

$$\delta z = A\ \delta O + \frac{1}{T} \int \delta O\ dt \tag{3.5}$$

where A is the proportional gain and T is the sampling time constant. This is known as the proportional-integration (PI) feedback loop, and there are two parameters (A and T) to adjust. Proportional gain A can be tuned like the proportional controller. The integrator helps to increase the gain at low frequencies. The integrator gain decreases as frequency increases. At a certain cutoff frequency f_I, equal to $1/(2\pi AT)$, the integrator gain is so small that the controller is essentially proportional again. It is customary to require f_I (by adjusting T) to be about 10% of f_c. If f_I is too close to f_c, the phase margin will be reduced. If f_I is too small, the settling time of the feedback loop will be increased.

A complete feedback control often involves a differentiate gain in addition to the proportion and integration gains, called PID controller. The effect of the differentiator is to increase the loop gain at higher frequencies (like what the integrator does to low frequencies). Since most SPMs scan at relatively low frequencies, the effect of differentiation is not very significant. Some commercial SPM controllers use only a PI feedback loop for simpler operation.

3.3. SCANNING TUNNELING MICROSCOPE

The scanning tunneling microscope was the first SPM and truly caught the attention of scientists and engineers. When two conductors are placed very close together without touching, it is still possible for electric current to pass through the gap between them. This phenomenon is called tunneling and is a quantum mechanical effect. The tunneling current depends on two major factors: (i) the distance between the two electrodes, and (ii) the electronic properties of the conducting electrodes. We will discuss these two aspects in more detail here.

In quantum mechanics, a free electron can also be considered as a wave with wavelength λ proportional to $1/mv$, where m is the mass and v is the velocity. However, if the "particle" really behaves like a plane wave, it will not look like a particle anymore. For example, we cannot know exactly where the particle is, at least up to an uncertainty of λ. This essentially puts a limit on the resolution if we use electrons like free traveling waves as in a scanning electron microscope. A similar situation occurs in an optical microscope also, in which we are using wave optics and the resolution is limited by the wavelength of the light source.

Classically, a free particle of kinetic energy $K = mv^2/2$ cannot enter and pass through a barrier of potential energy V that is larger than K. However, in quantum mechanics, it is possible. The penetration of the free particle through this forbidden region is called the tunneling phenomenon. In the forbidden region, the particle is becoming "virtual" and the conservation of energy will not be strictly followed. The STM makes use of this virtual particle to break the resolution limit of a free electron. In quantum mechanics, conservation of energy can be written in the form of a wave equation, known as the Schroedinger equation:

$$-\frac{\hbar^2}{2m}\frac{d^2}{dx^2}\Psi(x) + V(x)\Psi(x) = E\Psi(x) \tag{3.6}$$

The electron is represented by the wave function $\psi(x)$. The first term represents the kinetic energy, the second term is the potential energy, and the right hand side is the total energy term. The probability of finding an electron between positions x and $x + dx$ is given by $\psi^*(x)\psi(x)dx$. This yields a normalization condition $\int_{-\infty}^{\infty} \Psi^*(x)\Psi(x)dx = 1$. This will ensure the electron to be found somewhere in the allowed space.

This equation can be solved for an electron of energy E with the potential barrier as shown in Figure 3.9. Assume the electron is initially traveling in the $+x$ direction. The solution for the wave function in region I ($x < 0$) will be $\psi(x) = \mathrm{A}e^{\mathrm{ikx}} + \mathrm{B}e^{-\mathrm{ikx}}$, where k is the wave number ($k = \frac{2\pi}{\lambda} = \frac{\sqrt{2mE}}{\hbar}$), $\mathrm{A}e^{\mathrm{ikx}}$ is the incident wave, and $\mathrm{B}e^{-\mathrm{ikx}}$ is the wave reflected by the barrier. The solution for region III ($x > d$) includes only a wave traveling in the $+x$ direction, i.e., $\psi(x) = Ee^{\mathrm{ikx}}$. In the barrier region ($0 \le x \le d$), the solution is not a plane wave, but exponential $\psi(x) = \mathrm{C}e^{\mathrm{kx}} + \mathrm{D}e^{-\mathrm{kx}}$. The property of the differential equation requires continuity of ψ(x) and dψ/dx at all values of x, including the boundaries

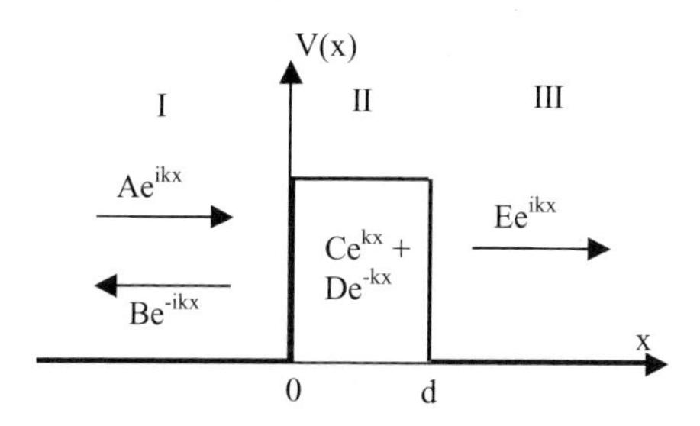

FIGURE 3.9. Tunneling of a single electron through a potential barrier.

at $x = 0$ and $x = d$. Together with the normalization condition, all coefficients A, B, C, D, and E can be calculated. The ratios $|B|^2/|A|^2$ and $|E|^2/|A|^2$ are known as the reflectance R^2 and transmittance T^2, respectively. Careful calculation can show that

$$|T|^2 \approx \exp(-2\,\kappa d) \quad \text{where } \kappa = \sqrt{\frac{2m}{\hbar^2}(V - E)} \tag{3.7}$$

and this can be generalized to a potential of general form $V(x)$:

$$|T|^2 \approx \exp\left\{-2\int\sqrt{\frac{2m}{\hbar^2}[V(x) - E]\,dx}\right\} \tag{3.8}$$

The current density for a single electron can be calculated as

$$j(x) = \frac{\hbar}{2mi}\left[\Psi^*(x)\frac{d\Psi(x)}{dx} - \Psi(x)\frac{d\Psi^*(x)}{dx}\right] \tag{3.9}$$

and the tunneling current density at $x > d$ will be proportional to $|T|^2$. Since $|T|^2$ decreases exponentially with d, the tunneling current will decay exponentially with the barrier thickness *d*. This explains why STM has such a high resolution in the z-direction, vertical to the surface. For the tunneling current to be large enough to be measurable, d has to be of the order of $1/\kappa \approx \lambda$.

In the case of STM, the electron energy is often less than one electron Volt. This corresponds to a wavelength of ten to hundred Angstroms. It is good enough to provide atomic resolution, because we are not using it like a free particle to probe the surface as in the case of SEM. The above analysis applies specifically to a single electron in vacuum. For electrons in a metal, there are many of them moving in a periodic crystal potential. Even if we assume they do not interact with each other directly, we still need to consider the fact that they are fermions and the Pauli exclusion principle applies.

We can consider electrons in a metal as electrons trapped in a box of dimensions L_x, L_y, and L_z. To fit an integral number of waves inside the box, the wave length in a particular direction has to be in the form $\lambda = L/n$ $(n = 1, 2, 3, \ldots.)$ and hence $k = 2\pi/\lambda = 2n\pi/L$. Electrons are fermions, so each particular combination of $\mathbf{k} = (k_x, k_y, k_z) = (2n_x\pi/L_x, 2n_y\pi/L_y, 2n_z\pi/L_z)$ can hold only two electrons (one spin up and the other spin down). Since n_x, n_y, and n_z are integers, each k state has a "volume" of $(8\pi)^3/V$ in the k-space. V is the real volume of the box (i.e., sample size). The number of states enclosed by the two spherical surfaces of radii k and $k + dk$ is $4\pi k^2 dk/[(8\pi)^3/V] = Vk^2dk/2\pi^2$. Including spin, the number of electrons you can pack within this shell is $= Vk^2dk/\pi^2$. The N electrons will fill up the states from the lowest energy one with $k = 0$ up to a sphere of radius called Fermi radius k_F. Hence, we have

$$\int_0^{k_F} \frac{Vk^2dk}{\pi^2} = N \Rightarrow k_F = \left[\frac{3N\pi^2}{V}\right]^{1/3} \tag{3.10}$$

and the corresponding energy is called Fermi energy E_f:

$$E_f = \frac{\hbar^2 k_F^2}{2m} \tag{3.11}$$

The number of states available per energy range per sample volume is called the density of states, $n(E)$.

The number of states between k and $k + dk$ is given as

$$dn = \frac{Vk^2\, dk}{2\pi^2} \tag{3.12}$$

But $E = \frac{\hbar^2k^2}{2m}$ and $\mathrm{dE} = \frac{\hbar^2 k}{m}dk$. Substituting this to the above above equation, we have:

$$n(E) = \frac{1}{V}\frac{dn}{dE} = \frac{mk}{2\pi^2\hbar^2} = \sqrt{\frac{m}{2}}\frac{\sqrt{E}}{\pi^2\hbar^4} \tag{3.13}$$

It is clear that the density of states *n(E)* depends on the energy. It is proportional to $E^{1/2}$ in the present case. Our discussion so far assumes zero potential inside the box. In other words, these electrons are still free within the box. Realistically, the electrons are moving in an effective potential due to the nuclei and other electrons. *n(E)* for real materials will not be as simple as the form given above. In most cases, it can be calculated by a more accurate model or it can also be measured experimentally. This is an important piece of data because it determines many electronic properties of the material. The two common techniques that can be used to measure density of states are photoemission spectroscopy and tunneling spectroscopy. We will discuss the latter case here.

Electrons will fill the quantum states from the lowest energy level to the Fermi level. The difference in energy between the Fermi level and the vacuum at a point outside the sample is the energy needed to remove an electron from the sample surface, equivalent to the work function of the sample. When two conductors are placed very close together, the electrons will redistribute themselves between the two conductors until the Fermi energy is equal at both sides. This is similar to the situation when the two conductors make direct contact with each other. The tunneling current not only depends on the barrier height and width, but also on the number of electrons available for the process and the number of empty states for the electrons to populate in the other metal. In other words, the tunneling current depends on the density of states of the two conductors also.

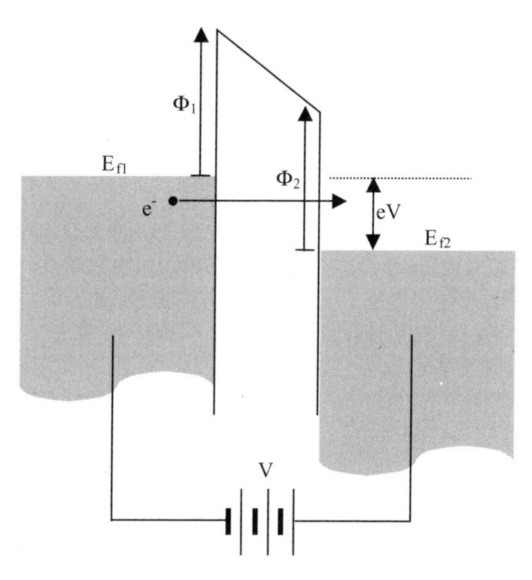

FIGURE 3.10. Tunneling of electrons between two metals.

When there is no potential difference between the two conductors, the tunneling current from one side to the other side equals that in the opposite direction, and the net tunneling current will add up to zero. To create a net tunneling current, a voltage bias V must be applied across the conductors. As shown in Figure 3.10, the Fermi level on side 1 will be raised by eV with respect to the Fermi level on side 2, where e is the charge of an electron. The electrons on side 1 will find more empty sites on the other side of the barrier to tunnel into than the electrons on side 2. Consequently, there is a net tunnel current from side 2 to side 1. Quantitative argument will give the tunnel currents as:

$$I \propto \int_{-\infty}^{\infty} n_1(E)n_2(E+eV)[f(E)-f(E+eV)]\,dE \tag{3.14}$$

$f(E)$ in the above equation is the Fermi-Dirac distribution, which is the average occupation number of a state of energy E at temperature T. k_B is the Boltzman constant. The functional form of the Fermi-Dirac distribution is sketched in Figure 3.11. It can be seen that its

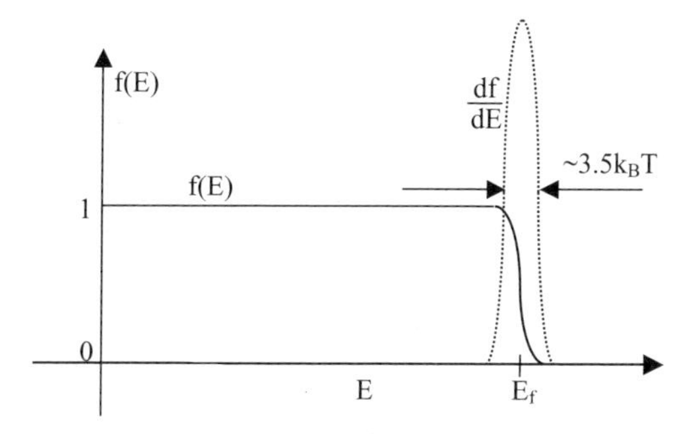

FIGURE 3.11. Fermi-Dirac distribution and its derivative.

derivative is Gaussian like, with a width of about 3.5 $k_B T$. As temperature goes down, the derivative will be sharpened and gradually becomes a delta function. With $[f(E + eV) - f(E)]/\Delta(eV)$ replaced by delta function $\delta(E)$, the derivative of the tunneling current (dI/dV) in the above equation will give the convolution of the density of states of the conductors on both sides.

dI/dV (also known as conductance) can be measured by two methods. The first method is simply to measure a sequence of tunneling currents I at different bias voltages V and then calculate dI/dV numerically. This method is very vulnerable to noise, since any noise component, no matter how small it is, can cause a very steep derivative. Most researchers use a lock-in amplifier to measure the derivative. A small oscillation $\delta V \sin \omega t$ is superimposed on the constant bias V, where ω is the frequency. With I as a function of V, the small oscillation will also cause oscillation in I at the same frequency ω. Since $\delta I = (dI/dV)\delta V$, the amplitude of the oscillating current will be proportional to dI/dV if δV is constant. With the reference frequency set to ω, a lock-in amplifier can be used to detect the oscillating component in the tunneling current. It will output a voltage that is proportional to δI, and this voltage will be taken as data proportional to dI/dV. The lock-in amplifier serves as a very narrow band filter, and the noise level can be limited and reduced significantly. Since the feedback control of the STM will interpret a changing current as variation in sample-tip distance, it is important to shut off the feedback whenever the bias voltage is varied.

If the density of states on one side is roughly constant, then the derivative of the tunnel current will give the density of states of the other side. This is how STM can be used to measure the local density of states of a metal. Note that the density of states can only be determined up to a proportionality constant, which depends on many parameters like the sample-tip distance and the area of the tunneling spot. In most cases, these parameters are not well known in STM measurements, hence the absolute value of density of states is difficult to determine. Nevertheless, the information obtained is enough for most purposes. Furthermore, tunneling is the most direct method to measure the density of states. It can also measure the density of both occupied and unoccupied states by simply reversing the polarity of the bias. STM finds many applications because of this capability. One most common example is superconductivity.

Recall that the derivative of the tunneling current involves the convolution of two densities of states, and we have assumed one of them to be constant. The measurement will be very effective if the density of states of the other side has some pronounced features in it. This is indeed the case for superconductivity. According to the conventional theory, the density of states of the superconducting electrons is given by

$$n_s(\xi) = \begin{cases} n_n(0)\dfrac{|\xi|}{\sqrt{\xi^2 - \Delta^2}} & \text{if } |\xi| \geq \Delta \\ 0 & \text{if } |\xi| \leq \Delta \end{cases} \tag{3.15}$$

where $n_n(0)$ is the normal density of states and ξ is the energy measured from the Fermi level E_f as reference zero. 2Δ is the energy gap resulting from electron pairing in the superconductor. Typical tunneling conductance (dI/dV) of a superconductor is shown in Figure 3.12. Note the singularity at $E = \Delta$. For this reason, STM is often used to measure the energy gap of a superconductor. STM can also be used to measure the density of states of a semiconductor, from which the band gap can be determined.[10]

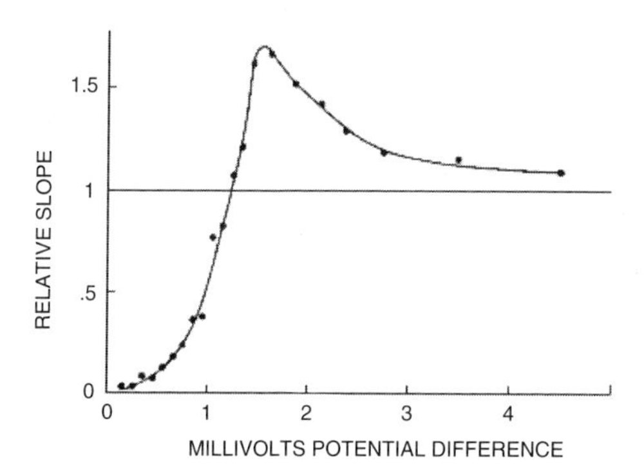

FIGURE 3.12. The density of states of superconducting Al. One of the first obtained by tunneling spectroscopy. The peak position roughly estimates Δ. (From Ref. 9 by permission of American Physical Society.)

Note that STM is a local probe, the major difference between STM and a tunnel junction is that it can probe the density of states locally. STM is often used to image superconducting vortices because of this capability. For most superconductors (called type II superconductors), an external magnetic field can penetrate the material partially. When this happens, the external field lines will be bundled together into filaments inside the superconductor. These filaments are actually becoming normal (rather than superconducting) because of the penetrating magnetic field (Figure 3.13). These normal regions will appear as spots (called vortices) like islands on the surface, and they will arrange themselves in a hexagonal structure. The low bias conductivity of the superconducting region will be much less than the normal region, because of the energy gap 2Δ. If we can scan and measure the local conductivity with a bias less then Δ, we can assemble an image showing how these vortices look (Figure 3.14).

There are many other situations where electronic properties vary spatially on sample surfaces. Impurities, defects, or even fabricated nanostructures will modify electronic properties significantly in the proximity area. For this reason, STM is a very powerful tool in nanotechnology. A well-known example is the observation of electron wave properties near fabricated structures (Figure 3.15). These structures can be fabricated by moving and arranging

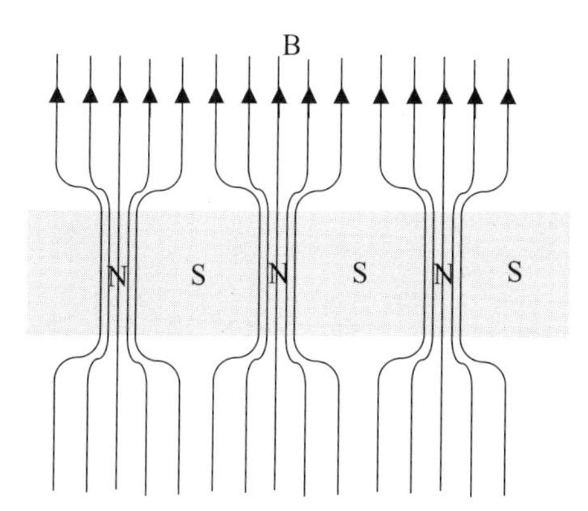

FIGURE 3.13. Formation of vortices in a superconductor. S region is still superconducting, but the area where the field penetrates (N) has now become normal (vortex core).

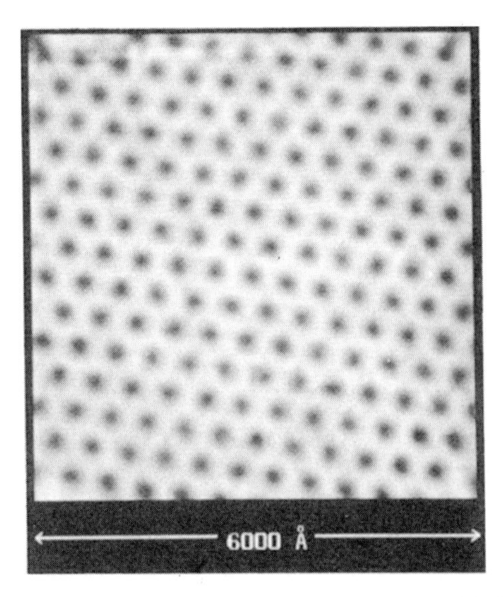

FIGURE 3.14. STM vortex image of $NbSe_2$ taken at 1.8 K, with an external field of 1T. (From Ref. 11 by permission of American Physical Society.)

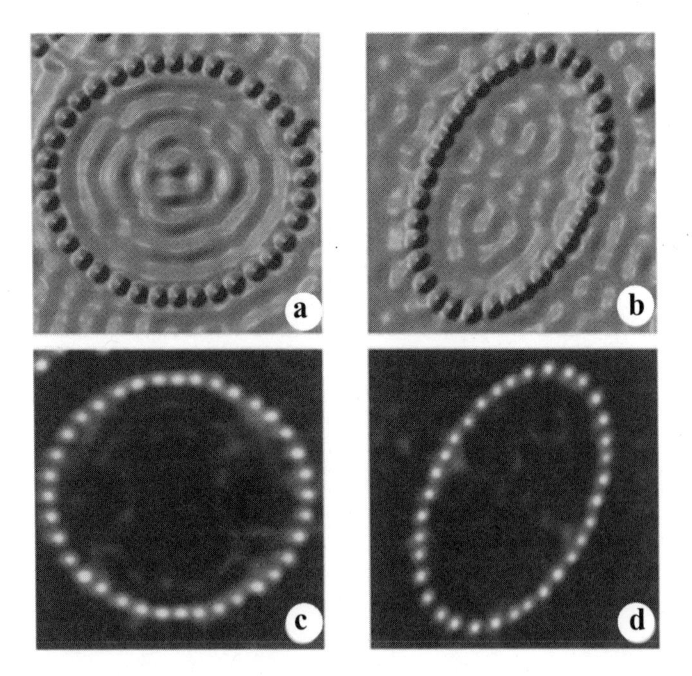

FIGURE 3.15. Co atoms on smooth Cu(111) surface. The Co atoms are moved to the desired pattern by carefully manipulating the position and voltage of the STM tip. Note how the electron waves in the background are being focused by the boundary. Lower pictures are dI/dV images of the top ones. (From Ref. 12 by permission of Macmillan Magazines Ltd.)

atoms into the desired pattern with an STM tip.[12] Loose atoms on a smooth surface can be picked up or released by careful manipulation of the tip voltage. Surface electron waves will interact with these structures and result in small spatial variation of conductivity. This will appear in the STM image as small variation in tunneling current and display diffraction patterns as expected by the wave property of electrons.

It is clear that when we try to interpret STM images as a map of local conductivity, like the vortex imaging or the present case, the underlying surface has to be perfectly flat. Otherwise, it is impossible to tell if the features we see in the image are an electronic effect or a result of surface topography. We have to limit the scanning range for surface flatness. If the surface is not so flat, or if we want to confirm the flatness of the surface, we can try to take the image of the same area at several different bias voltages. This may allow us to de-convolute the electronic component from the topographic one. For example, in vortex imaging, we can image with a bias voltage $V \gg \Delta$. Since the density of states at high energy is about the same between the superconduting state and the normal state, we can assume this to be a topographical image. If we use this image to normalize (i.e., divide) the image obtained by low bias at $V < \Delta$, the result will be a closer resemblance of the real vortex image.

Besides the density of states, STM can also be used to measure the local work function. As we have already discussed, the tunneling current I is proportional to $e^{-2\kappa e}$, where κ is given by

$$\kappa = \frac{\sqrt{2m(V-E)}}{\hbar} \cong \frac{\sqrt{2m\Phi}}{\hbar} \tag{3.16}$$

In obtaining the previous expression, we have assumed the Fermi level to be zero energy, and E is small (this is the case as long as the bias voltage is small). V is replaced by the average work function of the two electrodes Φ. Taking the logarithm of tunneling current, we have

$$I = Ke^{-2\kappa s} \Rightarrow \ell n I = \ell n K - 2\kappa s \tag{3.17}$$

where K is a proportionality constant. To obtain the work function Φ, we just need to measure I with different z-piezoelectric voltages (proportional to s). If we plot Log I versus V_z, the work function Φ can be determined by the slope of this plot.

3.4. OTHER SCANNED PROBE MICROSCOPES

A tunneling current is not the only method that can be used to probe and measure the surface. For topographical imaging, another type of microscope called atomic force microscope (AFM) or scanning force microscope (SFM) is commonly used. In this type of microscope, the interacting force between the probe and the sample is measured as an indication of sample-probe distance. Since no current is involved, it can image both conducting and insulating surfaces. This is its major advantage over STM. However, its resolution is in general not as high as that of STM, although atomic resolution is still possible under very favorable conditions. Compared to a scanning electron microscope, it

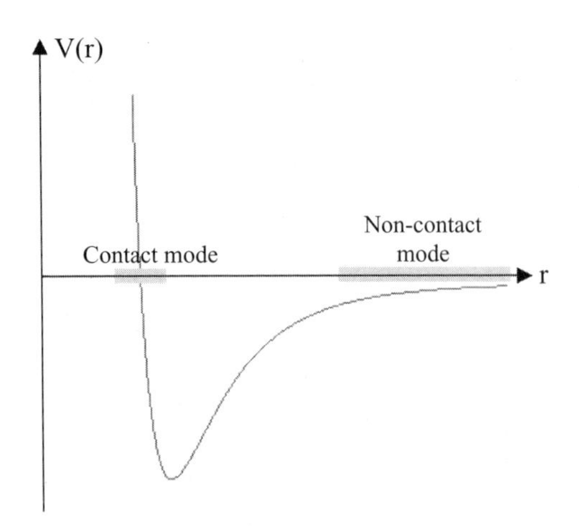

FIGURE 3.16. Potential energy between tip and sample as a function of the distance between them. The potential is attractive when they are far apart (non-contact), but it will become strongly repulsive when they are close together (contact).

has a simpler instrumentation setup, it can be operated with the sample in ambient air, and it is much cheaper in price. For these reasons, it is more and more commonly found in many laboratories.

In an atomic force microscope, the probe is scanned on the surface and the deflection of the probe is measured as the interaction force between the probe and the sample. The probe of AFM is often called a cantilever because of its special design, but here we will also use the more general term “probe”. The interaction force between the tip and sample can be modeled by the Lennard-Jones potential, which describes the interaction between two neutral atoms. An atom is not a solid point particle, but a small nucleus surrounded by a larger electron cloud that can be deformed. When two atoms are placed close together, they will induce a dipole in each other by deforming the electron cloud of the other atom. The force between the dipoles is called the Van der Waals force and it is attractive with a potential in the form $-A/r^6$ when the atoms are relatively far apart. When the atoms are very close together, another force due to the Pauli exclusion principle comes into effect. The principle does not allow two electrons to occupy the same state at the same place. The result is a large repulsive force within a very short range when the atoms are close to each other. The potential of this repulsive force is in the form of $+B/r^{12}$. This repulsive force is needed to prevent the collapse of two atoms together by the attractive dipole force. Figure 3.16 shows schematically the Lennard–Jones potential between two atoms. Between the attractive (far) and the repulsive (near) regions, there is an equilibrium point at which the potential is minimum and the interaction force between the atoms is zero. This is the point where the two atoms will settle.

A similar situation occurs when a probe is placed close to the sample surface. Starting from far away, the probe will first experience an attractive force and then a repulsive force after passing through the equilibrium point. This can be used to define the two operation modes of AFM. When the tip of the probe is kept far from the sample, the AFM is operating in non-contact mode and the force between the probe and the sample is attractive. In this regime, the tip is on the order of several tens to several hundreds of angstroms from the

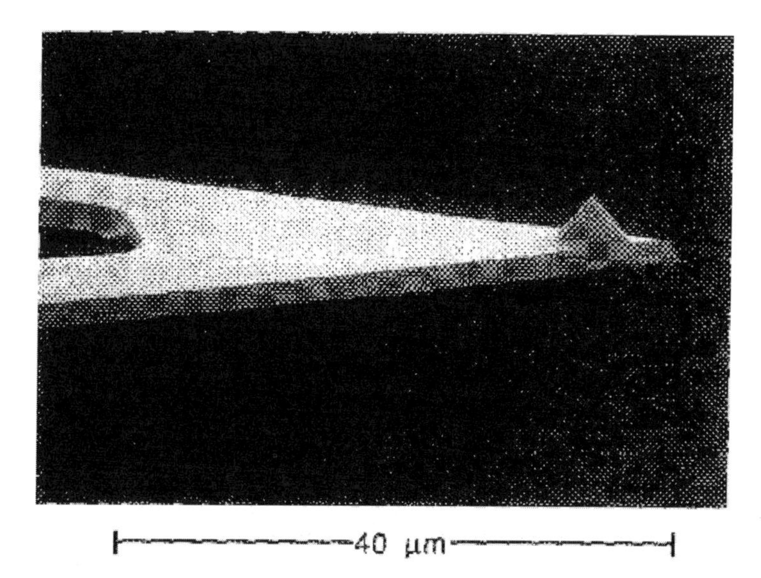

FIGURE 3.17. A SiO_2 AFM cantilever fabricated by photolithography. (From Ref. 13 by permission of American Vacuum Society.)

surface and the force is on the order of 10^{-12} N. In the repulsive regime, the tip in considered as making direct contact with the surface and the AFM is operating in contact mode. The force varies dramatically over a small distance because of the steep potential curve in the repulsive region. The typical operating range is between 10^{-6} to 10^{-7} N, but it can be as small as 10^{-9} N.

In any case, the cantilever is used as a spring balance to measure the interacting force. The deflection of the cantilever is measured and the force can then be calculated if the spring constant is known. For maximum sensitivity, the spring constant has to be as small as possible so that a small force can produce a large deflection. The spring constant of a solid plate can be estimated as

$$k = \frac{E}{4}\frac{wt^3}{\ell^3} \tag{3.18}$$

where w, t, and ℓ are the width, thickness, and length of the plate respectively. E is the Young's modulus of the material that makes the cantilever. A real cantilever can have a more complicated shape, often made by photolithography (Figure 3.17). We want the spring constant to be small for higher sensitivity, but the mass of the cantilever must also be small for a lower resonance frequency. We need to compromise between a small Young's modulus and a small density for the cantilever material. Common materials used are silicon, silicon oxide, or silicon nitride (Si_3N_4).[13] For example, the Young's modulus for silicon nitride is about 250 GPa. With a silicon nitride plate 200 μm long, 20 μm wide, and 2 μm thick, the spring constant k is about 1.25 N/m. With a density of 3190 kg/m^3, the mass of the cantilever is about 2.5×10^{-11} kg. The resonance frequency can be estimated from the equation $\omega_0 = (k/m)^{1/2}$ to be about 220 kHz. This resonance frequency is reasonably high since it is comparable to that of the other parts of microscope, such as the piezoelectric tube. With the spring constant known, the interacting force can be measured by the deflection of the cantilever. With the spring constant on the order of N/m, a 10^{-10} N force can produce a deflection of one Angstrom, which is measurable by methods discussed below.

The major advantage of non-contact mode is clearly that the tip does not make direct contact with the surface, and hence it is suitable for soft surfaces. However, the resolution of non-contact mode is not very high, since the force does not vary much over a long distance. For higher sensitivity, the shift in resonance frequency of the cantilever is often measured instead of measuring the deflection directly. The cantilever is vibrated at its resonance frequency, typically several hundred kHz. The frequency can be easily measured to high accuracy. Since the spring constant is defined as $k = \Delta F/\Delta z$, it will be offset by the force gradient $F' = dF/dz$ if the force itself is a function of position z. In other words, $k_{\text{eff}} = k - F'$ where k is the nominal spring constant of the cantilever. When the cantilever is far away from the surface, $F' = 0$ and the natural frequency is $\omega_0 = (k/m)^{1/2}$. As the cantilever is brought closer to the surface, the natural frequency will become $\omega = (k_{\text{eff}}/m)^{1/2} = [(k - F')/m]^{1/2} \sim \omega_0[1 - 2F'/k]$ and there is a fractional shift of $2F'/k$. The shift in resonance frequency measures the force gradient F', which in turn indicates the sample-tip distance. The shift can be either measured by tracking the resonance frequency of the cantilever directly, or by the change in vibration amplitude. Depending on the Q factor of the cantilever, the change in amplitude can be very dramatic at resonance frequency. A feedback loop can be set up to adjust the sample-tip distance to maintain the vibration amplitude at a constant value. In this case, the atomic force microscope is operating in constant force gradient, or simply constant force, mode. Without the feedback, the AFM will be in constant height mode. These two operating modes are similar to the constant current and constant height modes in STM. Constant force mode is recommended for large scanning area.

When the AFM operates in contact mode, the tip of the cantilever makes direct contact with the surface in the repulsive force regime. The repulsive force is much stronger and can vary over several orders of magnitude within a short distance. The cantilever will deflect in order to follow the surface profile smoothly. In contact mode, the deflection is large enough for it to be measured directly. Resolution is improved because of the greater force gradient and deflection. For some materials, atomic resolution is possible when the AFM is operated in contact mode. The most serious drawback of contact mode is scratching of the sample surface by the tip. The scratching is caused by the shear (lateral) force between the probe and the sample. Contact mode is thus not suitable for soft and delicate surfaces, which includes most biological samples. Many AFM manufacturers have developed special techniques to reduce the problem, without losing a lot of resolution.

One common technique used is the so-called tapping mode operation. Tapping mode is in the intermediate regime between contact and non-contact modes, near the valley of the potential curve in Figure 3.16. The force involved is slightly less than the low end of contact mode, on the order of 10^{-10} N. The probe is set to vibrate at an amplitude on the order of 10 nm. As the probe scans along the sample surface, the amplitude of vibration will change in accordance to the surface topography. The vibration amplitude can be taken as the distance between the tip equilibrium position and the surface. In this regime, the tip is considered as hitting the surface gently and hence given the name of "tapping mode". Although the tip is

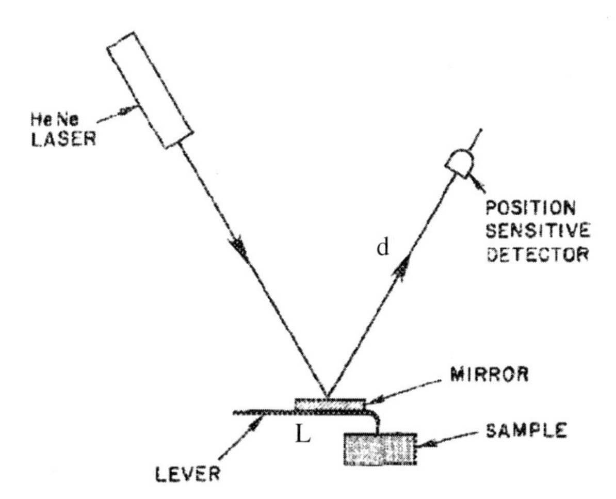

FIGURE 3.18. A laser optical system used to measure the deflection of the cantilever. This method is commonly used in many AFMs. (From Ref. 14 by permission of American Institute of Physics.)

slightly tapping the surface, this operating mode can reduce the lateral force and scratching significantly. This is very useful for imaging soft materials and biological samples.

Depending on the operation mode, we may want to measure the tip position directly as in contact mode, measure the resonance frequency as in non-contact mode, or the vibration amplitude of the probe as in the tapping mode. In all these cases, the measurement is obtained from the deflection of the probe. There are different ways to detect the probe deflection. For example, in the early days when AFM was first invented, STM circuitry was used to measure the tunneling current between a conducting AFM cantilever and another fixed tip above it. The distance of the gap between them will change as the cantilever is deflected along the surface. Since two junctions are involved in this method of detection, the operation is difficult and the structure is not very stable. The tunneling result depends on the surface condition of the AFM probe. Furthermore, the interaction between the STM tip and AFM cantilever will interfere with the cantilever deflection. This method was replaced soon by better ones, such as the optical method we will discuss in the following.[14]

In the optical method, the back of the probe is coated with metal like a mirror. A beam from a laser diode shines on this surface, either by mirrors or by optical fiber (Figure 3.18). After being reflected by the coated surface, the beam is detected by a position sensitive detector (PSD). A PSD is an electronic device that is readily available commercially. It is just two photodiodes constructed side by side. The photocurrent from each of the diodes can be converted to a voltage and the difference can be magnified by a differential amplifier. If the beam is equally split between the two diodes, there will be no imbalance in the photocurrent and the output voltage will be zero. If there is a small shift in the beam spot because of the deflection of the cantilever, the photodiode that receives more light will generate more current. For this reason, the output voltage from the differential amplifier can be used as a measurement of the cantilever deflection.

The actual shift of the beam spot δx is given by $d\delta\theta$ where d is the distance between the photo detectors and the cantilever and $\delta\theta$ is the angular deflection of the laser beam. This deflection of the laser beam is caused by the deflection of the cantilever of the same angle

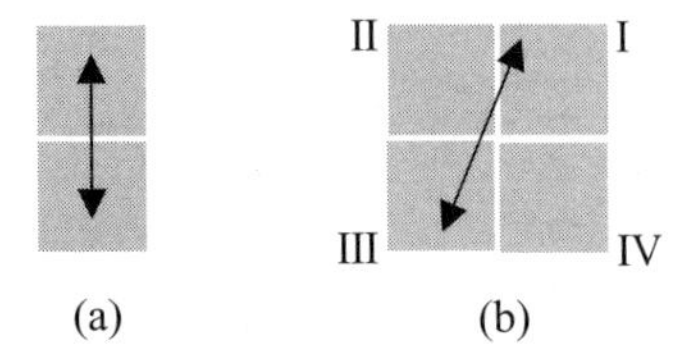

FIGURE 3.19. Schematics for PSD. Arrows indicate displacement of the laser spot. (a) A simple PSD can only measure vertical displacement. (b) A quad-PSD can measure both vertical and lateral displacement.

(or more correctly, $\delta\theta/2$). To the first approximation, we can estimate the deflection of the cantilever as $\delta z = L\delta\theta$, where δz is the vertical displacement of the tip and L is the length of the cantilever. From these, we have $\delta x = (d/L)\,\delta z$. In other words, there is a geometric magnification of d/L. This magnification can be as large as a few hundred, depending on the actual size of the AFM.

There are also quad-PSDs available, in which four photodiodes are fabricated in the four quadrants (Figure 3.19). While the normal force will bounce the laser spot "up and down" on the photodetector, the lateral force will twist the cantilever and move the laser spot in the "left and right" direction. A quad-PSD can thus measure both the normal force and lateral force. To measure the lateral force, we just need to add the photocurrents from the diodes in quadrants II and III and compare it with the sum of the currents from diodes in quadrants I and IV. A lateral force microscope is important in some applications, like studies of friction or the mechanical properties of some polymers and biological samples.

The photodiode method discussed above is quite commonly used in commercial AFMs. This is because of the ease in operation and minimal readjustment needed when the cantilever is replaced. However, there are other techniques to detect the deflection of the cantilever. For example, interferometry and capacity methods have been used to detect the nano-displacement of the cantilever. Some AFM companies also fabricate cantilevers with piezoresistive material so that the deflection can be easily measured in terms of the cantilever resistivity. More recently, there is also effort in using a tuning fork to hold the scanning tip directly. The tuning forks used for this purpose have a frequency of 32.768 kHz. They can be easily found in crystal oscillators for timing circuits. As the tip is scanned along the surface, the frequency of the tuning fork will be shifted according to the atomic force. This is a convenient approach since it is quite often for AFM (like the tapping mode and non-contact mode) to use frequency detection in measurement. This method also helps to make the AFM more compact and less vulnerable to mechanical noises.

STM and AFM are two of the most well known members in the SPM family. However, other sensing techniques can also be used to probe the surface. Some of these techniques may not be able to achieve atomic resolution, but they provide a means to measure some interesting properties of the surface. For example, the capacitance C between the tip (or an electrode) and the sample can be measured. Since C is proportional to the dielectric constant, scanning capacitance microscopy can provide local dielectric information on the sample surface. Unlike STM, the signal is inversely proportional to the sample tip distance, The resolution cannot be as good as STM in this case. Besides electronic and dielectric properties, optical and magnetic properties are some other important natures of the material under study. A Hall-effect sensor or SQUID (for very weak fields) can be used to measure the local magnetic field. These are called scanning Hall probe and scanning SQUID respectively.

If the material displays ferromagnetic behavior, an AFM can be used to measure the magnetic force between a magnetized cantilever and the sample. A cantilever can be magnetized by either coating with a magnetic material, or attaching a magnetic particle at the tip end. This is known as the magnetic force microscopy (MFM). Sometimes STM can be used to image magnetic domains in a ferromagnetic material with a magnetic tip (spin-polarized STM). Most efforts in MFM or spin-polarized STM are in the fabrication of the magnetic tips. The measurement is actually carried out by a regular AFM or STM. For some other sensing techniques, additional instrumentation is needed. An example is the near field scanning optical microscope, which we will discuss in the next section.

3.5. NEAR-FIELD SCANNING OPTICAL MICROSCOPE (NSOM)

It is well known that the resolution power of all optical instruments is limited by the wavelength of the electromagnetic wave used. More precisely, the minimum size of an object that can be resolved by a microscope is about 0.61λ. This is known as the Abbé limit. The only way to improve microscope resolution in wave optics (or far field) is to decrease the wavelength of the source. Scanning electron microscopes are used for higher magnification, for example, because high energy (~keV) electrons have much shorter wavelengths than visible light.

We can consider the scattering of monochromatic light of wavelength λ by the sample surface. Only a scattered wave of wavelength λ can propagate through space over a long distance (far field) to the objective lens of the microscope. These propagating scattered waves are formed by surface diffraction of long length scale ($> \lambda$). They carry only information of length scales longer than λ from the surface. This explains why the resolution is limited by the wavelength of the propagating waves. However, since there are many atoms within the length of λ, diffraction at shorter length scales will also occur. Short-range information is hidden in the so-called "evanescent waves" produced by the diffraction. These evanescent waves are not physical waves. They have imaginary wave vector, and they cannot propagate through space like a real wave. Their wave intensity decreases rapidly over distance, as s^{-4}. In the case of STM, if we consider electrons in a metal as waves, then the wave function under the barrier can be considered as evanescent waves. In this sense, near-field scanning optical microscopes (NSOMs) and STMs (both near-field microscopes) have similar analogy as optical microscopes and electron microscopes (far-field microscopes). Both NSOM and STM make use of the rapidly decaying wave function to probe the surface. The difference is that NSOM uses photons and STM uses electrons.

From the experience of STM, we know that the detector has to be placed very close to the surface in order to catch the evanescent wave component. This possibility to circumvent the Abbé limit had been proposed long time ago[15] in 1928, but became reality only after the invention of STM. The reason behind such a long delay is mostly in the knowledge in fabricating a fine probe with an extremely small aperture. Other techniques used in scanners and force feedback loops are also required. They were not fully developed until the invention of STM and AFM.

Detection of the near field can be performed either reflectively (source and detector on the same side), or transmissively (source and detector on opposite sides of a thin sample). Since an electromagnetic wave is being detected, the detector used in NSOM is usually a

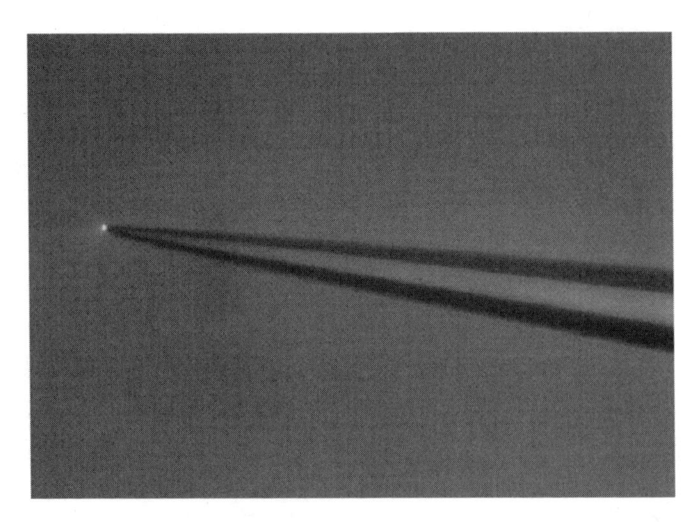

FIGURE 3.20. A fiber optical tip used as a light source. The tip end is placed very close to the sample surface. (From Ref. 16. Reproduced with kind permission of L. Goldner and J. Hwang.)

photomultiplier tube (PMT) or charge-coupled device (CCD), which convert light intensity into a voltage that can be recorded. The light detector is optically coupled to the flat end of the optical probe. The optical probe is actually a glass or optical fiber. The other end is a very sharp tip end that helps to bring the probe as close to the surface as possible to pick up the evanescent waves. The whole probe is coated with metal, except the very tip end (the aperture). The size of the aperture will essentially determine the resolution of the NSOM. Typical aperture size can range from a few hundred nanometers down to about 50 nm. The resolution of NSOM is not as high as STM or even AFM. Similar to STM, the piezoelectric tube holds this optical probe and scans it along the surface. The output from the light detector will be taken as data of the image. An NSOM image is associated with the optical properties of the surface (i.e., surface that we see with our eyes). This makes it different from STM or AFM. NSOM can probe both conducting and insulating surfaces.

There are more and more NSOMs that use a reverse approach in design. The fiber tip serves as a light source (Figure 3.20). When the source is placed very close to the surface, it is the evanescent waves that will interact locally with the surface. The signal will then be picked up by a regular optical system (such as an optical microscope) and eventually measured with a PMT or CCD to form the image.

As pointed out before, the size of the aperture is the ultimate factor that determines the resolution. However, we cannot indefinitely reduce the size of the aperture, because the light intensity through the aperture (throughput) will decrease dramatically. It is critical to make a high quality tip to remove other factors (like defects and impurities) that can reduce the throughput. The tip can be formed either by chemical etching or by mechanical pulling. In mechanical pulling, a laser beam is used to soften the middle part of the fiber and then the two ends are pulled apart until separation occurs. The cleaved end is rather flat by this method, and the aperture is well defined. Pullers specially designed for this process are available commercially. Reproducibility of the tip by mechanical pulling is not as good as chemical etching. Glass fiber can be chemically etched with hydrogen fluoride (HF) solution with a toluene layer at the top. If a fiber is dipped into the solution, a tip will be formed at the HF/toluene interface. Depending on the concentration of the solution, the whole process will take about 30 minutes to 1 hour to complete. After the tip is formed, the whole fiber has to be coated with metal except the very end. Thermal evaporation in a vacuum of 10^{-6} to 10^{-7} Torr is good enough for this process. The coating has to be deposited on the tip at an angle so that the aperture is in the shadow of the flux. The fiber is rotated continuously so that the deposition is even on the cylindrical surface. Gold or aluminum are commonly used for the tip coating. The coating has to be thicker than the penetration depth, about a few hundred nanometers. Though not as common as AFM cantilevers, NSOM tips are available commercially.

The tip of the NSOM has to be kept at a distance of about 10 nm to 50 nm (in general, a factor of the aperture size) from the sample, a distance less than the aperture size and the limiting resolution. Furthermore, variation of local optical properties within the relatively large scan range of NSOM should be expected. If the sample-tip distance is not fixed properly, it is difficult to ensure the signal reflects the optical properties of the surface. For these reasons, NSOMs need an independent but simultaneous measurement of sample tip distance. This is another major difficulty in the development of NSOM in becoming a useful tool. Tunneling between the tip and sample was used at the beginning, but this method is limited only to conducting samples. The most common method nowadays is to use the shear force feedback loop to control the sample tip distance and keep it constant.[17] This is effectively an AFM, using the NSOM probe as a "cantilever". A laser beam and PSD can be used, but it is more common now to use a tuning fork as we discussed in the AFM section. The optical tip is actually mounted to the tuning fork. A "dithering" piezoelectric tube is used to excite the tuning fork so that the tip oscillates parallel to the sample surface at the fork resonance frequency. The amplitude and phase of oscillation depends on the shear force between the tip and the sample, which in turn depends on the sample tip distance. The oscillation amplitude or phase is measured and maintained at the reference level by adjusting the z-control in the feedback loop. In this way, the sample-tip distance is kept constant.

3.6. SUMMARY

- Most scanning probe microscopes have three major parts: the sensor, the scanner, and the coarse approach mechanism.
- The scanner is made of piezoelectric material, a material that contracts or expands when an electric field is applied to it. It can provide motion with sub-Angstrom resolution, but not over a long distance.
- The coarse approach mechanism places the sensor close to the sample (before scanning) for a detectable signal. Its design is critical to the performance of the instrument. It affects the rigidity and stability, and ultimately the resolution, of the microscope
- Different sensing techniques can be used to probe the local surface properties. In this chapter, we have reviewed the three most commons ones. They are STM (electronic properties), AFM or SFM (interacting force), and NSOM (optical properties).
- Scanning probe microscopes can also be used to modify surfaces and construct atomic scale devices or structures.
- STM can only image conducting surfaces. AFM and NSOM can image both conducting and insulating surfaces, but it is more difficult to achieve atomic resolution.

QUESTIONS

1. The scanner of an STM is constructed with a piezoelectric PZT-8 tube that is 0.25 inch in length. The outer diameter and wall thickness of the tube are 0.25 inch and 0.020 inch, respectively. How much will the tube contract if a voltage difference of 100V is applied across the inner and outer electrodes? In an atomic image from the STM, the atoms have a corrugation of 0.2 Angstrom, what is the corresponding change in the voltage applied to the piezoelectric tube to maintain constant height?
2. For a metal with one valence electron per atom, calculate the Fermi energy using the free electron model. Estimate a typical sample-tip distance by calculating the wavelength of an electron at the Fermi level.

REFERENCES

1. Gerd Binnig and Heinrich, *Rev. Mod. Phys.* **59**, 615 (1987).
2. D. P. E. Smith, G. Binnig, and C. F. Quate, *Appl. Phys. Lett.* **49**, 1166 (1986).
3. D. W. Pohl, W. Denk, and M. Lanz, *Appl. Phys. Lett.* **44**, 651 (1984).
4. A. M. Chang, H. D. Hallen, L. Harriott, H. F. Hess, H. L. Kao, J. Kwo, R. E. Miller, R. Wolfe, J. van der Ziel, and T. Y. Chang, *Appl. Phys. Lett.* **61**, 1974 (1992).
5. John Kirtley, *IEEE Spectrum*, **33**, 40 (1996).
6. G. Binnig and D. P. E. Smith, *Rev. Sci. Instrum.* **57**, 1688 (1986).
7. J. W. Lyding, S. Skala, J. S. Hubacek, R. Brockenbrough, and G. Gammie, *Rev. Sci. Instrum.* **59**, 1897 (1988).
8. P. K. Hansma and J. Tersoff, *J. Appl. Phys.* **61**, R1 (1987).
9. Ivar Giaever, *Phys. Rev. Lett.* **5**, 147–148 (1960).
10. R. M. Feenstra, Joseph A. Stroscio, J. Tersoff, and A. P. Fein, *Phys. Rev. Lett.* **58**, 1192 (1987).
11. H. F. Hess, R. B. Robinson, R. C. Dynes, J. M. Valles, Jr., and J. V. Waszczak, *Phys. Rev. Lett.* **62**, 214 (1989).
12. H. C. Manoharan, C. P. Lutz, and D. M. Eigler, *Nature* **403**, 512 (2000).
13. T. R. Albrecht, S. Akamine, T. E. Carver, and C. F. Quate, *J. Vac. Sci. Techno.* A**8**, 3386 (1990).
14. Gerhard Meyer and Nabil M. Amer, *Appl. Phys. Lett.* **53**, 1045 (1988).
15. E. H. Synge, *Phil. Mag.* **6**, 356 (1928).
16. The picture is obtained from http://physics.nist.gov/Divisions/Div844/facilities/nsom/nsom.html.
17. E. Betzig, P. L. Finn, and J. S. Weiner, *Appl. Phys. Lett.* **60**, 2484 (1992).

II

Nanomaterials and Nanostructures

4

The Geometry of Nanoscale Carbon

Vincent Crespi
Department of Physics and Materials Research Institute
Pennsylvania State University, State College, PA

Since later chapters will cover the synthetic, electronic, and transport properties of carbon nanostructures in depth, here we take a different point of view and focus on the theory of their structure and geometry. The well-defined covalent bonding geometries of graphitic carbon lead to a simple set of geometrical rules that relate the global shape of a carbon nanostructure to the types of carbon rings within it.

4.1. BONDING

Carbon is an unusual element. The isolated carbon atom has filled $1s$ and $2s$ states and two electrons in the $2p$ state for a configuration of $(1s^2 2s^2 2p^2)$. Since carbon is a first-row element, the atom is very small and the Coulomb potential felt by the valence electrons is correspondingly high (remember the Coulomb potential energy varies as $1/r$). When carbon atoms are assembled into a larger structure, the potentials from nearby atoms perturb the $2s$ and $2p$ atomic orbitals and create bonding, nonbonding, and antibonding molecular orbitals formed from linear combinations (i.e., sums and differences) of the $2s$ and $2p$ states. Bonding occurs when the charge density of the electronic wavefunction occupies favorable areas, wherein the attractive atomic potentials of neighboring atoms overlap. Normally, the bonding orbitals between neighboring atoms pile up electron charge in the space that lies directly between the atoms, since this is the region where the attractive atomic potentials overlap most strongly. Such bonds are called σ states. However, for carbon, an accident of the fundamental constants (i.e., the mass of the electron, Planck's constant, the charge of an electron) implies that two neighboring atoms can also bond strongly by piling up charge in the regions above and below the line of intersection between the atoms, the so-called π states (Figure 4.1). Because carbon can bond "sideways" in this manner, using the

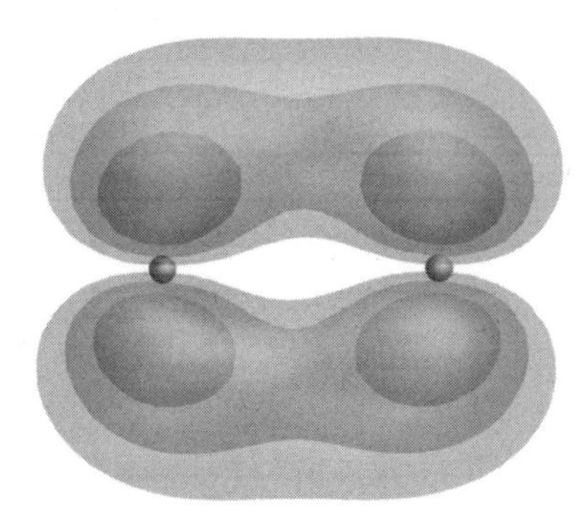

FIGURE 4.1. Charge density of a carbon dimer π state.

p states that point perpendicular to the line connecting the neighboring atoms, it can form highly anisotropic and stable two-dimensional layered structures.

The lowest-energy state of elemental carbon at ambient pressure and temperature is graphite. Graphite consists of individual graphene layers, each composed of interlinked hexagonal carbon rings tightly bonded to each other, stacked loosely into a three-dimensional material.

Within a single graphene layer (Figure 4.2), oriented in the x-y plane, each carbon atom is tightly bonded to three neighbors within a plane; these planes are then very weakly bonded to each other. The in-plane sp^2 bonding is best understood by first considering graphene, a single layer of graphite. The $2s$, $2p_x$, and $2p_y$ orbitals are recombined to form three new linear superpositions (thus the name sp^2). These three new linear combinations form three lobes of charge reaching outward from the carbon atom at 120 degree angles to each other, all within the x-y plane. These lobes form σ bonds to three neighboring carbon atoms. The leftover $2p_z$ orbital, which points perpendicular to the plane of the sp^2 bonds, overlaps with $2p_z$ orbitals on neighboring atoms to form an extended sheet-like bonding state that covers the upper and lower surfaces of the graphene sheet. The $2p_z$ orbitals of neighboring atoms overlap most effectively if they point in the same direction; therefore the sheet has its lowest energy when it is perfectly flat.

Since the overlap of the atomic potentials is strongest along the line between the atoms, the electronic bands arising from the σ states are lower in energy than those arising from

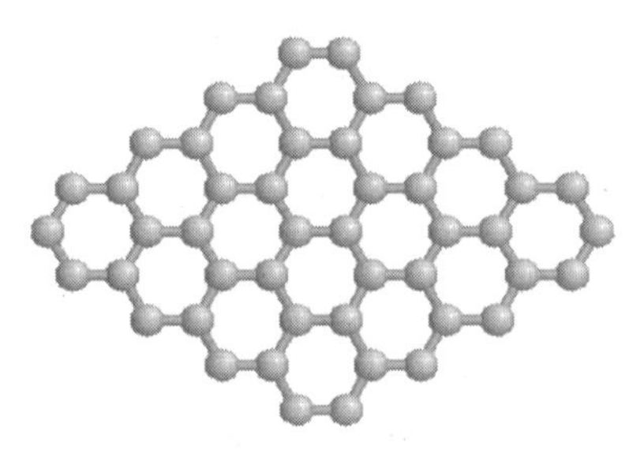

FIGURE 4.2. Graphene.

the π states. In fact, the π states, plus their antibonding cousins the π^* states, span the Fermi energy. (The antibonding π^* states are just higher-energy combinations of the same $2p_z$ orbitals. They are unoccupied in a pure graphene sheet). The interactions between the layers are very weak, arising from a combination of van der Waals interactions and electron delocalization in the $\hat{z}$ direction. The subtleties of the electronic overlap in the $\hat{z}$ direction and the exact patterns of stacking for graphite will not be covered here.

Normally, strong atomic potentials and covalent bonding imply the existence of a large bandgap, since strong potentials give well-separated atomic energy levels and large separations between the electron bands. However, nature again conspires to make carbon in the graphene structure special: sp^2-bonded atoms prefer three-fold coordination, so they naturally assemble into hexagonal sheets with two atoms in each unit cell. The potential arising from this structure can't distinguish between the π and π^* states at the Fermi level, so they remain equal in energy and these two bands formed from the $2p_z$ orbitals actually touch each other at E_{Fermi}. Since the system can't lower its energy by creating a bandgap, it does the next best thing and instead necks down the highest-energy filled electron bands (the π states) to the smallest possible number of states, a set of isolated single points, at the Fermi energy. In addition, the carbon atoms draw closer together to increase the interatomic overlap. This overlap spreads out the π and π^* bands over a wide range of energy and thereby pushes the occupied π states as low as possible in energy. As a side effect, the velocity of the electrons at the Fermi energy becomes rather large. A graphene sheet lives on the borderline between metallic and semiconducting behavior: it is both a metal with a vanishing Fermi surface and a semiconductor with a vanishing bandgap. Chapter 6 describes this electronic structure in more detail.

Carbon's neighbors to the right and left in the first row of the periodic table (boron and nitrogen) can also make strong π bonds. However, only carbon occupies the favored position of having exactly four electrons and requiring exactly four bonds to make a closed shell; therefore only carbon is stable as an extended covalently-bonded elemental two-dimensional structure. Nitrogen, in contrast, requires only three bonds, so can form a highly stable triple-bonded N_2 molecules in preference to an extended sheetlike structure. Boron, in contrast, lacks the fourth electron that stabilizes the π-bonded sp^2 sheet; instead it forms complex structures with multi-centered bonding.

4.2. DIMENSIONALITY

What exactly do we mean that a graphene sheet is two-dimensional? Like anything else, it's really a three-dimensional object, with a non-zero extent in the x, y and z directions. The *effective* two-dimensionality of graphene is fundamentally a question of quantum mechanics and energies. For directions within the sp^2 plane, the structure extends long distances; therefore one can form quantum states for electrons with many different finely-grained wavelengths, very closely spaced in energy. It's easy to squeeze in one more node into the wavefunction when one has so much space. However, perpendicular to the plane, the graphene sheet is quite thin, about 0.3 nanometers. Adding another node to the wavefunction in this direction (which means essentially, creating an excitation to the $3p$ atomic level) requires putting a very high curvature into the wavefunction and consequently implies a very large energy, far beyond the thermal energies available.

The ability of carbon to form these highly stable effectively two-dimensional structures is fundamental to their great promise in nanoscience and technology. Why? Because we live in a three-dimensional world, so we can distort this two-dimensional graphene sheet in the third dimension to form a very rich family of structures. The energy cost to perform these distortions is relatively small: graphite is a single atomic layer, so it can be bent without changing the in-plane bond lengths significantly. Since the direction of bending is perpendicular to the direction of the in-plane bonds, the energy to create a bending distortion is quadratic in the magnitude of the distortion, rather than linear.[a]

How can we exploit the third dimension to bend and distort a two-dimensional graphene sheet into interesting structures? First off, to be stable, any dangling bonds at the edges of such a distorted sheet must be eliminated. There are two ways to do this: either cap off dangling bonds with chemical groups such as hydrogen atoms or wrap the carbon structure around onto itself so that it forms a closed sheet with no edges. In the first case, one obtains an open structure; in the second, a closed structure. We begin with a discussion of the closed structures, known as fullerenes, since the geometrical rules that govern that situation can be easily extended to consider open structures as well.

4.3. TOPOLOGY

The fundamental requirement in a closed graphene-like structure is that every carbon atom have three bonds to neighboring atoms, and that the entire structure folds back on itself without any dangling bonds. These conditions are questions of topology, meaning the connectivity of an interlinked network of bonds. Topology establishes the constraints that must be satisfied when sp^2-bonded atoms bond together seamlessly into a closed structure. When analyzing the topology, we can imagine the bonding network to be infinitely deformable, so long as we don't break any bonds. For the moment forget what you might know about bond angles and bond distances; topology concerns itself only with the presence or absence of a connection between atoms.

Think of any closed sp^2-bonded carbon structure as a polyhedron, where each carbon atom is a vertex, each bond is an edge, and each closed ring of atoms forms a face. The condition that this polyhedron closes back onto itself imposes a universal mathematical relationship between the number of edges, faces and vertices. To find this relationship, we can begin with the simplest possible closed polyhedron: a tetrahedron, and successively extend the structure by adding new atoms. Remember that we're concerned for now only with topology: carbon atoms don't really form tetrahedra, since the bond angle distortions are too great. Nevertheless, the tetrahedron is the natural starting point for the mathematical construction of larger, more chemically plausible carbon polyhedra. The tetrahedron has four faces ($F = 4$), four vertices ($V = 4$), and six edges ($E = 6$). Notice that $F + V = E + 2$. We can extend the tetrahedron to form more complex polyhedra in any of the three ways depicted below.

Adding the bond connecting a vertex and an edge (left-hand side of Figure 4.3) creates one new vertex, one new face, and two new edges. Alternatively, adding the thick line

[a] This is why the acoustic phonon of graphite that is polorized in the $\hat{z}$ direction has a nearly quadratic dispersion at low wavevector.

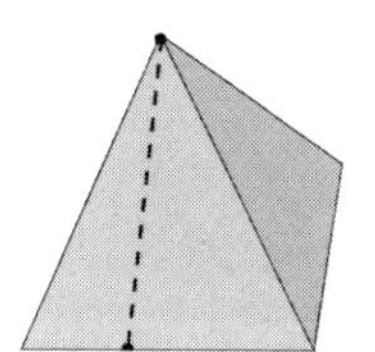
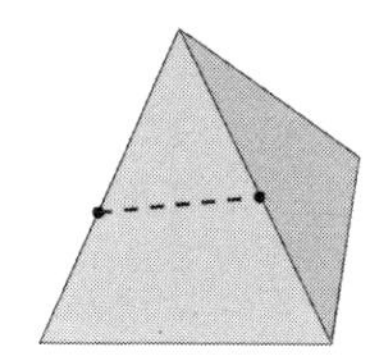
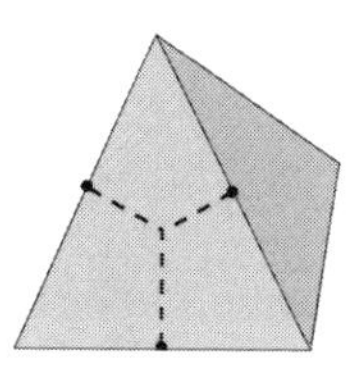

FIGURE 4.3. Euler's rule in a tetrahedron.

connecting two edges (middle) creates two additional vertices $V \rightarrow V+2$, one additional face $F \rightarrow F+1$ and three additional edges $E \rightarrow E+3$. Finally, adding a new vertex in the middle of a face and connecting it to n edges and m vertices (right-hand side) creates $n+1$ new vertices, $n+m-1$ new faces, and $2n+m$ new edges. By successive action of these operations we can construct any polyhedron, starting from the tetrahedron. How can we convince ourselves that we can make any polyhedron this way? Just think in the reverse: start from the polyhedron that you want to reach, and successively remove vertices and bonds; eventually only four vertices will remain, and the structure at that point must be a tetrahedron.[b]

Notice a very interesting fact: each of the operations

Left	Middle	Right
$V \rightarrow V+1$	$V \rightarrow V+2$	$V \rightarrow V+n+1$
$F \rightarrow F+1$	$F \rightarrow F+1$	$F \rightarrow F+n+m-1$
$E \rightarrow E+2$	$E \rightarrow E+3$	$E \rightarrow E+2n+m$

preserves the validity of the relation $F+V=E+2$ which we wrote down initially for the tetrahedron.

Actually, we are missing one more subtle operation: we could deform a sufficiently large polyhedron by bending it around onto itself and fusing together two faces, both which have the same number of sides, call it s, to give a donut-like shape with a hole in the center. This operation eliminates two faces $F \rightarrow F-2$, s edges $E \rightarrow E-s$, and s vertices $V \rightarrow V-s$. To retain our relationship $F+V=E+2$, we must subtract 2 from the right-hand side for each of the G times that we perform this operation: $F+V=E+2-2G$. G is called the genus of the polyhedron, the number of donut-like holes that it contains.

OK, enough abstract topology. Let's introduce the chemistry of the bonding. Can we make a closed polyhedron, with no dangling bonds, from sp^2-bonded carbon atoms? For sp^2 bonded carbon we need a new rule: each vertex has three edges emanating from it and each of these edges is shared between two vertices: $3V=2E$. This rule immediately requires that V be even: there are no closed carbon fullerenes with an odd number of atoms. Now graphene is made of hexagonal rings, so let's try to impose another condition: every ring of carbon atoms must have six edges. A polyhedron with purely hexagonal faces has 6 edges per face, each edge being shared by two faces: $2E=6F$. If we plug these two requirements ($E=\frac{3}{2}V$, $F=\frac{1}{3}E=\frac{1}{2}V$) into our rule $F+V=E+2-2G$, we obtain $G=1$. Our carbon structure with graphene-like bonding and hexagonal faces has a hole

[b] The operations that attach a new bond to a vortex are not relevant to sp^2-bonded structures, since they produce atoms with more than three nearest neighbors. However, they are necessary to create an arbitrary polyhedron, and including them here doesn't change any of the rules derived below.

FIGURE 4.4. A ring.

in it, like a donut! That is only a theorist's idealized picture of a nanotube (Figure 4.4). Usually, the theorist will think of the nanotube as perfectly straight, wrapping around onto itself at infinity. That's a nice idealization, but in real life, of course, no real nanotube is infinitely long (nor is it likely to wrap back onto itself seamlessly), so how does the end of a nanotube terminate without creating dangling bonds? One way is to cap the ends with metal particles or hydrogen atoms—we'll cover that possibility when we discuss open structures. Right now, we want to know how to make a closed polyhedral structure, with three-fold sp^2 bonding, but with no holes: $G = 0$.

Such a structure must contain fewer edges per face overall than our pure-hexagonal starting point. We can reduce the number of edges per face by using z-gons where $z < 6$ (for example, pentagons or squares). How many z-gons must we add to obtain a closed polyhedron with $G = 0$? A closed structure with N hexagons and M z-gons has $N + M$ faces, $\frac{6N+zM}{2}$ edges, and $\frac{6N+zM}{3}$ vertices. Plugging into $F + V = E + 2 - 2G$, we obtain a very simple answer: $(6 - z)M = 12(1 - G)$. The number of hexagons N is irrelevant, but for $G = 0$ the total depletion below purely hexagonal faces must be 12. Twelve pentagons suffice ($M = 12$, $z = 5$), as do 6 squares. The curvature from these z-gons bends the sp^2 sheet into a closed surface.

What z-gon does nature choose, and how are these z-gons arranged amongst the hexagons? The distortions in bond angles around a square embedded in a hexagonal network are about twice as large as the distortions surrounding a pentagon. Like any deviation from equilibrium, the energy cost is approximately quadratic in the distortion. Compared to two pentagons, the square then imposes a four times greater cost per atom across about half as many atoms. Therefore nature chooses pentagons. Similar arguments determine whether the pentagons are fused or separated by intervening hexagons. The two atoms shared by a pair of fused pentagons have about twice the local distortion in bond angle as have the atoms in separated pentagons. The fused pentagons contain 2 fewer atoms than a pair of separated pentagons, but they pay quadruple cost for the two shared atoms. Therefore nature prefers separated pentagons. Longer-range distortions then favor a uniform distribution of the twelve pentagons.[c]

Therefore a carbon nanotube can pinch off into a close polyhedral structure by incorporating six pentagons into a roughly hemispherical cap on each end of the tube. One could also subtract off the belly of the tube, since it contains only hexagons, and connect the two endcaps into a roughly spherical closed cage, such as the C_{60} molecule, to which we return later.

4.4. CURVATURE

The nanotube that we constructed above, with a straight belly composed entirely of hexagons plus two endcaps, each with six pentagons mixed into the hexagonal matrix, nicely illustrates the two mathematically distinct kinds of curvature that can be imposed on a graphene sheet. The cylindrical belly of the tube possesses mean curvature. This is the sort of curvature that one can impose on a sheet of paper without creating wrinkles or tears. Each of the pentagons acts as a point-like source of Gaussian curvature. Gaussian curvature is most easily thought of as the curvature of a sphere, the type of distortion that would wrinkle or tear a flat sheet of paper. Both types of curvature impose an energetic cost, since they weaken the overlap between the p_z orbitals of neighboring atoms.[d]

4.5. ENERGETICS

Now that we have covered the abstract geometrical requirements for a closed sp^2-bonded structure, let's consider the relative energies of various structures to get insights into why and how they form. In particular, we will examine why closed sp^2 bonded structures form in the first place. Later, when discussing kinetics, we will consider the distinction between ball-like structures such as C_{60} and long thin cylindrical structures such as carbon nanotubes.

Later chapters will give a more detailed description of fullerene synthesis; here we need know only that carbon clusters are produced in a high-temperature and low-density environment. The carbon source generally provides single carbon atoms or dimers that extend the growing structure, often in the midst of an unreactive buffer gas that helps encourage thermal equilibration. Tubes often grow while attached to a surface, such as that of a small metallic particle, whereas smaller ball-like fullerenes typically grow entirely in the gas phase.

If we want to use the relative energies of different structures to shed light on which ones are preferred during synthesis, we are restricted to situations where the system is near a thermal equilibrium. Only then does the system have time to explore the whole range

[c] These bond distortions are just a discrete analog to the continuum elasticity theory result that a spherical surface has minimal curvature energy.

[d] If one defines radii of curvature along the two principle axes passing through a given point on a surface, then the mean curvature at that point is the arithmetic mean of the inverses of the two radii of curvature, while the Gaussian curvature is the geometric mean of the inverses of these two radii. Since the radius of curvature of a cylinder is infinite along the axial direction, the Gaussian curvature of a cylinder is zero, while the mean curvature is finite.

of possible structures accessible to it; the lowest energy structure is then selected out as the system cools. As always in thermodynamics, one has to be careful about which degrees of freedom in the system are fast enough to become equilibrated and which ones are sufficiently slowed down (usually by large activation barriers or large phase spaces to explore) to prevent the system from accessing all possible configurations on the experimental timescales. Fullerenes typically grow very quickly in a highly transient environment. Therefore, if we want to keep things simple and consider all of the degrees of freedom to be thermalized, then we are restricted to considering only small clusters of atoms that have less configuration space to explore.

The smallest clusters of carbon atoms (those below about $N = 20$ atoms) do not form sp^2 bonds at all. Instead, they form linear chains. In this regime of very small sizes, where edges are very important, the decreased edge-to-interior ratio in one dimension favors chain-like structures (which have only two edge atoms at the exposed ends) over two-dimensional graphene-like structures. As the number of atoms in the cluster increases, the one-dimensional chains eventually became long enough that the reward for eliminating the two edge atoms outweighs the cost of bending, so the chains close into rings. However, one-dimensional structures make inefficient use of the strong carbon nuclear potential, since a linear structure has weaker overlap between the atomic potentials. In a double-bonded carbon chain, the binding energy for an interior atom is about 6 eV. The binding energy in a flat two-dimensional sp^2 sheet is larger, about 7.5 eV/atom. As the number of atoms in a cluster increases, the binding energies of the interior atoms begin to dominate; the edge atoms and curvature become less important and the system transitions from one-dimensional chains with sp bonding to two-dimensional sheets with sp^2 bonds.[e]

Are these sp^2-bonded sheets open like a bowl or closed like a ball? When an sp^2-bonded sheet is bent away from a perfectly flat geometry, the energy per atom goes up proportional to $1/R^2$, where R is the radius of curvature.[f]

For a patch of graphene with N atoms, the characteristic linear dimension (which sets the size of the radius of curvature in a closed structure) is proportional to $\sqrt{N}$, so the bending energy per atom is proportional to $1/N$, and the total energy of bending is independent of the number of atoms. The energetic penalty for the edge atoms along the perimeter of an open graphene sheet goes up as $\sqrt{N}$. Since the energy cost of curvature in the closed structure is roughly constant, while the energetic cost of dangling bonds in the open structure goes up as $\sqrt{N}$, for large enough clusters a closed structure is lower in energy.

As we showed in the section on topology, every closed sp^2-bonded structure has twelve pentagons and the most favorable such structures are those that separate out the twelve pentagons as evenly as possible and avoid having the edges of any two pentagons

[e] The transition from closed one-dimensional rings to closed two-dimensional sheets is perhaps best thought of as the favorable interior bonding energy in two dimensions overpowering the increased curvature energy which arises from the reduction in the radius of curvature from $R \sim N$ in one dimension to $R \sim \sqrt{N}$ in two dimensions.

[f] Why $1/R^2$ and not $1/R$? The curvature-induced change in the potential felt by the electrons is sensitive to the sign of R and hence proportional to $1/R$ (not $1/R^2$). The curvature introduces σ character into the π states and π character into the σ states. This hybridization is also sensitive to the sign of R and therefore is also proportional to $1/R$. Since the perturbation in the potential is odd, it only has a finite matrix element between perturbed and unperturbed portions of an electronic wavefunction. The two factors of $1/R$ then yield an energetic change proportional to $1/R^2$.

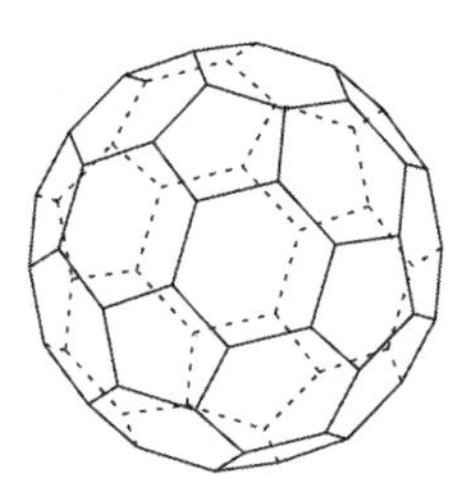

FIGURE 4.5. C_{60}.

fused together (Figure 4.5). In addition, since thermal equilibrium in the transient environment of fullerene synthesis can only be obtained for small structures, the best bet for being able to exploit thermodynamics to select out the most stable accessible structure is to choose the smallest structure that keeps the pentagons isolated from each other. That structure is C_{60}. The next smallest isomer that also has isolated pentagons provides the second most common fullerene, C_{70}. C_{70} consists of two C_{60} caps with an extra row of hexagons in between. For larger-scale carbon polyhedra such as nanotubes, nanocones, or giant onion-like fullerenes, the system no longer has time to explore all possible configurations, so the types of structures produced are determined by a mixture of energetics and kinetics.

4.6. KINETICS

In ever-larger sp^2-bonded carbon structures the curvature and/or edge atoms generally become less and less important and the energies of nearly all structures approach that of planar graphite. These larger structures do not have enough time to thermally explore all possible configurations, particularly since the strongly directional covalent bonding produces many metastable minima in the energy surface. Kinetics, meaning the non-equilibrium exploration of only a fraction of the possible structures, becomes more important. This dominance of kinetics over thermodynamics allows for a rich variety of large-scale structures.

The detailed microscopic mechanisms by which carbon nanostructures nucleate and grow remain largely mysterious, since growth is a fleeting and high-energy process that is difficult to characterize. However, the geometry of sp^2-bonded carbon imposes certain topological constraints that help us classify the possibilities. The first step toward creating a large-scale fullerene structure is to nucleate a small seed structure; the geometry of this seed then defines a growth zone, the part of the structure that incorporates the new carbon atoms as the structure grows. The geometry of the growth zone has a very strong influence on the shape of the final structure. These large-scale structures grow predominately through the addition of hexagonal rings, since the ring geometries in an active growth zone are reasonably well-thermalized and hexagonal rings are usually the lowest-energy sp^2-bonded rings.

We can consider three distinct geometrical possibilities for the seed, depending on how many pentagonal rings it contains: from one to five pentagons, from seven to eleven

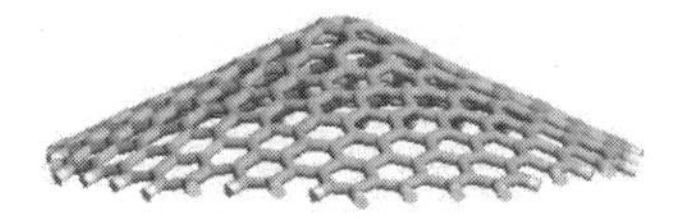

FIGURE 4.6. Cone.

pentagons, or exactly zero, six, or twelve pentagons. A seed with one to five pentagons has less than half the Gaussian curvature necessary to wrap into a closed fullerene, so it defines an open expanding cone (Figure 4.6). These sp^2-bonded carbon cones have been made with all degrees of acuteness from one pentagon to five pentagons. Assuming that exclusively hexagons are added to the growing edge, the length of the cone's open edge expands as the square root of the number of atoms in the cone. Since the growth edge is ever-expanding, one expects that it eventually becomes difficult to maintain satisfactory growth conditions across this entire perimeter.

During growth, such a cone may occasionally add a pentagonal ring to the mostly hexagonal structure. Should the system accumulate seven or more pentagons, then the Gaussian curvature is strong enough to curl the seed structure from an outwards expanding cone into an inwards tapering cone (Figure 4.7). As additional hexagons extend the structure, the open edge shrinks and eventually closes up upon itself, when it accumulates a total of twelve pentagons. The subtle interaction of bond angles and dangling bonds actually favors pentagons over hexagons once the growth edge tapers down into a sufficiently small opening. Eventually, the structure forms a large, somewhat lumpy closed fullerene. This sort of structure is favored in conditions that allow for the occasional creation of a pentagonal ring during growth.

What of the seed structures with exactly zero, six, or twelve pentagons? These structures are special: they have the ability to grow while maintaining a steady, unchanging growth zone. How? By extending the structure only in one dimension, the growth zone can retain a constant shape. A seed with zero pentagons is a sheet of graphene wrapped around into a belt with open edges at both ends (the other option for zero pentagons is a flat sheet, which forms a simple graphene flake). Such a cylindrical seed could extend along its axis

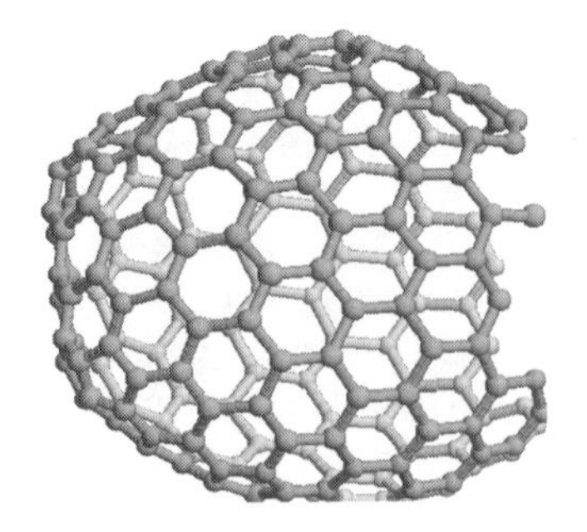

FIGURE 4.7. Taper.

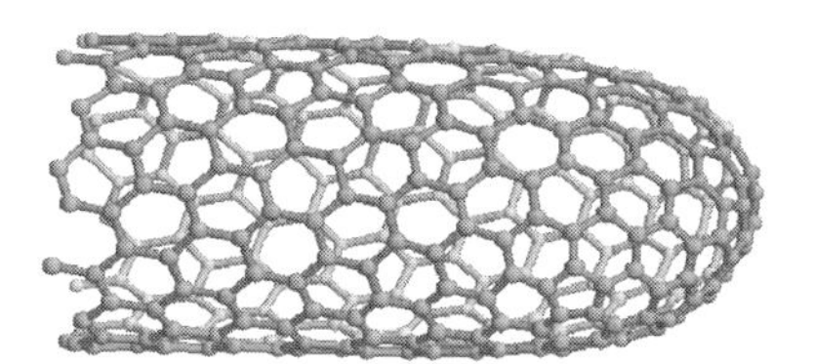

FIGURE 4.8. Tube closed on one end.

by adding hexagonal rings to the edges while maintaining an open edge of constant size. A seed with six pentagons forms a hemisphere. Adding hexagons to the open edge of such a seed will extend the edge to form a long thin cylinder (Figure 4.8). Once a such cylindrical extension begins to form, it becomes very difficult to insert new hexagons into the hemispherical cap, since that would expand the cap and destroy the match in diameter between the cap and the cylindrical extension; (any such hexagons would tend to migrate into the cylindrical extension to re-match the diameters and thereby minimize strain). These zero and six pentagon seeds have one or two open edges; these edges could be either plugged by a metallic (or metal carbide) nanoparticle or left open to the environment. In either case, these highly reactive edge regions are likely to be the growth zones.

A seed with twelve pentagons is a closed fullerene with two hemispherical caps joined together (Figure 4.9). Such a structure would act similarly to the closed hemispherical end of a six-pentagon seed. It is not yet clear experimentally if and when nanotubes grow from seeds with zero, six or twelve pentagons. However, the fundamental geometrical mechanisms favoring one-dimensional structures are identical in these three cases. These seeds can extend into a one-dimensional cylindrical shape whose growth zone maintains constant size and shape as the number of atoms increases. Since the growth edge retains its shape, the kinetics of growth remain constant and such a structure can grow very long.

The energy of a long carbon nanotube is dominated by the $1/R^2$ curvature energy of the walls. Energetics plays a role in constraining the possible diameters: nanotubes with diameters smaller than about 0.7 nanometers suffer from reduced stability.

4.7. OTHER RINGS

Pentagons provide an efficient way to close up the structure and thereby cap dangling bonds. Heptagons, in contrast, open up the structure. Since expanding the edge of an open structure is normally energetically unfavorable, heptagons are less common in low-density synthesis conditions where dangling bond energy is more important.

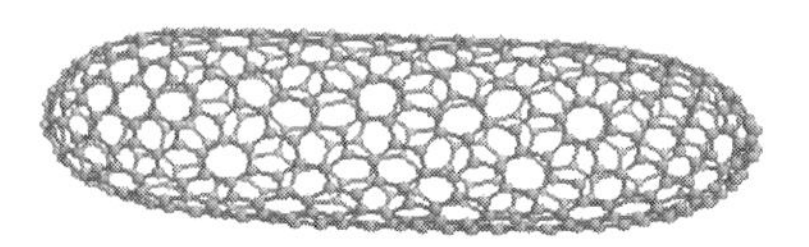

FIGURE 4.9. Tube closed on both ends.

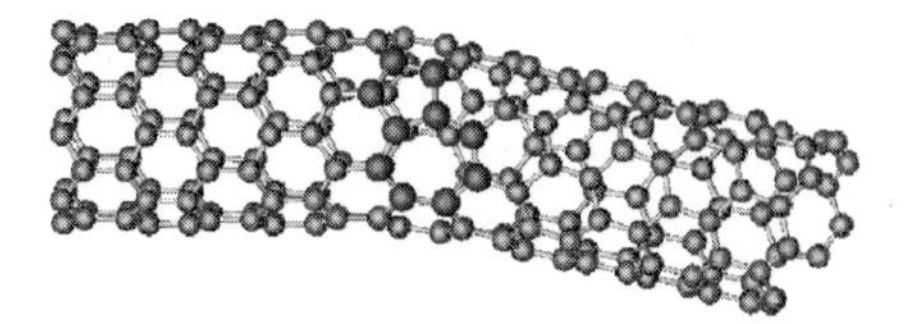

FIGURE 4.10. Junction.

This expansion of the open edge can be avoided by pairing each heptagon with a pentagon. Just as one can add an arbitrary number of hexagons to a closed sp^2-bonded structure without disturbing the rule $F + V = E + 2 - 2G$, one can also add equal numbers of heptagons and pentagons.[g]

Such pentagon-heptagon pairs are observed: when the pentagon and heptagon are separated by intervening hexagons along the axis of a tube, the nanotube tapers (at the pentagon) and flares (at the heptagon). When the heptagon and pentagon are close together, the tube diameter does not change much, but the tube may bend abruptly (Figure 4.10).

4.8. SURFACES

So far we have been considering only the energy of the σ and π bonding within a single graphene layer. However, these graphene layers also prefer to stack one atop another in graphite: Graphene sheets attract each other; pulling them apart and exposing free surfaces costs energy. This surface energy is very small, but can become important for large surfaces.

Sometimes the surface energy is important enough to open new kinetic pathways, wherein new atoms stick onto an existing sp^2-bonded surface and form another layer. The attractive interaction between two curved graphene layers is opposed by the $1/R^2$ curvature energy required to bend an outer incomplete sp^2-bonded patch into contact with the curved inner sheet. Therefore for sufficiently large R, an outer layer can form (Figure 4.11). The crossover radius is small, so most all fullerene structures would prefer to accrete additional

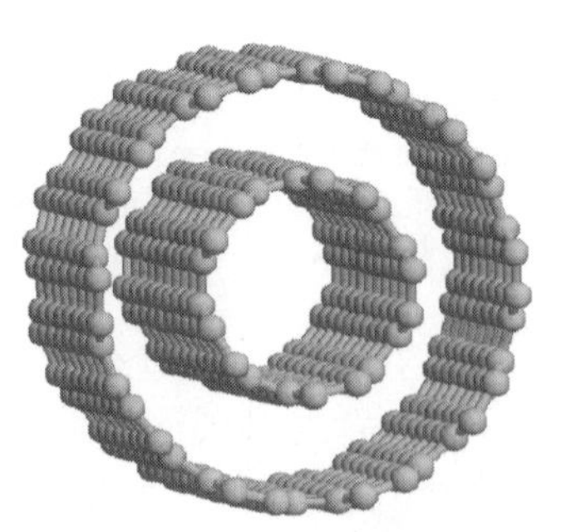

FIGURE 4.11. Two-walled tube.

[g] The Gaussian curvatures of the pentagon and heptagon cancel each other.

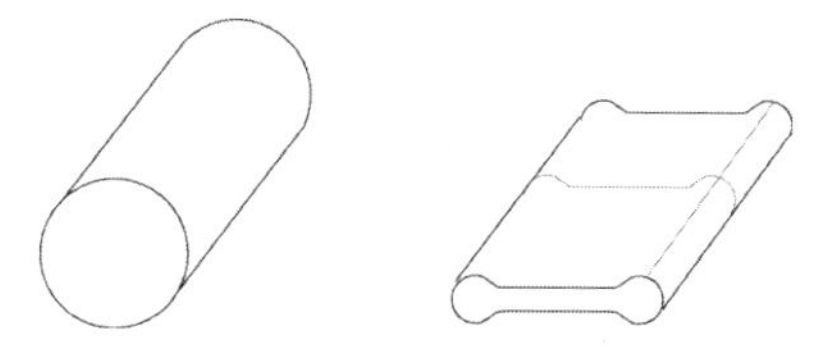

FIGURE 4.12. Flattened tube.

outer layers, if the outermost layer is exposed to a source of new carbon atoms and synthesis conditions allow graphene-like patches to form on an exposed graphene surface. Multiwalled nanotubes and onion-like fullerenes result.

Nanotube synthesis can be catalyzed by metallic particles that plug the open end(s). Formation of outer layers is then suppressed (although an outer layer of amorphous carbon may form instead) and the tubes are predominately single-layered. However, the surface energy still plays a role: the growing tubes attract each other, aligning into bundles with the constituent tubes arranged into a roughly triangular lattice transverse to the bundle axis.

The surface energy can also change the cross-section of an individual nanotube. Tubes with a large diameter can flatten into ribbons which take advantage of the attraction between opposing interior faces (Figure 4.12). The energetic gain due to the intersheet attraction is proportional to the tube's diameter. The energetic cost of curvature is determined by the shape of the bulbs along either edge of the ribbon; this is essentially independent of the tube diameter. Therefore, the lowest-energy state of a large-enough diameter tube is flattened into a ribbon. Since the energetic curvature cost of any distortion from a circular cross-section is immediate while the gain in surface energy is short-ranged, there is a kinetic barrier against collapse. Experimentally, tubes with a particularly large internal radius can be flattened by a moderate uniaxial compression transverse to the tube axis. Single-walled tubes of diameter $\sim$1 nm are more stable when inflated than when collapsed.

4.9. HOLES ($G \neq 0$)

The topological rule $F + V = E + 2 + 2G$ implies that a structure with $G > 1$ must have an excess of heptagons over pentagons (Figure 4.13). Since heptagons are disfavored in low-density conditions where dangling bond energies dominate, holey carbon structures should be preferred only in high-density synthesis conditions. The best such example is nanoporous carbon. Nanoporous carbons are formed by pyrolysis: at high temperatures the material decomposes into pure carbon plus various gaseous species that must escape

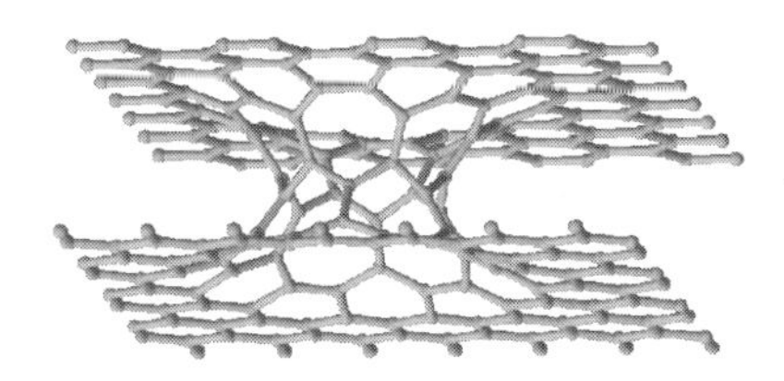

FIGURE 4.13. Wormhole.

from the still-forming disordered sp^2-bonded structure. These gases induce the formation of a disordered network of interconnected escape channels. Since the density of carbon in a pyrolyzing sample is much higher than that during a gas-phase fullerene synthesis, heptagons can form more readily. The resulting structure is complex, with a mixture of five-fold, six-fold and seven-fold rings, an unknown admixture of sp^3 bonds, and a very large number of holes. If we assume for simplicity that there are no sp^3-bonded carbon atoms, then the excess of heptagonal rings over pentagonal rings should be $12(G - 1)$ where G is the number of holes in the structure. The number of holes is equivalent to the number of times that the sp^2-bonded surface must be mathematically cut before one can contract the surface into a single giant closed fullerene ball (perhaps with some rather large rings).

4.10. CONCLUSIONS

Carbon's rich variety of two-dimensional structures arises not only from kinetics, but also because three-dimensional sp^3-bonded diamond-like structures are actually slightly less stable than graphite at zero pressure. The development of large, complex two-dimensional structures is not arrested by another transition in dimensionality, as happened for one-dimensional structures. In contrast to carbon, clusters of nearly all other elements are essentially always three-dimensional, with edge effects imposing at most a local surface reconstruction.

We end this section with the parable of the squirrel and the ant. A student asks a teacher whether a carbon nanotube is one dimensional, since it is long and thin, or two-dimensional, since it is composed of an sp^2 bonded sheet. The teacher responds, "Consider the squirrel and the ant. The squirrel, crawling on a telephone line, declares that the telephone line is a one-dimensional object, since the squirrel can scamper only back and forth along it. The ant, however, declares the telephone line to be two-dimensional, since it can happily crawl both along the length and around the circumference of the wire. So it is with the nanotube."

QUESTIONS

1. Describe how one could make extended sp^2-based solids from boron and/or nitrogen by combining more than one element into the structure. How might one expect this material to differ from graphene?

2. Show that bending a graphite sheet perpendicular to the plane of the σ bonds changes the carbon-carbon bond-length by an amount that is quadratic in the upwards shift of a carbon atom away from the original flat plane.

3. The transition metal dicalcogenides also form nanotubes. The fundamental structural subunit of these materials is a sheet with a triangular (not hexagonal) lattice. Derive the topological rules that determine the geometries of closed surfaces formed from sheets of transition metal dicalcogenides.

4. As of yet, there are no known closed sheet-like structures formed from materials that prefer square lattices. Can you think of any candidate materials that might form such structures? Discuss the strengths and weakness of your candidates as regards the energetic and kinetic aspects of their putative syntheses.
5. Describe why heptagons are energetically preferable to octagons as defects within a hexagonal graphene sheet.
6. Define the circumference of a nanotube as (n, m) in graphene lattice coordinates. How are the indices (n, m) changed when a pentagon/heptagon pair, is added to the structure of a growing tube? Treat only the special case where the pentagon and the heptagon share a bond in common.
7. Write the surface energy per atom of a graphene sheet as ϵ. Write the mean curvature modulus (i.e., energy-length^2 per atom) of a graphene sheet as κ. The radius of the bulb on the edge of a flattened nanotube can be written as a function of a certain combination of ϵ and κ. What is this combination?

ACKNOWLEDGMENTS

The author thanks P. Lammert, D. Stojkovic, and S. Jordan for providing graphics, and also thanks P. Lammert for a helpful critical reading of the manuscript.

REFERENCES

Reviews

1. J. R. Heath, S. C. O'Brien, R. F. Curl, H. W. Kroto, and R. E. Smalley, "Carbon Condensation" *Comments Cond. Mat. Phys.* **13**, 119 (1987).
2. L. D. Lamb and D. R. Huffman, "Fullerene production" *J. Phys. Chem. Sol.* **54**, 1635 (1993).
3. *Buckminsterfullerenes* ed. W. E. Billups and M. A. Ciufolini (VCH Publishers, Inc. 1993).
4. G. E. Scuseria, "Ab initio calculations of fullerenes" *Science*, **271**, 942 (1996).
5. M. S. Dresselhaus, G. Dresselhaus, and P. C. Eklund, *Science of Fullerenes and Carbon Nanotubes* (Academic Press, San Diego, 1996).

Small fullerenes, experiment

6. H. W. Kroto, J. R. Heath, S. C. O'Brien, R. F. Curl, and R. E. Smalley, "C_{60}: Buckminsterfullerene" *Nature*, **318**, 162 (1985).
7. W. Kratschmer; L. D. Lamb, K. Fostiropoulos, D. R. Huffman, "Solid C_{60}: a new form of carbon" *Nature*, **347**, 354 (1990).

Large fullerenes (including nanotubes), experiment

8. S. Iijima, "Helical microtubules of graphitic carbon" *Nature*, **354**, 56 (1991).
9. S. Iijima, T. Ichihashi, and Y. Ando, "Pentagons, heptagons and negative curvature in graphite microtubule growth" *Nature*, **356**, 776 (1992).
10. L. D. Lamb, D. R. Huffman, R. K. Workman, S. Howells, *et al.* "Extraction and STM imaging of spherical giant fullerenes" *Science*, **255**, 1413 (1992). (C_{60}—C_{330})
11. S. Iijima, "Growth of carbon nanotubes" *Mater. Sci. and Eng.* **B19**, 172 (1993).
12. S. Iijima and T. Ishihashi, "Single-shell carbon nanotubes of 1-nm diameter" *Nature*, **363**, 603 (1993).

13. D. S. Bethune, C. H. Klang, M. S. de Vries, G. Gorman, R. Savoy, J. Vasquez, and R. Beyers, "Cobalt-catalysed growth of carbon nanotubes with single-atomic=layer walls" *Nature*, **363**, 605 (1993).
14. N. G. Chopra, L. X. Benedict, V. H. Crespi, M. L. Cohen, S. G. Louie, and A. Zettl, "Fully collapsed carbon nanotubes" *Nature*, **377**, 135 (1995).
15. A. Thess, R. Lee, P. Nikolaev, H. Dai, D. Petit, J. Robert, C. Xu, Y. H. Lee, S. G. Kim, A. G. Rinzler, D. T. Colbert, G. E. Scuseria, D. Tománek, J. E. Fischer, and R. E. Smalley "Crystalline ropes of metallic carbon nanotubes" *Science*, **273**, 483 (1996).
16. A. Krishnan, E. Dujardin, M. M. J. Treacy, J. Hogdahl, S. Lynum, and T. W. Ebbersen, *Nature* (*London*) **388**, 451 (1997).

Large fullerenes (including nanotubes), theory

17. D. H. Robertson, D. W. Brenner, and J. W. Mintmire, "Energetics of nanoscale graphitic tubules," *Phys. Rev.* B **45**, 12592 (1992).
18. D. J. Srolovitz, S. A. Safran, M. Homyonfer, and R. Tenne, "Morphology of Nested Fullerenes" *Phys. Rev. Lett.* **74**, 1779 (1995).

5

Fullerenes

Harry C. Dorn
Department of Chemistry, Virginia Tech, Blacksburg, VA
James C. Duchamp
Department of Chemistry, Emory & Henry College, Emory, VA

5.1. FAMILIES OF FULLERENES: FROM C_{60} TO TNTs

5.1.1. Discovery

Although carbon represents just one of over one hundred known chemical elements, it plays a vital role in nature and represents a very important building block of nanomaterials and nanostructures. The reason for this critical role is the propensity for carbon to bond in so many different ways. Many of nature's most important biological compounds, for example, carbohydrates, proteins, lipids, and DNA are largely based on carbon uniquely bonded to other atoms, such as, nitrogen, oxygen, and hydrogen. The latter case of carbon bonded to hydrogen (hydrocarbons) is of course the important class for all petroleum related products (natural gas, gasoline, diesel fuel). These carbon compounds are extensively studied by sophomore level chemistry students in organic chemistry courses.

Even as late as 30 years ago, carbon was thought to exist in only two allotropic forms, that is, carbon bonded only to other carbon atoms, as represented by graphite and diamond. For diamonds, the sparkling and extremely hard solid crystalline structure is chemically described by interlocking sp^3 hybridized carbon atoms bonded to four other carbon atoms (Figure 5.1) in a perfect tetrahedral (C—C—C bond angles, 109.5°) 3-D lattice. Whereas, as explained in the previous chapter, the structure of graphite (common pencil lead) is quite different and is best described as stacked hexagonal sheets of sp^2 hybridized with π bonding orbitals (C—C—C bond angles, 120°) as shown in Figure 5.1 and corresponding hexagonal crystal structure. The very different properties of these two materials are more clearly understood based on their "nanolevel architecture." For example, the highly delocalized π bonding network for graphite accounts for the higher electrical conductivity of graphite

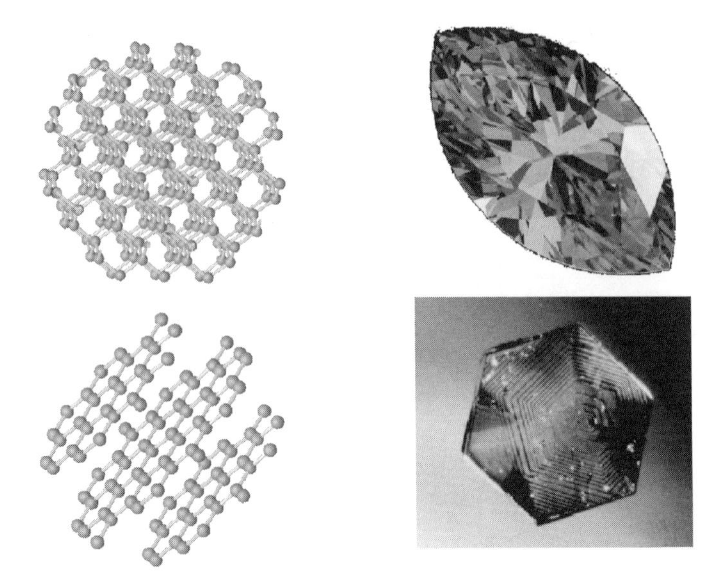

FIGURE 5.1. Forms of Carbon. Top. Diamond's sp^3 crystalline structure (left) and a crystalline diamond sample (right). Bottom. Graphite's sp^2 crystalline structure (left) and a crystalline graphite sample. (Part of this Figure was reproduced with kind permission of D. Bethune.)

relative to diamond. Whereas, the strong covalent interlocking network of sp^3carbons in diamond is responsible for the characteristic hardness of diamonds.

Although Jones in 1966 had suggested from the perspective of Daedulus in the *New Scientist* the possibility of hollow all-carbon cage molecules, it was almost another twenty years before a team at Rice University headed by Rick Smalley in the mid 1980's experimentally verified these predictions. This team was studying atomic clusters formed using a laser-supersonic cluster beam apparatus (Figure 5.2) and had studied cluster chemistry for several elements (e.g., silicon). In 1984, Harry Kroto (visiting Professor from University of Sussex) suggested looking at carbon clusters because of his interest in the conditions for formation of these species in the atmosphere of giant red stars. When the team started analyzing carbon clusters, they observed by analytical mass spectrometry even numbered clusters of two to thirty carbon atoms. However, under certain conditions they observed a preponderance of a peak at higher mass corresponding to a mass of 720 (in units of the proton mass) which correspond to exactly 60 carbon atoms (Figure 5.3). This was invariably also followed by a peak with a mass of 840 corresponding to 70 carbon atoms.* The team quickly recognized that the peak with a mass of 720 (60 carbons) was consistent for an all-carbon molecule with a well known polyhedra shape, a truncated icosahedron, (I_h symmetry). This well known spheroidal shape is the football to most of the world except the United States where it is better known as the soccer ball (Figure 5.3). The soccerball shape was also well known in architecture and especially by Buckminster Fuller who was

* The Rice team maximized these two masses at 720 and 840. During these studies, the team named these new peaks the "Lone Ranger and Tonto" because the former was always more prominent and the latter always followed behind.

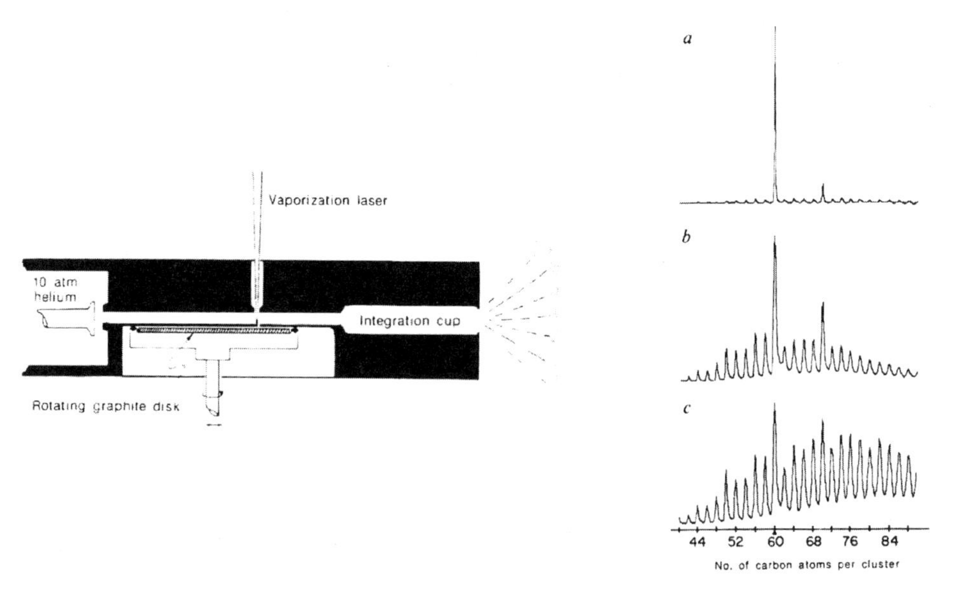

FIGURE 5.2. Laser generated fullerenes. Left. Laser ablation apparatus. Right. Mass spectra of generated carbon clusters under different conditions. (From Ref. 1 by permission of Nature Publishing Group.)

one of the great philosopher-scientists of the past century. Buckminster Fuller "Bucky" was a strong proponent of building with polyhedra and the corresponding advantages for architecture, such as, geodesic domes. With this chemistry-architecture connection established, the Rice team named the new spherical carbon C_{60} molecule buckminsterfullerene which has also been shortened on occasion to buckyballs and/or fullerenes.

5.1.2. *Production*

Although the discovery of C_{60} and subsequent publication by the Rice group is a very important discovery, it remained a curiosity in the late 1980's and was even doubted by

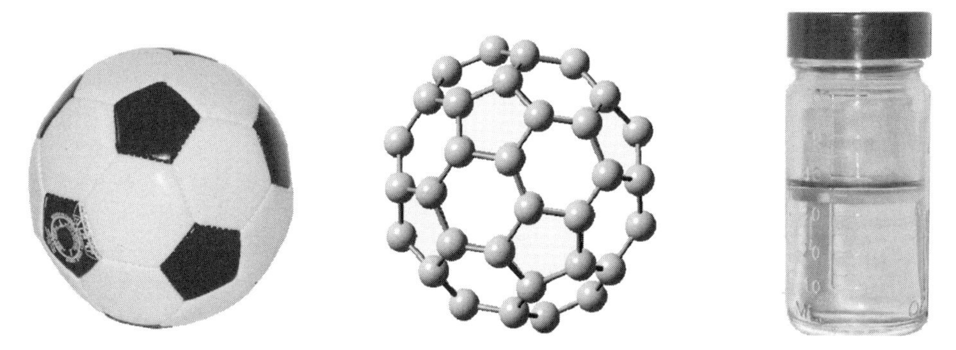

FIGURE 5.3. Buckminsterfullerene. Left. Soccerball showing the pentagons and hexagons that make C_{60}. Center. C_{60} with pentagons highlighted. Right. C_{60} in toluene.

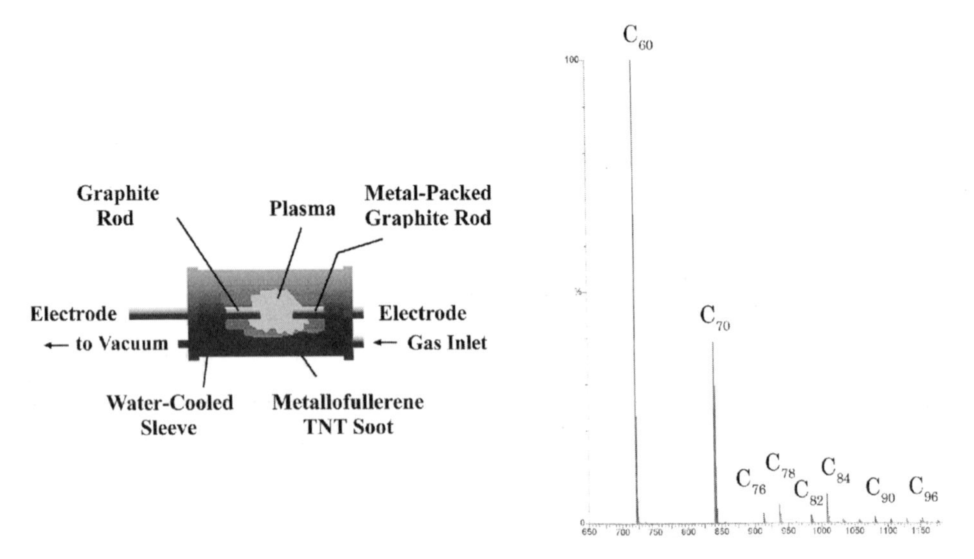

FIGURE 5.4. Plasma generation of fullerenes. Left. Krätschmer–Huffman apparatus. Right. Mass spectrum showing the different fullerenes generated.

some workers. Furthermore, studies were hampered because only limited amounts of these new fullerenes were produced in the laser experiments at Rice. In 1989, Krätschmer and Huffman (at the University of Arizona) were trying to produce laboratory carbon species that would perhaps be material they had been studying in interstellar dust. In 1990, they reported for the first time that fullerenes and especially C_{60} can be produced in an electric-arc apparatus (Figure 5.4). In this preparation, graphite rods are vaporized in an inert gas atmosphere. The role of the buffer gas is critical because it cools the plasma by collisions with the vaporized carbon. The highest yields are observed with helium as the buffer gas and at pressures ranging between 100–200 torr. Under optimized conditions, the Kratschmer-Huffman apparatus provides 5–15% yields of the fullerene product (mainly C_{60} and C_{70}). It was recognized early on that this fullerene mixture is soluble in various non-polar solvents, such as, toluene, carbon disulfide, and benzene. At this stage, the soluble fullerene product still contains a complex mixture of fullerenes (C_{60}, C_{70}, C_{84}, C_{88}, C_{92}, and even higher fullerenes) as shown in Figure 5.4. The next step is the separation of fullerenes from each other—usually by high performance liquid chromatography (HPLC). When purified, the fullerenes are black crystalline powders, but in solution (e.g., toluene) each of the fullerenes exhibit unique colors, for example a purple color for C_{60} (see Figure 5.3) and a red-wine color for C_{70}.

With the addition of certain metals (e.g., Ni and Co) into the core of the graphite rods, it was found independently by Ijima (NEC) and Bethune (IBM) that single walled nanotubes (SWNTs) are also produced in the Kratschmer–Huffman apparatus as described in Chapter 6. On the other hand, the inclusion of other metals and/or metal oxides especially from Group III and rare earths (e.g., Sc, Y, La, Gd) into cored graphite rods produces endohedral metallofullerenes as described below. As previously noted, laser methods can be used for the preparation of fullerenes. Solar generators can be used to produce fullerenes by focusing sunlight on a graphite carbon target, but usually the yields are not very high by this method. More recently, fullerenes have been prepared by lower temperature processes. For example, at lower temperatures (1800 K) fullerenes are formed in a sooty flame (limited oxygen) using benzene or other hydrocarbon precursors. and also pyrolysis of aromatic polycyclic aromatic hydrocarbons. The latter method has the advantage of providing fullerenes in a continuous process at lower formation temperatures. Also, the pyrolysis of polycyclic aromatic hydrocarbons (e.g., naphthalene) has also been used to make fullerenes at 1300 K in an argon stream.

5.1.3. Formation Process

Although fullerenes, such as C_{60}, can be produced in a variety of ways, it is surprising that these beautiful molecules are commonly produced in a chaotic electric-arc process at temperatures in excess of 3000 K. This leads to several key questions: What are the features of a formation process where graphitic carbon materials are vaporized, yet the recovered products are not just graphite (or even diamond), but rather significant quantities of fullerenes or even endohedral metallofullerenes or nanotubes with the addition of metals and/or metal oxides to the graphite rods in the electric-arc process *vide supra*? Although the mechanistic pathway(s) are not totally understood, it begins with the recognition that C_{60} is formed under a kinetically controlled process since the thermodynamic stability of graphite is significantly greater than C_{60}. Furthermore, C_{60} usually dominates C_{70} in the recovered fullerene mixture in a Kratschmer–Huffman electric arc generator (ratio of $\sim 5/1$) as shown in Figure 5.4 (right), yet it is well known that the higher fullerene C_{70} has greater thermodynamic stability. It is also well recognized that the first mechanistic step in the vaporization of graphite (Figure 5.5) is the formation of carbon atoms and/or small atomic carbon clusters C_n $(n = 1-7)$. These results have been verified by several groups using $^{12}C/^{13}C$ isotopic scrambling experiments. In the next mechanistic step, the small clusters form small linear chains which also progress to monocyclics in the range of C_n $(n = 7-10)$. These monocyclic rings progress to larger rings, polycyclic rings and even bicyclic carbon structures. With growth to the range of thirty carbon atoms and beyond, fullerenes are formed. Of critical importance in the mechanistic view is the annealing process by which the precursor high energy fullerene carbon species undergo collisions with other gaseous species and subsequently isomerize to fullerenes. Thus, a key role is the inert buffer gas, typically helium, at relatively high pressures 100–200 torr that leads to energy deactivation in fullerene formation.

Another important factor in the final annealing process is the formation of pentagons in the growing fullerene precursor which provides curvature instead of a flat hexagonal graphitic sheet. The role of pentagons in the fullerene formation mechanism and closure to fullerene spheres has been described as . . . going down "the pentagon road." It is also known that the formation of pentagons instead of hexagons minimizes the number of dangling bonds in a given growing structure and closure reduces the number of dangling bonds to zero. This model also takes account of the extra stability provided by isolated pentagons in the growth process. This has been stated as the well known isolated pentagon rule (IPR) which is known to provide extra stability for fullerenes that minimize the number of adjacent or fused pentagons in their structure. This is illustrated by the three structural motifs illustrated in Figure 5.6. The most common motif (prominent in C_{60}) is the one in the center (meta) which has a so-called (6,6) bond linking two pentagon rings, but flanked between two

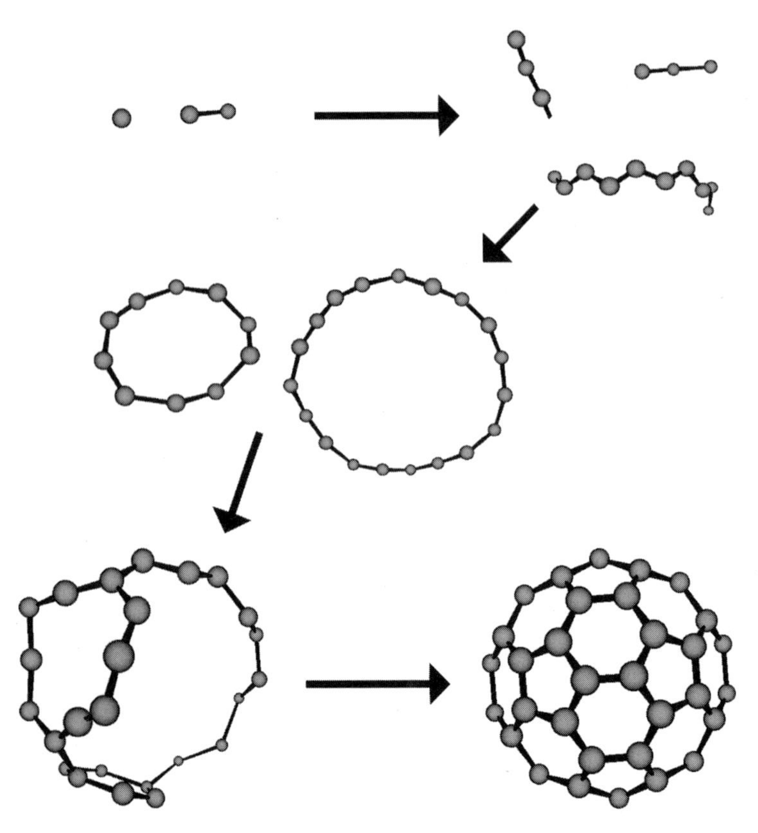

FIGURE 5.5. Formation of fullerenes as they cool in a Krätschmer–Huffman apparatus.

hexagon rings (only one hexagon is shown in the motif). This motif is also extremely important in understanding the chemical reactivity of fullerenes *vide infra*. Thus, soccer ball shaped C_{60} can be formally described as a [5,6]–fullerene-60-I_h because it consists of five and six membered rings containing sixty carbons with icosahedral symmetry. That is, for a given IPR allowed fullerene C_n, it is known from Euler's theorem that there are always 12 pentagons and $(n/2)$-10 hexagon faces which corresponds to 20 hexagon faces for [5,6]–fullerene-60-I_h. The icosahedral symmetry is also unique with only C_n IPR fullerenes of the series $n = 60, 80, 140, 200, 260\ldots$ carbons having the possibility of I_h symmetry. Although there are literally thousands of possible C_{60} fullerene structural isomers that violate the IPR, [5,6]–fullerene-60-I_h is the smallest fullerene that obeys the isolated pentagon rule. The para motif has the pentagon rings but the pentagon rings are isolated from each other and do not have the (6,6) bond juncture. This motif is illustrated by an icosahedral C_{80}-I_h cage, however, this cage is an electronic open-shell structure. Open shell molecules are generally less stable than closed shell molecules (see below) and C_{80}-I_h has not been isolated to date *vide infra*. However, metal stabilized examples of this cage are well recognized $Sc_3N@C_{80}$ (see below) and $La_2@C_{80}$. For the motif on the left, this is an example of a violation of the IPR and examples with this case are also known for certain metal-stabilized endohedral metallofullerenes ($Sc_3N@C_{68}$, $Sc_2@C_{66}$).

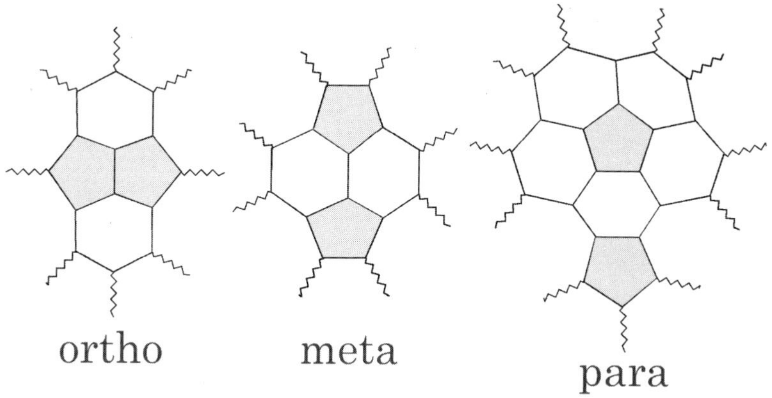

ortho meta para

FIGURE 5.6. Fullerene ortho, meta, and para motifs.

For the case of the isolated fullerene C_{70}, and the higher fullerenes C_{76}, C_{78}, C_{84} that have been isolated to date, the IPR has been observed with a corresponding total of twelve pentagons in each fullerene structure and a corresponding increase in the number of hexagons faces (32 for C_{84}). It is also interesting to note that many of the higher fullerenes typically exhibit lower symmetry. For the case of C_{70}, the isolated stable isomer has ellipsoidal shapewith D_{5h} symmetry. For the higher empty-cage fullerenes, examples are illustrated, D_2 for one of the C_{76} isomers, and D_{2d} for one of the C_{84} isomers (Figure 5.7). Furthermore, the number of IPR allowed isomers increases dramatically for the higher fullerenes with the fullerenes C_{70}, C_{76}, C_{78}, and C_{84}.

5.1.4. Properties

A necessary starting point in understanding fullerenes is the electronic structure of these spherical closed carbon cages. Starting with a rather simplistic approach, one can construct the fullerene C_{60} molecule from atomic carbon atoms. Specifically, we start with single carbon atoms with localized core ($1s$) electrons and four valence electrons consisting of one $2s$ and three $2p$ electrons. If sixty isolated carbon atoms, collapse together and form a sphere bonded with sp^2 hybridized orbitals, three of the four valence electrons on each carbon atom form a σ bonding network (three sp^2 hybridized orbitals) with 60 remaining

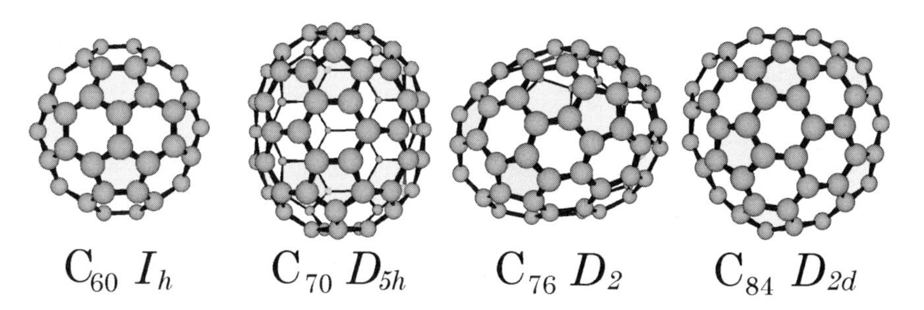

FIGURE 5.7. Fullerene structures with pentagons shaded.

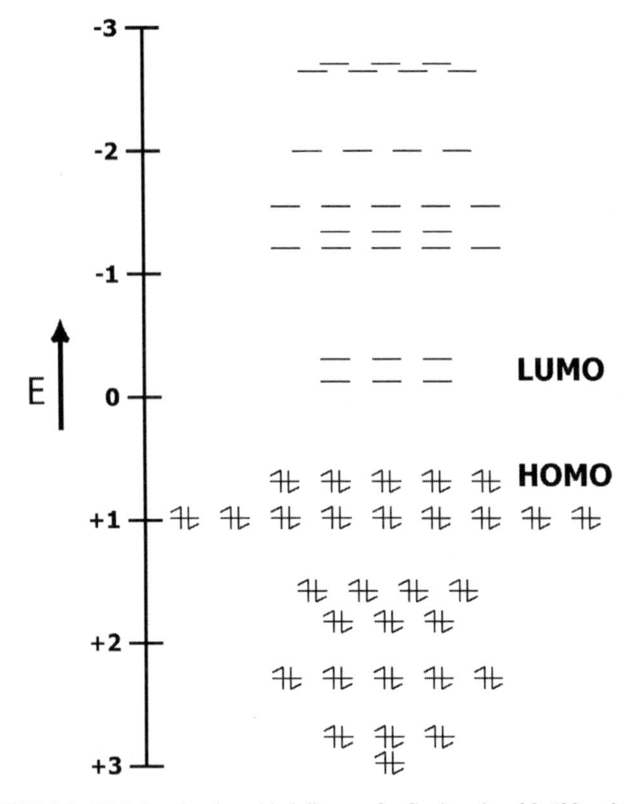

FIGURE 5.8. Hückel molecular orbital diagram for C_{60} in units of β. (2β ~ 36 kcal)

p orbitals. Thus, there will be a single radial *p* orbital centered on each of the 60 carbon atoms.

If we now employ a simple Hückel molecular orbital (HMO) approximation, then the σ bonding network is treated separately and attention focuses on the *p* orbitals and corresponding π bond formation. This simple picture leads to a HMO representation for the molecular level diagram as illustrated in Figure 5.8. Since we started in this approach with 60 *p* orbitals, we have a total of sixty molecular orbitals. As illustrated, there are thirty highest occupied molecular orbitals (HOMO's) and thirty unoccupied molecular orbitals (LUMO's) in the ground state for [5,6]–fullerene-60-I_h. The electron configuration of the neutral molecule for [5,6]–fullerene-60-I_h is a closed shell (all 60 electrons occupy bonding MO's) and this represents one measure of the stability of this molecule. For this level of theory, the HOMO-LUMO gap can also be shown to be fairly large. In the analysis above, we have ignored a very important feature of the hybridization process, namely, because of the local curvature at each carbon site the hybridization in C_{60} is not pure sp^2 as is the case for an infinite graphite sheet. The local curvature at each carbon site leads to a mixing of the carbon 2*s* orbital with the π-orbitals leading to MO's with intermediate rehybridization between pure sp^2 (graphite) and sp^3 for diamond. This rehybridization leads to enhanced electron affinity for C_{60} with a measured value of 2.65 eV and has higher electronegativity than most hydrocarbons.

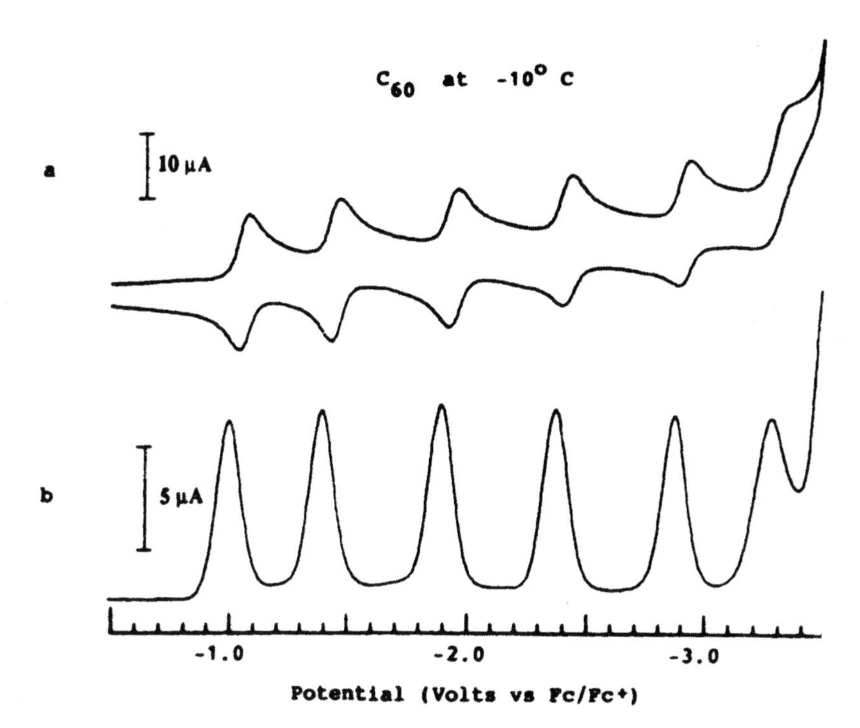

FIGURE 5.9. Cyclic voltammogram for C_{60}. (From Ref. 7 by permission of the American Chemical Society.)

It is also apparent from the energy level diagram that under reductive conditions, C_{60} can readily accept six additional electrons that can go into the three degenerate LUMO's. This is clearly observed in the solution acetonitrile/toluene electrochemistry of C_{60} with the formation of the mono- to $(C_{60})^{-1}$ to hexaanion $(C_{60})^{-6}$ as shown by the cyclic voltammetry and differential pulse voltammetry as shown in Figure 5.9. Also, the addition of three electrons to (half-filling) the degenerate LUMO by exohedral metal doping with potassium leads to a face-centered cubic (fcc) K_3C_{60} structure (see Figure 5.10 right) that is an electrical conductor at higher temperatures, but is a superconductor at temperatures below 19 K. Various other A_3C_{60} superconductors have been prepared with even higher superconducting transition temperatures (e.g., $RbCs_2C_{60}$ has a superconducting transition of 33 K). Whereas, the addition of six electrons by external doping yields a body-centered cubic (bcc) structure K_6C_{60} (see Figure 5.10 left) and has much poorer electrical conductivity.

Another interesting feature of fullerene chemistry is their unique colors when dissolved in nonpolar solvents (e.g., hexane, carbon disulfide, xylene, benzene). The beautiful purple and red-wine colors associated with C_{60} and C_{70}, respectively, in solution are well known and are due to the moderate electronic absorption in the visible region (400–600 nm). C_{60} exhibits strong absorption in the UV range (below 300 nm) but absorption extends well beyond the visible to nearly 700 nm.

An interesting structural feature of C_{60} is the fact that all 60 carbon atoms are chemically equivalent. This is easily recognized from the icosahedral symmetry and also experimentally based on the single peak observed in the ^{13}C NMR at ~143 ppm. The ^{13}C NMR technique

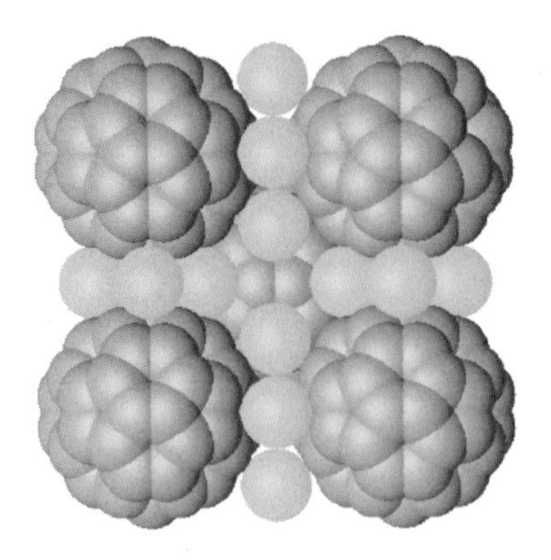
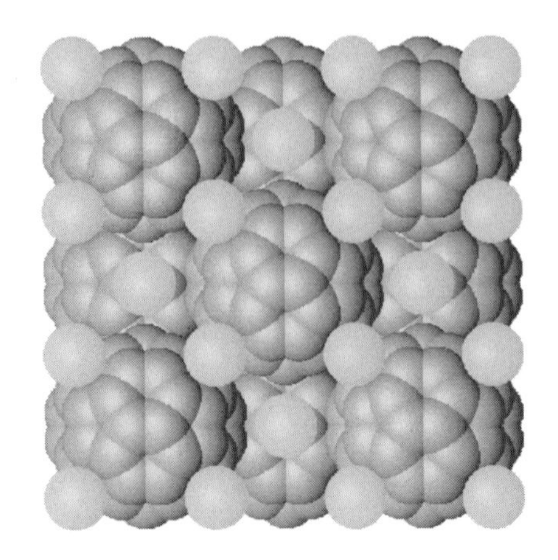

FIGURE 5.10. Structures of C_{60} doped with potassium. Left. K_6C_{60}—body centered cubic structure (bcc). Right. K_3C_{60}—face centered cubic structure (fcc). Structural data from Ref. 3.

is a powerful probe of local structure around a given nucleus. Although there is only one unique carbon atom in C_{60}, there are two different bond lengths in C_{60}. One bond length is for the hexagon-hexagon edges (30) and the other is hexagon-pentagon edges (60). The latter edge is the (6,6) bond described above and represents increased double bond character with a carbon-carbon bond length of $\sim$1.38 A. In contrast, the bond length of the hexagon-hexagon edge (5,6) is somewhat longer $\sim$1.45 A. This supports the view that the double bonds in C_{60} are more localized at the hexagon-hexagon (6,6) bond and are not significant inside the pentagon rings.

The X-ray crystalline C_{60} data indicates a center to center distance of 10 Å between adjacent C_{60} molecules and a molecular diameter of 7.1 Å which is consistent for a Van der Waals interatomic distance of 2.9 Å. It is also known from X-ray and ^{13}C NMR studies that the C_{60} molecule rotates rapidly in the solid state crystalline lattice ($\sim 10^{10}$ s^{-1} at room temperature). This rapid molecular rotation of solid C_{60} at room temperature is "liquid-like" motional behavior and very unusual for a solid. In addition, at 249 K the solid progresses through a phase transition from a simple cubic structure to a fcc structure. The change in phases is characterized by free rotation above 249 K and "ratched phase" motion below 249 K.

5.2. REACTIVITY

5.2.1. *Fullerenes*

To understand the reactivity of fullerenes, it is important to remember that most common organic chemistry reactions involve either front-side and/or back-side attack of a carbon atom in a molecule. A good example from organic chemistry is the well known electrophilic reactions of benzene (C_6H_6) derivatives at either face of the benzene ring. This is representative of most simpler molecules that do not have a closed carbon surface shell. The closed surface fullerenes only allow addition of attacking agents to the exohedral surface of the fullerene. In addition, the relatively high electron affinity of

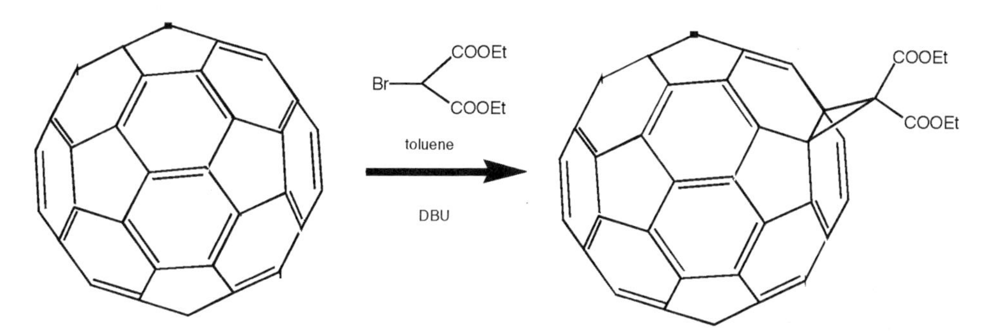

FIGURE 5.11. Bingel-Hirsch reaction for C_{60}.

the fullerenes provide a predictive description of the reactions that fullerenes should undergo as nucleophilic addition reactions with a number of nitrogen, carbon, sulfur, and oxygen based nucleophiles. Although electrophilic reactions do occur with fullerenes, they are less common and can lead to reactions involving opening the fullerene cage. These can include oxygenation, halogenations, and reactions with strong oxidizing acids. It should also be noted that fullerenes also readily undergo reactions with free radicals.

A second important feature of the reactions of fullerenes is the higher reactivity at the (6,6) carbon junction between two pentagon rings (see the meta motif). Thus, one of the most common reactions of fullerenes is nucleophilic addition reactions of the Bingel–Hirsch type as shown in Figure 5.11 for C_{60}. The corresponding reaction for C_{70} is shown in Figure 5.12. In the latter case, the greater number of more reactive (6,6) carbon bonds at the polar endcap of this fullerene leads to a dominance of products formed from this addition with multiple isomers. In addition, multiple additions (di- and tri-) occur to give other products that exploit the regio-chemistry of the fullerenes surface. These functionalization reactions are very important in the development of new fullerene products. In one example, the fullerenes require functionalization to convert their hydrophobic surfaces to provide more hydrophilic character for ultimate diagnostic and therapeutic medical applications (e.g., MRI contrast agents see below).

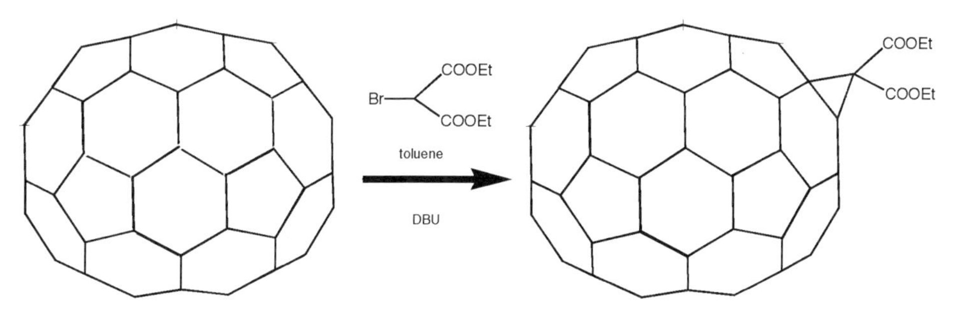

FIGURE 5.12. Bingel–Hirsch reaction for C_{70}.

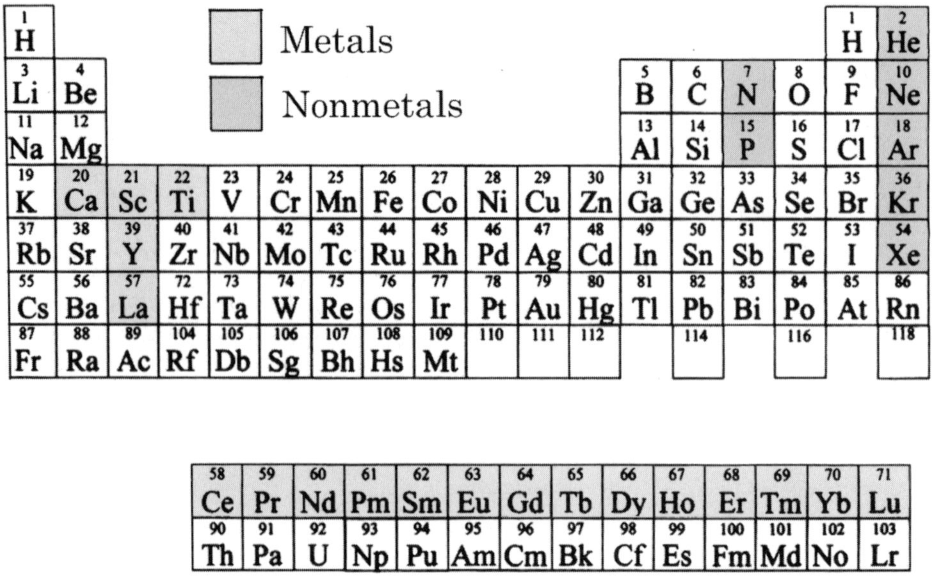

Metals

Nonmetals

1 H																1 H	2 He
3 Li	4 Be											5 B	6 C	7 N	8 O	9 F	10 Ne
11 Na	12 Mg											13 Al	14 Si	15 P	16 S	17 Cl	18 Ar
19 K	20 Ca	21 Sc	22 Ti	23 V	24 Cr	25 Mn	26 Fe	27 Co	28 Ni	29 Cu	30 Zn	31 Ga	32 Ge	33 As	34 Se	35 Br	36 Kr
37 Rb	38 Sr	39 Y	40 Zr	41 Nb	42 Mo	43 Tc	44 Ru	45 Rh	46 Pd	47 Ag	48 Cd	49 In	50 Sn	51 Sb	52 Te	53 I	54 Xe
55 Cs	56 Ba	57 La	72 Hf	73 Ta	74 W	75 Re	76 Os	77 Ir	78 Pt	79 Au	80 Hg	81 Tl	82 Pb	83 Bi	84 Po	85 At	86 Rn
87 Fr	88 Ra	89 Ac	104 Rf	105 Db	106 Sg	107 Bh	108 Hs	109 Mt	110	111	112		114		116		118

58 Ce	59 Pr	60 Nd	61 Pm	62 Sm	63 Eu	64 Gd	65 Tb	66 Dy	67 Ho	68 Er	69 Tm	70 Yb	71 Lu
90 Th	91 Pa	92 U	93 Np	94 Pu	95 Am	96 Cm	97 Bk	98 Cf	99 Es	100 Fm	101 Md	102 No	103 Lr

FIGURE 5.13. Periodical table showing elements successfully incarcerated inside fullerene cages to make endofullerenes.

5.2.2. Endofullerenes

One of the first questions after the discovery of C_{60} and the rest of the family of fullerenes was whether atoms or molecular clusters could be trapped inside of fullerene cages. This question was quickly answered in subsequent years with reports that a variety of different elements (see Figure 5.13) can indeed be encapsulated in the fullerene sphere. It should be noted that most metals encapsulated to date have +2 to +4 oxidation states with +3 being very common (Group III and rare earths). Because these atoms are trapped inside of fullerenes, they have been called endohedral fullerenes or endofullerenes. If metals are encapsulated, then the description has often been endohedral metallofullerenes. More properly these molecules represent examples from the class of jailed or incarcerated fullerenes or simply, *incar*-fullerenes. Nevertheless, the development of this exciting class of fullerenes is approximately 10 years behind the development of empty-cage fullerenes because of low yields and purification difficulties.

Various noble gases (He, Ar, Xe) have also been encapsulated in carbon cages at elevated temperatures and pressures (600–1000°C and 40,000 psi). A particularly interesting example is the case of radioactive tritium ^{3}He (obtained as a nuclear processing by-product) and can be incorporated in fullerene cages, He@C_{2n}. Since this isotope can be monitored by ^{3}He NMR, this provides a very valuable analytical approach for determining the number of isomers in a fullerene mixture. Another important example is the encapsulation of N or P atoms, the formula for these molecules is represented by the case of a single N atom inside a C_{60} cage as N@C_{60}. It is interesting to note that these molecules are paramagnetic since there is an odd unpaired electron centered on the N or P atom in the carbon cage. The incarceration of "naked nitrogen atoms" in a carbon cage leads to very interesting

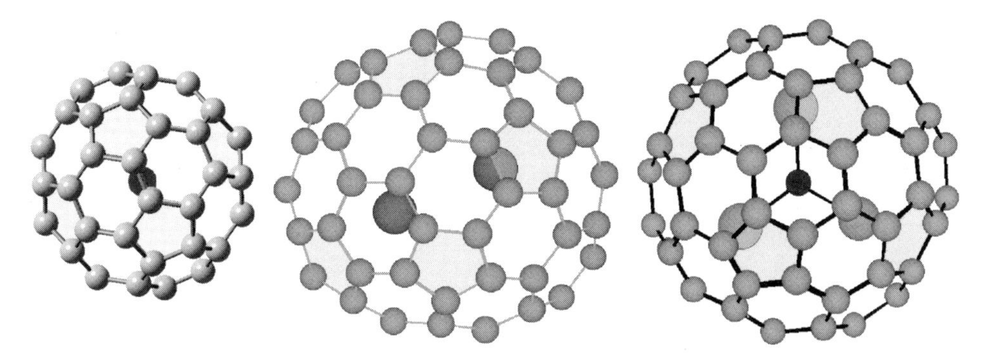

FIGURE 5.14. Three endofullerenes. Left. $N@C_{60}$. Center. $Sc_2@C_{82}$. Right. $Sc_3N@C_{80}$.

properties for these endofullerenes and potential use in future quantum computing applications.

As illustrated in Figure 5.14, representative endohedral metallofullerenes (Group III) include examples, such as, $Sc@C_{82}$, $Sc_2@C_{82}$, $Sc_3@C_{82}$, and even $Sc_4@C_{82}$ as illustrated for the case with two scandium atoms encapsulated ($Sc_2@C_{82}$). It should be noted that transition metal endohedral metallofullerenes (e.g., Co, Fe, and Ni) have not been commonly prepared, but these metals are common catalysts for preparing nanotubes in the Kratschmer–Huffman electric-arc approach. One of the important features of cages with at least 80 carbon atoms is the internal size (~0.8 nm) of these cages is large enough to accommodate four atom molecular clusters. For example, a molecular cluster Lu_3N can be encapsulated in a carbon cage of 80 carbons, $Lu_3N@C_{80}$.

The latter C_{80} carbon cage has icosahedral symmetry and is an unstable open shell molecule in the absence of the Lu_3N cluster. However, the A_3N cluster donates six electrons to stabilize the carbon cage $(Lu_3N)^{+6}@(C_{80})^{-6}$ and this enhanced stability has led to the isolation and characterization of various other trimetallic nitride endohedral metallofullerenes, $A_3N@C_{80}$ (A = Sc, Y, Gd, Ho, Er).

Little is known about endohedral metals' effects on reactivity and regiochemistry compared to our understanding of empty fullerenes. Recently, endohedral metallofullerenes have been functionalized to enhance solubility in aqueous media especially for biological applications. The $A_3N@C_{80}$ family lacks the (6,6) pyracylene sites (see above) that are the reactive positions in C_{60} and C_{70}. Initial results indicate that for the case of $Sc_3N@C_{80}$ is less reactive than C_{60} in typical addition reactions, but can be functionalized under more strenuous conditions. For example, treatment of $Sc_3N@C_{80}$ with an excess of 6,7-dimethoxy-isochroman-3-one in refluxing 1,2,4-trichlorobenzene (see Figure 5.15) yields the adduct shown with the corresponding highly oriented crystal structure shown on the right in Figure 5.15. The results of this study illustrate a wide range of new TNT endohedral metallofullerenes can be prepared in reasonably high yields and purity. The unique structural, chemical, and reactivity features for these new endofullerenes will clearly provide new directions in host-guest chemistry. This opens new vistas for development of these unique materials in a wide range of electronic, optoelectronic, magnetic, catalytic, nanomechanical, and medical applications.

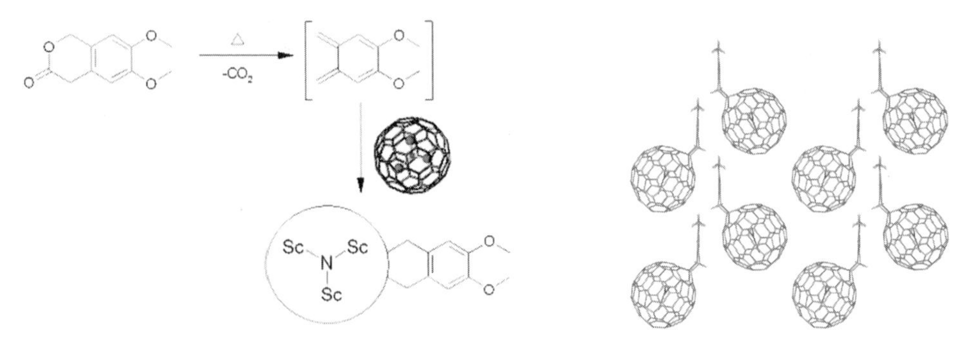

FIGURE 5.15. Diels–Alder reaction scheme for the functionalization of $Sc_3N@C_{80}$ (left) (From Ref. 4 by permission of the American Chemical Society). Crystal structure diagram for $Sc_3N@C_{80}C_{10}H_{10}O_2$. Note the ordered packing in the solid-state resulting from the functionalization. Structural data from Ref. 5.

5.3. POTENTIAL APPLICATIONS

Progress toward fullerene applications and commercial products has been slow to date. Only recently have macroscopic quantities of the fullerene, C_{60}, been available and even smaller quantities available for the endofullerenes. However, the recent announcement of the construction of large-scale fullerene production facilities in Japan by Mitsubishi, will undoubtedly increase the availability of this key nanomaterial. The early potential application of C_{60} or fluorinated C_{60} ("fuzzyball") as a new class of lubricants has been tempered by the relative instability of the latter $C_{60}F_n$ molecules and the relatively high cost of C_{60}. Interest in the magnetic properties of fullerenes stems from early reports of [TDNE]C_{60} as a new organic magnet material, but further reports were less promising. Likewise, the use as a new generation of conductors or superconductors (A_3C_{60}) would appears to require extensive further exploration.

However, as new materials for various electronic, photovoltaic, and other opto-electronic applications, the picture appears more promising. For example, single wall nanotubes (SWNT's) provide a framework upon which functional nanoscale systems can be built with encapsulated fullerenes and endofullerenes "peapods" as illustrated in Figure 5.16 on the right for encapsulated $La_2@C_{80}$. An important variable for carbon nanotube properties is the charge transfer between the guest molecules themselves and between the guest molecules and the SWNT. Hybrid materials that exhibit controllable interactions between the molecular encapsulated fullerene or endofullerene and the SWNT could be basic components of molecular electronic devices. These nanoscale materials are already viewed as important new components for batteries, fuel cells, and hydrogen storage.

Already scientists at IBM have succeeded in fabricating field-effect transistors (FETs) based on individual carbon nanotubes (see above left and Chapter 6) and the properties of this new generation of nanoscale materials can be tuned by encapsulation of fullerenes and endofullerenes as in the peapods illustrated above.

Perhaps the most promising applications will be in the area of new diagnostic and therapeutic pharmaceuticals. The spherical shape of fullerenes, C_{60} can be used for molecular recognition, for example, the inhibition of the HIV protease (HIVP). In addition, it has been shown that fullerenes and endofullerenes can efficiently photosensitize the conversion of

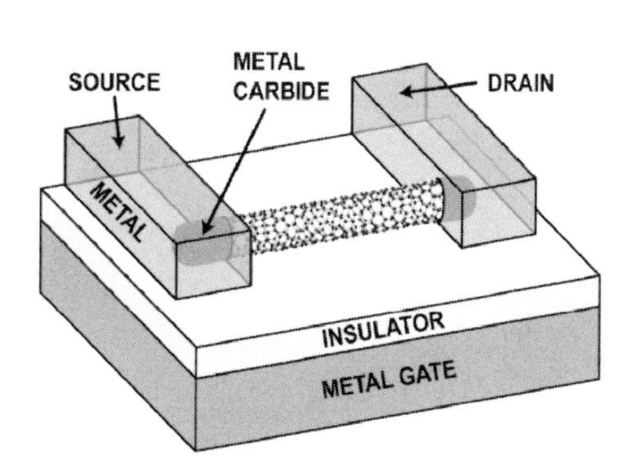

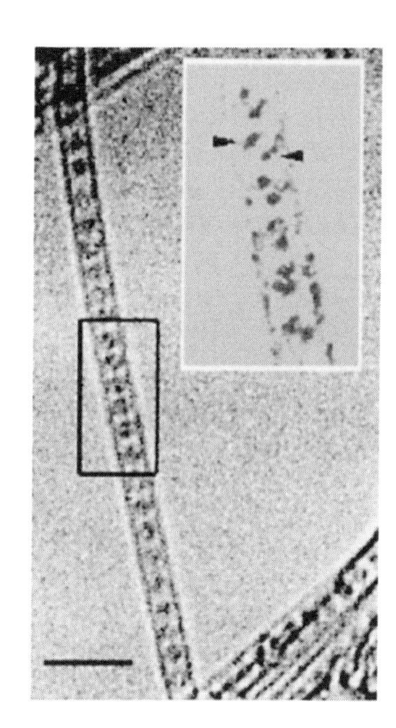

FIGURE 5.16. Nanoscale electronic device connected with a nanotube (left). (Reproduced with kind permission of Ph. Avouris.) $La_2@C_{80}$ trapped inside a single walled carbon nanotube. a.k.a PEAPODS (right). (Reproduced with kind permission of D. E. Luzzi.)

oxygen *in vivo* to singlet oxygen. The singlet oxygen which is well known to cleave DNA can then be used as a directed therapeutic agent.

In the area of new diagnostic pharmaceuticals, new gadolinium based endofullerenes show great promise as new MRI contrast agents. In all these cases, the ability of the carbon cages to more efficiently isolate the metal from biological tissue has discrete advantages over current MRI contrast agents. This advantage also extends to radiopharmaceuticals as well. However, new methodologies for functionalizing fullerenes and endofullerenes are clearly needed to maximize the desired properties of these new nanoscale materials. A hydroxylated endofullerene is shown in the Figure 5.17. The hydroxylation of the endofullerene dramatically improves its solubility in water.

Although new nanotechnology industries based on fullerenes and endofullerenes are still only at their infancy, the train has clearly left the station but the destination is still unclear. It should be remembered that it has been less than twenty years since the discovery of fullerenes. As stated by professor Kroto, "five hundred years after Columbus reached the West Indies, flat carbon has gone the way of the flat earth."

5.4. FURTHER READING

Discovery of Fullerenes

C_{60}-buckminsterfullerene the heavenly sphere that fell to earth, H. W. Kroto, Angewandte Chemie, **31**, 1992, 111–129.

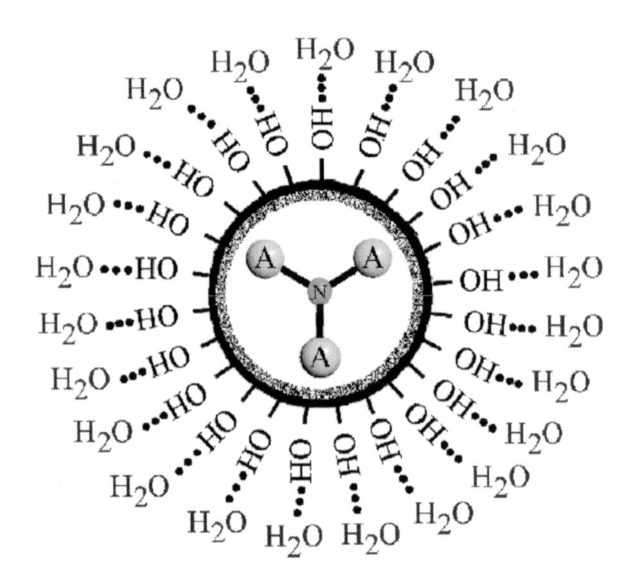

FIGURE 5.17. Hydroxylated endohedral metallofullerene. The addition of OH groups improves water solubility.

Solid C_{60}: a new form of carbon, W. Kraetschmer, L. D. Lamb, K. Fostiropoulos, and D. R. Huffman, *Nature (London)* **1990**, *347*, 354–358.
Probing C_{60}, Robert F. Curl and Richard E. Smalley, *Science*, New Series, Vol. 242, No. 4881. (Nov. 18, 1988), pp. 1017–1022.

Properties of Fullerenes

An Atlas of Fullerenes, P. W. Fowler and D. E. Manolopoulos, Oxford University Press, 1995.
The Chemistry of the Fullerenes, Andreas Hirsch, Organic Chemistry Monographs, Thieme Medical Publishers Inc., New York, 1994.
The March 1992, *Accounts of Chemical Research* thematic issue on Fullerenes.
Endofullerenes A New Family of Carbon Clusters, T. Akasaka and S. Nagase, Kluwer Academic Press, 2002.

Applications of Fullerenes

The May 1999, *Accounts of Chemical Research* thematic issue on Nanomaterials.
The July 1999, *Chemical Reviews* thematic issue on Nanostructures.

QUESTIONS

1. Since carbon arcs have been used as high intensity lights (e.g., search lights) for over 75 years, what was different about the Krätschmer–Huffman electric-arc procedure that provided the production of macroscopic quantities of fullerenes for the first time in 1989. Or, why were fullerenes not discovered until 1985 by the Rice team?
2. For the fullerene C_{60} there is one type of carbon, but two different bond lengths; whereas, for the simple monocyclic hydrocarbon benzene (C_6H_6) there is only one unique carbon and only one bond length, explain these results.

3. What factors are important in predicting which elements or molecular clusters can be encapsulated in fullerenes?

REFERENCES

1. H. W. Kroto, J. R. Heath, S. C. O'Brien, R. F. Curl, and R. E. Smalley, *Nature*, **318**, 165 (1985).
2. A. Hirsch, *The Chemistry of Fullerenes* (Thieme Medical Publishers, 1994).
3. K. M. Allen, W. I. F. David, J. M. Fox, R. M. Ibberson, and M. J. Rosseinsky, *Chem. Mater.* **7**, 764 (1995).
4. E. B. Iezzi, J. C. Duchamp, K. Harich, T. E. Glass, H. M. Lee, M. M. Olmstead, A. L. Balch, and H. C. Dorn, *J. Am. Chem. Soc.* **124**, 524 (2002).
5. H. M. Lee, M. M. Olmstead, E. B. Iezzi, J. C. Duchamp, H. C. Dorn, and A. L. Balch, *J. Am. Chem. Soc.* **124**, 3494 (2002).
6. B. W. Smith, M. Monthioux, and D. E. Luzzi, *Nature*, **396**, 323 (1998).
7. Q. Xie *et al.*, *J. Am. Chem. Soc.* **114**, 3978 (1992).

6

Carbon Nanotubes

Brian W. Smith and David E. Luzzi
Department of Materials Science and Engineering,
University of Pennsylvania, Philadelphia, PA

Few molecules have acclaim equaling that of the carbon nanotube. Perhaps no other chemical structure has garnered so much attention since the double-helix of DNA was introduced to the world. It is unusual within its cohort of famous molecules because it is non-biological, and therefore it exists at the confluence of physics, chemistry, and molecular biotechnology. In many ways the nanotube has exemplified the era of nanoscale science. While it is true that Richard Feynman spoke of nanotechnology as early as 1959,[1] and contemporary figures like K. Eric Drexler introduced the concept of molecular manufacturing to the masses,[2] the scientists behind the carbon nanotube have done much to advance the field of the exceptionally small.

Why all the fuss? One reason is that the name 'nanotube' is descriptive: a carbon nanotube is, in fact, a *nano* tube, and its structure is not veiled by a vexing IUPAC name. Most everyone can imagine a tiny cylinder, which makes tangible an otherwise esoteric field. This visual metaphor serves as a common denominator between technical and non-technical people, and so the carbon nanotube has become one of the hallmarks of nanotechnology in the popular press. A second and more important reason is that a carbon nanotube is wonderfully complex in its simplicity. Its seemingly insipid structure is a single sheet of carbon atoms wrapped into a cylinder of perfect registry, yet this gives rise to a host of tantalizing and unparalleled properties. Whether as molecular wires or as delivery vectors for drug molecules, the prominence of carbon nanotubes in the nanotechnology revolution is secure.

6.1. HISTORY

As explained in the previous two chapters, carbon nanotubes belong to the fullerene family of carbon allotropes. Like their spheroidal brethren, they are composed entirely of covalently bonded carbon atoms arranged to form a closed, convex cage. The first of these molecules, C_{60}, was reported in *Nature* in 1985[3] by a team of researchers at Rice University. The molecule's distinctive soccer ball structure resembled architect Buckminster Fuller's geodesic domes, winning it the now-renown name *buckminsterfullerene*. The structure of C_{60} was theorized as early as 1970 by E. Osawa,[4] but it was the investigators from the Rice team—Robert Curl, Harold Kroto, and Richard Smalley—who shared the 1996 Nobel Prize in Chemistry for its experimental discovery.

Although carbon nanotubes are proper fullerenes, their discovery did not stem directly from the discovery of C_{60}. Instead, nanotubes have their roots in the pyrolytic and vapor phase deposition processes by which conventional carbon fibers are historically grown. In 1960 scientist Roger Bacon of Union Carbide produced structures of graphitic basal layers (e.g., graphene sheets) rolled into scrolls, supporting his discovery with both diffraction and microscopy data.[5] Nearly two decades later, Peter Wiles, Jon Abrahamson, and Brian Rhoades of the University of Canterbury reported finding hollow fibers on the anode of a carbon arc-discharge apparatus.[6] These fibers consisted of concentric layers of wrapped graphene, spaced by essentially the usual graphitic interlayer separation (c.a. 0.34 nm).

The actual discovery of carbon nanotubes is credited to Sumio Iijima who, while working at the NEC Corporation in 1991, provided a rigorous structural solution to fibers created by arc-discharge and bearing resemblance to those found at Canterbury. In a seminal paper published in *Nature*,[7] Iijima described "helical microtubules of graphitic carbon" having outer diameters of 4–30 nm and lengths of up to 1 μm. Because of their Russian-doll-like coaxial packing, these soon became known as multiwall carbon nanotubes (MWNTs). Subsequently, papers published in 1993 by Iijima et al. and Bethune et al. reported similar fibers, each composed of a single graphene cylinder only 1.37 nm in diameter.[8,9] The appearance of these so-called single wall carbon nanotubes (SWNTs) was the true genesis of the modern field.

N.b. In the scientific literature, it has become customary to omit the "C" from acronyms that refer to carbon nanotubes. For example, it is often implied that "SWNT" refers to single wall *carbon* nanotubes. But because other types of nanotubes do exist, the context of the usage is important.

6.2. MOLECULAR AND SUPRAMOLECULAR STRUCTURE

The basic structure of both a SWNT and a MWNT is derived from a planar graphene sheet. An isolated sheet is composed of sp^2 hybridized carbon atoms arranged with D_{6h} point group symmetry. Overlap of the unhybridized p_z orbitals yields a π complex both above and below the plane containing the atoms, which is related to the high electron mobility and high electrical conductivity of graphene.

A SWNT can be imagined to be a sheet that has been wrapped into a seamless cylinder. There are many different ways this wrapping can be accomplished—in principle, the ends of any vector connecting two crystallographically equivalent points in the graphene sheet can be joined to form a nanotube. Any such vector $\vec{c}$ is a linear combination of the two lattice translation vectors $\vec{a}_1$ and $\vec{a}_2$ such that:

$$\vec{c} = n\vec{a}_1 + m\vec{a}_2 \tag{6.1}$$

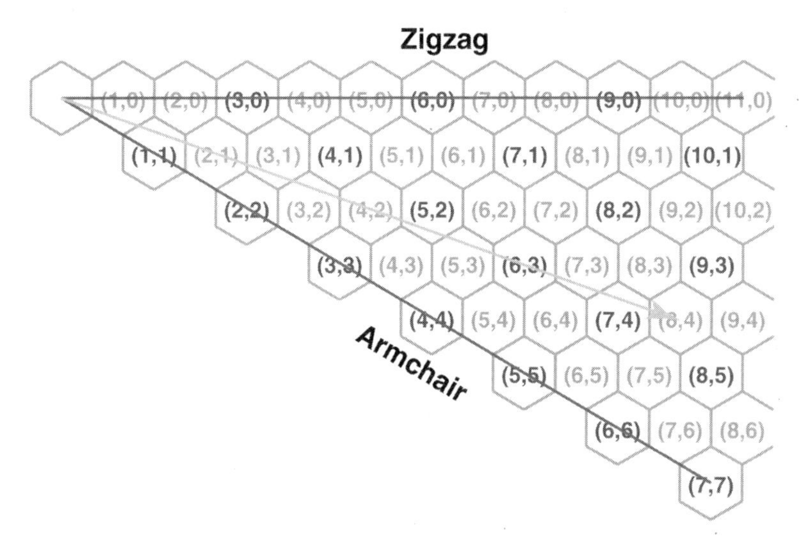

FIGURE 6.1. Diagram explaining the relationship of a SWNT to a graphene sheet. The wrapping vector for an (8,4) nanotube, which is perpendicular to the tube axis, is shown as an example. Those tubes which are metallic have indices shown in red. All other tubes are semiconducting.

where the integer indices (n, m) provide a complete crystallographic description of the tube. This is shown diagrammatically in Figure 6.1. Many different nanotube diameters and chiralities (both left- and right-handed enantiomers) are possible, although a typical tube might be 1.5 nm in diameter. Note that chirality is sometimes called helicity because in certain cases, tracing the carbon-carbon bonds in a systematic way yields a helical path that winds around the tube.

Of course, there are restrictions on the types of tubes that can exist. Wrapping graphene into a cylinder distorts the preferred planar orientation of the three equivalent sp^2 molecular orbitals located at each carbon atom. There is strain energy associated with this distortion, and for a very small radius of curvature it becomes energetically unfavorable to roll up the graphene sheet. For this reason, isolated SWNTs having diameters less than approximately 1 nm are less commonly observed, and larger tubes are generally more stable than smaller ones. Two types of tubes, the $(n, 0)$ and (n, n) tubes, are achiral. These are sometimes called zigzag and armchair tubes, respectively, because of the configurations of the carbon-carbon bonds along the direction of the wrapping vector $\vec{c}$. True to the structure-property principle of materials science, the specific wrapping of a SWNT directly determines many of its properties. For example, as is explained below, certain nanotubes are metallic while others are semiconducting. The electronic characters of various SWNTs are also indicated in Figure 6.1. To date, no diameter- or chirality-monodisperse nanotubes have been synthesized. A typical expectation for a synthesis yield might be a diameter range of 1.4 ± 0.5 nm, although these numbers vary dramatically depending on growth parameters.

Ideal SWNTs might be many hundreds of nanometers long (the aspect ratio is on the order of 1000) and are closed at both ends by hemispherical caps, each formed by introducing six pentagonal defects into the otherwise hexagonal graphene lattice. Half of a

C_{60} molecule is the correct cap for a (5,5) nanotube; the caps of larger tubes have similar structures. A carbon atom that belongs to a pentagonal ring has even more strain energy than one belonging only to hexagons, which has important consequences with respect to the chemical stability of nanotubes. In fact, the strain energy of an sp^2 bonded pentagon is so high that fullerenes almost always satisfy the isolated pentagon rule, which is described in Chapter 4.

It is interesting to ask whether a SWNT has the structural perfection that is expected of smaller molecules. While this is sometimes claimed to be the case, the effect of entropy must be carefully considered. Any defect (e.g., a lattice vacancy) has an equilibrium concentration that depends on the defect's energy and the temperature.[a] Although they vary greatly, defect energies are typically on the order of 1 eV. The concentration of such defects is vanishingly small at 300 K but may be as high at a few ppm at 1000 K. Since a SWNT might encompass more than 100,000 individual carbon atoms, it is likely that it would contain at least a few defects (on the order of tens) at high temperatures. However, it is doubtful that such a small defect density would have a measurable effect on properties. More catastrophic defects like partially severed tubes are sometimes detected in raw material and might be attributable to stochastic 'errors' that occur during synthesis.

The delocalized π complex of the inner and outer surfaces of the nanotube wall means that nanotubes easily experience fluctuating and induced dipole moments. Consequently, they exhibit excellent van der Waals adhesion to other molecules and to each other. This affinity results in the spontaneous aggregation of SWNTs into crystalline bundles, which are commonly called ropes. In most cases, ropes are the expected gross morphology of SWNTs when they are recovered from a synthesis reactor. Ropes may contain from a few to hundreds of agglomerated nanotubes, and rope cross-sections sometimes have irregular shapes. The packing has been determined by X-ray diffraction to be triangular with a lattice parameter of $\sim$1.7 nm for 1.4 nm diameter nanotubes,[10–12] although packing defects are common, and even small ropes cannot be considered to be perfect crystals. Only tubes of similar diameters cocrystallize, suggesting that the packing of as-synthesized SWNTs depends on thermodynamic considerations during growth and/or aggregation as much as the kinetics of entanglement. Despite this fact, the chirality distribution within a given rope can be broad. The parallel orientation of tubes in a bundle maximizes the interfacial surface area over which van der Waals forces can act, resulting in an energetically stable structure. Exfoliating and untangling these ropes is a daunting obstacle that, as of this writing, stands in the way of the bulk utilization of SWNTs. Figure 6.2 summarizes the molecular and supramolecular structures that have been discussed.

Multiwall nanotubes are essentially multiple SWNTs of different sizes that have formed in a coaxial configuration, although MWNTs are typically tens of nanometers in diameter. Spacing between the layered shells in the radial direction of the cylindrical nanotube is approximately 0.34 nm, which matches the c-axis interlayer separation in bulk graphite. There is not necessarily a correlation between the wrapping vectors of the individual shells that compose a MWNT, although in certain cases a systematic variation in the chiralities of successive layers has been found.[7] This is in marked contrast to multiwall BN nanotubes, which form nested shells of nanotubes with matched chirality due to the slight puckered structure of each individual shell.

[a] This can be derived from Boltzmann's definition of statistical entropy.

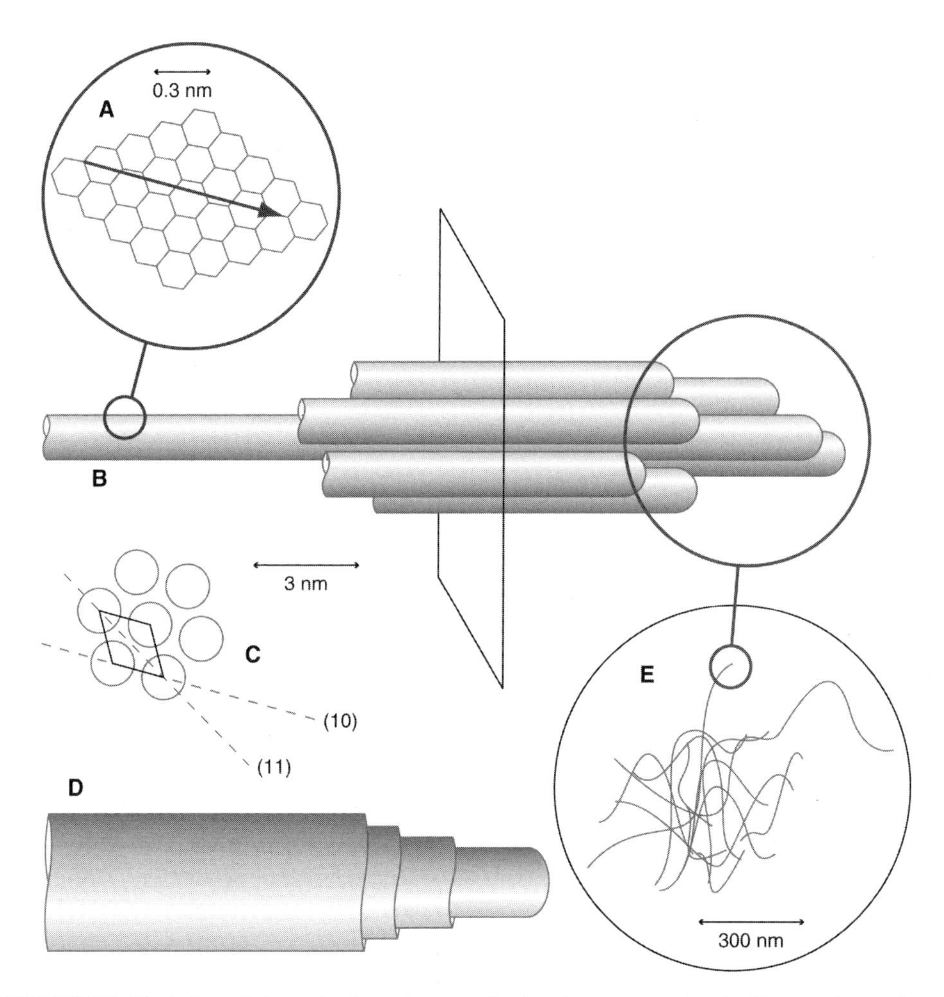

FIGURE 6.2. Illustration of the molecular and supramolecular structures associated with nanotubes at three different length scales. (a) shows the wrapping of a graphene sheet into a seamless SWNT cylinder. (b) and (c) show the aggregation of SWNTs into supramolecular bundles. The cross-sectional view in (c) shows that the bundles have triangular symmetry. (d) A MWNT, another nanotube polymorph composed of concentric, nested SWNTs. (e) At the macromolecular scale, bundles of SWNTs are entangled.

6.3. INTRINSIC PROPERTIES OF INDIVIDUAL SINGLE WALL CARBON NANOTUBES

The scientific literature is replete with manuscripts documenting the observed and theoretical behaviors of nanotubes and their supramolecular assemblies. The vast amount of work that has been spent to characterize these materials is due in part to the difficulty in making accurate property measurements on nanoscale materials and in part to the enormous potential embodied in these properties. Presented here is an overview of the most germane aspects of this considerable topic. Nanotube ensembles (junctions, bundles, MWNTs) are not included because a discussion of the relevant inter-tube and inter-layer

TABLE 6.1. Selected characteristics of single wall carbon nanotubes (ranges indicate the spread of commonly reported values)

Typical diameter	1–2 nm
Typical length	100–1000 nm
Intrinsic bandgap (metallic/semiconducting)	0 eV/~0.5 eV
Work function	~5 eV
Resistivity, 300 K (metallic/semiconducting)	10^{-4}–10^{-3} Ω cm/10 Ω cm
Current density	10^{7}–10^{8} A cm^{-2}
Typical field emission current density	10–1000 mA cm^{-2}
Longitudinal sound velocity	~20 km s^{-1}
Thermal conductivity, 300 K	20–3000 W m^{-1} K^{-1}
Thermoelectric power, 300 K (bulk sample)	200 μV K^{-1}
Elastic modulus	1000–3000 GPa

interactions is beyond the scope of this chapter. For similar reasons, the effect of defect structures on transport properties is not addressed. However, this does not belie the importance of understanding the properties of individual SWNTs, because it is from these that the more complex behaviors are derived. Some of the characteristics of SWNTs appear in Table 6.1.

6.3.1. Chemical and Physical Properties

Because of the non-polar nature of their bonds, carbon nanotubes are insoluble in water. This hydrophobicity stems from both the more positive enthalpy of forming weak SWNT-water hydrogen bonds as compared to strong water-water hydrogen bonds, and the decreased entropy of water molecules at the non-polar SWNT surface. SWNTs can be made to form stable suspensions in certain organic solvents like toluene, dimethyl formamide (DMF), and tetrahydrofuran (THF), but they are generally insoluble in *any* medium without chemical modification or coordination with surfactant.

Nanotubes are subject to the rules of carbon chemistry, which means that they can be covalently functionalized. While they are not especially reactive, SWNTs respond well to strong acids and other chemical oxidizers that are believed to attach functional groups (e.g., hydroxyl and carboxyl groups) to side walls. Direct fluorination of side walls has also been achieved. Subsequent substitution reaction with alkyllithiums or Grignard reagents, for example, provides a route to more complex derivatization. As might be expected, the more highly strained caps at the ends of SWNTs contain the most reactive carbon atoms, and smaller diameter SWNTs are more reactive than larger ones.

Related to derivatization is the fact that each carbon atom on a nanotube is accessible to both the interior and exterior chemical environments. In this way, a SWNT has an extremely good surface area to volume ratio (where the volume of the lumen must be excluded because it contains no atoms belonging to the molecule). When coupled with its excellent van der Waals physisorption properties, this attribute makes a nanotube a natural candidate for gas filtration, sensing, and energy storage applications. A more detailed discussion of covalent and non-covalent interactions appears in a following section.

SWNTs exhibit excellent thermal stability in inert atmospheres. They are routinely vacuum annealed at temperatures up to 1200°C and can survive temperatures in excess of 1500°C, although in the latter case the coalescence of several small tubes into larger diameter SWNTs has been reported. Like graphite, nanotubes burn when heat treated in air or a similar oxidizing environment. Small diameter tubes burn at lower temperatures than large diameter tubes due to the difference in strain energy. Thermogravimetric data show that typical SWNTs burn in the range 450–620°C.

6.3.2. *Electronic Properties*

6.3.2.1. Electronic Band Structure Even though the chemical composition of all carbon nanotubes is fundamentally the same, not all SWNTs are identically wrapped. For this reason, it makes sense to think of nanotubes not as individual molecules but instead as an entire class of molecules capable of having myriad properties. Nowhere is this more manifest than in the electronic properties of SWNTs.

The electronic properties of a molecule with discrete energy states are strongly related to the gap between the highest occupied molecular orbital (HOMO) and the lowest unoccupied molecular orbital (LUMO). The analog for a solid state material is the shape of the energy dispersion in the neighborhood of the Fermi level. Now imagine creating a SWNT by taking individual sp^2 hybridized carbon atoms and assembling them together one at a time. When there are only two atoms, the HOMO and LUMO states are the π and π^* molecular orbitals produced by overlap of the p_z atomic orbitals. As more atoms are added, the these states are split as a consequence of the Exclusion Principle. A SWNT has many individual atoms, so ultimately this splitting is extensive. In this way, overlap of the p_z orbitals produces HOMO and LUMO *bands*, i.e., the π and π^* bands, which are separated by the Fermi level. Thus it is the p_z orbitals that are most important in determining a nanotube's electronic character. (The σ and σ^* bands, which are formed by splitting of the sp^2 molecular orbitals, lie so far below and above the Fermi level that they do not factor in. Mixing of the σ and π orbitals is small and can be similarly omitted.)

Fortunately, the band structure of a SWNT can be easily obtained from a simple quantum mechanical derivation that begins with the p_z orbitals of a planar graphene sheet. The methodology is essentially that of both Saito[13] and Hamada.[14] A tight binding calculation (which usually excludes multi-body interactions and assumes the mixing of only nearest-neighbor wavefunctions) applied to these orbitals gives the two-dimensional energy dispersion around the Fermi level:

$$E_{2D}\left(\vec{k}\right) = \pm\gamma_0 \sqrt{1 + 4\cos\left(\frac{\sqrt{3}k_x a}{2}\right)\cos\left(\frac{k_y a}{2}\right) + 4\cos^2\left(\frac{k_y a}{2}\right)} \tag{6.2}$$

where γ is the nearest-neighbor overlap integral, k_x and k_y are components of the electron wavevector $\vec{k}$, and $a = |\vec{a}_1| = |\vec{a}_2|$. The details of this equation are far less important than its qualitative implications. In the reduced zone scheme, the energy dispersion maps into the first Brillouin zone of graphene, which is shown in Figure 6.3 in relation to the unit cell. The dispersion itself appears in Figure 6.4. The high symmetry points are labeled Γ, M, and K. Since there are two π electrons per (hexagonal) unit cell of graphene, the lower π band is completely filled. The π band is tangent to the π^* band at each K point. This means that the Fermi level exists at the degeneracy between the valence band maximum and the

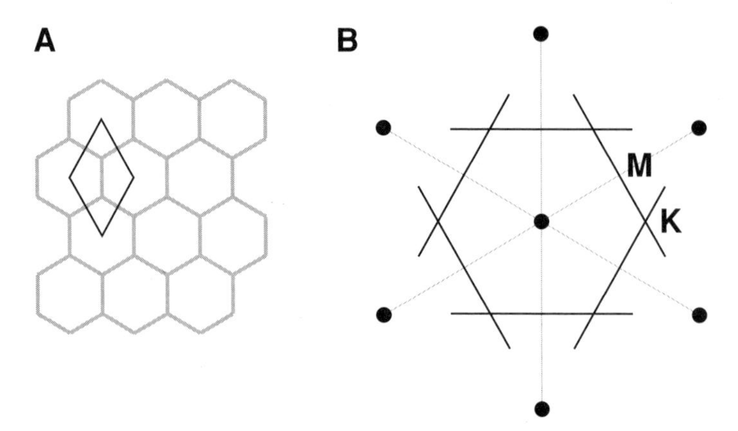

FIGURE 6.3. (a) The unit cell of graphene, and (b) the corresponding reciprocal lattice and Brillouin zone construction by the perpendicular bisector method. Dimensions are not to scale, but orientation between the real and reciprocal lattice is preserved. Important locations within the Brillouin zone are Γ at the zone center, K at the zone corner, and M at the midpoint of the zone edge.

conduction band minimum—the density of states at the Fermi level is zero. For this reason, graphene is a semimetal.

This treatment incorporates the usual Born-Von Karman periodic boundary condition $\vec{k} \cdot \vec{r} = 2\pi j$ where $\vec{r}$ is a lattice translation vector and j is an integer. The spacing of k-states in the reciprocal space direction parallel to $\vec{r}$ is given by:

$$\frac{dk_r}{dN} = \frac{2\pi}{|\vec{r}|} \tag{6.3}$$

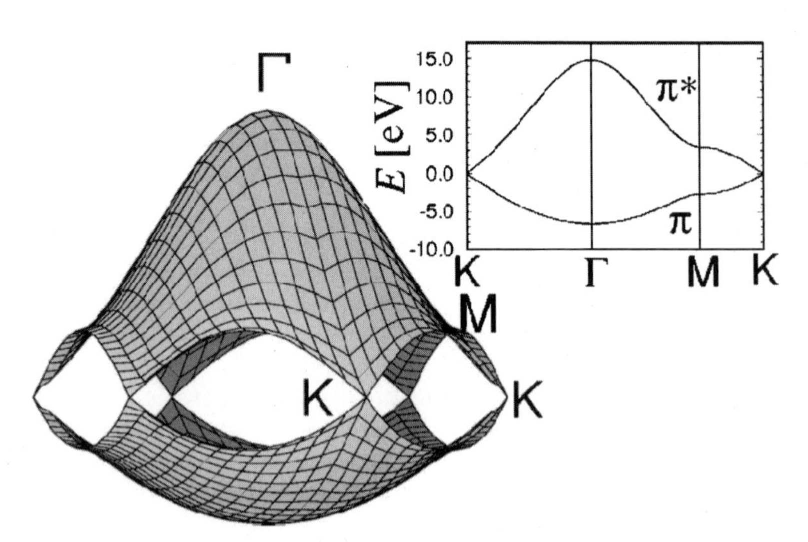

FIGURE 6.4. The dispersion surface of two-dimensional graphene in proximity to the Fermi level. The valence and conduction bands are tangent at each K point. (From Ref. 48 by permission of the American Physical Society.)

N refers to the number of states, so (6.3) tells us the change in k per change in N in one dimension. Because the graphene sheet is assumed to be infinitely extended, the largest allowed $|\vec{r}|$ is also infinite, so the k-states are spaced infinitely close together in any direction. A continuum of states is allowed within the first Brillouin zone of graphene.

Imagine slicing a SWNT parallel to its axis so that it could be unrolled into a flat, contiguous strip. This strip looks like an infinitely extended graphene sheet along the direction of the axis, but it is a *finite* sheet along the direction of its wrapping vector. For this latter case, the periodic boundary condition becomes $\vec{k} \cdot \vec{c} = 2\pi j$. Since $|\vec{c}|$ is only a few nanometers, equation (6.3) indicates that $dk_c/dN \gg 0$. The k-states form a continuum along the tube axis but are discrete along $\vec{c}$.

It follows immediately that the allowed k-states of a SWNT are restricted to lines within the first Brillouin zone of graphene. They lie parallel to the tube axis and are separated from each other by dk_c/dN. The energy dispersion is obtained by restricting equation (6.2) to these k-states. Examples are shown in Figure 6.5. Although the lines appear discontinuous in the reduced zone scheme, they are actually continuous when viewed in the extended zone scheme.

As seen in Figure 6.4, there is a gap between the π and π^* bands of graphene everywhere except at the K points of the Brillouin zone. This means that a SWNT will have a bandgap unless $\vec{K}$ (the vector connecting Γ and K) is an allowed wavevector. To figure out which

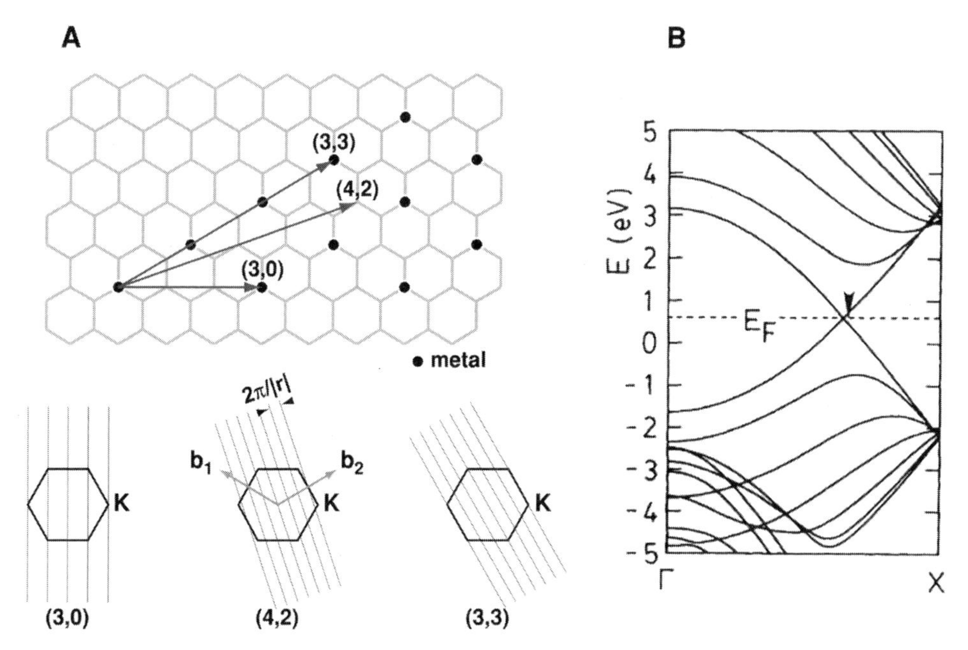

FIGURE 6.5. (a) Wrapping vectors and allowed k-states for (3,0) (zigzag), (4,2), and (3,3) (armchair) SWNTs. The degeneracy at the K point is allowed only for the (3,0) and (3,3) tubes, which behave like metals. The (4,2) tube does not contain the degeneracy, so it has a band gap. Note that the lines of allowed k-states are perpendicular to the wrapping vector for each tube. (From Ref. 49 by permission of Annual Reviews.) (b) The band structure of a (6,6) SWNT. The presence of many overlapping subbands is typical for SWNTs. (From Ref. 14 by permission of the American Physical Society.)

tubes have an allowed k-state at the K point, $\vec{K}$ is simply inserted into the periodic boundary condition along the circumference:

$$\vec{K} \cdot \vec{c} = 2\pi j \tag{6.4}$$

In terms of the reciprocal lattice vectors, $\vec{K} = (\vec{b}_1 - \vec{b}_2)/3$. To evaluate $\vec{K} \cdot \vec{c}$, we invoke the definition of the reciprocal lattice: $\vec{a}_i \cdot \vec{b}_j = 2\pi\delta_{ij}$. Thus:

$$\begin{aligned}\vec{K} \cdot \vec{c} &= \left(\frac{\vec{b}_1 - \vec{b}_2}{3}\right) \cdot (n\vec{a}_1 + m\vec{a}_2) \\ &= \frac{n}{3}\vec{b}_1 \cdot \vec{a}_1 - \frac{m}{3}\vec{b}_2 \cdot \vec{a}_2 \\ &= \frac{2\pi}{3}(n - m)\end{aligned} \tag{6.5}$$

Notice that the cross terms disappear since $\delta_{12} = \delta_{21} = 0$. Substitution of (6.5) into (6.4) gives $(n - m) \bmod 3 = 0$. This is the criterion for a SWNT to have an allowed k-state at the K point. All other SWNTs will be semiconductors with a gap on the order of 0.5 eV.

But are the tubes that satisfy this rule metallic? To answer this question, we must consider the curvature of the graphene wall. The distortion of planar graphene into a cylinder considerably complicates the tight binding method, affecting the overlap integrals. The result is a shift of the K point degeneracy in the direction of $\vec{K}$ and slightly off the corner of the Brillouin zone. Thus, many tubes having $\vec{K}$ as an allowed wavevector will still have a small bandgap. Only those tubes having lines of allowed states that superimpose with $\vec{K}$ are truly metallic because the shifted degeneracy is still allowed, and no bandgap appears in the dispersion. It is trivial to show that these tubes are the armchair tubes having $n = m$.

One-dimensional solids are subject to a lattice instability known as a *Peierls distortion* whereby certain bonds are lengthened and others are shortened, yielding a doubled unit cell. This bond alteration occurs because the atomic displacements change the energy dispersion of the material in such a way that when the electrons are rearranged into these new states, the result is to lower the total energy of the system. An additional consequence is the opening of a band gap at the Fermi level. Therefore, the Peierls distortion can transform a 1D metal into a semiconductor or an insulator. Because of this, it might be expected that metallic carbon nanotubes should not exist. However, the energy cost of repositioning carbon atoms around the entire circumference of the tube is so high that the Peierls distortion is suppressed, e.g., the energy loss associated with elastic energy is large in comparison to the gain in electronic energy for a large Peierls distortion. Effectively, the Peierls gap in a SWNT is sufficiently small that it is negligible in the presence of finite temperatures or fluctuation effects. The physical basis for the large strain energy that ultimately inhibits the Peierls distortion is explained in the section on mechanical properties of SWNTs.

6.3.2.2. Electronic Density of States In general, the density of states (DOS) is given by:

$$\frac{dN}{dE} = \frac{dN}{dk}\frac{dk}{dE} \tag{6.6}$$

The reciprocal of (6.3) is dN/dk_r, the number of k-states between k and $k + dk$ along a line parallel to $\vec{r}$ (where each state can accommodate two electrons), and dk/dE can be obtained by manipulation of (6.2) subject to the constraints of the allowed wavevectors. By making these substitutions into (6.6), the density of states for any given SWNT can be obtained from the energy dispersion.

Examples of the DOS for several SWNTs are shown in Figure 6.6. An explicit derivation is straightforward so it is omitted for simplicity, although some important features merit discussion. First, the DOS of each nanotube contains singularities called *van Hove singularities*. These arise because the equation for the DOS contains a $1/\sqrt{E - E_0}$ term, where E_0 is the energy at a band minimum. Thus, the singularities appear at each band minimum, even where the bands overlap. This is a common feature to all 1D energy bands. Second, it follows directly from the band structure derivation that the energy separation between band edges, and therefore between singularities, depends on $\vec{c}$. This has important consequences

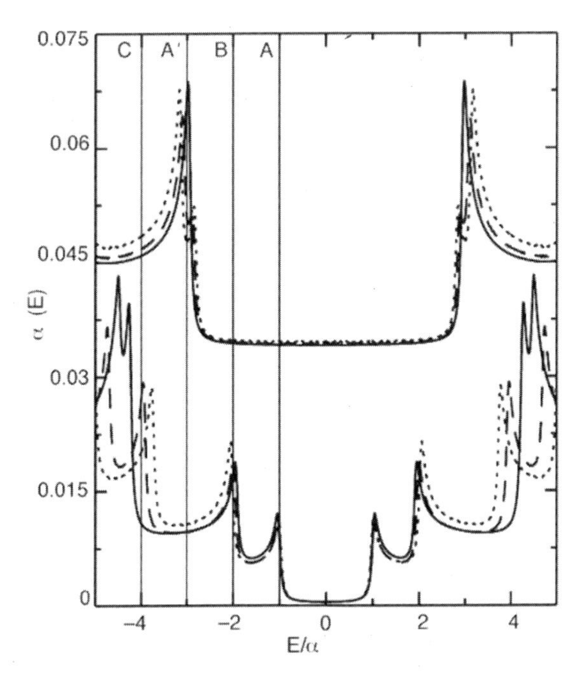

FIGURE 6.6. DOS for three metallic and three semiconducting SWNTs. The Fermi level is at 0, and the DOS for the metallic tubes are offset by 0.03 along the vertical axis. Notice that the DOS is zero at E_F for the semiconducting tubes but finite for the metallic tubes. For all SWNTs, singularities arise at each band edge due to the low dimensionality of the material. (See Question 1.) (From Ref. 50 by permission of Macmillan Publishers.)

with respect to the spectroscopy of SWNTs. Interestingly, it has been recently shown that SWNTs can be induced to fluoresce via the de-excitation of promoted electrons across the gap between singularities.

6.3.3. *Vibrational Properties*

6.3.3.1. Phonon Spectrum and Density of States The symmetry of a carbon nanotube results in a unit cell that has many atoms. This gives rise to a large number of vibrational degrees of freedom and, in turn, to a large number of possible phonon modes. Even a high-symmetry (10, 10) SWNT has a unit cell containing 160 atoms, resulting in 120 vibrational modes. Usually some of these modes are degenerate, and the number of distinct phonon branches is smaller than the number of modes. Nevertheless, a chiral tube may have more than 100 branches in its phonon spectrum. An exhaustive listing of the phonon modes of a SWNT seldom appears in the literature.

Because this lexicon can be confusing, it is worthwhile to be pedantic. A phonon mode is simply a coherent oscillation of atoms about their lattice points. In general, each phonon mode has its own dispersion curve $\hbar\omega(\vec{k})$. In the case of a SWNT, some of these modes are degenerate, i.e., their dispersion curves are identical. Modes that are degenerate with each other are called a branch, and all the branches taken together form the spectrum. Optical modes are often designated by the wavenumber of the vibration at the Γ point, $\overline{\nu} = \omega(0)/2\pi c$. Acoustic modes are those which have $\omega \rightarrow 0$ as $k \rightarrow 0$; all others are denoted optical modes.

Each branch of the graphene phonon spectrum has a 2D dispersion. Because rolling graphene into a cylinder imposes quantization on the circumferential wavevector, it might be expected that the phonon dispersions of a SWNT are obtainable by zone folding of the phonon dispersions of graphene into 1D subbands. This treatment is essentially the same as the one by which the electronic band structure was derived and more or less agrees with what is known to be physically true: each branch of the graphene spectrum is split into multiple 1D subbands (each of which corresponds to one branch in the SWNT spectrum), and each subband has a singularity in the phonon density of states (PDOS) at the band edge. Furthermore, the dispersion for each subband is sensitive to the direction (chirality) and magnitude (diameter) of the wrapping vector $\vec{c}$. However, zone folding fails to reproduce the experimentally measured low-frequency branches. Alternate tight binding and force-field approaches applied directly to the molecular structure more accurately reflect these modes. Figure 6.7 shows calculated phonon spectra and PDOS for three different SWNTs having approximately the same diameter.

The modes illustrated in Figure 6.8, which include four acoustic modes and three optical modes, are especially important. The acoustic modes are the longitudinal acoustic (LA) mode, which has atomic displacements parallel to the tube axis; two degenerate transverse acoustic (TA) modes, which have displacements perpendicular to the tube axis; and the twisting (TW) mode, which has torsional displacements. Some of these are derived from similar modes in graphene. The LA mode, for example, is exactly the same as the LA mode of graphene. Graphene also has a TA mode, corresponding only to in-plane displacements, and a ZA mode, corresponding only to out-of-plane displacements. Therefore, the doubly-degenerate TA modes of a SWNT are produced by mixing the normally uncoupled TA and ZA modes of graphene (c.f. Figure 6.8). The TW mode of a SWNT has no graphene analog.

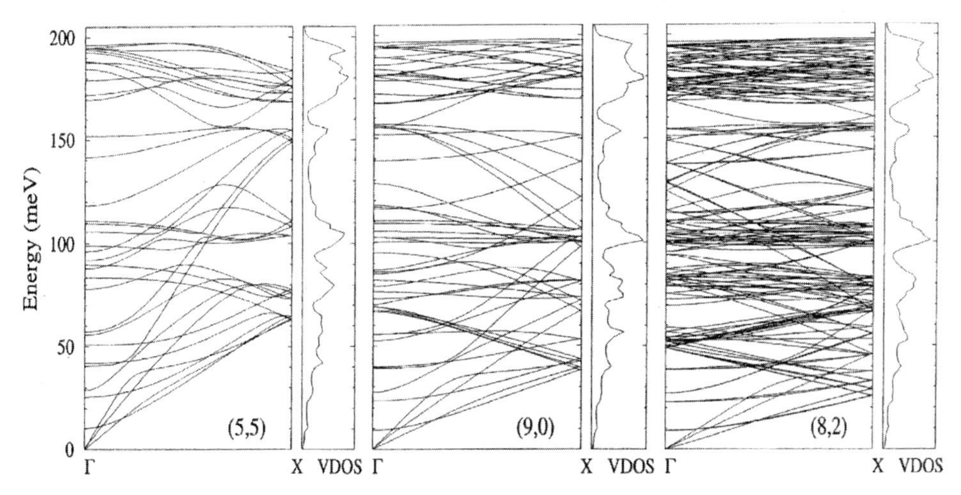

FIGURE 6.7. Calculated phonon spectra and PDOS for (5,5), (9,0), and (8,2) SWNTs. The large number of subbands is due to the complexity of the unit cell. The (8,2) tube has the lowest symmetry, corresponding to the largest number of vibrational degrees of freedom and the most complicated phonon spectrum. (From Ref. 15 by permission of Elsevier Science Ltd.)

Two optical branches have special significance. The first is the radial breathing mode (RBM), which is a low-frequency oscillation derived from the pure ZA mode of graphene. For a (10, 10) tube, this occurs at 20.4 meV (165 cm^{-1}). It is called the 'breathing' mode because the atomic displacements produce an axially-symmetric, respiratory-like expansion and contraction of the tube. The RBM is sometimes denoted the A_{1g} mode, which is the point group that refers to the symmetry of the vibration. The second is the doubly-degenerate $E_{2g(2)}$ mode which, for a (10, 10) tube, occurs at 196.5 meV (1585 cm^{-1}). This mode is similar to the LA mode, but it involves antiparallel instead of parallel displacements and so occurs at higher energies. Both the A_{1g} and $E_{2g(2)}$ modes are important because they have strong Raman activity.

The diameter dependence of the RBM has been studied extensively. Sauvajol and coworkers have generated a mastercurve, shown in Figure 6.9, that incorporates both empirical and calculated results.[15] The best fit to these data is:

$$\overline{\nu} = 238/d^{0.93} \tag{6.7}$$

where $\overline{\nu}$, the wavenumber of the RBM, is in cm^{-1} and d, the SWNT diameter, is in nm. This has proved to be an important equation in the characterization of SWNTs by Raman spectroscopy. (An alternate equation, $\overline{\nu} = (224/d) + 14$, sometimes appears in the literature. Both equations are valid and give nearly identical results.)

6.3.3.2. Specific Heat and Thermal Conductivity When heat is applied to a material, internal energy is increased because electrons are promoted to excited states and higher energy phonon modes are made accessible. Therefore, there are both electron (C_{el}) and phonon (C_{ph}) contributions to the specific heat (C) such that $C = C_{el} + C_{ph}$. For a SWNT

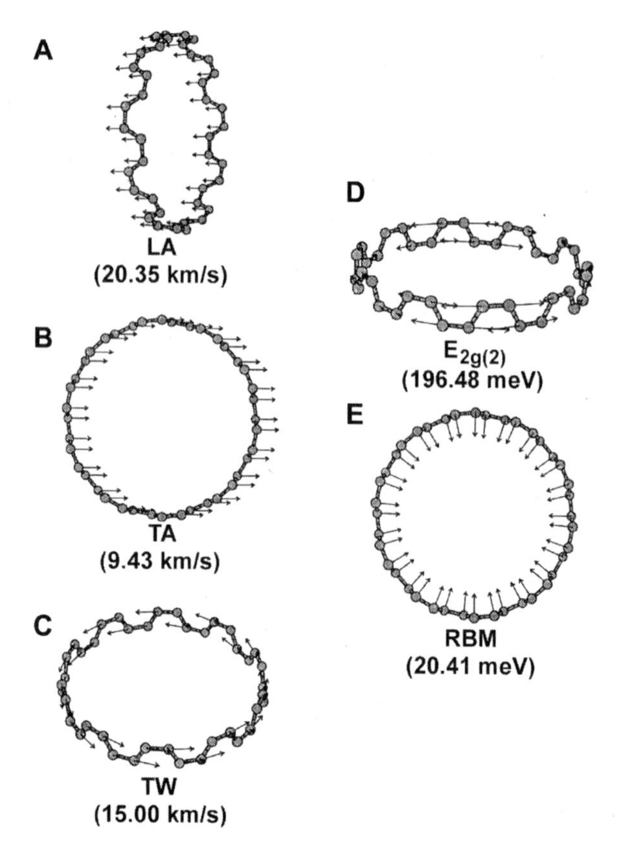

FIGURE 6.8. Important vibrational modes in SWNTs, illustrated for a (10,10) SWNT. (a) Longitudinal acoustic mode. (b) Transverse acoustic mode (doubly degenerate). (c) Twisting (acoustic) mode. (d) $E_{2g(2)}$ mode (doubly degenerate). (e) A_{1g} mode (radial breathing mode). Calculated sound velocities are indicated for the acoustic modes, (a–c). (d–e) are Raman active optical modes. (From Ref. 52 with kind permission of Z. Benes.)

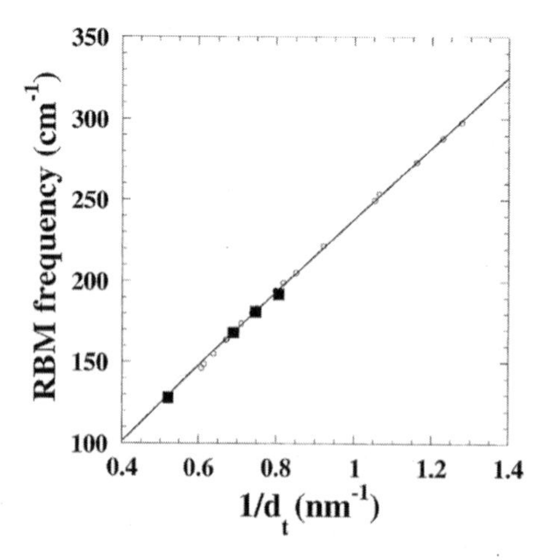

FIGURE 6.9. Mastercurve showing the relationship between RBM frequency and diameter for both calculated (hollow symbols) and empirical (solid symbols) data. (From Ref. 15 by permission of Elsevier Science Ltd.)

it has been shown that $C_{ph}/C_{el} \sim 100$ so the phonon contribution dominates and $C \sim C_{ph}$. This is true all the way down to $T = 0$.

One definition of specific heat is the temperature derivative of the energy density u. In turn, u can be obtained by integrating over the PDOS with a weighting factor to account for the energy ($\hbar\omega$) and statistical occupation (given by the Bose distribution in the case of phonons) of each state[b]. Quantitatively:

$$C(T) = \frac{du}{dT} = \frac{d}{dT}\int d\omega D(\omega)\frac{\hbar\omega}{\exp(\hbar\omega/kT) - 1} \tag{6.8}$$

where $D(\omega)$ is the PDOS and k is Boltzmann's constant. This commonly appears in the literature with the temperature derivative already evaluated such that the integrand is expressed as the product of $D(\omega)$ and a "heat-capacity convolution factor." Rarely is an analytical solution possible since $D(\omega)$ is a complicated function. Nevertheless, it is clear from Figure 6.7 that at moderate temperatures, many of the overlapping subbands of a SWNT will be occupied, but at low temperatures (<5K) only the acoustic branches are accessible. In this regime, (6.8) can be evaluated analytically, and the result is a linear dependence of C on T. This is a consequence of the quantized phonon spectrum of a SWNT and has been empirically confirmed by direct measurement of the low-T heat capacity.

Sound propagates though a material primarily via longitudinal acoustic phonons, and the sound velocity is determined by the slope of the corresponding dispersion at $k = 0$. Furthermore, phonon thermal conductivity (κ) is approximately $\kappa = Cv_s l$, where l is the mean free path. Most solids have $v_s \sim$ 3–6 km s^{-1}. For graphene and SWNTs, the largest v_s corresponds to the LA mode and is ~20 km s^{-1}. This anomalously high value stems from the stiffness of the carbon-carbon covalent bond and is directly responsible for the remarkable thermal conductivity of carbon-based materials: diamond and graphene display the highest measured κ of any known material at moderate temperatures. Because of the sensitivity to mean free path length, it has been postulated that the inherent long-range crystallinity of a SWNT might result in even better thermal conduction. Theory predicts a room temperature thermal conductivity of 6000 W m^{-1} K^{-1} in an ideal, defect-free SWNT, although conflicting experimental data place this value somewhere in the range 20–3000 W m^{-1} K^{-1}. For comparison, Cu metal has room temperature $\kappa = 400$ W m^{-1} K^{-1}.

6.3.4. Mechanical Properties

6.3.4.1. Elasticity One might imagine that the strain energy associated with placing a SWNT in axial tension is somehow related to the dispersion of the longitudinal acoustic phonon because both involve atomic displacements along the tube axis. In fact, there is a very simple relationship between the elastic modulus (Y) and the longitudinal sound velocity: $v_s = \sqrt{Y/\rho}$, where ρ is the mass density. There is a range in calculated moduli for a SWNT due in part to varying estimates of the thickness of the graphene monolayer that makes up a tube's wall. Also, since the acoustic phonon dispersions are sensitive to the wrapping vector, both v_s and Y should depend on diameter and chirality. Despite these caveats, the calculated

[b] A useful but unrelated point: the probable occupancy of a particular *electronic* quantum state at a particular temperature is determined by the Fermi distribution.

elastic modulus of a SWNT (by both molecular dynamics and *ab initio* methods) is on the order of 1000 GPa (1 TPa). For comparison, the moduli of alumina, carbon steel, Kevlar, and titanium are approximately 350, 210, 130, and 110 GPa, respectively. Direct measurement of Y has proved elusive due to the length scale that must be probed, although it can be carefully inferred from spectroscopic and transmission electron microscopy data. These experiments have consistently produced values ranging from 1–3 TPa, making SWNTs among the stiffest materials known.

6.3.4.2. Plasticity 'Resilient' is a term sometimes applied to nanotubes when referring to their mechanical properties. Certainly there is empirical evidence that SWNTs can suffer high bending angles, kinks, and other tortuous deformations, and that these deformations are fully reversible. The remarkable ability of nanotubes to accommodate large elastic strains is attributed to their atomic monolayer shell thickness, which leaves no room for stress-concentrators, and to the ability of sp^2 carbon to rehybridize when subjected to an out-of-plane distortion. A coarse analogy is that a nanotube is like a plastic drinking straw which buckles under bending stress but snaps back to its original shape when the stress is released. However, no buckling deformation is possible when a SWNT is placed in pure tension. Just as a drinking straw will snap under sufficient tensile stress, a nanotube also has a mechanism of plastic deformation.

The primary strain release mechanism in a SWNT is believed to be the Stone-Wales transformation. It has been calculated that at a certain critical value of tension, the strain energy becomes so high that it becomes unfavorable to maintain a perfect hexagonal lattice. The transformation to a lower energy state involves the rotation of one carbon-carbon bond by 90°, giving rise to a pair of topological defects, each of which is composed of a pentagon fused to a heptagon and is called a 5–7 defect. The two defects are initially conjoined (i.e., a 5–7–7–5 arrangement) but have opposing stress fields such that further relaxation is obtainable if the defects glide apart through successive bond rotations as shown in Figure 6.10.[16] The formation of 5–7 defects becomes energetically favorable at 5–6% strain (for a (5,5) tube, the formation energy at 0% strain is ~2 eV), but the activation barrier is so high (and the formation kinetics so slow) that much larger elastic strains can be realized before the onset of ductile flow. The in-plane yield strain of a SWNT might be as high as several dozen percent depending on the strain rate. The plasticity of a single SWNT has yet to be studied empirically.

6.4. SYNTHESIS AND CHARACTERIZATION OF CARBON NANOTUBES

The previous section has shown that there is fantastic promise in the carbon nanotube, whose repertoire of useful material properties is unequalled. A SWNT has the highest elastic modulus that has been measured and it can withstand large bending and tensile strains without yielding, yet it is intrinsically lightweight due to its low-density structure and carbon composition. It is expected to have the highest thermal conductivity of any material known, exceeding that of even diamond. The low dimensionality of a SWNT generates an unusual electronic density of states whose singularities could provide the basis for intrinsic transistor-like behavior. Indeed, as integrated electronics devices shrink

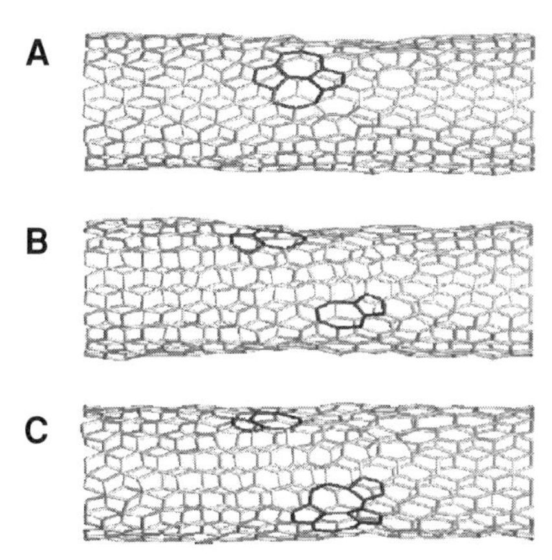

FIGURE 6.10. Time evolution of plasticity in a (10,10) SWNT under 10% uniaxial strain after (a) 1.5 ns, (b) 1.6 ns, and (c) 2.3 ns. The 5–7–7–5 defect is seen to glide apart into separate 5–7 pairs. In (c), it can be seen in the lower 5–7 pair that a second bond rotation has led to the formation of a 5–7–5–8–5 defect. (From Ref. 16 by permission of the American Physical Society.)

towards the nanoscale, SWNTs are certain to play a vital role as molecular wires. And amazingly, all of these solid state-like properties are packaged within an organic molecule that can be functionalized. The attachment of ligands to a SWNT could conceivably facilitate self-assembly into more complex, functional structures. Finally, one important 'property' has received much less acclaim: nanotubes are an unusually tractable system for both theorists and experimentalists to study one-dimensional phenomena and certain aspects of fundamental physics.

Unfortunately, the engineering side of the story is not nearly as unambiguous as the scientific side. In order to unlock their potential, nanotubes must first be controllably manufactured. In this section, we explore the ways in which nanotubes are synthesized and studied. Many of the tangible challenges associated with carbon nanotechnology will be made evident.

6.4.1. Synthesis and Processing

6.4.1.1. Methods: The Black Art that is Nanotube Synthesis There is no lack of creativity in the study of carbon nanotubes, and surprisingly many techniques have proved successful at producing these molecules. Most methods are permutations of one of four primary approaches which are illustrated in Figure 6.11: (1) arc discharge, (2) laser ablation, (3) catalyzed decomposition, or (4) chemical vapor deposition.

The serendipitous discovery of MWNTs was made in material produced by the arc discharge process. In this method, a dc voltage is applied between two closely spaced graphite electrodes under an inert atmosphere (typically 500–600 mbar of flowing He). The voltage is sufficiently large to induce dielectric breakdown of the gas molecules between the electrodes, causing a current of ~100 A to flow in the form of an electric arc whose peak temperature is approximately 3000°C. The anode is gradually consumed in this process as

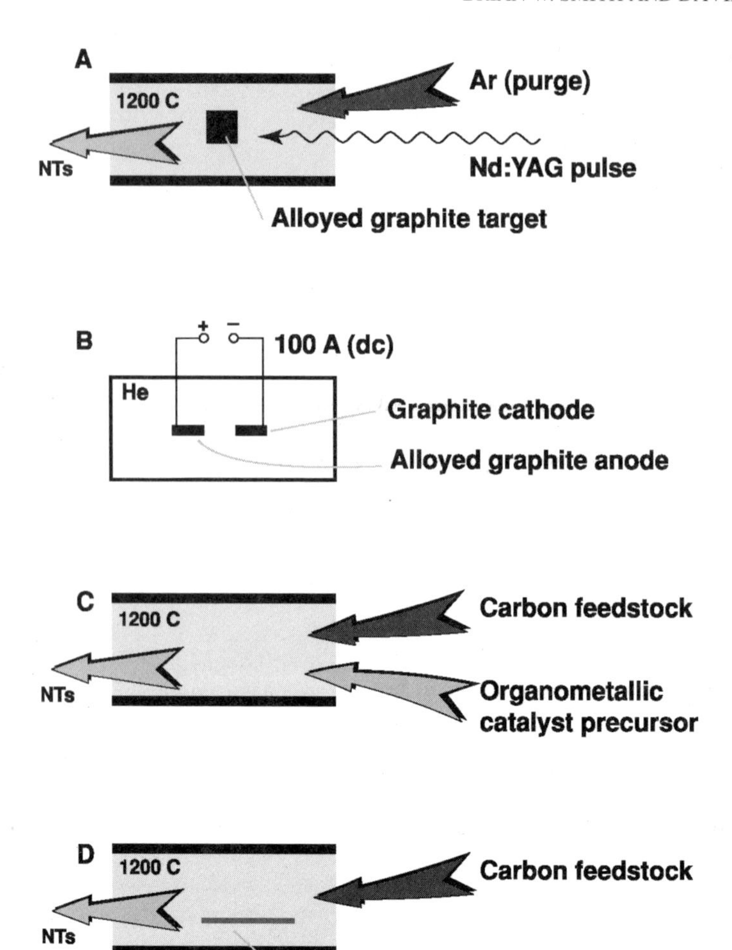

FIGURE 6.11. Illustration of the four primary approaches for synthesizing carbon nanotubes: (a) pulsed laser vaporization, (b) arc discharge, (c) catalytic decomposition, and (d) chemical vapor deposition.

carbon atoms are vaporized from its surface. Ultimately the evaporated atoms redeposit on the cathode and chamber walls in the form of amorphous carbon, graphitic nanoparticles, fullerenes, and MWNTs in lesser abundance. The same process can be adapted to synthesize SWNTs if the anode is cored and filled with a mixture of graphite and certain metal catalysts. Transition metals (Co, Ni, Fe) and rare earth elements (Y, Gd) alloyed into the anode at a few atomic percent have all successfully catalyzed SWNT growth. In what has become a seminal paper, Journet and coworkers experimented with Co/Ni, Co/Y, and Ni/Y mixtures, achieving the best production with 1 at.% Y and 4.2 at.% Ni. Most of the nanotubes were located in a small collar-like deposit around the cathode. This so-called 'collarette' comprised 20 wt.% of the total recovered product and contained 70–90 vol.% SWNTs.[11] Interestingly, SWNTs are essentially *never* obtained in the absence of catalyst. Proposed mechanisms for catalysis are discussed in the following section.

Laser ablation is another means for vaporizing a graphitic target to create SWNTs. In fact, the pulsed laser vaporization (PLV) technique was the first by which SWNTs could be grown efficiently. The apparatus consists of a tube furnace maintained at c.a. 1200°C under flowing argon and a high power pulsed laser (typically Nd:YAG operated at 532 nm, 30 Hz, and ~500 mJ per pulse) directed down the furnace tube. A graphite target impregnated with transition metal catalyst is ablated inside the furnace, forming SWNTs that are then collected downstream. The beam can be rastered and the target rotated to maximize utilization of the starting material. When they first introduced the PLV method, Guo and coworkers reported 15% conversion of all vaporized carbon atoms into SWNTs[17] using Co/Ni (0.6/0.6 at.%) and Co/Pt (0.6/0.2 at.%) catalysts. Follow-on work by other members of this research team increased the yield to 70–90 vol.% by modification of certain parameters.[10,12]

The PLV and arc discharge methods were the first by which gram quantities of reasonably pure SWNTs could be obtained. They have in common reaction products with impressively narrow diameter distributions averaging ~1.4 nm. Unfortunately, both also generate large quantities of unwanted byproduct and require high temperatures (3000–4000°C) to evaporate solid carbon sources, although the nanotubes form at a much lower temperature within the chamber. Attempts to circumvent these inefficiencies have evolved into a pyrolytic technique, which uses gaseous feedstocks for both carbon and catalyst. In this simple approach, organometallic precursors are sublimed or evaporated at low temperatures (200–300°C) and delivered into a furnace held at 900–1200°C by an inert carrier gas. As the precursors flow into the heat zone of the reactor, they decompose. The 'metallic' part of the organometallic then coarsens into metal clusters or nanoparticles that can catalyze nanotube growth from the 'organic' part. Care must be taken to prevent the metal particles from forming oxides, which, according to some, poison SWNT growth.[c] Argon can be flowed continuously to purge the system of air, and hydrogen (an effective reducing agent) can replace the inert carrier gas to reduce any oxides to elemental metal. It is common to supply a separate hydrocarbon gas or vapor for the main carbon source. Acetylene, methane, hexane, xylene, and benzene have all been successfully pyrolyzed into SWNTs and MWNTs by metallocene catalysts, and especially by ferrocene. The so-called HiPco (high pressure carbon monoxide) method uses $Fe(CO)_5$ to catalyze the formation of SWNTs by CO disproportionation. There are many possible variations on this theme.

The difference between chemical vapor deposition (CVD) and catalyzed decomposition is largely semantical. The two terms are commonly used interchangeably, although a distinction is made to reflect the fact that CVD often refers to growth on a surface (although this does not necessarily mean that the interface between a growing nanotube and a catalytic substrate is coherent as in epitaxy). CVD processes use a supported catalyst to grow nanotubes along, or off of, a surface. There are numerous strategies for affixing catalyst particles to substrates. The simplest involves spin-coating a metal salt like $Fe(NO_3)_3$ onto a silicon wafer (the native SiO_2 layer is often left intact). Alternatively, metal salts can be infused into the pore network of alumina or silica. Calcination (heating in air) transforms the metal salts into discrete metal oxide clusters. Growth is subsequently accomplished in the same manner as in the decomposition method, i.e., in a reaction furnace with flowing hydrogen for catalyst reduction and a hydrocarbon feedstock. The direct evaporation of Fe

[c] There are recent reports that certain oxides can catalyze SWNT growth at unexpectedly low temperatures. For example, see Ref. 18.

through a shadow mask onto porous Si has also proved effective, enabling the patterned growth of MWNTs.

6.4.1.2. Growth Models There are many ways to make nanotubes. Moreover, any one method is more or less adaptable to produce the same material as any other method. Journet and coworkers were the first to make an important point: the similarities between nanotube samples prepared by different techniques mean that growth depends less on the details of the experimental conditions than on the thermodynamic conditions created by the experiment.[11] For this reason, we forgo discussing the many parametric details of each method and instead focus on the probable requirements for producing good nanotubes.

There are some general trends relating catalyst size, temperature, and product. Nanoparticulate catalyst (1–5 nm in diameter) seems to be necessary for SWNT growth, whereas MWNTs are formed in the presence of larger particles or in the absence of catalyst altogether. High temperature favors the formation of SWNTs over MWNTs. Possibly this is because at low temperature, the activation barrier to grow a tube with such a small radius of curvature is too high even in the presence of catalyst. (Why, then, is catalyst not required to synthesize fullerenes, which have an even smaller radius of curvature? It could be that the *nucleation* of SWNTs and fullerenes is possible without catalyst, but the sustained *growth* is energetically unfavorable. Perhaps many fullerenes are incipient SWNTs that were prematurely closed.) Perversely, high temperature also causes the rapid coarsening of catalyst particles, thereby quenching SWNT production as the particles grow. This is the great paradox of SWNT synthesis: how does one maintain the proper catalyst size at the proper reaction temperature? One option is to find a more efficient catalyst so that SWNTs can be grown at low temperatures where the driving force for coarsening is smaller. A second (and possibly more challenging) option is to prevent known catalysts from coarsening at high temperature. This is the rationale behind infusing catalyst into ceramic substrates for CVD growth. A third option is to find a method for making SWNTs that does not require an extrinsic catalyst. This may not be such a pipe dream: the catalyst-free synthesis of SWNTs on the (001) face of SiC was recently reported. An interesting corollary: under conditions that produce SWNTs, higher reaction temperatures yield larger average diameters. This suggests a positive correlation between catalyst size and SWNT diameter, although substantial evidence for this is largely absent.

What atomistic mechanisms are at work during growth? There are at least as many thoughts as there are researchers working in the field. A common hypothesis is that nanotubes are formed by a 'root growth' mechanism at the surface of a catalyst particle. The catalyst serves as a nucleation site for a hemispherical cap and stabilizes the cap against spontaneous closure. Carbon atoms are then added by direct adsorption at the interface. (It is less likely that synthesis occurs in the traditional manner of vapor-grown carbon fibers, i.e., by supersaturation and reprecipitation of carbon by the catalyst particle.) An alternate possibility is that nanotubes are closed at *both* ends during growth, and the catalyst facilitates the incorporation of atomic carbon (in the form of C_1, C_2, and C_3) at the strained pentagonal sites in the caps. In this case, the mechanistic role of the catalyst is unclear. A third hypothesis is that nanotubes grow from open ends by the direct adsorption of carbon species to dangling bonds. This has become known as the 'scooter' mechanism because the tube is presumably kept open by a chemisorbed metal atom that continuously scoots

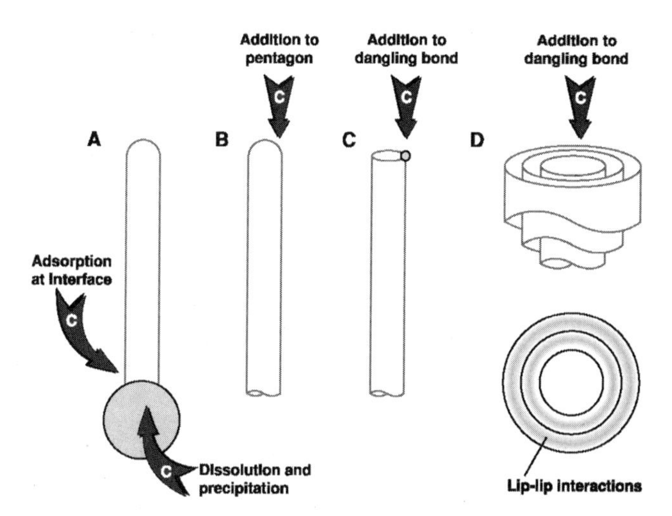

FIGURE 6.12. Illustration of proposed, atomistic growth mechanisms. (a) Root growth. Growth occurs at the nanotube-catalyst interface, and the tube remains closed at its distal end. (b) Direct addition to strained sites (pentagons) at end caps. (c) Open-ended growth by the 'scooter' mechanism. Carbon addition is to dangling bonds, and a migrating, chemisorbed metal catalyst particle (or atom) inhibits spontaneous closure. (d) Illustration of the lip-lip interactions that stabilize MWNT against closure during growth. Carbon addition is to dangling bonds.

around the circumferential edge, rearranging pentagons and other lattice defects into regular hexagons. It has been proposed that MWNTs grow by the addition of carbon to open ends, where outer shells are epitaxially nucleated. Conceivably the open ends are stabilized by 'lip-lip' interactions between the exposed circumferential edges, which might explain why catalyst is not necessary for MWNT production. Each of these possibilities is illustrated in Figure 6.12. There are many other, more exotic hypotheses.

It is worth reiterating that these atomistic processes are mere speculations, and that they are not exclusive of each other. In fact, for any proposed mechanism there exists empirical data that has been interpreted as supporting evidence. The most important point is that the thermodynamic conditions established by a particular experimental setup and the choice of catalyst and carbon source ultimately determine the nature of the product. Dependant variables include whether the product is single wall or multiwall, the diameter and chirality distribution, and the yield. At present, there is no coherent explanation for the synthesis of SWNT ropes. It remains an open question whether these are formed by the correlated growth of many SWNTs, by the assembly of discrete SWNTs into bundles after growth, or by some combination of the two.

6.4.1.3. Composition and Purification of the Reaction Product So far we have ignored the fact that the conversion of carbon to nanotubes is an inefficient process. Progress towards synthesizing analytically pure nanotubes continues to be made, but most methods currently in use produce a raw reaction product that is a multi-component solid, only a small fraction of which consists of nanotubes. The balance of the material is composed of residual catalyst particles (some of which are encased in concentric graphitic shells that resemble onions),

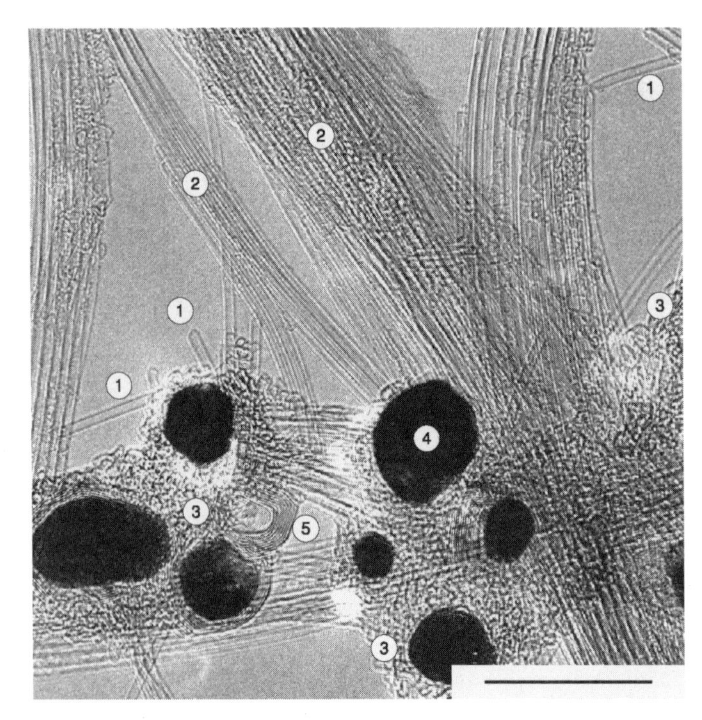

FIGURE 6.13. Transmission electron micrograph (see section 6.4.2.1) of raw nanotube material produced by arc discharge. The multicomponent reaction product includes (1) isolated SWNTs, (2) SWNT ropes, (3) amorphous or uncatalyzed carbons, (4) residual catalyst particles, and (5) polyaromatic graphitic shells. (From Ref. 53.)

fullerenes, other graphitic materials, and amorphous carbon. Many of these components are visible in the electron micrograph of raw nanotube material shown in Figure 6.13. Quantitative yield determination remains a controversial topic because there is no easy way to distinguish nanotubes from carbonaceous impurities on the macroscopic scale, although the weight fraction of residual catalyst can be easily quantified by energy dispersive X-ray fluorescence or atomic absorption spectroscopy. Reported yields are best regarded with an appropriate level of skepticism depending on the method of assay.

If the carbon nanotube is ever to fulfill its promise as an engineering material, large, high purity aliquots will be required. Ideally, improved synthesis methods will reduce or eliminate the need for purification, but the current reality is that SWNTs must be methodically extracted from their byproducts. One significant challenge is the separation of carbon tubes from carbon impurities which exhibit similar chemistry. As of this writing, traditional electrophoretic and chromatographic separation techniques have been disappointingly ineffective, although this could be the result of the lack of good methods for solubilizing nanotubes or for producing stable suspensions of isolated nanotubes. The most commonly employed purification techniques are categorized into two major methods: oxidation and size-exclusion.

Oxidation schemes derive their selectivity from the fact that carbon atoms found in impurities are often more reactive than those found in SWNTs. Small fullerene isomers and amorphous carbons are easily digested by concentrated acids and acid mixtures (HNO_3,

H_2SO_4) and other strong oxidizers (H_2O_2, $KMnO_4$). Refluxing (mixing while evaporating and recondensing the solvent) at moderate temperatures (70–120°C) results in the addition of functional groups, improving the solubility of the impurities. To ensure good intermixing of the nanotubes with the oxidizing agent, these steps are sometimes assisted by dispersion in an ultrasonic bath. The reflux decomposition products can then be separated by repeated washing in slightly basic NaOH solutions, resulting in the removal of many byproducts. Unfortunately, SWNTs also have reactive sites at defects and their strained caps, and the tubes themselves are also functionalized. The nanotubes remain largely insoluble due to steric constraints, but it is worrisome that this chemical modification could considerably alter their properties. Aggressive oxidation has been shown to completely sever tubes, leaving their open ends terminated primarily with —COOH groups. It is likely that such treatment also nucleates holes at any defect sites in the graphene shell itself.

An alternative to wet chemical oxidation is burning in air. This has proved particularly useful for removing amorphous carbon since it burns at such a low temperature, but the burn temperature of many graphitic impurities is precariously close to that of certain SWNTs. For this reason, the efficacy of air oxidation is limited.

The most obvious route to pure nanotubes is by size-exclusion. SWNTs are mesoscopic in one dimension, while most of their impurities are no larger than tens of nanometers in size. Filtration, by both lamellar and tangential flow, is often used in conjunction with wet chemical oxidation to recover purified nanotubes. Without prior oxidation, dissolution or suspension followed by filtration is generally less effective because the impurities tend to adhere to the tubes in lieu of entering into solution. SWNTs deposit onto a filter membrane in the form of an entangled film that can be peeled away as freestanding paper. This mat is a common starting point for certain characterization studies.

Another purification issue relates to the removal of metallic catalyst particles. Their presence prevents the high temperature utilization of SWNTs because upon heating, they could begin to pyrolyze the tubes themselves. Furthermore, certain catalytic metals have unpaired nuclear spins, which would render ^{13}C nuclear magnetic resonance (NMR) and electron spin resonance (ESR) studies of the nanotubes immensely difficult. At least partial removal is easily accomplished if the catalysts are exposed: treatment with HCl will dissolve the catalyst particles into soluble metal chlorides but does not react appreciably with carbon materials. The intact tubes can then be recovered by filtration. If the catalysts are encased in graphitic onions that cannot be opened by oxidation, purification is a greater challenge, but magnetic separation might be possible. This involves eluting a well-dispersed suspension of nanotubes through an inhomogeneous magnetic field. Magnetic catalyst particles will sense the magnetic field gradient in the form of a force and are attracted, while the SWNTs (which are only weakly paramagnetic) can pass through. Unfortunately, magnetic chromatography has been only marginally successful, and the efficient removal of catalyst remains a nontrivial problem.

The various purification steps that are sequentially applied to nanotubes can be damaging, leaving holes and chemisorbed functional groups on the tubes' graphene walls. Wet oxidation typically destroys the crystallinity of bundles, possibly due to intercalation of acid molecules into the rope interstices. For these reasons, purification is sometimes followed by a high temperature (1100–1200°C) vacuum anneal to remove certain chemisorbed groups and residual physisorbed reflux decomposition products. Annealing recovers rope crystallinity and ostensibly heals defects in the nanotubes' walls, although

TABLE 6.2. Characterization techniques commonly applied to carbon nanotubes

Technique	Information typically obtained	References
Transmission electron microscopy (TEM)	Atomic and supramolecular structure	[19]
Scanning electron microscopy (SEM)	Macrostructural information, sometimes used for purity assay	[12]
X-ray diffraction (XRD)	Supramolecular structure	[10, 11]
Raman spectroscopy	Diameter distribution	[20]
Fluorescence spectroscopy	Electronic DOS	[21]
Solid-state nuclear magnetic resonance (NMR)	Electronic structure	[22]
Electron paramagnetic resonance spectroscopy (EPR)	Structure of intercalated SWNTs, catalyst content	[23]
Temperature programmed desorption (TPD) with mass spectroscopy	Binding energy, amount, and compositions of physisorbed and chemisorbed species	[24]
Thermogravimetric analysis (TGA)	Burn temperatures of SWNTs, catalyst content	[25]
Inelastic neutron scattering	Phonon DOS	[26]
Scanning tunneling microscopy (STM)	Atomic structure, electronic DOS	[27, 28]
Atomic force microscopy (AFM)	Supramolecular structure, direct manipulation	[29]
Transport measurements	Electronic and thermal conductivity, thermopower	[30–32]

it has been evidenced that this healing is incomplete even after high temperature treatment.

6.4.2. *Characterization*

Materials scientists usually deploy characterization techniques to determine or evaluate an unknown material. Perhaps we have preempted ourselves by discussing the qualities of SWNTs prior to the methods by which they were determined. Nevertheless, knowing the structure and properties of nanotubes in advance will aid in understanding the interpretation of characterization data. Nanotubes have been examined by a large battery of techniques. Many of these are summarized in Table 6.2 with key literature sources cited for each, although this is not an exhaustive list. Some techniques are so fundamental to the study of nanotubes that they deserve further discussion. As has been true throughout this chapter, our focus in this section is on SWNTs.

6.4.2.1. Diffraction Transmission electron microscopy (TEM), powder X-ray diffraction (XRD), and elastic neutron scattering are three of the most common methods for investigating the structure of carbon nanotubes.[d] These are complementary techniques,

[d] Most people would categorize TEM as an imaging technique. However, it has been sagaciously said that the TEM is a phase-conserving diffractometer. Anyone who is familiar with a TEM will appreciate that 'diffractometer' is a much better description of the function of the instrument than 'microscope.'

but they are fundamentally different in that electrons and neutrons interact primarily with nuclei, while X-rays interact with bound electrons. Therefore electron microscopy and neutron diffraction probe the specimen potential, while XRD probes the electron density. Of course, the techniques ultimately give the same structural information because the electron density is determined by the specimen potential. One challenge in the analysis of carbon nanotubes by diffraction is that carbon has a small scattering factor due to its low atomic number. The scattering factor for electrons is still large enough to give measurable intensities from small sample volumes (perhaps tens of unit cells). However, the scattering power for X-rays is 10^{-4} that for electrons, which means that measurable intensities are obtainable only from macroscopic volumes.[e] The situation is more dire for neutrons, which have a smaller scattering factor than X-rays due to their neutral charge.

For carbon nanotubes, a TEM is usually used to obtain high-resolution images of the molecules. (Resolutions of 0.1 nm are possible with a good microscope.) In this mode, contrast is derived from mass-thickness, e.g., the integrated atomic number of the atoms intercepted by a collimated electron beam as it passes through the sample. If the 3D sample has sufficiently small mass-thickness, it can be treated as a 2D projection in the direction of the beam, and therefore the image is a direct projection of the specimen potential. This is called the *weak phase object approximation* (WPOA). While this is a technical detail, we bring it up for the following reason. An isolated SWNT satisfies the WPOA, so its image can be directly interpreted: contrast is greatest where the beam is tangent to the tube's walls (because the beam intercepts the most carbon atoms), so the image is two parallel lines separated by the tube's diameter. The graphene sheet usually cannot be directly resolved. However, even a medium-sized bundle of SWNTs is too thick to be a weak phase object. The image is more complicated, giving rise to interference patterns that cannot be directly interpreted. Even experienced scientists can fail to make this distinction, sometimes resulting in the misinterpretation of microscopy data. One disadvantage to TEM is that nanotubes quickly degrade when exposed to typical illumination conditions (both electron flux and energy are important), limiting the time that a sample can be non-destructively observed.

With imaging, it is possible to measure directly SWNT diameters and MWNT interlayer separations; and to get a general sense of a sample's composition and purity. In principle, quantitative structure determination is possible using electron diffraction. For the purpose of this chapter, the details of diffraction are less important than understanding the strengths and limitations of the technique. The usual Ewald criterion holds for electron diffraction: the observed pattern is the image of the intersection of the Ewald sphere with the reciprocal volume. Because the reciprocal volume is obtained by the Fourier transform of the real structure, and because the Fourier transform of a cylinder is a cylindrical Bessel function, SWNT patterns are characterized by (1) nodal cylinders concentric with the tube axis, and (2) spots that reveal the helicity of the constituent graphene lattice. The mathematics are omitted here, but a definitive treatise on the electron diffraction of nanotubes has been prepared by Amelinckx and coworkers.[19] A simulated pattern of an isolated (10,10) tube is shown in Figure 6.14a. Typically the reflections in experimental patterns of isolated SWNTs are weak due to the smallness of both the scattering factor and the sample volume. For this

[e] Of course, measurable intensities could be obtained from small volumes if the scattering intensity were integrated over a much longer time than is practical in a typical experiment. The real issue is signal-to-noise ratio, not sample volume.

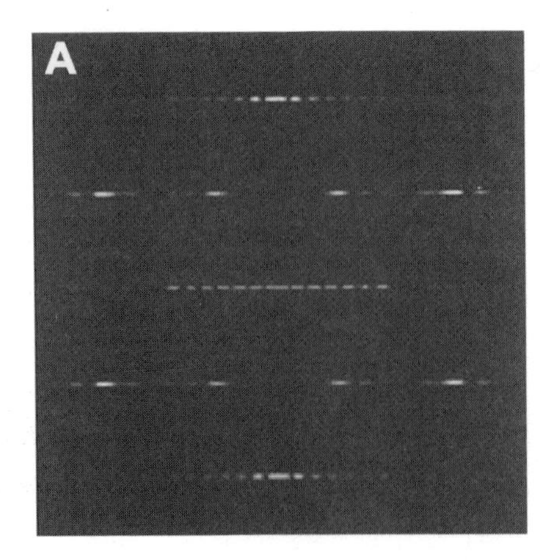

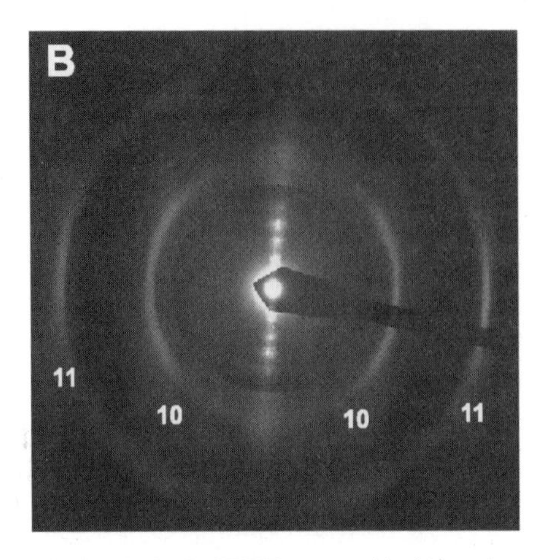

FIGURE 6.14. (a) Simulated electron diffraction pattern of an isolated (10,10) SWNT at normal incidence (e.g., with the electron beam perpendicular to the tube axis). (From Ref. 19 by permission of IOP Publishing Ltd.) (b) Empirical electron diffraction pattern from a SWNT bundle. Reflections from both the rope lattice and the graphene lattice are indicated. The circle delineates a region of the pattern outside of which the contrast has been enhanced for clarity. Therefore, the circle is an analysis artifact and is not intrinsic to the pattern. (From Ref. 53.)

reason, essentially no effort has been spent determining the (n, m) indices of a SWNT from empirical diffraction data.[f]

The packing of SWNTs in a rope can be easily evaluated by electron diffraction. The pattern of a rope, an example of which appears in Figure 6.14b, looks like the superposition of multiple SWNT patterns with an extra row of systematic reflections perpendicular to the rope axis. This extra equatorial row is due to the two-dimensional lattice of tubes, and the spacing of spots along this row is related to the spacing of lattice planes that contain the directions of both the tube axis and the zone axis of the pattern. Because this spacing is larger than the spacing of atomic planes, the rope lattice reflections appear at low Q. (The diffraction pattern of a MWNT has a similar row from which the interlayer separation can be determined.) Furthermore, the mosaic of chiralities within a bundle means that the superposition of the discrete graphene lattice spots associated with each individual tube forms continuous arcs in the rope pattern. These 10 ($Q = 2.94$ Å) and 11 ($Q = 5.09$ Å) arcs are centered on the meridian, which is parallel to the real space direction of the nanotube axis. The angle subtended by these arcs is partially due to the chirality distribution in the bundle. Both the equatorial and graphene lattice reflections are angularly broadened due to any bending distortion or misalignment of the bundle.

While electron microscopy can provide a local probe of the sample, the poor sensitivity of X-rays and neutrons necessitates large sample volumes. Since no large nanotube single crystals exist, XRD and neutron diffraction are intrinsically powder-averaged methods using polycrystalline samples (i.e., multiple ropes). For this reason, the anisotropy of the graphene *hk*0 reflections is lost, giving rise to a Warren lineshape (a sharp onset with a gradual tail

[f] Structure determination is also complicated by the fact that the relative phases of the diffracted beams are lost when the diffraction pattern is recorded on film. Clever solutions to this phase problem have been developed in the field of structural biology.

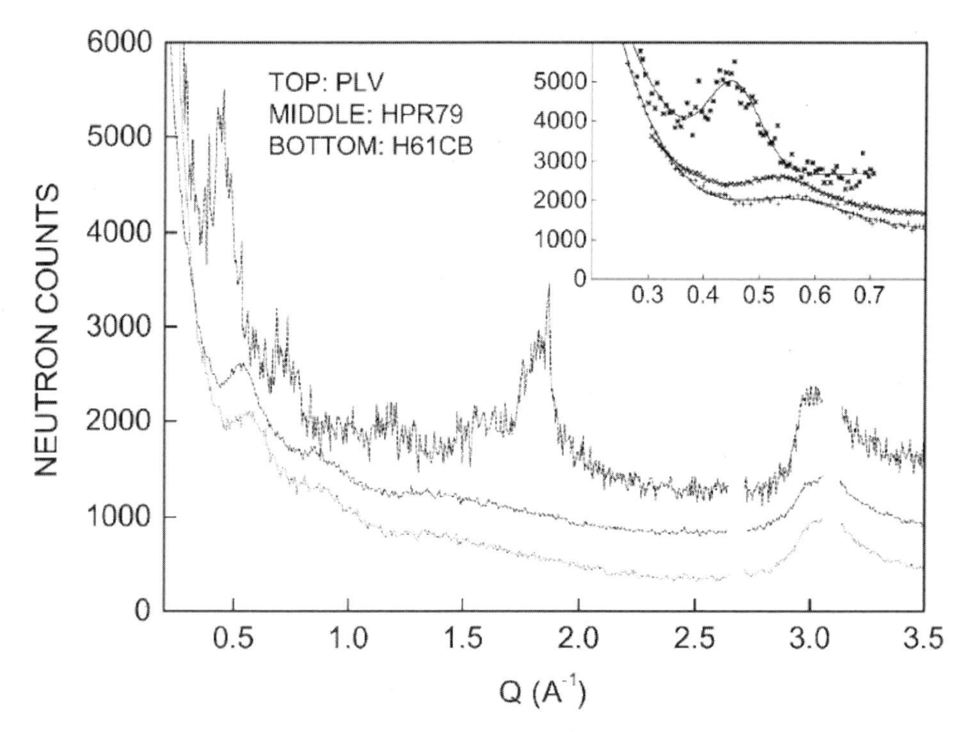

FIGURE 6.15. Powder neutron diffraction patterns of three different nanotube samples. The inset shows Gaussian fits to the strongest rope lattice peak. The feature at Q $\sim$ 1.8 Å^{-1} in the PLV pattern is attributed to the presence of graphitic shells that survived purification. (From Ref. 25 by permission of Elsevier Science B.V.)

as a function of increasing scattering angle) for the graphene diffraction peaks. XRD has become the technique of choice for the routine investigation of the crystallography of the rope lattice. The large lattice constant of a nanotube bundle ($\sim$1.7 nm) means that the pertinent reflections lie at small scattering angles, very close to the origin of diffraction space and well inside the graphene 10 and 11 peaks. As was true of an electron diffraction pattern, the intensity in both XRD and neutron diffraction patterns is modulated by a cylindrical Bessel function due to the shape of each SWNT. A typical neutron diffraction pattern of polycrystalline SWNT material is shown in Figure 6.15, with the most important features indicated. XRD has also proved useful in evaluating rope diameter (i.e., coherence length) from peak breadth and lattice dilation during interstitial doping of the channels between SWNTs.

6.4.2.2. Raman Spectroscopy The most common application of Raman spectroscopy to nanotubes is the determination of the diameter distribution within a bulk sample. The technique involves the probing by laser light of the intramolecular vibrational and electronic states of the material. Incident monochromatic radiation promotes a bound electron into a 'virtual' excited state. Because this virtual state does not exist in the energy dispersion, it immediately decays into an available real state within the same electronic subband, resulting in the emission of a photon. Sometimes this event is inelastic such that the emitted (scattered) photon has more or less energy than the incident photon. This energy difference is due to a concomitant vibrational transition during the electronic excitation-decay process and is called the Raman shift. Thus, the Raman shift corresponds to the energy between allowed vibrational modes in the sample. Typically, Raman probes transitions from acoustic modes (which are always populated) to unpopulated optical modes. The electronic wavevector $\vec{k}$ of the involved electron does not change in the scattering process.

As we previously explained, the RBM of a SWNT is Raman active because of its symmetry. Therefore, the Raman spectrum (i.e., intensity versus Raman shift) of a sample of nanotubes is a direct probe of the allowed RBMs and therefore of the diameter distribution, c.f. equation (6.7). There is one caveat: the separation of the van Hove singularities in the electronic DOS also depends on diameter, and for certain diameter SWNTs this separation will equal the energy of the incident photon. Promotion can be to a real state instead of to a virtual state, which increases the excitation probability by a factor of 100–10,000. Accordingly, those tubes that resonate with the incident wavelength are more likely to result in a Raman shift of this radiation. In this way, the Raman intensity of a particular RBM depends on the incident wavelength. This *resonance enhancement* of only certain tubes makes it nearly impossible to correlate Raman intensity with the number of tubes at a particular diameter. Raman is most useful for determining the endpoints, and not the shape, of the diameter distribution. In addition to the RBMs, there is another band of peaks called the D-band whose intensity relates to the fraction of disordered carbons in the sample. The D-band has been sometimes used as a qualitative metric for sample purity. Some of the salient features of typical Raman spectra are indicated in Figure 6.16.

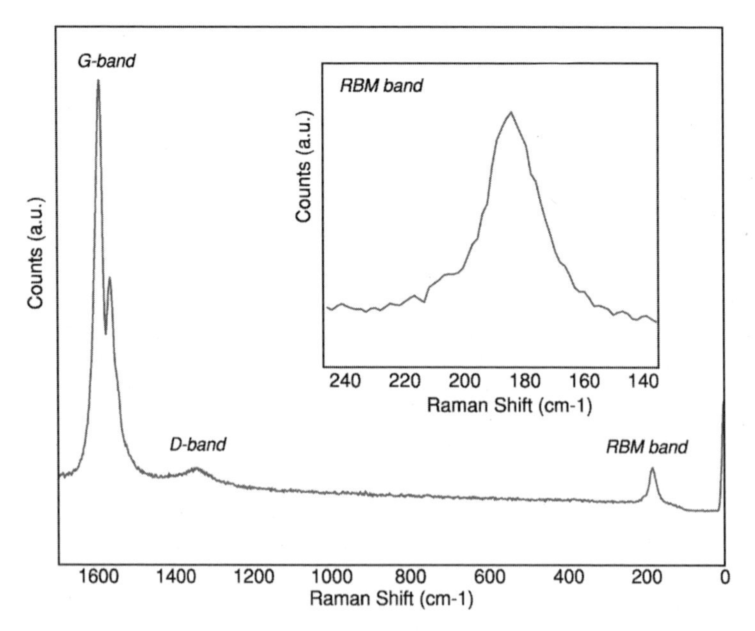

FIGURE 6.16. Raman spectrum of acid purified nanotube material. The important spectral features are (a) the $E_{2g(2)}$ modes (G-band), (b) the D-band, and (c) the A_{1g} modes (RBM band). Inset: detail of the RBM band. (From Ref. 53.)

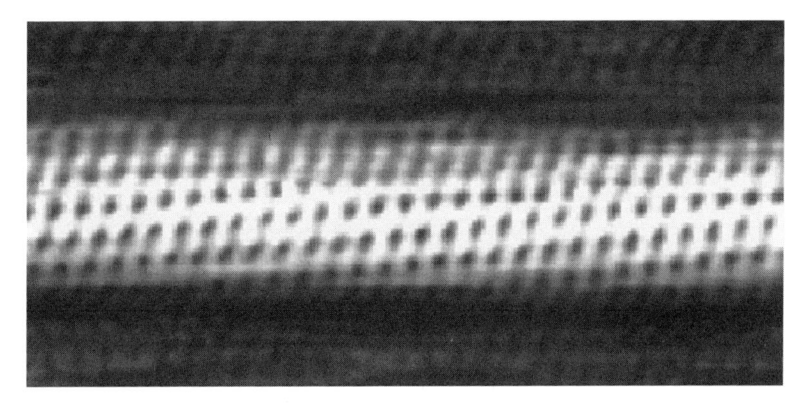

FIGURE 6.17. Atomic resolution scanning tunneling microscopy image of an individual (11,7) SWNT. The indices are determined by direct measurement of the chiral angle (7°) and diameter (1.3 nm). (From Ref. 28 by permission of Macmillan Publishers Ltd.)

6.4.2.3. Scanning Tunneling Microscopy Scanned probe microscopy is so essential to the examination of nanoscale systems that an entire chapter of this book is devoted to the topic (see Chapter 3). For carbon nanotubes, scanning tunneling microscopy (STM) in particular has been the centerpiece of many seminal experimental studies. Samples can be prepared for STM by spin-coating SWNTs suspended in an organic solvent onto a conducting substrate such as Au(111). Imaging is generally accomplished at low temperature ($\leq$77 K) under ultra-high vacuum. An STM tip is biased and brought close to the sample so that a tunneling current is generated across the vacuum barrier. As the tip is scanned, its height is varied to keep the tunneling current constant. This displacement of the tip maps out the electronic topology of the sample, and an atomically resolved image is generated. An STM image of a SWNT is shown in Figure 6.17. The hexagonal symmetry of the graphene lattice is obvious, and the (n, m) indices of the tube can be determined by observation.

The potency of STM lies in its ability to determine simultaneously the atomic structure and the electronic density of states of a SWNT. The latter is obtained by scanning tunneling spectroscopy (STS), in which both scanning and feedback to the tip are turned off. With the tip fixed in position, the bias is swept, and the tunneling current is collected as a function of voltage. The differential conductance dI/dV is proportional to the density of states, so differentiation of the collected data gives the spatially localized DOS. A positive bias to the tip draws electrons out of the SWNT, mapping the occupied states, while a negative bias injects electrons into the SWNT, mapping the unoccupied states. By using STM to resolve both the atomic structure and the DOS of the *same* tube, the suspected correlation between the chirality and electronic properties of a SWNT has been empirically verified.[27,28] The same methodology has been used to confirm the localized perturbation of the DOS of a SWNT by an extrinsic dopant.[33]

6.4.2.4. Transport Measurements The electrical conductivity, thermal conductivity, and thermopower of a SWNT mat can be easily measured by affixing the appropriate leads to a macroscopic sample. However, measurement of the transport properties of an individual

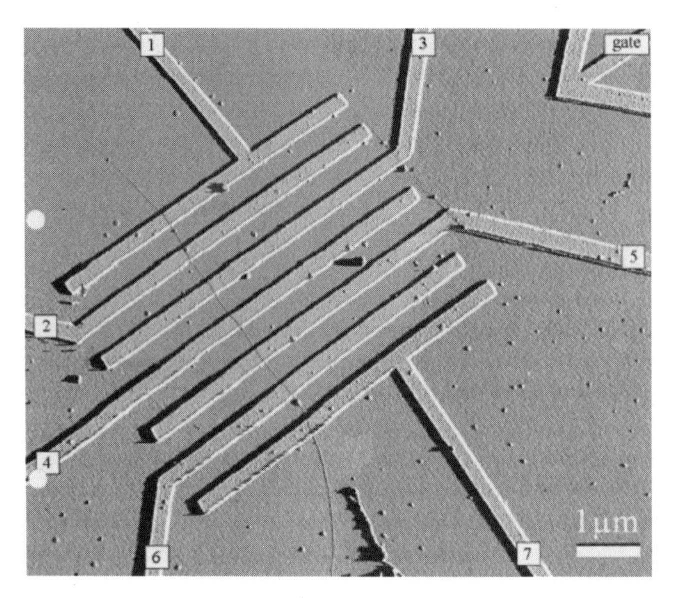

FIGURE 6.18. Atomic force microscopy image of an isolated SWNT deposited onto seven Pt electrodes by spin-coating from dichloroethane solution. The substrate is SiO_2. An auxiliary electrode is used for electrostatic gating. (Reproduced with kind permission of C. Dekker.)

SWNT requires a fair amount of ingenuity. At present, only the electrical conductivity of an individual SWNT has been directly measured. The main difficulty concerns the attachment of reliable electrical contacts in a configuration that allows the contact resistance to be accounted for. (Some believe that a four probe configuration is optimal for this purpose.) The simplest method is to spin-coat SWNTs onto an insulating substrate upon which electrodes have already been patterned. The sample is imaged by atomic force microscopy (AFM) to locate any tubes serendipitously lying across the correct electrode configuration. For example, Figure 6.18 shows an AFM image of a SWNT spanning seven Pt electrodes on a SiO_2 substrate. A nearby auxiliary lead can be used as a capacitively coupled gate electrode, as can the substrate itself if the silicon below the insulating oxide layer is heavily doped. An STM tip can also be used as a movable electrode. An alternate approach involves the use of a focused ion beam (FIB) to deposit leads directly on top of a nanotube, although this more complicated method is less commonly employed. Similar experiments have been performed on SWNT bundles and MWNTs. While various other techniques exist for measuring transport phenomena through molecules, many of these have yet to be applied to nanotubes.

Multiprobe measurements on SWNTs such as the one shown in Figure 6.18 exhibit plateaus of non-zero current in the I–V spectrum.[34] This evidences ballistic transport (i.e., without scattering), which is believed to be the dominant mode of electronic conduction in a SWNT.

6.5. MODIFICATION

In the construction of bridges, airplanes, and even microchips, it has become possible to tailor the properties of various materials to suit a particular purpose. Steel can be alloyed and tempered if more toughness is required, and silicon can be doped to change its conductivity. This ability to engineer a material to do something specific is possibly even more important in the field of nanotechnology, where there may be few alternatives available if a given molecular component doesn't quite suit its intended purpose. Part of the attraction of carbon nanotubes is that their structure and composition makes them amenable to chemical modification. While the properties of nanotubes cannot yet be tuned systematically, there are many incontrovertible examples of changes in properties due to functionalization. In this section, we describe some of these methods and the nanotube-based materials that they effect. There are three principal approaches: covalent modification, physisorption, and filling.

6.5.1. Covalent Modification of Carbon Nanotubes

The first forays into covalent functionalization were intended to improve the solubility of SWNTs, which are extremely resistant to wetting. SWNTs lack surface functional groups that would be prime sites for derivatization, and their gently curved graphene walls are so energetically stable as to render the constituent carbon atoms unreactive. However, it was previously described that the strong oxidizers used in purification can attach —COOH groups primarily to the termini of SWNTs but also to strained or defective sites on their sidewalls. Early attempts at solubilizing SWNTs in organic solvents aimed to replace this acid moiety with long-chain hydrocarbons. In fact, conversion of the carboxylic acid to the amide of octadecylamine via an acid chloride intermediate was successful at increasing the solubility of SWNTs in chloroform, dichloromethane, and various aromatic solvents.[35] This scheme has become a general way to create amide or ester linkages to SWNTs. It has since been exploited to assemble SWNTs together either through covalent amide bridges or through the attachment of DNA oligonucleotides that can then base pair with complementary strands on other tubes.

The preferential addition of —COOH groups to nanotube ends means that an alternate scheme must be found for efficient sidewall addition. Some electrophilic compounds, including dichlorocarbene, can be added directly to SWNT sidewalls, as can hydrogen up to ~9% coverage by the Birch reduction. While these may provide a starting point for further derivatization, the most common route to sidewall addition has been through fluorination. Nanotubes can be easily fluorinated by flowing F_2 gas diluted with He at low temperatures (150–325°C). Stoichiometries of nearly C_2F can be obtained without decomposition of the tubes.[36] Efficient alkyl substitution for the fluorine is then possible using sterically unhindered alkyllithium reagents (RLi, R = methyl, *n*-butyl, *n*-hexyl, phenyl), Grignard reagents (alkylmagnesium bromide), or alkoxides (sodium methoxide). These reactions are carried out simply by bath sonication of the fluorotubes dispersed in an organic solvent with excess reagent.[37] Because the hybridization of carbon atoms in a nanotube must change to accommodate sidewall modification, these derivatized SWNTs might be expected to have markedly different electronic structures than pristine SWNTs or SWNTs with only terminal groups. Solution spectroscopy of alkylated SWNTs confirms this: no van Hove singularities can be detected.[37] Interestingly, fluorotubes can be defunctionalized by reaction with anhydrous hydrazine to recover the original starting material. Pyrolysis of fluorotubes under argon at ~1000°C also gives the curious result of severely shortened, defluorinated SWNTs. The mechanism responsible for this latter observation remains unknown.

In summary, two primary methods are commonly used for the covalent derivatization of nanotubes. These are (**I**) carboxylation followed by conversion to an ester or amide

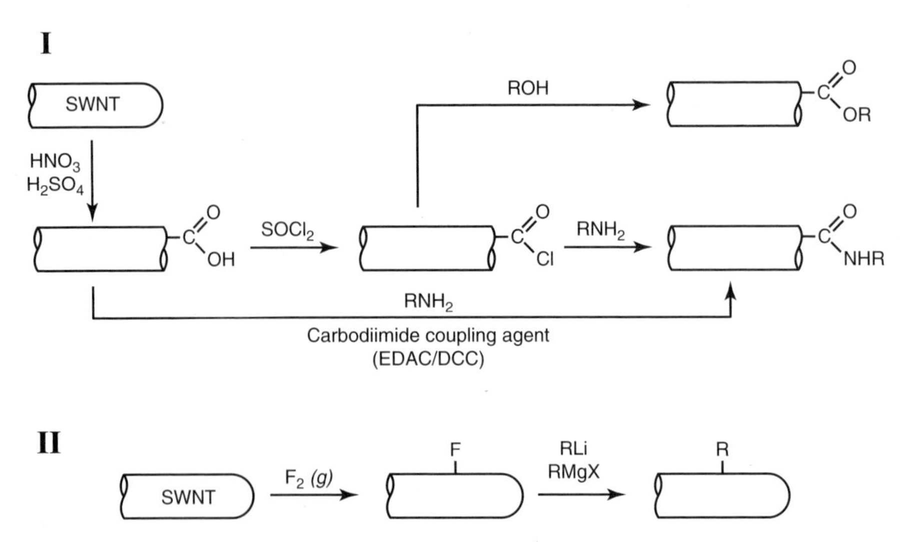

FIGURE 6.19. Two common reaction schemes for the covalent derivatization of SWNTs: (**I**) carboxylic acid derivatization, and (**II**) fluorination. Many variations on these schemes are possible.

linkage via an acid chloride intermediate, and (**II**) fluorination followed by substitution reaction with an organometallic. **I** favors terminal modification because of the preferred attachment of carboxylic acid groups to strained sites, while **II** is an efficacious pathway to sidewall derivatization. The general schemes are illustrated in Figure 6.19 and enable the solubilization, directed assembly, and property modification of SWNTs. The covalent chemistry of nanotubes is a nascent field. Chemists have hesitated to study reactions with SWNTs largely because of the absence of analytically pure starting materials. As this problem is surmounted, it is likely that a rich field will be uncovered.

It is also noteworthy that substitutional doping of SWNTs is possible. The usual dopants are boron and nitrogen, which are predicted to contribute acceptor and donor states, respectively, to the band structure. Certain $B_xC_yN_z$ stoichiometries can be attained either *in situ* during arc discharge synthesis or by gas phase pyrolysis.

6.5.2. Physisorption to Carbon Nanotubes

Physisorption includes both surface adsorption and intercalation. The one-dimensional trigonal channels between SWNTs in a bundle are obvious sites for the insertion of extrinsic dopants. Such intercalants could transfer charge to or from the nanotubes while minimally perturbing the tubes' intrinsic DOS. Both potassium and bromine are archetypal ionic intercalants for graphite, residing in the galleries between the constituent graphene sheets as ions (K^+ and Br_2^-). Similarly, both have been intercalated into SWNT rope channels at stoichiometries of approximately KC_8 and Br_2C_{52}. Doping occurs by direct exposure of nanotubes to K and Br_2 vapors. Despite the fact potassium is an electron donor while bromine is an electron acceptor, intercalation with *either* results in a decrease in bulk resistivity by a factor of ~30[38]. This suggests charge exchange, displacing the Fermi level (either up or down) out of the band gap in the case of semiconducting nanotubes and/or to where the number of charge carriers is large in the case of metallic nanotubes. This strategy opens the possibility of tuning the Fermi level to coincide with a singularity in the DOS, which could dramatically enhance the conductivity of a SWNT rope. Non-covalently doped nanotubes are not limited to intercalation compounds. Isolated SWNTs (which of course lack intercalation channels) have been successfully n-doped via charge exchange with surface adsorbed potassium at temperatures below 260 K[39].

It is also possible to form nanotube intercalation compounds from alkali metals by an electrochemical route. In this method, the alkali to be inserted serves as a sacrificial anode, while SWNTs compose the cathodic half-cell. In the presence of electrolyte, galvanostatic charging of the cell causes intercalation of the metal ion. Both lithium and potassium have been intercalated electrochemically, and lithium has a particularly large storage capacity due to its small size. Furthermore, electrochemical doping is at least partially reversible by flipping the polarity of the cell. Combined with the fact that a large volume fraction of a rope is occupied by interstices, these results suggest that nanotubes could outperform even the best known anode materials for lithium ion batteries. SWNT ropes can be charged to a stoichiometry of $Li_{1.23}C_6$ (corresponding to a reversible capacitance of 460 mAh/g)[40], while a state-of-the-art intercalated graphite anode can be charged to only LiC_6. Unlike vapor phase intercalation, the electrochemical approach results in an irreversible loss of rope crystallinity. This curious behavior has been attributed to co-intercalation of the solvent in the electrochemical case.

Although covalent functionalization has been shown to increase the solubility of SWNTs in various media, non-covalent alternatives have been pursued because they are perhaps less likely to modify the tubes' intrinsic electronic properties. Dissolution has been reported in water using a range of surfactants including octyl phenol ethoxylate (syn. Triton X-100), sodium dodecyl sulfate (SDS), and sodium dodecylbenzene sulfonate (NaDDBS). Each of these has a hydrophobic, long-chain alkane tail that coordinates with the SWNT and a hydrophilic head group that mediates the interaction with water. Recent motivation to introduce nanotubes into biological systems has resulted in the discovery that even common starch (i.e., amylose) wraps SWNTs via hydrophobic interactions, stabilizing them in aqueous solutions.

Some of the best surfactants also contain benzene rings, which are expected to have a high affinity for a nanotube's surface due to π–π interactions. The same principle enables SWNT bundles to be coordinated or wrapped by certain conjugated polymers such as poly(aryleneethynylene) and poly(m-phenylenevinylene). The solubility of polymer-coordinated SWNTs could be enhanced by selective modification of the polymer's side chains. Similarly, pi stacking facilitates the assembly of various molecules on the outer walls of nanotubes using molecular linkers that contain conjugated moieties. For example, proteins have been immobilized on SWNTs through amide bonding to a succinimidyl ester with a pyrene group.

6.5.3. Filling Carbon Nanotubes

A carbon nanotube has one characteristic that virtually no other molecule duplicates: it has an interior channel, separated from the exterior environment by an essentially impervious graphene shell. Furthermore, the lumen of most cage molecules like fullerenes and macrocycles is zero-dimensional—the available space is confined in all directions—while the lumen of a nanotube is extended in one-dimension. Therefore, if the core of a nanotube could be filled with some other atom, ion, or molecule, it would enable the creation of an entirely new class of 1D heterostructured materials. Because so few known materials have such low dimensionality, these entirely synthetic structures could have unexpected properties just as the linear nature of a SWNT is responsible for its many enticing qualities. One benefit of filling is obvious: isolated SWNTs cannot be doped by intercalation. Moreover, even though the surface adsorption of potassium has been shown to dope isolated SWNTs, these exposed dopants are vulnerable to chemical reaction and thermal desorption. It is not likely that this method will be robust under real conditions. However, an encapsulated dopant is sterically protected from chemical attack or spontaneous de-doping by the surrounding nanotube sheath. In this way, filling may be a viable means for tuning the electronic structure (or other properties) of an individual SWNT.

6.5.3.1. Filling MWNTs How exactly do you fill a nanometer-sized vessel? Substantial effort has been devoted to the filling of MWNTs with low surface tension liquids by capillarity. This was first accomplished by a one-step process, in which capped MWNTs were opened and filled with an unidentified lead compound by annealing MWNTs decorated with evaporated Pb particles in air at 400°C (above T_m for Pb) for 30 minutes.[41] Other one-step processes have relied upon wet chemical methods, in which acids and/or chemical precursors of the filling medium were reacted directly with MWNTs for a number of hours. MWNTs have also been filled by a two-step process, in which tubes were first opened by annealing at 700–850°C in air or under flowing CO_2, and the open tubes were then filled by direct immersion in a liquid or molten filling medium. Often, the encapsulated compound assumes a crystalline structure. This can be seen in Figure 6.20, which shows a TEM micrograph of a filled MWNT.

These techniques have in common the opening of MWNTs, either by air oxidation (possibly catalyzed by residual metal in the case of the one-step process) or chemical oxidation, and the subsequent uptake of a liquid. An upper limit of 100–200 mN/m has been determined for the surface tension of the filling medium above which capillary filling will not occur, although this value is greatly affected by the actual diameter of the MWNT cavity.[42] For these reasons, filling by capillarity is limited to those compounds having low melting points and low surface tensions. Examples of filled MWNTs include tubes which contain Ag, Au, Bi, Pd, and Re metals; V_2O_5, La_2O_3, Pr_2O_3, CeO_2, Y_2O_3, Nd_2O_3, Sm_2O_3, $FeBiO_3$, and other Ni, Co, Fe, and U oxides; $AgNO_3$, $CoNO_3$, $CuNO_3$, AuCl, and other

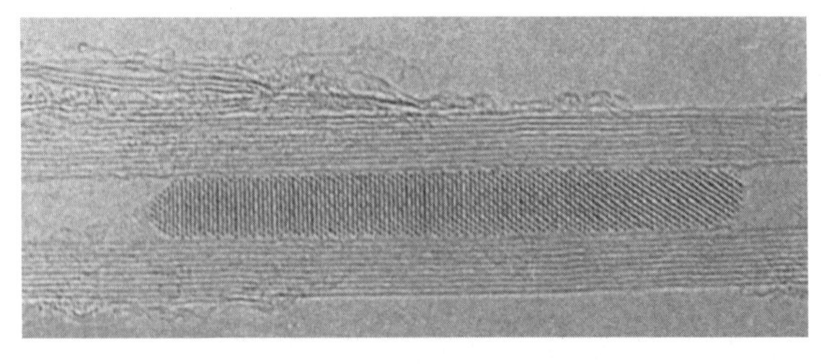

FIGURE 6.20. Transmission electron micrograph of a MWNT filled with Sm_2O_3. The interlayer separation in the MWNT is c.a. 0.34 nm. Lattice planes in the oxide are clearly seen. (From Ref. 55 by permission of The Royal Society of Chemistry.)

nitrate and chloride salts; and the eutectic systems KCl–$CuCl_2$ and KCl–UCl_4. Sometimes reduction of the filling medium and removal of the MWNT template are possible, yielding free-standing nanowires. (It is also possible to template the synthesis of nanowires by coating MWNTs with oxides. This method has been used to prepare WO_3, MoO_3, Sb_2O_5, MoO_2, RuO_2, and IrO_2 nanowires.) Yields of these filling methods range greatly, perhaps from 10–90%.

Arc discharge based techniques have also been developed for the direct synthesis of MWNTs filled with various poly- and monocrystalline transition metal (Cr, Ni), rare earth (Dy, Gd, La, Sm, Yb), and covalent (Ge, S, Sb, Se) compounds and carbides. In this case, the graphite anode is drilled and packed with the filling medium or its corresponding precursor, and filled MWNTs are collected from the resulting cathode deposit.

6.5.3.2. Filling SWNTs Mass transport into SWNTs is a fundamentally different challenge due to their smaller diameters. No techniques have been developed for the direct synthesis of filled SWNTs. The capillarity method used to fill MWNTs has proved successful at filling SWNTs with Ru, Au, Ag, Pt, Pd, various KCl–UCl_4 and AgCl–AgBr compounds, and KI. Unfortunately, the surface tension of most liquids prevents uptake into the sub-nanometer lumen of a SWNT, limiting the scope of capillarity as a technique for producing filled SWNTs in high yield. However, gases are free of this physical limitation and can easily permeate even nanoscopic pores. This suggests that where possible, a better method for delivering extrinsic species to the inside volume of a SWNT is through the vapor phase.

The canonical example of a filled SWNT is the 'peapod' structure, $(C_{60})_n$@SWNT (where the 'at' symbol should be read 'non-covalently contained within'), which consists of a close-packed arrangement of C_{60} molecules collinear with the axis of a surrounding single wall nanotube.[43] A TEM micrograph of a typical peapod is shown in Figure 6.21a. These supramolecular materials are formed by vaporizing C_{60}, which exists as an fcc molecular

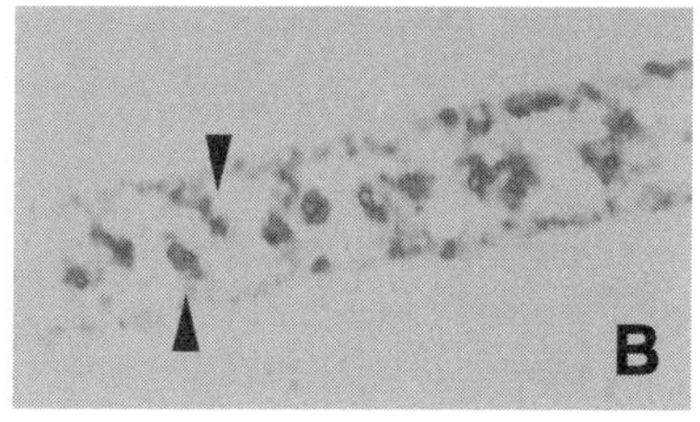

FIGURE 6.21. (a) Transmission electron micrograph of C_{60}@SWNT. The nanotube is surrounded by vacuum and does not lie on a substrate. The encapsulated fullerenes form a one-dimensional chain with a lattice periodicity of c.a. 1.0 nm. It is possible to obtain diffraction signatures from these structures. (b) False-color transmission electron micrograph of La_2@C_{80}@SWNT. Each C_{80} cage contains two point scattering centers which are the individual La atoms contained within.

solid under standard conditions, in the presence of SWNTs that have been opened by oxidation. Vaporization is typically accomplished by vacuum annealing at 300–600°C. The vaporized fullerenes stochastically collide with the SWNTs, in some cases physisorbing for a short time due to favorable pi stacking. Directed diffusion of adsorbed C_{60} along the outside of a tube is possible. These processes lead C_{60} molecules to openings in the SWNTs' walls or ends, admitting the fullerenes into the lumens. Fortuitously, C_{60} is exactly the right size to fit inside a typical SWNT with the two molecules separated by the graphitic interlayer separation all the way around the circumference. This is energetically favorable, so the fullerenes are retained inside. Linear diffusion enables the encapsulated molecules to assemble into periodic chains, increasing their stability even further.[44]

Conceivably, hierarchical materials could be created from SWNTs and quantities of any molecule fitting the steric and thermodynamic constraints suggested by the vapor phase filling method. Generality has been demonstrated. SWNTs have been filled with metallofullerenes including $La_2@C_{80}$ (Figure 6.21b), $A_nB_{3-n}N@C_{80}$ ($A =$ rare earth metal, $B =$ Sc or Y), and $Gd@C_{82}$; with various metallocenes; and with *o*-carborane. Some, but not all, of these form a periodic array when encapsulated. The emphasis has been on filling with molecules that have some three-dimensional structural features. This is because a planar molecule, an atom, or an ion may not be large enough to benefit from the three-dimensional coordination afforded by the inside of a nanotube. In this case, the driving force for filling is removed because the interior and exterior surfaces offer equivalent energetic stability.

The ability to produce one-dimensional molecular crystals inside SWNTs is a powerful tool in the creation of molecular materials, potentially begetting quantization conditions that yield novel electronic and thermal conductors. Moreover, the perturbing effect of encapsulated molecules on the intrinsic electronic properties of a SWNT has been convincingly demonstrated by STM. Figure 6.22 shows STM differential conductance spectra obtained from the same spot on a C_{60} peapod before and after the C_{60} molecules have been shuttled to an empty part of the tube. In both spectra, the occupied DOS (obtained at negative sample bias, e.g., tunneling out of the sample) is essentially that of an empty SWNT. However, the unoccupied DOS (positive sample bias) is seen to have spatially localized features only when the C_{60} molecules are present. These features evidence the creation of a narrow band of unoccupied states due to overlap of the C_{60} and SWNT molecular orbitals.[33] It is not yet known how these dopant states affect conductivity, but it is interesting to speculate that the insertion of molecules with a tendency to withdraw or donate charge could displace the Fermi level in addition to forming hybrid electronic states. The possibility of doping isolated SWNTs in a systematic way is fantastic.

6.6. APPLICATIONS OF NANOTUBES

The extraordinary structure and properties of carbon nanotubes include a high aspect ratio; a one-dimensional form; the ability to assemble into ropes; the presence of an internal lumen; high mechanical strength, resilience, and thermal conductivity; either electrical conductance or resistance; the ability to be modified by chemical reactions; and chemical stability in a nanoscale form. These impart a great potential for use in a number of applications. In this section, the current and projected applications of carbon nanotubes are discussed.

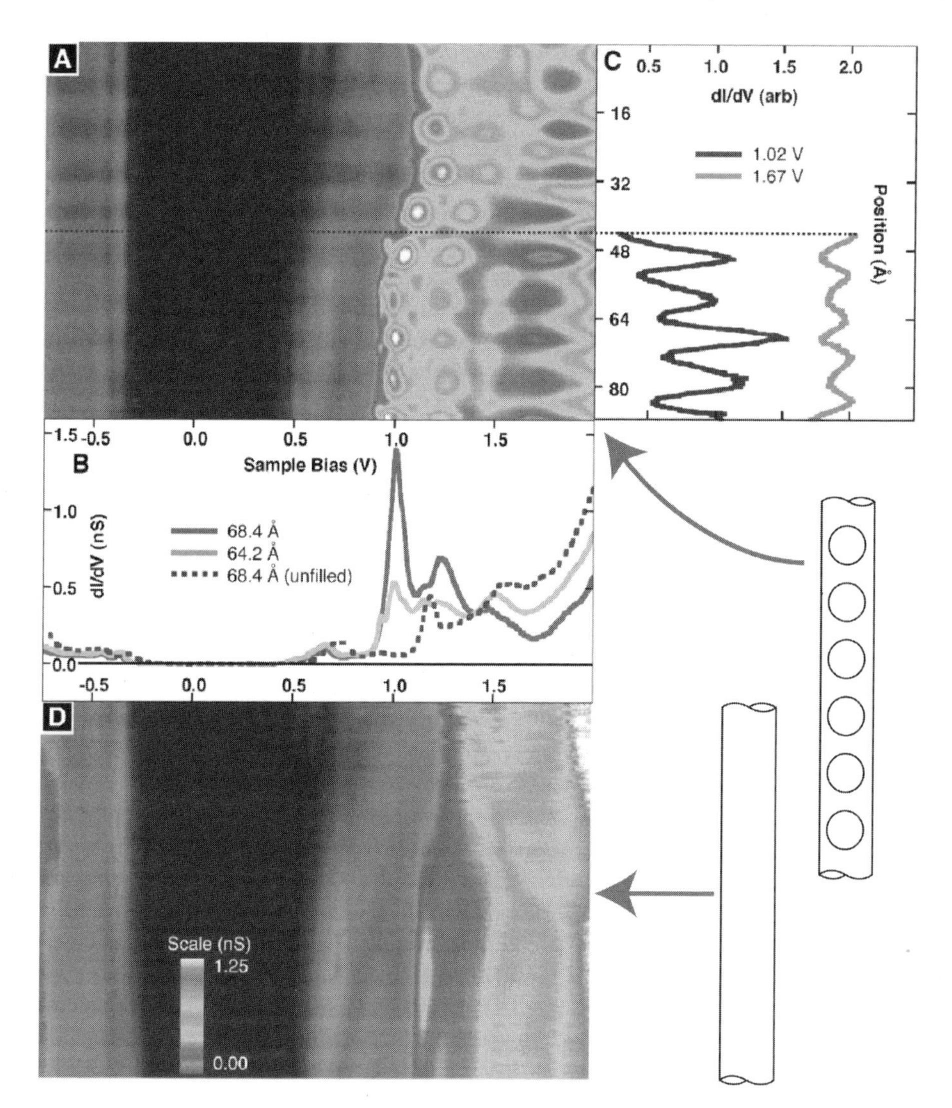

FIGURE 6.22. Differential conductance spectra of a C_{60} peapod. (a) Conductance versus position (Å) and sample bias (V) for the peapod. Spatially localized modulations are observed only for positive sample bias, i.e., in the unoccupied density of states. The periodicity of these modulations matches the periodicity of the encapsulated fullerenes. (b) and (c) show conductance at constant position and at constant sample bias. (d) Conductance versus position for the same location on the SWNT after the C_{60} molecules have been shuttled into an empty part of the tube by manipulation with the STM tip. No periodic modulations are observed. (From Ref. 33 by permission of The American Association for the Advancement of Science.)

6.6.1. Nanotubes for Storage Applications

The large uptake of Li ions into mats of SWNTs, mentioned above, provided initial excitement about the use of the material for batteries with much higher capacity than has been previously available. However, it was soon discovered that as-synthesized and purified

SWNT materials suffer from two significant barriers to this technological application: they do not discharge at a constant voltage, and a large component of the Li uptake is irreversible such that efficiency on a per-weight basis is undermined. Even so, it has been found through empirical development in industry that the addition of some content of nanotubes to a conventional carbon-based anode—up to 20%—produces a significantly higher storage capacity. Therefore, nanotubes are currently an important constituent of the most-advanced lithium ion batteries in industrial production.

Nanotubes are also being considered for hydrogen storage applications. The technological driver is the need for a system that can store sufficient H_2 for fuel cell powered automobiles. Theoretical studies have predicted, and experimental desorption studies have confirmed, that the interaction of hydrogen with nanotubes is through physisorption of H_2 on the exterior, and possibly the interior, surfaces. This process yields a level of hydrogen uptake that is insufficient to make pure nanotubes practical as a medium for hydrogen storage. However, experimental studies of a number of groups have produced a wide range of values for hydrogen uptake in nanotube materials produced by various methods. Although no systematic trends have resulted from these studies, the largest values are well above the levels needed for a practical storage material. It remains a possibility that either imperfect nanotubes, or nanotubes intermixed with other carbonaceous or non-carbonaceous materials, may provide a solution to the hydrogen storage challenge. These inconclusive results and the potential enormous economic impact of a successful storage material ensure that extensive, continued research into the development of carbon-nanotube-based hydrogen storage materials will continue.

6.6.2. Nanotubes for Environmental Remediation and as Catalyst Supports

A related application of nanotubes is as filter materials for the purification of water, air, or other substances. For this application, the greatest attribute of the nanotube is the large specific surface area inherent to its structure. If one considers the lumen of the nanotube as well as its exterior surface, this property can have values as high as 3000 $m^2\ g^{-1}$ for a typical 1.4 nm diameter SWNT. This extraordinarily large value suggests that the nanotube could also be a valuable material for use as the support structure for catalytically active materials. Additional nanotube properties that make them potential candidates for such applications are high aspect ratio and mechanical strength, which allow tubes to form self-supporting filter and substrate structures with high mechanical integrity. For post-filtering processing, the pure carbon composition of the nanotube creates an ideal construct for incineration, especially useful for the conversion of volatile organic compounds into more benign compounds.

6.6.3. Nanotubes for Field Emission

The sharp tip formed by the end of a nanotube produces a strong, local enhancement of electric field when the nanotube is placed in a potential. If the nanotube is a cathode and is surrounded by vacuum, it can emit electrons through a process known as field emission. In essence, field emission occurs when an applied electric field reduces the local barrier to electron emission, allowing electrons to tunnel out of the material. Once the electrons have escaped the attractive atomic potential of the nanotube tip, the applied electric field draws them to the anode. The process of electron emission is governed by the Fowler-Nordheim equation:

$$J \propto F^2 \exp\left(\frac{-\phi^{3/2}}{\beta F}\right) \tag{6.9}$$

where J is the emission current density, F is the applied electric field, β provides a measure of the local field enhancing effect and ϕ is the work function of the material. The work function of the nanotube ~5 eV, which is not especially low. However, the molecular sharpness of the nanotube creates a strong local electric field enhancement, promoting field emission. The value of the nanotube as a field emitter also arises from its inherent ability to carry large electric currents for its size with minimal resistance, allowing its use as high efficiency, bright electron sources.[45]

Nanotubes as field emitters are being contemplated for use in flat panel displays. In this application, the high brightness and good energy efficiency are expected to allow the nanotube-based display to be used under conditions (e.g., bright sunshine) that hinder the use of current liquid-crystal-based displays (LCDs). Due to the entrenched market position of LCDs and the large industry investment in production capacity for these products, nanotubes are first expected to be used in larger flat panel displays of up to 70″ diagonal, only moving to the smaller sizes (laptop computers) after economies of scale and prior investment make such a move economical. Both energy efficiency and absence of significant waste heat generation are expected to make nanotube-based displays strongly competitive against traditional cathode-ray-tube-based displays.

Nanotubes could be used as efficient electron sources for diagnostic tools and in niche areas such as fast switches. Nanotube emitters will likely find early application as electron sources in portable X-ray generators and electron microscopes. For X-ray applications, the high energy efficiency and high total current from an electron source using multiple nanotubes in parallel makes it possible to produce even a hand-held X-ray source. Thus one can conceive of the possibility for paramedics to quickly produce X-ray films at the scene of an accident, increasing the ability of rescue personnel to tailor treatments for trauma. Yet another application pertains to homeland security: a portable X-ray source facilitates the screening of packages and shipping containers in the field, thereby increasing the possibility for early interception of illicit materials. For microscopy applications, nanotube emitters enable high current density (brightness), small diameter electron probes while maintaining the good energy stability (temporal coherence) inherent in a field emission electron source. This might improve the resolution of the modern microscope for material and biological studies at the nanoscale.

6.6.4. Nanotubes for Sensor Applications

Nanotube-based sensor devices are being developed by several companies. These devices make use of the strong electronic response of nanotubes to changes in the local environment. In this application, the nanotube is the active sensing component of a small device that includes a power source, logic, basic signal processing, and perhaps a wireless transmitter. As has been described, nanotubes undergo a large change in electrical conductivity when they are exposed to certain types of gases. This property allows the nanotube to be used as a sensitive, low-power device for detecting the presence, or concentration, of an active gas, such as carbon monoxide, ammonia, etc. Another embodiment of a nanotube sensor utilizes mechanical transduction. The small mass, large elastic modulus, and relative perfection of a SWNT make it a high frequency, high quality factor resonator. If a nanotube is functionalized to selectively bind a specific antigen, it is possible that this would change the mechanical resonance in a predictable way, thereby sensing the binding (and perhaps the nature of the ligand itself). The simple architecture of such devices would enable the cost per sensor to fall drastically from that of today's sensors.

This lower cost is expected to produce a large increase in the number of sensors in use, thereby increasing the control and efficiency of industrial products and processes. For example, a better knowledge of carbon monoxide levels in building heating, ventilation, and air conditioning systems, will allow engineers to vary the amount of outside air that is brought into a building, keeping the CO levels within acceptable limits. This control would prevent the unexpected build-up of unhealthy air during peak building occupancy, while allowing better system efficiency by reducing the need to heat or cool unneeded excess volumes of outside air during low demand periods. A second benefit is expected in the chemicals and petroleum industries, where an inexpensive, wireless sensor network can be used to provide leak detection at valves and pipe flanges. Certainly this sensor network would provide increased worker safety. It would also reduce the costs of plant maintenance by allowing components to be replaced as-needed rather than according to a fixed schedule that incorporates a large engineering safety margin and the associated shorter useful component life.

6.6.5. Nanotubes for Structural Applications

Nanotubes have the ideal combination of mechanical strength and aspect ratio for use as fiber reinforcing elements in structural composites. Therefore, it is not surprising that this was the first envisaged application. Initially, nanotubes were used as fine probes for atomic force microscopy. Individual nanotubes were affixed to the tip of a conventional AFM probe and then used to image the surface of a material mechanically, in much the same method as a record stylus. It was shown that the fine diameter of the nanotube provided higher resolution images in test experiments such as the characterization of trenches in semiconducting integrated circuits and the imaging of the helical structure of DNA. Nanotube AFM tips are now grown directly by a CVD process.

The continued work on nanotubes for structural applications has spawned a worldwide effort devoted to the development of techniques to align nanotubes and to mate them with a variety of polymer and non-polymer matrix materials. This has confirmed the potential of nanotubes as reinforcing members in future composite materials, demonstrating that the addition of nanotubes yields improved composite mechanical properties such as modulus and strength.[46] However, there are a number of challenges that must be overcome in the synthesis and processing of the constituents in order to produce superior, and economical, material solutions.

One of the unusual properties of the carbon nanotube (and of other nanotubes such as BN, MoS_2, etc.) is that the bonds on the surface are fully satisfied; there is no need to terminate the exterior of the walls of the nanotube with hydrogen or other elements. This is in contrast to the surface of diamond or the edge of graphite, which is saturated with hydrogen. Similarly, nanowires of silicon are typically terminated with a silicon dioxide layer. Although this inherent chemical stability of the exterior of a nanotube is useful for many applications, it poses a problem for the development of composite materials. In composites, unless sufficient load transfer can be produced between the matrix and the reinforcing fiber, the mechanical properties will not approach the theoretical values. Therefore, significant effort has been placed on modifying the chemistry of the nanotube wall to promote interfacial bonds with the matrix material that are stronger than the natural van der Waals interactions that occur. This is a bit of a dichotomy: modifications to the nanotube wall for improved interfacial bonding could conceivably affect the tube's mechanical integrity. In multiwall nanotubes, similar issues apply with the additional problem of providing sufficient load transfer between the exterior and interior shells—without the participation of the inner shells, the specific reinforcement provided by the nanotube is compromised.

An additional challenge arises from the processing of the nanotubes. An ideal composite would provide reinforcement from a set of isolated nanotubes within the matrix. Due to their inherent perfection and long length, SWNTs exhibit a strong driving force to produce bundles (c.f. Section 6.2). For composites and many other applications, separating nanotubes from these bundles is an important enabling step. Methods for solubilizing nanotubes were discussed in Section 6.5.2. Improvements in such processing methods are likely to translate directly into better composite materials.

Optimal positioning of nanotubes for load bearing is yet another important issue. Ideally, tensile stresses should be borne along the axis of the nanotube, necessitating the need for alignment. While only slight progress has been made in the production of pure, aligned nanotube fibers, significant progress has been made in the production of composite fibers, typically composed of a polymer matrix with reasonably dispersed nanotubes and nanotube bundles. In similarity to other fibers, nanotubes are found to align in shear fields when dispersed in a medium like a viscous polymer above the glass transition temperature. At low nanotube density and high extensions in spun fibers, excellent nanotube alignments have been reported. Particular success has been achieved with poly(vinylalcohol) (PVA) and poly(methyl methacrylate) (PMMA). These materials have demonstrated the best mechanical performance improvements, underscoring the importance of using aligned, dispersed nanotubes within the composite. In the production of aligned nanotube fiber composites, the stiffness and aspect ratio of the nanotube provide the mechanical strength, but also negatively affect the rheological properties by strongly increasing the viscosity of a melt or suspension with nanotube content. This compromises the ability to obtain the large draw ratios that provide excellent alignment and aid fiber production methods.

A final challenge to the production of nanotube-based composites is economic. Structural applications require the largest amount of nanotubes. At the time of writing, the typical cost of commercially available, purified SWNTs is $750 per gram. While progress is being made in the economical mass production of nanotubes, at the present time the cost of these materials limits their application in composite materials.

6.6.6. Nanotubes for Thermal Management Applications

The strength of the sp^2 carbon–carbon bond, responsible for the outstanding modulus and strength of the nanotube, also provides excellent one-dimensional thermal conduction. This opens the possibility for using nanotubes in thermal management or high thermal conductivity applications. Only small loadings of nanotubes in a material are required to produce large increases in thermal conductivity, and imperfect alignment can actually improve the connectivity between the nanotube dispersion by yielding good percolation for thermal conduction. An additional benefit is that different levels of nanotube content are required to achieve good material thermal conductivity and good electrical conductivity (for electrically conductive nanotubes). Thus, it may be able to produce nanotube-based thermal management materials that are either electrically insulating (e.g., for electronic devices in computers) or electrically conductive (e.g., for shielding materials in satellites) just by altering the loading levels. These properties also suggest that nanotubes could be used as contact materials in thermoelectric devices. Such contacts are required to conduct heat to active thermoelectric components while also providing a low resistance contact for electrical connections.

6.6.7. *Nanotubes for Electronics Applications*

The electronic properties of nanotubes, detailed earlier in this chapter, have spurred an intense interest in the potential of these materials for computer logic and memory circuits. In principle, nanotubes could compose both the functional semiconducting elements as well as the metallic interconnects in such circuits. Nanotubes provide a host of useful properties for such applications: high electron mobility, small size, stable non-interacting surfaces, and configurability into field-effect transistor geometries. This major worldwide effort has made notable progress in crossing the technical hurdles associated with nanotube-based computing. Reactions with metals such as titanium have yielded ohmic contacts between nanotubes and electrical leads, which are preferred over the tunnel junctions created by simple deposition of nanotubes onto prefabricated electrodes. Synthesis methods have been developed for growing nanotubes on a surface, between metal contacts, and vertically in a cavity as conducting vias. By controlling current, processing procedures are now available to selectively remove metallic or semiconducting nanotubes that span two metal leads. Simple component circuits such as "And" and "Nor" gates have been created, as well as more complex devices like voltage inverters.[47] While this progress is promising, significant challenges remain. Not the least of these is the economic barrier faced by any material system that will displace silicon technology for electronic applications. For this reason, the first application of nanotubes in electronics will likely be for sensing, where a nanotube sensor element will be married to a traditional silicon-based circuit.

6.6.8. *Nanotubes for Medical Applications*

Through functionalization of the exterior surface, and through use of the lumen, researchers are seeking medical applications for carbon nanotubes. In principle, the chemically-inert nature of the nanotube suggests that it will be non-toxic to mammals. However, there have been no detailed studies examining the toxicity and immunogenicity of nanotubes. One concern relates to a nanotube's high aspect ratio: other high aspect ratio materials, the most famous being chrysatile asbestos, are known carcinogens that lodge in the alveoli of the lungs. However, a nanotube is much smaller than these fibrous materials, so a direct analogy is not defensible.

One application in which nanotubes can be used without regard to sterically-induced toxicity is in their use as probes for medical research. Several groups are working to develop nanotube-based needles that can be inserted non-destructively through a cell membrane. The operating assumption is that the combination of mechanical resilience and small diameter will enable the nanotube to penetrate the lipid bilayer of the cell with minimal disruption. Once inside, a functionalized nanotube could deliver or extract material from a cell. It might also inject electrical charge, local magnetic fields, or light for diagnostics or manipulation. The ability to operate on this small scale is expected to yield insights into the behavior of cells in response to a variety of stimuli.

For in vivo work, nanotubes are envisioned as carriers for therapeutic or diagnostic agents. With the demonstrated success in the basic functionalization of the exterior of the nanotube, it is clear that specific antigens or biofunctional ligands can be covalently attached to the nanotubes, a necessary ingredient to deliver site specificity for the targeting of nanotube-based medical agents. A challenge for this application is the need to control precisely nanotube length and diameter since steric monodispersity is an important factor in providing reproducible and predictable behavior of in vivo agents. This has not yet been accomplished for carbon nanotubes.

6.6.9. Directed and Patterned Growth of Nanotubes for Applications

A number of the envisioned products rely upon nanotubes that are located precisely within a device with a correct orientation. These nanotubes may extend as cantilevers (e.g., field emission cathodes or medical probes), or they may be contacted by device connections (e.g., sensors). Techniques have been under development with which to control the location and orientation of nanotubes. The first successes were brute force methods in which nanotubes were teased out of mats onto tips using fine control and real time imaging at high magnification, e.g., in a scanning electron microscope, or by affixing clumps of nanotube mat onto particular locations on a surface of a device. In the past few years, synthesis and fabrication methods have become more sophisticated. CVD in particular has played an important role in the production of material geometries for filter, electronic, sensor and medical probe applications, while advances in the chemical processing of nanotubes have enabled progress in the alignment and dispersion of nanotubes for composites and fibers.

Using CVD on a variety of metallic and non-metallic, solid and porous substrates, conditions have been developed in which a wide range of nanotube-based material morphologies can be produced. In one manifestation, vertical forests of parallel MWNTs are grown perpendicular to a substrate, much like a well-groomed lawn. By controlling the location of catalyst islands on the substrate material, these forests can be patterned to grow only in certain locations. Analyses of the quality of these nanotubes tend to show that they have a higher defect density than nanotubes produced by high energy processes or by CVD at low densities. It has also proved possible to pattern the deposition nanotubes grown by catalytic decomposition onto an oxidized silicon substrate. In this method, the nanotubes (synthesized with unsupported catalyst) deposit selectively on SiO_2 but not on the native silicon.

Another manifestation is the growth of nanotubes parallel to a substrate from the side surface of metal catalyst. This tactic has been successfully employed to produce well-formed nanotubes with long lengths and is quickly becoming the preferred method of production for groups studying the electronic properties of nanotubes. In this method, the flow of gas through the reaction chamber can be used to influence the alignment of the nanotubes, which tend to orient themselves in the direction of gas flow. Nanotubes have been grown from one metal island to another, a useful strategy for electronic circuit and sensor creation.

Using their inherent (weak) paramagnetic susceptibility and (strong) electric polarizability, nanotubes have been aligned to varying degrees using magnetic and electric fields. In these cases, alignment occurs parallel to the field lines. The need for very large magnetic fields or closely-spaced electrodes has tended to limit the application of these methods. Nevertheless, the directed CVD growth of nanotubes in electric fields of <0.25 V μm^{-1} has proved possible.

6.7. CONCLUSIONS

In this chapter, we have detailed the structure, properties, synthesis, characterization, modification, and applications of carbon nanotubes. This is a fantastically rich area of research: essentially no other material exhibits such a range of extraordinary fundamental properties that merits consideration for an equally broad multitude of applications. For this reason, nanotubes have drawn interest from both basic scientists and engineers. Despite the large scope of this chapter, it is important to remember that the content presented herein is only the proverbial tip of the iceberg. The field is both expansive and embryonic. We have covered only the basics of what is state-of-the-art for today, and the rate at which nanotube research progresses is staggering. Nevertheless, the knowledge we have provided will enable the inclined reader to examine the primary literature and uncover more of the evolving nanotube story.

ACKNOWLEDGEMENT

B.W.S. and D.E.L. thank Dr. B. C. Satishkumar for his valuable comments.

QUESTIONS

1. The appearance of van Hove singularities in the DOS is a common feature to all one-dimensional electronic energy bands. Show why this is the case. (Hint: begin with the simplest form for a 1D dispersion, $E - E_0 = \hbar^2 k^2/2m$.)
2. Raman spectroscopy provides a good metric for the types of SWNTs in a sample. Provide as much information as you can about the diameter distribution in the sample whose spectrum appears in Figure 6.16. Be quantitative where possible. Can you determine qualitative chirality information from these data? If not, propose a way to study chirality by Raman spectroscopy.
3. The energetically preferred graphitic van der Waals gap is approximately 0.34 nm, which means the effective interior lumen inside a 1.4 nm diameter nanotube is about 0.7 nm. Nevertheless, it has proved possible to insert C_{80} fullerenes, which are nominally 0.8 nm in diameter, inside such SWNTs. How can this be the case?
4. It is sometimes observed during transport experiments that isolated, semiconducting SWNTs behave as though they were *p*-doped. Give a possible explanation for this observation.

REFERENCES

1. R. P. Feynman, in *lecture to the American Physical Society*, 1959).
2. K. E. Drexler, *Engines of creation* (Anchor Books, 1986).
3. H. W. Kroto, J. R. Heath, S. C. O'Brien et al., *Nature* **318**, 162 (1985).
4. E. Osawa, Kagaku **25**, 850 (1970).
5. R. Bacon, *Journal of Applied Physics* **31**, 283 (1960).
6. J. Abrahamson, P. G. Wiles, and B. L. Rhoades, Proceedings of the 14th Biennial Conference on Carbon, 254 (1979).
7. S. Iijima, *Nature* **354**, 56 (1991).
8. S. Iijima and T. Ichihashi, *Nature* **363**, 603 (1993).
9. D. S. Bethune, C. H. Kiang, M. S. de Vries et al., *Nature* **363**, 605 (1993).
10. A. Thess, R. Lee, P. Nikolaev et al., *Science* **273**, 483 (1996).
11. C. Journet, W. K. Maser, P. Bernier et al., *Nature* **388**, 756 (1997).
12. A. G. Rinzler, J. Liu, H. Dai et al., *Appl. Phys.* A **67**, 29 (1998).
13. R. Saito, M. Fujita, G. Dresselhaus et al., *Appl. Phys. Lett.* **60**, 2204 (1992).
14. N. Hamada, S. Sawada, and A. Oshiyama, *Phys. Rev. Lett.* **68**, 1579 (1992).
15. J.-L. Sauvajol, E. Anglaret, S. Rols et al., *Carbon* **40**, 1697 (2002).
16. M. B. Nardelli, B. I. Yakobson, and J. Bernholc, *Phys. Rev.* B **57**, 4277 (1998).
17. T. Guo, P. Nikolaev, A. Thess et al., *Chem. Phys. Lett.* **243**, 49 (1995).
18. A. R. Harutyunyan, B. Pradhan, U. J. Kim et al., *Nano Letters* **2**, 525 (2002).
19. S. Amelinckx, A. Lucas, and P. Lambin, Reports on Progress in Physics **62**, 1471 (1999).
20. A. M. Rao, E. Richter, S. Bandow et al., *Science* **275**, 187 (1997).
21. M. J. OConnell, S. M. Bachilo, C. B. Huffman et al., *Science* **297**, 593 (2002).
22. X. P. Tang, A. Kleinhammes, H. Shimoda et al., *Science* **288**, 492 (2000).
23. A. S. Claye, N. M. Nemes, A. Janossy et al., *Phys. Rev.* B **62**, 4845 (2000).
24. H. Ulbricht, G. Moos, and T. Hertel, *Phys. Rev.* B **66**, 075404 (2002).
25. W. Zhou, Y. H. Ooi, R. Russo et al., *Chem. Phys. Lett.* **350**, 6 (2001).
26. S. Rols, Z. Benes, E. Anglaret et al., *Phys. Rev. Lett.* **85**, 5222 (2000).
27. T. W. Odom, J.-L. Huang, P. Kim et al., *Nature* **391**, 62 (1998).
28. J. W. G. Wildoer, L. C. Venema, A. G. Rinzler et al., *Nature* **391**, 59 (1998).
29. J. Lefebvre, J. F. Lynch, M. Llaguna et al., *Appl. Phys. Lett.* **75**, 3014 (1999).
30. J. Hone, M. C. Llaguno, N. M. Nemes et al., *Appl. Phys. Lett.* **77**, 666 (2000).
31. M. Bockrath, D. H. Cobden, P. L. McEuen et al., *Science* **275**, 1922 (1997).
32. Z. Yao, C. Dekker, and P. Avouris, in *Carbon nanotubes: synthesis, structure, properties, and applications*, edited by M. S. Dresselhaus, G. Dresselhaus and P. Avouris (Springer-Verlag, Berlin, 2001), Vol. 80.
33. D. J. Hornbaker, S.-J. Kahng, S. Misra et al., *Science* **295**, 828 (2002).
34. S. J. Tans, M. H. Devoret, H. Dai et al., *Nature* **386**, 474 (1997).
35. J. Chen, M. A. Hamon, H. Hu et al., *Science* **282**, 95 (1998).
36. E. T. Mickelson, C. B. Huffman, A. G. Rinzler et al., *Chem. Phys. Lett.* **296**, 188 (1998).
37. V. N. Khabashesku, W. E. Billups, and J. L. Margrave, Accounts of Chemical Research **35**, 1087 (2002).
38. R. S. Lee, H. J. Kim, J. E. Fischer et al., *Nature* **388**, 255 (1997).
39. C. Zhou, J. Kong, E. Yenilmez et al., *Science* **290**, 1552 (2000).
40. A. S. Claye, J. E. Fischer, C. B. Huffman et al., *J. Electrochem. Soc.* **147**, 2845 (2000).
41. P. M. Ajayan and S. Iijima, *Nature* **361**, 333 (1993).
42. E. Dujardin, T. W. Ebbesen, H. Hiura et al., *Science* **265**, 1850 (1994).
43. B. W. Smith, M. Monthioux, and D. E. Luzzi, *Nature* **396**, 323 (1998).
44. B. W. Smith and D. E. Luzzi, *Chem. Phys. Lett.* **321**, 169 (2000).
45. J.-M. Bonard, H. Kind, T. Stockli et al., Solid State Electronics **45**, 893 (2001).
46. R. Haggenmueller, H. H. Gommans, A. G. Rinzler et al., *Chem. Phys. Lett.* **330**, 219 (2000).
47. P. Avouris, Accounts of Chemical Research **35**, 1026 (2002).
48. R. Saito, G. Dresselhaus, and M. S. Dresselhaus, *Phys. Rev.* B **61**, 2981 (2000).
49. J. Bernholc, D. Drenner, M. B. Nardelli et al., *Annual Review of Materials Research* **32**, 347 (2002).
50. C. T. White and J. W. Mintmire, *Nature* **394**, 29 (1998).

51. S. Rols, Ph.D. thesis, (University of Montpellier II, Montpellier, France, 2000).
52. Z. Benes, Ph.D. thesis (University of Pennsylvania, Philadelphia, PA, USA, 2001).
53. B. W. Smith, Ph.D. thesis (University of Pennsylvania, Philadelphia, PA, USA, 2001).
54. A. Bezryadin, A. R. M. Verschueren, S. J. Tans et al., *Phys. Rev. Lett.* **80**, 4036 (1998).
55. Y. K. Chen, A. Chu, J. Cook et al., *Journal of Materials Chemistry* **7**, 545 (1997).
56. B. W. Smith and D. E. Luzzi, *Chem. Phys. Lett.* **331**, 137 (2000).

7

Quantum Dots

A. B. Denison, Louisa J. Hope-Weeks, Robert W. Meulenberg, and L. J. Terminello
Lawrence Livermore National Laboratory
Livermore, CA

7.1. INTRODUCTION

The advent of reliable production of nano-structures has opened a frontier in material science. As the size of these structures or devices approaches the nano-meter scale (1 nm = 10^{-9} meter) the laws of quantum mechanics come into play. Quantum dot structures are being considered for a variety of technological applications ranging from semiconductor electronics to biological applications including optical devices, quantum communications and quantum computing. The understanding of the quantum structure's electronic properties and quantum confinement is of paramount importance. The basic concept of quantum confinement comes from the interplay of two fundamental principles of quantum mechanics; namely the electronic system must obey the Schrödinger equation and also follow the de Broglie momentum-wavelength relationship.

7.2. QUANTUM MECHANICAL BACKGROUND

Before discussing quantum confinement itself it is necessary to make a short review of the Schrödinger equation and the theory of electrons in solids. In bulk crystalline lattices the electron states are guided by the Schrödinger equation (see standard texts for quantum mechanics, e.g., Richtmyer, Kennard and Cooper, Schiff, Eisberg)

$$H\psi(r) = [-(h/8\pi m)^2\Delta + V(r)]\psi(r) = E\psi(r), \tag{7.1}$$

where H is the Hamilton operator, Δ the kinetic energy operator $(\partial^2/\partial p^2 + \partial^2/\partial q^2 +$

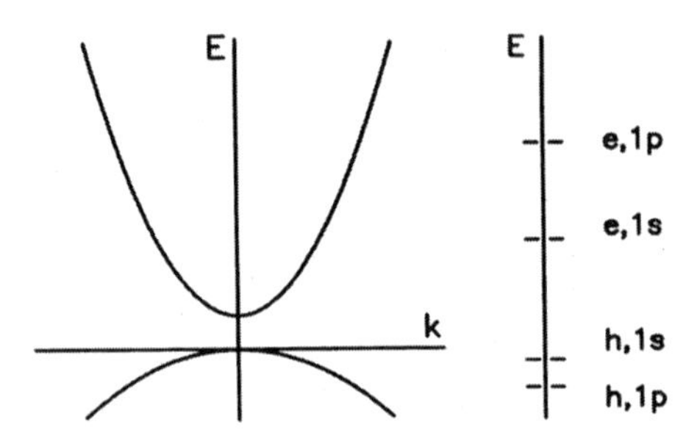

FIGURE 7.1. Schematic plot of the single particle energy spectrum in a bulk semiconductor for both the electron and hole states on the left side of the panel with appropriate electron (e) and hole (h) discrete quantum states shown on the right. The upper parabolic band is the conduction band, the lower the valence.

$\partial^2/\partial r^2$) and $V(r)$ is the periodic potential representative of the inter-atomic lattice spacing a; i.e., $V(r) = V(r+a)$. h is Planck's constant and m is the mass of the electron. The eigenfunctions, or solutions to the Schrödinger equation, are the well known Bloch functions which are also periodic with the lattice spacing (a) (see standard texts on Solid State Physics, e.g., Kittel, Ashcroft and Mermin);

$$\psi(r) = e^{ikr}u(r) \tag{7.2}$$

where $u(r)$ represents the interatomic potential well and is also periodic; i.e., $u(r) = u(r+a)$. k is the wave vector determined by the symmetry and spacing of the lattice. The solution to the above *wave equation* for a periodic potential predicts energy bands with gaps. The simplest approximation, yet which yields sensible results and provides insight into more complicated solutions, is the so called *parabolic band approximation*. The eigenvalues for this approximation are:

$$E(k) = (hk)^2/(8\pi m_{e,h}), \tag{7.3}$$

where $m_{e,h}$ is the effective mass of the electron or hole (missing electron in an allowed energy state). Figure 7.1 shows a schematic plot of the single particle energy spectrum in a bulk semiconductor for both the electron and hole states. The gap between the two parabolas is the so-called *band gap* found in all semiconductors and is directly related to the optical properties of a given material. The region below the lower parabola is typically called the valence band (VB) and energy states in this region are typically filled in semiconductors. The region inside the upper parabola is typically called the conduction band (CB) and is normally empty. However, an electron may be excited into the conduction band by, for example, a photon which then leaves a *hole* in the valence band region. This electron-hole pair exists as an *exciton*. The exciton can, in fact, be considered an *exotic atom* in its own right with its own atomic or *Bohr radius*.

7.3. QUANTUM CONFINEMENT—3D QUANTUM DOT

As a first example of quantum confinement we consider a spherical crystal with a diameter $D(=2R)$. In order to rightly be called a quantum structure, the diameter, D, must be smaller than the de Broglie wavelength, λ;

$$\lambda = h/p, \tag{7.4}$$

where h is, again, Planck's constant and $p = hk/2\pi$ is the momentum of the electron in the sample at the temperature T. If we take $E = (3/2)kT = p^2/2m$, E the kinetic energy, k Boltzmann's constant and m the mass of the electron (or hole) we find a λ at room temperature ($\sim$300 K) to be $\sim$6 nm. This means that for the spherical crystal of diameter $D < 6$ nm the electrons (or holes) wave packet is squeezed somewhat unnaturally into a space smaller than it normally likes to have. The result is the electron takes a higher energy than it would normally have in its host structure. Another way of looking at this effect is to say the natural exciton Bohr radius is larger than the host crystal structure. Such small host crystals are called *quantum dots* and the resulting physical state is called *quantum confinement*.

Now let us look in more detail for the spherical quantum dot case. For simplicity we take an infinite spherical well, which actually turns out not to be a bad approximation and allows insight into the consequences of quantum confinement. We now take the Hamilton operator (neglecting the Coulomb interaction to be discussed later) as:

$$H = -(h^2/8\,\pi\, m_e)\Delta_e \; - \; (h^2/8\,\pi\, m_h)\Delta_h \; + \; V_e(r_e) + V_h(r_h) \tag{7.5}$$

and the potential:

$$V_i(r_i) = \begin{matrix} 0 & \text{for } r_i < R \\ \infty & \text{for } r_i > R. \end{matrix} \qquad i = e, h \tag{7.6}$$

For the case of a single exciton we can take

$$\Psi(r_e, r_h) = \varphi_e(r_e)\,\varphi_h(r_h) \tag{7.7}$$

The Schrödinger equation for a spherical infinite well has been well studied and its solutions can be found in most standard texts on quantum mechanics. The normalized solution to Eq. (7.5) is given through spherical harmonics (Y_{lm}) and Bessel functions (J_l) (see e.g., Abramowitz and Stegun, *Handbook of Mathematical Functions*),

$$\phi^i_{nlm}(r) = Y_{lm}(2/R^3)^{1/2}(J_l(\chi_{nl})/J_{l+1}(\chi_{nl})) \tag{7.8}$$

The χ_{nl} are the zeroes of the Bessel functions. The Bessel function $J_l(\chi_{nl}r/R)$ must also vanish as the surface of the quantum dot sphere. With these conditions the energies of the electrons or holes are given by

$$E^{e,h}_{nl} = (h^2/8\pi m_{e,h})(\chi_{nl}/R)^2 \tag{7.9}$$

FIGURE 7.2. Solutions of quantum dots of varying size. Note the variation in color of each solution illustrating the particle size dependence of the optical absorption for each sample. Note that the smaller particles are in the red solution (absorbs blue), and that the larger ones are in the blue (absorbs red).

What we now have is an additional term added on to the band gap energy E_g predicted for the bulk, namely $E_{nl}^{e,h}$ due to the quantum confinement of the electron or hole. Thus the photon energy required to produce the electron-hole pair (exciton) is

$$h\nu = E_g + (h^2/8\pi m_{e,h})(\chi_{nl}/R)^2 \tag{7.10}$$

This produces an effective band gap shift that is proportional to $1/R^2$ in the quantum regime. This $1/R^2$ has been confirmed numerous times experimentally (e.g., Ref. 1, Figure 1) and gives rise to the beautiful array of photoluminescence colors in quantum dots of different sizes dissolved in solution (see Figure 7.2). This can be understood by the following argument. By shrinking the size of a quantum particle one widens the band gap. In so doing, the optical transition—either absorption (promotion of an electron from the occupied valence band to the empty conduction band) or the inverse process of emission, or photoluminescence, blue shifts compared to the bulk material or a larger sized quantum dot. Although this simple model ignores many of the more complicated interactions that clearly must be considered it has been remarkably successful in predicting many of the experimental results and clearly gives a firm basis from which to start.

7.4. OTHER INTERACTIONS

In order to predict in detail the exact energy levels and shape of the band gap other interactions need to be considered. We will not go into detail, but will mention a few of the more important ones. The masses m_e and m_h are not the free space masses with which we are familiar, but are the *effective masses*. The effective masses are related to the curvature of the band gaps where the transition occurs between the valence band and the conduction band. In fact, the effective mass goes as $\partial^2 E/\partial p \partial q$ at the transition point. This means accurate band calculations must be made which represents much of the theoretical work in this area. The electronic states also possess angular (L) and spin (S) momentum so interactions such as spin-orbit ($\lambda L.S$) effects occur which cause modifications of the energy levels. The interaction of multiple excitons in quantum structures so excited, also modify and shift the levels. As seen in the Coulomb interaction discussed previously the electric susceptibility of the host nanocrystal, ε, modifies the levels. Of particular importance is the condition and constitution of the surface where at these small sizes the surface consists of a large fraction of the atoms in the quantum structure. The material with which the quantum dot is covered and surface reconstruction different than the bulk structure, make easily measured changes in the optical spectroscopy.

From our simple example of the spherical quantum dot we have seen how shrinking the size of a host crystal into the quantum region ($<$6 nm) modifies the energy levels of the electrons, causing in general a shift of the band gap with $1/R^2$. The production of an exciton, the electron-hole pair, introduces a further small contribution to the energy level which goes as $1/R$. The shift in the visible light observed from the photoluminescence of quantum dots of various sizes in solution (Figure 7.2) testifies to the confinement effect on the band gap ($E_g \sim 1/R^2$). The photoluminescence observed arises from the recombination of the electron and hole as the electron falls back into the valence band across the existing band gap. It should be pointed out that even without the existence of an exciton there is a band gap shift. Although we considered in some detail the spherical quantum dot, it was noted that quantum confinement can also occur in 1D (wire), 2D (planar) and 3D (sphere or other 3D geometry). The approach to modeling these systems follows the same procedure of solving a Schrödinger equation and for the case of excitons including the Coulomb interaction. We further discussed briefly other interactions which determine the energy levels and band gaps in quantum systems where confinement occurs. All of these effects are currently topics of active research and become even more important as device applications are developed. The future of quantum devices is very promising and an understanding of the effects of quantum confinement is of vital importance.

With this theoretical motivation for how we can vary and control the electronic and optical properties of a material by varying its particular size, it drives the need to have some means of varying the size through some synthesis method. It is important to point out that most quantum dot materials that exhibit interesting or useful electronic or optical properties are semiconductor nanocrystals. One should note that a whole class of metallic (e.g., gold) nanoparticles also exhibit interesting electro-optic properties but do so through a somewhat different mechanism than the one described above. We will focus on the synthesis of semiconductor quantum dots in the remainder of this chapter.

The most common methods of manufacturing semiconductor quantum dots are colloidal growth, epitaxial growth, and ion implantation. Each method has its own strengths and weaknesses with respect to particle size, order, and distribution, but represent the majority of the useful ways of making quantum dots. It is important to note that the development of characterization tools that can "see" down to the nanometer scale, and below are critical for rigorous study and understanding of these materials. While this is not the focus of this chapter (see Chapter 1 for a detailed discussion), it is useful to mention several of these tools for observing nanoparticles. Most notable are Transmission Electron Microscopy—which gives atomic resolution images of quantum dots, and Scanned Probe Microscopy (most notably Scanned Tunneling Microscopy—for atomic resolution images of dots on surfaces, and Atomic Force Microscopy—for near atomic resolution images of dots).

7.5. COLLOIDAL GROWTH OF NANOCRYSTALS

The novel electronic and optical properties of semiconductor and metal nanocrystals are highly dependant on the crystals size. To permit studies of these novel properties, nanocrystals of a monodisperse nature need to be synthesized, with a well-defined crystalline core. A monodisperse sample would be one where nanocrystals would be identical to each other in terms of size, shape, internal structure and surface chemistry. In the case of nanocrystals the definition of monodisperse is slightly relaxed and is used to describe a sample with a mean deviation of less than 5% in diameter. This need for monodispersity has resulted in synthetic procedures that result in a size control with rational adjustments of the synthetic protocol. Thus, it is not surprising that colloidal, or chemical synthesis, methods for producing semiconductor quantum dots is the most common technique.

The principle for colloidal synthesis of semiconductor nanocrystals is based on a study by Le Mur and Dinegar, which showed that a temporally short cluster nucleation event followed by controlled slow growth on the existing nuclei results in the formation of monodisperse colloids (Figure 7.3).[2] In practice, reagents are rapidly injected to a vessel charged with hot coordinating solvent (Figure 7.4), thus raising the concentration above the nucleation threshold. This is possible when the temperature is sufficient to decompose the reagents, forming a supersaturated solution of species. Supersaturation is followed by a short nucleation period that partially relieves supersaturation. This results in a drop in concentration of species below the critical concentration for nucleation, and the clusters "bang" out of solution. As long as the rate of addition of precursor does not exceed the rate at which it is consumed by the growing nanocrystals, no additional nuclei form. Since the growth of each of the nanocrystals is similar, the size distribution if mainly governed by the time over which the nuclei are formed and continue to grow. This simply means that for increasing reaction time the larger and more uniform the nanocrystals become and thus gives one considerable control over size. This trend is often referred to as focusing of the size distribution. A size series of nanocrystals can be produced via removal of aliquots from the vessel over periodic intervals.

A second approach to nanocrystals synthesis involves the mixing of the precursor in a coordinating solvent below the temperature of reaction. The reaction then undergoes a

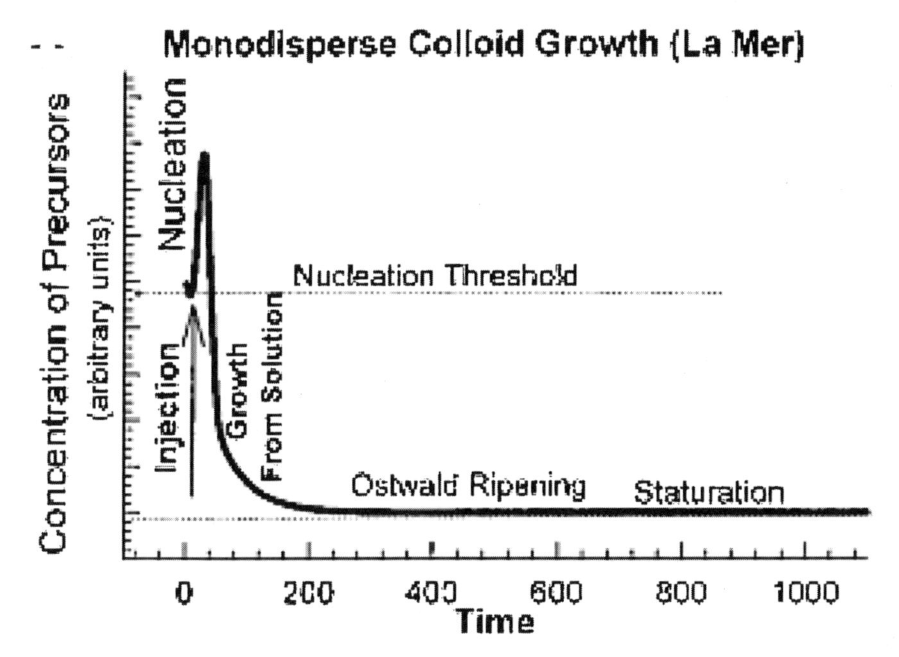

FIGURE 7.3. La Mer model for the growth stages of nanocrystals.

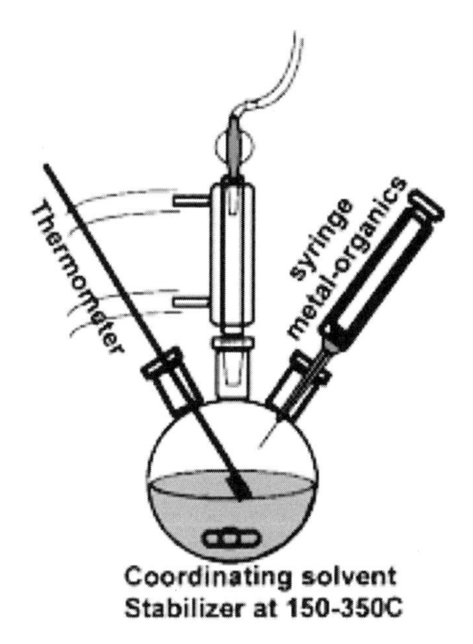

FIGURE 7.4. Synthetic apparatus for the preparation of nanocrystals.

controlled increase in temperature that accelerates the reaction to the point of supersaturation. Supersaturation is again relieved by a discrete nucleation burst. The temperature is then strictly controlled to ensure the rate of reaction of the precursors is less than or equal to the rate of addition of material to the surface of the nanocrystals, thus avoiding a second supersaturation state.

Many nanocrystals synthesis systems exhibit a second growth phase referred to as Ostwald ripening. During Oswald ripening the smaller nanocrystals due to their high surface energy are dissolved and the material is redeposited on the larger nanocrystals. This process results in a decrease in nanocrystals number, with an increase in nanocrystals size. Oswald ripening also aids in producing samples of monodisperse nanocrystals in a size series, with initial size distributions of 10–15% in the nanocrystals diameter. Nanocrystal size is limited mainly by the short time period during which nuclei form and begin to grow. In addition higher solution temperatures enhance Ostwald ripening therefore also leading to larger average nanocrystals sizes.

Colloidal reactions to form nanocrystals can be systematically modified to have more control over the size of nanocrystals formed by changing the variables of the reaction such as concentration, temperature and reactant choice. The ratio of reactants to surfactants allows addition control over size, for example a high ratio results in the formation of more small nuclei during nucleation that subsequently result in the formation of smaller nanocrystals. The choice of surfactant can also strongly influence nanocrystals size. Surfactants adsorb reversibly to the surface of the nanocrystals to form a shell that acts to stabilize the nanocrystals in addition to moderating their growth. The higher the binding constant of the surfactant groups to the surface, the lower the rate of addition of material to the forming nanocrystals. This therefore results in a smaller average size of nanocrystals being produced. Pairs of surfactants can also be used to moderate growth and increase size control. For example one that binds tightly to the surface with one less tightly bound that permits rapid growth results in a combination where growth rate can be tuned and size more strictly controlled.

Final narrowing of the size distributions of nanocrystals is achieved via a precipitation induced by the addition of a second solvent to the reaction solution. The second solvent is often referred to as a nonsolvent as it has an unfavorable reaction with the surface groups of the nanocrystals, but is miscible with the reaction solvent. Introduction of the nonsolvent results in aggregation of the nanocrystals causing their flocculation. This is much the same process that takes place when an organic is recrystallized from solution by the addition of a solvent in which it is insoluble. Centrifugation to the nanocrystal suspension allows the solvent to be decanted and nanocrystal powders to be isolated. Slow titration of a nonsolvent allows only partial flocculation in which the largest nanocrystals aggregate first results in powders enriched with larger nanocrystals.

Group II–VI Semiconductor Nanocrystals

The synthesis of group II–VI semiconductors ME, where M = zinc, cadmium, and mercury and E = sulfur, selenium and tellurium, are prepared using the principles described previously. This is perhaps the most widely synthesized and studied class of nanocrystals mostly due to their ease of synthesis. Supersaturation and subsequent nucleation are triggered by rapid injection of metal-organic precursors into vigorously stirred very hot coordinating solvent (150–350°C). The group II sources that have been widely employed are

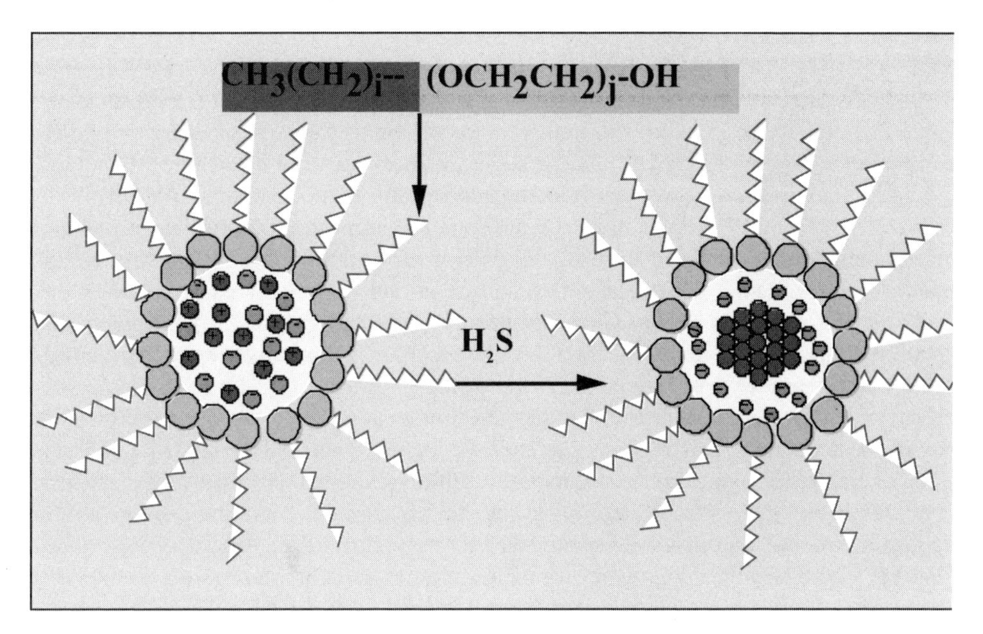

FIGURE 7.5. Schematic for how cluster precipitation can occur in solution from an inverse micel. Cluster size is regulated by micel size (concentration).

metal alkyls such as dimethylcadmium, diethylcadmium, diethylzinc and dibenzylmercury. The sources of group VI precursors are often organophosphine chalcogenides (R_3PE) or bistrimethylsilylchalogenides (TMS_2E), where E = S, Se or Te. TMS_2S is often selected as a sulfur source as it is more reactive than R_3PS, however organophosphine chalcogenides are easy to prepare and are widely used as source for both Se and Te. The coordinating solvents that are used for these reactions are long chain alkylphosphines (R_3P), alkylphosphine oxides (R_3PO), alkylphosphites, alkylphosphates, pyridines, alkylamines and furans. See Figure 7.5 for example.

Reactions involving the use of mixed precursors such as combinations of TMS_2S and R_3PSe results in the formation of an alloy. However the resulting stoichiometry of the nanocrystals does not directly reflect the ratio of the initial precursors but rather the differential rate of precursor incorporation. For each reaction it is important to choose the appropriate precursors and solvent. For example, R_3PO strongly interacts with Zn to such an extent that it retards growth but replacement of the alkylphosine oxide with an alkylamine results in an enhanced growth rate.[3]

Group III–V Semiconductors

The synthesis of group III–V semiconductor nanocrystals InE, where E = phosphorus and arsenic, are produced via similar procedure as described for II–VI semiconductors. Except the In precursor [$InCl(C_2O_4)$] is already present in the coordinating solvent [R_3P/R_3PO] prior to injection of the TMS_2E, where E = As or P. For this group of nanocrystals Ostwald ripening over 1–6 days is required to achieve crystals of the desired size, shape, internal structure and surface chemistry.[4]

7.6. EPITAXIAL GROWTH

Another method for the synthesis of quantum dots is through epitaxial growth of clusters on a substrate. There are many different variants of epitaxial growth of materials and can be done in a variety of ways. While the subject of how epitaxial growth occurs is much too broad for this book, it essentially is the condensation of a crystalline film (or film of crystallites ranging in particle size from atoms to many microns) from a vapor or liquid onto some substrate. Some of the epitaxial growth methods from the vapor phase include molecular beam epitaxy, chemical vapor deposition, laser ablation—each with their own spectrum of sub-methods. From a liquid, epitaxial growth can occur electrochemically (i.e.: "plating"), through crystallization, and through Langmuir-Blodgett ("dipping") processes. These methods are routinely used for the production of films ranging in thickness from atoms to millimeters. However, for the production of quantum dots on a substrate from any of these methods, it is necessary to exercise greater control of the growth conditions (flux of material to the substrate, substrate itself, crystallinity or order of the substrate itself, temperature of substrate, and other ambient conditions). Through this more precise approach to depositing material on a given substrate, clusters, or dots can be produced. One key ingredient on the growth of quantum dots instead of dense large crystallite or single crystal films is the appropriate match (or mismatch) of the free energies of the depositing material and substrate. Details of this can be obtained in any of many epitaxial growth method texts.

The epitaxial growth method allows for a wide range of control of principally the order of the quantum dots on a substrate since through this method a regular array is achievable through selective growth conditions. If we take one of these sub-methods as an example of how this method occurs, we can illustrate some of the strengths of the technique. In molecular beam epitaxy, a stream of atoms or molecules produced by a number of methods in a high, or ultra-high vacuum environment, is impingent on a substrate where these atoms condense and "cluster" on the surface in a two dimensional disordered array of particles. In a common example, the flux of material can be produced by evaporating bulk material in a Knudsen cell, or crucible of some kind. By varying of the ambient gas pressure and type (e.g., use of some inert gas such as argon or helium), pre-condensation of clusters in the vapor phase can be achieved. This is a useful method since size-selection of clusters can be done by varying the gas conditions, and thus, nanoclusters can be deposited directly on a given substrate. In Figure 7.6 we see an example of an evaporated and condensed film of Ge nanocrystals on two types of surfaces (graphite in the left two panels, and Si substrate on the right). The image shown was collected using an atomic force microscope (AFM) which gives not only cluster size information, but also film morphology.

While not strictly epitaxial, thicker films of clusters can be produced on surfaces with a broad size distribution, but higher number density. With thicker films, disordered arrays are fairly common and size can be regulated by carrying the conditions of the atom delivery method to the surface, as well as the annealing conditions of the substrate (if desired). This subsequent anneal of the clusters on the substrate is what can yield the largest clusters in this class of quantum dot films.

Another method of producing this material flux is laser ablation—which for alloy semiconductors (e.g., metal oxides) preserves the stoichiometry of the starting material in the stream. Sputtering of the starting bulk material is yet another method for producing a

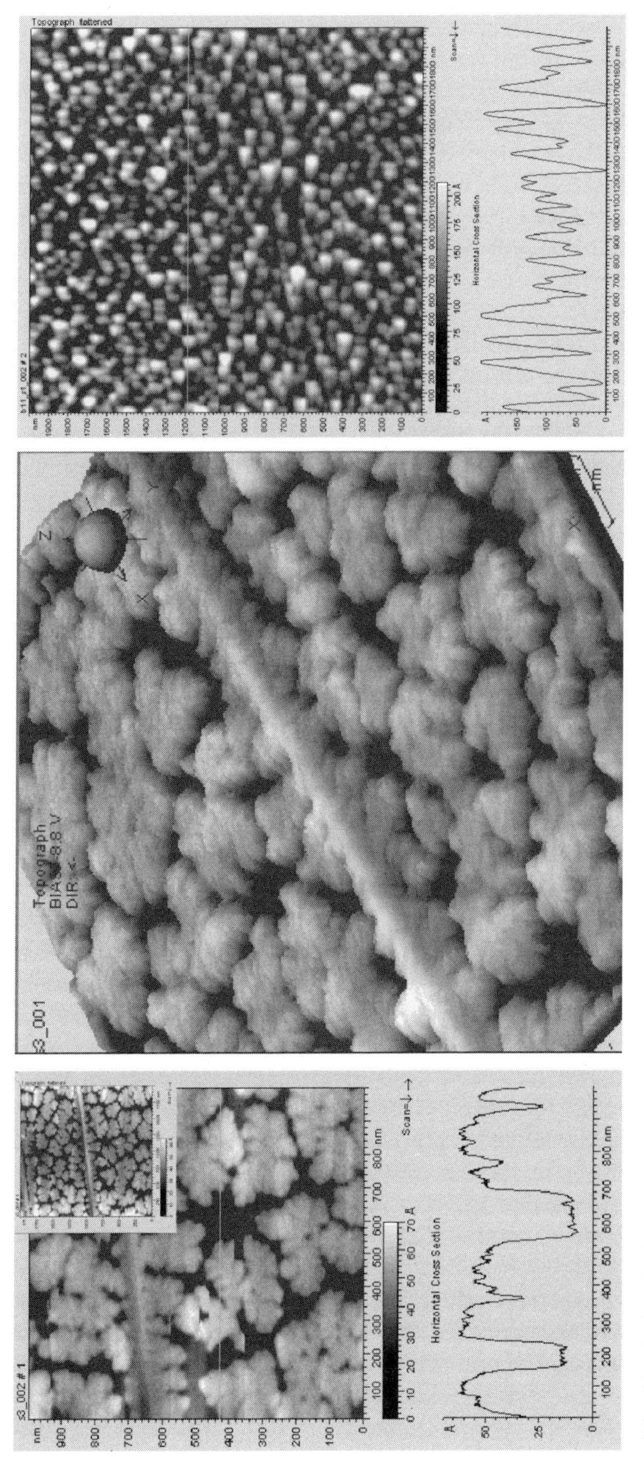

FIGURE 7.6. Atomic Force Microscope images of Ge clusters on two types of surfaces. Graphite in the left two panels, and SiO_2 in the right. The line plots on the figure give vertical profiles of line cuts through the AFM images directly above and give the quantitative size information.

material flux (or flood in this case owing to the high flux of atoms this method can produce) to deposit on a given substrate.

Once the material gets to the substrate several things can happen. If there is sufficient energy and number density, the atoms on the surface can move in 2 dimensions across the surface and agglomerate into either: a dilute array of well ordered small clusters (number of atoms per cluster <20) or a random agglomeration of clusters that can range in size from <1 nm to many microns (see the rightmost panel of Figure 7.6). In the first case, a well ordered array of small clusters on a substrate requires especially stringent growth conditions, such as single crystal substrates with clean surfaces, and the appropriate matching of surface free energies between the cluster material and the substrate) in order to achieve the dilute film morphology. Quantum dot size control is achieved by keeping the amount of material on the substrate low and the ambient conditions pristine. These are particularly interesting quantum dot arrays because they hold great potential as extremely high-density memory or recording substrates and lend themselves nicely to quantitative nanoscale characterization. The random ordered films of clusters are more typical of epitaxial growth methods and hold great promise because they are usually cheaper to make and require less stringent growth conditions in some (but not all) cases. In both cases, annealing of the cluster films can change either the film morphology or particle size through a two dimensional diffusion and cluster ripening process (Ostwald as well as others).

While we have focused on only one variant of epitaxial growth of quantum dots, it is clearly illustrative of the general method. It essentially produces films of clusters that can be regulated in size and stoichiometry through various means. One distinct advantage this method has is it produces quantum dots in direct contact with a macroscopic material (substrate) that would lend itself to directly connect the dot with some electronic source, thus lending itself to further processing in the manufacture of some electro-optic device.

7.7. QUANTUM DOTS FORMED BY ION IMPLANTATION

Alternative routes to chemically and epitaxially prepared quantum dots are found in quantum dot composite (QDC) materials. QDC materials are typically produced using high energy (several kV) ion implantation into a solid (Figure 7.7). Ion implantation is the preferred method of producing QDC materials as it allows control of the profile of the implanted ion, such as spatial and dose control. Because of the high degree of selectivity in the ion implantation process, the QDC materials can be used in applications ranging from light emitting diodes to semiconductor electronics. Understanding how the ion implantation process produces well formed quantum particles, as well as how these quantum dot particles interact with light are the focus of this section.

Producing the QDC materials via ion implantation is quite straightforward. An intense ion beam of an appropriate semiconductor species is bombarded into a glass target, typically, silicon dioxide (SiO_2). For example, the growth of Si quantum dots in SiO_2 matrices has been achieved by implanting Si ions (dose $\sim 10^{17}$ ions cm^{-2}) at energies of 200 keV into a 0.65–0.75 μm thick SiO_2 layer grown on a Si(100) wafer (Figure 7.8). After implantation, the Si/SiO_2 substrate is annealed at temperatures as high as 1100°C. Depending on the Si^+ beam dose and energy, well formed QDC materials with various sized Si quantum dots (20–60 Å in diameter) can be fabricated (Figure 7.8). This general scheme can be applied

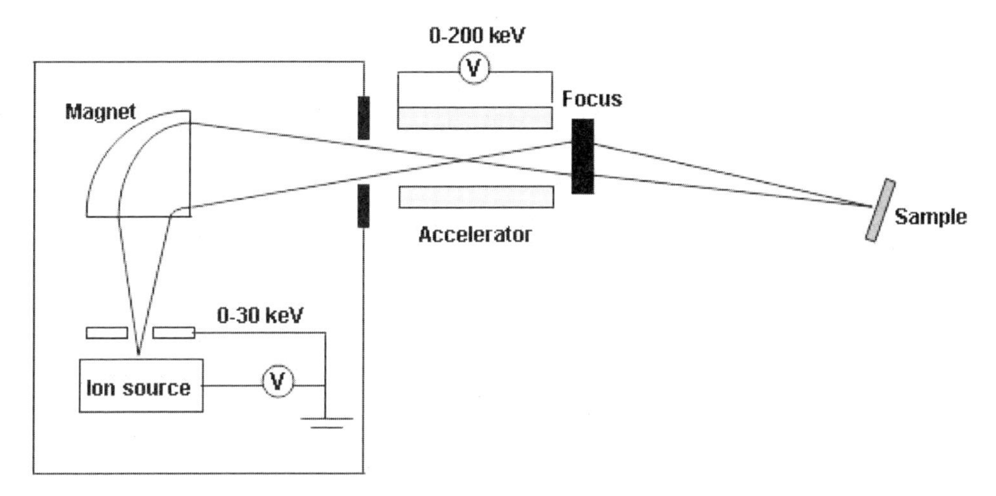

FIGURE 7.7. Illustration of high-energy ion implantation process to fabricate quantum dots.[5]

for a variety of other systems including Ge, Au, and Ti and some binary systems such as GaAs, CdSe, and GeSi.

How does a high energy ion that is implanted into a glass substrate actually form a quantum dot? When an energetic ion is implanted into the substrate, it will lose energy very quickly by interacting with the substrate. The ions will collide with substrate atoms in a random fashion until they slow down to an energy that is comparable with the thermal energy ($kT = 0.025$ eV). Because of the extreme conditions of the initial implantation event, it is possible to have the substrate *supersaturated* with the ion dopant. Supersaturation refers to the situation in which a larger quantity of solute (the ions) is dissolved in a solution (the matrix) than would normally be possible. The supersaturated substrate is then annealed slowly to promote the first step of quantum dot formation, *nucleation and growth*. This

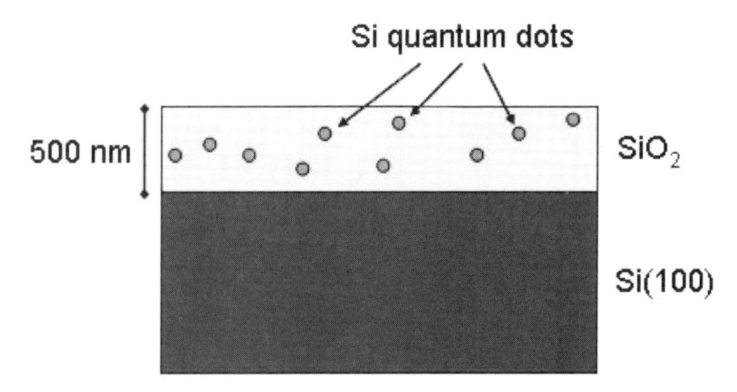

FIGURE 7.8. Illustration of a cross sectional view of Si quantum dots formed in a glass matrix via ion implantation. Note that the random arrangement and spherical shape of the quantum dot particles is expected for quantum dots implanted in an amorphous media.[6]

supersaturation of ionic species in the matrix is relieved by nucleation of the ions into small quantum sized (<20 Å) particles. Upon nucleation, the concentration of these species in solution drops below the critical concentration for nucleation, and further material can only add to the existing nuclei. This initial nucleation and growth phase happens at temperature of greater than 500°C.

The second phase of quantum dot growth occurs via a process called *Ostwald ripening* that was introduced earlier in this chapter. Ostwald ripening is the process in which many small crystallites are initially formed but slowly disappear except for a few that grow larger at the expense of the small crystals. In other words, the smaller crystals act as "seeds" for the bigger crystals. As the larger crystals grow, the area around them is depleted of smaller crystals. This is a spontaneous process that occurs because larger crystals are more energetically favored than smaller crystals. While the formation of many small crystals is kinetically favored, (i.e., they nucleate more easily) large crystals are thermodynamically favored (i.e., they have a lower surface energy). Thus, from a kinetic standpoint, it is easier to nucleate many small crystals. The nuclei, however, have a large surface energy and can attain a lower energy state if transformed into large crystals. For diffusion-limited Ostwald ripening, the change in average crystal size is proportional to the cube root of time and can be expressed as $R \propto (At)^{1/3}$, where R is the particle radius, t is the time, and A represents a combination of constants that are related to the type of material in question. For long annealing times, Ostwald ripening tends towards an asymptotic region, meaning that there exits some absolute limit to the maximum size a quantum dot can grow via Ostwald ripening (see Figure 7.3).

The interplay between ion dose and QDC annealing to determine particle size is also very important. Increasing the ion dose in the implantation process will increase the level of ion supersaturation. As the ion supersaturation percent increases, the average particle size will also increase, as there are more ions per area to initiate nucleation and growth. In addition, higher annealing temperatures will also produce large quantum dot sizes, as Ostwald ripening is the growth process at higher temperatures. It is interesting to note that changing the two variables accordingly can produce a broad range of quantum dot sizes. For instance, researchers have found that Ge quantum dots grown from a Ge^+ dose of 3×10^{17} ions cm^{-2} have an average quantum dot size of ~50 Å when annealed at 600°C, while annealing temperatures of 1000°C shift the average size closer to 80 Å with dot sizes as large as 200 Å. However, a Ge^+ ion dose of 6×10^{16} ions cm^{-2} with a subsequent annealing at 1000°C leads to well formed Ge quantum dots with an average size of 45 Å and low size dispersity.

As a result of quantum confinement, semiconductor materials tend to show a quantization of the bulk band structure as well as a blue shift of the optical band gap. This means, for example, Si and cadmium selenide (CdSe), two famous quantum dot materials, which have bulk band gaps of 1.12 and 1.74 eV (at 298 K), respectively, can have their band gap energies shifted toward the visible region of the electromagnetic spectrum. This unique property makes these materials very useful in technological applications. Si QDC materials show strong quantum size effects with annealing temperature, which is expected due to the quantum dots growing larger as the annealing temperature is increased (Figure 7.9). After heating at 950°C, the Si quantum dots have a photoluminescence (PL) energy of ~1.82 eV. On the other hand, heating at 1100°C leads to a PL energy of ~1.65 eV. This difference corresponds to a change in Si quantum dot size from less than 20 Å at 950°C to about 30 Å at 1100°C.

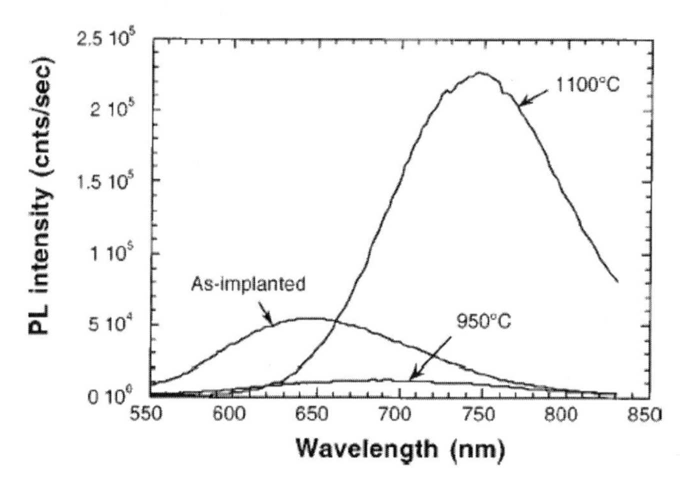

FIGURE 7.9. Photoluminescence spectra from Si (400 keV, $1.53 \times 10^{17} \text{cm}^{-2}$) implanted SiO_2 as implanted and after annealing at 950 and 1100°C. (From Ref. 4 by permission of the American Institute of Physics.)

Ion implantation can also provide a resource to examine the structure of QDC materials. Researchers at Delft University have used ion implantation to produce Li quantum dots in MgO substrates. Upon implantation with 30 keV Li ions and subsequent annealing at 700°C, Li quantum dots with an average size of less than 100 Å can be formed. These quantum dots adapt an unusual fcc structure, although Li exists in a bcc structure in the bulk lattice. These

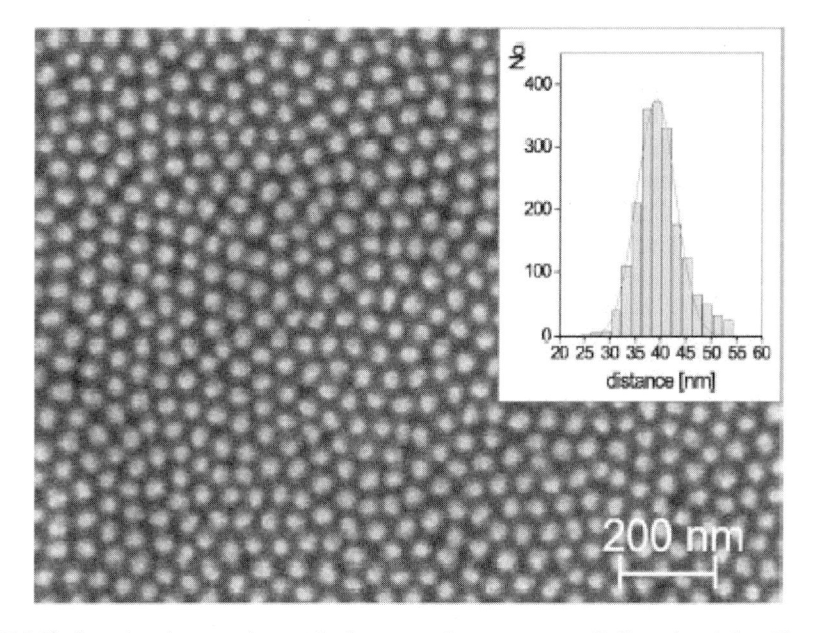

FIGURE 7.10. Scanning electron micrograph of quantum dot patterns on a GaSb surface induced by Ar-ion sputtering with an ion energy of 500 eV. The dots show a hexagonal ordering with a characteristic wavelength that depends on ion energy. The insets show the corresponding distribution of the nearest-neighbor distance. (From Ref. 5 by permission of the American Physical Society.)

types of QDC materials will allow further understanding into the physics of metastable materials.

Another form of ion implantation called *ion sputtering* can also be used to make well formed quantum dot materials. Ion sputtering is a low energy variant of ion implantation. The main difference between implantation and sputtering is that the sputtered materials are not composite materials; rather they are well formed quantum dot arrays on the surface of a substrate. Bombarding a substrate surface with 1 keV Ar ions makes it possible to produce hexagonal arrays of quantum dots (Figure 7.10). This occurs due to surface instability caused by the energetic Ar ion beam. Surface instability causes atoms on the surface of the substrate to undergo ion-induced surface diffusion. The surface diffusion can be related to the characteristic wavelength, l_c, of the hexagonal array given by, $l_c = 2E\sqrt{(2D/E)}$, where D is the diffusion constant and E is ion energy.

7.8 FURTHER READING

L. I. Schiff, *Quantum Mechanics* (3rd edition) (McGraw-Hill, 1968).

F. K. Richtmyer, E. H. Kennard, and J. N. Cooper, *Introduction to Modern Physics* (6th edition) (McGraw-Hill, 1969).

C. H. Kittel, *Introduction of Solid State Physics* (John Wiley & Sons, Inc., 1996).

N. W. Ashcroft and N. D. Mermin, *Solid State Physics* (Holt, Rinehart and Winston, 1976).

U. Woggon, *Optical Properties of Semiconductor Dots,* Springer Tracts in Modern Physics (Springer-Verlag, 1997).

L. Banyai and S. W. Koch, *Semiconductor Quantum Dots* (World Scientific, 1993).

L. Jacak, P. Hawrylak, and A. Wojs, *Quantum Dots* (Springer-Verlag, 1998).

P. Harrison, *Quantum Wells, Wires and Dots* (John Wiley & Sons, Ltd., 2000).

J. G. Zhu, C. W. White, J. D. Budai, S. P. Whitrow, and Y. Chen, *J. Appl. Phys.* **78**, 4386 (1995).

C. V. Falub, P. E. Mijnarends, S. W. H. Eijt, M. A. van Huis, A. van Veen, and H. Schut, *Phys. Rev. B* **66**, 075426 (2002).

L. E. Brus, *J. Chem. Phys.* 80, 4403 (1984); 90, 2555 (1986).

Y. Kayanuma, *Phys. Rev.* **38**, 9797 (1988).

QUESTIONS

1. Using Figure 7.10, use the experimentally derived characteristic length and the Ar ion energy to calculate the surface diffusion rate.

REFERENCES

1. C. B. Murray, D. J. North, and M. G, Bawendi; *J. Am. Chem. Soc.* 115, 8706 (1993).
2. V. K. La Mer and R. H. Dinegar, *J. Am. Chem. Soc.* **72**, 4847 (1950).
3. M. A. Hines and P. Guyot-Sionnest, *J. Phys. Chem. B* **102**, 3655 (1998).
4. A. A. Guzelian, U. Banin, A. V. Kadavanich, X. Peng, and A. P. Alivisatos, *Appl. Phys. Lett.* **69**, 1432 (1996).
5. S. P. Withrow, C. W. White, A. Meldrum, J. D. Budai, D. M. Hembree, Jr., and J. C. Barbour, *J. Appl. Phys.* **86**, 396 (1999).
6. S. Facsko, H. Kurz, and T. Dekorsy, *Phys. Rev. B* **63**, 165329 (2002).

8

Nanocomposites

Robert C. Cammarata
Department of Materials Science and Engineering
Johns Hopkins University
Baltimore, MD

8.1. INTRODUCTION

Nanocomposites can be defined as multiphase materials where one or more of the phases have at least one dimension of order 100 nm or less. Most nanocomposites that have been developed and that have demonstrated technological importance have been composed of two phases, and can be microstructurally classified into three principal types (see Figure 8.1): (a) Nanolayered composites composed of alternating layers of nanoscale dimension; (b) nanofilamentary composites composed of a matrix with embedded (and generally aligned) nanoscale diameter filaments; (c) nanoparticulate composites composed of a matrix with embedded nanoscale particles. As with conventional composites, the properties of nanocomposites can display synergistic improvements over those of the component phases individually. However, by reducing the physical dimension(s) of the phase(s) down to the nanometer length scale, unusual and often enhanced properties can be realized. An important microstructural feature of nanocomposites is their large ratio of interphase surface area to volume. For example, in dispersions of layered clay (aluminosilicates) in nanocomposite polymers, this ratio can approach 700 m^2/cm^3, which is of order the area of a football field within the volume of a raindrop.[1] This large surface area can result in novel and often enhanced properties that can be exploited technologically.

In terms of their engineering applications, nanocomposites can be classified either as functional materials (based on their electrical, magnetic, and/or optical behavior) or as structural materials (based on their mechanical properties). An example of a functional nanocomposite material is a nanolayered semiconductor (generally called a "semiconductor superlattice") composed of alternating layers of single crystal GaAs and $GaAl_xAs_{1-x}$. When the layer thicknesses are reduced below the electronic mean free path of the bulk

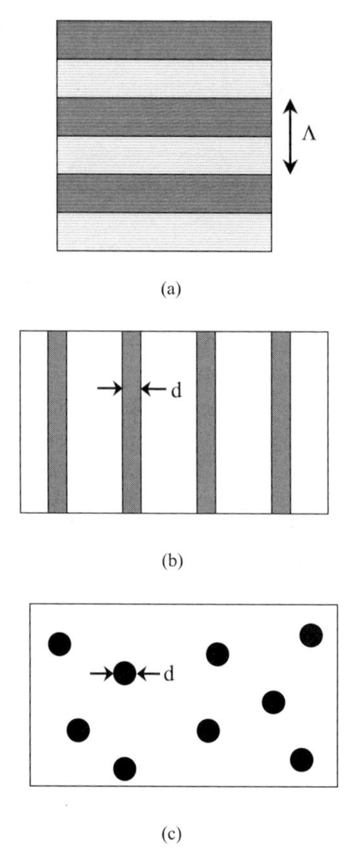

FIGURE 8.1. Schematic representations of nanocomposite materials with characteristic length scale: (a) nanolayered composites with nanoscale bilayer repeat length Λ; (b) nanofilamentary (nanowire) composites composed of a matrix with embedded filaments of nanoscale diameter d; (c) nanoparticulate composites composed of a matrix with embedded particles of nanoscale diameter d.

three-dimensional materials, novel electronic and photonic properties can be realized owing to quantum confinement effects. Figure 8.2 shows a schematic (real space) energy band diagram for such a GaAs/$GaAl_xAs_{1-x}$ superlattice. An electron in the GaAs layer can be considered as partially confined in a "quantum well" of barrier height ΔE equal to the difference in the energies of the bottom of the conduction band E_c for the two layer materials. In the limit where $\Delta E \to \infty$, all of the electron energy levels are quantized and these levels can be expressed as[2]

$$E_n = n^2 h^2 / 8m^* w^2 \tag{8.1}$$

where n is a quantum number $= 1, 2, 3, \ldots$, h is Planck's constant, m^* is the effective mass of the electron, and w is the width of the quantum well (GaAs layer thickness). By altering

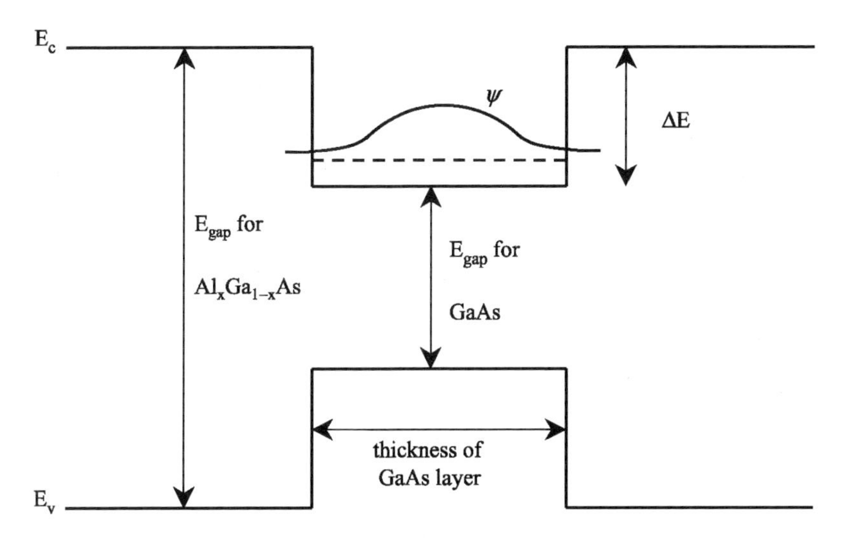

FIGURE 8.2. Schematic energy band diagram of GaAs/GaAl$_x$As$_{1-x}$ quantum well. An electron (represented by its wavefunction ψ) can be considered as partially confined in the quantum well of width equal to the GaAs thickness. The barrier height ΔE is equal to the difference in the energies of the bottom of the conduction band E_c for the two layer materials. E_v is the energy of the top of the valence band and E_{gap} is the band gap energy.

the width of the well, the electron energies can be tuned for certain electronic and photonic applications.[3–7]

As an example of a structural nanocomposite material, consider a ductile metal matrix embedded with a reinforcing second phase composed of hard ceramic nanoparticles. The yield strength of a material is often governed by the stress necessary for mobile dislocations to overcome obstacles to their motion. If the average distance between the particles in the nanocomposite (assumed significantly smaller than the diameter of the particles) is the smallest microstructural length scale, then the yield strength will be determined by the stress needed for dislocations to overcome the particles by bowing around them. This process, known as the Orowan bowing mechanism[8] is schematically illustrated in Figure 8.3. The stress σ needed to bow around the particles is approximately given by the expression

$$\sigma \approx Gb/\lambda \tag{8.2}$$

where G and b are the shear modulus and Burgers vector, respectively, for the matrix material and λ is the average interparticle spacing. As long as the particle diameter does not become too small so that it is possible for dislocations to cut through the particles, it is the particle density that determines the bowing stress. Therefore, by using nanoscale particles, it is possible to achieve significant strengthening with a relatively small particle volume fraction.[9,10]

This chapter will survey artificially produced nanocomposite materials with emphasis on advances made during the past two decades. Examples of metallic, ceramic, polymer, and

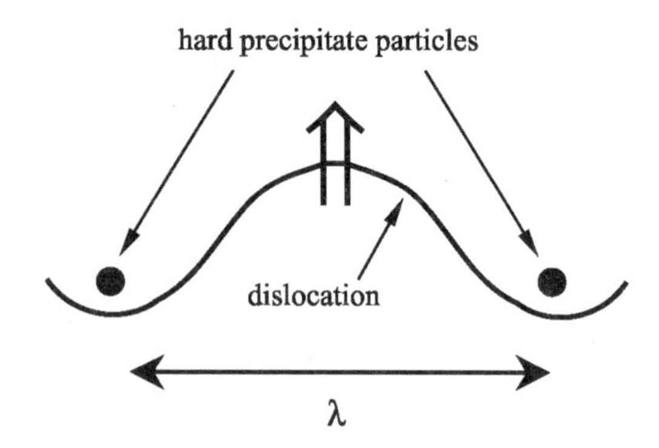

FIGURE 8.3. Precipitate particles of spacing λ acting as obstacles to dislocation motion.

semiconductor systems, as well as hybrid systems, will be presented. For each principal type of nanocomposite (nanolayered, nanofilamentary, and nanoparticulate), a discussion of the major synthesis and processing methods are discussed, as well as important technological applications.

8.2. NANOLAYERED COMPOSITES

The production of technologically useful multilayered materials with very thin layers dates back to the forging of Damascus[11] and Japanese[12] swords composed of layers such as soft wrought iron and hardened steel that were repeatedly folded and rolled. The ability to fabricate high quality artificially multilayered materials coincides with the advent of advanced thin film deposition methods. These methods allow the production of materials with precise control of the composition and thickness of the layers. Compared to conventional bulk laminate composites, the individual layer thicknesses can be reduced to atomic dimensions, resulting in the epitome of microstructural engineering.

Referring to Figure 8.1a, the characteristic microstructural length scale for nanolayered materials composed of periodically alternating two phase layers is the bilayer repeat length (or bilayer period) Λ, equal to the combined thickness of two adjacent layers. Artificially mulilayered materials composed of layers of different phase are generically known as heterostructures, whether they consist of only a few layers or of many layers. Multilayers composed of many single crystal layers that possess the same crystal structure and where there is perfect lattice matching at the surfaces (interphase interfaces) are called superlattices. An important type of this material is the semiconductor superlattice, such as the $GaAs/GaAl_xAs_{1-x}$ superlattice mentioned in the introduction. Figure 8.4 is a high resolution transmission electron micrograph showing a cross-sectional view of an InAs-GaSb (100) superlattice grown by molecular beam epitaxy.[13]

In the case of metallic nanolayered composites, the requirement that the layers be single crystals is often relaxed, so that a large grained, highly textured metallic multilayer is often referred to as a superlattice. In addition, a metallic multilayer composed of layers with different crystal structures but with a well-defined epitaxially relationship at the interfaces is also called a superlattice.

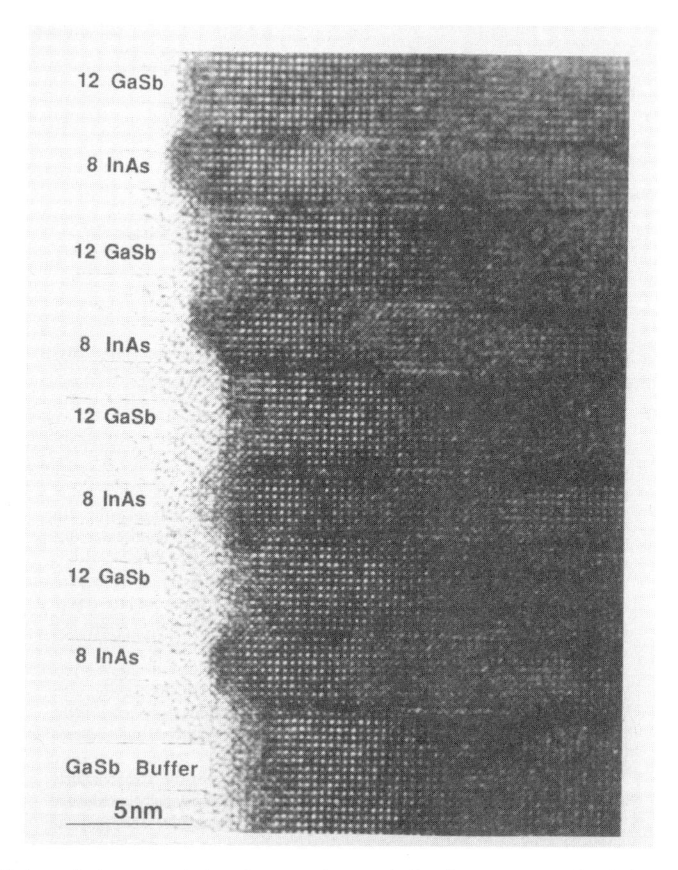

FIGURE 8.4. High resolution transmission electron micrograph showing a cross-sectional view of an InAs-GaSb (100) superlattice. (Reproduced with kind permission of M. Twigg.)

8.2.1. Synthesis and Processing

The most common method of producing inorganic nanolayered materials involves standard thin film deposition methods that have modified to allow for the alternate deposition of two (or more) different materials.[13] Because these approaches have been used extensively in making other types of nanocomposites as well as nanolayered materials, it is worthwhile to discussing them in some detail. Physical vapor deposition (PVD) methods, such as evaporation and sputtering, have been widely used to produce metallic, ceramic, and semiconductor artificially layered thin films.[13–15] These methods involve deposition in an evacuated chamber, typically from sources that are alternately shuttered in order to produce the layered film. Sputtering involves the collision of ions, usually of an inert gas such as argon, with the surface of a target material, resulting in the ejection of target atoms that are collected onto a substrate to form the thin film. In the case of ceramic multilayers where the layers are oxides or nitrides, it is possible to directly sputter from ceramic targets (using a neutralized beam or by radio frequency sputtering[14]). Alternately, it is possible introduce oxygen or nitrogen into the chamber and reactively sputter using metal targets. Sputtering is useful when attempting to produce films composed of alternating metal and ceramic layers such as Al-Al_2O_3. This can be performed either by alternate sputtering from two different targets composed of the layer materials, or by sputtering from a single metal target (Al) while alternatingly opening and closing a valve that bleeds in oxygen to reactively sputter the ceramic layer.[14] Also, sputtering has been used to produce amorphous metallic and amorphous ceramic nanolayered materials.

In the case of semiconductor films, such as GaAs/$GaAl_xAs_{1-x}$ superlattices mentioned in the introduction, the need to fabricate high quality single crystal materials for most applications has required the use of molecular beam epitaxy (MBE).[16,17] An MBE system is an evaporation system where deposition occurs under ultrahigh vacuum conditions, and generally a variety of *in situ* characterization methods, such as reflection high energy electron diffraction and Auger electron spectroscopy are available to monitor film deposition. In the case of an MBE system fabricating a GaAs/$GaAl_xAs_{1-x}$ superlattice, deposition of the elemental constituents (Ga, Al, As) occurs by evaporation or sublimation from effusion cells. When depositing high melting temperature materials such as silicon, and electron beam is used to evaporate the material.

Pulsed laser deposition (PLD)[18] involves the use of a short pulse from a focused output of a laser to vaporize a material from a target and collect it onto a substrate inside a vacuum chamber. An important feature of PLD is that it allows deposition of a multicomponent material from a target of the same composition, and therefore has become a popular method to produce ceramic systems such as high temperature superconductors and ferroelectrics. Multiple targets are used to deposit multilayered films.

In addition to PVD methods, chemical-based approaches have also been used to fabricate nanolayered materials. Chemical vapor deposition (CVD) is a common method for producing a variety of metallic, ceramic, and semiconductor systems.[6,13] A variety of reaction types can be utilized during CVD, such as pyrolysis, reduction, and oxidation. Metallo-organic chemical vapor deposition (MOCVD) is an important method for producing compound semiconductor superlattices. It involves reactions of metal alkyls with a hydride of the nonmetal component to form the compound semiconductor layer materials. As an example, the following reactions can be used to fabricate a GaAs/$GaAl_xAs_{1-x}$ superlattice:

$$Ga(CH_3)_3(g) + AsH_3(g) \rightarrow GaAs(s) + CH_4(g) \tag{8.3}$$

$$x Al(CH_3)_3(g) + (1-x)Ga(CH_3)_3(g) + AsH_3 \rightarrow Al_xGa_{1-x}As(s) + 3CH_4(g). \tag{8.4}$$

Precise control of the temperature, gas pressures, and gas compositions allows for the production of high quality superlattices.

Electrochemical deposition is a relatively inexpensive method that has allowed for the producing of a variety of metallic, semiconductor, and conducting ceramic multilayers.[19,20] Generally the process involves alternate deposition between two different plating potentials using a single electrolyte. However, this restricts the number of multilayered systems that can fabricated, as the electrochemistry of the two layer materials may not be compatible with the use of a single electrolyte. For these systems, it is possible to alternately deposit from two different electrolytes. In addition to the low cost, other advantages of electrodeposition includes the ability to uniformly coat non-planar substrates, to plate over wide areas, and to deposit a nanocomposite to an overall thickness much larger than can be generally achieved with vacuum deposition methods.

Langmuir–Blodgett[21,22] and self–assembly[23] methods for producing organic films composed of monomolecular layers have been investigated for many years. The fragility of the Langmuir–Blodgett films and the inability to produce high quality self–assembled films with overall thicknesses greater than about 100 nm have limited their technological usefulness. Multilayered polymer composites have been produced by coextrusion of alternatingly layered thermoplastics.[24,25] This approach can fabricate materials composed of thousands of layers with individual layer thicknesses as small as a few tens of nanometers. In addition, multilayered conducting polymer films have been produced by electrodeposition.[19]

8.2.2. Functional Materials

As has previously been discussed, semiconductor heterostructures in general, and superlattices in particular, display a variety of interesting and tunable microelectronic and photonic properties owing to quantum confinement effects, and these systems represent the most important technological application of nanolayered materials. It is not possible to adequately discuss in detail these applications here; the interested reader is referred to the reviews and books cited in the references.[3–7]

The most heavily studied behavior of metallic superlattices has been their magnetic properties.[14,26,27] A variety of interesting phenomena can occur based on magnetic coupling between ferromagnetic layers in films where the alternating layers are both ferromagnetic as well as when the layers are alternating ferromagnetic and nonmagnetic materials. One of the most interesting phenomena displayed in the latter type of superlattice is the magneto-electrical effect referred to as giant magnetic resistance (GMR).[28] GMR refers to the large change in electrical resistance in certain systems such as Co/Cu when an external magnetic field H is applied. The effect occurs in superlattices where the ferromagnetic layers display an antiparallel alignment when $H = 0$. When a sufficiently strong field is applied, the magnetic moments of the ferromagnetic layers assume a parallel alignment, resulting in a large reduction in the electrical resistance. This phenomenon has the potential for many technological applications, and is currently employed in computer hard disk drives for data storage.

Because of the periodic electron density variation in multilayers, it is possible to use them as Bragg reflectors, particular for x-ray optics elements.[29] By adjusting the bilayer period, the wavelength of the reflected radiation can be sensitively tuned. Optimal materials for the alternating layers will have a large electron density difference, and Mo/Si has become a common system for this application.

Nonmetallic multilayers have also been used as Bragg reflectors. For example, high reflectivity GaAs/AlAs layered heterostructures have been used a mirrors in solid state layers.[6] Polymer multilayers where the layers have different refractive indices have been produced that reflect ultraviolet or near-infrared radiation as well as displaying iridescent color effects.[24,25]

8.2.3. Structural Materials

Layered nanocomposites have displayed a wide variety of enhanced mechanical behavior that make them useful for protective coating applications.[12,29,30] Large hardness enhancements have been observed in both ceramic and metallic multilayers when the bilayer repeat length is reduced below about 100 nm. This can be understood in a general way as resulting from dislocation pinning effects associated with the interfaces. A significant difference in the elastic moduli of the layers will result in dislocation "image forces" that can act as a barrier to dislocation generation and motion. In addition, if the layer materials have different bulk lattice spacings, lattice matching at the interfaces creates a stress field that can impede dislocation mobility. When the layers display different crystal structures (whether or not there is interfacial lattice matching), the two materials may not possess compatible slip systems that allow transmission of dislocations at the interface.

In addition to hardness enhancements, multilayered metallic films have displayed improved wear resistance and fracture strength.[12] Recently, Cu/Ni multilayers electrodeposited on Cu rods have been observed to significantly increase the fatigue lifetime.[31,32] Coextruded polymer multilayers composed of alternating brittle and ductile materials have displayed increased fracture toughness that makes them attractive for heavy-duty wrapping and packaging materials.[24,25]

8.3. NANOFILAMENTARY AND NANOWIRE COMPOSITES

Two types of nanocomposites composed of a matrix embedded with aligned second phase filaments will be discussed. The first has been termed a nanofilamentary composite, and is generally associated with mechanical processing methods to produce materials with enhanced mechanical strength as well as other properties. The second type is often referred to as an embedded nanowire array, and is generally produced by electrodeposition in a compliant matrix. The interest for these materials is not in the mechanical behavior but in the functional properties of the nanowires.

8.3.1. Nanofilamentary Composites

Nanofilamentary composites can be considered as wires that are composed of a metal matrix with aligned second phase metal filaments.[33–36] They are produced by starting with a bulk two-phase ingot produced using conventional metallurgical fabrication methods such as casting or powder processing. In general, the two phases are two metallic elements that display little to no solid solubility. The most common system that has been investigated is Cu/Nb where the Nb volume percent is typically of order a few tens of percent. In order to produce the nanocomposite, the two-phase ingot is subjected to large scale deformation, for example, by drawing and swaging, forming the filamentary microstructure where, in the case of Cu/Nb, the Nb wire diameters can be reduced to tens of nanometers. Although the filaments produced in this manner are not generally continuous, the aspect ratio is large, typically of order 10^3 to 10^6. Since this process can result in wires with very small overall diameters, it has become common to bundle and bond several wires that can then be further swaged and drawn.[35,36]

Nanofilamentary composites often display significantly enhanced tensile strength (approaching the theoretical limit) as well as a high electrical conductivity. These properties are maintained down to very low temperature down where, in the case of Cu/Nb, the niobium is superconducting. Because of this behavior, these nanocomposites have application as windings for high field pulsed magnets.[37]

8.3.2. Nanowire Composites

In contrast to the nanofilamentary composites just discussed, nanowires are functional materials whose behavior are generally associated with quantum confinement effects. A variety of novel approaches have been developed to produce arrays of nanowires.[39] With regard to producing composites involving a matrix with embedded and aligned nanowires, electrodeposition has become an extremely important synthesis method.[19,40,41] This approach uses a nanoporous membrane such polycarbonate. The pores are produced by exposing the membrane to a radioactive source that creates nuclear tracks that are subsequently etched. In this way, it is possible to form high aspect ratio pores with diameters down to tens of nanometers. Prior to electrodeposition, a thin layer of an electrical conductor such as gold is deposited by, e.g., sputtering, on one side of the membrane. This conductive layer is used as an electrode in an electrochemical cell. During the plating, the pores are filled with the deposited material, resulting in an array of nanowires within the membrane. Figure 8.5 shows a scanning electron micrograph of FeCo nanowires produced by this method after the membrane has been dissolved.[41]

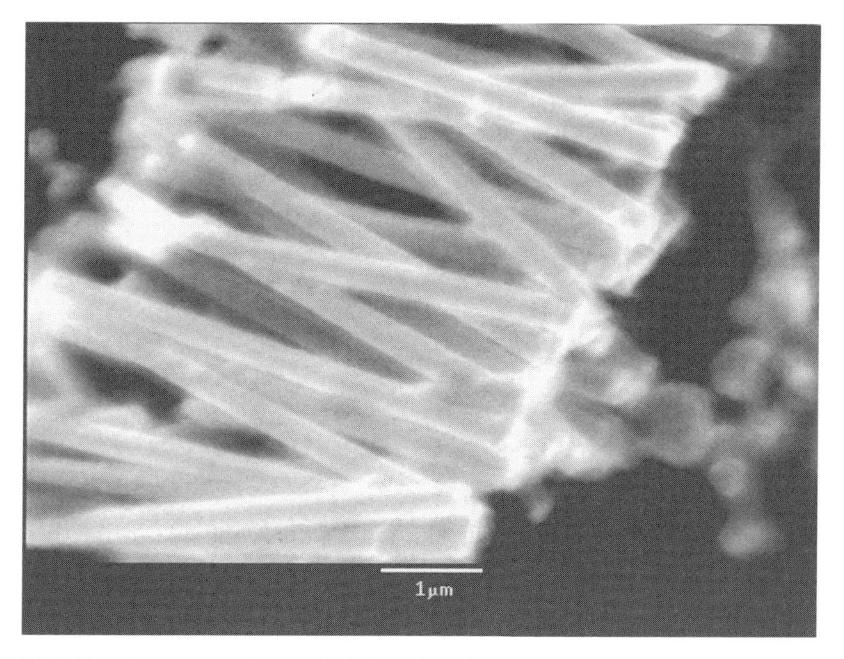

FIGURE 8.5. Scanning electron micrograph of electrodeposited FeCo nanowires (the polycarbonate matrix in which the wires were embedded has been completely dissolved).

Using this approach, nanowire composites with arrays of metals and alloys, conducting polymers, and semiconductors can be produced.[18] Examples of potential applications for these systems include field emitter areas for flat panel displays and magnetic data storage.

8.4. NANOPARTICULATE COMPOSITES

The fabrication and study of artificial nanoparticulate composites in the form of granular metals can be dated back to the 1960s.[42,43] These materials are composed of nanoscale metal particles embedded in an immiscible metal, ceramic, or semiconductor matrix. Since that time, there has been an explosion in the number and types of systems that have been developed and investigated, involving a wide variety of processing approaches.

8.4.1. Synthesis and Processing

As with other types of nanocomposites materials, thin film processing methods have been used extensively to produce nanoparticulate composites. Granular metals have been generally produced by simultaneous thin film deposition of two immiscible phases by, for example, evaporation or sputtering. Figure 8.6 show bright field transmission electron micrographs of Ni/SiO_2 granular metal films.[30] The ceramic phase is amorphous and the metal is in the form of nanoscale single crystal particles that are close to spherical in shape. In these films, the metal phase displays a percolation threshold at a composition between

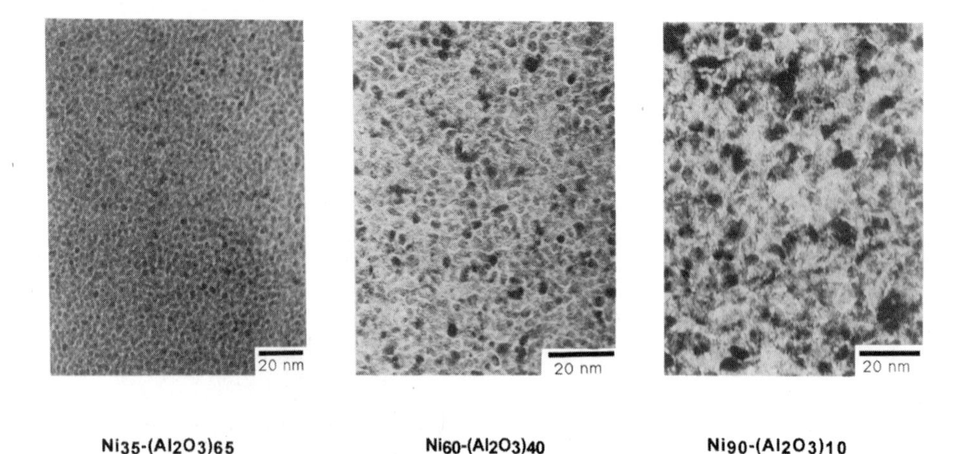

FIGURE 8.6. Bright field transmission electron micrographs of Ni/SiO_2 granular metal films. (From Ref. 29 by permission of Elsevier Science B.V.)

50 and 60 volume percent. Above this composition the metal is an interconnected network, and below this composition the metal is in the form of distinct particles. This percolation effect has a major influence on the properties, as discussed in the next section.

Electrodeposition has been used to deposit a nanocomposite in the form of a thin film matrix embedded with reinforcing second phase particles.[9,18,29,40] In this approach, the electrolyte contains a uniform suspension of nanoscale particles. The plating occurs on a rotating disk electrode that creates a hydrodynamic boundary layer that controls mass transport-limited processes. The kinetics of particle codeposition has been modeled and successfully used to characterize the deposition of Ni/Al_2O_3 films.[44] By varying the particle concentration in the electrolyte, the deposition current, and the electrode rotation rate, the volume fraction of particles can be sensitively controlled.

Recently, other novel processing routes to producing metal-based nanoparticulate composites have been developed. One approach involves partially devitrifying bulk metallic glasses.[45] This can be performed either by annealing the glassy precursor or by direct quenching from the liquid state. The materials produced are composed of nanoscale metal crystals embedded in an amorphous metal matrix. Another processing approach involves self-assembly of nanoscale particles.[46] This method has been used to produce three-dimensional $FePt/Fe_3O_4$ nanocomposite assemblies. A hexane dispersion of FePt and Fe_3O_4 nanoparticles is mixed under ultrasonic agitation, and then self-assembly is induced by evaporation of the hexane or by addition of ethanol.[46] Figure 8.7 shows transmission electron micrographs of such assemblies.

A variety of processing routes has been developed to fabricate ceramic-based nanocomposites, such as Al_2O_3/SiC, where particles of the carbide are used to reinforce the alumina for structural material applications.[47] One common approach involves powder processing, where mixtures of ultrafine powders of the two phases are homogenized by, for example, wet ball milling in an organic or aqueous medium. After drying of the slurries, the nanocomposite is consolidated; this has been generally performed by hot-pressing. Because of certain limitations associated with particle agglomeration and dispersion problems, alternate processing approaches have been developed. One of these involves pyrolysis of a silicon-containing polymer precursor.[48] Polycarbosilane is coated onto a surface-modified alumina powder and pyrolysed to produce nanoscale SiC particles. These powders are then hot-pressed into the fully dense composite. Sol gel processing has also been investigated. Boehmite gels are used as a source of the alumina that is either coated onto SiC particles[49,50] or mixed with the SiC precursor polysilastyrene.[51] After drying and calcination, the powder is consolidated by hot-pressing.

Polymer-based nanocomposites for structural[52] and functional[53,54] material applications have received a great deal of recent attention. In terms of commercial importance, structural nanocomposites produced using layered clay minerals, such as montmorilonite and hectorite are the most significant.[55] The layered clay (aluminosilicates) materials are synthesized by mixing the layered clay with the monomer, followed by polymerization, by melt mixing the layered clay with the polymer, or by mixing the layered clay with a solvated polymer followed by solvent removal. Another type of commercially important nanocomposite are thermoplastics reinforced with carbon nanotubes formed by extrusion and injection molding, The high modulus and tensile strength of carbon nanotubes make them extremely attractive filler materials for nanocomposite polymers.

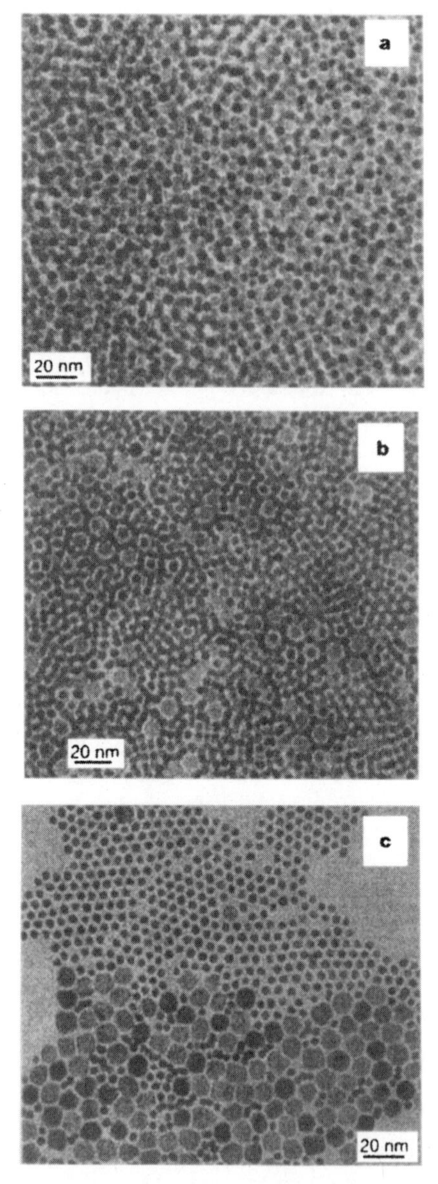

FIGURE 8.7. Transmission electron micrographs of binary nanoparticle assemblies. (a) Fe_3O_4 (4 nm)-$Fe_{58}Pt_{42}$ (4 nm) assembly; (b) Fe_3O_4 (8 nm)-$Fe_{58}Pt_{42}$ (4 nm) assembly; Fe_3O_4 (12 nm)-$Fe_{58}Pt_{42}$ (4 nm) assembly. (From Ref. 46 by permission of Macmillan Publishers Ltd.)

8.4.2. *Electrical and Magnetic Properties*

Granular metal films have displayed a variety of interesting electronic and magnetic properties.[56] As mentioned previously, there is a percolation transition when the volume fraction of the metal is varied, with the transition generally occurring around 50 to 60 volume percent of the metal. In the case of a metal/ceramic film above the percolation threshold (where the metal forms an interconnected network), the nanocomposite is an electrical conductor. Below the percolation threshold (where the metal is in the form of single crystal granules) the nanocomposite displays the conductivity of a dielectric material. Thus, associated with the percolation transition is a several orders of magnitude change in conductivity.

If the metal is a ferromagnet, the nanocomposite will behave like a bulk ferromagnetic when the metal volume fraction is above the percolation threshold. When the metal volume fraction is below the percolation threshold and the metal granule size is smaller than the bulk equilibrium domain size, the assembly of single domain particles can lead to novel magnetic behavior. One of these effects is giant magnetoresistance (GMR), similar to that found in magnetic superlattices, where the application of an external magnetic field leads to a significant reduction in the electrical resistance. This discovery demonstrated that GMR is not associated solely with a layered structure as was once thought, but can be induced in a nanoparticulate composite structure as well.

8.4.3. *Structural Materials*

As discussed in the introduction, reinforcing second phase particles in a nanoparticulate composite can result in significant mechanical property enhancements using a relatively small volume fraction of the second phase. Such enhancements have been observed in, for example, electrodeposited Ni/Al_2O_3[9,10] and chemical vapor deposited Si_3N_4/TiN[57] films. This behavior makes these materials attractive as protective coatings. Ceramic nanocomposites such as Al_2O_3/SiC have shown promise for enhanced wear and creep resistance applications.[47] Polymer nanocomposites using layered clay and carbon nanotube reinforcing particles are already used for automobile material applications such as timing chain covers and mirror housings.[55]

8.5. SUMMARY

A wide variety of nanocomposite materials have been synthesized that display a spectrum of interesting and technologically useful functional and structural material properties. Processing methods have been developed that allow for very precise control of the microstructure that in turn allows for sensitive tuning of these properties. Several nanocomposites have already been produced that are currently being used in commercial applications, and there is great promise that many more technologically useful systems will be available in the near future.

REFERENCES

1. R. A. Vaia and E. P. Giannelis, *MRS Bulletin*, **26**, 394 (2001).
2. C. Kittel, *Introduction to Solid State Physics* (Wiley, New York, 1996) Chapter 6.
3. L. Esaki, in *Synthetic Modulated Structures*, L. L. Chang and B. C. Geissen, eds. (Academic, Orlando, 1985) Chapter 1.
4. G. Burns, *Solid State Physics* (Academic, San Diego, 1985) Chapter 18.
5. D. A. B. Miller, *Opt. Photon*. News, February, 257 (1990).
6. K. N. Tu, J. W. Mayer, and L. C. Feldman, *Electronic Thin Film Science for Electrical Engineers and Materials Scientists* (Macmillan, New York, 1992) Chapter 8.
7. *Semiconductor Superlattices: Growth and Electronic Properties*, H. T. Grahn, ed. (World Scientific, 1995).
8. G. E. Dieter, *Mechanical Metallurgy* (McGraw-Hill, Boston, 1986) p. 218.
9. R. R. Oberle, M. R. Scanlon, R. C. Cammarata, and P. C. Searson, *Appl. Phys. Lett.* **66**, 46 (1995).
10. I. Shao, P. M. Vereecken, C. L. Chien, P. C. Searson, and R. C. Cammarata, *J. Mater. Res.* **17**, 1412 (2002). G. Slayter, *Sci. Amer.* **206**, 124 (1962).
11. M. Chikashige, *Alchemy and Other Chemical Achievements of the Ancient Orient, Engl. trans.* Sasaki (Tokyo, Rokakuho Uchide, 1934) p. 84.
12. R. C. Cammarata, in *Nanomaterials: Synthesis, Properties and Applications*, A. S. Edelstein and R. C. Cammarata, eds. (Institute of Physics, Bristol, 1998) Chapter 6.
13. A. L. Greer and R. E. Somekh, in *Materials Science and Technology: A Comprehensive Treatment-Volume 15*, R. W. Cahn, P. Haasen, and E. J. Kramer (VCH, Weinheim, 1991).
14. M. Ohring, *Materials Science of Thin Films* (Academic, Boston, 2001).
15. J. Y. Tsao, *Materials Fundamentals of Molecular Beam Epitaxy* (Academic, Boston, 1993).
16. *Molecular Beam Epitaxy: Applications to Key Materials*, R. F. C. Farrow, ed. (Noyes, 1995).
17. *Pulsed Laser Deposition of Thin Films*, D. B. Chrisey and G. K. Hubler, eds. (Wiley-Interscience, New York, 1994).
18. P. C. Searson and T. F. Moffat, *Crit. Rev. Surf. Chem.* **3**, 171 (1994).
19. C. Ross, *Annu. Rev. Mater. Sci.* **24**, 159 (1994).
20. J. D. Swalen, D. L. Allara, J. D. Andrade, E. A. Chandross, S. Garoff, J. Israelachvili, T. J. McCarthy, R. Murray, R. F. Pease, J. F. Rabolt, K. J. Wynne, and H. Yu, Langmuir, **3**, 932 (1987).
21. G. G. Roberts, *Langmuir-Blodgett Films* (New York: Plenum, 1990).
22. N. Tillman, A. Ulmann, and T. L. Penner, Langmuir, **5**, 101 (1989).
23. W. J. Schrenk and T. Alfrey, in *Polymer Blends*, Volume 2, D. R. Paul and S. Newman, eds. (Academic, New York, 1978) p. 129.
24. E. Baer, A. Hiltner, and H. D. Keith, *Science*, **235**, 1015 (1987).
25. D. Altbir and M. Kiwi, in *New Trends in Magnetism, Magnetic Materials, and Their Applications*, J. L. Moran-Lopez and J. M. Sanchez, eds. (Plenum, 1994).
26. E. E. Fullerton, in *Handbook of Thin Film Process*, D. A. Glocker and S. Ismat Shah, eds. (Institute of Physics Publishing, Bristol, 1997).
27. *Magnetic Multilayers and Giant Magnetoresistance: Fundamentals and Industrial Applications*, U Hartmann, ed. (Springer Verlag, 2000).
28. Ph. Houdy and P. Boher, *J. Physique III*, **4**, 1589 (1994).
29. R. C. Cammarata, *Thin Solid Films*, **248**, 82 (1994).
30. S. A. Barnett and M. Shinn, *Annu Rev. Mater. Sci.* **24**, 481 (1994).
31. M. R. Stoudt, R. C. Cammarata, and R. E. Ricker, *Scripta Mater.* **43**, 491 (2000).
32. M. R. Stoudt, R. E. Ricker, and R. C. Cammarata, *Int. J. Fatigue* **23**, S215 (2001).
33. J. Bevk, J. P. Harbison, and J. L. Bell, *J. Appl. Phys.* **49**, 6031 (1978).
34. J. Bevk, *Annu. Rev. Mater. Sci.* **13**, 319 (1983).
35. S. I. Hong, *Scripta Mater.* **39**, 1685 (1998).
36. S. I. Hong and M. A. Hill, *J. Mater. Sci.* **37**, 137 (2002).
37. K. Han, V. J. Toplosky, R. Walsh, C. Swenson, B. Lesch, and V. I. Pantsyrnyi, *IEEE Trans. Appl. Supercond.* **12**, 1176 (2002).
38. S. M. Prokes and K. L. Wang, *MRS Bulletin*, **24**(8), 13 (1999).
39. T. M. Whitney, J. S. Jiang, P. C. Searson, and C. L. Chien, *Science*, **261**, 1316 (1993).

40. P. C. Searson, R. C. Cammarata, and C. L. Chien, *J. Electronic Mater.* **24**, 955 (1995).
41. I. Shao, M. W. Chen, C. L. Chien, P. C. Searson, and R. C. Cammarata, to be published.
42. B. Abeles, P. Sheng, M. D. Coutts, and Y. Arie, *Adv. Phys.* **24**, 407 (1975).
43. B. Abeles, in *Applied Solid State Science: Advances in Materials and Device Research*, Volume 6, R. Wolfe, ed. (Academic, New York, 1976) p. 1.
44. I. Shao, P. M. Vereecken, R. C. Cammarata, and P. C. Searson, *J. Electrochem. Soc.* **149**, C610 (2002).
45. A. L. Greer, *Materials Sci. Eng.* A **304–306**, 68 (2001).
46. H. Zeng, J. Li, J. P. Liu, Z. L. Whang, and S. Sun, *Nature* **420**, 395 (2002).
47. M. Sternitzke, *J. Eur. Ceram. Soc.* **17**, 1061 (1997).
48. C. E. Borsa and R. J. Brook, in *Ceramic Transactions, Volume 51, Ceramic Processing and Science*, H. Hausner, G. L. Messing, and S-I. Hirano (American Ceramic Society, Westerville, OH, 1995) p. 653.
49. Y. Xu, A. Nakahira, and K. Niihara, *J. Ceram. Soc. Jpn.*, **102**, 312 (1994).
50. R. J. Conder, C. B. Ponton, and P. M. Marquis, *BR. Ceramic Proc.* **51**, 105 (1993).
51. R. S. Haaland, B. I. Lee, and S. Y. Park, *Ceram. Eng. Sci. Proc.* **8**, 879 (1987).
52. R. A. Vaia and R. Krishnamoorti, in *Polymer Nanocomposites: Synthesis, Characterization, and Modeling*, R. Krishamoorti and R. A. Vaia, eds. (Oxford University Press, New York, 2001) p. 1.
53. R. Gangopadhyay and A. De, *Chem. Mater.* **12**, 608 (2000).
54. E. Vasilu, C.-S. Wang, and R. A. Vaia, *Mat. Res. Soc. Symp. Proc.* **703**, 243 (2002).
55. J. Collister, in *Polymer Nanocomposites: Synthesis, Characterization, and Modeling*, R. Krishamoorti and R. A. Vaia, eds. (Oxford University Press, New York, 2001) p. 7.
56. K. M. Unruh and C. L. Chien, in *Nanomaterials: Synthesis, Properties and Applications*, A. S. Edelstein and R. C. Cammarata, eds. (Institute of Physics, Bristol, 1998) Chapter 14.
57. S. Veprek, S. Reiprich, and L. Shizhi, *Appl. Phys. Lett.* **66**, 2640 (1995).

III

Nanoscale and Molecular Electronics

9

Advances in Microelectronics—From Microscale to Nanoscale Devices

Jan Van der Spiegel

Department of Electrical and Systems Engineering
University of Pennsylvania
Philadelphia, PA

9.1. INTRODUCTION

Microelectronic devices have evolved rapidly in terms of size, cost and performance. Scaling of device dimensions has been the engine of the semiconductor industry[1] allowing manufacturers to produce consecutive generations of integrated circuits of ever decreasing dimensions and increasing transistor densities. This trend has resulted in feature sizes with nanometer dimensions. Current physical gate lengths of transistors used in high performance integrated circuits are around 50 nm and will go down to 18 and 9 nm by 2010 and 2016, respectively, according to projections made in the 2003 International Technology Roadmap of Semiconductors (ITRS).[2] Prototype transistors with gate lengths as small as 15 nm have already been fabricated in research labs around the world.[3,4] Clearly, the microelectronics industry has entered the nanotechnology era and is manufacturing millions of nanoscale transistors on a phenomenal scale.

The semiconductor industry has produced one of the most sophisticated manufacturing processes known to mankind. The key productivity drivers have been the scaling of the transistors, the device switching speed, and the reduced cost per function. To get an appreciation of the magnitude it is instructive to mention that the number of transistors produced in 2002 in DRAMs alone exceeds the number of grains of rice produced yearly. In addition, for each grain of rice one can buy 100's of transistors.[5]

Microlectronic devices will play a key role in the future of nanoelectronics. Traditional microelectronic devices (CMOS transistors) are entering the nanoscale regime and are giving rise to extremely inexpensive but extraordinary powerful circuits and systems. Any new (nano) technology will have to reckon with this powerful force in order to become a viable alternative. Also, CMOS circuits and microelectronic technology will be used as a substrate on which to build future nanoelectronic structures. It is not unthinkable that a hybrid between microelectronic devices/technology and nanotechnology will give rise to powerful structures and systems. Certain novel nano devices, such as carbon nanotube transistors for instance, have the same structure as traditional CMOS transistors. Understanding the operation and the limitations of CMOS transistors is important to understand these new device structures.

The goal of this chapter is to review the basics of microelectronic devices (CMOS). We will discuss the structure and operation of CMOS transistors, the concept of scaling and its limitations. The challenges associated with nanoscale CMOS transistors will be reviewed. Finally, we will look at non-classical nanoscale CMOS devices and structures.

9.2. BRIEF HISTORY OF MICROLECTRONIC DEVICES AND TECHNOLOGY

The invention of the bipolar transistor in 1947 by J. Bardeen, W. Brattain, and W. Shockley at Bell Labs, was one of the milestones that made the microelectronics revolution possible. Although the first transistor was made of germanium material, silicon quickly became the material of choice since it could be easily grown as single crystal material. In addition, silicon has a high quality silicon oxide layer that can be used for insulating layers (e.g., gate oxide) and for passivating and masking purposes, which are key steps in the fabrication of a today's integrated circuit.

Another breakthrough was the introduction of the planar process at Fairchild Semiconductor. This process still forms the basis of the fabrication of current-day integrated circuits. It makes use of the masking properties of SiO_2 to define a region through which impurities can penetrate during the gas phase diffusion, as illustrated in Figure 9.1. Silicon was also superior over germanium in this respect due to its ability to form a stable oxide layer.

The next major invention occurred in 1959 when J. Kilby of Texas Instruments and R. Noyce of Fairchild Semiconductor introduced, independently, the concept of the

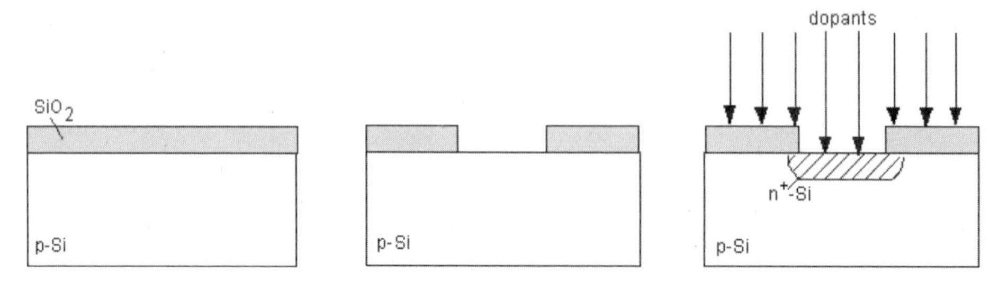

FIGURE 9.1. Use of silicon oxide as a masking layer during diffusion of dopants.

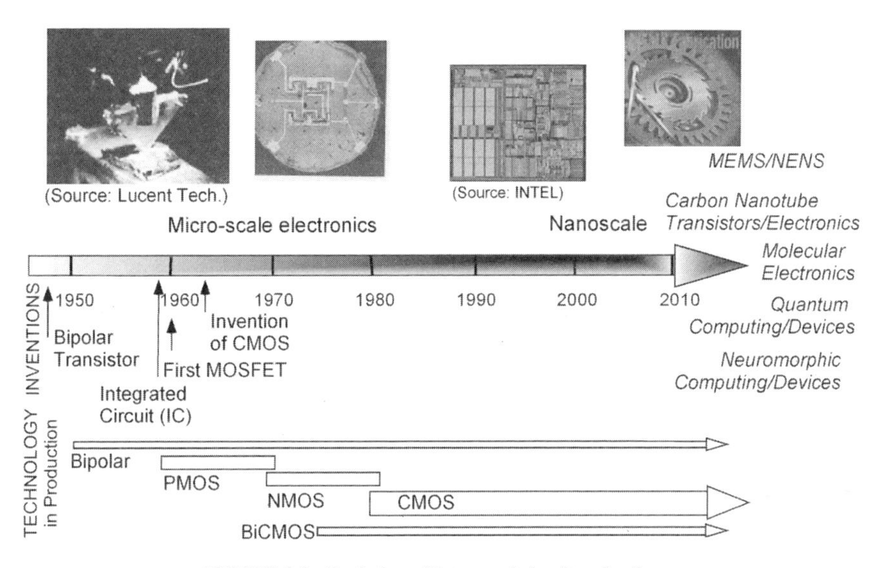

FIGURE 9.2. Evolution of integrated circuit technology.

integrated circuit. This allowed the fabrication of multiple devices and their interconnection on the same wafer. Subsequent improvements in materials, devices and the planar process have allowed the mass-fabrication of millions of transistors on a single chip. Figure 9.2 summarizes the evolution of the integrated circuit technology since the invention of the transistor.

Another milestone was reached after the successful fabrication of Metal Oxide Semiconductor (MOS) field effect transistors. The idea of the MOS transistor dates back to 1927 when Lilienfeld patented the field effect transistor, but it was not until the early 60's that one was able to overcome the technical difficulties associated with the fabrication of the MOS transistor.[6] One of the main challenges was the quality of the silicon-oxide interface and associated interface states and oxide charges. The advantage of the MOS transistor over a bipolar transistor is its simpler device structure, fewer fabrication steps, the absence of a DC input current and the suitability for mixed-mode circuits.

The first MOS transistors were p-channel devices (PMOS) in which the current consists of positive charge carriers. The introduction of the NMOS transistor was another step forward since electrons in a NMOS transistor move faster than the holes in a PMOS transistor. The combination of both NMOS and PMOS transistors on one substrate gave rise to the Complementary MOS or CMOS. The main advantage of CMOS over NMOS is that CMOS gates do not consume standby power, except for a small leakage current (which can become significant in sub-micron and nanoscale transistors). Advanced CMOS provides high performance (speed) at relatively low power, which has made it the dominant technology of choice for the fabrication of large-scale integrated circuits since the early 80's. Bipolar transistors are mainly used for high-speed, low-noise analog and microwave applications. The integration density of bipolar ICs is considerably lower than that of CMOS ICs.

Many subsequent improvements in the fabrication have allowed the realization of ultra-small feature size transistor. Shrinking transistor dimensions (scaling) combined with larger

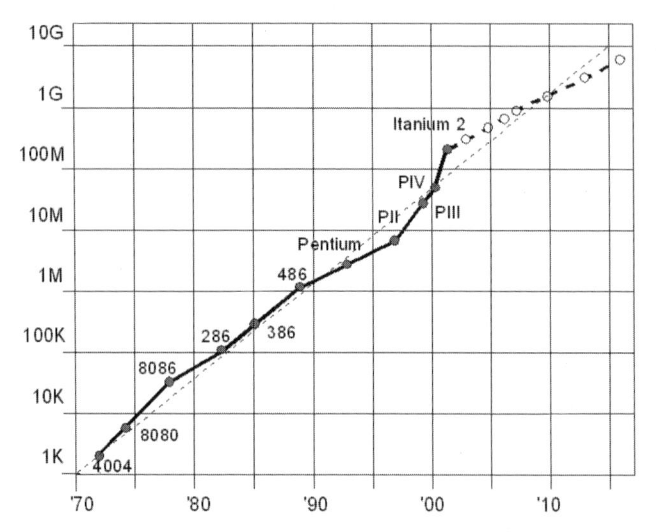

FIGURE 9.3. Moore's law (data points are for INTEL's microprocessors; the projections are based on the 2003 Technology Roadmap ITRS03).[2]

chip sizes has given rise to ever more complicated integrated circuits. Over the last 40 years, the transistor count per chip has doubled about every 18 months, as is illustrated in Figure 9.3. This trend was originally observed by G. Moore of Intel Corp and has been called Moore's law.[7,8] This has brought us from the SSI (small scale integration), to MSI (medium scale), to VLSI (very large scale), to ULSI (ultra large scale) integrated circuits. Current integrated circuits contain up to tens of millions of transistors, giving rise to sophisticated *systems* on a chip.

A different way to look at Moore's law is to plot the minimum feature size being used to fabricate integrated circuits. Figure 9.4 illustrates the trend over the last 40 years.

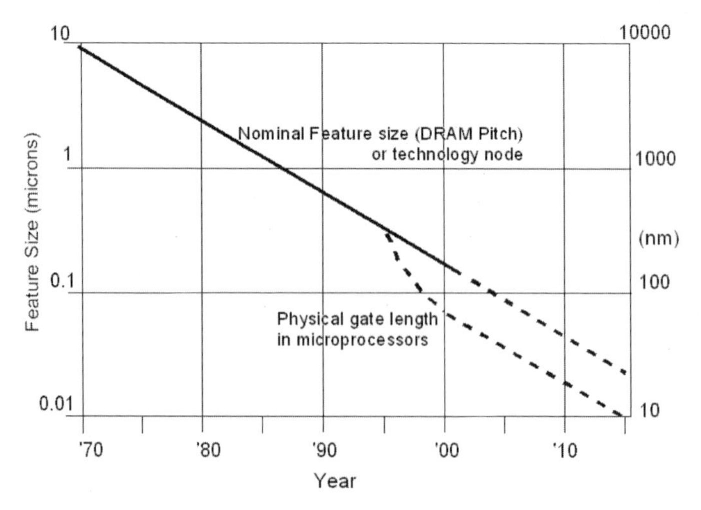

FIGURE 9.4. Trend of the minimum feature size (2003 Technology Roadmap ITRS03).[2]

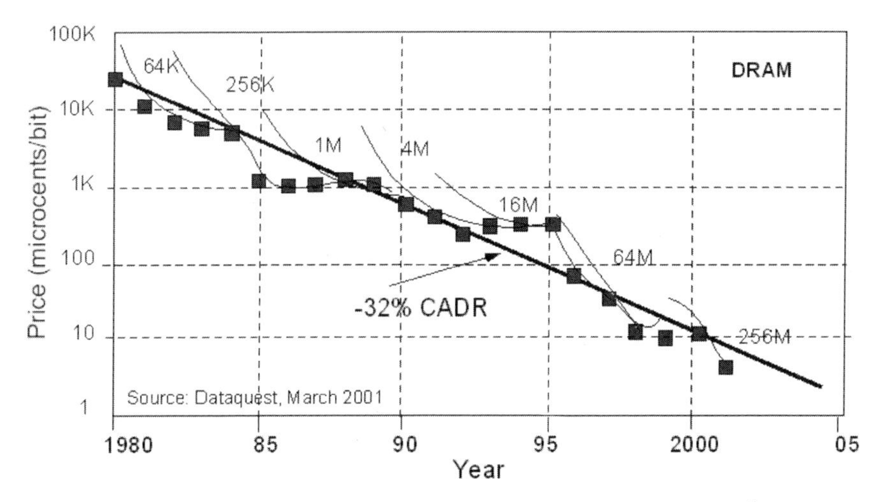

FIGURE 9.5. Average selling price per bit of DRAM memory since 1980.[5]

Dimensions have been shrinking by 12–14% per year. As can be seen, continued scaling at this pace will eventually lead to devices on a molecular and atomic scale that operate at very different principles than those we are using today.

The reduced dimensions, increased chip and wafer size, and improvement in process technology have allowed the cost per function to go down by 25–32% per year. Figure 9.5 illustrates this trend for the average selling price per bit of DRAM memory which has been decreasing at a cumulative annual rate of 32%. This remarkable trend has been an important driving force behind the semiconductor industry.

Figure 9.6 shows schematically the breakdown of the factors contributing to the cost reduction. About 12–14% of the cost reduction has become possible due to shrinking of minimum feature size, 4% is due to increased wafer size, 2% is the result of yield improvement, and another 7–19% is due to other innovations in technology and design efficiency. Feature size reduction accounts for almost 50% of the cost reduction.

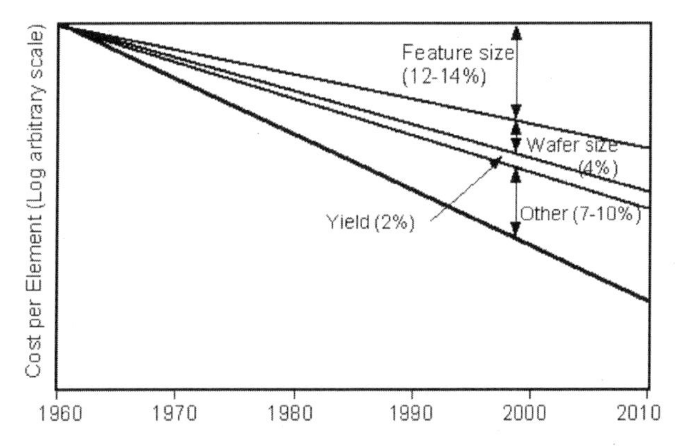

FIGURE 9.6. Trend of the cost per element over the years indicating the main contributing factors to the cost reduction.[9]

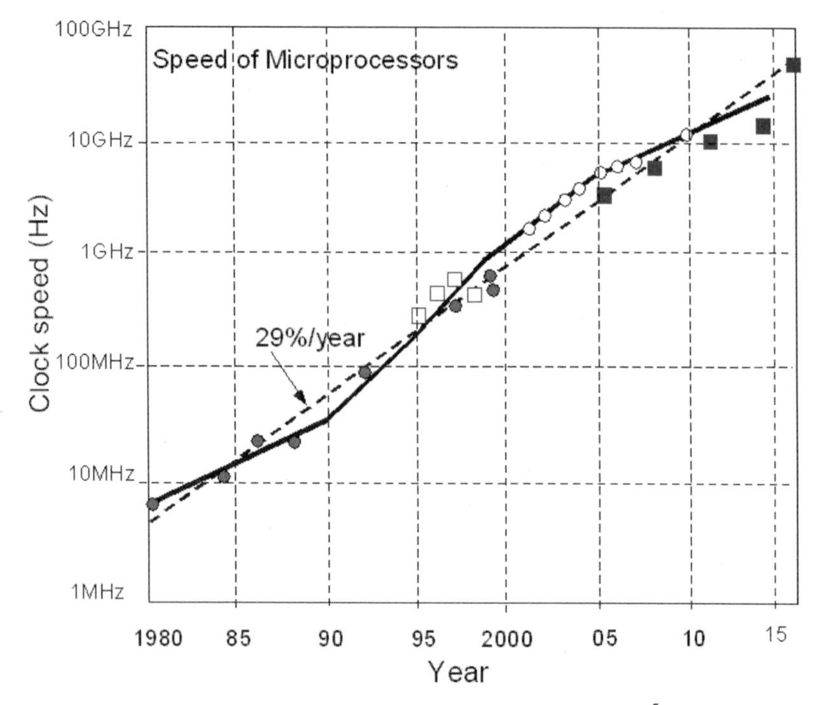

FIGURE 9.7. Clock speed of microprocessors since 1980.[5]

At the same time the cost per function has been decreasing, the performance of devices has increased exponentially. This is illustrated by the clock frequency of microprocessors that have been increasing at an average rate of 29% per year, shown in Fig. 7.

9.3. BASICS OF SEMICONDUCTORS

9.3.1. Semiconductor Model and Energy Band Structure

A majority of microelectronic devices and circuits are fabricated with silicon (Si). Devices for specialized applications are sometimes built with germanium (Ge), gallium arsenide (GaAs) and other III–V or II–VI compounds. As can be seen from the periodic table in Figure 9.8, these atoms belong either to column IV (elemental semiconductors of Si, and Ge), or columns III and IV (compound semiconductors of Ga and As). The atoms in column IV have the common property that they have 4 electronics in their outer shell, while the atoms in columns III and V have 3 and 5 electrons, respectively. The electrons in the outer orbit are the valence electrons and determine to a large extent the chemical and electrical properties of the material.

We will use silicon as an example to discuss the properties of semiconductors. Silicon has a total of 14 electrons, as illustrated schematically in Figure 9.9.[10,11,12] These electrons circle around the atom nucleus. The closer the electrons are to the nucleus of the atom the larger the attractive force will be and the tighter they are bound to the nucleus. This

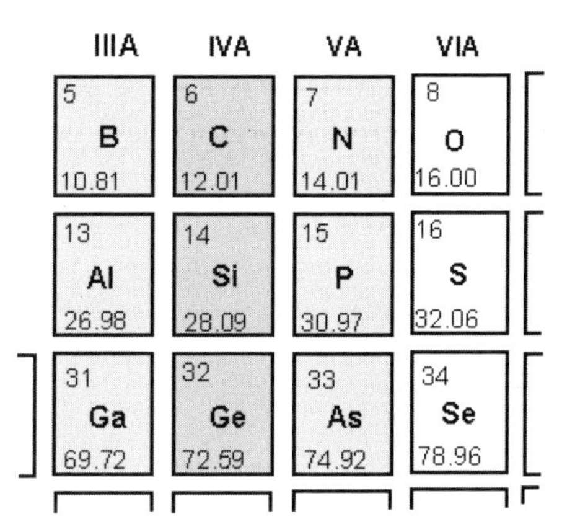

FIGURE 9.8. Part of the periodic table showing the III, IV, and V columns. The elemental semiconductors Si and Ge belong to column IV.

can be conveniently expressed by the amount of energy of each electron. In contrast to the macroscopic word, only certain energy levels are allowed. It was Niels Bohr who predicted in 1913 that the electrons are restricted to certain orbits or that the energy of the electrons is quantized. This is the result of the quantum mechanical nature encountered in systems of atomic scale.

We notice that in Figure 9.9 only two electrons occupy the same energy levels. This is the result of Pauli's exclusion principle which states that each energy level can accommodate only two electrons, corresponding to two energy states. The outer shell (valence shell) has 4 energy levels of which the lowest ones are occupied by two valence electrons each. A similar picture is valid for the other atoms in column IV of the periodic table since all these atoms have 4 valence electrons. The energy quantization and Pauli's exclusion principle have important consequences for the electronic structure of materials as will become clear shortly.

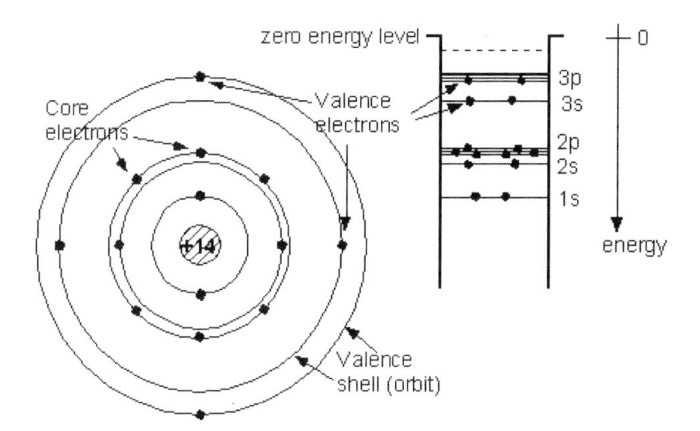

FIGURE 9.9. Schematic representation of the orbital model of a single silicon atom, with its 14 electrons of which the 4 valence electrons reside in the outermost shell. The energy levels are shown to the right (not to scale).

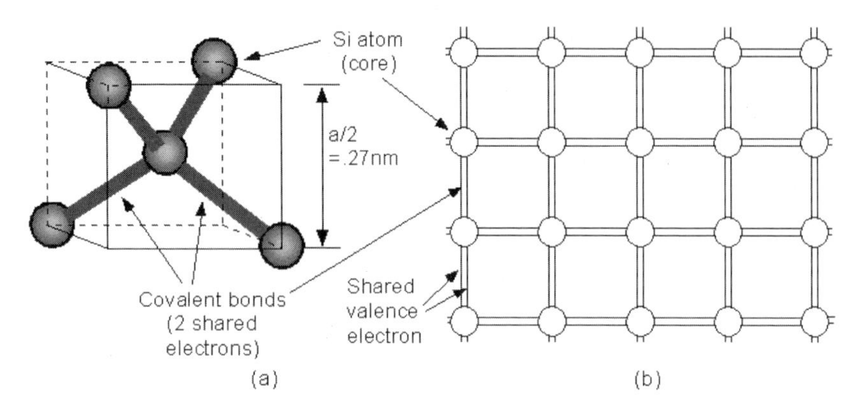

FIGURE 9.10. (a) Tetrahedral bonding of silicon in the diamond structure showing the four nearest neighbors connected through covalent bonds; (b) 2-D bond model illustrating the sharing of the valence electrons between nearest-neighbor silicon atoms. Each line between the silicon core represents one valence electron.

Let us now consider a silicon crystal that consists of many silicon atoms. Silicon material can be in the form a single crystal, poly-crystalline or amorphous. For the fabrication of semiconductor devices, single crystal, high purity semiconductors are used in which each atom occupies a precise location. Silicon crystal has a diamond lattice in which each silicon atom has four neighbors as shown in Figure 9.10. In this configuration each silicon atom shares a valence electron with its four neighbors. As a result of the electron sharing, the eight allowed energy states in the outer shell of the silicon atom are filled with electrons. The neighboring atoms form covalent bonds through the sharing of its valence electrons which gives rise to a stable structure. By repeating the structure of Figure 9.10a, one can build up the silicon crystal. This is schematically illustrated by the two-dimensional bond model of Figure 9.10b.

As a result of the close proximity of silicon atoms, the energy states of the atoms in a crystal will change slightly from those of a single atom. Let us assume that there are N silicon atoms, which will give rise to 4N valence electrons, since each atom in the crystal contributes four valence electrons. According to Pauli's exclusion principle, only two electrons can occupy the same energy level. In order to accommodate the 4N electrons, the energy levels of the single atom have to change slightly so that there will be a total of 4N energy levels spread out over two energy bands, illustrated in Figure 9.11. The bottom energy band contains 2N closely spaced energy levels and the top band contains another 2N energy levels. Since there are 4N valence electrons, the bottom energy band will be able to accommodate all the electrons. As a result, at a temperature of 0 K, the lower band will be completely filled and the top one will be empty. We call the lower energy band the valence band and the upper one the conduction band. The energy difference between the two bands is called the band gap E_G. The band gap in silicon is 1.12 eV at room temperature.

9.3.2. Charge Carriers in Semiconductors

The model of Figure 9.10b indicates that all the valence electrons are used to form the covalent bonds between neighboring atoms and as a result, there are no free electrons

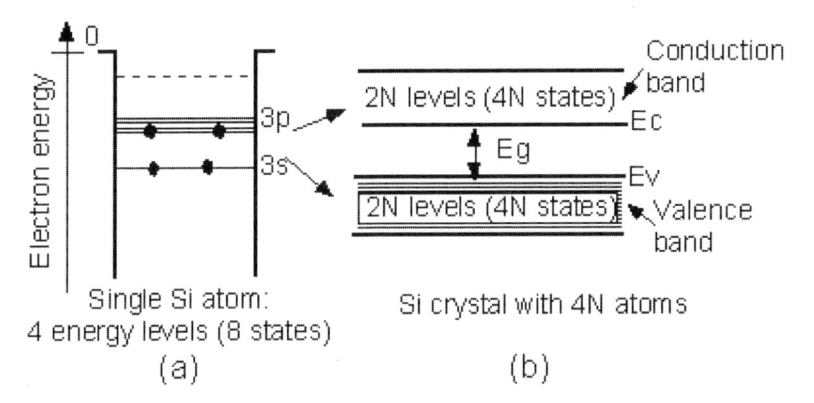

FIGURE 9.11. (a) Energy levels in an isolated silicon atom and (b) in a silicon crystal of N atoms, illustrating the formation of energy bands. The valence band contains 4N states and can accommodate all 4N valence electrons.

available that can give rise to conduction. This is confirmed by the energy band diagram of Figure 9.11b in which the valence band is completely filled with electrons, leaving no room for the electrons to move around. One can compare the situation of a filled valence band to a bottle completely filled with grains of salt. It is going to be very hard to move the grains around, even after shaking it, since there is no space for the grains to move. For the electrons to move there should be some empty energy levels to which they can jump. The above situation is true only at zero absolute temperature. However, at room temperature some electrons will have sufficient thermal energy to break away from the silicon atom. This occurs when the electron jumps across the band gap, as illustrated in Figure 9.12. Once these electrons are in the conduction band, they encounter plenty of empty energy states which make it easy to move around. The higher the temperature the more electrons will have crossed the band gap. We call these *mobile* electrons since they give rise to conduction in the semiconductor. What is interesting is that for each electron that jumps from the valence band to the conduction band, there will be an empty state created in the valence band. Thus, the electrons in the valence band will be able to move around as well. Since

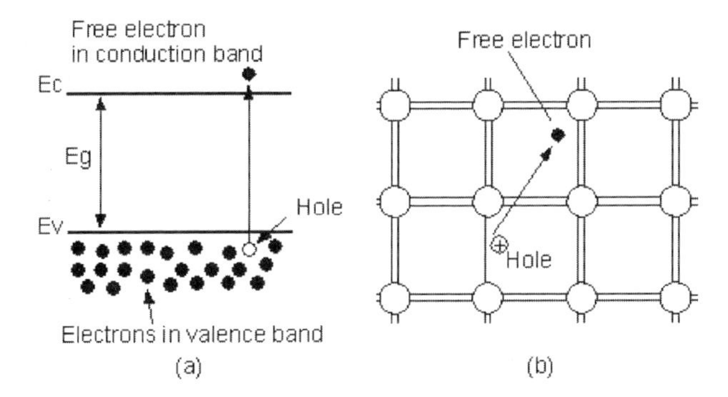

FIGURE 9.12. A valence electron jumping across the energy gap in pure silicon resulting in the generation of a free electron and hole in the crystal: (a) energy band model, (b) bond model.

the conduction in the valence band is the result of the removal of an electron, which leaves a positive charge behind, we call these carriers conveniently *holes*. Thus, the conduction in a pure semiconductor is the result of both electrons and holes. Since each electron that jumps across the band gap creates a hole, the concentration of the mobile electrons, n, and holes, called p, is equal. We call this concentration the intrinsic carriers concentration n_i.

It is clear that the width of the band gap is an important characteristic of a semiconductor and of materials in general. For instance, it allows us to explain the difference between insulators, conductors and semiconductors. An insulator is a material whose valence band is completely filled and whose energy gap is so large that no electrons can jump across it. An example is SiO_2 whose large energy gap of 8–9 eV makes it an excellent insulator. On the other hand, a conductor is a material whose valence band is only partially filled so the electrons can easily jump to the empty states in the valence band and move around freely. A semiconductor acts as an insulator at zero temperature and becomes a poor conductor at higher temperature since the energy gap is small enough for some electrons to free themselves by jumping across the energy gap at room temperature.

For silicon, the fraction of valence electrons that jump across the energy gap is less than 10^{-12}. There are 5×10^{22} silicon atoms per cubic centimeter so the total amount of mobile electrons and holes n_i in a pure silicon semiconductor is about 1.4×10^{10} cm^{-3} at room temperature (T = 300 K). Since the thermal energy allows the electrons to jump across the band gap, it is not surprising the intrinsic career concentration n_i is a strong function of the temperature, as expressed by the Eq. (9.1) below[11]

$$n_i = 3.1 \times 10^{16}\, T^{1.5} \times e^{-0.603\ \mathrm{eV}/kT}\ \mathrm{cm}^{-3} \tag{9.1}$$

in which T is the absolute temperature and k the Boltzmann constant. For pure semiconductors one can write that

$$n = p = n_i \quad \text{and} \quad \text{thus also} \quad np = n_i^2. \tag{9.2}$$

The last expression will be proven later on.

9.3.3. Intrinsic and Extrinsic Semiconductors

The semiconductor material we discussed in the previous section is called intrinsic since it consists of pure semiconductor material without added dopants. One of the key properties of semiconductors is that one can modify its characteristics by adding dopants or impurities. Adding relatively tiny amounts of dopants has a very strong effect on the electrical properties of semiconductors. These impurities have one more or one fewer electron than silicon in their outermost shell. Looking at the periodic table in Figure 9.8 one notices that phosphorous P sits next to silicon and is located in column V, indicating that it has five valence electrons. One can now dope a silicon crystal with phosphorous atoms using a process called ion implantation or diffusion. Since P has a similar size as silicon, it is relatively easy to replace some of the silicon atoms with phosphorous ones, as shown in Figure 9.13.

Since only four of the five valence electrons are needed for the valence bonds to fill the outer shell around the silicon atom, the fifth electron supplied by the phosphorous atom is very weakly bound to its atom core. Actually, at room temperature it has enough energy

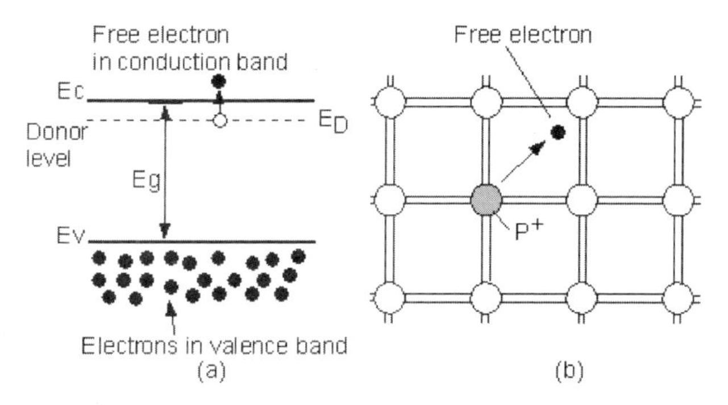

FIGURE 9.13. Extrinsic n-type silicon doped with P donor atoms. (a) Energy band diagram and (b) Bond model.

to break away and move freely inside the silicon crystal, giving rise to conduction. In the energy band model of Figure 9.13a, the energy level of the valence electron associated with the phosphorous atom is very near the energy of the bottom of the conduction band E_C. As a result, the electron will jump into the conduction band leaving behind a fixed positive phosphorous ion.

At room temperature, each added phosphorous atom will donate one mobile electron. For that reason we call the phosphorous atoms *donors*. The concentration of the mobile electron concentration n will be equal to the sum of the concentration of the donor atoms N_D and the intrinsic concentration n_i.

$$n = N_D + n_i. \tag{9.3}$$

A typical doping concentration N_D is 10^{15} cm^{-3} or higher so that at room temperature, $n \cong N_D$. Notice that the hole concentration is no longer equal to the electron concentration since the donor only contributes a free electron and no hole, resulting in a larger electron concentration than hole concentration, $n > p$. We call this an n-type semiconductor in which electrons are called *majority* carriers and holes are called *minority* carriers. Other atoms, such as As (arsenic) and Sb (antimony) can be used as donors.

By doping silicon with an atom from column III in the periodic table, such as boron, we can obtain p-type silicon. Since boron has only three valence electrons it will try to capture one electron so that the silicon atoms have 8 electrons to fill its outer shell. This results in the generation of a free hole, as schematically illustrated in Figure 9.14.

Each boron atom contributes an energy level E_A close to the top of the valence band E_V. This allows an electron to jump easily from the valence band to the acceptor level, leaving behind a free hole and creating a negative B ion. Since boron accepts an electron, we call them *acceptors*. At room temperature, all the acceptor energy levels will be filled with electrons from the valence band, contributing a number of holes equal to the number of acceptors. As a result, at room temperature the hole concentration p will be equal to

$$p = N_A + n_i. \tag{9.4}$$

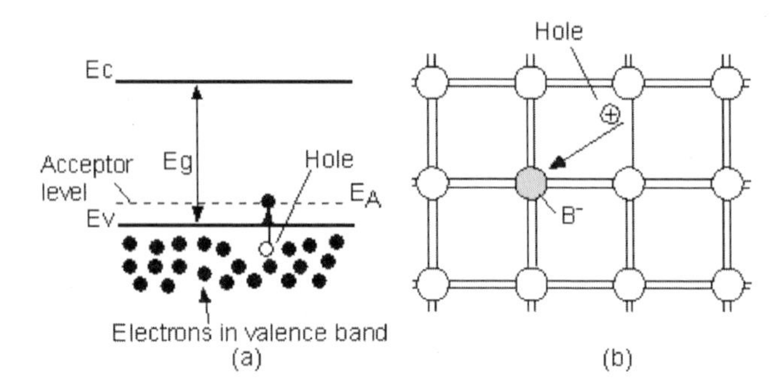

FIGURE 9.14. Extrinsic p-type silicon doped with B acceptor atoms. (a) Energy band diagram and (b) Bond model.

For typical acceptor concentration used in semiconductors $N_A > n_i$ at room temperature making $p \cong N_A$. As will be seen later on, the product of the electron and hole concentration np at equilibrium is equal to,

$$np = n_i^2. \tag{9.5}$$

So far we have qualitatively described that a certain amount of electrons will jump across the bandgap and fill up states in the conduction band, leaving states empty in the valence band, giving rise to mobile electrons and holes. In order to quantitatively describe the number of electronics or holes, we introduce an important function called the Fermi function $F(E)$. The Fermi function gives the probability that an energy state E will be occupied by an electron. Assuming discrete energy levels in which Paul's exclusion principle allows only two electrons per energy level, the function $F(E)$ is given by,

$$F(E) = \frac{1}{1 + e^{(E-E_F)/kT}} \tag{9.6}$$

in which E_F is the Fermi level that is defined as the energy level at which the probability of finding an electron is equal to 0.5. Notice that when E is much larger than the Fermi energy E_F, the function $F(E) \cong 0$, and when E is much smaller than E_F, the function $F(E) \cong 1$. Assuming that we know the density of states $N(E)$ in the conduction and valence bands, we can then use the Fermi function to calculate the concentration of the electron and holes in the semiconductor.[13] When the energy E of the state is a few times kT larger than the Fermi energy E_F, the function $F(E)$ can be approximated as,

$$F(E) \cong e^{-(E-E_F)/kT} \tag{9.7}$$

We call this the Boltzmann approximation. This allows us to write the electron and hole concentration as follows.

$$n = \int_{E_C}^{\infty} F(E)N(E)dE \cong N_C e^{-(E_C-E_F)/kT} = n_i e^{(E_F-E_i)/kT} \tag{9.8}$$

$$p = \int_{-\infty}^{E_V} [1 - F(E)]N(E)dE \cong N_V e^{-(E_F-E_V)/kT} = n_i e^{(E_i-E_F)/kT} \tag{9.9}$$

in which N_C and N_V are the effective densities of states in the conduction and valence bands, respectively, and E_i is the intrinsic energy level corresponding to the Fermi level in an intrinsic semiconductor. From Eqs. (9.8) and (9.9) one can easily prove that the product $np = n_i^2$. For n-type semiconductors the Fermi level will lie above the middle of the energy band gap E_i and for p-type semiconductor E_F will lie in the bottom half of the band gap.

The conductivity σ of the semiconductor can be expressed as a function of the electron and hole concentrations and carrier mobility.

$$\sigma = 1/\rho = q\mu_n n + q\mu_p p \tag{9.10}$$

in which ρ is the resistivity, q the electron charge, and μ_n and μ_p the electron and hole mobility, respectively. By adding dopants one can change the conduction over a very large range. This is a key property of semiconductors that will become important for the operation of semiconductor devices as discussed later on. The mobility of the electrons is typically 2–2.5 times higher than that of the holes. This will imply that devices in which electrons cause the current will be faster than devices in which holes are responsible for the current flow. The mobility is a function of both doping levels and temperature. The higher the temperature and doping concentration, the lower the mobility will become due to increased scattering of the charge carriers with the atoms.

An interesting aspect is the effect of elastic strain on the mobility of mobile charges.[14] This phenomenon has recently been used to improve the performance of field-effect transistors. By implanting germanium in the silicon, the lattice of the $Si_{1-x}Ge_x$ alloy will expand in the plane of the surface as compared to that of the pure silicon crystal. By growing a thin single crystal film (epitaxial layer) on top of the $Si_{1-x}Ge_x$ ($x = 0.25$–0.30) alloy, the silicon lattice expands up to 1.2% which is sufficient to increase the mobility considerably. It has been reported that electron mobility enhancement up to 110% and hole mobility enhancement up to 45% has been achieved in n and p-type MOS transistors.[15]

9.4. STRUCTURE AND OPERATION OF A MOS TRANSISTOR

The metal-oxide-semiconductor field effect transistor (MOSFET) has been the workhorse of the semiconductor industry. The basic structure is conceptually simple and is the reason why the overwhelming majority of integrated circuits are fabricated with MOS transistors. The active element of a MOS transistor consists of a MOS capacitor, whose operation is briefly described in the next section.

9.4.1. MOS Capacitor

A schematic cross section of a MOS capacitor is shown in Figure 9.15a. The top electrode consists of a conductor that can be a metal such as aluminum or can be doped polysilicon or a silicide. The insulator has traditionally been SiO_2 but could also be Si_3N_4, oxinitride, or a high-dielectric material. The bottom electrode consists of a semiconductor. It is the semiconductor that makes this MOS capacitor different from a traditional parallel plate capacitor. The corresponding energy band diagram across the metal-oxide-silicon structure is shown in Figure 9.15b. In this diagram the electron energy is positive upwards and the potential is positive downwards. Notice that the Fermi level E_F lies in the bottom

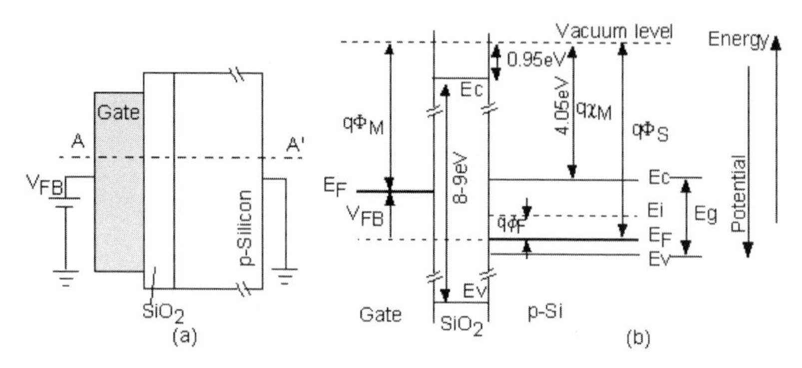

FIGURE 9.15. Schematic cross-sectional view of a MOS capacitor; (b) Energy band diagram (not to scale) of the MOS capacitor under flat band conditions along the cross section AA′.

half of the band gap for p-type silicon. The symbol ϕ_F is sometimes called the Fermi potential and is defined as the voltage difference between the intrinsic and the Fermi levels. This voltage can be found from Eqs. (9.8) or (9.9) knowing that $p \cong N_A$,

$$\phi_F = \pm(E_i - E_F)/q = \frac{kT}{q}\ln\left(\frac{N_B}{n_i}\right). \tag{9.11}$$

in which N_B is the doping concentration of the substrate. For p-type material, $N_B = N_A$, and for n-type, $N_B = N_D$. The Fermi potential is positive for p-type and negative for n-type silicon. For a substrate doping level $N_A = 10^{17}$ cm^{-3}, the Fermi potential will be equal to +0.41 V at room temperature.

As can be seen from Figure 9.15b, the band gap of silicon dioxide is much larger (8–9 eV) than that of silicon (1.12 eV) as we would expect from a good insulator. The difference between the vacuum level and the Fermi level is called the work function. The value of silicon workfunction $q\Phi_S$ can be found from Figure 9.15b and written as,

$$\Phi_S = \chi + \frac{E_G}{2q} + \phi_F \tag{9.12}$$

in which the quantity $q\chi$ is called the electron affinity which corresponds to the energy that an electron at the bottom of the conduction band must acquire to break loose from the crystal. The workfunction of the metal gate is called $q\Phi_M$. For an aluminum gate the value of the workfunction is 4.1 eV.

To explain the behavior of the MOS capacitor, we will vary the voltage V_G over the capacitor from a negative value to a positive one. A negative gate voltage will induce an electric field over the insulator that will attract positive charges at the semiconductor-insulator interface. These are the majority carriers, i.e., holes, which will accumulate at the surface. When one makes the voltage V_G slightly positive, the induced electric field will repel the holes from the surface region which results in the creation of a depletion region near the surface, as schematically shown in Figure 9.16. It is convenient to look at the corresponding energy band diagram near the silicon surface. The width x_d of the depletion

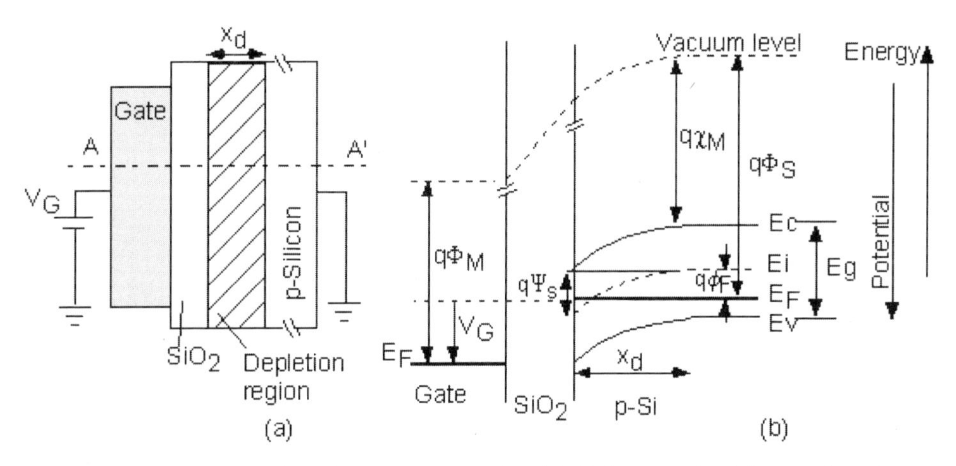

FIGURE 9.16. MOS capacitor in depletion: (a) cross section and (b) the energy band diagram along the cross section AA′.

region can be found by solving the Poisson's equation and is given by[16]

$$x_d = \sqrt{\frac{2\varepsilon_{\text{Si}}\,|\psi_S|}{qN_B}} \tag{9.13}$$

in which ε_{Si} is the dielectric constant of silicon ($= 1.04 \times 10^{-12}$ F/cm), ψ_S is the surface potential defined as the energy band bending, as indicated in Figure 9.16b, and N_B is the doping concentration of the substrate. Notice that the depletion width narrows when the doping level increases.

When one keeps increasing the gate potential, the energy diagram bends further downward. At a certain point the surface potential ψ_S will become equal to $2\phi_F$. From Eqs. (9.8) and (9.9) one sees that at this point the concentration of electrons at the surface will be equal to the concentration of the holes in the substrate. We call this point the onset of inversion. The corresponding gate voltage is called the threshold voltage V_T. Increasing the gate voltage beyond the threshold voltage will result in a fast increase of the electrons in the inversion layer without a significant increase in the surface potential due to the exponential dependency of the electron concentration on the surface potential according to Eq. (9.8). We can assume that once inversion has been reached, the value of the surface potential ψ_S will remain equal to $2\phi_F$. This implies that the depletion width has reached its maximum value equal to,

$$x_{d\,\max} = \sqrt{\frac{2\varepsilon_{\text{Si}}.2\,|\phi_F|}{qN_B}} = \sqrt{\frac{4\varepsilon_{\text{Si}}kT\,\ln(N_B/n_i)}{q^2N_B}} \tag{9.14}$$

The charge per unit area in the depletion region is given by,

$$Q_D = \mp qN_Bx_d \tag{9.15}$$

The minus sign is for p-type material and the plus sign is for n-type since the ionized acceptors and donors are negative and positive, respectively. One of the important technological parameters of a MOS capacitor and transistor is the threshold voltage V_T. The threshold voltage is equal to the gate voltage at the point that strong inversion is reached. We can write the gate voltage V_G using the diagram of Figure 9.16b.

$$V_G = \Phi_M + V_{\text{ox}} + \psi_S - \Phi_S \tag{9.16}$$

The voltage over the insulator V_{ox} is a function of the charge Q_D in the depletion layer and oxide charge Q_{ox}. The oxide charge is the result of impurities and defects in the insulator. We will assume that Q_{ox} is the equivalent charge located at the insulator-silicon interface. These charges are the result of the non-ideality of the oxide and often play a major role in the operation and reliability of MOS devices. The oxide charges can be grouped into mobile charges, oxide-trapped charges, fixed oxide charges and interface-trapped charges. High quality SiO_2 layers should have a defect charge density in the range of 10^{10} cm^2 or less. At the onset of inversion the surface potential ψ_S is equal to $2\phi_F$. This allows us to write the threshold voltage V_T as follows,

$$\begin{aligned} V_T &= \Phi_M - \Phi_S + 2\phi_F - Q_{\text{ox}}/C_{\text{ox}} - Q_D/C_{\text{ox}} \\ &= \Phi_{MS} - Q_{\text{ox}}/C_{\text{ox}} + 2\phi_F - Q_D/C_{\text{ox}} \\ &= V_{FB} + 2\phi_F - Q_D/C_{\text{ox}} \end{aligned} \tag{9.17}$$

in which Φ_{MS} is the work function difference between the gate and substrate material, V_{FB} is called the flat band voltage. Control over the threshold voltage will be very important to ensure proper operation. Notice that the threshold voltage is a function of the doping level through the charge in the depletion region Q_D and, to a lesser extent, through the Fermi potential $2\phi_F$. Also, the gate material plays a role through the workfunction Φ_M. Typical gate materials used are highly doped n-type and p-type polysilicon in NMOS and PMOS devices, respectively. The threshold voltage can be adjusted by ion implantation near the silicon-oxide interface. This can be modeled by adding a term, $Q_{\text{Impl}}/C_{\text{ox}}$ in the expression for the threshold voltage. Applying a gate voltage V_G larger than the threshold voltage will result in a build up of minority carriers at the surface. The charge in the inversion layer per unit area can then be written as,

$$Q_n = -C_{\text{ox}}(V_G - V_T) \tag{9.18}$$

9.4.2. MOS Transistor

The metal-oxide-semiconductor field-effect transistor consists of a MOS capacitor as described above and two adjacent diodes called source and drain. The structure is schematically shown in Figure 9.17. It has four terminals: the gate, source, drain and substrate terminal. The region underneath the MOS gate is called the channel of the transistor.

9.4.2.1. Long Channel Transistor and I–V Characteristics in Strong Inversion As we discussed in the previous section, the voltage applied to the gate terminal of a MOS

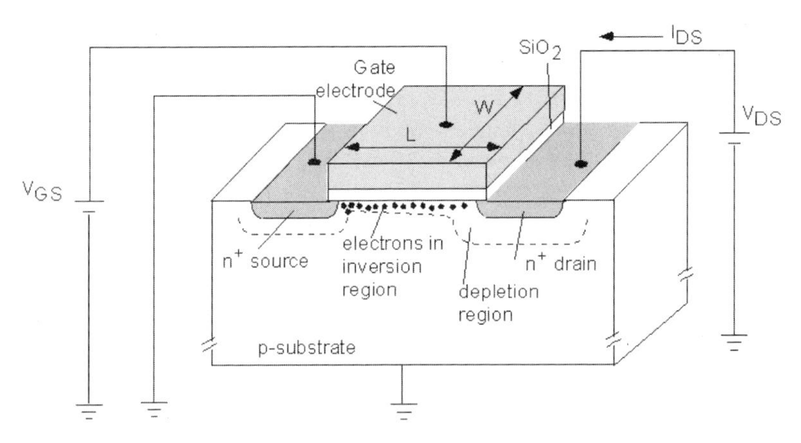

FIGURE 9.17. Schematic view of a n-type MOSFET.

capacitor determines the amount of mobile charges in the inversion layer. When using a p-type substrate, the inversion charge, also called the channel charge, will consist of electrons. We call this a NMOS transistor. Similarly, a p-type transistor (PMOS) is built from an n-type substrate with holes as the charges in the inversion layer. The operation of a transistor can be explained as follows. Let us assume that we connect both the substrate and the source to the ground terminal, as shown in Figure 9.17. When we apply a gate voltage V_{GS} between the gate and substrate terminals which is larger than the threshold voltage V_T, an inversion layer will be created. In contrast to the MOS capacitor where the charges in the inversion layer are supplied from the substrate through thermal generation of minority carriers, the electrons in the channel region in a transistor will be supplied by the adjacent n^+ source. The source-substrate diode will be slightly forward biased at the surface near the channel region. Since there are plenty of electrons in the source, the build-up of the inversion layer will be very fast. The amount of charges is given by expression (9.18), repeated below, in which V_{GS} is the voltage between the gate and source terminals.

$$Q_n = -C_{\text{ox}}(V_{GS} - V_T) \tag{9.19}$$

If we now apply a positive voltage V_{DS} at the drain terminal, the electrons underneath the gate will flow towards the positive drain terminal. For each electron that is taken away from underneath the gate region, another will be supplied by the source. This causes current to flow from drain to source (electrons flow from source to drain), shown in Figure 9.18. For small V_{DS} voltages, we can consider the channel region as a linear resistor whose resistance is a function of the amount of mobile charges in the channel and thus linearly function of the gate to source voltage V_{GS} is expressed in Eq. (9.20).

$$I_{DS} = \mu_n C_{\text{ox}} \frac{W}{L}(V_{GS} - V_T)V_{DS} \quad (V_{GS} \gg V_{DS}) \tag{9.20}$$

in which μ_n is the effective electron mobility at the surface, C_{ox} the oxide capacitor per unit area, and W/L the width to length ratio of the channel region (Figure 9.17).

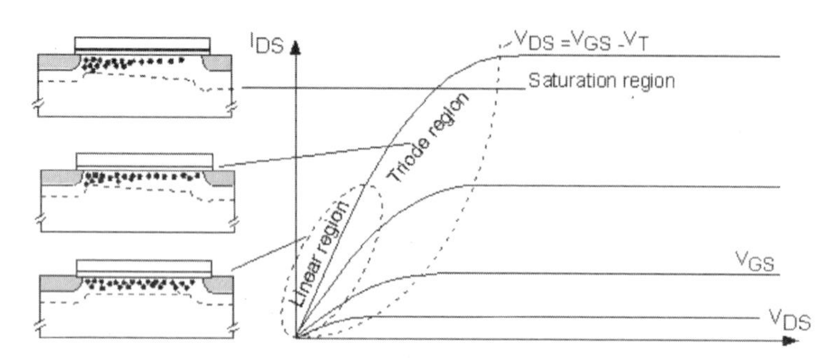

FIGURE 9.18. Current flow in a NMOS transistor, illustrating the three operating regions (linear, triode and saturation).

If we continue our experiments and keep increasing the drain voltage, the concentration of electrons in the channel region will not be uniform anymore. This can be easily understood as follows. Close to the source region the voltage V_{GS} determines the charges while near the drain region and the voltage V_{GD} determines the amount of charges in the channel. Since $V_{GD} = V_{GS} - V_{DS} < V_{GS}$, the amount of charges near the drain end of the channel will have decreased as compared to those near the source. This will cause the current to level off as a function of the drain voltage as shown in Figure 9.18. Assuming a gradual-channel approximation, a first order model of the current flow can be derived.[13,16] The current in the triode region can be expressed as,

$$I_{DS} = \mu_n C_{\text{ox}} \frac{W}{L} \left((V_{GS} - V_T) V_{DS} - m \frac{V_{DS}^2}{2} \right) \quad \text{for } V_{GS} - V_T > V_{DS} \tag{9.21}$$

The parameter m is called the body effect coefficient and can be expressed as

$$m = 1 + \frac{C_{d\max}}{C_{\text{ox}}} = 1 + \frac{\varepsilon_{\text{Si}} t_{\text{ox}}}{\varepsilon_{\text{ox}} x_{d\max}} = 1 + \frac{1}{C_{\text{ox}}} \sqrt{\frac{\varepsilon_{\text{Si}} q N_B}{4\,|\phi_F|}} \tag{9.22}$$

The value of m depends on the substrate doping level and the oxide thickness t_{ox}. One tries to keep m as close to 1 as possible. Typical values range between 1.1 and 1.5. If we continue to increase the drain voltage, we reach a point where the channel charge is reduced to zero near the drain. The charge along the channel can be expressed as,

$$Q_n = -C_{\text{ox}}(V_{GS} - V_T - mV) \tag{9.23}$$

in which the voltage V varies from $V = 0$ V near the source end and $V = V_{DS}$ near the drain end of the channel. By increasing the drain voltage, the inversion charge near the drain will become equal to zero when $V = V_{DS} = V_{DSat} = (V_{GS} - V_T)/m$. We say that the channel is pinched off and the transistor goes into saturation. The current becomes basically independent of the drain voltage in this region as shown in Figure 9.18. By substituting V_{DS} by $(V_{GS} - V_T)/m$ in Eq. (9.21), we find that the current in saturation is equal to,

$$I_{DS} = \mu_n C_{\rm ox} \frac{W}{L} \frac{(V_{GS} - V_T)^2}{2m} \quad \text{(Saturation region: } V_{DS} > (V_{GS} - V_T)/m) \tag{9.24}$$

When the gate length L is made small, the effect of the drain voltage will not be negligible anymore. Increasing the drain voltage will increase the depletion layer of the drain and cause a reduction of the effective gate length. Since the current is proportional to $W/L_{\rm eff}$, the current will increase with V_{DS}. This can be modeled by a channel length modulation parameter λ. The expression of the drain current can now be written as,

$$I_{DS} = \mu_n C_{\rm ox} \frac{W}{L} \frac{(V_{GS} - V_T)^2}{2m} (1 + \lambda V_{DS}) \tag{9.25}$$

The above current-voltage expressions are based on first-order models that are only valid for long channel transistors. For small dimension transistors other effects need to be taken into account. The model gets easily complicated whose parameters in many cases are based on fitting of measured transistor characteristics.

When we discussed the threshold voltage of a MOS capacitor we assumed that the substrate was shorted to ground. In the case of a MOS transistor, the substrate and source terminals can be at different potentials. As a result of the V_{SB} voltage between the source and substrate, the voltage over the depletion region during inversion will increase from $2\phi_F$ to $2\phi_F + V_{SB}$. This will affect the threshold voltage through the value of the depletion charge Q_D increases,

$$Q_D = \mp q N_B x_d = \mp \sqrt{2q\varepsilon_{\rm Si} N_B (2\,|\phi_F| \pm V_{SB})} \tag{9.26}$$

The threshold voltage V_T is equal to

$$V_T(V_{SB}) = {\rm V}_{FB} + 2\phi_F \pm \frac{\sqrt{2q\varepsilon_{\rm Si} N_B (2\,|\phi_F| \pm V_{SB})}}{C_{\rm ox}}$$

$$V_T(V_{SB}) = V_{VT}(0) + \gamma[\sqrt{(2\,|\phi_F| \pm V_{SB})} - \sqrt{2\,|\phi_F|}]$$

$$\gamma = \pm \frac{\sqrt{2q\varepsilon_{\rm Si} N_B}}{C_{\rm ox}} \tag{9.27}$$

in which the top signs apply for NMOS and the bottom signs for PMOS transistors. This effect is called the *body effect* or the substrate bias effect. The coefficient γ is related to the parameter m defined in Eq. (9.22). Notice that for a NMOS, the threshold voltage will become more positive with V_{SB} and for a PMOS, the threshold voltage becomes more negative for a more negative V_{SB}. The larger the body effect coefficient γ or the substrate doping concentration, the larger the effect of the substrate-source voltage will be. Since the body effect increases the threshold voltage, the smaller the value of γ the better.

9.4.2.2. Subthreshold Characteristics In the above discussion we have assumed that the drain current is cut-off when the gate voltage V_{GS} is less than or equal to the threshold voltage V_T. However, the transition between on and off state of the transistor is not that abrupt and there is a transition region, called *weak inversion*, when mobile minority carriers start to build up at the silicon-oxide interface. These mobile electrons (for a NMOS transistor) will give rise to a diffusion current. The concentration of the electrons in weak inversion is exponentially dependent on the surface potential, Ψ_S, according to Eq. (9.8). We can also expect that the drain current will be an exponential function of the gate voltage, as long as the transistor is in weak inversion. The expression for the current in weak inversion is,

$$I_{DS} = \mu_n C_{\text{ox}} \frac{W}{L} (m-1) \left(\frac{kT}{q}\right)^2 e^{\frac{q(V_G - V_T)}{mkT}} \left(1 - e^{\frac{-qV_{DS}}{kT}}\right) \quad (9.28)$$

For a drain-source voltage a few times larger than kT/q, the transistor is in saturation and the current-voltage relation becomes,

$$I_{DS} = \mu_n C_{\text{ox}} \frac{W}{L} (m-1) \left(\frac{kT}{2}\right)^2 e^{\frac{q(V_G - V_T)}{mkT}} = I_o e^{\frac{q(V_G - V_T)}{mkT}} \quad (9.29)$$

The subthreshold slope S is defined as the gate voltage required to change the drain current by one decade. From the above equation we find the slope to be

$$S = mkT/q.\ln 10 = 2.3\,mkT/q = 2.3\,kT(1 + C_d/C_{\text{ox}})/q \quad (9.30)$$

The subthreshold slope is typically 60 to 100 mV/decade, depending on the value of m. This is an important parameter of a transistor since it is a measure of how easy it is to switch off a transistor. The factor $(1 + C_d/C_{\text{ox}})$ indicates what fraction of the gate voltage V_{GS} is used to control the charge at the silicon-oxide interface. The smaller the slope, the more efficient the gate voltage will be in controlling the current flow. It should be noted that the actual value of the subthreshold slope can be somewhat larger than given by expression (9.30) due to the presence of interface states. The effect of these states can be modeled by a capacitor C_{it} in parallel with the depletion capacitor C_d.

9.5. SCALING OF TRANSISTOR DIMENSIONS

Over the years, transistor dimensions have been scaled down from about 50 micrometer in the early 1960s to nanoscale dimensions in today's technology. The main driving forces for reducing the dimensions have been increased circuit density per unit chip area, increased speed of operation, and lower cost per function. The result is that systems on a chip (SOC) have become a reality with today's technology.[17] When reducing transistor dimensions one needs to make sure that the characteristics do not change and that reliability does not become a problem. This was accomplished by a set of guidelines called *scaling*. Scaling refers to reducing both horizontal and vertical dimensions by the same factor. Let us assume that the scaling factor is called α. Besides scaling the gate length and width, we also need to scale the vertical dimensions such as the oxide thickness t_{ox} and the depletions width x_d. To scale x_d we can increase the substrate doping level N_B and/or reduce the voltage. Let us

TABLE 9.1. Transistor scaling assuming long channel device operation

	Multiplicative scaling factor		
Parameter	Constant Field	Constant Voltage	Generalized Scaling
Transistor dimensions (t_{ox}, W, L, x_j)	$1/\alpha$	$1/\alpha$	$1/\alpha$
Voltage	$1/\alpha$	1	k/α
Doping concentration (N_B)	α	α	$k\alpha$
Electric Field	1	α	k
Capacitance ($C = \varepsilon A/t$)	$1/\alpha$	$1/\alpha$	$1/\alpha$
Current (I)	$1/\alpha$	α	k^2/α
Current per gate width (I/W)	1	α^2	k^2
Delay (CV/I)	$1/\alpha$	$1/\alpha^2$	$1/k\alpha$
Power dissipation ($P = IV$)	$1/\alpha^2$	α	k^3/α^2
Power density (P/A)	1	α^3	k^3

assume that the voltage scales with the same factor as the dimensions. We call this *constant-field* scaling. According to the expression (9.13) of x_d the doping level should increase by a factor α to keep the depletion width constant, assuming that the voltage scales with the same factor. We can now find the effects of scaling on the current, propagation delay and the power consumption. The current will reduce by the factor α according to Eq. (9.24). The capacitance will also decrease by the same factor. This implies that the delay, which is proportional to CV/I, will be reduced by a factor α. The total power consumption, which is proportional to the product of current and voltage, will decrease by a factor of α^2. On the other hand, the power per unit area will remain constant. The result of the device scaling is given in the Table 9.1.

An alternative scaling method is *constant-voltage* scaling where the voltage is kept constant. The advantage of constant-voltage scaling is reduced circuit delay, which scales with the square of the scaling factor. The main drawback is the large electric fields that can cause reliability problems for the devices. In reality, one often takes an approach between constant voltage and constant field scaling, called *generalized scaling*. This is done to optimize the speed, power dissipation and device reliability. The scaling rules for the different scaling approaches are given in Table 9.1, assuming that the transistor does not suffer from velocity saturation and operates in strong inversion.

The scaling trend of the nominal feature size and physical gate length of transistors in high-performance circuits is shown in Figure 9.4. As can be seen from the figure, the scaling of the transistor's gate length has become more aggressive since the middle of 1990's than that of the features used in DRAMs, indicating that microprocessors have been driving the technology. Figure 9.19a illustrates the scaling trends of the delay, power density and current drive per unit gate width as found in industry, as well as the trend predicted by constant field scaling (dashed lines).[18] The delay follows nearly the constant field scaling law but the power density has increased significantly faster, mainly because the power supply has not been scaled with the same factor as the gate length. Also, the current drive has increased more than predicted by the constant field scaling, while the gate capacitance has decreased more than predicted. This is due in part to the fact that the gate length has scaled more aggressively than the oxide thickness and also due to the subscaling of the power supply.

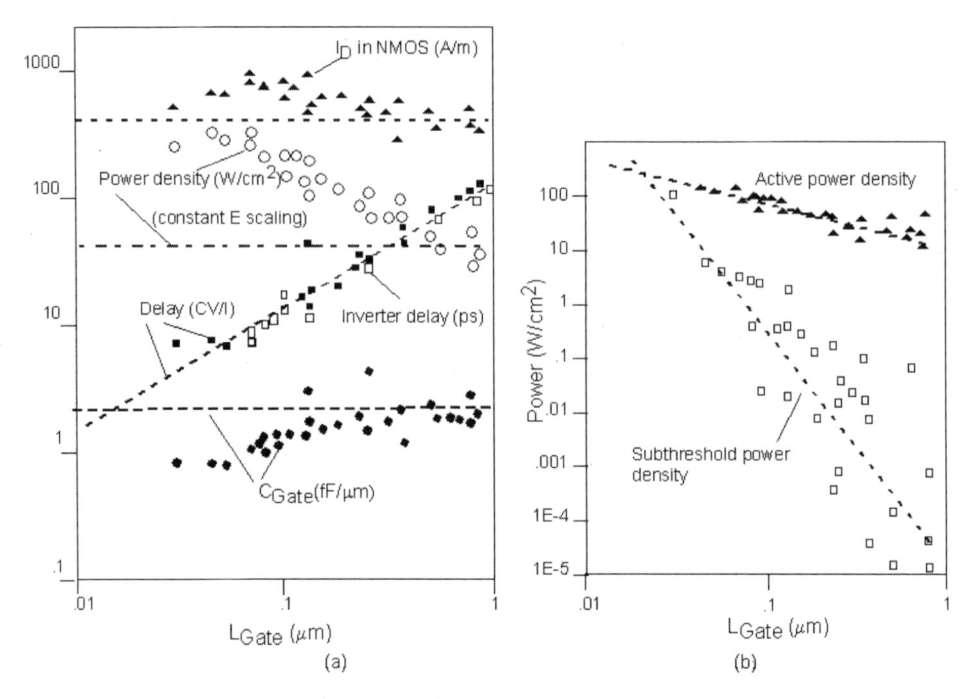

FIGURE 9.19. Scaling trends in industry (data points) as compared to the trends for constant field scaling (dashed lines): (a) current per unit gate width, delay, power density, and gate capacitance; (b) active and subthreshold power densities vs. the gate length. (From Ref. 18, courtesy of International Business Machines. Unauthorized use not permitted.)

With the rise of mobile computing applications and the increased integration density, the need for low power devices and power efficient circuit design has grown considerably. Table 9.1 illustrates that the power density increases as α^3 for constant voltage scaling and as k^3 for the generalized scaling method. This problem is highlighted in high performance microprocessors that consume up to about 100 Watt.[19] If the trend in scaling continues, the power consumption of microprocessors is going to become excessively large, approaching power densities comparable to those found in a nuclear reactor! Clearly, power dissipation is going to become a major limitation that needs to be addressed. Reducing both the active power and the subthreshold power dissipation by improving the technology, lowering the power supply aggressively, dynamically adjusting the threshold voltage, switching off part of the circuits when not in use, new device structures, more efficient architectures, etc., are becoming hot research topics.

9.6. SMALL-DIMENSION EFFECTS

The goal of scaling is to adjust the device dimensions in such a way that the transistor characteristics will remain basically the same as the long channel devices. Although scaling has provided a successful method to reduce the device dimensions without decreasing its performance or reliability, some parameters do not scale well which has caused specific problems associated with small-dimension transistors. These include increased off-current, short-channel and narrow-width effects, increased current density and hot electrons.[16]

9.6.1. Scaling of the Subthreshold Current

When one switches a transistor off by making the gate voltage $V_G = 0$, the transistor is not completely off but will be in the subthreshold regime. A certain amount of *off-current* will flow, as is given by the expression (9.29) with $V_G = 0$. Since the capacitance C_{ox} increases with scaling, the off-current or the subthreshold leakage current also increases with scaling. In addition, the exponential dependency of the leakage current on the threshold voltage has caused the power density associated with the subthreshold current to increase *exponentially* with scaling. In order to limit this increase in power dissipation, one has reduced the threshold voltage to a lesser extent than is prescribed by the constant field scaling. Nevertheless, according to industry trends, the subthreshold power density has increased much faster than the active power density. Unless this trend changes, the subthreshold power density will become equal to the active power for a gate length of about 20 nm, as illustrated in Figure 9.19b. The situation gets even worse at higher temperatures since the subthreshold current increases faster with temperature than does the active power dissipation. Techniques such as using two threshold voltage levels, or biasing the substrate to reduce or increase the threshold voltage (through the body effect) are being applied to overcome the tradeoffs between leakage current and delay, active and passive power dissipation. For low power applications, one often has to sacrifice performance by selecting processes with a higher V_T and thus lower off-currents.

9.6.2. Hot Electrons

Due to the lack of scaling of the subthreshold current, constant electric field scaling is not feasible for high performance circuits. Thus, the electric field across the channel in small dimension transistors increases significantly over long channel transistors. An electron that travels from the source to the drain can gain enough kinetic energy and move into higher energy levels in the conduction band. Such electrons are called "hot" electrons. A few of these electrons will have gained enough energy to overcome the energy barrier ($E =$ 3.1 eV) between the silicon and the oxide (Figure 9.15). The majority of these electrons will be collected at the gate electrode but some of them will be trapped inside the oxide. These charges contribute to the effective oxide charge Q_{ox} and will increase the threshold voltage V_T. Also, the amount of interface states will increase due to the damage caused at the interface, increasing the body-effect parameter m and the subthreshold slope S. Careful design of the device parameters and the use of a lightly doped drain (LDD) help alleviate this problem.

9.6.3. Short-Channel Effects and Drain-Induced Barrier Lowering

For short channel transistors, the charge in the depletion region is shared between the source/drain and the gate. As a result, the amount of depletion charge that determines the threshold voltage is reduced for short length transistors, causing a roll-off of the threshold voltage with decreasing gate length. Figure 9.20 illustrates that the depletion region under the gate is in part determined by the voltage of the drain and source terminals.

When the gate length becomes very small as compared to the source and drain junction depth and the depletion width, the gate may not be able to fully control the potential in the channel region. The drain voltage will influence the voltage in the channel near the

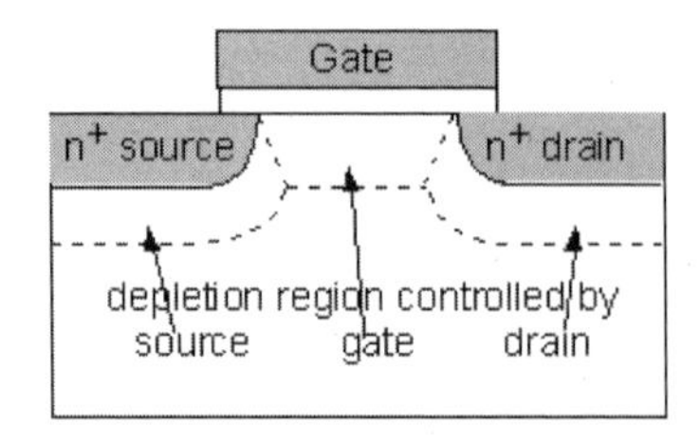

FIGURE 9.20. Cross section of a MOSFET showing the depletion region under the gate, source and drain. The depletion region under the gate is in part a function of the depletion region controlled by the source and drain, causing short channel effects.

source. This is called Drain-Induced barrier lowering (DIBL). This results in a reduction of the threshold voltage, an increase in the subthreshold current with drain voltage, and in punch-through between the source and drain. The latter gives rise to a large leakage current independent of the gate voltage. Minimizing the distance between the channel region and the quasi-neutral bulk can reduce the DIBL effect. This requires a small substrate depletion width (and high channel doping), shallow source and drain junctions, or a non-uniform channel implant in the lateral and vertical direction (superhalo implant). DIBL is one of the major limitations to scaling of conventional planar FETs.

9.6.4. *Narrow-Width Effects*

For narrow width channels the depletion width under the gate will extend along the width direction of the channel. This causes an increase of the threshold voltage since the total charge Q_D in the depletion region has increased (Eq. 17). As was the case for the short channel effect, the narrow width effect can also be reduced by a smaller depletion region width (or higher substrate doping). On the other hand higher substrate doping levels degrades the mobility of the carriers in the inversion layer. Careful engineering of the substrate doping profile has become essential to ensure proper operation of the scaled devices.

9.6.5. *Velocity Saturation*

The velocity of the mobile carriers increases proportionally with the electric field. However, when the field gets large ($E = 2.10^4$ V/cm for electrons), the carriers lose their energy quickly and the mobility decreases. The drift velocity of the carriers reaches a maximum value $v_{\text{sat}} = 7-8 \times 10^6$ cm/s for electrons and $6-7 \times 10^6$ cm/s for holes in the channel region. This phenomenon is called velocity saturation. In short channel devices, velocity saturation can occur due to the large electric field. As a result, the current will not increase with the drain voltage anymore and be independent of the channel length L. The current for very short channel devices then becomes equal to,

$$I_{D\text{sat}} = C_{\text{ox}} W v_{\text{sat}} (V_{GS} - V_T) \tag{9.31}$$

This has an effect on the scaling behavior of the transistor since the current is independent of the gate length and linearly dependent on $(V_{GS} - V_T)$. The scaling factor for the generalized scaling of the current in the velocity-saturation regime is equal to k/α instead of k^2/α (Table 9.1).

9.6.6. Source-Drain Resistance

Scaling of the device dimensions also includes the junction depths. This causes the sheet resistance of the source and drain to increase. This resistance R_S gives rise to a voltage drop $R_S I_{DS}$ that reduces the available voltage between the gate and the internal source terminal. This parasitic series resistance can give a considerable current reduction in short channel transistors. Techniques such as an amorphization implant for shallow junctions or laser annealing of the junctions will become important to obtain low sheet resistance in shallow junction.

9.6.7. Doping Fluctuations

For small dimension transistors the amount of dopants in the channel region that controls the threshold voltage becomes very small. The randomness in the number and location of these dopants will give rise to significant variations in the threshold voltage from device to device.[20] As an example, a transistor with a gate length of 100 nm and width of 400 nm has about 1000 dopants in its depletion region. When the gate length is reduced to 25 nm, the number of dopants decreases to about 120. The number fluctuation from device to device is given by the standard deviation that is equal to the square root of the number of dopants. For a 25 nm gate length this amounts to about 11 out of 120 dopants, which is a very significant fluctuation. This has a direct impact on the threshold voltage variations from device to device since the number of ionized dopants determines the value of the threshold voltage. These variations can be reduced by carefully tailoring the doping profile in the channel region, such as in a retrograde-doped channel. A retrograde-doped channel consists of a low-high doping profile. Doping fluctuations are now moved away from the channel to reduce their effect on the threshold voltage.

Transistors with a 15 nm physical gate length have been demonstrated recently.[3,4] These transistor dimensions are well ahead of the ITRS target (Figure 9.21). One paper describes a transistor that was fabricated with a 1.4 nm (0.8 nm effective oxide) thickness nitride/oxinitride gate dielectric stack with a poly-silicon gate electrode, with ultra-shallow source/drain junctions, and compact halo profiles in the channel region to reduce short-channel effects. The NMOS and PMOS devices obtained gate delays (CV/I) as small as 0.29 ps and 0.68 ps, respectively, at a supply voltage of 0.8 V. This illustrates that planar CMOS technology has the potential to stay the mainstream technology for the next decade. However, continued efforts will be needed to overcome the device design and process barriers to meet the ITRS target.

9.7. NANOSCALE MOSFET TRANSISTORS: EXTENDING CLASSICAL CMOS TRANSISTORS

Transistor dimensions have been reduced by about 30% every 2–3 years, allowing the semiconductor industry to follow Moore's Law which states that the number of functions per chip doubles every 18 months. According to the International Technology Roadmap of the

TABLE 9.2. Key technology requirements for high-performance logic circuits, based on the ITRS2002 roadmap[2]

	Near Term				Long Term	
Year	2001	2003	2005	2007	2010	2016
DRAM Half-pitch or Node (nm)	130	100	80	65	45	22
Printed gate length of MPU (nm)	90	65	45	35	25	13
Physical gate length of MPU (nm)	65	45	32	25	18	9
Equivalent oxide thickness, t_{ox} (nm) (high performance applications - MPU)	1.3-1.6	1.1-1.6	0.8-1.3	0.6-1.1	0.5-0.8	0.4-0.5
Wiring levels	7	8	9	9	10	10
Clock Frequency MPU on chip (GHz)	1.7	3.1	5.1	6.7	11.5	28.7
Trans. per MPU chip (Millions)	276	439	697	1106	2212	8848
Power supply, *Vdd* (V)	1.2	1.0	0.9	0.7	0.6	0.4
Power dissipation MPU (W)	130	150	170	190	218	288
Bits per chip on DRAM (Gb)	0.54	1.07	2.15	4.29	8	64

Manufacturable solutions exist and are being optimized
Manufacturable solutions are known
Manufacturable solutions are not known

Semiconductor Industry Association, downscaling will continue in the foreseeable future, with the printed and physical gate lengths of transistors reaching 13 and 9 nm, respectively, by 2016.[2] The following table gives some of the key requirements and parameters, projected for the next 15 years. The technology *node* in the table refers to the dimension of the half-pitch of a DRAM cell that has historically been the technology driver. However, since the 90's the technology to manufacture high-performance microprocessors (MPUs) has accelerated and is currently driving the most advanced processes. The gate length of transistors used in microprocessors has dimensions that are considerably smaller than the DRAM half-pitch, as can be seen in Table 9.2.

The transistor physical gate lengths will decrease from 45 nm in today's advanced processes to about 9 nm in the next decade. Figure 9.21 illustrates this trend for the physical and printed gate length of transistors used in MPUs. As can be seen, the gate length has decreased by a factor of 0.7 every 2 years over the past 5 years, which is faster than what was originally projected. It is anticipated that this acceleration will continue until about 2005 after which the scaling will be done on a 3-year cycle instead of a 2-year cycle. As a result of the aggressive scaling, the small-dimension effects discussed in the previous section will become more pronounced and the limits imposed by the physics and the materials will pose serious challenges. The traditional solutions involving complex doping profiles of the channel and drain/source regions will become inadequate to keep the small-dimension effects from degrading the transistor performance.

As discussed earlier, among the main challenges posed by the limits of scaling are short channel effects, the thin gate oxides, in particular the *leakage currents* including the gate leakage due to quantum-mechanical tunneling, the subthreshold leakage, junction leakage, band-to-band tunneling between the reverse biased drain and the highly doped substrate, and direct tunneling between the source and drain through the channel potential barrier. Replacing the traditional SiO_2 insulator by alternative materials with high K dielectrics will be required to reduce some of the leakage problems. Since subthreshold leakage current is

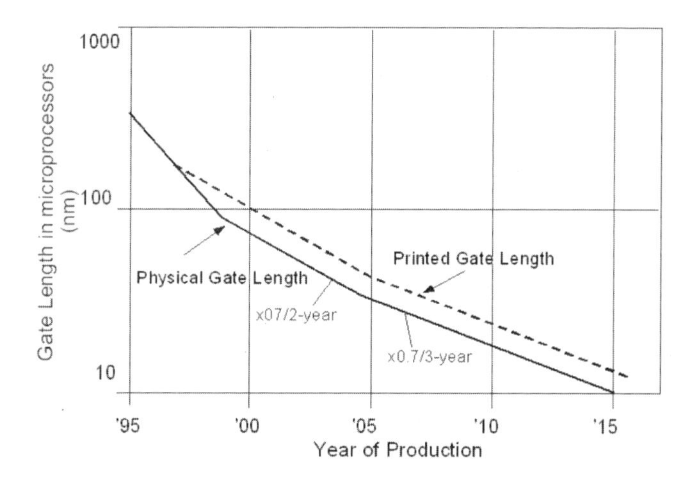

FIGURE 9.21. Projection of the printed and physical gate lengths of transistors in MPUs according to the ITRS2002.[2]

not suppressed by introducing new materials, it is going to be one of the ultimate limitations to scaling. This implies that the limits to scaling are going to be *application dependent*, with low power applications imposing stricter limitations on leakage currents than high-performance applications.

In addition, the polysilicon gate electrode has the limitation associated with the depletion region and boron out-diffusion, reducing the beneficial effects of device scaling. As a result, new metal gate materials will be needed in conjunction with the high K dielectrics. Innovations in both the device structures and the materials will ensure high-performance operation of nanoscale electronic devices. Some of the more promising approaches are discussed below.

9.7.1. Silicon on Insulator (SOI)

The conventional way to fabricate CMOS circuits has been on bulk silicon wafers as discussed above. The need for continued improvement in performance of scaled transistors has put considerable constraints on the devices, such as low parasitic capacitances and reduced threshold voltage, in order to obtain good drive current. On the other hand the requirement for low power operation demands low off-currents and thus a large threshold voltage. Fulfilling these contradictory requirements becomes more and more difficult in conventional bulk CMOS. Silicon on insulator (SOI) technology provides an attractive alternative since SOI CMOS transistors have minimal junction capacitance, they have no body effect, do not suffer from latch-up, and have a good subthreshold slope. This makes SOI particularly attractive for low-power applications and radiation hard circuits.

SOI CMOS transistors are built on a thin layer of single crystal silicon that is separated from the substrate by a silicon dioxide film, as schematically shown in Fig. 22. There are two main methods to obtain the SOI wafer. In the first method, oxygen atoms are implanted beneath the silicon surface at about 500°C to form a buried oxide (SIMOX—Separation by

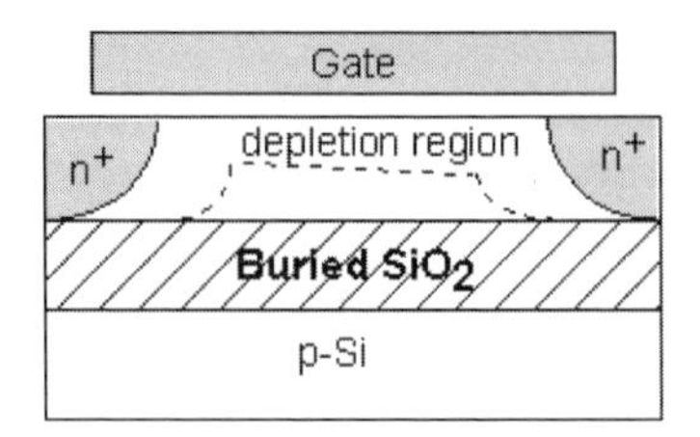

FIGURE 9.22. Schematic cross section of a SOI NMOS.

IMplantation of OXygen) of about 400 nm thickness. The other method, called SOITEC, makes use an oxidized wafer implanted with hydrogen atoms at a certain depth and a second wafer that is bonded to the first one. After heating, the two wafers split at the place of high hydrogen concentration, leaving a thin layer of crystalline silicon on top of the oxide.

The SOI devices can be grouped into two categories, depending on the thickness of the silicon layer and doping level. For very thin silicon layers, the depletion width underneath the channel of the transistor is usually larger than the silicon thickness, giving rise to a fully depleted (FD SOI) transistor. The partially depleted (PD) SOI transistor makes use of a thicker silicon layer and has a quasi-neutral substrate region.

9.7.1.1. Partially Depleted SOI When the silicon layer is thicker than the maximum gate depletion width, the transistor is called partially depleted. The PD SOI is currently the most used SOI technology since the processing is similar to that of bulk CMOS, with the additional advantages of low junction capacitance and the lack of body effect. However, PD CMOS poses a problem as a result of a floating-body. This leads to the charging of the body due to the generation of carriers by impact ionization in the drain region. These carriers are stored in the floating substrate and will change body potential and thus the threshold voltage.[21] Also, the source-substrate junction can become forward biased causing larger leakage currents. Using a substrate contact can reduce this effect at the expense of larger area and the loss of the body-effect advantage in SOI. In many cases, circuits fabricated in PD SOI technology need to be redesigned in order to reduce the detrimental effects of the floating body. When done properly, PD SOI is an attractive alternative over bulk CMOS for low-voltage and low-power CMOS circuits.[22]

9.7.1.2. Fully Depleted SOI In a fully depleted CMOS transistor the subthreshold slope can be near ideal since the body effect coefficient m (see Eq. 22) is almost equal to 1. This is because the effective depletion width can be considered to be very large. A steep subthreshold slope allows one to use a lower threshold voltage for the same leakage current and thus lower supply voltages. However, a thick buried oxide reduces the beneficial effects of PD SOI for short channel transistors. This is because the drain-source field penetrates through the oxide and thus results in poor short channel effects.[23] It has been reported that by using an ultra-thin silicon film of 2 nm and a local buried oxide of 20 nm, the short channel effects can be suppressed in devices with channel lengths of 20 nm.[24] The unique feature of this approach is that the buried oxide is not continuous as in the other SOI approaches but is limited to the gate regions. This has the advantage of SOI combined with the deep

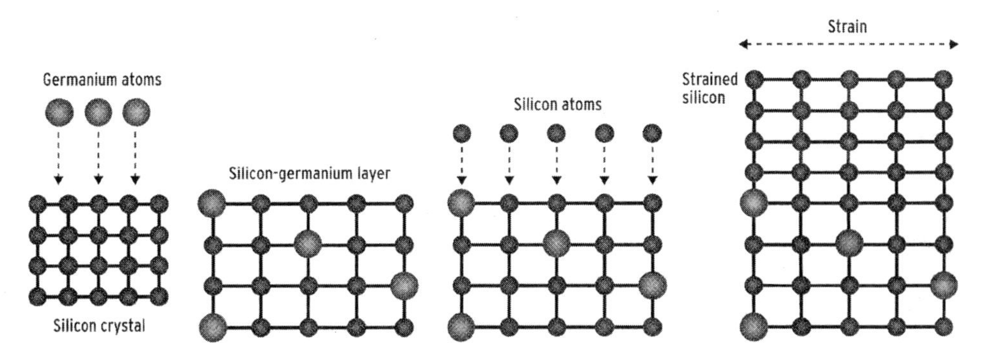

FIGURE 9.23. Schematic illustration of the effect of Ge atoms, implanted in a Si crystal, on the strain of Silicon. (From Ref. 25 by permission of the Institute of Electrical and Electronics Engineers.)

source/drain regions giving rise to reduced series resistance of the source and drain. The silicon film and the buried oxide are defined by epitaxy on a bulk substrate and can thus be very accurately controlled, allowing the growth of very thin and uniform layers. These thin layers suppress the short channel effects and drain induced barrier lowering effects (DIBL). It is expected that PD SOI will become the dominating technology in the near future for CMOS for the 50 to 30 nm nodes.

9.7.2. Strain (Silicon-Germanium)

As mentioned in section 9.3.3, the mobility of the carriers increases significantly in strained silicon. Transistors built on strained-silicon wafers have shown mobility improvements up to 110% in NMOS and 45% in PMOS transistors.[15] The strained silicon is obtained by growing an epitaxial silicon layer on top of a silicon-germanium layer (Figure 9.23). When the larger germanium atoms replace the silicon atoms, the distance between the silicon-germanium atoms is larger than in a pure silicon crystal. The change in the crystal geometry reduces the scattering and decreases the effective mass of the carriers. This technology has the advantage that the conventional device structure can be maintained. A drawback is the increased junction leakage and junction capacitance (source and drain to substrate diodes) due to the energy gap in SiGe layer being smaller and the dielectric constant being larger than that of silicon. To eliminate the excessive leakage and junction capacitance, silicon on insulator (SOI) wafers will be used for the silicon-germanium process.[25]

9.7.3. Hi-K Gate Dielectrics

Reducing the thickness of the gate dielectric is important in order to reap the advances of scaling in terms of reduced device delays (*CV/I*). Indeed, scaling the thickness is necessary not only to obtain a large transistor drive current, but to also reduce the short-channel effects as discussed earlier. The gate oxide thickness in state-of-the art devices for high-performance applications (e.g., microprocessors) is currently in the range of 1–1.5 nm and it is projected that the equivalent oxide thickness will be reduced to 0.4–0.5 nm by 2016

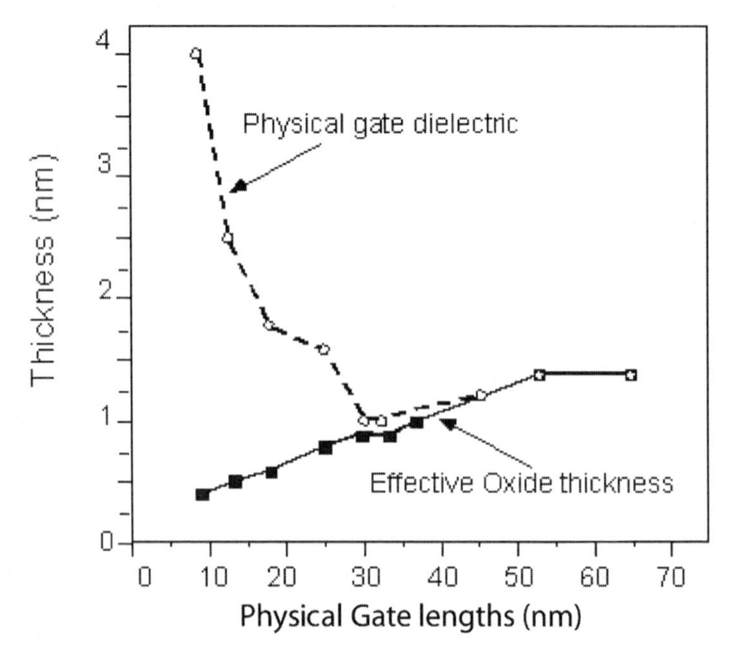

FIGURE 9.24. Scaling of the physical gate dielectric thickness and effective oxide thickness (EOT), based on data from the ITRS2002. (From Ref. 4 by permission of the Institute of Electrical and Electronics Engineers).

(Figure 9.24 or Table 9.2). The thickness will have been reduced to only a few atomic layers which causes large gate leakage currents due to direct tunneling through the oxide. Leakage current densities for a 1.5 nm thick SiO_2 is about 1 A/cm^2 at 1 V. Since the main leakage mechanism through these thin layers is direct tunneling of electrons, the leakage current is an exponential function of the layer thickness. Reducing the oxide thickness to 1 nm would increase the leakage current to 100 A/cm^2 at 1 V.[26] For high performance applications it is generally assumed that a silicon oxide can be used down to a thickness as small as 0.8 nm.[20] Also, oxinitride, nitride films or a nitride/oxinitride stack will be near-term solutions for high performance devices. However, for low power applications where smaller leakage currents are required, silicon oxide films below 1.5 nm will give excessive leakage currents and alternative materials will be required as soon as the 80 nm node (Table 9.2).[2]

One solution for the increased leakage current is to replace the traditional SiO_2 dielectric layer with a material that has a larger dielectric constant K than that of silicon oxide. This allows one to use a layer that has a larger physical thickness $t_{\text{Hi-K}}$ while maintaining a smaller equivalent physical SiO_2 thickness t_{ox}(or EOT),

$$t_{ox} = (K_{SiO_2}/K_{\text{Hi-K}})t_{\text{Hi-K}} \tag{9.32}$$

in which $K_{SiO_2}(=3.9)$ is the dielectric constant of SiO_2 and $K_{\text{Hi-K}}$ is the dielectric constant of the actual gate dielectric. One notices that by using a high-K material, one obtains a small equivalent oxide thickness t_{ox}, even if the actual physical thickness $t_{\text{Hi-K}} = t_{phys}$ is considerably larger.

Materials being investigated as potential candidates belong to the family of binary metal oxides such as Al_2O_3 (K = 10), HfO_2 (K = 20), Ta_2O_5, (K = 25), ZrO_2 (K = 23) and others.[27] Introducing these high-K materials poses serious challenges to make the materials compatible with conventional CMOS processing. Other requirements are thermal stability and crystallization temperature, thermal expansion coefficient that matches that of silicon, low interface state traps at the dielectric-silicon surface, large energy bandgap and large silicon-dielectric energy barrier (see Figure 9.15b). The latter is important since the tunneling current is a strong function of the barrier height. These films suffer from a significant amount of traps and fixed charges that give rise to shifts in the flat band voltage, cause reliability problems, and degradation of the carrier mobility. Also, in many cases a thin interfacial SiO_2 layer (t_{SiO2}) is formed between the silicon substrate and the high-K material that increases the equivalent thickness of the dielectric,

$$t_{ox} = t_{SiO_2} + (3.9/K_{Hi\text{-}K})\, t_{Hi\text{-}K} \tag{9.33}$$

Thus, the minimal achievable EOT (t_{ox}) will never be less than the interfacial oxide thickness. This poses a serious limitation on the maximum gate capacitance of a high-K dielectric. Processing of these materials poses equally challenging problems since compatibility with existing CMOS processing is important. They can be deposited by various means, such as metallo-organic chemical vapor deposition (MOCVD), atomic layer CVD (ALCVD), or sputtering. Transistors of 50 nm gate length with HfO_2 dielectric of 1.3 nm deposited with ALCVD have been demonstrated recently.[28]

9.7.4. Metal Gate Electrode

Current day transistors are fabricated with a highly doped poly-silicon gate. The advantage of poly-silicon is the value of its work function, high thermal stability, compatibility with high-temperature processing steps, and the self-aligned feature that reduces the overlap capacitance between the gate and the source/drain. However, a small depletion layer (about 0.5 nm) is formed in the poly-silicon material near the gate oxide interface. This causes a depletion capacitance that is in series with the gate oxide capacitance. For oxide thicknesses in the range of 1.5 nm, the depletion capacitance associated with the poly-silicon gate starts to become important and reduces the effect of the gate capacitance.[25,27] As a result, the coupling between the gate and the channel of the transistor is reduced. Another potential problem with the poly-silicon gate material is the high processing temperatures that are required. This can cause difficulties with the stability and integrity of the high-K gate dielectric. It is expected that poly-silicon gates will need to be phased out beyond the 65 nm technology node.[2]

A solution to the above problems is to replace the traditional poly-silicon gate with a metal gate. Since the carrier concentrations in a metal are much higher than in poly-silicon, the depletion layer is virtually non-existent. Metals can also be processed at considerably lower temperatures than poly-silicon. Among the criteria to select the proper metal material are the process compatibility, thermal stability, stability of the metal-dielectric interface and the work function. As discussed earlier, the work function influences the value of the threshold voltage (see Eq. 17). By selecting a material that has the proper work function, one can tailor the threshold voltage without having to rely on modifying the channel doping level. This gives some leeway to optimize the channel doping in order to reduce short-channel effects without compromising the threshold voltage. Ideally, one likes to have dual metal gates with work functions that are comparable to those of n^+ and p^+poly-silicon.

Several candidate materials have been investigated—both single metal and alloys. One aspect that complicates the choice of metal is the fact that one likes to have a different work function of the NMOS and PMOS transistors in order to tailor the value of the threshold voltage for each. This would require two different materials or modifying a single metal gate. Dual metal gates have been used such as Titanium for NMOS and Molybdenum for PMOS transistors. Molybdenum has also been used as a metal gate whose work function has been modified by nitrogen implantation. Also, metal mixtures such as Ti-Ni and Ru-Ta have shown to be promising. By changing the metal mix one can adjust the work function to obtain the right value of the threshold voltages for the NMOS and the PMOS. These approaches are still under development and more research is needed to validate their success.

9.7.5. Cooled CMOS

Scaling has been the main factor driving up the performance of devices. However, the lack of full voltage scaling and the increased tunneling through the gate oxide will limit the amount of scaling of conventional bulk CMOS transistors to gate lengths of about 20 nm.[29] One way to extend the devices beyond the 20 nm gate length is to cool them. At lower temperatures the scattering of the carriers is reduced resulting in higher mobility. In addition, the subthreshold slope S is proportional to kT/q, Lowering the temperature will give a steeper slope, which reduces the off-current considerably. This allows one to work with a lower threshold voltage and power supply. As a result, it may become feasible to extend the CMOS scaling down to the 10 nm regime.

9.7.6. Double-Gate MOSFET

Device scaling continues to pose more demanding constraints on the device design that makes optimization more and more difficult. As discussed above, one of the limitations of device scaling is the drain-induced barrier lowering, giving rise to reduced threshold voltage and increased leakage currents. By shielding the electric field of the drain from the source through the gate, the effect of drain-induced barrier lowering can be significantly reduced. This can be obtained with a smaller substrate depletion region and thus higher substrate doping levels. On the other hand, a small depletion region of the substrate reduces the coupling of the gate voltage to the channel potential and thus gives lower drive current, a larger body coefficient m and worse subthreshold slopes. These contradictory requirements on the substrate doping as a result of the need to reduce short channel effects and obtain good drive currents, pose serious limitations on the effectiveness of further scaling of conventional CMOS transistors.

The double-gate field effect transistor (DG FET) basically decouples these two requirements for the substrate doping.[30,20] It has been predicted that the DG FET can be scaled down to 50% of that of a bulk FET, making gate lengths of about 8–10 nm feasible for devices operating at room temperature.[31] These predictions and recent successes with the fabrication of such devices make the DG FET a strong candidate to become the dominant device structure in the last era of CMOS scaling before one reaches atomic levels.[32]

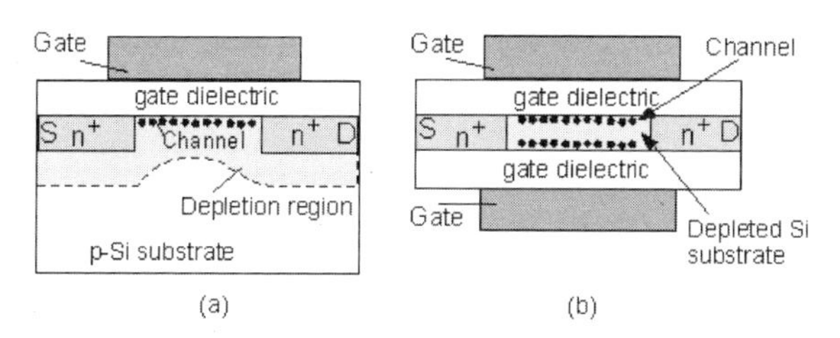

FIGURE 9.25. Schematic cross sections: (a) conventional bulk MOSFET and (b) dual gate FET.

A schematic cross-sectional representation of the DG FET is shown in Figure 9.25 together with a bulk MOSFET. A DG FET is a non-planar transistor with a top and bottom gate on both sides of a thin silicon channel. This allows one to control the potential of the channel much tighter than in the case of a conventional planar transistor where the effect of the substrate played a major role. The short-channel effects are now basically a function of the device geometry and not of the substrate doping. As a result, the substrate can be lightly doped or even undoped. The negligible depletion capacitance of the substrate allows for an effective coupling of the gate voltage to the channel potential which gives a steep subthreshold slope of about 60 mV/dec and a good drive current. This is an important advantage of the DG FET since it allows for a larger gate voltage over-drive and makes the DG FET particularly attractive for low power applications.

The low channel doping is also advantageous for the carrier mobility. However, it has been found that in a silicon channel thickness below 20 nm, the carrier mobility degrades when the inversion carrier density is low. For large inversion carrier densities (above $10^{13}/\text{cm}^2$), which is usually the case in nanoscale CMOS, the degradation is much less pronounced. In addition, in undoped substrates the carrier transport is considerably better than in conventional devices. This is mainly due to the reduced scattering in the channel and the lower surface electric field. Both of these effects increase the mobility in DG FET by a factor of about two as compared to that in conventional bulk FET.[27]

As can be seen from Figure 9.26, a DG FET can be fabricated with any of the three topologies shown.[33] The type I structure most closely resembles the conventional planar bulk FET. It allows for uniform control of the channel thickness but poses difficulties to fabricate the bottom gate and a high quality gate dielectric. Also, making contact to the bottom gate is not easy and requires extra real estate which reduces the overall device density.

Type II and III DG FETs have their channel perpendicular to the substrate. Access to the top and bottom gates is easier than in the planar structure. On the other hand the substrate thickness is determined by lithographic techniques that make the uniformity control critical. For proper operation, the substrate (channel) thickness should be considerably smaller (3–4$\times$) than the gate length, further complicating the dimensional control of the substrate for small length devices.

Type III structure has been most successful to date and is often called a FinFET.[34,35,36] A schematic view of the FinFET is shown in Figure 9.27. The channel substrate sits on top of a silicon-on-insulator (SOI) substrate in the form of a fin. The thickness of the fin corresponds to the thickness of the silicon channel that needs to be tightly

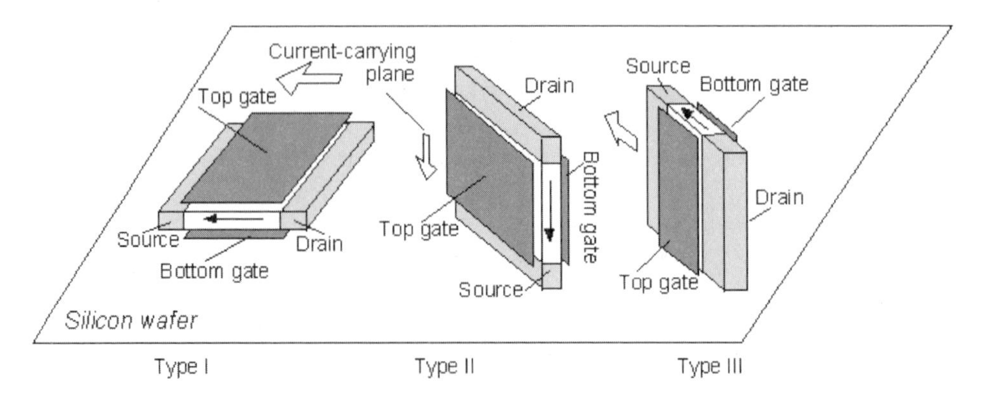

FIGURE 9.26. Conceptual representation of double-gate field effect transistors (Adapted from H.-S. Wong, K. Chan, and Y. Taur, *IEDM Tech. Digest*, 427 (1997)).

controlled, using advanced patterning and etching techniques. A gate dielectric is grown or deposited on top of the fin before the poly-silicon gate is deposited on top of it, forming a saddle-like structure. The source and drain regions are adjacent to the fin. The distance between the source and drain regions determines the gate length while the height of the fin determines the gate width. In order to reap the advantages of the double gate structure, one must make sure that the parasitic overlap capacitances between the gate and source/drain regions are kept to a minimum. Process innovations will be necessary to reduce these and other parasitic effects, and to take full advantage of the reduced short channel effects and increased drive currents in these nanoscale devices.

The structures and materials discussed in this section will be needed to extend CMOS technology towards the end of the time horizon of the ITRS. Although no manufacturable solutions are known for several of the materials and structures discussed, it is believed that these materials and manufacturing challenges will be solved in order to keep the ITRS projections on track. The question that arises is, what will happen at the end of the roadmap? Will CMOS be replaced by more advanced technologies or will it co-exist with other, non-conventional technologies? The next section gives some insights to what lies beyond the traditional CMOS devices.

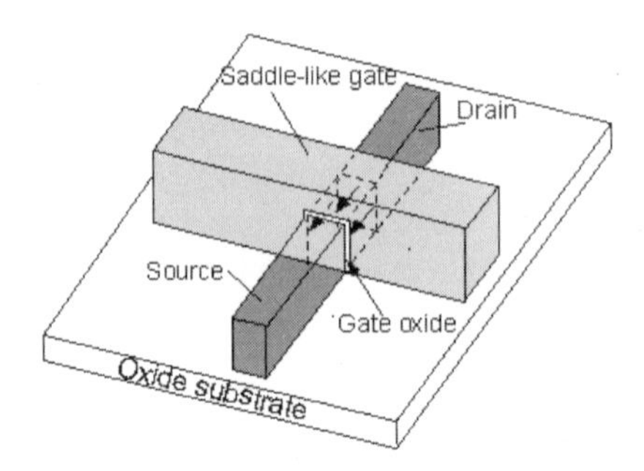

FIGURE 9.27. Schematic view of a FinFET, type III double-gate FET.

9.8. BEYOND TRADITIONAL CMOS

As discussed above, the technology roadmap (ITRS 2003) projects that transistors will have physical gate lengths of 8–9 nm by 2016. Initial experimental results obtained with non-traditional MOSFETs, such as the DG FET, indicate that the projections of the ITRS will most likely stay on its trajectory. However, manufacturing multi-billion-transistor circuits with nanoscale dimensions will become exceedingly difficult and expensive. These circuits will be extremely fast but will also consume a lot of power, requiring careful circuit design and choice of both high- and lower-performance devices on the same chip to manage the excessive power dissipation. In addition, the dimensions of the devices are reaching molecular and atomic levels ushering us into a different regime dominated by quantum mechanical effects. Alternative technologies to CMOS need to be explored that will most probably become complementary to the existing CMOS technology. Among the candidate technologies are quantum devices, molecular electronics, organic (plastic) transistors, nano-electromechanical structures (NEMS), carbon nanotube transistors, and optical devices (see the next three Chapters).[2,37,38]

Figure 9.28 depicts the various technologies in a three-dimensional cube such as axes speed, size and cost. The energy per consumption is indicated by the intensity. Each

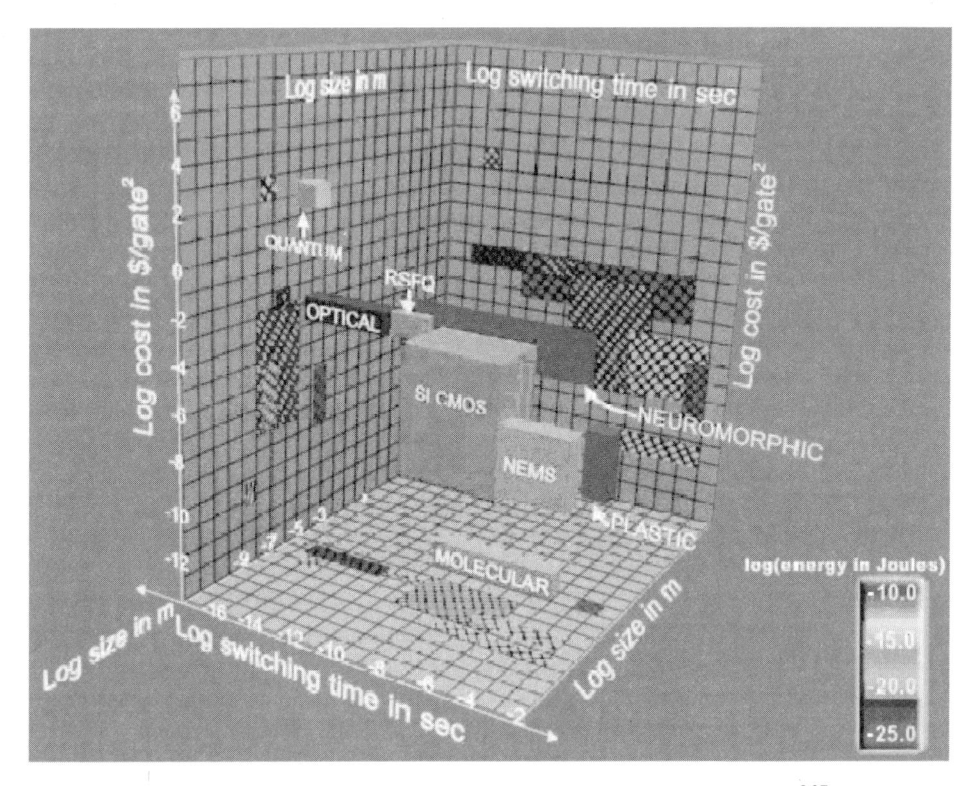

FIGURE 9.28. Representation of emerging technologies in a 3-D space of speed, size and cost.[2,37] (From Ref. 37 by permission of the Institute of Electrical and Electronics Engineers.)

TABLE 9.3. Estimated parameters for the emerging technologies of Figure 9.28[37]

Technology	Tmin (s)	Tmax (s)	CD_{min} (m)	CD_{max} (m)	Energy J/op	Cost (min) $/gate	Cost (max) $/gate
Si CMOS	3E-11	1E-6	8E-9	5E-6	4E-18	4E-9	3E-3
Molecular	1E-8	1E-3	1E-9	5E-9	1E-20	1E-11	1E-10
Plastic	1E-4	1E-3	1E-4	1E-3	1E-24	1E-9	1E-6
Optical	1E-16	1E-12	1E-7	2E-6	1E-12	1E-3	1E-2
NENS	1E-7	1E-3	1E-8	1E-7	1E-21	1E-8	1E-5
Quantum computing	1E-16	1E-15	1E-8	5E-7	1E-21	1E3	1E5

technology occupies a certain volume in this three-dimensional representation.[2] Each volume is projected on the three planes in order to make it easier to determine the value of the three parameters for each technology. As can be seen for CMOS, the cost lies in the range of $4 \times 10^{-9} - 3 \times 10^{-3}$ \$/gate, the switching time varies between 30 ps–1 μs, and the size lies in the range from 5 μm to 8 nm. Several of the technologies shown in the graph are still in their infancy and lack experimental data. In that case, the data shown is based on assumptions and physical principles.[2,37] Table 9.3 summarizes the parameters for the devices shown in Figure 9.28. T_{min} and T_{max} refer to the switching tie, and CD is the critical dimension.

It is interesting to notice that the applicability of several of the technologies shown in Figure 9.28 is application specific. As an example, *Quantum computation* provides an immense speed improvement for solving factoring algorithms such as encountered in encryption, turning an impossible computation into a practical one. However, quantum computers are not expected to be useful for general purpose computing.[39] Quantum computing is in its infancy stage and devices that are suitable for realizing quantum computers are still under investigation. One such promising device is called a spin-resonance transistor that consists of Si-Ge multilayers.[40]

Plastic or *organic* transistors are thin film transistors that are fabricated on plastic substrates instead of silicon. One application of organic electronics is the Organic Light Emitting Diodes or OLEDs. This technology offers the potential of very inexpensive electronics and displays that can be printed on flexible substrates. This opens up opportunities for new applications such as digitized newspapers, product labels, RF tags, printable electronics on clothing and other products. These applications do not require the high performance and densities of CMOS circuits but should be inexpensive and flexible.[41] This technology is well advanced and several commercial products are being developed.[42] The main disadvantages are the low speed (kHz) and large size (about 10 μm) of these devices.

Molecular electronics makes use of molecules that act as switching devices. The small size of these molecules would make it possible to build computers and memories of extraordinary density, a million times denser than today's memories. Although the realization of molecular electronics is still far from reality, several groups have succeeded in developing molecules that act as electronic switches and memory. Simple nano-circuits built from molecular switches and wires are being developed.[43] Of course, developing a circuit with a complexity similar to those of current day CMOS chips is still far away and many problems need to be overcome before the potential of molecular electronics can be realized.

Carbon Nanotube Field Effect Transistors (CNFET)

Carbon nanotubes are large, all carbon molecules which belong to the family of fullerenes. These tubes can be thought of as an expanded buckyball in which the mid-section has been expanded to form cylinders. These tubes have diameters ranging from one to several nanometers and lengths of hundreds of micrometers. The carbon atoms in the tube contain no dangling bonds, making them extremely inert and stable. The tubes can be metallic or semiconducting depending on how the tube has been wrapped (see chapter on Carbon Nanotubes). Metallic nanotubes are very good electrical and thermal conductors, exceeding conductivity of copper. Semiconducting nanotubes are promising for the fabrication of high performance nanoscale transistors. Several research groups have fabricated carbon nanotube field effect transistors (CNFET). The advantage of these transistors is that their structure is very similar to silicon based field effect transistors and do not need a completely new fabrication technology.[44]

IBM recently announced a high performance single carbon nanotube transistor that was fabricated with standard CMOS technology.[45] A schematic cross section of a CNFET based on the conventional MOSFET structure is shown in Figure 9.29. The transistor has a top gate similar to a MOSFET. The carbon nanotube of only 1.4 nm in diameter has been carefully deposited by spinning the tubes from a solution of decholoroethane (1.2 M) on top of a silicon oxide layer. Titanium source and drain electrodes were patterned. By annealing the structure, good contact between the titanium and nanotube was obtained through a thin titanium-carbide interfacial layer. A gate oxide of 15 nm was deposited at 300°C on top of the tube. A top electrode consisting of titanium or aluminum was deposited.

The mechanisms of the CNFET operation are not fully understood yet but recent results have indicated that CNFET operate as Schottky barrier transistors rather than conventional bulk transistors. This implies that one does not need to dope the carbon nanotubes. By

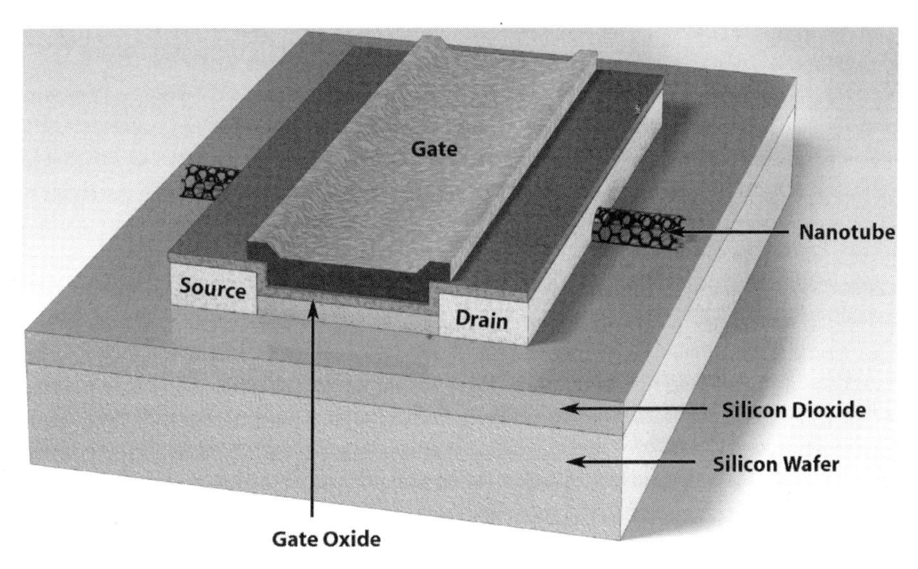

FIGURE 9.29. Schematic cross-section of a carbon nanotube field effect transistor (CNFET). (Adapted from Ph. Avouris, *Acc. Chem. Res.* **35**, 1026 (2002) with kind permission of Ph. Avouris).

modifying the barrier height one can obtain *n*-type and *p*-type devices, making it feasible to fabricate CMOS devices based on carbon nanotubes.

The CNET has a steep subthreshold slope and a transconductance which are better than that found in silicon transistors. Also, the speed of the electrons along the tubes is considerably better than in silicon due to the ballistic transport. Carbon nanotubes can be packed as close as 3 nm apart, giving rise to extraordinary densely packed devices. Among the improvements that are needed are thinner gate oxide of about 1.5 nm, and shorter gate lengths in the range of a few nanometers. It is expected that these changes will give a few orders of magnitude improvements in performance. The current results and the further improvements in performance are very promising and give a glimpse of how high performance nanoelectronic circuits may be fabricated in the post ITRS roadmap era.

These and other emerging technologies are discussed in different chapters in this book. We expect that these technologies will co-exist with CMOS technologies in a hybrid fashion. They will enhance microelectronics by adding functionality. Systems consisting of various subsystems manufactured with different technologies will offer the most economical and reliable solution. Information processing systems will not be limited to electronic systems but will include microfluidics, biosensors and systems on a chip, and other complex information processing systems that go beyond conventional CMOS systems.

9.9. SUMMARY

- Microelectronics is one of the most sophisticated manufacturing methods devised by mankind.
- CMOS is the dominant technology.
- The transistor count and functionality per chip have doubled every 18 months for the last 40 years (Moore's law).
- Scaling of device dimensions has allowed the microelectronics industry to stay on the trajectory of Moore's law.
  - Feature size reduction of 30% every 2 to 3 years.
  - Generalized scaling laws most popular in industry
  - Improved performance: 30% speed increase per year.
  - Reduced cost per function: 25–32% per year.
  - Devices are entering the nanoscale. Scaling is expected to continue until physical transistor gate lengths of about 8–10 nm are obtained.
- Limits of scaling:
  - Subthreshold current (off-current) does not scale.
  - Hot electrons and tunneling in the small-dimension transistors can become excessive.
  - Doping fluctuations cause large threshold voltage variations.
  - Power density as a result of the leakage current increases more rapidly. than the active power and will be one of the major limitations to scaling.
- Improved device structures and materials are required to continue the scaling for the next decade. Promising approaches are:
  - Silicon on insulator
  - Strained silicon

- Double gate FET
- Hi-K gate materials replace the traditional SiO_2.
- Metal gate electrodes will replace the traditional poly-Si gate.
- Approaches beyond the conventional CMOS transistors:
  - Plastic electronics
  - Molecular electronics
  - Quantum electronics
  - Carbon Nanotubes
- It is expected that these non-conventional technologies will not replace CMOS silicon but will co-exist.
  - CMOS will form a substrate and be integrated with non-traditional devices
  - Silicon technology will form the basis for new nanoscale structures
  - Powerful systems on a chip (SoC) consisting of various technologies such as nanoscale microelectronics, microfluidics, biosensors, novel nano-devices and complex information processing systems will emerge.

QUESTIONS

1. Consider a *p*-type Silicon wafer with a resistivity of 25 Ωcm. Assume that the hole and electron mobility in bulk silicon at 300 K is 400 cm^2/Vs and 1,300 cm^2/Vs, respectively.
   a. What is the doping concentration N_A of the acceptor atoms in the bulk silicon?
   b. What is the concentration of minority carriers in the bulk?
   c. Calculate the Fermi voltage ϕ_F of the bulk silicon at room temperature.
   d. At what temperature will the intrinsic carrier concentration n_i be equal to the doping level?

2. Consider a MOS capacitor whose gate dielectric stack consists of SiO_2/Si_3N_4 of thicknesses of 0.8 nm and 0.5 nm, respectively. The substrate doping level is $N_A = 10^{18}$ cm^{-3}. The gate is a n^+-poly silicon layer. The effective oxide charge $Q_{ox}/q = N_{ox} = 2 \times 10^{11}$ cm^{-3}.
   a. Calculate the equivalent oxide thickness EOT and the dielectric capacitance C_{eq} per unit area.
   b. What is the maximum width of the depletion layer when the surface is inverted?
   c. Calculate the flat band voltage V_{FB}, assuming that no ion implantation has been performed to adjust the threshold voltage.
   d. Calculate the value of the threshold voltage V_T.
   e. Assume that one needs a threshold voltage of 0.65 V. What is the implantation dose needed to adjust the threshold voltage? Do you implant donors or acceptors?

3. Consider the following modified MOS capacitor (gated diode). The capacitor has the same parameters as in problem 2. Explain qualitatively how the following parameters will be affected: $x_{d\max}$, V_{FB} and V_T.

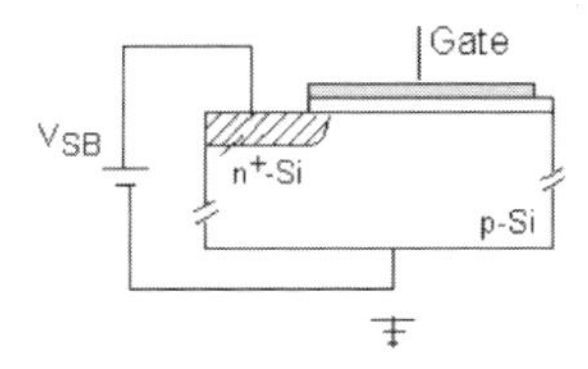

4. A NMOS transistor is fabrication on a silicon substrate with doping level $N_A = 1 \times 10^{18}\ \text{cm}^{-3}$. The oxide thickness is 6 nm and the threshold voltage at room temperature is equal to 0.45 V.
   a. Calculate the body effect coefficient m and the subthreshold slopes of the NMOS and PMOS transistors at 300 K, 375 K and 77 K. Assume that the contribution of the interface states to the body effect coefficient is 0.15.
   b. Calculate the leakage current in the NMOS at 300 K, 375 K and 77 K when $V_G = 0$ V, for a transistor with a W/L = 10. Assume that the mobility of the electrons at the respective temperatures is equal to: 500, 300, 1800 cm^2/Vs, respectively. The threshold voltage temperature coefficient is −0.8 mV/°C.
   c. Also calculate the current in saturation for $V_G = 2$ V for the three temperatures.
   d. In order to keep the leakage current at 77 K the same as at 300 K, how much smaller can one make the threshold voltage? Give the value of the threshold voltage at 77 K and 300 K.
   e. In order to keep the current in saturation at 77 K the same as at room temperature, how can one reduce the power supply $V_{GS} = V_{DD}$ at 77 K?. Assume that the V_{DD} at room temperature is 2 V.

5. Assume an oxide thickness $t_{\text{ox}} = 1.1$ nm, and a depletion layer thickness in the poly-silicon gate of $t_p = 0.2$ nm.
   a. What is the combined capacitance of the oxide and poly-silicon depletion layer?
   b. Assume that you need to keep the contribution of the poly-silicon depletion capacitor to the effective oxide less than 10% of the oxide capacitance. What is the maximum depletion thickness t_p of the poly-silicon layer one can tolerate?

6. Using the rules of constant field scaling, prove that the scaling of the transistor current I, the delay and power dissipation behave as given in Table 9.1.

7. Using the rules of constant voltage scaling, prove that the scaling of the transistor current I, delay and power dissipation behave as given in Table 9.1.

8. Using the rules of constant-field scaling, find the scaling rule for the subthreshold leakage current.

9. The scaling rules given in Table 9.1 assume that the transistor does not suffer from velocity saturation. What would the scaling rules be for the current, delay and power dissipation in case of velocity saturation and generalized scaling?

10. The transit time τ_{tr} in a transistor is defined as the ratio Q/I_{DS}, in which Q is the charge in the inversion region.
    a. Find the expression of the transit time assuming that the transistor is operating in the linear region. How does the transit time scale for constant voltage scaling?
    b. Find the expression of the transit time when the transistor is operating in the saturation region. You can assume a long channel transistor. The expression of the charge Q in the channel is two-thirds of the charge when the transistor is in the linear region.

11. Consider the following gate dielectric:
    a. A dielectric layer has a K-value of 16. What is the thickness of the layer in order to obtain an EOT of 1 nm? Assume that the K value of SiO_2 is 3.9.
    b. The dielectric stack consists of a 0.5 nm SiO_2 interfacial layer. What is the thickness required of the high-K layer in order to obtain an overall EOT of 1 nm? Assume that the high-K material has a K value of 16.

APPENDIX A. Useful Constants and Materials Properties:

Electronic charge, $q = 1.6 \times 10^{-19}$ C
Permittivity, ε_0 (free space) $= 8.85 \times 10^{-14}$ F/cm

$$\varepsilon_{SiO_2} = 3.45 \times 10^{-13}\ \text{F/cm}(K = 3.9)$$

$$\varepsilon_{Si3} = 6.64 \times 10^{-13}\ \text{F/cm}(K = 7.5)$$

$$\varepsilon_{Si} = 1.06 \times 10^{-12}\ \text{F/cm}(K = 11.9)$$

Boltzmann's constant: $k = 1.38 \times 10^{-23}$ J/K $= 8.62 \times 10^{-5}$ eV/K
Planck's constant: $h = 6.63 \times 10^{-34}$ *J-s* $= 4.14 \times 10^{-15}$ eV-s
kT at room temperature: $kT = 0.0259$ eV
kT/q at room temperature: $kT = 25.9$ mV
Intrinsic carrier concentration of Si at room temperature $n_i = 1.5 \times 10^{10}$ cm^{-3}.
Mobility of carriers in bulk silicon at room temperature: $\mu_n \approx 1.500$ cm^2/V-s
$\mu_p \approx 450$ cm^2/V-s

Density of Si atoms: 5×10^{22} atoms/cm^3
Bandgap at room temperature:

Si:	$E_g = 1.12\,\text{eV}$
Ge:	$E_g = 0.67\,\text{eV}$
GaAs:	$E_g = 1.43\,\text{eV}$
SiO_2:	$E_g \approx 8-9\,\text{eV}$

Breakdown field of Si, $E_c \approx 3 \times 10^5$ V/cm
Dielectric strengths of SiO_2, $E_d \approx 5-10 \times 10^6$ V/cm

*APPENDIX B. Typical Si Process Parameters**

	0.25 μm Process		0.18 μm Process	
	nMOS	pMOS	nMOS	pMOS
t_{ox} (nm)	5.5	5.5	4	4
N_{sub} (cm^{-3})	10^{17}	10^{17}	1.5×10^{17}	2×10^{17}
V_{TO}	0.42	−0.55	0.36	−0.41
μ (Low fields) (cm^2/V-s)	400	120	400	120
V_{max}(m/s)	1.4×10^5	1.0×10^6	1.4×10^5	1×10^6
$\gamma(V^{1/2})$	0.55	0.62	0.68	0.64

* Actual process parameters may vary.

REFERENCES

1. P. Gargini, The global route to future semiconductor technology, *IEEE Circuits and Devices Magazine*, 13 (March 2002).
2. International Technology Roadmap for Semiconductors (ITRS), Semiconductor Industry Association (ISA), 2003 Update (San Jose, 2003); http://public.itrs.net/.
3. F. Boeuf, T. Skotnicki, S. Monfray, C. Julien et al., *Technical Digest, IEEE International Electron Devices Meeting*, 637 (2001).
4. B. Yu, H. Wang, A. Joshi, Q. Xiang et al., *Technical Digest, IEEE International Electron Devices Meeting*, 937 (2001).
5. R. Goodall, D. Fandel, A. Allan, P. Landler, and H. R. Huff, *Proceedings Electrochemical Society* **2**, 125 (2002).
6. D. Kahng and M. M. Atalla, Silicon-silicon dioxide field induced surface devices, *IRE-AIEE Solid-State Device Research Conference* (Pittsburgh, 1960).
7. G. Moore, Electronics Magazine **39**, 114 (1965).
8. R. Schaller, IEEE Spectrum, 53 (June 1997).
9. M. Pinto, *Proceedings IEEE International Conference, Solid State Circuits,* 26 (2000).
10. R. Turton, *The Quantum Dot, A Journey into the Future of Microelectronics*, (Oxford University Press, New York, 1995).
11. J. Plummer, M. D. Deal, and P. B. Griffin, *Silicon VLSI Technology*, (Prentice Hall, Upper Saddle River, NJ, 2000).
12. R. Pierret, *Semiconductor Fundamentals: Volume I, 2/E*, (Prentice Hall, Upper Saddle River, 1988).
13. B. G. Streetman and S. Banerjee, *Solid State Electronic Devices,* 2nd ed., (Prentice Hall, Upper Saddle River, 2000).
14. R. W. Keyes, *IEEE T. Electron Dev.* **33**, 863 (1986).
15. K. Rim et al., Symposium, VLSI Technology, 98 (2002).
16. Y. Taur and T. Ning, *Fundamentals of Modern VLSI Devices*, (Cambridge University Press, New York, 1998).
17. D. Buss, *Digest of the 2002 IEEE International Solid-State Circuit Conference*, 3 (2002).
18. E. J. Novak, IBM J. Res. Dev. **46**, 169 (2002).
19. P. Gelsinger, *Digest of the 2001 IEEE International Solid-State Circuit Conference,* 3 (2001).
20. D. Frank, R. Dennard, E. Nowak, P. Solomon, Y. Taur, and H. S. Wong, *P. IEEE* **89**, 259 (2001).
21. S. Krishnan and J. G. Fossum, *IEEE Circuits Device*, 32 (July 1998).
22. M. Palella and J. Fossum, *IEEE T. Electron Dev.* **49**, 96 (2002).
23. H. S. Wong et al., *IEEE IEDM Technical Digest,* 407 (1998).
24. M. Jurczak et al., *IEEE T. Electron Dev.* **47**, 2179 (2000).
25. L. Geppert, *IEEE Spectrum,* 28 (October 2002).
26. R. M. Wallace and G. Wilk, MRS Bulletin, 192 (March 2002).
27. P. H. Wong, IBM *J. Res. Dev.* **46**, 133 (2002).
28. J. M. Hergenrother et al., IEDM, 3.11 (December 2001).
29. Y. Taur, IBM *J. Res. Dev.* **46,** 213 (2002).
30. F. G. Pikus and K. K. Likharev, *Appl. Phys. Lett.* **71**, 3661 (1997).
31. C. Svensson, *Technical Digest IEEE International Solid-State Circuits Conference*, S28 (February 2003).
32. E. J. Nowak, IBM J. Res. Dev. **46,** 169 (2002).
33. P. H. Wong et al, P. IEEE, **87,** 537 (1999).
34. X. Huang et al, *Technical Digest IEDM*, 67 (1999).
35. D. Hisamoto et al, *IEEE T. Electron Dev.* **47,** 2320 (2000).
36. Y. K. Choi et al., *Technical Digest IEDM*, 421 (2001).
37. J. Huchby, G. Bourianoff, V. Zhirnov, and J. Brewer, Extending the Road beyond CMOS, *IEEE Circuits & Device Magazine*, 28 (March 2002).
38. R. Compano, Ed., *Technology Roadmap for Nanoelectronics*, 2nd ed., European Commission Information Society Programme, (2000). (ftp://ftp.cordis.lu/pub/ist/docs/fetnidrm.zip).
39. A. Steane and E. Rieffel, *IEEE Computer*, 38 (January 2000).
40. D. DiVincenzo, *Technical Digest IEDM*, 12 (2000).
41. S. Forrest et al, *IEEE Spectrum*, 29 (August 2000).

42. Ten technologies that will change the world, *MIT Technology Review*, 97 (February 2001).
43. D. Rothman, Molecular Computing, *MIT Technology Review*, 53 (May/June 2000).
44. P. Colling and P. Avouris, *Scientific American*, 62 (December 2000).
45. J. Appenzeiler, Ph. Avouris, V. Derycke, R. Martel, and S. Wind, *39th Proceedings of the Design Automation Conference*, 94 (2002).
46. Nanotube Industry Means Business, *Smalltimes* **2**, 32 (July/August 2002).

10

Molecular Electronics

Michael Zwolak[1] and Massimiliano Di Ventra[2]

[1] *Physics Department, California Institute of Technology, Pasadena, CA*

[2] *Department of Physics, University of California, San Diego, La Jolla, CA*

As discussed in the previous chapter, the limits of silicon-based computer technology (microelectronics) are fast approaching. Alternative technologies are thus being investigated. Molecular electronics is one of these alternatives. Molecular electronics can be loosely defined as a subfield of nanotechnology that envisions the use of single molecules, or small groups of molecules, as components in electronics applications. Along this line, molecular electronic devices could form the next generation of transistors, sensors, and circuits. Molecules can have feature lengths as small as 1 nm, making it possible to extend the validity of Moore's law for many years to come, should we be able to integrate them into circuits. In addition, although the envisioned operation of many molecular devices mimics traditional devices, the quantum world can open up novel possibilities for the way devices work.

The idea of using molecules as components in electronics was suggested more than two decades ago.[1] However, it is only in the last decade that we have witnessed an increased interest in the field. This is partly due to both our growing ability to fabricate contacts with nanometer size, which can accommodate a small number of molecules in between them, and the development of self-assembly. The methods used in making prototype devices are, however, often difficult to control and, as of now, not quite ready to make the necessary step into commercialization. Nevertheless, the field of molecular electronics is experiencing tremendous growth, with new ideas being generated at an amazing rate.

This chapter will provide a limited overview of the field, not a review. This simply means that we will select a few examples from the literature and discuss them in the broader context of the physical phenomena we observe at the nanoscale. In particular, we will discuss ways to assemble and probe molecular devices, report on their electrical characteristics, and outline the various factors that influence their properties. We will also discuss possible charge transport mechanisms in these devices, and finally we will outline integration strategies and related difficulties associated with working at such small length scales. Our focus will be on the transport properties of organic molecules and small fullerenes. We will also briefly discuss the possibility of using DNA in electronics. Nanotube-based electronic devices and their transport properties have been covered in the previous chapter and in Chapter 6.

10.1. TOOLS AND WAYS TO BUILD AND PROBE MOLECULAR DEVICES

Two ingredients are necessary to make a molecular device: an aperture of nanoscale dimensions and a way to arrange one or more molecules within it. Two main methods have been used in recent years to make nanoscale apertures: the break-junction technique[2] and electromigration.[3] Combined with the property of certain molecules to self-assemble on metallic surfaces, these two methods can, in principle, be used to create single-molecule nanojunctions. Self-assembly has been discussed at length in Chapter 2. We just recall that its basic principle is to exploit chemical interactions to create a structure on the nanoscale. In the context of molecular electronics, an example of such an interaction is the binding of a thiol group (—S—H) with a gold surface. The use of a Scanning Tunneling Microscope (STM) to probe the conductance of molecules self-assembled on a metal surface will be discussed later on in this section.

10.1.1. Break-Junction Technique

A break-junction is formed by literally breaking a thin metal wire to form a very small gap in the wire, in which a molecule or a group of molecules can be placed. This method was developed in the late 1980's and early 1990's.[4] Typically, a thin metal wire is fabricated using standard techniques like optical or e-beam lithography (see Chapter 1), on top of a flexible substrate, e.g., polyimide. A thin wire can also be attached to the substrate by gluing it on using an epoxy. A sharp object (or e-beam lithography) is then used to notch the wire. The wire is subsequently broken at the point of the notch by bending the substrate, which, due to its flexibility, will remain intact. The bending is then slightly relaxed to bring the two parts of the wire back into contact. A piezoelectric actuator is used to both break the wire and to adjust the width of the gap between the two parts of the broken wire. The piezoelectric actuator uses a piezoelectric material, typically a ceramic, to precisely adjust the distance between the two wires. A piezoelectric material deforms under application of an external voltage. This material is placed below the substrate, and when it expands, it bends the substrate. The more the substrate is bent, the larger the gap between the two wires. A typical experimental setup is shown in Figure 10.1. This is the so-called mechanically controllable break-junction.

There are several major advantages to the break-junction technique. It works in a range of conditions from ambient to low temperature/high vacuum, and in conjunction with other experimental setups. It also has a relatively high experimental "success rate," in the sense that the formed nanojunction can be reproduced fairly well. Additionally, since the substrate is bent to adjust the size of the electrode gap, only small fluctuations in the gap size are expected. This is because the size of the electrode gap changes by a lower amount than the amount of expansion of the piezoelectric element (called the reduction factor) due to the setup geometry. In this way, the effects of fluctuations in the piezoelectric element expansion are reduced. This allows for accurate control of the size of the electrode gap.

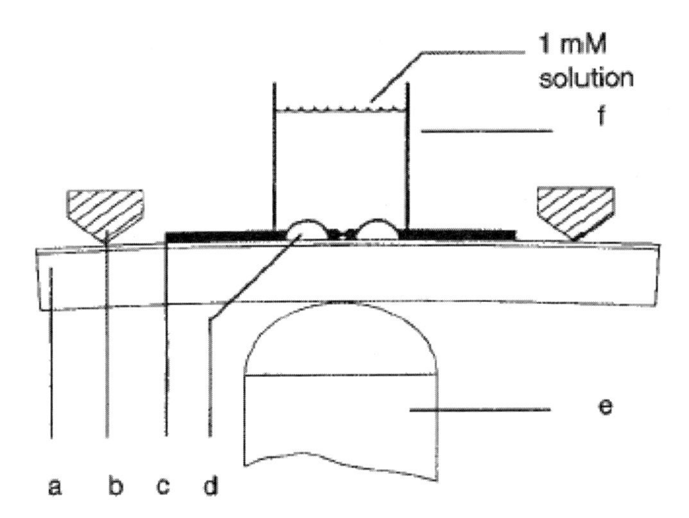

FIGURE 10.1. Experimental setup of a mechanically controllable break-junction with (a) the flexible substrate, (b) the counter supports, (c) the notched wire, (d) the glue contacts, (e) the piezoelectric element, and (f) the solution containing the molecule of interest. (From Ref. 2 by permission of the American Association for the Advancement of Science.)

The conductance of benzene-1,4-dithiolate molecules was first measured using this technique.[2] In this experiment, a notched gold wire was glued onto a substrate. Benzene-1,4-dithiol molecules were self-assembled onto the gold surface via adsorption of the thiol group on the gold surface to form a self-assembled monolayer (SAM). The gold wire was then broken in the solution, which causes the SAM formation around the two newly formed gold tips (note that by "tip" we do not necessarily mean a single atom termination of the wire). The solution was subsequently evaporated, and the two sides of the wire were brought together until conduction was observed. This is thought to be the point when a single molecule bridges the two gold tips. The spacing of the two gold tips was estimated to be about 0.8 nm, i.e., enough to accommodate a benzene-1,4-dithiolate molecule. A schematic of the break-junction used in this experiment is shown in Figure 10.2. The results of the experiment are discussed below in Section 10.2. We just note here that even though the spacing of the nanojunction can be controlled with some accuracy, it is not at all clear how many molecules bridge the gap and what the actual bonding configuration is to the gold tips.

10.1.2. Forming Nanogaps with Electromigration

An alternative way of creating nanoscale junctions is by using a physical phenomenon known as electromigration. Electrons diffusing in a conductor can scatter atoms away from their equilibrium position. If the electric current density is large enough, i.e., if a lot of these scattering events takes place, the atoms can be moved along the conductor, leaving holes of missing atoms behind. Electromigration is still a major concern in microelectronics since it is the major failure mechanism in solid-state circuits. However, in the context of molecular electronics it can be used advantageously.[5] A nanowire is first made using

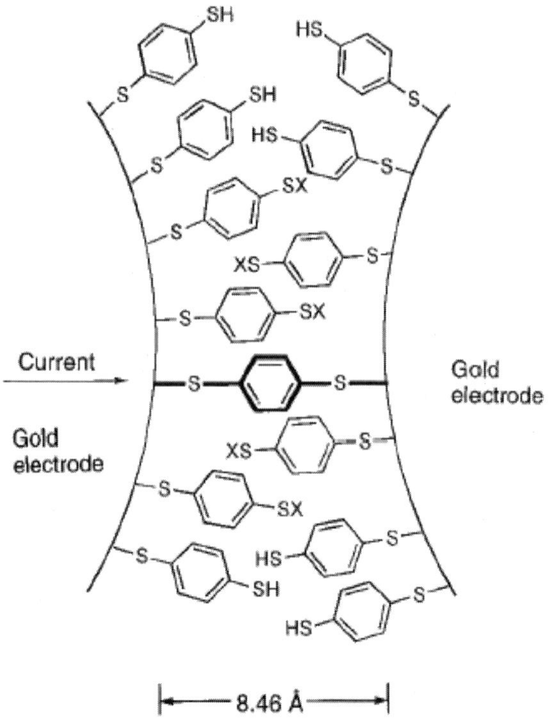

FIGURE 10.2. Schematic of a break-junction with a SAM of benzene-1,4-dithiol molecules. (From Ref. 2 by permission of the American Association for the Advancement of Science.)

e-beam lithography. A nanometer-size gap is then created by controlling the amount of current that passes in the wire. While increasing the current, the conductance of the wire is monitored with a voltage probe. For small current values, the conductance of the wire remains unaffected. With further increase of the current, however, a change in the resistance is observed that indicates the onset of electromigration. By increasing the current further, the conductance eventually drops almost to zero, indicating a gap has formed. The width of this gap is estimated to be about 1 nm from the value of the tunneling resistance between the two newly formed electrodes. The process seems to be highly reproducible as long as the diameter of the wire is a few nanometers.[5] Figure 10.3 shows a Scanning Electron Microscope (SEM) image of a gold wire before [Figure 10.3(a)] and after [Figure 10.3(b)] the break-up process due to electromigration. This technique can be combined with self-assembly to create a molecular junction in a similar manner to the break-junction technique. As in the case of the mechanically controllable break-junction, it is uncertain how many molecules bridge the nanogap and what the bonding configuration is.

10.1.3. Probing Individual Molecules

Information on the transport properties of molecules can also be obtained using a Scanning Tunneling Microscope (STM) [see Chapter 3]. The molecules are first self-assembled

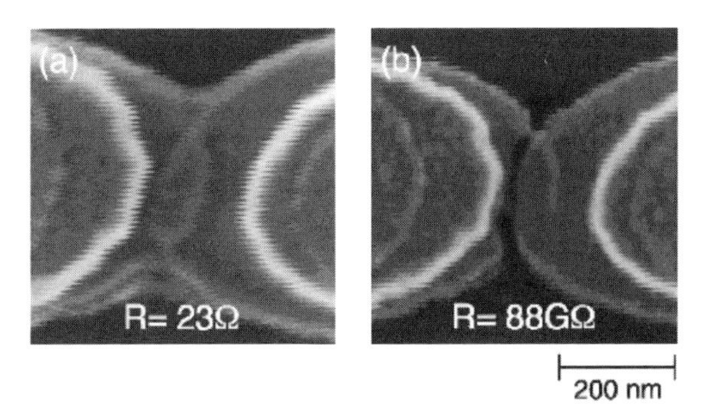

FIGURE 10.3. SEM image of a gold wire (a) before and (b) after break-up by electromigration. (From Ref. 5 by permission of the American Institute of Physics.)

on a metal surface and then the STM tip is used as a second contact. This is schematically shown in Figure 10.4, where a single molecule is shown between an STM tip and an electrode surface.

As discussed in Chapter 3, the operation of the STM relies on what is called the tunneling current, i.e., on the fact that a particle can penetrate into a classically forbidden energy region. In Figure 10.4, the tunneling current flows between the tip and the SAM to be scanned when the tip is sufficiently close to the sample (a distance generally less than 1 nm). The tip is moved toward the surface similarly to the way the break-junction gap is adjusted: a piezoelectric element, to which a voltage is applied, is used to adjust the tip height until the current flows. The whole sample can be scanned by, for instance, changing the tip height in order to keep a constant tunneling current, in this way obtaining a map of the surface height (surface topography). To measure the conductance of single molecules, the feedback loop, which adjusts the height of the STM, is shut off. This keeps the tip at a constant position while measuring the current-voltage (I-V) characteristics, i.e., changing the external voltage and recording the current through the sample. It is important to note that the measured conductance can correspond to a small group of molecules, not necessarily to a single one. Another way to measure the I-V characteristics of the sample is to use an alternating current STM. By applying an alternating voltage across the tip and the metal surface, a frequency-dependent current can be measured.

Using an STM to measure the I-V characteristics of molecules has actually been one of the first methods to obtain information on their transport properties. However, this approach has some limitations. The magnitude of the current can strongly depend on the tunneling barrier between the STM tip and the molecule, introducing an additional and undesirable "contact" resistance. This can be partly corrected for by, e.g., adding a thiol group to the end of the molecule and then attaching it to a gold nanoparticle.[6] The nanoparticle can then make contact with the STM tip. However, this introduces additional uncertainties in the current due to the particle size and how well it is contacted to the STM tip. Another disadvantage is that it is unlikely that lone molecules will be perpendicularly aligned to the surface. A possible solution to this problem is to use an SAM of "insulating" molecules to prop up the small group of "conducting" molecules that need to be probed, as shown in Figure 10.4.[7] Here we call "insulating" a particular molecule whose resistance has been measured to be much larger than the molecule of interest in a similar experimental set-up.

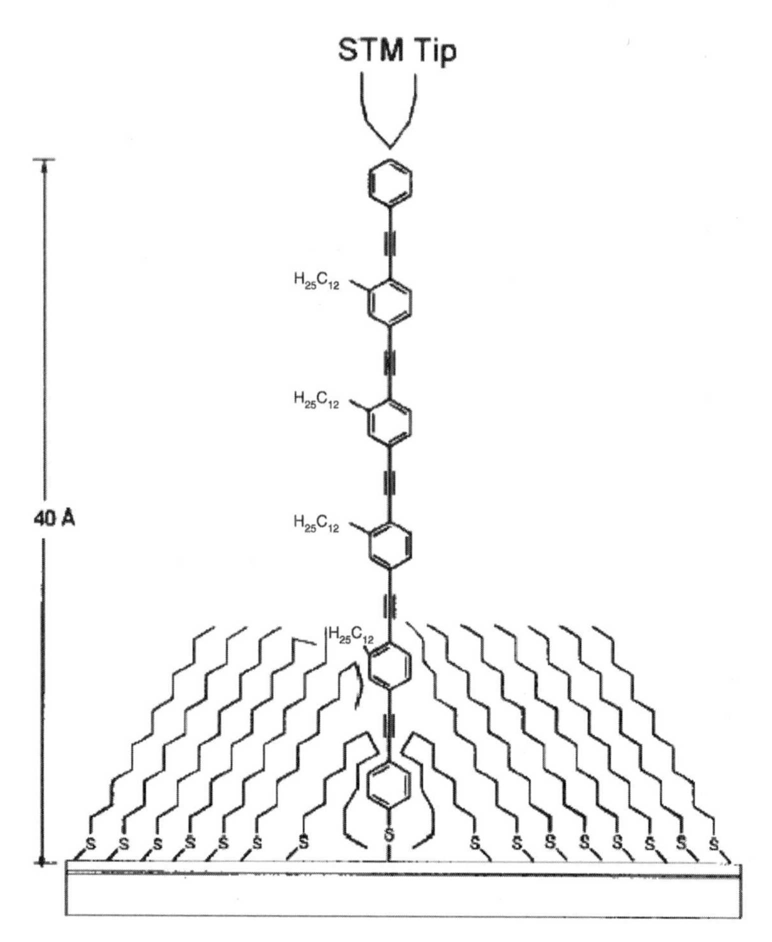

FIGURE 10.4. Schematic of a SAM used to prop up a single molecule so that charge transport along the length of the molecule can be probed using an STM. (From Ref. 7 by permission of the Institute of Electrical and Electronics Engineers.)

For instance, an alkanethiol can be used to form the support SAM, while π-conjugated molecules (which must be longer so they stick out far from the support SAM and can be located by the STM) are deposited afterward.[8] It has been shown that the molecules of interest will insert into defect regions in the support SAM. These are regions where the directionality of the molecules in the SAM change. In this way, the molecules of interest will be propped up by the SAM and located by the STM, and then their conductance can be measured.

Similarly, the conductance of single molecules or small groups of molecules can be measured by using an Atomic Force Microscope (AFM) [see Chapter 1]. Here, a tip is held at the end of a cantilever (or arm). The tip is dragged across the sample (or the sample moved) using a piezoelectric element. The tip will move up and down with the height of the local area of the sample. Generally, a laser beam is reflected off the back of the cantilever, and a multi-segment photodiode detects the movement of the beam. There are two modes that the AFM can be run in, either a mode with no gain or a mode with high gain (constant force). In the former, the tip is just dragged across the sample, but the cantilever will bend and exert additional force where the sample is higher. As long as this is calibrated for, then the surface can be scanned and the height extracted. In high gain mode, either the sample height or cantilever height is adjusted so that there is no additional bending of the cantilever, which gives a constant force. The measurement of the conductance of a SAM with an AFM (conducting probe AFM) can be done by applying a voltage across the cantilever (and tip) and the electrode on which the sample is assembled.

10.2. CONDUCTANCE MEASUREMENTS

In the previous section, we discussed ways to measure and probe the I-V characteristics of individual molecules or small groups of molecules. We are now ready to discuss some of the measurements reported in literature. Instead of reviewing all measurements done so far (which is beyond the scope of this chapter), we will select a few that show different features. This will allow us to introduce the different transport mechanisms that are believed to occur in these systems. The reader is warned that the field is rapidly changing, with experiments being performed in an increasingly controlled manner. Many of the physical interpretations we report can therefore change in the near future.

10.2.1. Contact Resistance and Quantized Conductance

A prototype molecular device that has received much attention both experimentally and theoretically is one formed by a benzene-1,4-dithiol molecule between two gold electrodes. In the previous section, we discussed how this device can be made using self-assembly and a mechanically controllable break-junction. The I-V characteristics of this device are shown in Fig. 10.5(a) and a schematic of its possible atomic configuration in Fig. 10.5(b). It is assumed that the hydrogens of the thiol groups (—S—H) at each end of the molecule desorb and the sulfur atoms bind strongly to the surface of the two gold tips. The I-V characteristics show nonlinear behavior with peaks and valleys in the conductance as a function of bias. A molecule like this has several electronic states that are filled with electrons up to the highest occupied molecular orbital (HOMO), which is a few electron volts below the lowest unoccupied molecular orbital (LUMO). When the molecule makes contact with the gold electrodes these states broaden and the Fermi level lies between the HOMO and the LUMO [see Figure 10.6(a)]. By increasing the bias, one of the electronic states of the molecule can align with the left or right chemical potentials giving rise to a peak in the conductance [see Figure 10.6(b)]. This phenomenon is known as "resonant tunneling" and has been invoked to explain the peaks and valleys observed in this experiment.[9,10] Negative differential resistance, a phenomenon also associated with resonant tunneling (see below), was not observed in this experiment. What theory has not been able to explain yet is the large resistance observed in this and many other experiments with organic molecules. This brings us to the question of what actually determines the resistance in a nanoscale junction.

The electrodes have many current-carrying modes (essentially infinite), while the junction, specifically the molecule, has only a few. Thus, as current flows from one electrode into

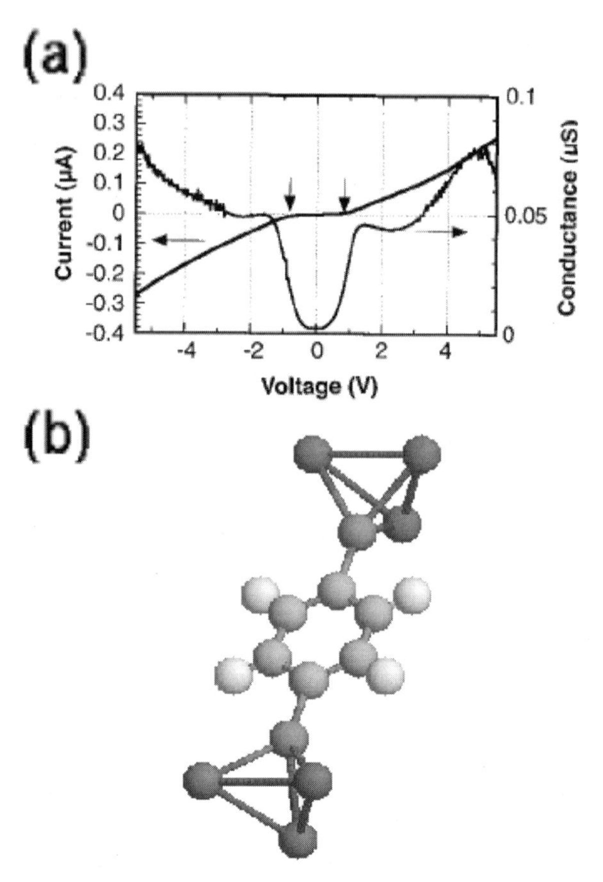

FIGURE 10.5. (a) *I-V* characteristics for Benzene-1,4-dithiol between gold electrodes. (From Ref. 2 by permission of the American Association for the Advancement of Science.) (b) Schematic of the molecule in contact with gold electrodes, represented here as three gold atoms.

the junction, it must redistribute itself into the fewer modes available. This redistribution introduces a *contact resistance* which is inversely proportional to the number of modes in the junction and inversely proportional to the probability per mode that an electron can cross the junction (transmission probability). The larger the number of modes in the junction, the less resistance this redistribution will introduce. If only one mode is accessible in the junction and its transmission probability is one, then a conductance (inverse of resistance) of $2e^2/h$ should be observed, where e is the electron charge and h is Planck's constant. This phenomenon is referred to as *quantized conductance*, and implies that even an ideal wire will show quite a large resistance of $\sim$12.9 kΩ. This resistance is due to the contact resistance associated with each mode. That is, the resistance is not in the wire itself, but is instead associated with redistribution of the current carrying modes in the contact where it connects to the conductor (for a more extensive discussion the reader is encouraged to look at Ref. 11). Large contact resistance can affect the device performance in many ways. For instance, it can decrease the operational speed of the device or facilitate inelastic scattering (which will heat the device).

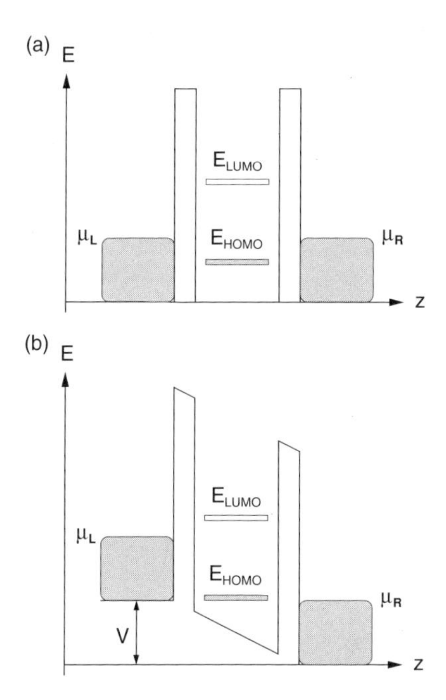

FIGURE 10.6. (a) Schematic of the energy configuration of a molecular junction with the Fermi level located in the HOMO-LUMO gap of the molecule. The HOMO and LUMO states are depicted as broadened energy levels due to their interaction with the electrodes. (b) As the voltage is increased, eventually the HOMO or LUMO will align with the Fermi level of the right or left electrode. This will cause an increase in the magnitude of the current.

Quantized conductance has been observed, for instance, in break-junction experiments, where the conductance of a gold wire was measured as a function of time during break-up of the wire.[12] In this case, it is believed that quantized conductance is a direct consequence of the formation of linear chains of a small number of gold atoms, each contributing a small number of channels.[12] On the other hand, in molecular devices quantized conductance is rarely observed. Actually, their resistance is quite large. This can be explained as follows. The number of channels accessible at zero temperature is determined by the number of states with energies between the Fermi levels of the two electrodes. When the molecule is in contact with the electrodes there is generally only a small amount (density) of states between the HOMO and the LUMO, which is due to the presence of the electrodes. Thus, the molecule offers a large resistance to current flow. The amount of resistance in molecular junctions is linked to the amount of overlap between the electronic states of the molecule and the conduction states in the bulk electrodes. This overlap determines the probability that an electron can actually transmit across the molecule. Poor overlap can be caused by, e.g., the spatial distribution of the molecular orbitals or by poor physical contact (like in STM experiments), which causes the electronic states of the electrode and the molecule to be spatially separated.

The resistance due to poor overlap of electronic states is well illustrated by the break-junction experiment with benzene-1,4-dithiolate molecules. A possible atomic configuration of this system is the one depicted in Fig 10.5(b). If this is the case, the sulfur atoms at each end of the molecule strongly bind to the two gold electrodes. By strongly we mean that the energy required to break the sulfur-gold bond is substantial (of the order of few electron volts). However, the sulfur-gold contact is a poor contact in terms of transmission probabilities: this contact induces a bad overlap of the electronic states of the molecule responsible for transport with the conduction states of the electrodes.[10] If the contact is instead through a single gold atom bound to the sulfur, there is even worse overlap of the conduction states of the junction and the electrode, which gives an even higher resistance.

That the contact geometry can change the resistance by significant amounts has fostered an intense debate on the actual atomic configuration of these systems. Theoretical work has been in fairly good agreement with the experimental results regarding the shape of the I-V curve but disagrees with the absolute value of the current.[9,10,13] Part of this disagreement has been attributed to the role of the contact geometry in changing the resistance of the junction. This has been generally true with many of the molecular devices reported in literature. To better understand this issue we need to recall that molecular devices are generally fabricated with either mechanically controllable break-junctions or electromigration. Both techniques yield atomic-scale contacts of unknown geometry: the atomic regions where the molecules are supposed to bind to the electrodes are not necessarily smooth nor symmetric on opposite sides of the junction. In addition, the molecules can bind on just one side of the junction and not the other, or many molecules can bridge the two electrodes allowing current to flow along different paths.[14]

Recent theoretical reports have also shown that unintentional adsorption of atomic species, e.g., oxygen, on the electrode surfaces close to (but not necessarily binding to) the molecule can completely change the I-V characteristics of these devices.[15,16] The latter effect is partly due to the electrostatic interaction between the charge on these species and the states of the molecular device. Similar effects have been experimentally demonstrated in carbon nanotube field-effect transistors.[17] While these findings show that the I-V characteristics of molecular devices can be easily tuned (much more than in conventional microelectronic devices), they also suggest that in order to have reproducible and reliable operation, atom-by-atom control of their configuration is necessary.

10.2.2. Molecular Switches, Transistors and the Like

Despite the above difficulties, several groups have succeeded in making fundamental circuit elements, like diodes, switches and transistors.[2,3,18,19,21,22] Here we give a few examples of how these basic elements can be realized at the molecular level. However, it is important to stress again that the details of the electronic and structural properties of these systems are not completely known and, therefore, the interpretation of the following experimental results could change in the future.

A molecular switch is a device which will be activated by an external event to change its state from "on" to "off". An electrical switch uses a change in voltage to change a molecular junction from a conducting state (on) to a non-conducting state (off). A transistor is somewhat similar, where a gate voltage changes the current from a low to a high value, or vice versa, but the transistor does not maintain its state.

A molecule that can be used as a switch is shown in Figure 10.7; it is called a catenane. This molecule changes its internal structure when different voltages are applied across it. For details on how these structures have been made, the reader is referred to the original papers

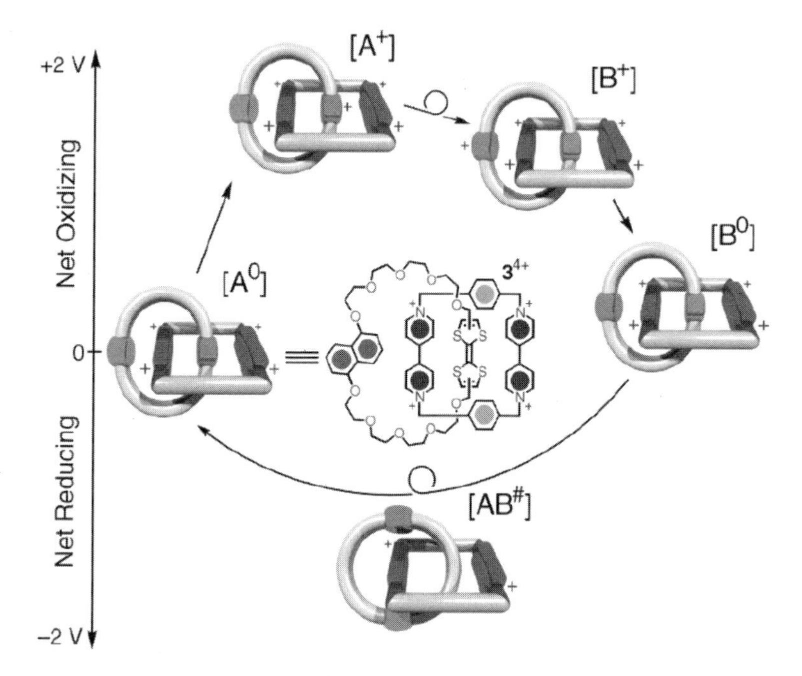

FIGURE 10.7. Structure of the [2]catenane 3^{4+} is shown in the middle. Mechanochemical mechanism that is thought to be responsible for the hysteresis observed in the electrical current. This mechanism is described in the text. (From Ref. 18 by permission of the American Chemical Society.)

(see, e.g., Ref. 18 and references therein). Here we just report that the catenanes were first synthesized and then deposited on a series of polysilicon wires using the Langmuir–Blodgett technique (see Chapter 2). A second layer of perpendicular wires was deposited on top of that to form the circuit. The current through the device was measured by applying a series of high voltage pulses, and then probed with a low applied voltage after each pulse. Hysteresis was observed in the current-voltage characteristics (see Figure 10.8), which means that the current assumes a different value when it is scanned by increasing the magnitude of voltage

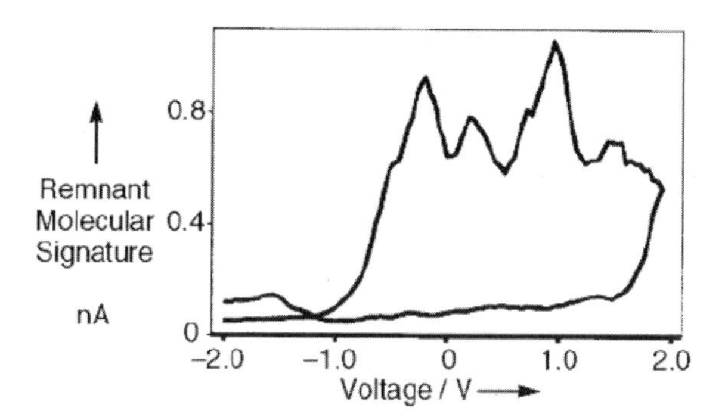

FIGURE 10.8. Electrical current hysteresis observed in catenane switches. (From Ref. 18 by permission of the American Chemical Society.)

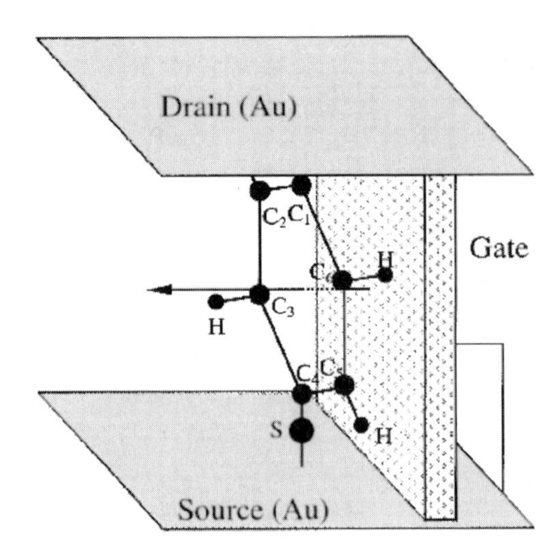

FIGURE 10.9. Three terminal setup for a molecular device made with a benzene-1,4-dithiolate molecule. (From Ref. 15 by permission of the American Institute of Physics.)

pulses compared to decreasing the magnitude of voltage pulses. Hysteresis is necessary in devices like computer memory.

A possible mechanism for such a behavior is what has been called "mechanochemistry". This mechanism is shown in Figure 10.7. The catenane has two states: "open" ($[A^0]$) and "closed" ($[B^0]$). The "open" state being a high current state and "closed" being a low current state. Application of a positive voltage brings $[A^0]$ to the state $[A^+]$ which will rearrange due to repulsion of the positive charge on the two rings. When the voltage is decreased, this will bring the catenane to the closed state $[B^0]$. The catenane can be brought back to the $[A^0]$ state by the application of a negative voltage. Similar results have been reported with other molecules, like rotaxanes (see, e.g., Ref. 18).

One of the most important elements in microelectronics is the field-effect transistor (FET). Using nanotubes to make FETs has been described in the previous chapter and in Chapter 6. Single-electron transistors made with quantum dots will be discussed in the following chapter. Realizing such an element out of single molecules is certainly an important achievement. However, making an FET from small molecules presents a greater challenge, partly due to the limits of the fabrication techniques described above, and partly to the difficulty of placing a third terminal in close proximity to the molecules. The prototype benzene-1,4-dithiolate molecule was one of the first proposals of a molecular FET.[20] A possible setup of such device is shown in Figure 10.9. Here, the molecule plane faces an insulating surface across which a gate voltage is applied to modulate the current that flows from source to drain. It was found theoretically that the source-drain current can be modulated by an order of magnitude by increasing the gate voltage. The effect of the gate voltage is simply to displace the energy of the electronic states of the system with respect to the chemical potentials of the electrodes until resonant tunneling occurs.

An alternative transistor has been demonstrated using C_{60}.[3] In this case, electromigration was used to create two gold electrodes with a nanoscale junction in between. C_{60} was deposited from a dilute solution in order to ensure that multiple C_{60} molecules would not

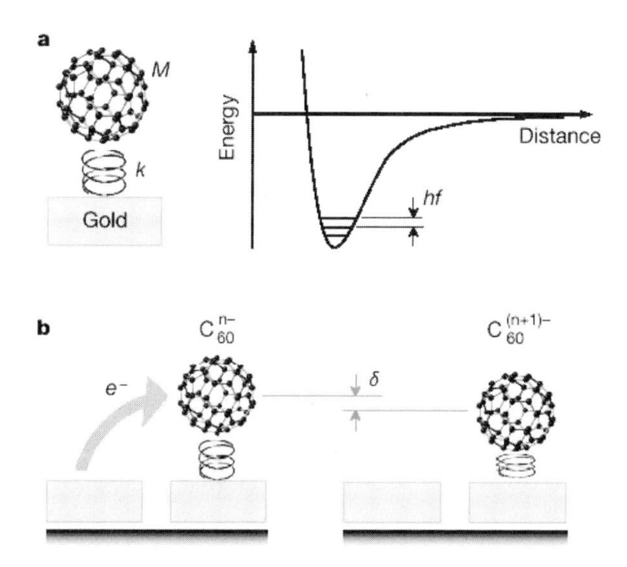

FIGURE 10.10. (a) Schematic drawing of C_{60} as an oscillator on a gold surface, with an interaction potential shown on the right. (b) As an electron goes onto C_{60}, the molecule is attracted toward the surface due the attraction between the electron and its image charge. This gives C_{60} mechanical motion. (From Ref. 3 by permission of Nature Publishing Group.)

end up in the junction. By changing the gate voltage, the magnitude of the current could be changed. Stairsteps were observed in the *I-V* curve. The steps were attributed to coupling of the tunneling through the junction with the vibrational mode of the whole C_{60} molecule with respect to the gold surface (see Figure 10.10).[3]

Single-molecule transistors were also demonstrated recently using both $[Co(tpy-(CH_2)_5-SH)_2]^{2+}$ [21] and divanadium.[22] A schematic of the first one is shown in Figure 10.11. The device was made by first self-assembling the molecules onto a gold wire. Then electromigration was used to create the junction as discussed above. For certain samples, it was believed that a single molecule bridged the electrodes. In this case, the current as function of both gate voltage and applied source-drain voltage displayed characteristics similar to that of single-electron transistors (see next chapter). Since such characteristics are related to the ability of the molecule to accommodate single electrons at a time, the cobalt atom is likely to play a central role in the operation of this device.

10.2.3. Electronics with DNA

Charge transport in biological molecules is of interest to several disciplines and is now the subject of intense research. In particular, DNA and its transport properties have received much attention in the last decade in view of its possible use in molecular electronics.[24,25] Since this topic is increasing in importance and is not covered in any part of this book, we give here a brief overview of the status of this area.

DNA is a double helix formed by a sequence of base pairs with a phosphate-sugar backbone. There are four possible bases: thymine, cytosine, adenine, and guanine.

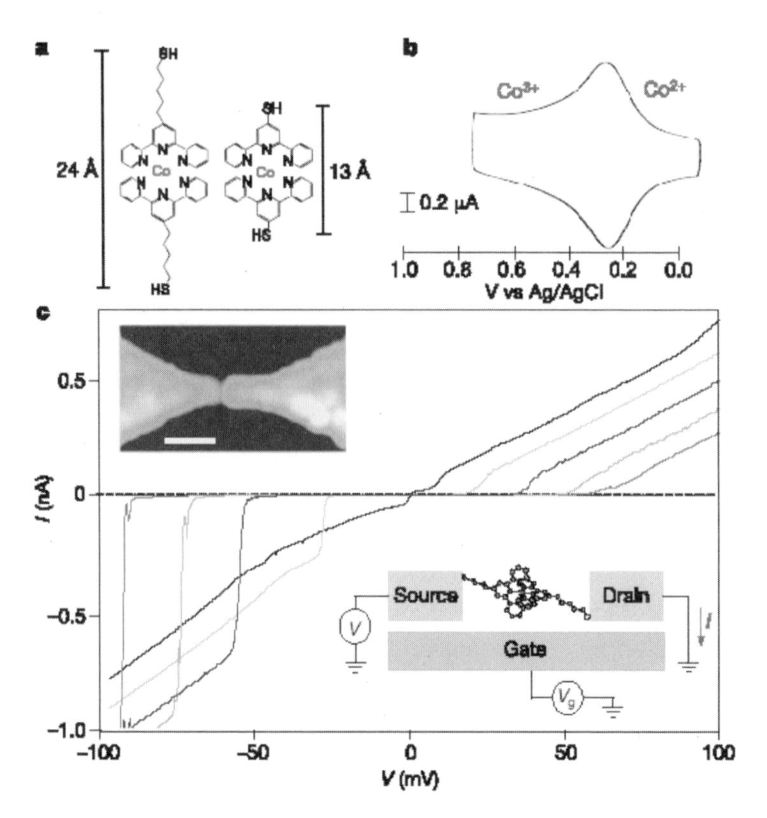

FIGURE 10.11. (a) Structure of the $[Co(tpy\text{-}(CH_2)_5\text{-}SH)_2]^{2+}$ and $[Co(tpy\text{-}SH)_2]^{2+}$ molecules. (b) Cyclic voltammogram of $[Co(tpy\text{-}SH)_2]^{2+}$. (c) *I-V* curve at different gate voltages for $[Co(tpy\text{-}(CH_2)_5\text{-}SH)_2]^{2+}$, the upper inset shows an AFM image of the experimental setup and the lower inset shows a schematic of the single molecule transistor made with $[Co(tpy\text{-}(CH_2)_5\text{-}SH)_2]^{2+}$. (From Ref. 21 by permission of Nature Publishing Group.)

Many experiments have examined the charge transport properties of DNA between electrodes, but there have been seemingly contradictory conclusions. Experiments show DNA has metallic, semiconducting, insulating, or even superconducting properties. This wide range of characteristics can be attributed to both the complexity of the DNA structure and the variety of experimental conditions under which the transport properties of this molecule can be measured. For example, base sequence, length, orientation, counterions, temperature, electrode contact, adsorption surface, structure fluctuations, and so on, can all affect its conducting properties. Therefore, it is still not clear at all how DNA (more precisely, one of its many forms) *intrinsically* conducts. In some experiments, DNA is believed to conduct by tunneling of holes between the guanine bases. Guanine has the lowest oxidation potential, so it is the most favorable place for holes to be located. Between guanines there will be some sequence of bases that acts as a tunneling barrier. Holes can then tunnel between the guanine bases, and transverse the length of DNA (see Section 10.3 below for details on transport mechanisms).

DNA is particularly versatile, and even though it might not turn out to have the desired transport properties for certain applications, it can still be of great use in molecular electronics. For instance, DNA templated wires have been demonstrated.[26] To make these wires, DNA was used to connect two electrodes and the counterions (which bind along the DNA backbone to neutralize the negative charge on the phosphate groups) were replaced with ions of, e.g., silver or gold. Further metal was then deposited along the DNA to create very thin wires, ten's of nanometers thick. These wires have been found to conduct very well and could potentially be used to serve as interconnects in molecular circuits. The fact that DNA could serve to connect up molecular devices has been shown in recent experimental work, which used DNA self-assembly to perform molecular lithography.[27] Here, a group of researchers exploited DNA interactions with proteins to selectively metallize (i.e., create a metal wire around) lengths of DNA. In this way, the researchers were able to create DNA chains that had a bare DNA region between two metallized regions. This bare region could provide a location for a molecular device. There are numerous other possibilities to use DNA's self-assembly properties to create well structured nanoscale templates. The interested reader is referred to Ref. 25 for a review on DNA's possible use in molecular electronics. Also contained there is an overview of the possible mechanisms by which charge transport could occur in DNA, some of which are reviewed below in a general context.

10.3. TRANSPORT MECHANISMS AND CURRENT-INDUCED EFFECTS

After outlining the fabrication methods and some of the characteristics of molecular devices, we can now get into more of details of the transport mechanisms and current-induced effects occurring in these systems. This will give us a better feeling of the physics behind their operation. We will not be able to discuss all possible mechanisms, in particular those related to many-body effects (e.g., the Kondo effect), as they would require more advanced knowledge. We refer the reader to the specialized literature for these topics (see, e.g., Ref. 28). Single-electron tunneling phenomena will be discussed in the next chapter.

10.3.1. Resonant Tunneling: Coherent and Sequential

The laws of quantum mechanics allow a particle to overcome a large energy barrier even if the particle does not have the required energy. This quantum mechanical effect is known as tunneling. This is schematically shown in Figure 10.12. This effect arises because a particle is described by a wavefunction, $\Psi(\vec{x})$. The absolute value squared of this function

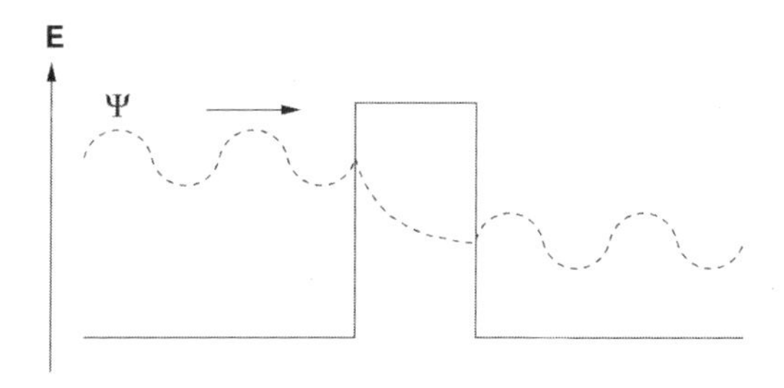

FIGURE 10.12. Particle tunneling through an energy barrier. The real part of the right moving wavefunction $\Psi(x)$ is schematically shown overlaid on top of the energy barrier (note that the wavefunction is not given in units of energy).

$|\Psi(\vec{x})|^2 = \Psi^*(\vec{x})\Psi(\vec{x})$, i.e., the function times its complex conjugate, gives the probability that the particle will be found at some position $\vec{x}$. This wavefunction decays exponentially through an energy barrier (schematically shown in Figure 10.12), such that the probability that the particle will be found on the other side of the barrier depends on its width and height.

What we have been considering so far is a particle that can be in any of the (continuum of) states originating from one side of the barrier and ending on the other side. To make more connection with the experiments described in the previous sections, let us consider instead a double barrier system, where an electron comes toward the first barrier as a free particle, tunnels through the barrier into the middle region (quantum well) and eventually tunnels through the second barrier. In many cases the quantum well can accommodate just a few (quasi-)discrete states which are available for the electron to be in. The situation is depicted in Figure 10.13(a), where for simplicity, we show one energy state, E_R, in the junction. If the electron impinges first, say, on the left barrier with energy E not equal to the discrete state E_R in the middle of the junction, the probability that the electron can be

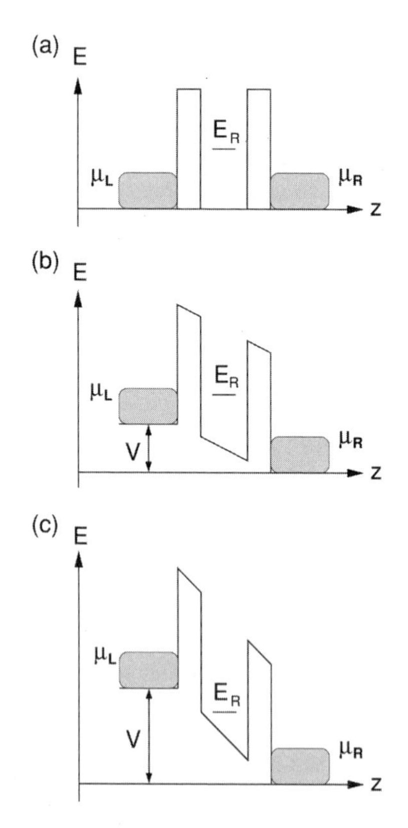

FIGURE 10.13. (a) A double energy barrier shown with a (quasi-)discrete state in the junction, which represents a molecular state. (b) Applying a bias V to the double barrier system causes the (quasi-)discrete state to line up between filled states in one electrode and empty states in the other. (c) Increasing V further brings the system out of resonance.

found on the other side of the junction is very small. However, if the electron has an energy equal to the quantum well energy level E_R, then the probability that it can cross the whole double-barrier junction is high. This is due to enhanced probability that the electron can be located in the middle of the junction, with a correspondingly higher chance of making it past the second barrier. In this case, we say that the particle is *resonant tunneling* through the junction and a large peak in the current would be observed. If there are multiple energy levels in the junction, one condition for the peak in the current to be observable is that the levels in the quantum well be far apart compared to the thermal energy $k_B T$. The effect of the thermal energy is to spread out and lower the absolute value of the peak due to the smearing out of the electronic distribution which is described by the Fermi function

$$f(E) = \frac{1}{\exp(E - E_f)/k_B T + 1} \tag{10.1}$$

where E_f is the Fermi energy.

To clarify further the phenomenon of resonant tunneling let us step through what happens when the voltage is increased in the case where there is only one energy level in between the junction (see Figure 10.13). There is a continuum of states filled up to the chemical potential in the left electrode μ_L, and likewise in the right electrode with chemical potential μ_R. When no bias is applied (and at zero temperature) there are either no electrons with an energy high enough to be equal to E_R or there are no empty states for electrons to go into at that energy, the former is shown in Figure 10.13(a). However, when a bias is applied (high voltage on right side), there will eventually be electrons on the left electrode with energy E_R and empty states on the right electrode for those electrons to tunnel into. An increase in current will be observed at this point. However, when the bias is increased further, eventually no electrons will have energy low enough to be in resonance with the energy level in the junction. In this case, the current decreases (as shown in Figure 10.14). The bias window in which the current decreases while the voltage increases provides a region with negative differential resistance. The differential resistance is simply

$$R_d = \left(\frac{\partial I}{\partial V}\right)^{-1} \tag{10.2}$$

or, in other words, the inverse derivative of the curve in Figure 10.14. There can be resonant tunneling without a region of negative differential resistance. In this case, only a peak in conductance, not current, is observed.

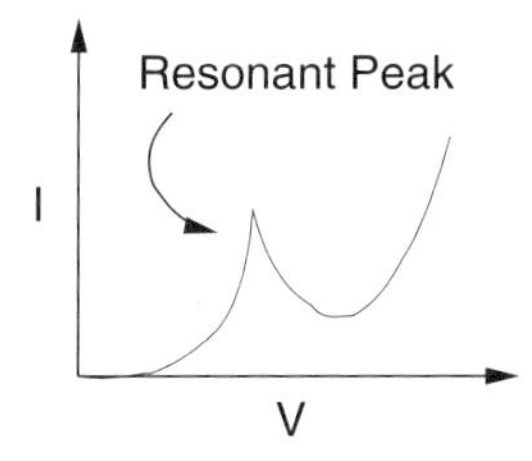

FIGURE 10.14. *I-V* curve with a resonant peak.

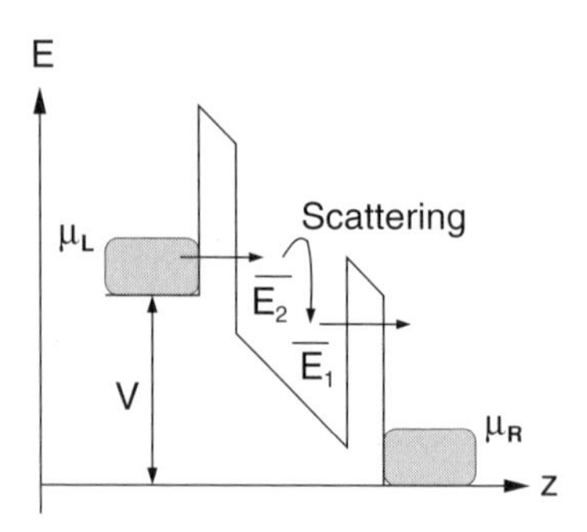

FIGURE 10.15. Schematic of sequential tunneling with two energy states in the junction. The charge carrier comes from the left electrode into the upper state, scatters into the lower state, and finally tunnels into the right electrode.

So far we have neglected any scattering effect that would change the energy of the electron and/or its localization properties inside the quantum well. If such effects take place then we can think of the electron as first resonant tunneling from, say, the left electrode into the quantum well, then loosing part of its energy (or becoming localized) in the quantum well, and finally tunneling into the right electrode. This phenomenon is called *sequential tunneling* as opposed to coherent resonant tunneling described above. It is depicted in Figure 10.15. Sequential tunneling can occur when the electron scatters off an impurity or with a vibrational state in the junction. The scattering effect can be elastic or inelastic. In the former case, the electron maintains its initial energy after scattering, in the latter, the electron looses or gains energy. Both inelastic or elastic scattering effects can contribute to sequential tunneling.

The rate (or probability) of coherent tunneling drops off dramatically with the distance between the electrodes. However, sequential tunneling has smaller distance dependence since the rate to transverse a series of barriers is just a product of the rates to transverse each individual barrier. Charge transport in DNA, for instance, is believed to occur by sequential tunneling of holes. Charge transport in the benzene-dithiolate molecule described above is believed to occur by coherent tunneling.

In the case of coherent tunneling, the current is given by

$$I = \frac{2e}{h} \int T(E)[f_L(E) - f_R(E)]\, dE \tag{10.3}$$

where e is the electron charge, h is Plank's constant, $T(E)$ is the total transmission coefficient for an electron with energy E, and $f_L(E)$ ($f_R(E)$) is the Fermi distribution function for the left (right) electrode. If there is resonant tunneling, the transmission coefficient will have a peak at the resonant energy E_R.

We conclude by noting that another possible transport mechanism is thermal hopping which occurs when the Fermi level lies below a low but wide energy barrier. The tunneling probability across the barrier is considerably suppressed due to the width of the barrier. However, at higher temperatures, the electron can raise its energy with the assistance of a vibrational mode (phonon mode of the structure). The electron is said to "hop" from one side of the barrier to the other via an intermediate (phonon-assisted) state. While thermal hopping has been invoked as a possible transport mechanism in DNA, it is unlikely it plays a major role in the molecular devices described in the previous sections.

10.3.2. Current-Induced Mechanical Effects

When current flows in a device it can affect its atomic structure by literally moving the atoms (electromigration), as discussed in Section 10.1, or by exciting vibrational modes and heating up the system. Both effects have tremendous consequences in electronics and are still not fully understood at the nanoscale. While in Section 10.1 we reported that electromigration can be used to build molecular junctions, here we refer to the consequences of this effect on the fully-formed device.

Electromigration is observed when a large current density flows into the device. In this case, some of the momentum of the charge carriers is transferred to the ions, which are consequently moved. In traditional microelectronics devices, this effect increases when the dimensions of the devices become smaller. It is natural to ask, then, if such an effect is detrimental for molecular devices. While more work is needed to understand this issue, it has been experimentally demonstrated that carbon nanotubes can sustain current densities much larger than normal microelectronic devices without current-induced break-up.[29] Similarly, theoretical work has shown that the benzene-dithiolate molecule described above should be quite resistant to electromigration effects.[30] Both results can be attributed to the strong carbon-carbon bond.

Even less work has been done on the heating effects of molecular devices. Theoretical results[31] indicate that even though nanoscale junctions can heat up quite substantially when current flows, the majority of such heat is dissipated into the bulk electrodes if the contacts easily allow for heat conduction. Clearly, in a real structure both effects are present at the same time. However, the interrelation between current-induced forces and heating is still unclear.

10.4. INTEGRATION STRATEGIES

Up to this point, we have discussed the fabrication and operation of molecular devices. However, we need to recognize that most likely the major challenge of creating an electronics from single molecules is not the fabrication of individual devices (though difficult), but to *integrate* them into complex circuits. In traditional microelectronics, individual devices are etched onto a larger microprocessor and connected by wires. A similar concept is envisioned for molecular electronics where billions of single molecule devices should be connected in a reliable (and desirably not too expensive) way. We conclude this chapter by discussing some ideas and efforts toward this goal.

10.4.1. Defect Tolerance and New Molecular Architectures

As the size of traditional microelectronics devices decreases, the fabrication costs increase dramatically mainly because every component (or at least a large percentage) has to work properly. In other words, the number of defects in the circuit has to be very small

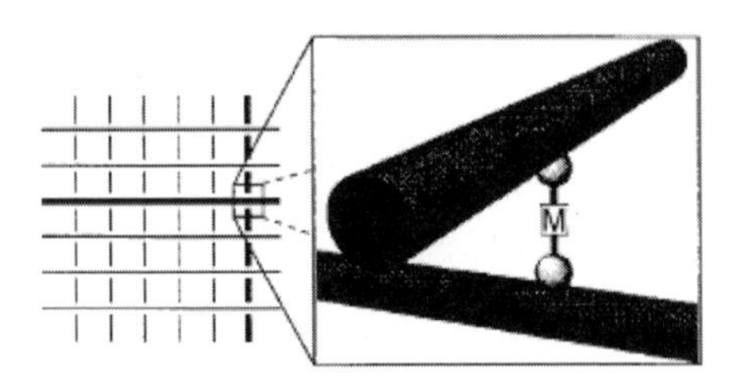

FIGURE 10.16. Schematic of a molecular memory built with a crossbar architecture. (From Ref. 18 by permission of the American Chemical Society.)

for the circuit to function. If self-assembly will be the method of choice to build molecular devices, then a large number of defects will likely be part of the circuit. With self-assembly, the molecules are not as neatly aligned on a metal surface as one would like, and many of them might not interact with the electrodes at all; thus creating domains with totally different *I-V* characteristics. Since it is conceivable that such defects will be unavoidable, then *defect-tolerant* architectures will be necessary so that the circuit keeps functioning even with a large number of defects. Incidentally, this could lower the fabrication costs as less control on the quality of each component is necessary.[37] Clearly, there has to be some amount of defect tolerance built into both the computer hardware and software (in the latter case, the term "fault tolerance" is used). Here we just focus on the hardware.

Redundancy is one way to create a defect-tolerant architecture: By making many of the same devices, i.e., devices with same functionalities, even if some of them do not work, the circuit can still operate because it has access to the ones that still work. A practical example of how to create such redundancy is provided by an experimental computer based on a crossbar architecture. The latter is simply a series of horizontal lines (wires) crossing a series of parallel lines (see the schematic in Figure 10.16). The points where the wires intersect are the locations of the circuit elements. This computer, called Teramac and built at Hewlett-Packard, used traditional microelectronic components. However, the computer contained many (with respect to regular computer architectures) fatal hardware defects. Many of the same components were hooked up in a crossbar architecture. This allowed for software rerouting around the defective components, and thus, the computer was able to work with very high percentage of defects.

A crossbar architecture was used, as well, to make a molecular memory. Previously, we discussed molecular switches made of catenanes. These switches can be made by depositing the molecules on parallel wires, and then depositing perpendicular wires on top (see Figure 10.16).[18] Here, each molecular switch (which is comprised of a number of molecules) is accessed by a bias across the two wires which cross at that point. The array of switches can then act as a memory device, with each bit stored at the points where the wires cross. This type of architecture seems promising for future computing applications.

10.5. CONCLUSIONS

We have given just a limited overview of the fundamental concepts and current research in molecular electronics. We refer the reader to the reviews suggested below and original papers to get a more detailed account. The field is expanding at a fast rate and many of the problems we encounter today in making and integrating molecular devices might be overcome in the future. However, irrespective of whether this electronics will become a commercial reality or not, it is important to mention that we have learned, and will learn, a big deal regarding the fundamental transport properties of nanoscale systems, and knowledge is always at the heart of any technological revolution.

10.6. FURTHER READING

- For more advanced readers, see Ref. 11 and Ref. 32 for a more complete discussion of charge transport on the mesoscopic and molecular scale. Included in these two references are details on how the transmission coefficient is calculated.
- For reviews of molecular electronics, see Ref. 13 and Refs. 33–36.
- For a review of catenanes and rotaxanes as molecular switches, see Ref. 18. Transport in DNA and its application to nanoscale electronics are reviewed in Ref. 25.

QUESTIONS

1. Show that at zero temperature, expression 10.3 can be written as

$$I = \frac{2e}{h} \int_{E_{f,R}}^{E_{f,L}} T(E)\, dE \tag{10.4}$$

   where $E_{f,L}$ ($E_{f,R}$) is the Fermi energy of the left (right) electrode.

2. Show that if the transmission coefficient, $T(E)$, is energy independent and equal to one, the conductance of a junction assumes the value $2e^2/h$, called the quantum of conductance.

3. For resonant tunneling, the total transmission coefficient will have a sharp peak at the resonant energy E_R. Consider the system shown in Figure 10.13, where there is one discrete level in the junction above the Fermi level E_f^0 at zero bias. Suppose the transmission function $T(E)$ is proportional to a sharply peaked Gaussian function plus a constant. Take the standard deviation to be 10^{-2} eV, the Gaussian to be centered around the energy $E_R = 10^{-1}$ eV (with $E_f^0 = 0$ eV for both electrodes at zero bias), and the constant to be 1. Let the proportionality factor be 10^{-4} to keep the magnitudes within a reasonable range. Assume the bands in the electrodes are large and that there is a linear drop in the voltage across the junction. Plot $T(E)$ on the interval $[-1,1]$ eV. Now calculate and plot the I-V curve using Equation 10.3 at zero temperature from a bias of 0 V to a bias of 1 V. Assume $2e/h = 1$.

4. Now plot the differential resistance from the I-V curve of Question 3. Why does the differential resistance have a broader peak than the transmission coefficient, when the Fermi function is just a step function? Also, note that there is no region of negative differential resistance.

5. Increase the temperature and plot the I-V curve. What happens when you increase the temperature to room temperature? If you increase the temperature further? Why does this happen? Now plot the differential resistance, and note the difference between the differential resistance in this case and the result of Question 4.

REFERENCES

1. A. Aviram and M. A. Ratner, *Chem. Phys. Lett.* **29**, 277 (1974).
2. M. A. Reed, C. Zhou, C. J. Muller, T. P. Burgin, and J. M. Tour, *Science*, **278**, 252 (1997).
3. H. Park, J. Park, A. K. L. Lim, E. H. Anderson, A. P. Alivisatos, and P. L. McEuen, *Nature*, **407**, 57 (2000).
4. C. J. Muller, J. M. van Ruitenbeek, and L. J. de Jongh, *Physica*, C **191**, 485 (1992).
5. H. Park, A. K. L. Lim, A. P. Alivisatos, J. Park, and P. L. McEuen, *Appl. Phys. Lett.* **75**, 301 (1999).
6. R. P. Andres, T. Bein, M. Dorogi, S. Feng, J. I. Henderson, C. P. Kubiak, W. Mahoney, R. G. Osifchin, and R. Reifenberger, *Science*, **272**, 1323 (1996).
7. M. A. Reed, *Proceedings of the IEEE*, **87**, 652 (1999).
8. L. A. Bumm, J. J. Arnold, M. T. Cygan, T. D. Dunbar, T. P. Burgin, L. Jones, D. L. Allara, J. M. Tour, and P. S. Weiss, *Science*, **271**, 1705 (1996).
9. E. G. Emberly and G. Kirczenow, *Phys. Rev.* B **58**, 10911 (1998).
10. M. Di Ventra, S. T. Pantelides, and N. D. Lang, *Phys. Rev. Lett.* **84**, 979 (2000).
11. S. Datta, *Electronic Transport in Mesoscopic Systems* (Cambridge University Press, 1995).
12. See, e.g., C. J. Muller, J. M. Krans, T. N. Todorov, and M. A. Reed, *Phys. Rev.* B **53**, 1022 (1996).
13. M. A. Ratner, *Introducing molecular electronics*, Materials Today, Feb. 20–27 (2002).
14. E. G. Emberly and G. Kirczenow, *Phys. Rev.* B **64**, 235412 (2001).
15. Z. Q. Yang, N. D. Lang, and M. Di Ventra, *Appl. Phys. Lett.* **82**, 1938 (2003).
16. N. D. Lang and Ph. Avouris, *Nano. Lett.* **2**, 1047 (2002).
17. V. Derycke, R. Martel, J. Appenzeller, and Ph. Avouris, *Appl. Phys. Lett.* **80**, 2773 (2002).
18. A. R. Pease, J. O. Jeppesen, J. Fraser Stoddart, Y. Luo, C. P. Collier, and J. R. Heath, *Acc. Chem. Res.* **34**, 433 (2001).
19. S. J. Tans, A. R. M. Verschueren, and C. Dekker, *Nature*, **393**, 49 (1998).
20. M. Di Ventra, S. T. Pantelides, and N. D. Lang, *Appl. Phys. Lett.* **76**, 3448 (2000).
21. J. Park, A. N. Pasupathy, J. I. Goldsmith, C. Chang, Y. Yaish, J. R. Petta, M. Rinkoski, J. P. Sethna, H. D. Abruña, P. L. McEuen, and D. C. Ralph, *Nature*, **417**, 722 (2002).
22. W. Liang, M. P. Shores, M. Bockrath, J. R. Long, and H. Park, *Nature*, **417**, 725 (2002).
23. R. M. Metzger, B. Chen, U. Hopfner, M. V. Lakshmikantham, D. Vuillaume, T. Kawai, X. L. Wu, H. Tachibana, T. V. Hughes, H. Sakurai, J. W. Baldwin, C. Hosch, M. P. Cava, L. Brehmer, and G. J. Ashwell, *J. Am. Chem. Soc.* **119**, 10455 (1997).
24. C. Dekker and M. A. Ratner, *Electronic properties of DNA*, *Phys. World*, **14**, 29 (2001).
25. M. Di Ventra and M. Zwolak, *DNA Electronics* in "Encyclopedia of Nanoscience and Nanotechnology," H. S. Nalwa (American Scientific Publishers, 2004).
26. J. Richter, *Physica*, E **16**, 157–173 (2003).
27. K. Keren, M. Krueger, R. Gilad, G. Ben-Yoseph, U. Sivan, and E. Braun, *Science*, **297**, 72 (2002).
28. P. Phillips, *Advanced Solid State Physics* (Westview Press, 2002).
29. P. J. de Pablo, E. Graugnard, B. Walsh, R. P. Andres, S. Datta, and R. Reifenberger, *Appl. Phys. Lett.* **74**, 323 (1999).
30. M. Di Ventra, S. T. Pantelides, and N. D. Lang, *Phys. Rev. Lett.* **88**, 046801 (2002).
31. M. J. Montgomery, T. N. Todorov, and A. P. Sutton, *J. Phys. Cond. Mat.* **14**, 1 (2002).
32. V. Mujica and M. Ratner, *Molecular Conductance Junctions: A Theory and Modeling Progress Report* in "Handbook of Nanoscience, Engineering, and Technology," W. A. Goddard III, D. W. Brenner, S. E. Lyshevski, and G. J. Iafrate (CRC Press, 2002).
33. K. S. Kwok and J. C. Ellenbogen, *Moletronics: future electronics*, Materials Today, Feb., 28 (2002).
34. R. Lloyd Carroll and Christopher B. Gorman, *The Genesis of Molecular Electronics*, Acta Cryst. B **41**, 4379 (2002).
35. M. A. Reed and J. M. Tour, *Computing with molecules*, Scientific American, June, 86 (2000).
36. C. Joachim, J. K. Gimzewski, and A. Aviram, *Electronics using hybrid-molecular and mono-molecular devices*, *Nature*, **408**, 541 (2000).
37. J. R. Heath, P. J. Kuekes, G. S. Snider, and R. S. Williams, *Science*, **280**, 1716 (1998).

11

Single Electronics

Jia Grace Lu
Department of Electrical Engineering and Computer Science &
Department of Chemical Engineering and Materials Science
University of California, Irvine, CA

11.1. SINGLE ELECTRON TUNNELING

11.1.1. Introduction

With the advances of fabrication techniques in recent years, it has become possible to fabricate tunnel junctions of increasingly small dimensions. This opens the realm of mesoscopic physics which enables the study of a wide range of novel phenomena. It involves samples that combine both the characteristics of macroscopic world (i.e., classical effects) and those of microscopic world (i.e., quantum effects). The characteristic length scale of mesoscopic samples varies from nanometers to tens of micrometers.

In mesoscopic devices, one of the surprising results is that the electron wave interference effects produce quantum conductance fluctuation and weak localization.[1–3] In addition, it has also been shown to give rise to the Aharonov-Bohm effect in ring geometries.[4] In contrast, the ballistic transport of electrons through short and narrow channels results in conductances that have quantized values (see Chapter 10).[5] The discreteness of the electron energy spectrum has been probed in semiconductor heterostructures[6,7] and small metal particles.[7,8]

Here, we will focus on the study of systems in which the discrete nature of the electronic charge is important. This discreteness is usually not evident in conventional electronic devices, in which the current is regarded as a continuous charge flow. However, if electrons are confined to small isolated regions (islands) that are weakly coupled to the external circuit, the discreteness of the electronic charge can greatly influence the electrical transport properties of the system. The capacitance of the island to the external circuit can be so small that the charging energy required to add a single electron becomes the dominant energy. This phenomenon is called the "single electron charging effects".

Single electron charging effects have been widely studied on double tunnel junction systems, also called single electron transistors (SET). Such sample is conventionally fabricated by electron beam lithography and shadow evaporation techniques (see Chapter 1). It consists of a metallic island which is weakly coupled to two bias leads through small capacitance and high resistance tunnel junctions, and capacitively coupled to a gate electrode. The islands and the leads are separated by thin oxide layer tunnel barriers, through which electrons can only pass by quantum mechanical tunneling. The gate is used to control the average number of electrons on the island. In this chapter, we will emphasize on the current transport mechanisms in SETs with normal metal island or superconductor island, and describe the application of SETs.

11.1.2. Theoretical Background

11.1.2.1. Electron Tunneling The field of single charge tunneling was formulated in the mid-80's. The "orthodox" theory[10–13] of the correlated single charge tunneling was shown to be extremely successful in the description of most experimental results. The general operating principle of SET is to control the tunneling of a single charge. Such system must have small islands that connect to other metallic regions only via tunnel barriers, with a total tunneling resistance R_Σ that exceeds the quantum resistance $R_Q = h/e^2 \approx 25.8\ k\Omega$. This condition is necessary by an uncertainty relation argument: for an excess charge to tunnel onto the island, the energy uncertainty $\hbar/R_\Sigma C_\Sigma$ associated with the lifetime due to tunneling has to be much smaller than the Coulomb charging energy $E_c = e^2/2C_\Sigma$. Another requirement to observe single charge tunneling effects is that the islands must be small enough (small C_Σ, i.e., large E_c) and the temperature (T) low enough, so that the energy E_c required to add a charge carrier onto the island far exceeds the available energy of thermal fluctuations, i.e., $E_c \gg k_B T$. These two conditions ensure that the transport of charges through the island is governed by the Coulomb charging energy.

To understand the electron transport through the SET, let us start by considering the tunneling between two normal metal electrodes separated by an insulating barrier, as illustrated by Figure 11.1 and outlined in Chapter 10. The Fermi energies of the two electrodes are offset from each other by an amount eV.[a] The basic idea is that there is a nonzero probability of charge transfer by quantum mechanical tunneling of electrons between two metals separated by a thin insulating barrier, which classically forbids tunneling. This probability falls exponentially with the distance of separation and it depends on the properties of the insulating material. Electrons can tunnel across the barrier via energy conserving horizontal transitions, from the filled states in electrode 1 to the empty states in electrode 2. We can determine the tunneling rate from electrode 1 to electrode 2 as (refer to Appendix A):

$$\Gamma_{1\to 2} = \frac{eV}{e^2 R_T(1 - e^{-\beta e V})} \tag{11.1}$$

with R_T being the normal state resistance of the tunnel junction.

[a] The symbol e represents the magnitude of the electron charge, i.e., $e = |e|$.

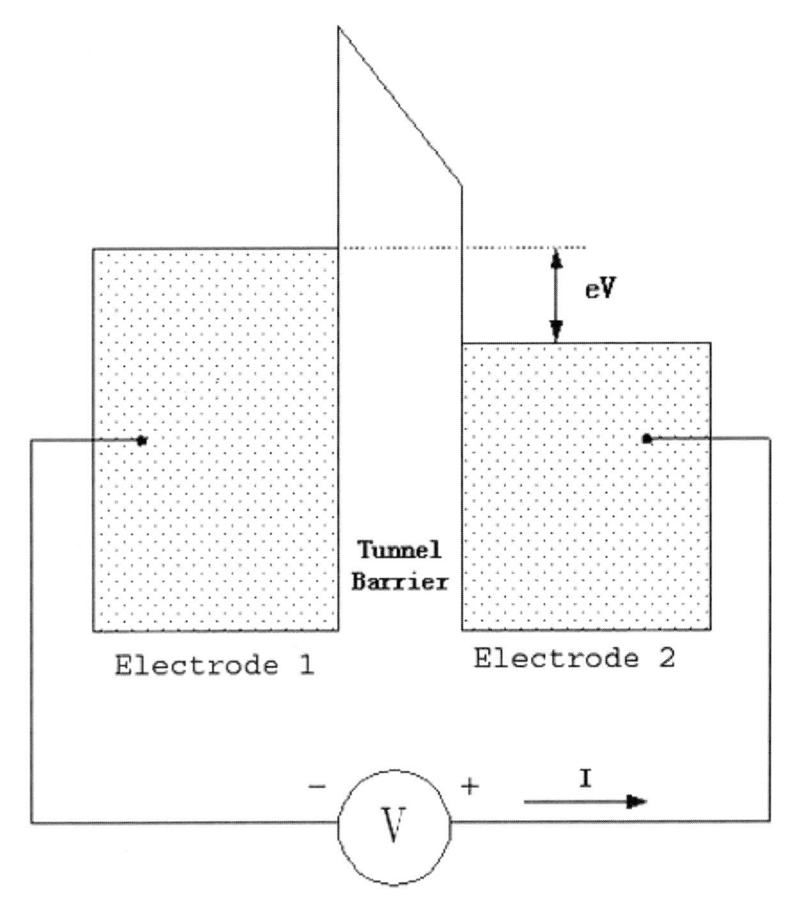

FIGURE 11.1. Schematic depicting tunneling across a normal-insulator-normal junction at $T = 0$.

The tunneling rate in the reverse direction $\Gamma_{2\to1}$ is likewise obtained by reversing the sign of the bias voltage:

$$\Gamma_{2\to1} = \frac{eV}{e^2 R_T (e^{\beta eV} - 1)} \tag{11.2}$$

The net current through the junction is then obtained by subtracting the reverse tunnel current from the forward tunnel current:

$$I(V) = e(\Gamma_{1\to2} - \Gamma_{2\to1}) \tag{11.3}$$

After inserting the tunneling rates, we get

$$I(V) = V/R_T \tag{11.4}$$

recovering the ohmic relationship.

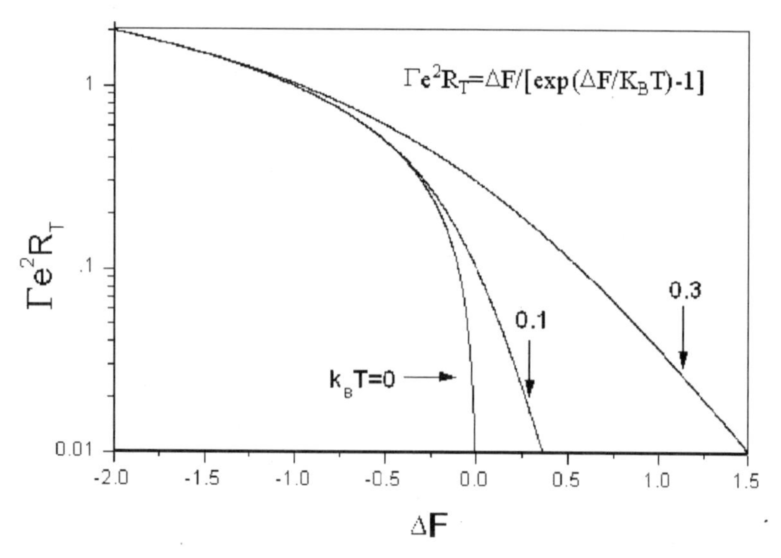

FIGURE 11.2. The electron tunneling rate through a NIN junction as a function of change in system free energy at different values of $k_B T$.

We can rewrite the tunnel rates in terms of a more general energy change $\triangle F$. $\triangle F$ is defined as the change in the system free energy in going from the initial state to the final state. In the case we have discussed, the electron travels from electrode 1 to electrode 2 in the direction favored by the applied bias, the bias source does an amount of work eV. Assuming that the tunneling electron rapidly relaxes to the Fermi level, it leads to a decrease in the system free energy by this amount since the tunneling process is essentially irreversible. Thus, $\triangle F = -eV$. So we can express the tunneling rate as a function of $\triangle F$.

$$\Gamma(\triangle F) = \frac{1}{e^2 R_T} \frac{\triangle F}{e^{\beta \triangle F} - 1} \tag{11.5}$$

Figure 11.2 plots the tunneling rate as a function of $\triangle F$ for different temperatures. This rate equation forms the basis of the orthodox theory of single electron tunneling.[10–13]

11.1.2.2. Coulomb Blockade As we have discussed before, when the capacitances of the junctions decrease sufficiently the Coulomb charging energy E_c of a single electron gets large enough, then the discreteness of the electronic charge becomes very important. Let us first look at a single tunnel junction with small capacitance. The electrostatic energy of an isolated capacitor C with charges $Q(>0)$ and $-Q$ on the two electrodes is $Q^2/2C$, or $CV^2/2$, where $V = Q/C$. If an electron tunnels from the negative electrode to the positive one, the charge on the capacitor becomes $\pm(Q - e)$, so that the energy of the capacitor becomes $(Q - e)^2/2C$.[b] This indicates an increase in the system energy unless the initial charge $Q \geq e/2$, i.e., $V \geq e/2C$. In other words, electron transfer is energetically forbidden

[b] Here we assume that the single junction is not in a closed circuit, so that the capacitor charges do not come to equilibrium with the rest of the circuit.

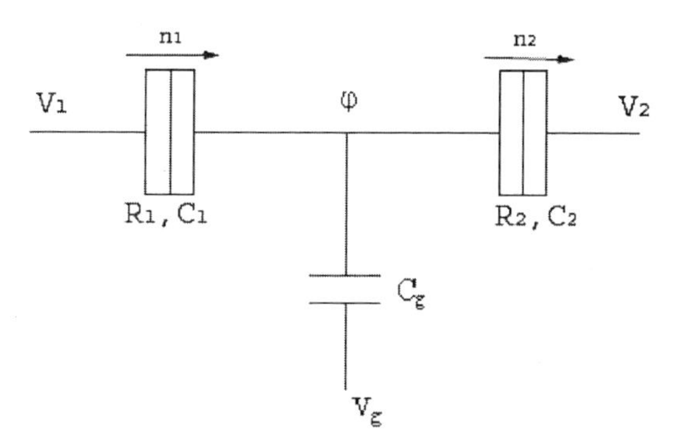

FIGURE 11.3. A schematic diagram of the single electron transistor showing two small capacitance tunnel junctions characterized by junction resistance R and capacitance C, and also the capacitively coupled gate.

for voltages $V < e/2C$. This regime of zero tunnel current despite a finite voltage across the junction is called the Coulomb blockade[10].

Since the relevant time scale for tunneling is set by the tunneling resistance and the capacitance: $\tau = RC$, which is on the order of $10^{-10}\,s$, the effective shunting resistance is determined by the high frequency properties of the leads near the junction, not the dc bias resistance. Hence, unless we insert ultracompact resistors right at the junction,[14] the effective impedance will be the high frequency impedance of the leads, around 100 Ω which is much less than R_Q. Thus the Coulomb blockade effect is hard to be observed in a single tunnel junction, no matter how low the temperature gets. To circumvent this problem, we need a double junction system in which each junction is effectively isolated from the low impedance environment by high tunnel resistance and low capacitance of the other junction[10,14] (see Figure 11.3).

11.1.2.3. Energy Relations in Double Junction Systems In this section, we consider the energetics of the double junction system made of only normal metals. The energetic considerations are important because once we know how to calculate the change in the system free energy ΔF for a tunneling event, then we can calculate the rate at which this particular process occurs. When the leads and the island are normal metals, this rate is determined by Eq. 11.5. Once all the tunneling rates are known, we can determine the tunneling current through the device.

As illustrated in Figure 11.3, n_1 and n_2 represent the number of electrons that tunnel forward across junctions 1 and 2, respectively. Let us define the integer $n = n_1 - n_2$ to be the excess number of electrons on the island; and the integer $m = n_1 + n_2$ to be the total number of electrons that have tunneled forward through either junction. The system free energy can be expressed as:

$$F_{\text{sys}}(n, m) = \frac{(Q_0 - ne)^2}{2C_\Sigma} - \left[m + n\frac{C_2 - C_1}{C_\Sigma}\right]\frac{eV}{2} \tag{11.6}$$

where C_Σ is the total capacitance of the island to the bias leads and the gate electrode, and $Q_0 = C_g V_g$ is the induced charge by the gate (refer to Appendix B for detailed derivation).

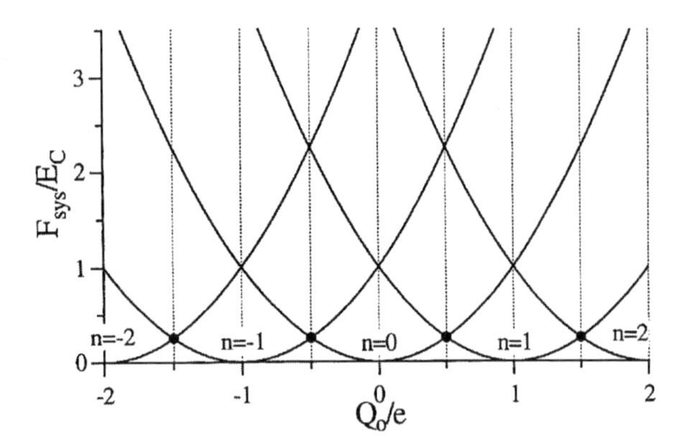

FIGURE 11.4. The Helmholtz free energy of the system as a function of Q_0/e for various charge states n at $V = 0$.

The first term in F_{sys} denotes the charging energy with an effective island charge of $Q_0 - ne$.[c] The second term represents the total work done by the bias voltage sources. Since n must be an integer, the minimum energy for given Q_0 is obtained if n is the integer closest to Q_0/e. That means, giving the lowest charging energy n must lie in the range

$$\frac{Q_0}{e} - \frac{1}{2} \leq n \leq \frac{Q_0}{e} + \frac{1}{2} \tag{11.7}$$

In Figure 11.4, the system energy is plotted as a series of parabolas for $V = 0$. Each curve corresponds to a different number of n excess electrons on the island. As we sweep along the gate voltage, i.e., changing the gate charge Q_0, we vary the system favorable energy level. The parabolas for n and $n \pm 1$ cross at $Q_0 = (n \pm \frac{1}{2})e$ at energy $E_c/4$. At each crossing point, one electron tunnels onto or off the island, changing the number of excess electrons on the island from n to $n \pm 1$. Since the system energy is periodic in Q_0, the current is also periodic in Q_0 with period e and current peaks occur at half-integer values of Q_0/e. In other words, in one modulation period of the current, i.e., gate voltage increases by $\pm e/C_g$, a single electron is added or removed from the island. The current through this device is a function of both the bias voltage V and the gate voltage V_g, exhibiting transistor action. Therefore, this system is given the name, single electron transistor.

11.1.2.4. Energy Diagrams Illustrating Single Electron Tunneling It is useful to draw simple energy diagrams to illustrate how the Coulomb blockade arises and how it can be overcome. Figures 11.5–11.7 show the energy diagram for tunneling in the normal metal SET (NNN system) with symmetric junctions, i.e., $C_1 = C_2$. The energy that can be supplied by the bias sources is represented by shifting the Fermi level of one of the leads with respect to the other by an amount eV. The change in the charging energy due to tunneling is represented by the distance between the Fermi level in the island and the solid

[c] Since the effective charge of the island influences the strength of the charging energy barrier, the gate voltage can be used to tune this barrier, and therefore modulate the current through the device.

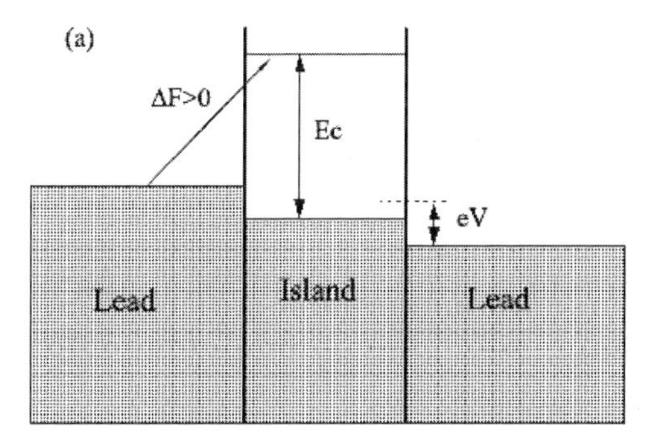

FIGURE 11.5A. Energy diagram of a SET with symmetric junction capacitances. Coulomb blockade exists when the tunneling process is energetically unfavorable.

line above it. This solid line represents the position of the Fermi level after the tunneling event.

Consider symmetrically biased junctions at $V_g = 0$, in order to overcome the blockade, the bias voltage must exceed a threshold value for which $\Delta F_{n \to n+1} = 0$. This yields

$$eV \geq (2n+1)\frac{e^2}{C_\Sigma} = 2(2n+1)E_C \tag{11.8}$$

Figure 11.5(a) shows the case for $n = 0$, tunneling from the left lead onto the island increases the charging energy by E_c. For $V < e/C_\Sigma$, this tunneling process is energetically unfavorable ($\Delta F > 0$) and occurs with an exponentially suppressed rate for $T \ll E_c$; so very little current will flow through the system, thus tunneling is blocked.

Figure 11.5(b) shows that when $V > e/C_\Sigma$, the bias source provides enough energy to overcome the charging energy barrier, and Coulomb blockade is lifted and current flows through the system.

Figure 11.5(c) shows that when $Q_0 = e/2$, the Coulomb blockade is completely tuned away, and current will flow through the system for any non-zero bias voltage.

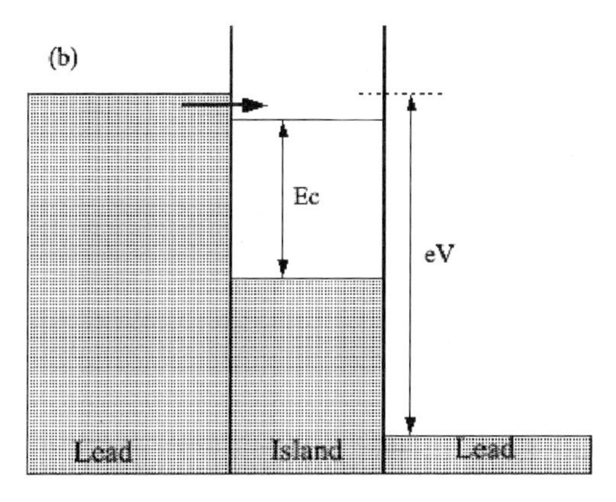

FIGURE 11.5B. When bias voltage sources provide enough energy to overcome the charging energy barrier, single electron tunneling occurs.

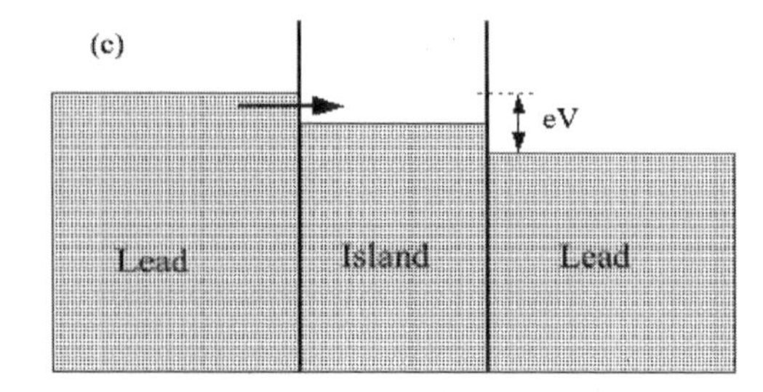

FIGURE 11.5C. At $Q_0 = e/2$, the potential of the island is lowered by E_C so that Coulomb blockade is absent at all bias voltages.

Figure 11.6 shows the $I - V$ characteristic for two junctions of equal tunnel resistances in the cases of $Q_0 = 0$ and $Q_0 = e/2$, respectively. The $Q_0 = 0$ curve shows a significant Coulomb blockade, and the $Q_0 = e/2$ curve exhibits a non-zero conductance at zero bias.

If we consider electron tunneling across junctions 1 and 2, respectively, we can calculate the energy changes in the transition from n to $n \pm 1$ state[15]:

$$\Delta E_1^{\pm} = \frac{e^2}{C_\Sigma}\left\{\left[\frac{1}{2} \pm \left(n - \frac{Q_0}{e}\right)\right] \pm \frac{(C_2 + C_g/2)V}{e}\right\} \tag{11.9}$$

$$\Delta E_2^{\pm} = \frac{e^2}{C_\Sigma}\left\{\left[\frac{1}{2} \pm \left(n - \frac{Q_0}{e}\right)\right] \pm \frac{(C_1 + C_g/2)V}{e}\right\} \tag{11.10}$$

For a transition to occur at $T = 0$, it is necessary that the relevant energy change ΔE is negative. For net current through the device, both tunneling onto the island across one junction (charging step) and then off across the other (discharging step) must be allowed. The sum of the two ΔE values for a through passage of charge is always exactly eV, so that the energy is always lowered by $e|V|$. The essence of the Coulomb blockade is that the energy must be lowered on each of the two successive transitions of charging and

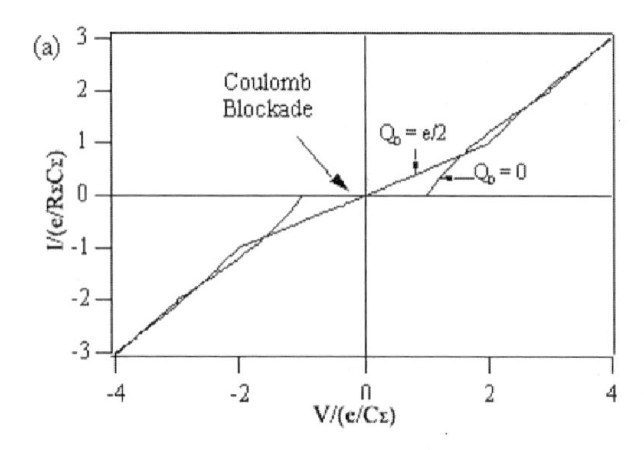

FIGURE 11.6. I–V curves at $T = 0$ for a NNN SET with identical junction resistances and capacitances for the cases of $Q_0 = 0$ and $Q_0 = e/2$.

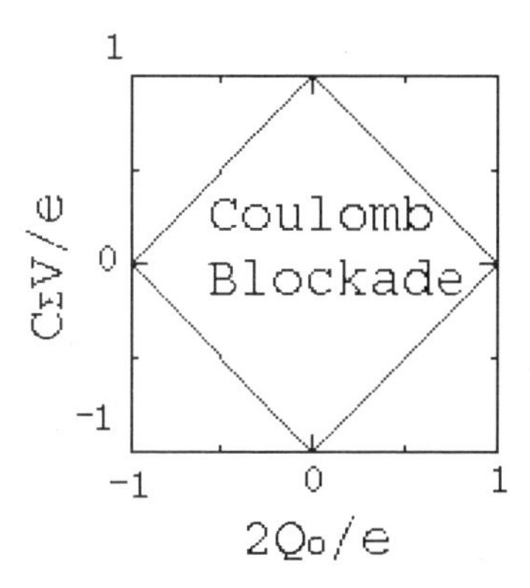

FIGURE 11.7. $V - V_g$ diagram showing the coulomb blockade for a symmetric SET in the $n = 0$ state.

discharging the island. In fact, as long as the first step becomes energetically favorable, the second step will always become favorable.[d]

From ΔE_1^+ and ΔE_2^-, we can determine the voltage thresholds for the charging step across junction 1 and for the discharging step across junction 2. The extent of the Coulomb blockade is determined by whichever is the lesser of the two thresholds:[e]

$$V_{1,\mathrm{th}}^+ = \frac{e}{C_2 + C_g/2}\left[\frac{1}{2} + \left(n - \frac{Q_0}{e}\right)\right] \tag{11.11}$$

$$V_{2,\mathrm{th}}^- = \frac{e}{C_1 + C_g/2}\left[\frac{1}{2} - \left(n - \frac{Q_0}{e}\right)\right] \tag{11.12}$$

Therefore, Coulomb blockade can be overcome by a combination of bias voltage V and gate voltage V_g. In the case of symmetric junctions (i.e., $C_1 = C_2$), when the following condition is fulfilled, tunneling is blocked:

$$C_\Sigma |V|/e < (2n + 1) - 2C_g |V_g|/e \tag{11.13}$$

For $n = 0$, we see that Coulomb blockade happens for the area within the diamond of Figure 11.7. If the junctions are asymmetric, then the diamond shape will be skewed. From the slopes, each junction capacitance can be extracted.

11.1.2.5. Single Electron Tunneling Current Calculation–Orthodox Theory In this section, we briefly describe how one calculates the tunneling current through a SET in which the leads and the island are normal metals. First we assume that the electrons can not tunnel across both junctions simultaneously. This process is referred to as co-tunneling

[d] For the case when an electron tunnels off from the island to the right electrode first, followed by an electron from the left onto the island, it is the same as considering an hole tunnels from the right electrode onto the island then off to the left electrode.

[e] ΔE_1^+ for the charging step becomes negative for positive bias voltage above the threshold $V_{1,\mathrm{th}}^+$, and ΔE_2^- becomes negative for positive bias voltage above $V_{2,\mathrm{th}}^-$.

which we will discuss in the next section. We calculate the $\Delta F's$ associated with transitions from charge state n to charge state n' using Eq. 11.6. Once $\Delta F's$ are known, all tunneling rates can be calculated using Eq. 11.5.

We define a steady-state probability $\sigma(n)$ to be the ensemble distribution of the number of electrons on the island. Assuming no charge accumulation on the island at steady state, we can determine $\sigma(n)$[f] by requiring the total probability of tunneling into a state to be equal to the total probability of tunneling out of it. The master equation to determine $\sigma(n)$ is

$$\sum_{n} \sigma(n)\Gamma(n \rightarrow n') = \sum_{n'} \sigma(n')\Gamma(n' \rightarrow n) \tag{11.14}$$

Here, $\Gamma(n \rightarrow n')$ includes both the rates $\Gamma_1(n \rightarrow n')$ and $\Gamma_2(n \rightarrow n')$ for going from state n to state n' by tunneling through either junction, and the summation runs over all possible island charges. In the case of normal metals, only adjacent n states are connected via tunneling. The distribution satisfies detailed balance:

$$\sigma(n)\Gamma(n \rightarrow n+1) = \sigma(n+1)\Gamma(n+1 \rightarrow n) \tag{11.15}$$

allowing us to solve for $\sigma(n)$ subject to the normalization condition $\sum_n \sigma(n) = 1$. When either the lead or the island are superconducting, the current mechanism changes to two-electron tunneling at low bias voltages, i.e., from n to $n \pm 2$. In these cases, the full matrix equation of 11.14 must be solved.

After determining the island charge distribution function $\sigma(n)$, we can calculate the steady state current through the transistor by considering transitions across a single junction for any possible island charge. For normal metals, taking account of the forward and reverse tunneling across each junction, the net current can be written as

$$I(V) = -e\sum_{n} \sigma(n)\,[\Gamma_1(n \rightarrow n+1) - \Gamma_1(n \rightarrow n-1)] \tag{11.16}$$

$$= -e\sum_{n} \sigma(n)\,[\Gamma_2(n \rightarrow n-1) - \Gamma_2(n \rightarrow n+1)] \tag{11.17}$$

Since there is no charge accumulation on the island in the steady state, the current in the two junctions are equal. This overall approach to calculate the current is referred to as the "orthodox theory" of single electron tunneling.[10–13]

11.1.2.6. Co-Tunneling Mechanisms Consider a Coulomb blockade region, where the sequential tunneling of an electron from the left electrode to the island is energetically unfavorable ($\Delta F > 0$) and thus forbidden even in the presence of a finite bias voltage. Therefore, from what we have previously discussed, it is impossible to transport an electron from the left to the right electrode by sequential single electron tunneling across the junctions. However, the co-tunneling processes make the transport possible. There are two types of co-tunneling processes: (1) an inelastic process, in which two different electron

[f] It can be viewed as the classical analog of the diagonal elements of the quantum mechanical density matrix.

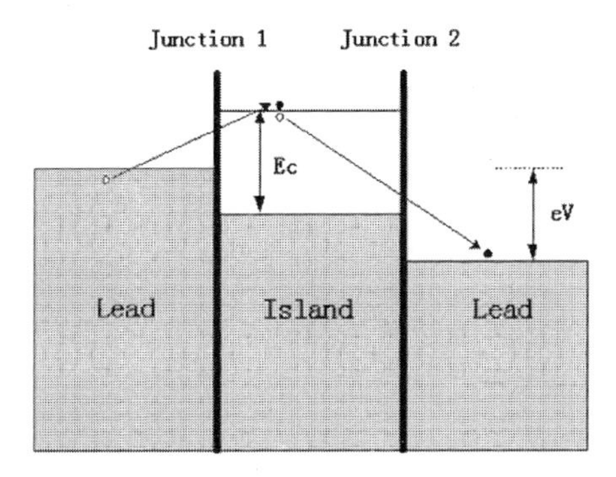

FIGURE 11.8. Schematic of an inelastic co-tunneling process. Although it is in the Coulomb blockade region, electron can still be transported across the device by simultaneous tunneling across both junctions, and leaving an electron and a hole excitation on the island.

states are involved, and there is an electron-hole excitation left on the island; and (2) an elastic process, in which the same electron state is involved in both tunneling processes and there is no excitation left over.

Figure 11.8 illustrates the inelastic co-tunneling process. The energy-time uncertainty relation of quantum mechanics results in the quantum fluctuation of electric charge on the island within a short period of time associated with the tunneling event. When one electron tunnels onto the island and a second electron tunnels off the island within time $\hbar/E_c$, the net result is that an electron has tunneled through the whole system. This process involves electrons tunneling simultaneously[g] through the two junctions via a virtual intermediate state with increased electrostatic energy, and leaving an electron-hole excitation on the island.[16] This inelastic co-tunneling process transfers electrostatic energy of the system into the energy of the electron-hole excitations. Since the process is of the second order, the rate of the co-tunneling is small compared to the rate of sequential tunneling, but nevertheless, this process is dominant in the Coulomb blockade regime, where the sequential tunneling is suppressed.

The rate for inelastic co-tunneling can be calculated by applying the Fermi's golden rule for the higher order transitions. At low temperatures, the current due to this process is:[17]

$$I_{\text{elastic}} = \frac{\hbar}{12\pi e^2 R_1 R_2}\left(\frac{1}{\Delta E_1} + \frac{1}{\Delta E_2}\right)^2 [(eV)^2 + (2\pi k_B T)^2]V \tag{11.18}$$

where ΔE_1 and ΔE_2 are the energy changes associated with the sequential tunneling in the first and the second junction as given by Eqs. 11.9 and 11.10. Here, $\Delta E_1, \Delta E_2 \gg eV$. The current varies as the third power of the voltage V, and it causes rounding in the $I-V$ curve right near the voltage onset of the single electron tunneling.

Co-tunneling is elastic if the same electron state is involved in both tunneling processes and the coherence of the tunneling electron is maintained in the tunneling event. Elastic

[g] Here, "simultaneous" is used to distinguish co-tunneling from sequential tunneling in which two tunneling events are separated by a time interval great than $\hbar/E_c$.

co-tunneling can also be viewed as a process in which an electron tunnels onto the island, virtually diffuses through it, and tunnels off across the opposite junction.[18]

Inelastic co-tunneling is an important charge transport mechanism in the SET. It was first observed experimentally by Geerligs et al.[19] in a lithographically-patterned SET. In contrast, elastic co-tunneling has only been observed in tunneling through extremely small metal particles ($\sim 1-10$ nm) in a double junction system using scanning tunneling microscope. Elastic co-tunneling occurs with a negligible rate in lithographically patterned SETs because it takes an electron much longer than $\hbar/E_c$ to virtually diffuse from one junction to the other.

11.2. SUPERCONDUCTING SINGLE ELECTRON TRANSISTOR

11.2.1. Sub-gap Charge Transport Mechanism in a NSN System

Let us first understand the charge transport mechanisms in a SET with normal metal leads and superconducting (Al) island (NSN system). Charge transport at low temperatures and at low bias voltages in the NSN system occurs by two sequential Andreev steps,[h] which we refer to as the Andreev cycle. The first is the charging step, in which one electron incident on junction 1 is Andreev reflected into a hole on the same side of the junction. It can be regarded as an event in which two electrons tunnel onto the island to form a Cooper pair.[20,21] The second is the discharging step, in which a hole incident on junction 2 is Andreev reflected into an electron. This can be regarded as an event in which a Cooper pair breaks up and forms two electrons tunneling off the island, and the island returns to the original state. The two steps of the Andreev cycle are analogous to the sequential single electron tunneling, except that two electrons now tunnel in each step. Andreev reflection is subject to the same energetic considerations as is single electron tunneling in the NNN system. Thus, it also exhibits a Coulomb blockade. To calculate the Andreev reflection rate, Hekking et al.[21] models each step of the Andreev cycle as a second-order process and break it down as follows. In the first transition of step 1 (step 2), an electron (hole) tunnels onto the island and takes the system from the initial state to an intermediate state in which a virtual quasiparticle exists on the island. In the second transition, another electron (hole) tunnels onto the island through the same junction to a state paired with the existing quasiparticles. This leads to immediate recombination and returns the system to the BCS ground state. By summing over appropriate initial, intermediate, and final states, the Andreev reflection rate across junction i is:

$$\Gamma^{i}_{\text{Andreev}}(\Delta E_i) = \frac{G^{i}_{\text{Andreev}}}{e^2}\frac{\Delta E_i}{e^{\beta\Delta E_i} - 1} \tag{11.19}$$

where ΔE_i corresponds to the energy change in transferring both electrons and G^{i}_{Andreev} is the Andreev conductance of junction i. Note that this rate equation is analogous to that of single electron tunneling in a NNN SET (refer to Eq. 11.5) except that the Andreev

[h] Andreev considered a normal metal and a superconductor in good metallic contact. But his physical picture can be generalized to tunnel junctions.

conductance of the junction appears in place of the normal tunnel conductance of that junction. Unlike the normal conductance, however, the Andreev conductance depends on the specific lead geometry and the electronic mean free path in the leads.[22–24] Since Andreev reflection is a second-order process and the junctions are tunnel barriers with very low transparency, the Andreev conductances are expected to be much smaller than the normal tunnel conductance. A first order estimate based on ballistic electron motion near the junctions and ignoring the effects of phase coherence between the tunneling electrons[21] gives Andreev conductances of the order $5 \times 10^{-12}\Omega^{-1}$, which are indeed much smaller than $1/R_\Sigma$. However, the experimentally determined Andreev conductances are approximately 10^3 times larger than this value. Since the elastic mean free path in the leads is smaller than the junction dimension, the ballistic picture assumed is not applicable, and one must consider directly the effects of phase coherence.[25] The magnitude of enhancement of the Andreev conductance by phase coherence is expected to depend on the precise geometry near the tunnel junctions as well as the location of impurities and other scattering sites.

11.2.2. Sub-gap Charge Transport Mechanisms in a SSS System

Figure 11.9 shows experimental data at low temperature and zero magnetic field for a sample when both the island and the leads are in the superconducting state (SSS system).[26] In this figure, the current is measured while slowly sweeping V and quickly sweeping V_g, forming the envelope of all possible $I(V, V_g)$ curves. The upper graph in Figure 11.9 shows a sharp rise at 960 μV. Since this is a SSS system, this rise should occur at $V = 4\Delta/e$, which yields the superconducting gap $\Delta = 240\,\mu$eV. The current peaks appearing at approximately 660 μV result from a "$2e - e - e$" Josephson-quasiparticle (JQP) cycle in which the tunneling of a Cooper pair in one junction is followed by two sequential quasiparticle tunneling events across the other junction.[27–30] Here, we focus on the behavior of the system for nearly zero bias voltages, shown enlarged in the lower graph of Figure 11.9.

In the SSS system, another energy scale, the Josephson coupling energy E_j, becomes important in addition to E_c and Δ. There are three distinguishable regimes: if $E_c \ll E_j$, the charging energy effects are small, and the system has essentially the classical Josephson effect. If $E_c \gg \Delta$, single electron charging energies dominate, observed phenomena show e periodicity, and Cooper-pair tunneling is unimportant. The most interesting regime is $E_c < E_j < \Delta$, in which we shall show that there is a supercurrent that is modulated by the gate charge Q_0 with period $2e$.[15]

If we consider junctions in which $R_\Sigma \gg R_Q$, then $E_j \approx (R_Q/R_\Sigma)\Delta \ll \Delta$, so there is a considerable range of E_c for which the two inequalities hold. The fact that Δ is larger than all other energies allows us to restrict our attention (at $T = 0$) to states of the island containing only an even number of electrons, which form Cooper pairs.[15] Let's define E_{j1} and E_{j2} to be the Josephson coupling energies of the two junctions which couple the island to the superconducting leads, φ_1 and φ_2 the phase differences across them, and the phase sum $\vartheta = \varphi_1 + \varphi_2$. For large C_Σ thus negligible E_c, the supercurrent is calculated to be:

$$I_s = \frac{2e}{\hbar}\frac{E_{j1}E_{j2}\sin\vartheta}{E_j} \tag{11.20}$$

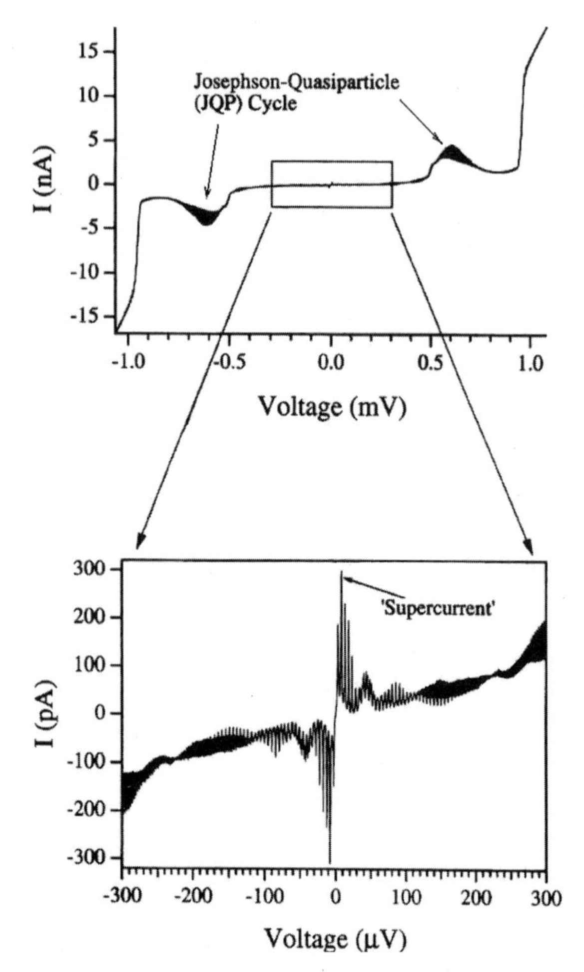

FIGURE 11.9. I (V, V_g) data for a SSS sample. Upper graph shows the Josephson-Quasiparticle peaks, and the lower graph shows the supercurrent which depends strongly on the value of Q_0.

where

$$E_j(\vartheta) = \sqrt{E_{j1}^2 + E_{j2}^2 + 2E_{j1}E_{j2}\cos\vartheta} \tag{11.21}$$

In the other limit, $E_c > E_j$, in which the charging energy term is dominant, it has been shown[31] that the supercurrent at the degeneracy points between even-n states (i.e., at odd integer values of Q_0/e) is:

$$I_s = \frac{2e}{\hbar}\frac{E_{j1}E_{j2}\sin\vartheta}{2E_j} \tag{11.22}$$

This is exactly half the value found in Eq. 11.20 for the case in which E_c is negligible. And in the midway between the degeneracy points, the supercurrent decreases to a minimum value of

$$I_s = \frac{2e}{\hbar}\frac{E_{j1}E_{j2}\sin\vartheta}{4E_c} \tag{11.23}$$

In summary, the Josephson current varies periodically with the gate charge. At odd integer values of Q_0/e, it reaches a maximum value given by Eq. 11.22, comparable to the classical values for the supercurrent in individual junctions. And in between the degeneracy points, the current is depressed by a factor of the order of $E_j/E_c \ll 1$. It should be noted that thermal and quantum fluctuations can result in phase diffusion, characterized by a finite resistance of the supercurrent. As a result, it leads to a switching current lower than the nominal critical current.

11.2.3. Parity Effects in a Superconducting Island

We know that thermodynamic properties of small systems depend on the parity of the number of particles N in the system. One example of such a dependence is provided by atomic nuclei, where the binding energy of nuclei with an even number of protons and neutrons is systematically larger than the binding energy of nuclei with an odd number of nucleons.[32] A similar phenomenon is expected to occur in small metallic clusters and one dimensional systems. Because of the spin degeneracy of the single particle states, the properties of clusters with even and odd number of electrons are quite different.[33]

In both of these examples, however, the even-odd asymmetry, specifically the ground state energy difference decreases with increasing number of particles and vanishes in the thermodynamic limit. It had been found,[34,35] that superconductivity amplifies the parity effect in such a way that it takes place even in a superconducting island with macroscopically large number (10^9) of conduction electrons. Before the experiments described in this section were performed, it was believed that the strength of even-odd effects in superconductors would be suppressed as $1/N$. Thus, it was generally agreed that a superconductor with $N \gg 1$ would not exhibit parity effects in its macroscopic properties. However, we will show that the strength of the parity effect at very low temperatures is not governed by $1/N$, but rather, is in principle independent of N. We show that its strength is directly related to the difference between the number of quasiparticle excitations in the island when N is even and when is N odd. Since this difference is in principle equal to 1 at $T = 0$ regardless of the size of the system, at very low temperatures the parity effect should not weaken as the system is made larger.

11.2.4. Experimental Observation of the Parity Effect

In this section, we will show the experimental temperature dependence of the parity effect in a NSN SET system. We have shown that when the bias voltage V is close to zero, the system free energy is simply the electrostatic energy

$$U = \frac{(-en + Q_0)^2}{2C_\Sigma} \tag{11.24}$$

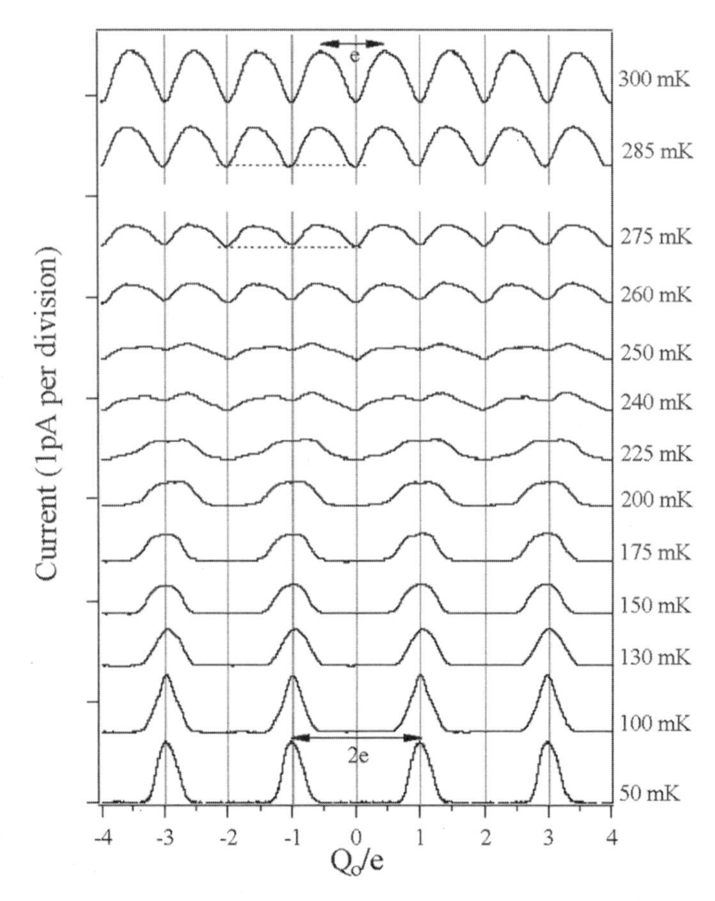

FIGURE 11.10. Experimental $I - Q_0$ curves for a NSN sample at a small bias voltage ($V = 125\,\mu V$) with temperatures ranging from 50 to 300 mK. The curves are displaced upward successively for clarity. At low temperatures, the curves are strongly $2e$-periodic. As temperature is gradually increased, the curves evolve until they become completely e-periodic above a crossover temperature $T^* \sim 285$ mK for this sample.

which depends on n, the number of excess electrons on the island. As V_g is swept, n changes by unity every time Q_0 passes through a half-integer value. This leads to a variation of the populations, and hence the current $I(V_g)$ at fixed bias V is e-periodic with peaks at half-integer values of Q_0/e, corresponding to single electron tunneling.

When the island is superconducting, it was found that the $I - Q_0$ curve taken under similar conditions becomes $2e$-periodic. Figure 11.10 shows a series of $I - Q_0$ curves for a NSN sample (Al island has a $T_c \approx 1.5$ K) at different temperatures and zero magnetic field.[36] At the lowest temperatures, the curves are strongly $2e$ periodic with two-electron tunneling (Andreev peaks) occurring at odd-integer values of Q_0/e. This $2e$ periodicity indicates that the number of electrons on the island changes by two as Q_0 is swept through one current peak. This in turn suggests that the island prefers to have a total number of conduction electrons of a certain parity. As the temperature rises, the Andreev peaks decrease in size, and gradually each Andreev peak splits apart into a subtle double maximum. The two maxima eventually move farther apart in Q_0 and develop into well separated peaks with the current slightly higher at odd-integer values of Q_0/e than at even-integer values. Finally, the current becomes completely e-periodic with equal peaks at all half-integer values of Q_0/e, and the magnitude of the current modulation grows with a further increase in the temperature. From observation, these $I - Q_0$ curves in Figure 11.10 are $2e$-periodic at temperatures up to and including 275 mK.[i] At 285 mK and above, only e periodicity can be distinguished in these curves. Thus, we estimate the crossover temperature from $2e$ to e periodicity, $T^* \approx 285$ mK, which is only about one fifth of the critical temperature T_c of the sample.

To thoroughly understand the parity effect, it is necessary to make a kinetic calculation,[37–39] solving a master equation to find the self-consistent steady state non-equilibrium populations of all relevant states, and the resulting current as a function of the bias voltage and the gate voltage. However, in the limit of low bias voltage, state populations will be close to the $V = 0$ equilibrium values. At sufficiently low bias voltages, we expect that the current through the device is proportional to V with a coefficient which is a function of V_g and T, depending on the equilibrium populations. Thus the period (e or $2e$) of the current will be determined by the period with which the populations vary with V_g. This allows one to simply use the periodicity of the equilibrium populations as a proxy for the periodicity of the current at low bias voltages. Such simple approach is very useful, even if it is limited to finding the period of $I(V_g)$, without finding the exact magnitude and function form.

11.3. IMPLEMENTATION OF SINGLE ELECTRON TRANSISTORS

Figure 11.11 gives an illustration of a SSS sample, which was fabricated in a $Al/Al_2O_3/Al$ three layer process. Since aluminum is a superconductor, this device can be operated either in the superconducting state or in the normal state. SETs can be made using a wide variety of metals, semiconductors, or conducting polymers. In this section, we will describe the type of SETs which use lithographically fabricated ultra-narrow thin film as the island.

Typically, this kind of SET devices is fabricated by electron-beam (e-beam) lithography followed by a shadow evaporation technique. This fabrication technique was pioneered by Dolan.[40] The fabrication procedure consists of a series of steps. The first one consists of growing a polished Si wafer, which has a thin layer ($\sim$ 30 Å) of native oxide. The Si wafers are first coated by a bilayer resist, which is an e-beam sensitive polymer. Scanning electron microscope (SEM) is used to write lithographic patterns on the wafer chips. Then they are developed so that the electron beam exposed areas are removed. After metal evaporation, the remaining resist gets removed in the liftoff step in acetone. Please refer to Chapter 1 of the book for more details on fabrication processes.

In order to observe the single electron effects, the SETs must be measured at low temperatures. This is because the energies of these phenomena are lower than 0.1 meV for a total capacitance on the order of 10^{-15} F, which approximately equals to the thermal energy at 1 K. Typically 4-probe current transport measurements are performed on samples cooled in dilution refrigerators.

[i] The dashed line on the $I - Q_0$ curve at 275 mK is drawn to assist viewing the existence of $2e$ periodicity.

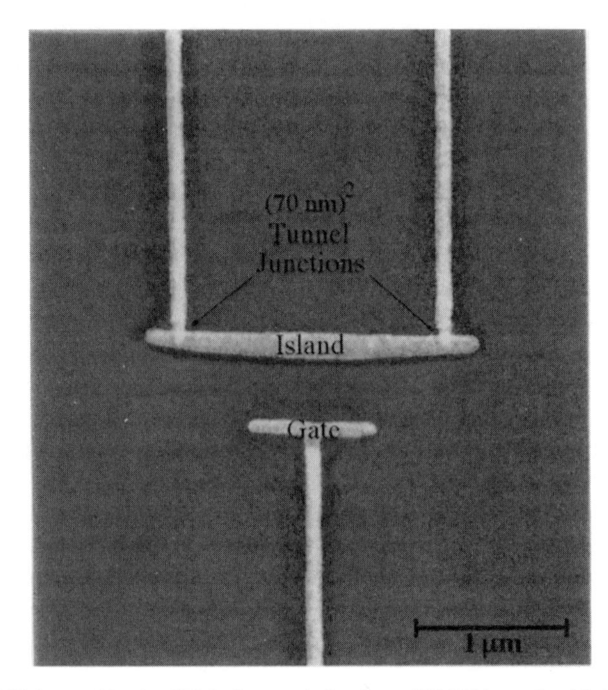

FIGURE 11.11. A high magnification SEM micrograph showing a SSS SET sample with 2 μm long island and 70 nm wide leads. It also shows the fabricated gate which allows the control of the number of electrons on the island. (Courtesy of Dr. J. Hergenrother.)

SETs are usually very sensitive to noise,[41] so measurement is often taken inside an electromagnetically shielded room, and sufficient lead filtering is needed. High frequency noise in the leads must be attenuated before it reaches the sample. Otherwise, the noise can lead to a rf heating of the sample, enable photon-assisted tunneling which can dramatically change the current transport behavior. And for superconducting SETs, it can change the phase dynamics of the Josephson junctions.

11.4. APPLICATION OF SINGLE ELECTRON TRANSISTORS

In semiconducting industry, it is now possible to produce MOSFETs and integrated circuits with gate oxides less than 10 atoms across (see Chapter 9). Such thin films are required to maintain the current response of the transistor to lower voltages at the gate electrode. Manufacturers need to lower the power supply to individual components in order to pack more devices onto a chip. The roadmap projects that by 2012 the gate oxide must be reduced to a thickness of 5 Si atomic layers as the device size miniaturizes.[42] Such scaling down of the gate oxide will eventually reach its fundamental physical limit, as the quantum mechanical tunneling of electrons will cause the break down of electrical insulation. As a result, nanoscale devices become an attractive option for the development of integrated circuits with dimensions and performances beyond the ultimate roadmap. Among them, SET presents promising features such as reduced dimension, extremely low power consumption, and high charge sensitivity.

Many measurements have been performed to elucidate the charge transport in the SETs. As we have discussed, by adding electrons to the island of a SET transistor, one can study the interaction of electrons on the island, the coupling of states in the leads to states in the island, and how superconductivity changes the system behavior. The remarkable capability to control current allows SETs to be used in metrological applications such as ultrasensitive electrometer and current standard. The low power dissipation makes them potentially useful in high density memory and logic circuits. SETs are also very sensitive to applied radiation. Photon assisted tunneling has been observed and the absorption of individual photons of microwave radiation can be detected. Furthermore, SETs have potential applications in magnetic recording industry. In this section, we will discuss some possible applications of SETs.

11.4.1. Metrological Applications

11.4.1.1. Precision Charge Measurements Because of their charge sensitivity, SETs are excellent for making precision charge measurements. A SET can be used to measure charge either in the normal state or in the superconducting state. Typically, it is voltage biased at a point where there is a large modulation of the current as a function of the gate charge Q_0. The charge that is to be measured is coupled to the gate, and the current modulated by the gate is monitored. Thus, charges much smaller than the electron charge e can be measured. The charge resolution that can be achieved is $8 \times 10^{-6} e/\sqrt{Hz}$ at 10 Hz.[44] SETs offer by far the best charge resolution among the available charge measurement devices.

SETs have also been capacitively coupled to a variety of systems so that the charge motion of those systems could be observed. Metallic SETs have been coupled to semiconductor quantum dots to monitor the charge fluctuations in the quantum dot.[45] They have been coupled to superconducting particles where it is possible to observe whether the particle has an even number or an odd number of electrons on it.[46] SETs have also been scanned over semiconductors to measure fluctuations in the dopant distribution.[47]

11.4.1.2. Current Standard One of the applications of SET devices is a fundamental current standard. In such a device, a known current is established by transferring individual electrons through the device with a frequency f. This results in a current $I = ef$. A number of different proposals have been made. They include modulating the gates coupled to the islands in an array of tunnel junctions,[48,49] modulating the tunnel barriers in a semiconducting quantum dot,[50] and transferring Cooper pairs in a superconducting circuit.[51] The most intensively studied current standard is called an electron pump. It consists of a number of tunnel junctions in series with a gate connected to each island between the junctions. By modulating the gates successively, one can draw a single electron through the array of tunnel junctions. The accuracy that has been achieved with this current standard is 15 parts per billion.[52]

11.4.2. Information Technology

11.4.2.1. SET Memories The small size and low power consumption of SETs make them a promising candidate for information technology industry. At present stage, SET memories seem more promising than SET logic circuits.[53] Here, we will discuss two memory schemes. The first is an offset-charge independent DRAM cell in which a bit is represented by the presence or absence of the charge of a few electrons which are stored on an island.[54] The charge on the island is monitored by a SET. When the memory cell is read, if there are charges on the island, the current through the SET undergoes oscillations as each electron tunnels off the island. The current oscillations through the device occur for any value of the offset charge. If there are no charges stored on the island, there will be no current oscillations.

Another SET memory type, called a single electron MOS memory, is also based on the motion of individual electrons. This device is very similar to a conventional floating gate MOS memory. The charge on a floating gate modulates the conduction through a channel nearby the gate. The gate is so small that even when one electron is added to the floating gate, the conduction through the channel significantly changes.[55,56]

11.4.2.2. SET Logic Quite a number of logic designs based on SET circuits have been proposed. Some of the designs are very similar to CMOS where bits are represented by voltage levels.[57,58] Some logic designs resemble superconducting single flux quantum logic.[10] In this case, bits are represented by the presence or the absence of individual electrons. Other logic designs contain elements that act like electron pumps for transferring charges.[59]

A CMOS-like SET inverter using two capacitively coupled single electron transistors has been successfully developed.[60] The inverter has a voltage gain of 2.6 at 25 mK and remains larger than 1 for temperatures up to 140 mK. Inverters can be used as fundamental building block of SET logic circuits and memory elements. NAND and NOR gates are realized by making slight variations of the inverter. With two inverters, a static RAM memory cell can also be constructed.

11.4.3. Ultrasensitive Microwave Detector

When the island is superconducting, the effects of photon assisted tunneling in the SET are dramatically enhanced.[43] Due to the parity effect in the superconducting island, it produces a secondary $I - Q_0$ peak, as shown in Figure 11.12.

This can be conspicuous in the presence of very small amounts of microwave radiation because each absorbed photon allows many electrons to tunnel through the system. Thus, the superconducting SET is an extremely sensitive microwave detector because it acts as a photon-activated switch from a low- to a high-current state.

11.4.4. Magneto-electronics

There has been a lot of interest in the past few years to study the spin transport and dynamics in ferromagnetic SET with ferromagnetic electrodes.[61–64] Here, let us consider a double junction system where the two electrodes are ferromagnets and the island is a normal

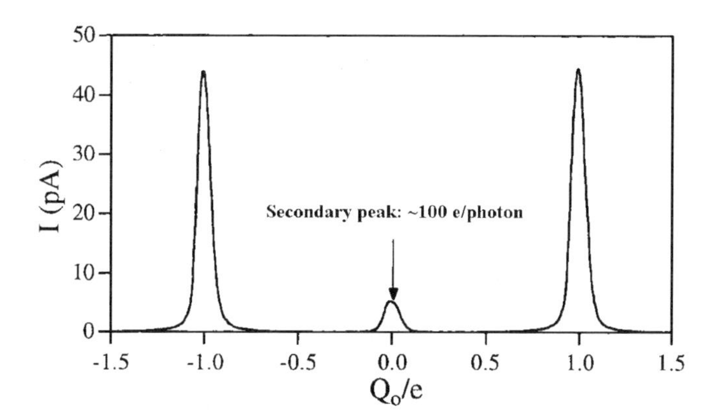

FIGURE 11.12. A SSS SET acts as a photon-activated switch. Near $Q_0 = 0$, the secondary current peak corresponds to the tunneling of many electrons for every absorbed photon.

metal. Such a tunnel junction can be used in ferromagnetic alignment (the magnetization of the two electrodes is parallel) and in anti-ferromagnetic alignment (the magnetization of the two electrodes is anti-parallel). Already for relatively large junctions (where the Coulomb blockade can be neglected), the anti-ferromagnetic alignment is interesting because a tunnel current generates a spin accumulation on the island. For nanoscale double junction systems with ferromagnetic leads, it has been calculated that spin accumulation caused by co-tunneling squeezes the Coulomb blockade in the anti-ferromagnetic configuration. And the tunnel magnetoresistance is strongly enhanced in the Coulomb blockade region by co-tunneling process.[65–67] This aspect will lead SET into new spin electronic devices useful for magnetic recording industry as read head sensors and magnetic memory elements.

11.5. SUMMARY

We have discussed the charge transport mechanisms in single electron transistors, with primary focus on the normal state SETs and superconducting state SETs. We have shown that superconductivity in the island leads to very interesting even-odd electron number effects even if there are as many as one billion conduction electrons in the island. Single charge tunneling effects play an important role in devices with very small dimensions as device miniaturization continues for dense integrated circuits. Because the phenomena of single electronics most groups have been studying manifest themselves at low temperatures, the main drawback of using SET for applications is the necessity for low temperature operation. However, the demonstration of silicon based SET operating at room temperature,[68] the recent demonstration of carbon nanotube and nanowire based logic circuits operating at room temperature,[69,70] and ongoing advances in nanofabrication to make even smaller feature sizes indicate that it will be possible to have SET circuits operating at room temperatures. Therefore, it is highly promising for single electron devices to realize their practical industrial applications.

QUESTIONS

1. In the chapter, we have shown the system free energy for a symmetric biased single-electron transistor (SET). Derive a form of the system free energy for a general biased SET.
2. We have discussed that for a symmetric SET, the $I-V$ curves for all values of Q_0, are enveloped between $Q_0 = 0$ (when there is a Coulomb blockade with a voltage threshold e/C_Σ) and $Q_0 = e/2$ (when it shows linear characteristic through origin).

   a. Determine what the extreme values of Q_0 are in an asymmetric SET with $C_1 \neq C_2$.
   b. Do the same for a SET with a superconducting island.
3. Find the energy change ΔE in tunneling transition from n to $n \pm 1$ state across junction 1 for a general bias.
4. For a superconducting island, find the rate of one *specific* electron tunneling *off* the island. Hint: Here, the density of states on the island $D_1(\varepsilon) = \delta(\varepsilon - \Delta)$.
5. Show the higher order approximation of N_{eff} as a function of T at H $= 0$ can be expressed as:

$$N_{\text{eff}} \approx V_I \rho_n(0)\sqrt{2\pi \Delta k_B T}\left[1 + \frac{3}{8}\frac{k_B T}{\Delta} + O\left(\frac{k_B T}{\Delta}\right)^2\right]$$

Where $\rho_n(0)$ is the density of states in normal state.

APPENDIX A—SINGLE ELECTRON TUNNELING RATE

Refer to Figure 11.1 in the text, let us assume that the coupling due to tunneling between the two electrodes is sufficiently weak. According to Fermi's golden rule, the transition rate from a single initial state k to a set of final states q is given by[j]

$$\Gamma_k = \frac{2\pi}{\hbar}\sum_q |T_{k,q}|^2\,\delta(\varepsilon_k + eV - \varepsilon_q) \tag{11.25}$$

The Dirac delta function ensures that energy is conserved in the tunneling process, i.e., $\varepsilon_q = \varepsilon_k + eV$. We can transform the sum over q states to an integral over the energy ε_q by substituting the density of states (DOS) of electrode 2, $D_2(\varepsilon_k + eV)/2$ at $\varepsilon_k + eV$, and obtain

$$\Gamma_k = \frac{2\pi}{\hbar}|T|^2\frac{D_2(\varepsilon_k + eV)}{2} \tag{11.26}$$

Note that the DOS of electrode 2 already includes the spin degeneracy factor of 2. The division by 2 reflects that only states with the same spin as the original can be tunneled into, i.e., no spin flips are possible. We have also assumed that the average tunneling matrix

[j] k and q are electron wave vectors, ε_k and ε_q are the electron energies measured with respect to the Fermi energies of electrodes 1 and 2, respectively.

amplitude $|T|$ is independent of the wave vectors k and q, therefore, it is independent of the energies ε_k and ε_q.

At thermal equilibrium, the occupation probability of the electron state is given by the Fermi-Dirac function

$$f(\varepsilon) = [1 + e^{\beta\varepsilon}]^{-1} \tag{11.27}$$

where $\beta = 1/k_B T$. The probability of the unoccupied state is $1 - f(\varepsilon)$. Tunneling can only occur if the initial state is occupied and the final state is unoccupied. Thus, summing over all initial occupied states ε_k, including the factor $D_1(\varepsilon_k)$ for the DOS of electrode 1, we obtain the rate for tunneling from electrode 1 to electrode 2:

$$\Gamma_{1\to 2} = \sum_k \frac{2\pi}{\hbar}|T|^2 \frac{D_2(\varepsilon_k + eV)}{2}[1 - f(\varepsilon_k + eV)] \tag{11.28}$$

Replacing the sum over initial states k by an integral over ε_k, we get,

$$\Gamma_{1\to 2} = \frac{\pi}{\hbar}|T|^2 \int_{-\infty}^{\infty} D_1(\varepsilon) f(\varepsilon) D_2(\varepsilon_k + eV)\,[1 - f(\varepsilon + eV)]d\varepsilon \tag{11.29}$$

Since the bias energies are much less than the Fermi energies in the normal electrodes, we can assume that the DOS is independent of energy, i.e., $D_1(\varepsilon) = D_1$ and $D_2(\varepsilon) = D_2$. Then we obtain the forward tunneling rate:

$$\Gamma_{1\to 2} = \frac{\pi}{\hbar}|T|^2 D_1 D_2 \int_{-\infty}^{\infty} f(\varepsilon)\,[1 - f(\varepsilon + eV)]d\varepsilon \tag{11.30}$$

Using mathematical identities for the Fermi functions:

$$f(\varepsilon)[1 - f(\varepsilon + eV)] = \frac{f(\varepsilon) - f(\varepsilon + eV)}{1 - e^{-\beta eV}} \tag{11.31}$$

We may express Eq.11.30 as

$$\Gamma_{1\to 2} = \frac{eV}{e^2 R_T(1 - e^{-\beta eV})} \tag{11.32}$$

with R_T being the normal state tunnel resistance:

$$R_T = \frac{\hbar}{\pi |T|^2 e^2 D_1 D_2} \tag{11.33}$$

APPENDIX B—DERIVATION OF THE SYSTEM FREE ENERGY

The total capacitance of the island to its surroundings is the sum of the capacitances to the three electrodes[k]

$$C_\Sigma = C_1 + C_2 + C_g \tag{11.34}$$

Since charge is quantized in units of e, the total charge on the island is always discrete. However, in calculating the system free energy, we will show that continuously increasing the gate voltage is mathematically equivalent to continuously adding charge to the island. Thus, the effective charge of the island is continuous. We will use this fact to illustrate how the gate can be used to tune the strength of the charging energy barrier.

To calculate the system free energy for the single electron transistor, we define n_1 and n_2 to be the number of electrons that have tunneled forward across junctions 1 and 2, respectively, and the integer $n = n_1 - n_2$ to be the excess number of electrons on the island. If we define ϕ to be the electrostatic potential of the island, then with the voltages and capacitances defined in the schematic diagram of Figure 11.3, we have (with $C_3 = C_g$, $V_3 = V_g$)

$$\sum_{i=1}^{3} C_i(V_i - \phi) = ne \tag{11.35}$$

From this, we can find the electrostatic potential of the island:

$$\phi = \frac{\sum_{i=1}^{3} C_i V_i - ne}{C_\Sigma} \tag{11.36}$$

where ϕ is measured with respect to the same reference as V_1, V_2, and V_g. The electrostatic energy U of the system is given by summing the field energies of the three capacitors. Thus,

$$U = \frac{1}{2}\sum_i C_i(V_i - \phi)^2 \tag{11.37}$$

If we use Eq. 11.36 and substitute for φ in Eq. 11.37, after some algebra, we obtain:[15]

$$\begin{aligned} U &= \frac{(-ne)^2}{2C_\Sigma} + \frac{1}{2C_\Sigma}\sum_i \sum_{j>i} C_i C_j (V_i - V_j)^2 \\ &= \frac{(-ne)^2}{2C_\Sigma} + \Lambda \end{aligned} \tag{11.38}$$

where Λ is a constant independent of n. Since we are only concerned with changes in the

[k] The self capacitance of this island is usually neglected because it is much smaller than the capacitances of the junctions and the gate.

system free energy associated with tunneling events, we do not need to know the exact value of Λ.

When considering the tunneling of an electron onto or off the island, we must include not only the change in the charging energy U from Eq. 11.37, but also the work done by the bias voltages. The total work done by the voltage sources when an electron tunnels onto the island across junction j is

$$W_j = e \sum_i \frac{C_i}{C_\Sigma}(V_i - V_j) \tag{11.39}$$

In the case of the symmetric voltage bias, i.e., $V_1 = -V/2$, and $V_2 = +V/2$, for an electron tunneling onto the island across junction 1, the voltage source does work:

$$W_1 = \frac{e}{C_\Sigma}[(C_2 + C_g/2)V + C_g V_g] \tag{11.40}$$

and similarly, for junction 2, the voltage source does work:

$$W_2 = -\frac{e}{C_\Sigma}[(C_1 + C_g/2)V - C_g V_g] \tag{11.41}$$

The system free energy can then be calculated by subtracting the work done by the voltage sources from the electrostatic energy, i.e.,

$$F_{\text{sys}}(n_1, n_2) = U - n_1 W_1 + n_2 W_2 \tag{11.42}$$

Substituting Eq. 11.40 and Eq. 11.41 into Eq. 11.42, we obtain

$$F_{\text{sys}}(n_1, n_2) = \Lambda + \left[\frac{(-ne)^2}{2C_\Sigma} - n\frac{eQ_0}{C_\Sigma}\right] - n_1\frac{e}{C_\Sigma}(C_2 + C_g/2)V - n_2\frac{e}{C_\Sigma}(C_1 + C_g/2)V \tag{11.43}$$

where we have introduced the induced gate charge $Q_0 = C_g V_g$,[l] which is a continuous variable as compared to the discrete change ne from tunneling events. If we define the integer $m = n_1 + n_2$ as the total number of electrons that have tunneled forward through either junction, then after completing the square of the term in brackets in Eq. 11.43, we find (up to a constant which is independent of n and m):

$$F_{\text{sys}}(n, m) = \frac{(Q_0 - ne)^2}{2C_\Sigma} - \left[m + n\frac{C_2 - C_1}{C_\Sigma}\right]\frac{eV}{2} \tag{11.44}$$

[l] In practice, naturally occurring random charged impurities near the island shift the polarization charge by an amount independent of V_g, and may drift or change discontinuously in time. This offset can be suppressed by shifting the zero of the gate voltage.

APPENDIX C—EVEN-ODD FREE ENERGY DIFFERENCE

The even-odd free energy difference F_0 is calculated using equilibrium statistical mechanics. The grand canonical partition function Z for quasiparticle excitations on the superconducting island is:[39]

$$Z = \prod_k (1 + e^{-\beta\xi_k}) \tag{11.45}$$

where ξ_k is the energy of a quasiparticle excitation in state k relative to the Fermi energy. In zero magnetic field,

$$\xi_k = \sqrt{\varepsilon_k^2 + \Delta^2} \tag{11.46}$$

where ε_k is the individual electron energy in the normal state relative to the Fermi energy. The grand canonical partition function can be separated algebraically into terms corresponding to an even or an odd number of quasiparticle excitations:

$$\begin{aligned} Z &= \left(1 + \frac{1}{2!}\sum_k \sum_{l\neq k} e^{-\beta\xi_k} e^{-\beta\xi_l} \ldots\right) \\ &\quad + \left(\sum_k e^{-\beta\xi_k} + \frac{1}{3!}\sum_k \sum_{l\neq k} \sum_{m\neq k,l} e^{-\beta\xi_k} e^{-\beta\xi_l} e^{-\beta\xi_m} \ldots\right) \\ &= Z_{\text{even}} + Z_{\text{odd}} \end{aligned} \tag{11.47}$$

If the number of electrons on the island is fixed, only Z_{even} or Z_{odd} is used, since quasiparticle excitations can only be created two at a time. Only the tunneling of a quasiparticle from the leads onto the island can change the parity. the even-odd free energy difference is then defined as the difference in free energy calculated for the two parity cases:

$$F_0 = k_B T \ \ln\left(\frac{Z_{\text{even}}}{Z_{\text{odd}}}\right) \tag{11.48}$$

Low Temperature Approximation of $F_0(T)$

At low temperatures, there are rarely any thermally excited quasiparticles. Consequently, the leading terms in Z_{even} and Z_{odd} dominate, yielding:

$$F_0 \approx -k_B T \ln\left[\sum_k e^{-\beta\xi_k}\right] \tag{11.49}$$

Making the continuum approximation and using the quasiparticle density of states $\rho_s(\xi)$ which includes the spin degeneracy, we get:

$$F_0 \approx -k_B T \ln\left[2V_I \int_0^\infty \rho_s(\xi)e^{-\beta\xi}d\xi\right] \tag{11.50}$$

V_I is the volume of the superconducting island, and the factor of two on the right-hand side is to count quasiparticles with $|k| < k_F$ and $|k| > k_F$. It is expected that $F_0(T=0, H) = \Omega_G(H)$, since at zero temperature the remaining quasiparticle will reside at the minimum excitation energy, i.e., the spectroscopic gap $\Omega_G(H)$, which is a decreasing function of the magnetic field. Thus, it is useful to rewrite Eq. 11.50 as:[m]

$$\begin{aligned}F_0 &\approx -k_B T \ln\left[e^{-\beta\Omega_G}2V_I \int_{\Omega_G}^\infty \rho_s(\xi)e^{-\beta(\xi-\Omega_G)}d\xi\right]\\ &= \Omega_G - k_B T \ln(N_{\text{eff}})\end{aligned} \tag{11.51}$$

with

$$N_{\text{eff}} = 2V_I \int_{\Omega_G}^\infty \rho_s(\xi)e^{-\beta(\xi-\Omega_G)}d\xi \tag{11.52}$$

We see that F_0 approaches zero at a $2e - e$ crossover temperature of $\Omega_G/k_B \ln(N_{\text{eff}})$.

Calculation of a General Form of $F_0(T)$

For T near or above T^*, the presence of thermally excited quasiparticle in the island becomes important. As a result, in order to accurately calculate the even-odd free energy difference in this temperature regime, we must consider higher-order terms in Eq. 11.47. To simplify the calculation, we model the quasi-continuum of excited states above the superconducting gap by a single level at energy Ω_G with degeneracy N_{eff}, then the partition sums of Eq. 11.45 can be written as:

$$Z = (1 + e^{-\beta\Omega_G})^{N_{\text{eff}}} \tag{11.53}$$

We assume that the island is in weak tunneling contact with a particle reservoir at $V = 0$, so that the island can contain either an even or odd number of electrons. By constructing forms in which either the even or odd terms cancel, we can then write down the partial partition sums as:[36]

$$Z_{\text{even}} = \frac{[(1 + e^{-\beta\Omega_G})^{N_{\text{eff}}} + (1 - e^{-\beta\Omega_G})^{N_{\text{eff}}}]}{2} \tag{11.54}$$

[m] Note that the magnetic field dependence enters into F_0 and N_{eff} through Ω_G and ρ_s. For a study of the field dependence of the parity effects, refer to Ref. 26.

and

$$Z_{\text{odd}} = \frac{[(1+e^{-\beta\Omega_G})^{N_{\text{eff}}} - (1-e^{-\beta\Omega_G})^{N_{\text{eff}}}]}{2} \tag{11.55}$$

At this point, it is helpful to make use of the approximation

$$\left(1+\frac{a}{b}\right)^b \approx e^a \tag{11.56}$$

which is valid for $b \gg 1$ and $|a| \ll b$. Applying this relation, for $e^{-\beta\Omega_G} \ll 1$ and $N_{\text{eff}} \gg 1$, one obtains:

$$(1+e^{-\beta\Omega_G})^{N_{\text{eff}}} \approx e^{N_{\text{eff}}}e^{-\beta\Omega_G} \tag{11.57}$$

and

$$(1-e^{-\beta\Omega_G})^{N_{\text{eff}}} \approx e^{-N_{\text{eff}}}e^{-\beta\Omega_G} \tag{11.58}$$

We can thus write

$$Z_{\text{even}} \approx \cosh\,(N_{\text{eff}}e^{-\beta\Omega_G}) = \cosh\,(\langle N_{\text{qp}}\rangle) \tag{11.59}$$

and

$$Z_{\text{odd}} \approx \sinh\,(N_{\text{eff}}e^{-\beta\Omega_G}) = \sinh\,(\langle N_{\text{qp}}\rangle) \tag{11.60}$$

Substituting Z_{even} and Z_{odd} into Eq. 11.48, we find that

$$F_0(T) \approx k_B T\,\ln[\coth(N_{\text{eff}}e^{-\beta\Omega_G})] = k_B T\,\ln\,[\coth\langle N_{\text{qp}}\rangle] \tag{11.61}$$

which asymptotically approaches zero. $\langle N_{\text{qp}}\rangle$ denotes the average number of thermally excited quasiparticles in the superconducting island.

REFERENCES

1. G. Bergmann, *Physics Reports* **107,** 1 (1984).
2. S. Washburn and R. A. Webb, *Advances in Physics* **35,** 375 (1986).
3. W. F. Smith, T. S. Tighe, G. C. Spalding, M. Tinkham, and C. J. Lobb, *Phys. Rev.* B **43,** 12267 (1991).
4. R. A. Webb, S. Washburn, C. P. Umbach, and R. B. Laibowitz, *Phys. Rev. Lett.* **54,** 2696 (1985).
5. B. J. van Wees, H. van Houten, C. W. J. Beenakker, J. G. Williamson, L. P. Kouwenhoven, D. van der Marel, and C. T. Foxon, *Phys. Rev. Lett.* **60,** 848 (1988).
6. P. L. McEuen, E. B. Foxman, U. Meirav, M. A. Kastner, Y. Meir, N. S. Wingreen, and S. J. Wind, *Phys. Rev. Lett.* **66,** 1926 (1991).
7. J. Kong, C. Zhou, E. Yenilmez, and H. Dai, *Appl. Phys. Lett.* **77,** 3977 (2000).
8. D. C. Ralph, C. T. Black, and M. Tinkham, *Phys. Rev. Lett.* **74,** 3241 (1995).
9. J. Petta, D. Salinas, and D. C. Ralph, *Appl. Phys. Lett.* **77,** 4419 (2000).
10. D. V. Averin and K. K. Likharev, in Mesoscopic Phenomena in Solids, eds., B. L. Al'tshuler, P. A. Lee, and R. A. Webb (Elsevier, 1991).
11. D. V. Averin and K. K. Likharev, *J. of Low Temp. Phys.* **62,** 345 (1986).
12. K. K. Likharev, *IBM J. Res. and Dev.* **32,** 144 (1988).
13. G.-L. Ingold and Yu. V. Nazarov, in Single Charge Tunneling, eds., M. H. Devoret and H. Grabert (Plenum, 1992).
14. D. B. Haviland and L. S. Kuzmin, *Phys. Rev. Lett.* **67,** 2890 (1991).
15. M. Tinkham, Introduction to Superconductivity, 2nd ed., New York: McGraw-Hill, (1995).
16. D. V. Averin and A. A. Odintsov, *Phys. Lett.* A **140,** 251 (1989).
17. D. V. Averin and Yu. V. Nazarov, in Single Charge Tunneling, eds., M. H. Devoret and H. Grabert (Plenum, 1992).
18. D. V. Averin and Yu. V. Nazarov, *Phys. Rev. Lett.* **65,** 2446 (1990).
19. L. J. Geerligs, D. V. Averin, and J. E. Mooij, *Phys. Rev. Lett.* **65,** 3037 (1990).
20. A. F. Andreev, *Sov. Phys. JETP* **19,** 1228 (1964).
21. F. W. J. Hekking, L. I. Glazman, K. A. Matveev, and R. I. Shekhter, *Phys. Rev. Lett.* **70,** 4138 (1993).
22. C. W. J. Beenakker, *Phys. Rev.* B **46,** 12841 (1992).
23. F. W. J. Hekking and Yu. V. Nazarov, *Phys. Rev.* B **49,** 6847 (1994).
24. B. J. van Wees, P. de Vries, P. Magnee, and T. M. Klapwijk, *Phy. Rev. Lett.* **69,** 510 (1992).
25. A. D. Zaikin, *Physca* B **203,** 255 (1994).
26. J. G. Lu, J. M. Hergenrother, and M. Tinkham, *Phys. Rev.* B **53,** 3543 (1996).
27. T. A. Fulton, P. L. Gammel, D. J. Bishop, and L. N. Dunkleberger, *Phys. Rev. Lett.* **63,** 1307 (1989).
28. A. Maassen van den Brink, G. Schön, and L. J. Geerligs, *Phys. Rev. Lett.* **67,** 3030 (1991).
29. M. T. Tuominen, J. M. Hergenrother, T. S. Tighe, and M. Tinkham, IEEE Trans. on Appl. Supercond. vol. 3, no. 1, p. 1972 (1993).
30. S. Pohlen, R. Fitzgerald, and M. Tinkham, *Physica* B, 284–288, 1812–1813 (2000).
31. K. K. Likharev and A. B. Zorin, *J. Low Temp. Phys.* **59,** 347 (1985).
32. P. J. Siemens and A. S. Jensen, Elements of Nuclei, Ch. 6 (Addison-Wesley, 1987).
33. W. P. Halperin, *Rev. Mod. Phys.* **58,** 533 (1986).
34. D. V. Averin and Yu. V. Nazarov, *Phys. Rev. Lett.* **69,** 1993 (1992).
35. M. T. Tuominen, J. M. Hergenrother, T. S. Tighe, and M. Tinkham, *Phys. Rev. Lett.* **69,** 1997 (1992).
36. M. Tinkham, J. M. Hergenrother, and J. G. Lu, *Phys. Rev.* B **51,** 12649 (1995).
37. G. Schön and A. D. Zaikin, *Europhys. Lett.* **26,** 695 (1994).
38. J. M. Hergenrother, M. T. Tuominen, J. G. Lu, D. C. Ralph, and M. Tinkham, *Physica* B **203,** 327 (1994).
39. M. T. Tuominen, J. M. Hergenrother, T. S. TIghe, and M. Tinkham, *Phys. Rev.* B **47,** 11599 (1993).
40. G. J. Dolan, *Appl. Phys. Lett.* **31,** 337 (1997).
41. G. Johansson, P. Delsing, K. Bladh, D. Gunnarsson, T. Duty, A. Kack, G. Wendin, and A. Aassime, Proceedings of NATO ARW "Quantum Noise in Mesoscopic Physics", Oct. 8 (2002).
42. M. Schulz, *Nature* **399,** 729 (1999).
43. J. M. Hergenrother, J. G. Lu, M. T. Tuominen, D. C. Ralph, and M. Tinkham, *Phys. Rev.* B **51,** 9407 (1995).
44. V. A. Krupenin, D. E. Presnov, A. B. Zorin, and J. Niemeyer, *J. Low Temp. Phys.* **118,** 287 (2000).

45. D. Berman, N. B. Zhitenev, R. C. Ashoori, H. I. Smith, and M. R. Melloch, *J. Vac. Sci. and Tech.* B **15,** 2844 (1997).
46. C. T. Black, D. C. Ralph, and M. Tinkham, *Phys. Rev. Lett.* **76,** 688 (1996).
47. M. J. Yoo, T. A. Fulton, J. F. Hess, R. L. Willett, L. N. Dunkleberger, R. J. Chichester, L. N. Pfeiffer, K. W. West, *Science* **276,** 579 (1997).
48. L. J. Geerligs, V. F. Anderegg, P. A. M. Holweg, J. E. Mooij, H. Pothier, D. Esteve, C. Urbina, and M. H. Devoret, *Phys. Rev. Lett.* **64,** 2691 (1990).
49. H. Pothier, P. Lafarge, C. Urbina, D. Esteve, and M. H. Devoret, *Europhys. Lett.* **17,** 249 (1992).
50. L. P. Kouwenhoven, A. T. Johnson, N. C. van der Vaart, and C. J. P. M. Harmans, *Phys. Rev. Lett.* **67,** 1626, (1991).
51. L. J. Geerligs, S. M. Verbrugh, P. Hadley, J. E. Mooij, H. Pothier, P. Lafarge, C. Urbina, D. Esteve, and M. H. Devoret, Zeitschrift für Physik B **85,** 349 (1991).
52. M. W. Keller, J. M. Martinis, N. M. Zimmerman, and A. H. Steinbach, *Appl. Phys. Lett.* **69,** 1804 (1996).
53. A. N. Korotkov, *J. App. Phys.* **92,** 7291 (2002).
54. K. K. Likharev and A. N. Korotkov, VLSI Design **6,** 341 (1998).
55. A. N. Korotkov, Coulomb blockade and digital single-electron devices, in Molecular electronics, eds. J. Jortner and M. A. Ratner, Blackwell, Oxford (1997), pp. 157–189; A. N. Korotkov, Single-electron logic and memory devices, *Int. J. Electronics* **86,** 511–547 (1999).
56. L. Guo, E. Leobandung, and S. Y. Chou, Science **275,** 649 (1997).
57. N. Yoshikawa, Y. Jinguu, J. Ishibashi, and M. Sugahara, *Jap. J. Appl. Phys.* **35,** 1140 (1996).
58. A. N. Korotkov, R. H. Chen, and K. K. Likharev, *J. Appl. Phys.* **78,** 2520 (1995).
59. M. G. Ancona, *J. Appl. Phys.* **79,** 526 (1996).
60. C. P. Heij, P. Hadley, and J. E. Mooij, *Appl. Phys. Lett.* **78,** 1140 (2001).
61. Y. Ootuka, R. Matsuda, K. Ono, and J. Shimada, *Physica* B **280,** 394 (2000).
62. J. Barnaś, J. Martinek, G. Machalek, B. R. Bulka, and A. Fert, *Phys. Rev.* B **62,** 12363 (2000).
63. H. Imamura, Y. Utsumi, and H. Ebisawa, *Phys. Rev.* B **66,** 054503 (2002).
64. S. Takahashi, T. Yamashita, H. Imamura, and S. Maekawa, J. Mag. and Mag. Mat. **240,** 100 (2001).
65. H. Imamura, S. Takahashi, and S. Maekawa, *Phys. Rev.* B **59,** 6017 (1999).
66. A. Brataas, Yu. V. Nazarov, J. Inoue, and G. E. W. Bauer, *Phys. Rev.* B **59,** 93 (1999).
67. J. Martinek, J. Barnaś, S. Maekawa, H. Schoeller, and G. Schön, *Phys. Rev.* B **66,** 014402 (2002).
68. Y. Takahashi, M. Nagase, H. Namatsu, K. Kurihara, K. Iwdate, Y. Nakajima, S. Horiguchi, K. Murase, and M. Tabe, *Electronics Lett.* vol. **31,** no. 2, p. 136 (1995).
69. Y. Huang, X. Duan, Y. Cui, L. Lauhon, K. Kim, and C. M. Lieber, *Science* **294,** 1313 (2001).
70. A. Bachtold, P. Hadley, T. Nakanishi, C. Dekker, *Science* **294,** 1317 (2001).
71. E. Merzbacher, Quantum Mechanics, New York: John Wiley & Sons, (1970).
72. J. M. Hergenrother, Ph.D. Thesis, Harvard University (1995).

IV

Nanotechnology in Magnetic Systems

12

Semiconductor Nanostructures for Quantum Computation

Michael E. Flatté

Department of Physics and Astronomy and Optical Science and Technology Center, The University of Iowa, Iowa City, IA

12.1. NANOSTRUCTURES FOR QUANTUM COMPUTATION

Computers have achieved a remarkable degree of sophistication, permeating almost all aspects of human activity. It is reasonable to ask, therefore, what a new paradigm of computation might offer to surpass the current one. Over the last decade the justification for hypothetical "quantum computers" has been constructed step by step. This justification includes the development of quantum algorithms, the construction of quantum error correction procedures, and the identification of model physical processes that could be employed in an actual quantum computer. The properties of electron and nuclear spin in semiconductor nanostructures now appear well suited for use in quantum computers. Control of the properties of an individual spin touches the central questions of quantum measurement theory, and from a practical perspective will require nanofabricated structures in which the spin can reside, and within which the spin can be manipulated.

In this chapter we shall begin with the desirable properties of quantum computers, to motivate the (considerable) effort that will be required for their construction. The focus will then shift to the physical requirements of any system for quantum computation, and the role of nanoscience in fulfilling these requirements. Turning then to the properties of spin, possible physical realizations of the elements of quantum computation will be described. These include the information itself, stored as "quantum bits" or "qubits", as well as manipulation of this quantum information using "quantum gates". The concluding sections describe two relatively complete proposals based on confinement of the qubit spin in nanostructured systems; one focused on electron spin in quantum dots, and the second emphasizing the nuclear spin of phosphorus atoms embedded in silicon. Although a commercial implementation of quantum computation is very far off, and may very well differ considerably from either of these two proposed architectures, their description will help clarify what one should expect from a total quantum computer architecture.

12.2. QUANTUM COMPUTATION ALGORITHMS

Despite the dramatic success of current computation technology, there remain well-defined problems of great interest that are impractical or impossible to solve. Several of these directly influence the sending of encoded information, either for secure economic transactions or for national security, as many cryptographic protections are constructed as difficult mathematical problems to solve. Two such problems, which to date have not yielded to efficient classical algorithms, are the factoring of large numbers into primes, and the finding of an entry in an unordered list. Current classical algorithms for factorization increase in computation time exponentially with the number of digits in the number to be factored. Finding an entry in an unordered list requires the examination of (on average) half the entries in the list.

The proposal by Shor of an algorithm for factoring large numbers in polynomial time in 1994 provided the first non-trivial potential application of quantum computation. Grover's algorithm, proposed in 1997, can find an entry in an unordered list with the examination of (on average) only the square root of the number of entries. Both of these algorithms take advantage of a key element of the quantum mechanical world, the quantum principle of superposition. Several variations of these algorithms have been suggested over subsequent years, and although no additional algorithms of equal generality and novelty have been developed, many continue to be optimistic that new interesting algorithms remain to be discovered.

A comparison of the properties of classical and quantum computers can clarify some of the challenges and opportunities of quantum computers. A brief summary of the similarities and differences between classical and quantum computers follows. In its essence, a classical computer consists of a machine to take information encoded in a set of bits, which take the values "0" and "1", and to manipulate individual bits in a controlled way depending on the values of other bits. The manipulation rules are the algorithm. Of course all practical classical computers used today rely on quantum mechanics to function properly. The term "quantum computer" here has a precise meaning, and refers to the ability to coherently manipulate quantum mechanical information. A quantum bit, which is the element of quantum information in the quantum computer, is the linear superposition of two "approximate" eigenstates denoted $|0>$ and $|1>$. We say "approximate" eigenstates because if they were precise eigenstates then the system's initial linear superposition would endure for all time. The qubit can be in any linear superposition of the two states, corresponding to a wavefunction $\phi = \alpha|0> + \beta|1>$, where $|\alpha|^2 + |\beta|^2 = 1$. In principle much more information is contained in the quantum bit than the classical one, but the limitations of quantum measurement prevent all from being accessed. Coherent manipulation of these qubits according to the particular algorithm of interest constitutes "quantum computation".

12.3. SUPERPOSITION AND QUANTUM PARALLELISM

To see how the principle of superposition can assist in the solution of problems, consider a quantum bit, ϕ, in the superposition $\alpha|0> + \beta|1>$. Now imagine a function of one variable, $F(\phi)$, where ϕ is this quantum bit. The result of applying the function to this quantum bit is an output bit, which can be (for example),

$$\alpha \exp[i\pi F(0)]|0> + \beta \exp[i\pi F(1)]|1>.$$

The output of the function evaluated a *single* time on this quantum bit contains the value of this function at *both* possible inputs. This principle of superposition can be extended by considering an input of n quantum bits in the superposition state described above with $\alpha = \beta = 1/2^{(1/2)}$. This superposition corresponds to inputs from 0 to $2^n - 1$. So one might imagine that this tremendous parallelism should provide tremendous speed-up for a variety of calculations. Unfortunately nature is not so kind.

According to quantum measurement theory, only a limited part of the information contained within such a quantum bit can be measured. In the measurement process the remaining information is forever lost. Simply put, one can measure the overlap of the quantum bit with a given superposition $\gamma|0> + \delta|1>$, but that means that out of two parameters describing the two-state system, one will be lost forever. Furthermore, the measurement process is a probabilistic one, so unless the qubit is either in the superposition $\gamma|0> + \delta|1>$, or is orthogonal to it, the result of measuring the qubit (and hence the result of the algorithm) cannot be predicted even if the initial state is completely known. This means that either the calculation must be run several times, or an algorithm must be constructed that places the desired information in an eigenstate of the final system.

So what is the use of this quantum parallelism if it cannot be fully used? It permits the shuffling of information around that is difficult to efficiently process classically. Consider an example of a coin, where each side can be either heads or tails. In order to determine classically if the coin is "fair", with one side heads and one side tails, or "false", with two heads or two tails, one would need to look at both sides of the coin. Yet this measurement process is inefficient—if you only wish to know whether or not the coin is fair. Classically in addition to this information you also find out which side is heads and which is tails on fair coins, and whether the false coin has two heads or two tails. This information is thrown away. In the quantum world it is possible to determine if the coin is fair without determining the face on each side—in fact with only a single examination. You perform a single measurement which is a coherent superposition of the examination of the face on one side and the examination of the face on the other side. The realization of this approach is called the Deutsch-Josza algorithm, and uses the output bit described above. If F is 1 for heads and is 0 for tails, and $\alpha = \beta = 1/2^{(1/2)}$, then for a fair coin the output bit is (ignoring an unimportant phase factor)

$$(|0> - |1>)/2^{(1/2)}$$

and for a false coin is

$$(|0> + |1>)/2^{(1/2)}$$

The overlap of the first result with $(|0> + |1>)/2^{(1/2)}$ is 0, and the overlap of the second result is 1. The desired information was efficiently extracted through a coherent measurement. It is a similar efficiency that is used in both the Grover and Shor algorithms, although in both of these more complex algorithms the computation process must be run several times anyway.

12.4. REQUIREMENTS FOR PHYSICAL REALIZATIONS OF QUANTUM COMPUTERS

Algorithms for quantum computers that are considerably faster than possible for classical computers are nice abstract ideas, but is it possible to make a real quantum computer? Only small realizations of quantum computers have been implemented, of the order of 5 qubits, and considerable progress will be required before larger quantum computers are possible. The number of qubits required for a very useful quantum computer is surprisingly small—merely 10^3–10^5 qubits would be enough to implement the prime factoring algorithm for a number bigger than practical for a classical computer. DiVincenzo has characterized five essential requirements for a quantum computer. They are:

(1) a well-defined physical representation of a quantum bit, which can be incorporated into a scalable architecture
(2) the ability to initialize the states of the qubits—to boot the quantum computer by starting with a given initial state such as $|000000000\ldots0000>$
(3) Long decoherence times (relative to gate times)
(4) A "universal" set of quantum gates, permitting any quantum operation to be performed on the set of quantum bits
(5) The ability to measure specific quantum bits

Some explanation of these five requirements is in order. Here a scalable architecture refers to a pathway or strategy to construct a significant number of quantum bits (say a few hundred) and to provide for the quantum gates that permit the calculational operations of interest. The decoherence described in requirement (3) is an essential problem for quantum computers that is not shared by digital classical computers. The quantum computer is fundamentally an analog computer—but with much more parallelism possible than a classical analog computer. In a digital classical computer typically a "1" and a "0" correspond to two different voltages. Fluctuations in the value of the voltage from one of these values can be corrected by setting a threshold voltage for a bit value. Once a determination is made of the correct value, a fresh copy of the classical bit can be made. Hence error correction is a straightforward process, although considerable sophistication must be employed in situations (like communications) where the signal corresponding to a bit value may be degraded to the point where it crosses the threshold.

Copies of quantum bits, however, cannot be made in principle because of the inability to measure all the parameters of the linear superposition of the two states constituting the quantum bit. Error correction here is a fundamentally more subtle procedure. Despite this complication, error correction algorithms have been constructed that may control decoherence

TABLE 12.1. Inputs and output for classical exclusive OR (XOR)

First bit	Second bit	Output
0	0	0
1	0	1
0	1	1
1	1	0

sufficiently to perform calculations of arbitrary length if the probability of a quantum bit error can be kept below (roughly) one error per million gate operations.

(4) is required for both classical and quantum computers. Exploration of the properties of classical computers has shown that the "exclusive or" operation (XOR) is a "universal" operation, for all possible classical algorithms can be constructed with only the XOR operation available for bit-bit interactions. Shown in Table 12.1 is the truth table for the classical XOR operation. An algorithm for a quantum computer can be thought of as a matrix which connects an initial state i of n quantum bits to a final state f of n quantum bits. Schematically the process is $f = Ai$. Here A is a complex $n \times n$ matrix, which must be unitary (AA^* is the identity matrix). From the rules governing unitary matrices it can be shown that a general unitary matrix can be constructed with a sequence of "single qubit" operations, where one of the qubits is transformed according to the rule

$$\alpha|0> + \beta|1> \Rightarrow \gamma|0> + \delta|1>,$$

and "two qubit" quantum XOR operations. In order to construct a quantum XOR operation a unitary operation must be constructed between the input bits and output bits. In Table 12.2 is one possibility. Here the first bit in the output is a "trash" bit, of no importance except in maintaining the unitary nature of the transformation. The second bit is the quantum XOR result. Further progress on the "universal" set of quantum gates has indicated that even without the single qubit operations, with just the quantum XOR any quantum computational algorithm can be constructed. Algorithms with significantly fewer operations, however, can be constructed if single qubit operations are used as well.

TABLE 12.2. Inputs and output for the quantum exclusive OR (XOR)

First bit	Second bit		Output
$\vert 0>$	$\vert 0>$	$\Rightarrow$	$\vert 00>$
$\vert 1>$	$\vert 0>$	$\Rightarrow$	$\vert 11>$
$\vert 0>$	$\vert 1>$	$\Rightarrow$	$\vert 01>$
$\vert 1>$	$\vert 1>$	$\Rightarrow$	$\vert 10>$

12.5. SPIN AS A PHYSICAL REALIZATION OF A QUBIT

A spin-1/2 particle forms a natural representation of a quantum bit. The two states $|0>$ and $|1>$ correspond to the spin-up and spin-down states along a given axis (let us call it the z-axis). A general unitary matrix connecting initial and final configurations of n quantum bits can be constructed using spin-dependent dynamics. A general unitary matrix can be constructed from a quantum Hamiltonian according to the following general expression:

$$A = \exp\left[-(2\pi i/h)\int H(t)dt\right].$$

Thus constructing a single or two qubit operation corresponds to constructing the physical Hamiltonian appropriate for the physical qubits chosen. The process of manipulation of one or two qubits corresponds to finding a physical situation where there is an energy difference between two or more states.

For example, a single qubit operation when the qubit is an individual spin has a straightforward physical implementation. A magnetic field B oriented along the z-axis applied to the spin-1/2 particle will make the energy of the spin-up state $|0>$ differ from that of the spin-down state $|1>$ by $-2\pi g S\mu_B B/h$, where g is the so-called g-factor, h is Planck's constant, and μ_B is the Bohr magneton. This energy difference, when allowed to persist for a time T, will produce a phase difference between α and β of $-gS\mu_B TB$. This permits a particular example of the single qubit transformation above, where $\gamma = \alpha$ and $\delta = \beta \exp(i2\pi g S\mu_B\, TB/h)$. For example, for a spin pointing along the x-axis, corresponding to $(|0> + |1>)/2^{(1/2)}$, if $2\pi g S\mu_B TB/h = \pi$, the orientation of the spin switches from $+x$ to $-x$, corresponding to $(|0> - |1>)/2^{(1/2)}$. All possible single qubit operations can be generated by applying a magnetic field along a general axis.

The quantum XOR requires somewhat more machinery, and a relatively complicated series of operations. Rather than describe the quantum XOR in detail, we will explore how one might construct such an operation. Imagine a spin-spin coupling between two qubits, describable with the "Heisenberg Hamiltonian"

$$H = J(t)S_1 \cdot S_2$$

The eigenstates of the two-spin system with this Hamiltonian are shown in Table 12.3. Note that the states of the individual qubits are not eigenstates of this Heisenberg Hamiltonian.

TABLE 12.3. Character of the eigenstates of the two-qubit system with the Heisenberg Hamiltonian

Eigenstate	Energy	Total spin
$|00>$	$J(t)$	1
$(|01> + |10>)/2^{(1/2)}$	$J(t)$	1
$|11>$	$J(t)$	1
$(|01> - |10>)/2^{(1/2)}$	0	0

If you started with an initial state $|01>$ and permitted $\int J(t)dt = \pi$, then the resulting state would be $|10>$. This operation is called the "swap", for it exchanges the value of qubit #1 with the value of qubit #2. The quantum XOR is composed of single qubit operations described above, along with two operations where $\int J(t)dt = \pi/2$.

12.6. QUANTUM COMPUTATION WITH ELECTRON SPINS IN QUANTUM DOTS

It should be apparent at this point that having stable isolated individual spins, with the capacity to apply magnetic fields to one and only one of these spins, and the ability to controllably couple pairs of spins, cannot even be contemplated without nanostructures. One popular proposed direction towards the quantum computer, suggested by Loss and DiVincenzo, recommends that electron spins in semiconductor quantum dots could be used as the quantum bits. Quantum dots are small regions of semiconductor (nanocrystals) embedded in a host. The energy for an electron to reside in the nanocrystal is lower than to reside in the surrounding material. Thus the quantum dot, which is typically from 2 to 10 nanometers in diameter, binds the electron within it. Other chapters deal in more detail with some of the optical and electronic properties of quantum dots. For our purposes, if the host crystal is seeded with a sufficient number of charge carriers it is possible for each quantum dot to contain one bound electron. The energy separation between the ground state of the electron in the dot and the first excited state can exceed 50 meV, which means that only the ground state is occupied even up to room temperature. Electron states in these dots are doubly degenerate, hence the two states of the qubit correspond to the spin-up and spin-down states of the conduction electron in the ground state of the dot.

Application of a magnetic field to one and only one such dot in order to perform a single qubit operation is a great challenge. Several clever ideas have been proposed, including using only quantum XOR operations (which can be performed as described below), or applying a uniform magnetic field to all the dots and changing the coupling to the magnetic field of an individual electron spin. The precession of a spin in a magnetic field proceeds at a frequency $gS\mu_B B/h$. The g-factor that enters this expression for the single-qubit operation depends on the spin-orbit interaction within a given material, and is different for dot and host. By applying an electric field to a quantum dot (as shown below in Figure 12.1) the average g-factor of the quantum dot's electronic ground state can be changed. This changes the precession frequency of this dot relative to other dots, and thus can be used to controllably rotate the relative value of the electronic spin.

Controlling the coupling between two dots presents fewer challenges. The Heisenberg Hamiltonian described above for coupling two qubits occurs naturally when electron tunneling between the dots is possible. In this situation a virtual process occurs where the electron tunnels from dot 1 to dot 2, then interacts with the electron in dot 2, then tunnels back. This process lowers the electron's energy. As the process is more likely when the spin in dot 1 is antiparallel to the spin in dot 2, there is an effective $J > 0$. In order to control J it is necessary to enhance or suppress the tunneling process. This can be done by applying an electric field to lower or raise the tunneling barrier between the two dots (as shown below in Figure 12.2).

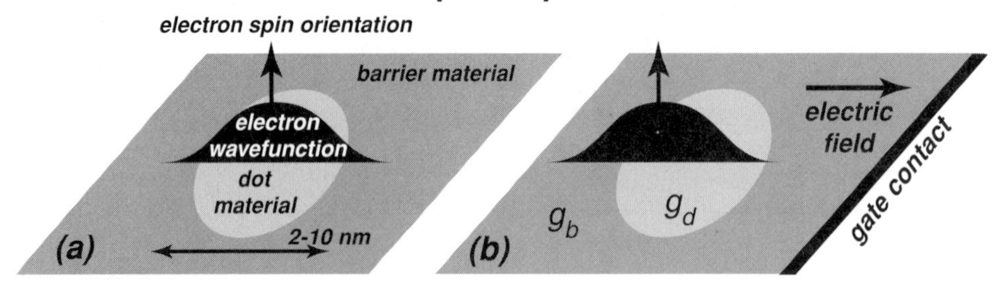

FIGURE 12.1. (a) Schematic of a quantum dot embedded in a host. The electron wavefunction is largely confined to the dot material, but does penetrate into the host (barrier) material. (b) Through the application of an electric field the wavefunction of the dot state is altered, including its amplitude in the barrier region. This produces an electric-field dependence of the g-factor, which can be used to rotate the electron spin relative to the spin in other dots.

12.7. QUANTUM COMPUTATION WITH PHOSPHORUS NUCLEI IN SILICON

An alternate suggestion, due to Kane, for scalable quantum computation with nanostructures, focuses on nuclear rather than electron spin. As there is a very common isotope of silicon with no nuclear spin, crystals of silicon could be grown entirely of this isotope. A few atoms with nuclear spin could be introduced into the crystal, and these could play the role of isolated qubits. For convenience in implementing the single and two qubit operations, the newly introduced atoms should not be isoelectronic to silicon, but should have a different charge. This suggests the use of phosphorus atoms, which have a nuclear spin 1/2, as these dopant atoms. As each phosphorus nucleus is well separated from all the other phosphorus nuclei, the decoherence of the nuclear qubit is expected to be extremely slow. This provides one of the major advantages of the nuclear qubit scheme.

Phosphorus has one additional proton in its nucleus, and one additional electron in its outer shell. When a phosphorus atom occupies a silicon site in the lattice, a so-called "shallow hydrogenic impurity" is formed. The extra positive charge of the nucleus attracts

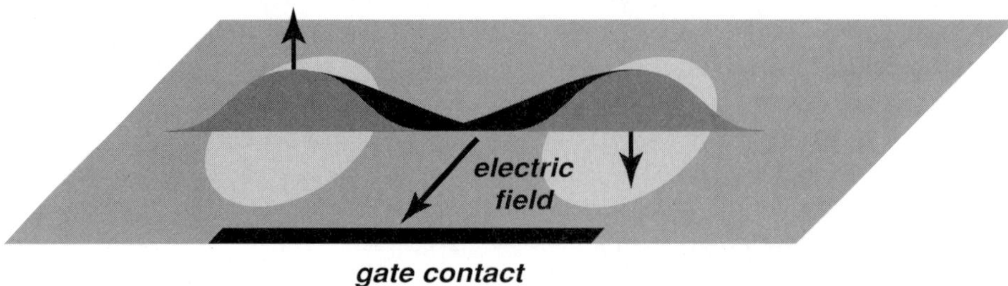

FIGURE 12.2. The wavefunctions in the individual dots (dark grey) do not overlap significantly most of the time. Under the application of an electric field the wavefunctions (black) leak more into the barrier region and overlap. This leads to a spin-spin interaction between the dots, which provides a controllable [$J(t)$] Heisenberg Hamiltonian.

electrons in its vicinity, and hence binds the extra electron into an extended (~10 nm) hydrogenic (1s) state. This electronic state bound to the nucleus provides a "handle" for manipulating the nuclear properties.

Single qubit operations are performed with traditional nuclear magnetic resonance (NMR) approaches, including radio-frequency (RF) waves tuned to the resonance frequency of the nucleus. Although in principle local variations of the crystalline structure, and weak couplings to neighboring nuclear spins, would provide a nearly individual resonance frequency for each nucleus in the solid, it would be far better to be able to tune the resonance frequency of a particular nucleus to bring it into or out of resonance with the driving field. This can be done through a phenomenon called the Knight shift. The electron circling the phosphorus atom interacts with the nuclear spin through the hyperfine interaction. This interaction has the form

$$H = CS \cdot I,$$

where C is the hyperfine constant and I is the nuclear spin. When a magnetic field is applied to the system the bound electron becomes slightly spin polarized. The expectation value of the electron spin parallel to the applied field, $<S_\alpha>$, is non-zero and is proportional to the applied magnetic field. As a result there is an additional field acting on the nuclear spin, called the hyperfine field. That additional effective field shifts the resonance frequency of the nucleus by an amount proportional to C. In standard NMR techniques using an RF field only the nuclear spins in resonance with the RF field will precess relative to the other spins.

The value of the constant C depends on the electron density at the nucleus. Thus a good approach to single qubit gates (shown in Figure 12.3) is to use an electric field to modify the electron density at the nucleus. That alters the constant C, which changes the resonance frequency of the nucleus, and can move the nuclear spin into or out of resonance with the external RF field.

Two qubit operations require nuclear spin-spin coupling. This might be performed indirectly, through coupling the bound electrons around the two nuclei. Pulling the electrons so that they overlap, as shown in Figure 12.4, tends to make the electrons prefer to align antiparallel. As a consequence of these anti-aligned electrons, and the hyperfine coupling

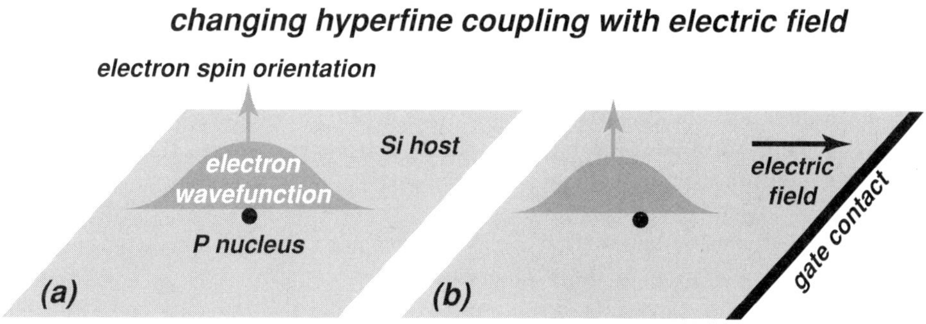

FIGURE 12.3. (a) Phosphorus atom embedded in a silicon host. The extra electron the phosphorus atom carries is bound to it in a highly extended state (~10 nm diameter). (b) An electric field can be used to modify the overlap of this electron with the phosphorus nucleus, changing the resonant frequency of its nuclear spin.

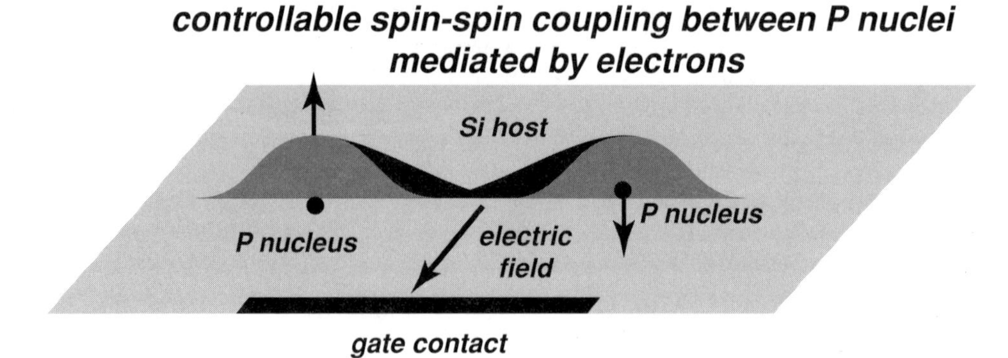

FIGURE 12.4. The wavefunctions of the individual bound electrons (dark grey) do not overlap significantly most of the time. Under the application of an electric field the wavefunctions (black) overlap more with each other. This leads to a spin-spin interaction between the electronic states, which is communicated to the nuclei through the hyperfine and produces a controllable Heisenberg interaction between the nuclear spins.

between the electrons and the nuclei, the nuclei experience a coupling which influences them to anti-align. We can write the nuclear spin-spin coupling that results as

$$H = J(t) I_1 \cdot I_2,$$

which has the same structure as the interacting Hamiltonian above for two electron spins. Thus by implementing two types of electrical gates to push the bound electrons around, both one and two qubit operations can be implemented.

12.8. CONCLUSIONS

Quantum computation is a new computational paradigm that has challenged the research community to find ways to control individual quantum elements. Perhaps the progress made in this direction will provide a foundation for construction of practical quantum computers. Even if not, manipulation and control of the smallest elements of information possible, these qubits, will lead to important applications in electronics, optics, and magnetics. Full control in the manipulation of qubits can only take place when the properties of materials and devices are engineered on the atomic scale, whether it is through quantum dots, or through the careful positioning of individual phosphorus atoms in silicon. Thus nanoscience must always play the central role in this new challenge.

- Certain problems are very hard for classical computers, yet quantum algorithms exist to dramatically speed up these calculations.
- Spin in nanostructures, either electron spin in quantum dots or nuclear spin in silicon, provides an excellent possible realization of a quantum computer
- Energy differences between states are used to manipulate qubits.

- Manipulation of qubits corresponds to manipulating the properties of nanostructures on the small scale, either by changing the g-factor of individual spins or by changing the Heisenberg coupling between two different spins.

QUESTIONS

1. Show how to construct, with n quantum bits, a superposition of all *even* numbers from 0 to 2^{n-1}.
2. Calculate the time that a field must be applied, and the axis along which the field should be applied to manipulate an initial state with $\alpha = 1, \beta = 0$ into a final state with $\gamma = \delta = 1/2^{(1/2)}$.
3. Construct the quantum XOR operation.
4. Express the value of the Heisenberg interaction between the nuclear spins in terms of the Heisenberg interaction between the bound electrons and the value of the hyperfine coupling C.

REFERENCES

Overviews

Introduction to Quantum Computation and Information, ed. H.-K. Lo, S. Popescu, and T. Spiller, World Scientific, Singapore, 1998.
Semiconductor Spintronics and Quantum Computation, ed. D. D. Awschalom, D. Loss, and N. Samarth, Springer, New York, 2002.
C. H. Bennett and D. P. DiVincenzo, "Quantum Information and Computation," Nature **404,** 247 (2000).
D. P. DiVincenzo, "Introduction to Quantum Computation and Information," Nature **399,** 119 (1999).

Algorithms

P. W. Shor, "Algorithms for Quantum Computation: Discrete Log and Factoring," in *Proc. 35th IEEE Symp. on Foundations of Computer Science*, ed. S. Goldwasser (IEEE Computer Society Press 1994).
L. K. Grover, "Quantum Mechanics Helps in Searching for a Needle in a Haystack," *Phys. Rev. Lett.* **79**, 325 (1997).
A. Ekert and R. Josza, "Quantum computation and Shor's factoring algorithm," *Rev. Mod. Phys.* **68**, 733 (1996).

Fault Tolerance

P. W. Shor, "Scheme for Reducing Decoherence in Quantum Computer Memory," *Phys. Rev.* A **52**, R2493 (1995).
J. Preskill, "Reliable Quantum Computers," *Proc. R. Soc. Lond.* A **454**, 385 (1998).

Implementations

D. Loss and D. P. DiVincenzo, "Quantum Computation with Quantum Dots," *Phys. Rev.* A **57**, 120 (1998).
D. P. DiVincenzo, D. Bacon, J. Kempe, G. Burkard, and K. B. Whaley, "Universal quantum computation with the exchange interaction," *Nature* **408**, 339 (2000).
B. E. Kane, "A silicon-based nuclear spin quantum computer," *Nature* **393**, 133 (1998).
D. P. DiVincenzo, "The Physical Implementation of Quantum Computation," Fortschritte der Physik—Progress of Physics **48**, 771 (2000).

13

Magnetoresistive Materials and Devices

Olle Heinonen
Seagate Technology, Bloomington, MN

13.1. INTRODUCTION

Conventional electronics rely on the ability to manipulate the electrical resistance or impedance (in AC applications) of a device. In virtually all such applications, we only care about the electron's (or hole's) charge, and ignore the fact that electrons come in two different spin species. Many materials, such as ferromagnetic metals and semiconductors, exhibit various magnetoresistive effects which have been known for many decades. The term magnetoresistance generally describes the manifestation of a magnetic-field dependent electrical resistance. This is obviously useful in applications in which we want to sense a magnetic field. More generally, *spintronics* refers to the idea that we should be able to use not only the charge in electron and hole transport, but also the spin of the charge carriers. The possibility of doing so potentially opens up new exciting applications in areas ranging from electronics to quantum computing. For spintronics to be useful, three criteria have to be satisfied: (a) we have to be able to create spin-polarized populations of charges, (b) we have to be able to manipulate the spin-polarization of these populations, and (c) we have to be able to detect the ensuing spin polarization, or the difference in spin polarization between the initial and final populations. Much effort has been directed recently on making spintronics work in semiconductor materials. The reason is that we can very carefully control materials properties of semiconductors, such as conductivity and electronic band gap, and there is already a very large semiconductor industry. The ability to make spintronic devices out of semiconductors has been hampered in large because of the difficulty to inject spin polarized populations of charge carriers into semiconductors. On the other hand, spintronic devices, using normal metals and metallic ferromagnets, have been in use for some time in industrial applications, in particular in the magnetic recording industry. In this chapter, we will discuss in some details the physics behind such devices and their applications in magnetic recording. We will close with an outlook that includes some aspects of semiconductor spintronics.

13.2. ELEMENTS OF MAGNETORESISTANCE

At the present, all read and write heads in magnetic hard drives use ferromagnetic materials. In the read head, the magnetic fields emanating from information stored on the medium somehow has to be translated into electric signals. One way to achieve this, is to have the read head consist of materials or structures such that the electrical resistance is sensitive to external magnetic fields. This so-called magnetoresistance is in fact a general property of ferromagnetic materials. However, the intrinsic sensitivity to magnetic fields is usually not large enough for the sensitivities required in modern read heads, and more sophisticated effects and designs, such as the spin valve, have to be used. In order to understand how they work, we must first understand something about ferromagnets and electric transport in ferromagnets.

13.2.1. Band Structure of Ferromagnets—Co, Ni, Fe

Individual atoms may acquire a net magnetic moment depending on the configuration its electrons. Each electron has a spin angular momentum of magnitude $\hbar/2$. Associated with the spin angular momentum is a magnetic moment μ_B. The z-component of the angular momentum can take the values $\pm\hbar/2$ ("up" and "down", respectively), and the z-component of the magnetic moment can correspondingly take the values $\mp \mu_B$. Depending on how the spin angular momenta of the electrons in an atom add up, the atom can end up with a net z-component of spin angular momentum, and a concomitant z-component of magnetic moment. There is of course also the orbital angular momentum of each electron, and the corresponding angular magnetic moment. However, the angular magnetic moment is usually much smaller than the spin magnetic moment and does not play any significant role in ferromagnetism, so we will ignore the orbital angular momentum here. Exactly how the spin angular momenta of the electrons are combined to form the total spin angular momentum of the atom is described by Hund's rules. These rules just articulate the fact that the electronic state of the atom tries to strike a balance between exchange interactions and kinetic energy. The exchange interactions originate in the Pauli principle, which states that two electrons cannot occupy the same place in the same state. This tends to keep electrons in the same spin angular momentum state apart, so that their wavefunctions do not overlap. As a consequence, the repulsive Coulomb energy is lowered from what it would have been, had the wavefunctions overlapped. The resulting reduction in Coulomb energy is the exchange energy, and it favors parallel alignment of the z-component of the electrons' spin angular momentum, and therefore parallel alignment of the z-components of the magnetic moment. On the other hand, keeping many electrons in the same state of z-component of spin angular momentum means that they have to have different states of orbital angular momentum and different principal quantum numbers, since each state can only be occupied by one electron. But the kinetic energy increases with principal quantum number (and also with orbital

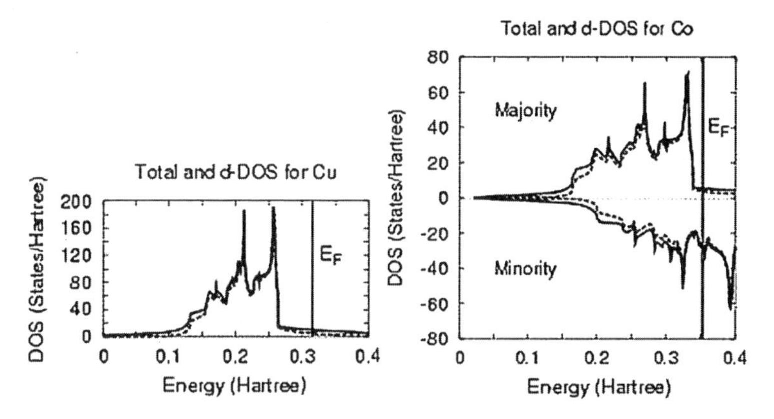

FIGURE 13.1. Densities-of-states for Cu (left panel) and for up- and down-spin states in Co (right panel). (From Ref. 12 by permission of Springer Verlag.)

angular momentum), so at some point, the cost in kinetic energy to keep the z-components of the electron spins parallel will be prohibitive. For most atoms, the final outcome is that the net z-component of the spin angular momentum, and hence the magnetic moment, is zero. However, for the transition metals, in particular we will here be concerned with Co, Fe, and Ni, the net result is a sizeable magnetic moment. In fact, each isolated transition metal atom has a z-component of magnetic moment which is an integer times μ_B. But when the atoms are combined to form a bulk metal, electronic energy levels from each atoms will combine to form bands, which have some finite width in energy. The result is that the kinetic energy cost to maintain the atomic magnetic moment with many electron spin angular momenta aligned on each atom increases, and the atoms will trade some z-component of electron spin angular momentum for kinetic energy, thus lowering the total energy per atom. The outcome is that each atom ends up with a z-component of magnetic moment which is some non-integer units of μ_B.

The net magnetic moment is clearly seen in the electronic density-of-states of the transition metals. Figure 13.1 shows the density-of-states for Cu, which is non-magnetic, and for up- and down-spin electrons in Co, which is a ferromagnet. Basically, the up- and down-spin bands and densities-of-states look pretty much alike in Co, except for the fact that the down-spin bands are shifted up by some amount of energy, which is called the exchange splitting. Since both up- and down-spin states are filled to the same Fermi energy, more up-spin states are filled than down-spin states, with the consequence that there will be a net magnetic moment per atom.

13.2.2. Drude Transport Theory

The exchange-splitting of up- and down-spin bands has two important consequences for magnetotransport. Only states at the Fermi surface participate in transport. Since up- and down-spin bands are exchange-split, the nature of the up- and down-spin states at the Fermi surface is different, and they will couple differently to scattering potentials. Consequently, and also because the densities-of-states are different, scattering rates $1/\tau_{(\vec{k},\sigma)}$ will be very different. Also, the dispersion, or the change in energy with wavevector, is different for up- and down-spin states at the Fermi surface. As a consequence, their group velocities $\vec{v}_{(\vec{k},\uparrow)}$, $\vec{v}_{(\vec{k},\downarrow)}$, defined as

$$\vec{v}_{(\vec{k},\sigma)} \equiv \left[\frac{1}{\hbar}\nabla\varepsilon_{(\vec{k},\sigma)}\right], \tag{13.1}$$

with $\varepsilon_{(\vec{k},\sigma)}$ evaluated at the Fermi energy ε_F, will also differ.

Let us for a moment first consider charge transport in a non-magnetic metal. In the Drude transport theory, the central quantity is the mean free path l of the charge carriers. This is roughly the average distance that a charge carrier travels between scattering events. As charge carriers travel at the Fermi velocity, we can think of a relaxation time τ of the carriers, which is something like the average time between scattering events. Theoretically, the scattering rate is the quantity that is more directly accessible to calculations by, for example, Fermi's Golden Rule. According to this rule, the scattering rate (the inverse scattering time) is proportional to square of the matrix elements of the scattering potential between initial and final states, times the density of final states. For impurity scattering, the scattering is elastic, which means that initial and final states must have the same energy, which confines all relevant states to the Fermi surface. In general, of course, the scattering rate depends on wavevector, so $1/\tau = 1/\tau_{\vec{k}}$.

Once we have obtained the scattering rate, we can relate the mean free path and the scattering rate:

$$l_{\vec{k}} = v_{F,\vec{k}}\tau_{\vec{k}}, \tag{13.2}$$

where we note that the mean free path, too, in general depends on wavevector. Usually, one uses some appropriate average of these quantities over the Fermi surface—since these are approximate and more phenomenological parameters, it doesn't serve any purpose to get too bogged down in details—in which case we then write $l = v_F\tau$.

The electrical conductivity can then be simply expressed in terms of the mean free path or the relaxation time. The model we use is that the electrons are accelerated by an external field $\vec{E}$ but suffers collisions at a rate $1/\tau$, which give rise to a friction force. If we balance the time-rate of change of momentum due to the acceleration by the external field with the loss due to collisions we get an expression of the steady-state average drift velocity $\vec{v}_d$ of the electrons. Since the electrical current density is just $\vec{j} = -en\vec{v}_d$, with n the electron number density, we can directly relate the current density to the electric field, with the following result for the electrical conductivity:

$$\sigma = \frac{ne^2\tau}{m} \tag{13.3}$$

13.2.3. Mott's Two-Current Model of Magnetotransport

It is relatively easy to extend the Drude model of transport to magnetic systems, under some reasonable assumptions. The main one is that spin-flip scattering events can be ignored, so that up-spin electrons remain up-spin, and down-spin electrons remain down-spin. Another way of saying this is that the z-component of spin angular momentum is conserved by the single-particle Hamiltonian. This is strictly not true in the presence of spin-orbit coupling, which mixes up spin angular momentum and orbital angular momentum. Furthermore, lose spins, e.g., Fe-impurities in a Cu matrix, can cause spin-flip scattering. In any case, for the systems in which we are interested here, the spin diffusion length, which is roughly the length an electron travels on the average before it suffers spin-flip scattering, is much larger than the relevant system dimensions, so we can for practical purposes consider electron spin conserved (or, more accurately, the z-component of electron spin angular momentum).

With this approximation we can then consider the spin-up and spin-down electrons constituting two separate parallel conducting channels, so that the total conductivity is just the sum of the conductivity in the up- and down-spin channels, $\sigma = \sigma_{\uparrow} + \sigma_{\downarrow}$. Now comes the crucial part. In general, the up- and down-spin conductivities are quite different. The reasons for this are the different characters of the up- and down-spin wavefunctions at the Fermi surface, the different group velocities, and the different densities-of-state. As we noted earlier, the scattering rate depends on the matrix element of the scattering potential between initial and final states, and on the final density-of-states. In general, both the wavefunctions and the scattering potentials are different for up- and down-spins at the Fermi surface. In addition (as is illustrated in Figure 13.1), the densities-of-state are quite different, with the down-spin density-of-state much larger than the up-spin one. As a consequence, down-spin states suffer much more scattering than up-spin ones, and the down-spin scattering rate is much larger than the up-spin one. Therefore, electrical transport tends to be dominated by the up-spin states. This opens the possibility to manipulating the conductivity of a magnetic system by manipulating the spins of the charge carriers, by, for example, subjecting them to external magnetic fields. This is the central thought in the area of spintronics.

13.2.4. Acronyms: AMR, GMR, TMR, BMR, and CMR

Let us first quickly survey the different magnetoresistive effects in use or of potential use in read heads. The first one is the Anisotropic Magnetoresistance (AMR). This one is in general observed in any magnetic system. If the magnetization density $\vec{M}$ and the current density $\vec{j}$ are at some angle θ with respect to one another, it is found that there is a component of the resistance of the system which depends on θ:

$$R = R_0 + \frac{\Delta R}{2}\left[1 - \cos^2\theta\right]. \tag{13.4}$$

The physics of this effect is rather involved and is not easily amenable to some simple picture. The basic ingredient is that spin-orbit scattering will cause anisotropy in the system and will cause up- and down-spin scattering to differ along different crystallographic directions.

The AMR effect is of the order of at the most a few percent in thin film systems. This was, however, enough to make thin film read heads based on the AMR effect superior to inductive read heads. In soft magnetic systems, such as Permalloy, the magnetization can respond to the field emanating from bits written on a medium. With the current running in a fixed direction, the resistance will then be modulated by the magnetization rotation. The most cumbersome part of AMR head design is that they have to be properly biased in order for the output signal to be linear in the applied field (for small enough fields). This is approximately achieved by biasing the reader so that current and magnetization density are at 45 degrees angle to each other in the quiescent state (in the absence of an external field).

The Giant Magnetoresistive effect (GMR) was discovered in 1988 by Baibich *et al.*[1] It was found in Fe-Cr multilayer systems, in which each layer had a thickness of the order of 10 A, that an external field applied in the plane of the layers gave rise to a magnetoresistance of up to 40% at room temperature with the current applied in the plane of the layers. This was observed if the Cr thickness was carefully chosen to promote an antiferromagnetic coupling between consecutive Fe layers. In the absence of an external field, consecutive Fe layers had antiparallel (AP) magnetization with the magnetization of each layer in the plane of the layer. By applying a sufficiently strong (of the order of a few thousand Oe) external field in the plane of the layers, the magnetization of all layers could be aligned in a parallel (P) configuration. The resistance was found to have a component which depended on the relative angle θ between the magnetization of consecutive layers:

$$R = R_0 + \frac{\Delta R}{2}\,[1 - \cos\theta]\,, \tag{13.5}$$

where R_0 is the minimum resistance (for P configuration). Note that this expression depends on the cosine, in contradistinction to the expression for the AMR effect, which depended on the square of the cosine. The potential commercial importance of the GMR effect in sensors in magnetorecording was rather quickly realized. One problem with the original discovery, however, was that the sensitivity to external fields was rather low—it took fields of the order of 10,000 Oe to appreciably change the resistance in the multilayer system. The invention of the spin valve by Dieny *et al.*,[2] solved that problem. We will look more closely at spin valves in the next section. A second improvement for the use in magnetic read heads was the discovery of the GMR effect in Co-Cu multilayers. Today, commercial spin valves use CoFe alloy ferromagnetic layers and Cu metal spacers. The advantage of the CoFe alloys is that the magnetic anisotropy can be made small.

Tunneling magnetoresistance is observed (in some cases) when two ferromagnetic layers are separated by an insulating barrier. If the barrier is thin enough, electrons can tunnel through the barrier when a potential difference is applied across it. If the relative orientation of the magnetizations in the ferromagnetic layers is varied, the magnetoresistance across the barrier can be written as

$$R = R_0 + \frac{1}{2}\Delta R\,[1 - \cos\theta]\,, \tag{13.6}$$

where θ is again the angle between the magnetizations in the two layers, and R_0 the minimum resistance in parallel configuration. This phenomenological equation describing the tunneling magnetoresistance looks very much like that describing the giant magnetoresitance, but the physics of the effects is quite different, and we will discuss some of those aspects later. Although tunneling magnetoresistance was discovered in the mid-1960s, it took a long time to demonstrate an appreciable tunneling magnetoresistance at room temperatures.[3] Currently, a magnetoresistive ratio exceeding 40% can be obtained at room temperatures using CoFe alloy magnetic layers separated by a tunneling barrier consisting of aluminum oxide. This very high magnetoresistive ratio obviously makes read heads based on tunneling magnetoresistance interesting and likely successors to read heads based on the giant magnetoresistance.

Ballistic magnetoresistance (BMR) was announced in 1999 by Garcia, Munoz, and Zhao.[4] The effect is observed when a very small contact, of the order of 10 nm or less, is made between two ferromagnets. Such contacts can be fabricated by electrodeposition, in which a contact is grown between a very sharp wire and a wider piece of material. By monitoring the resistance of the junction, the electrodeposition process can be turned off when a desired junction resistance has been reached. Another way of making BMR junctions is by mechanically joining two fine wires. The BMR effect can be very large indeed, with observed changes in the resistance of up to 3,000%,[5] and even reported to exceed 100,000%![6] Explanations put forward rest on the assumption that a domain wall can be confined to a very small space and be made very sharp in a nano-sized junction. If the magnetization of the two wires on each side of the junction are not parallel, a domain wall will be formed and confined to the junction. As a voltage is applied across the junction, electrons will be scattered by the effective spin-dependent potential caused by the domain wall. The domain wall must be very sharp compared to the spin diffusion length. Otherwise, the electrons will have time to relax their spins to the local direction of the magnetization in the domain wall, and no magnetoresistance will be observed. The BMR effect is obviously very interesting from the point of view of read heads in magnetorecording. Some of the problems which will have to be solved to use this effect include how to manufacture well-controlled junctions on wafers. Furthermore, the BMR effect typically decreases dramatically after cycling, and for this effect to be useful in read heads, it obviously has to be very stable over times of the order of several years.

The usual class of materials in which the Colossal Magnetoresistance (CMR) is observed are the doped perovskites ABO_3, where A is a rare earth, and B is a transition metal. The dopants can be elements such as Ba, Ca, Sr, or Pb, and substitute for the rare earth atoms. The best studied CMR perovskite is perhaps $LaMnO_3$. There are other compounds, such as EuO, which exhibit what is arguably CMR, but the term has come to be reserved almost exclusively for the transition metal oxide perovskites.

These materials are very complicated and typically exhibit rich phase diagrams as functions of dopant concentration and temperature, showing metal-insulator transitions and paramagnetic, ferromagnetic and antiferromagnetic phases.[7] The physics underlying this richness is very complicated and stems from the electronic structure that results when the transition metals are placed in the perovskite structure. We will not attempt to describe the details here, but refer the reader to more specialized texts on the subject.

The perhaps most studied compounds are the manganites, in which B is Mn. For small doping levels, around 30% or less, there is a high-resistivity high-temperature phase with a very small magnetoresistance (see Figure 13.2). The conductance is low and show thermally activated. As the temperature is lowered, the compound orders ferromagnetically at the Curie temperature T_C. As the Curie temperature is approached from above, the resistivity increases and peaks at a temperature T_m close to T_C, after which it decreases dramatically—the compound is a ferromagnetic metal below T_C. At T_m, the magnetoresistance has a large negative peak, and as the temperature is further reduced, the magnetoresistance decreases. The magnitude of the maximum magnetoresistance can be very large, up to four orders of magnitude—hence the term 'colossal magnetoresistance'. Unfortunately, the maximum

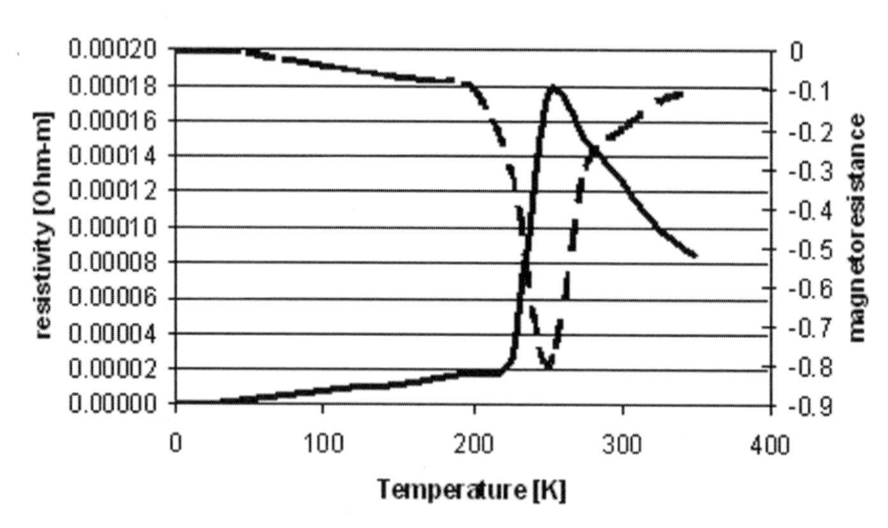

FIGURE 13.2. Schematic figure showing resistivity (solid line) and magnetoresistive ratio (dashed line) as a function of temperature for doped manganite.

magnetoresistance depends very sensitively on the Curie temperature of the compound, and increases exponentially with decreasing Curie temperature. For compounds with a Curie temperature at around room temperature, the maximum magnetoresistance in the range of 10%–20%. There was initially, after the discovery of the CMR effects in manganites, a great deal of excitement about the possible use of the CMR effects in magnetic read heads. This has not happened, and the industry appears not to pursue this. One big reason for this is that the operating temperature of read heads in disc drives exceeds the ambient temperature by 50 K–100 K. At these temperatures, there is no known CMR compound that shows any advantage over other technologies, such as GMR spin valves or tunneling magnetoresistive read heads.

The final entry on our list of magnetoresistive effects is the enormous magnetoresistance (EMR).[8] In contrast to all the other ones we have mentioned, the EMR effect does not rely on ferromagnetic materials and the rotation of magnetization in response to an external magnetic field. This immediately makes it interesting for several reasons. One is that it is very difficult to make stable read heads, in which the magnetization responds reversibly without jumps, so-called Barkhausen steps, or has open loops. This is obviously not a problem in a sensor which does not ferromagnetic in the first place. Another reason is that the maximum response frequency of a ferromagnetic sensor is ultimately limited by the ferromagnetic resonance frequency of the sensor. The magnetization responds by rotation, and the maximum rotation frequency is set by the ferromagnetic resonance frequency. This frequency depends on the effective anisotropy of the sensor, and is in the range of a few GHz to 10 GHz. As data rates in magnetic disc drives keep increasing with increasing areal density, we may in the near future enter a regime where the data rate is limited by the ferromagnetic resonance frequency of the sensor. The EMR effect is based on using the regular Hall effect in a clever way. In a rectangular bar with a current flowing perpendicularly to an external field, a voltage, the Hall voltage, will develop perpendicularly both to the current and the magnetic field. The reason is the electrons flowing along the current directions will

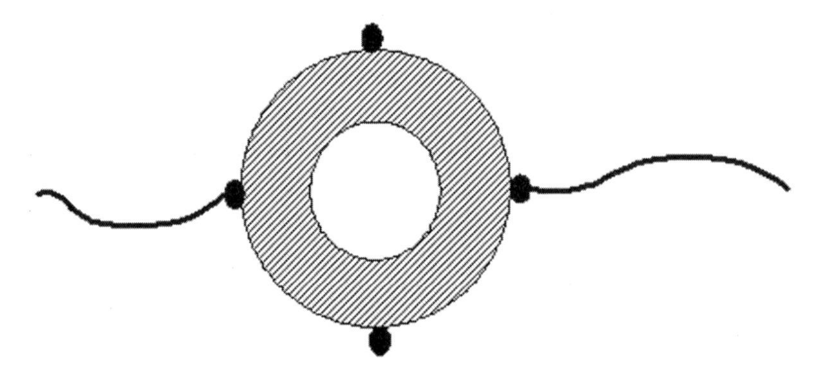

FIGURE 13.3. Cartoon of a four-terminal EMR device consisting of a semiconductor annulus (hatched) with the center cylinder filled with a high-conductivity metal.

experience a Lorentz force due to the magnetic field. The Lorentz force makes the electron trajectories curve and hit the sides of the device. As electrons are piling up along the sides, this charge accumulation gives rise to an electric field perpendicular to the current and the magentic field. Eventually, the field is so strong that the force due to the electric field precisely cancels the Lorentz force, $-e\vec{E} = -e\vec{v} \times \vec{B}$. The result is that a current I flows, but a Hall voltage $V_H = \frac{BI}{t}$, with t the thickness of the system, develops perpendicularly to the current and the magnetic field. Another way of stating this is that the Hall angle between the current density $\vec{j}(\vec{r})$ and the local electric field $\vec{E}(\vec{r})$ is 90 degrees. Now imagine that we use a circular geometry in which we have a semiconductor annulus with the center 'hole' filled with a very good conductor, such as Au (see Figure 13.3). In a semiconductor, the finite mobility μ gives rise to a Hall angle which depends on the external magnetic field, $\tan\theta = \omega_c\tau$, where ω_c is the cyclotron frequency $\omega_c = eB/m^*$, and τ is the relaxation time and m^* the carrier effective mass. We then run a current from one side to the other of the annulus, with an applied field perpendicular to the plane of the annulus. Since the Au center is basically a perfect conductor, the electric field lines are everywhere perpendicular to the Au boundary. In the absence of the magnetic field, the current flows along the electric field lines and a large part of the current flows through the high-conductance Au region. If we now apply a large magnetic field, the Hall angle tends to 90 degrees. As a consequence, the current is locally perpendicular to both the electric and magnetic field. This means that the current must flow around the Au center—the magnetic field effectively expels the current from the Au center. Since the semiconductor has a small conductivity compared to Au, this is a high-resistance state of the device. The magnetoresistance of such a device can reach several hundred percent, and it is in principle a high-speed device with very low noise. The drawbacks are first of all that the semiconductor has to be a very high-quality high-mobility semiconductor. These are very difficult to grow and typically have to be grown using special Molecular Beam Epitaxy (MBE) techniques, which are not easily transferrable to industry, since MBE processes have low throughput. Furthermore, the magnetoresistance is symmetric in the applied field, which means that the device has to be biased properly for a near-linear operation. Ideally, the bias should provide a Hall angle of about 45 degrees, which means that $\omega_c\tau \sim 1$. This again highlights the need for very high-mobility semiconductors with large τ, so that the biasing field can be reasonable.

13.2.5. *Giant Magnetoresistance in Multilayered Thin Films*

We will now consider in some detail the GMR effect in multilayered thin films. In multilayers, it can be arranged that consecutive magnetic layers have their magnetic moments aligned antiparallel in the absence of an applied magnetic field. This is accomplished by choosing the correct thickness of the non-magnetic (spacer) layer so that the exchange coupling between two magnetic layers across the spacer layer is negative. This exchange coupling is mediated by the conduction electrons in the spacer layer and oscillates with a period of the spacer layer thickness given by its Fermi surface. For Co-Cu layers stacked along the (111) direction, for example, the coupling is strongly antiferromagnetic for a Cu thickness of about 8.5 A. A second antiferromagnetic peak is observed for a Cu thickness of about 20 A.

By applying a sufficiently strong external field, the magnetization of all of the ferromagnetic layers can be aligned. With a current flowing either in CPP or CIP configuration, the resistance will change depending on the relative orientation of the magnetization in consecutive Co layers, with a concomitant change in current (for constant applied voltage).

13.2.5.1. CIP-GMR For current-in-plane (CIP), the electric field is uniform along the in-plane direction. One limit in which CIP GMR is relatively easy to understand is that in which the elastic mean free paths $l_\uparrow$ and $l_\downarrow$ for up- and down-spins, respectively, are both much larger then the thickness of the layers. The electrons will then, on the average, traverse many layers without being scattered as they drift along the in-plane direction. The resistance contributed by each spin channel is then proportional to the effective scattering rate for each channel. Let us denote the average scattering rate for each Co/Cu repeat unit by $1/\tau_\uparrow$ and $1/\tau_\downarrow$ for spins that are locally majority (up) and minority (down), respectively. In parallel (P) configuration, the resistance for the majority spin channel ($R_\uparrow$) and for the minority spin channel ($R_\downarrow$) are then proportional to $1/\tau_\uparrow$ and $1/\tau_\downarrow$ respectively. The current contributed by each spin channel adds in parallel so the total resistance of the system is then

$$R(P) = \frac{R_\uparrow R_\downarrow}{R_\uparrow + R_\downarrow} \tag{13.7}$$

In the antiparallel (AP) configuration on the other hand, the average scattering rate for each spin channel will be $\frac{1}{2}[1/\tau_\uparrow + 1/\tau_\downarrow]$, since each spin channel is a majority spin in half the Co layers, and a minority spin in the other half. Again, due to the absence of spin-flip scattering, the two spin channels conduct in parallel, and the total resistance of the system is $R(AP) = \frac{1}{4}[R_\uparrow + R_\downarrow]$. The change in resistance in going from P to AP configuration is

$$\Delta R = R(AP) - R(P) = \frac{\left(R_\downarrow - R_\downarrow\right)^2}{4\left(R_\downarrow + R_\uparrow\right)} \tag{13.8}$$

and the *GMR* ratio which is defined as $GMR = [R(AP) - R(P)]/R(P)$ is in the thin limit given by

$$GMR_{\text{thin-limit}} = \frac{1}{4}\frac{\left(R_\downarrow - R_\uparrow\right)^2}{R_\downarrow R_\uparrow} \tag{13.9}$$

Note that

$$\frac{R(AP)-R(P)}{R(P)}=\frac{G(P)-G(AP)}{G(AP)}=\frac{\Delta G}{G(AP)} \tag{13.10}$$

where *G(P)* and *G(AP)* are the conductances in P and AP configurations. This latter form tends to be more convenient to use in theoretical analyses, since by focusing on ΔG one can separate out shunting resistances which do not contribute to the GMR signal.

The opposite limit to the "thin" limit just described is the "thick" limit in which the mean free path is much shorter than the layer thicknesses. In the thick limit, the conductivity can be treated as local and there is no difference between the CIP conductance for parallel and antiparallel moment alignment. Thus $GMR_{\text{thick-limit}} = 0$.

13.2.5.2. CPP-GMR The theory for current flowing perpendicular to the planes (CPP) is more complicated than for CIP. For CPP, the current must be constant along the direction perpendicular to the planes. Maintaining a constant current across material boundaries generally requires charge and spin accumulations at the interfaces because of the different properties of the layers. This does not happen in the CIP case because there is no *net* current along the direction perpendicular to the planes. As a consequence, there will be discontinuities in the electrostatic potential at the interfaces for CPP and a concomitant non-uniform electric field, as well as chemical potential barriers due to the charge and spin accumulations.

We shall ignore such complications here, since they require rather sophisticated treatment. We can then describe CPP *GMR* rather simply using a model due to Valet and Fert.[9] It is relatively easy to show, for example, that within the semiclassical two current model, if all layers have the same electronic structure, the resistance in either spin channel is given by

$$R_\sigma = \frac{1}{A}\int dz\rho_\sigma(z), \tag{13.11}$$

where A is the area perpendicular to the current flow, and $\rho_\sigma(z)$ is the local resistivity which is proportional to the local scattering rate for electrons of spin σ. Thus consider a multilayer with N repeats of a ferromagnetic layer followed by a spacer layer. The resistance for parallel alignment of the magnetic layers will be given by

$$R(P) = \frac{R_\uparrow R_\downarrow}{R_\uparrow + R_\downarrow}, \tag{13.12}$$

where R_σ can be written in terms of the *local* resistivities of the ferromagnetic and non-magnetic spacer layers, $\rho_{F,\sigma}$, and ρ_N, and their thicknesses, t_F and t_N,

$$AR_\sigma = N[\rho_{F,\sigma}t_F + \rho_N t_N] \tag{13.13}$$

The resistance for antiparallel alignment will then be given by $R(AP) = \frac{1}{4}(R_\uparrow + R_\downarrow)$. This leads to a *GMR* of

$$GMR(CPP) = \frac{1}{4}\frac{(R_\uparrow - R_\downarrow)^2}{R_\uparrow R_\downarrow}. \tag{13.14}$$

So far we have only considered the effects of bulk scattering on the GMR. These effects enter through the different scattering relaxation times $\tau_\uparrow$ and $\tau_\downarrow$ into the up- and down-spin bulk resistivities. There is additional scattering which occurs at the interfaces between the different materials, as an electron from, e.g., a CoFe electrode impinges on the interface between the CoFe and Cu. If the electronic band structure is very similar on both sides of the interface, there is a state with the same momentum parallel to the interface at the same energy on both sides of the interface, and the electron easily crosses the interface. On the other hand, if the bands are poorly match, there may not be any available states for the electron to move into as it crosses the interface. The effect of such band mismatch is that the electron sees a potential step at the interface which makes the probability that the electron will cross the interface small, and the probability that the electron will be reflected at the interface large. Since up- and down-spin states at the Fermi surface are different in ferromagnetic materials, this opens up the possibility for a good bandmatch in one spin channel, and a poor bandmatch in the other. Material combinations which exhibit large *GMR* are such that there is a very good bandmatch in the spin channel which carries most of the electrical current, while there is a poor bandmatch in the other spin channel. In Co/Cu *GMR* systems, for example, with the growth direction in the (111) direction, there is an almost perfect bandmatch in the majority channel and a very poor bandmatch in the minority channel. Backscattering caused by the interface potential gives rise to increased resistance, since backscattering leads to reduced current as fewer charge carriers traverse the system. The net effect of the spin-dependent interface potentials due to the electronic structure can be approximately accommodated by adding spin-dependent interfacial resistances, r_σ to R_σ. Thus

$$AR_\sigma = N[\rho_{F,\sigma}t_F + \rho_N t_N + 2r_\sigma]. \tag{13.15}$$

In the usual Valet-Fert model, the interface resistance is expressed in terms of an interface spin-asymmetry parameter γ of that $r_{\pm\sigma} = r_0[1 \mp \gamma]$. As an example, in CoFe-Cu systems, this spin asymmetry parameter is about 0.7, which gives an appreciable contribution to the *GMR* effect.

Finally, we note that the simple models presented here do not include the effect of spin-flip scattering, which mixes up the spin channels and renders the two-current model useless. One can extend the Valet-Fert model to include spin-flip scattering, but we will not go through that exercise here.

13.3. READ HEADS AND MRAM

13.3.1. *Spin Valve Read Heads*

As we mentioned earlier, the CIP GMR effect in multilayers is not sensitive enough to make it useful for read heads. The spin valve is a design that uses CIP GMR in a useful device. The idea is to have only one layer respond to an external magnetic field and one other layer with its magnetization fixed as a reference, and to reduce the interlayer coupling between the magnetic layers. The basic structure is then as follows. First there is a reference layer (R), with the direction of its magnetization set in a fixed direction. Next is a metallic spacer layer (typically Cu), followed by the second ferromagnetic layer, the free layer (F). This latter is free to move in response to an in-plane external magnetic field. In order to increase the sensitivity, there should then not be a strong exchange coupling between the free and reference layers. Normally, this is not an issue in industrial applications. In these, the layers are grown by a sputtering process that gives the layer interfaces enough roughness that the interlayer exchange coupling is severely reduced. On the other hand, long wavelength roughness, or interface undulations, cause a so-called Ne'el orange peel coupling between the magnetic layers and care has to be taken to ensure that this coupling is not unacceptably large.

A typical spin valve has the following layer structure. On top of the wafer substrate is first deposited a seed layer. The layers usually develop a columnar grain structure along the growth direction. The purpose of the seed layer is to promote the growth of large grains, to provide the correct crystallographic texture (typically (111) texture along the growth direction). Ta is common seed layer, that gives a grain size of some tens of nm in the layers grown on top of it. More recently, a NiFeCr alloy is typically used as it promotes larger grains. The electrical transport is in the direction perpendicular to the grains, and grains boundaries do cause scattering which contributes to the resistivity in both spin channels. Larger grains are therefore desirable as any additional spin-independent scattering reduces the *GMR* signal. For the basic structure known as bottom spin valves (BSV), the next layer is an antiferromagnet, such as NiO or PtMn. In a simple BSV, there are only two ferromagnetic layers, the first of which is the reference layer, usually a CoFe alloy deposited on top of the antiferromagnet. Then come the Cu spacer, the free layer, and finally a cap, such as Ta, which protects the multilayer stack. The free layer is typically a CoFe alloy or a CoFe/NiFe bilayer. Another basic structure is the Top Spin Valve (TSV), in which the free layer is first grown on the seed layer, followed by the Cu spacer, reference layer, antiferromagnet, and cap layer.

The purpose of the antiferromagnet is to pin the direction of the magnetization in the reference layer. In general, a ferromagnet deposited on top of an antiferromagnet and then annealed in an external field will experience a net effective field in the direction of the field applied during the annealing. This effective field is called the pinning field. The physics of this exchange bias effect could be subject of a chapter of its own, and we will not discuss the details here.

In order to obtain a response that is as linear as possible, the free layer must be biased magnetically in a direction perpendicular to the reference layer in the absence of external fields. In addition, the device design has to be stable with a reversible response. Irreversible motion of the magnetization causes unacceptable noise and may also shift the electrical bias point. The magnetic bias is usually obtained by abutting permanent magnets to the free layer, with their magnetization in the desired free layer bias direction (cross-track). The strong magnetic field from the permanent magnets, and the demagnetizing field which tend to align the magnetization of the free layer along the edges, cause the response of the free layer in the device to be much stiffer than that of an unbiased sheet film structure. A figure of merit is the efficiency of the spin valve, which measures the average change

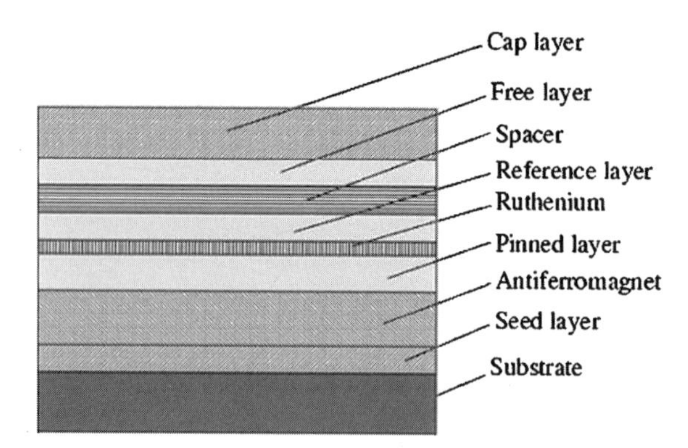

FIGURE 13.4. Cross-section of a SAF BSV stack.

in magnetization relative to saturation in the direction of the applied field. In typical spin valves, the efficiency may be about 20%–25%.

From a transport point of view, the seed, cap and antiferromagnetic layers simply constitute parasitic conductances that tend to shunt some of the current away from the active CoFe/Cu layers. Therefore, it is desirable for these parasitic conductances to be as small as possible. Fortunately, typical antiferromagnets, such as PtMn, and seed and cap layers, such as Ta, have resistivities of about hundred to a few hundred $\mu\Omega$-cm and do not shunt the current very much.

In a patterned simple spin valve, there is a large magnetostatic coupling between the free layer and the reference layer, as well as edge effects due to magnetostatic interactions within the layers. In addition, the reference layer does respond to some degree to the external field. Both of these effects tend to reduce the *GMR* since they reduce the degree to which the free layer magnetization can rotate relative to that of the reference layer. The Synthetic Antiferromagnetic (SAF) spin valve[10] is a design aimed at both reducing the response of the reference layer as well as the magnetostatic coupling between reference and free layers. The basic idea is to add another magnetic layer, the pinned layer (P), with a fixed magnetization direction and a very strong antiferromagnetic coupling to the reference layer. The magnetic field emanating from the pinned and reference layers then tend to form closed loops between the two layers, which substantially reduces the magnetostatic coupling to the free layer. In addition, the pinned and reference layer behave together as an antiferromagnet, with no net average response to an external field. This reduces the rotation of the reference layer in response to an external field.

The basic design of a SAF spin valve is shown in Figure 13.4. First there is the usual seed layer, followed by the antiferromagnet, which pins the pinned layer through the effect of exchange bias. After the pinned layer comes a very thin Ruthenium layer which separates the pinned and reference layers. Ruthenium has the property that it provides a very strong, of the order of 1 erg/cm^2, antiferromagnetic coupling between the pinned and reference layers for a Ruthenium thickness of about 8 A.[11] Following the reference layer come the usual Cu spacer, free layer and cap.

While the SAF provides a magnetically improved design over the simple spin valve, it appears at a first glance that it should have a *GMR* signal lower than that of a simple spin valve. The reason is that both Ru and the pinned layer are extra conducting layers which

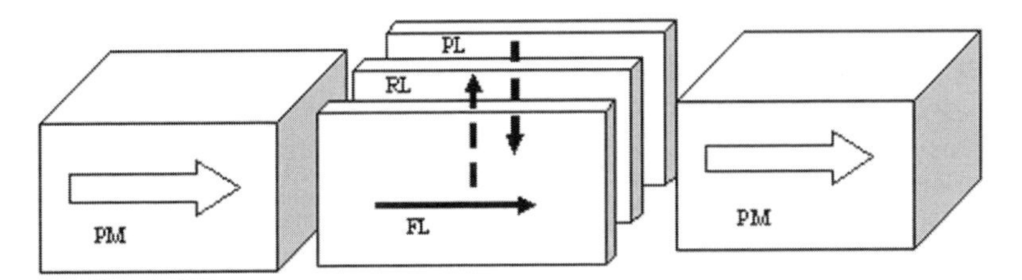

FIGURE 13.5. Cartoon of the magnetic layers in a permanent-magnet biased spin valve read head indicating the magnetization directions of the biasing magnets, of the free layer (FL), reference layer (RL), and pinned layer (PL). The $\hat{x}$-direction is horizontal, and $\hat{z}$ vertical.

shunt current. In addition, the pinned and reference layers are always in AP configuration, causing extra resistance for any electron no matter what the orientation of the reference and free layers. This tends to increase the resistance in P configuration more than in AP, so we would expect a net decrease in *GMR*, both due to the decrease in ΔG as well as the increase in $G(AP)$. However, in reality the *GMR* signal tends to be higher in SAF spin valves than in simple ones. Part of the reason lies in the electronic structure of the Ru relative to that of Co (or CoFe). Recall that at the CoFe/Cu interface, there is almost perfect bandmatch (for (111) texture) in the minority channel, while there is a large band mismatch in the minority channel. The resulting spin-dependent interface scattering contributes to the *GMR* effect as it further depresses the minority channel conductance relative to the majority one over what bulk scattering alone would do. At the CoFe/Ru interface the bandmatch/bandmismatch is reversed:[12] the majority spin electrons incident on the CoFe/Ru interface from CoFe have a large probability of being reflected, while the minority spins have a large probability of being transmitted into the Ru and the CoFe pinned layer, where they are quickly scattered. Therefore, the spin-dependent scattering at the CoFe/Ru interface also acts to enhance the asymmetry in conductance between majority and minority spins, which enhances the *GMR*.

13.3.2. Spin Valve Transfer Curves

We will now consider the output voltage from a well-designed, properly biased spin valve in response to media transitions in a longitudinal medium. If the spin valve is properly biased, the free layer magnetization is parallel to the plane of the medium, and the reference and pinned layer magnetizations are perpendicular to the plane of the medium. A cartoon of the device is shown in Figure 13.5. We will sense the signal voltage by applying a constant current across the spin valve and measuring the resulting voltage across it. The proper bias point is obtained by balancing the field in the magnetic layers due to the current in the device against the inter- and intralayer magnetostatic and interlayer exchange interactions acting on the layers.

Magnetostatic interactions confine the magnetization of the layers to the plane of the layers, so the free layer can only respond to $\hat{x}$- and $\hat{z}$-components of the magnetic field. If the track is wide enough compared to the reader and the reader is centered in the track we can ignore the cross-track component of the field, which we will do here. The vertical component of the field is not uniform but decays with z even in the absence of the reader. Shields in front of and behind the reader cause the field in the reader to decay even faster because of flux leakage from the reader to the shields. The actual strength of the field seen by the reader then depends sensitively on the medium magnetization, the head-to-medium separation, and the shield-to-shield separation but scales directly with the product of the medium magnetization M_r and thickness t. For a given reader design and head-to-medium separation, one can then put the reader right on top of a transition and plot the reader output voltage relative to the voltage at the quiescent point (in the absence of a transition) as a function of the medium $M_r t$. Alternatively, one can apply an external uniform field instead of using a medium transition. This field will also decay along the height of the reader, but one can plot the output voltage as a function of the applied field strength.

The transfer curve has the following shape. In the absence of an external field, the free layer and the reference layer magnetization are perpendicular and the signal voltage is zero, since this is measured relative to the quiescent state. We now subject the reader to a field pointing up along the z axis, either by moving the reader towards a transition with the magnetization on both sides head to head, or by applying a field. The field exerts a torque on the free layer magnetization, which rotates an angle θ up in response. As the angle between the free layer and reference layer magnetizations decreases, the resistance decreases as does the voltage across the reader and the signal voltage. For small fields, the rotation of the free layer magnetization is small and the resistance relative to the quiescent point is

$$R = R_0 + \frac{\Delta R}{2}\left[1 - \cos\left(\pi/2 - \theta\right)\right] \approx R_0 + \frac{\Delta R}{2}\left[1 - \theta\right]. \tag{13.16}$$

The transfer curve is linear as a function of excitation ($M_r t$ or applied field strength). If we continue to increase the excitation, the signal voltage becomes nonlinear in the excitation as the rotation angle becomes appreciable and we cannot approximate $\sin\theta \approx \theta$. For strong enough excitation, the free layer eventually becomes parallel to the reference layer, with a minimum signal voltage $-|V_-|$. Further increasing the excitation will eventually start rotating the pinned layer as the excitation field becomes stronger than the effective field coupling the pinned and reference layers, but this will not be manifested in the output signal.

If we instead start at the quiescent point and apply an excitation in the $-\hat{z}$-direction, θ will increase. At first, the signal voltage then increases linearly with the excitation, but starts to become nonlinear as θ increases further. At some point, θ reaches a maximum with a concomitant maximum V_+ in signal voltage. Further increasing the excitation will start to rotate the reference layer along the direction of the exciting field. This reduces θ and the signal voltage decreases. If we keep increasing the excitation, the reference layer will eventually become parallel to the free layer and the signal reaches its minimum level.

A properly designed and biased spin valve operates only in the linear region of the transfer curve. If $\theta < 0$ in the quiescent state, the maximum positive output voltage is greater than the absolute value of the minimum output voltage, and the spin valve is said to be overbiased. Conversely, if $\theta > 0$ in the quiescent state, the maximum output voltage is smaller than the absolute value of the minimum, and the reader is said to be underbiased. Over- or under-biased readers have nonlinear transfer curves. Nonlinear transfer curves are also the result of too weak biasing field from the permanent magnets, in which case the transfer curve can be nonlinear both at positive and negative signal voltages. Since the signal processing circuits expect a linear signal, nonlinearities give rise to noise, so-called nonlinear distortion noise. Even weaker biasing fields give rise to unstable readers with open loops in the transfer curve. Nonlinear distortion noise *vs.* sensitivity is a constant dilemma—as readers are made smaller, the sensitivity must increase. Increasing the sensitivity always runs the risk of yielding nonlinear transition noise or unstable readers.

13.3.3. Scaling and Extendability of Spin Valve Sensors

The signal voltage amplitude for a current-biased spin valve is

$$\Delta V = I_b R_S \frac{w}{h} \frac{\Delta R}{R} \eta \tag{13.17}$$

where I_b is the bias current, w and h the width and height, respectively, R_s the sheet resistance of the stack, $\Delta R/R$ the (sheet-film) magnetoresistive ratio, and η is the magnetic efficiency of the device. In order for signal-to-noise to remain constant as the areal density increases and the reader size decreases, the signal voltage of the reader has stay approximately constant. It is commonly believed that the use of spin valve readers may be extended to areal densities of 100–150 Gbit/in^2. We will now discuss what limits their extendability.

First of all, it appears that the magnetoresistive ratio is limited to about 20%. There have been great improvements in this quantity since the first spin valves. These improvements have been made through a substantial amount of hard work and research. Some improvements have come from materials research, which has lead to better seed layers and antiferromagnets. Others have come from improved processing techniques, which have allowed for thinner Cu layers, thinner free layers, better controlled magnetostriction and anisotropies, and better interfaces. Finally, improvements have also been made through new inventions, such as the use of "specular layers". These are oxide layers capping the free layer. Conduction electrons cannot enter the oxide and are reflected back into the free layer. If the interface between the free layer and the oxide cap is smooth enough, the reflection is near-specular. A specular reflection suffers no loss of momentum in the plane parallel to the interface and so gives rise to no resistance for current flowing parallel to the interface. The minority spin electrons have in general a very short mean free path in the free layer and are scattered many times as they travel through the free layer, so specular scattering does not impact the minority carriers much. The majority spin electrons, on the other hand, have a mean free path much longer than the free layer thickness. Therefore, specular scattering has a relatively large effect on the resistance of the majority spin electrons and tend to enhance the asymmetry between minority and majority spin resistivities, which increases the magnetoresistive ratio. In spite of all the significant (and expensive!) improvements, it seems unlikely that there will be spin valves with a magnetoresistive ratio exceeding 20%.

Another limiting factor is Joule heating due to the power dissipation in the spin valve. Elevated temperatures promote diffusion of atoms in the spin valve stack, in particular along grain boundaries which are predominantly perpendicular to the interfaces. This makes interfaces more diffuse which in turn causes more spin-independent scattering at interfaces. Diffusion of Mn also severely degrades the magnetic properties of the magnetic layers in the stack. Therefore, the bias current must be set at a level low enough that Joule heating does not degrade and cause the spin valve reader to fail during the expected lifetime (or at least during the warranty period!) of the disc drive. The heat generated in the spin valve primarily is dissipated through the top and bottom surfaces to the shields, which are large chunks of metal and good heat sinks. As the size of the spin valve is reduced, the resistance stays roughly the same while the area through which heat can be dissipated to the shields is reduced. Therefore, the bias current has to be reduced and the amplitude drops with spin valve size.

Finally, the efficiency of spin valves biased magnetically with permanent magnets decreases as the sensor size decreases. The reason is that for the field from the permanent magnets to be large enough to make the magnetization along the edges of the reader stable perpendicular to the edge, the field will be large enough to prevent the magnetization some distance in from the edge to respond. There is then an effective active area in the center of the sensor which is responsible for generating the signal. As the sensor shrink, the inert edge regions do not, so the active area must shrink.

The question is what to use once CIP spin valves run out of gas. One possibility is to use CPP spin valves, in which the current runs perpendicularly to the planes of the stack, rather than in the planes of the stack. On the one hand, the magnetoresistive ratio for CPP is much lower than for CIP spin valves. On the other hand, perhaps Joule heating will not limit the bias current as much as for CIP spin valves. This is because the shields themselves can be used as contacts. Eliminating the insulating layers between the spin valve and shields and putting them in direct contact with each other means that the heat can be dissipated more efficiently. So perhaps the bias current can be increased enough to more than compensate even if the CPP magnetoresistive ratio cannot be increased much further. This is complicated by the fact that large current densities can cause magnetic instabilities due to a phenomenon called spin-momentum transfer. CPP spin valves also must necessarily be targeting high areal densities. This is because the CPP resistance is very low. The product of resistance and area, which for CPP devices is an intrinsic quantity of the stack, is for CPP spin valves of no more than $0.1\ \Omega\mu\text{m}^2$. In order for the signal voltage to be acceptable, the device resistance must be at least of the order of $10\ \Omega$. This means that the area of the device must be no larger than $0.01\ \mu\text{m}^2$, or that the linear dimension of the device must be less than about 100 nm.

13.3.4. Tunneling Magnetoresistive Read Heads

In reality, most disc drive manufacturers don't count on CPP spin valves until areal densities of $200\ \text{Gbit/in}^2 - 400\ \text{Gbit/in}^2$ are reached. It appears that tunneling magnetoresistive read heads can bridge the gap between CIP and CPP spin valves. Tunneling magnetoresistance is observed when a thin insulating layer, typically aluminum oxide, separates two metallic ferromagnetic layers, and a current is flowing from one metallic layer to the other through the insulating barrier. The resistance of such a tunneling junction depends very sensitively on the barrier thickness, and increases exponentially with the thickness of the barrier. The intrinsic property is, just as for CPP spin valves, the product of resistance and area of the junction, usually measured in $\Omega\mu\text{m}^2$. A serious concern is that the frequency response of a circuit containing a tunneling junction depends very sensitively on the resistance and the capacitance of the junction. In order for the frequency response to have a large enough bandwidth to be useful in disc drive, at least 500 MHz, the resistance of the junction must be no more than a few hundred Ohms. If, as an example, the junction is $100\ \text{nm} \times 100\ \text{nm}$, this means that the RA-product must be only a few $\Omega\mu\text{m}^2$. This, in turn, translates into a barrier that is less than 10 A thick. How to reliably make such a thin barrier uniformly across a whole wafer at an acceptable yield is a formidable problem in high-tech manufacturing.

The magnetoresistance depends, just as for the *GMR* effect, on the angle θ between the magnetization of the two ferromagnetic layers, and can be written

$$R = R_0 + \frac{\Delta R}{2}\left[1 - \cos\left(\pi/2 - \theta\right)\right] \approx R_0 + \frac{\Delta R}{2}\left[1 - \theta\right], \tag{13.18}$$

where R_0 is the resistance in the parallel state. Although tunneling magnetoresistance was discovered quite a long time ago, it was not until very recently that an appreciable magnetoresistance was demonstrated at room temperature. It took some additional time to make available deposition systems which allow for manufacturable tunneling junctions in read heads. The tunneling magnetoresistive ratio is very sensitive to materials choices and processing conditions. The effect itself is entirely quantum mechanical and is poorly understood at the present. A simple model, due to Julliere, expresses the tunneling magnetoconductance as

$$\frac{G(P) - G(AP)}{G(AP)} = \frac{2P^F P^R}{1 - P^F P^R}. \tag{13.19}$$

In this equation, P^F and P^R are the spin polarizations of the free layer (F) and the reference layer (R), respectively. These are in turns defined by the product of tunneling probabilities $t_\uparrow$ and $t_\downarrow$ and densities-of-states at the Fermi surface $D_\uparrow$ and $D_\downarrow$ for each spin direction ($\uparrow$ and $\downarrow$) at each electrode (F or R):

$$P = \frac{t_\uparrow D_\uparrow - t_\downarrow D_\downarrow}{t_\uparrow D_\uparrow + t_\downarrow D_\downarrow}. \tag{13.20}$$

Although this is a very simple model and it can be shown that it makes sense only in some limiting cases, it is nevertheless frequently used to interpret experimental data. From this model, we can infer that the tunneling magnetoresistance will be very sensitive to the electronic structure in the barrier and in the electrodes, which includes the band gap in the barrier material and the band offsets between majority and minority spin electrons in the electrodes and the barrier. It is reasonable to suspect that interface states are very important as they may alter how electron states couple from the electrodes to the barrier, and careful experiments have also shown that this is the case. For example, inserting only a few angstroms of Cr between one electrode and the barrier can completely kill the magnetoresistance.

The standard design for a tunneling magnetoresistive read head is shown in Figure 13.6. The stack structure is the same as for a CIP spin valve, in which the Cu spacer has been replaced by the insulating barrier. The major difference is that the current flows in the direction perpendicular to the stack, so there are no leads and contacts on the sides, which have to be insulated from the shields. Instead, the shields themselves can be used as contacts. This has the potential advantage of placing the stack in better thermal contact with the shields, which will help heat dissipation from the stack.

Although the design and stack look deceptively similar to CIP spin valves, the tunneling junction reader raises several important questions which have to be answered before disc

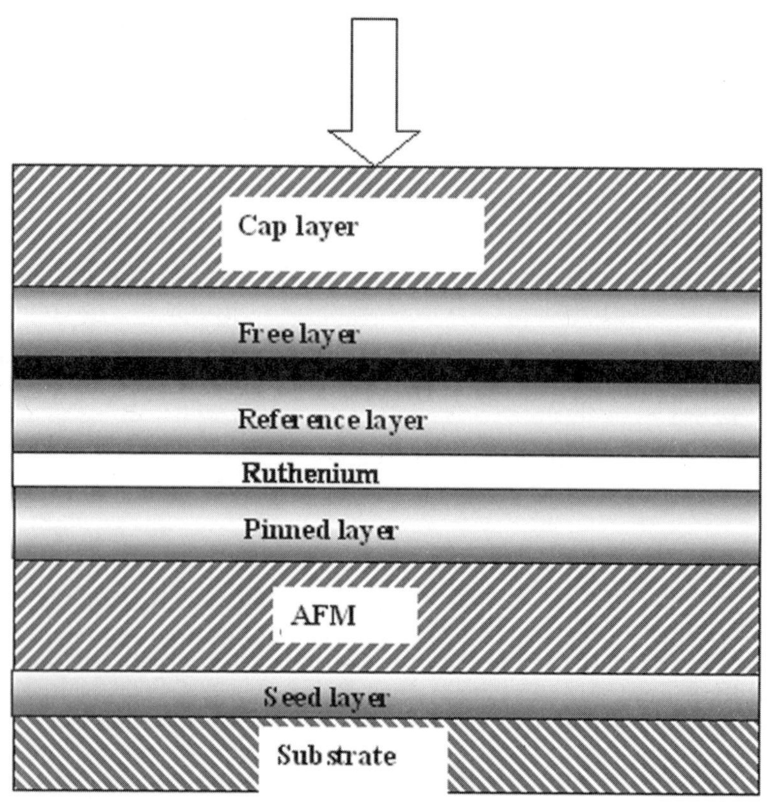

FIGURE 13.6. Cross-section of a tunneling magnetoresistive head stack. The tunneling barrier is indicated by the solid black layer. The current direction is CPP, indicated by the arrow.

drives with tunneling read heads can be sold. The first one of these is reliability. For spin valves, the reliability is limited by the power dissipation in the spin valve due to Joule heating. If the spin valve reader runs too warm, atoms will diffuse between different layers, degrading the electrical performance and the magnetoresistance. For tunneling readers, there may be other, more important, modes of degradation. One example is dielectric breakdown of the tunneling barrier. If too large a potential difference is applied across an insulating layer, dielectric breakdown occurs in which material is ionized with an ensuing large discharge. Clearly, tunneling readers have to be biased with a low enough bias that dielectric breakdown cannot occur. In addition, the tunneling barrier is ultra-thin, only a few atomic monolayers, with inhomogeneities and perhaps even pinholes. Such defects are expected to be weak spots were some kind of breakdown can occur before the regular dielectric breakdown. Furthermore, it is at the present not known how stable an ultra-thin barrier is to low bias voltages applied for long periods of time. Given the fact that the barrier is almost atomically thin, one can imagine that just a little rearrangement of atoms can alter the barrier, and hence reader, properties.

A second question concerns the electronic noise generated in tunneling readers. Normally, tunneling processes lead to shot noise, due to the discrete transmission of electrons across the tunneling barrier. The classical limit of such a process is well known to give rise

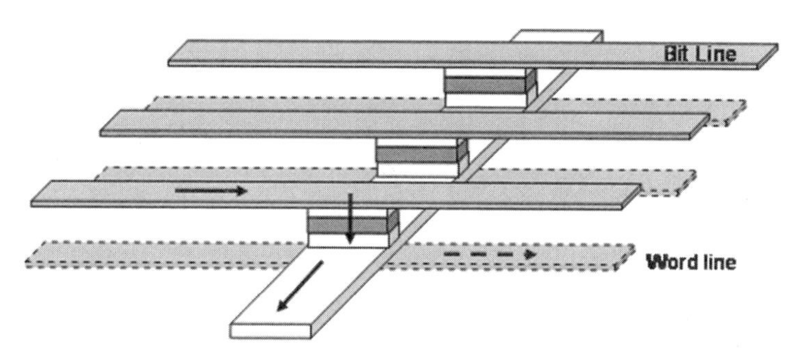

FIGURE 13.7. Cartoon of an MRAM element with bit lines (solid) and word lines (dashed). The current flow in the bit line and through the tunneling stack is indicated by the solid arrows. Current flows only in the bit lines when information is read. To write, current flows in the appropriate word and bit lines which intersect at the bit which is being written.

to a noise voltage which is

$$V_n = \sqrt{2e|I|\Delta f R} = \sqrt{2eVR\Delta f} = \sqrt{2eV(RA)\Delta f/A}, \tag{13.21}$$

where Δf is the bandwidth, $|I|$ the current, R the resistance, and A the area of the device. This clearly scales differently with device size compared to the Johnson noise of spin valves, for which

$$V_n = \sqrt{4k_BTR\Delta f} = \sqrt{4k_BTR_s w\Delta f/h}, \tag{13.22}$$

with R_s the sheet resistance, and w and h the reader width and height, respectively. To lowest order, the spin valve resistance is constant, since R_s is constant and the aspect ratio w/h is constant, while for tunneling readers RA and the bias voltage V are constant, while the area A decreases, so that the shot noise voltage increases. This raises the concern that electronic SNR will deteriorate with shrinking device sizes, and therefore with increasing areal density, for tunneling readers.

13.3.5. MRAM (Magnetoresistive Random Access Memory)

The *GMR* and tunneling magnetoresistive effects have opened the possibility to non-volatile random-access memory. Recall that regular RAM in a computer, such as SDRAM and DRAM, are volatile and the stored information is lost when the power is turned off. The basic idea in MRAMs is that the information is stored magnetically in a small structure like a spin valve or a tunneling reader, and so is non-volatile, provided the structure will maintain its magnetized state.

Most current MRAM designs are based on tunneling magnetoresistance.[13] The basic structure looks just like a tunneling reader, with a pinned layer, a tunneling barrier, and a free layer (see Figure 13.7). The crux of the matter is to design the shape of the device such that the free layer magnetization has to stable states. For example, the device can be shaped into a rectangle. This will provide a uniaxial shape anisotropy along the long side of the rectangle, with the bistable states with their magnetizations along the long side of the rectangle. These states can be further stabilized (and near-degenerate) by eliminating the stray fields from the pinned layer with a SAF structure (pinned and reference layers separated by a thin ruthenium layer). Bit lines are used to read out the state of the device—if the magnetization is parallel to the pinned (or reference, for a SAF structure) layer, the resistance is low and the device is a 0, while if the free layer magnetization is antiparallel to the device is a 1. The state of the device is written by using both word lines and bit lines. Currents running in these generate magnetic fields, but each one alone is too weak to alter the magnetization of any elements. But they both coalesce at the elements that are being written and the fields enhance each other, resulting in a total field large enough to change the magnetic state of the element.

The main difference between an MRAM stack and a tunneling reader stack is that barrier can be made much thicker in the MRAM—the voltage across the device is sensed in its 'high' or 'low' states, and the thicker the barrier, the higher the resistance, and the lower the power consumption during the read process. The problems with the MRAM have to do with processing difficulties: semiconductor-style architecture and line densities have to be integrated with magnetic and non-magnetic metals. In addition, the shape control of the magnetic devices and the placement of bit and word lines have to be extremely precise in order to ensure that only the elements intended to be written on switch during a write process. This all implies very different processes from what is currently used in the semiconductor industry and therefore different tools and large investments in tools. The question is then how long, if possible at all, it will take to recover the investments, both in research and development as well as in tools, necessary to take MRAMs to product.

A second area of concern is the so-called fan-out—the word and bit lines are low-resistance wires which are driven by some amplifiers. With many such lines in parallel, the power consumption by the word and bit lines increases and more amplifiers have to be used since each amplifier has a limited power capacity. In addition, the power consumption can be large, so heat has to be dissipated somehow.

13.3.6. What Comes Next?

13.3.6.1. Ballistic Magnetoresistance We mentioned earlier the BMR in nano-sized magnetic junction. Any magnetoresistive effect that has a magnetoresistive ratio of several thousand percent, or even a few hundred percent, is obviously very attractive from a device point of view. Add to this the facts that the BMR junctions are very small, so that devices based on this effect can conceivable read from very narrow tracks, and that the sensitivity of BMR junctions is very large, and the attraction becomes all the larger.

However, in order for there to be devices based on the BMR effect, there are several issues that have to be resolved. First of all, there has to be a manufacturable path to making devices. By this, we mean that there has to be a device design which allows for a very good process control in order for the yield to be acceptably high. At the present, the best BMR junctions (those with the highest magnetoresistive ratio), have been made by fusion a very sharp wire with a fatter one using electrodeposition. This is a slow and ill-controlled process which sometimes yields a good device and other times not, and there is basically no way of controlling the process so that junction with the same resistance and magnetoresistance are reproduced over and over again. However, there has been initial work aimed at developing processes to make BMR junction on wafers, basically by drilling nano-size holes in an oxide layer separating two magnetic layers.

In addition to a device design and a manufacturable path, the reproducibility and reliability of the effect and the devices have to be established. Currently, BMR junctions degrade with cycling and the thermal reliability is uncertain at the best. Finally, some theoretical work has argued that the conductance and magnetoconductance of ballistic junctions are multi-valued. This means that for an ensemble of junctions of the same junction area, the conductance and magnetoconductance can assume many different values. If this is true, it implies that even if the manufacturing process could be so well controlled that the variance in junction area can be ignored, the junction resistance and magnetoresistance will have huge variances, with seriously detrimental consequences for yield.

13.3.6.2. Enormous MagnetoResistance As I pointed out earlier, the EMR effect is very interesting as it can potentially allow for devices which are not based on magnetization rotation. This, in turn, completely eliminates the possible sources of instabilities and noise that can plague thin film ferromagnetic sensors. Furthermore, the noise in an EMR device is only thermal Johnson noise, so the noise should intrinsically be very small. The issues with EMR that have to be resolved mostly have to do with manufacturing. First of all, the EMR effect hinges on having a very high mobility semiconductor in a low-resistance contact with a very high-conductivity metal. At the present, only highly specialized MBE growth process can yield semiconductors with a high enough mobility. There has to be a relatively cheap, high-throughput way of growing them and pattering them without degrading the mobility. Second, not any semiconductor-metal combinations can do. The Schottky barrier between the semiconductor and the metal must be small. The Schottky barrier is an interface resistance that arises if the Fermi edges of the semiconductor and the metal are misaligned. If the Fermi level in the semiconductor is higher than that of the metal, electrons flow from the conduction band of the semiconductor into the metal. This leaves a depletion layer behind in the semiconductor and an associated potential barrier, which acts as an interface resistance for charges to flow from the semiconductor to the metal. Moreover, growing high-quality metals on semiconductors is a difficult materials science problem. Finally, present prototypical designs of EMR sensors have very precisely defined layouts of the leads to the device, with extremely narrow leads, of the order of 30 nm—the device performance changes if the lead resistances change or if the lead layout changes. This is clearly not a robust design which is easily manufacturable at a high yield.

13.3.6.3. Semiconductor Spintronics Spin valve heads and tunneling magnetoresistive head, and in fact any read head or sensor based on the GMR, TMR or BMR effects, are "spintronics" devices: they depend on spin-polarized populations of charges. Those populations are manipulated with an external magnetic field, and the results are detected using an electrical current or potential difference. However, the recent excitement about spintronics has been mostly due to the possibility of using semiconductor-based devices. From the perspective of magnetic recording, there are several potential advantages of semiconductor spintronic devices compared to those based on ferromagnetic materials. First of all the current, ferromagnetic-based devices contain an awkward soup of materials: soft and hard ferromagnets, non-magnetic metals, insulators, and antiferromagnets. The performance of the devices depends on the magnetic, electrical and mechanical properties of these materials, and how they interact with one another. Unfortunately, it is difficult at best to independently tune and optimize the materials properties such as electrical conductivity and magnetization for best device performance. For example, we can change, say, the magnetic properties of one ferromagnetic layer by doping it or changing its alloy composition, but then we are almost certainly also changing its electrical and mechanical properties, usually in undesirable ways. Furthermore, many of the critical materials combinations, such as CoFe/Cu in GMR spin valves or CoFe/aluminum oxide in tunneling junction, are dictated by delicate electronic structure matches (or mismatches), which severely limits the choice of materials from the outset. In addition, the materials soup poses some difficult challenges from processing and manufacturing perspectives. The different materials have to have proper adhesion to one another, corrosion has to be avoided, temperature expansion has to be avoided or at least controlled, the different materials have different etch and milling rates, which makes patterning difficult, and the different materials and their properties have to be able to withstand anneals (use to, e.g., set exchange bias at antiferromagnet-ferromagnet interfaces) and curing of photoresists and insulators. Semiconductor spintronics at least holds out the hope that much fewer materials combinations have to be used. In addition, an essential feature of semiconductors is that we can very carefully control, or "engineer", their properties, such as band gap and conductivity. We have also discussed the problems associated with using thin film ferromagnets to sense magnetic fields, such as the difficulty to produce a reversible, linear response and to avoid Barkhausen noise. Semiconductor spintronics devices would not necessarily be based on the rotation of the magnetization density in a ferromagnetic thin film, but could be based on the manipulation of individual electrons spins, which only weakly interact with one another.

So let us briefly survey the field of semiconductor spintronics and discuss some potential devices of interest. Currently, it has been demonstrated in numerous experiments that once created, spin polarized populations of charges can be transported over distances of the order of micrometers and manipulated with external magnetic fields.[14] In fact, semiconductors such as GaAs are very good spintronics materials in that the spin coherence length, the distance over which a spin packet maintains a coherent polarization, is quite long. The biggest problem is currently related to the creation of spin polarized charge carriers. Many experiments have used optical pumping techniques, in which the basic principle is that charges are excited to the conduction band from the valence band by polarized light. However, this is not suitable technique for magnetic recording. Charge injection from ferromagnetic metals is made difficult because of the large conductivity mismatch between the metal and the semiconductor, and this makes the spin injection efficiency very low. Ways to overcome this injection problem by using tunneling barriers or applied electric fields are currently under investigation. The most straightforward way of injecting spin polarized charges into semiconductors would be by using ferromagnetic semiconductors,[13] and this has indeed been demonstrated to work. The problem is that the to date most studied semiconductor, Mn-doped GaAs, has a maximum Curie temperature of about 100 K—this is clearly far too low for most applications. There have been reports of other magnetic semiconductors with Curie temperatures possibly exceeding room temperature, such as Mn-doped GaN, or Co-doped TiO_2. These magnetic semiconductors also have in common the fact that they are very difficult to produce. Mn-doped GaAs, for example, is a metastable solid solution of Mn in GaAs, and is usually made by cold MBE—if the growth temperature is too high, the Mn atoms diffuse too easily and the result is a phase separation of MnAs and GaAs. Similarly,

Co tends to form precipitates in TiO_2 and it is difficult to ascertain if the magnetism is just due to precipitates forming Co-cluster. However, bandstructure calculations indicate that Co-doped TiO_2 should be a half-metal, with a true energy dispersion in the majority band at the Fermi edge. In the end, given the resources dedicated to research in this area, and given the considerable progress that has been made within a few years, it is probably not too optimistic to assume that there will be useful room-temperature magnetic semiconductors before too long.

We close this section with a brief discussion of spintronics devices. The first one is the Monsma transistor, or spin valve transistor.[16] This one is a hybrid device, meaning it contains both metals and semiconductor. The basic design is to have a spin valve base wedged between semiconductor (usually Si) emitter and base. The emitter is forward biased and the collector is back-biased. At the Si-metal interfaces, there are Schottky barriers, the sum of which is the emitter-to-collector voltage. The emitter injects hot electrons into the base region. If they do not scatter appreciably in the base region, they will make it to the collector. If the spin valve magnetic layers are in parallel configuration, injected majority spin electrons do not scatter much and make it to the collector. In antiparallel configuration, both majority and minority spin carriers are scattered and are thermalized in the base without reaching the collector. The problem so far with the spin valve transistor is that it has low current gain—the spin valve has to be thick enough to discriminate between parallel and antiparallel configurations, but that severely reduces the current that reaches the collector. In addition, it is difficult to grow high-quality semiconductors on metals.

Tunneling devices based on semiconductors and magnetic semiconductors have been demonstrated. A Ga(Mn)As-AlAs-Ga(Mn)As tunneling device has reached a magnetoresistive ratio of about 70% at low temperatures. The good news is that using a semiconducting (AlAs) tunneling barrier offers the potential advantage of using a physically thicker barrier, which allows for better process control. The bad news is of course the low Curie temperature of Ga(Mn)As, but the device is a proof of principle, and there is no reason why it should not be operating at room or high temperature given magnetic semiconductors with a higher Curie temperature.

Finally, we mention the spin field-effect transistor.[17] This device is basically a field-effect transistor in which the source and drain are ferromagnetic. The key thing is that the source injects spin-polarized carriers into the channel, and the drain collects spin polarized carriers. For high enough quality channel, and high enough mobility, the carriers will travel fast enough that they will see a magnetic field component due to a Lorentz transformation of the electric field in the channel to the rest frame of the carriers. The magnetic field will affect majority and minority carriers differently. Therefore, and electrical signal applied to the gate voltage, or a magnetic field applied to the channel region, will change the source-to-drain conductance of the device.

13.4. SUMMARY

We have in this chapter discussed some magnetoresistive phenomena, and how some of them are applied, in particular in magnetic recording. The phenomena and their current applications form a fascinating and challenging mixture of research ranging from quantum mechanics to materials physics and state-of-the-art processing technologies. The main current applications of magnetoresistance are in magnetic field sensing, with perhaps some of the most advanced devices, such as spin valves and magnetic tunneling junctions, in the area of magnetic recording. However, current device applications are to some extent limited both by the magnitude and sensitivity of the magnetoresistive effect on the one hand, and by the complex sets of materials and concomitant complex and expensive processing technologies. Semiconductor spintronics opens up the possibility of both better materials and processing control, as well as the possibility of completely novel devices and applications. It remains to be seen when such devices and applications will actually be realized, but it is certain that research in semiconductor spintronics will remain an active and stimulating field at the intersection of basic physics, materials research, and engineering for a long time to come.

QUESTIONS

1. Using the Valet-Fert model, estimate the resistance in parallel and antiparallel states and the magnetoresistive ratio for Co(30 A)/Cu(20 A) multilayers consisting of (a) 5, (b) 15, and (c) 25 repeat units. Find reasonable values for resistivities from handbooks, use an interface asymmetry parameter $\gamma = 0.7$, and a resistivity asymmetry $\frac{\rho_\downarrow}{\rho_\uparrow} = 10$ for Co majority and minority bands. Do you think it is realistic to expect the magnetoresistive ratio to keep increasing monotonically with the number of repeats?
2. In a simple band model of a ferromagnet, the bands are free-electron like but are shifted by a constant energy Δ, the exchange potential. Thus the energy dispersions are

$$\varepsilon_{\vec{k},\uparrow} = \frac{\hbar^2 k^2}{2m} - \Delta$$

$$\varepsilon_{\vec{k},\downarrow} = \frac{\hbar^2 k^2}{2m} + \Delta$$

   Calculate the spin polarization of the densities-of-state at the Fermi surface assuming the exchange splitting is much less than the Fermi energy (use Eq. (13.20) with the tunneling probabilities set to unity). Use the result to calculate the tunneling magnetoresistive ratio using the Julliere formula.
3. Consider a tunneling junction with $RA = 4\ \Omega\mu\text{m}^2$, a magnetoresistive ratio of 30%, and a constant bias voltage of 150 mV. Assuming an efficiency of 20% and a bandwidth of 250 MHz, at what size does the electronic signal to noise power go below 20 dB if the noise is all classical shot noise?
4. In Ga(Mn)As, the Mn dopants occupy preferentially As sites, and each Mn atom has spin $(5/2)\hbar$. Assuming that all Mn spin contributes to the magnetic moment, and ignoring the expansion of the lattice constant with Mn doping, what is the magnetization density at 6% doping?

REFERENCES

1. M. N. Baibich, J. M. Broto, A. Fert, F. Nguyen Van Dau, F. Petroff, P. Etienne, G. Creuzet, A. Friederich, and J. Chaezelas, *Phys. Rev. Lett.* **61**, 2472 (1988).

2. B. Dieny, V. S. Speriosu, S. S. P. Parkin, B. A. Guerney, D. R. Wilhoit, and D. Mauri, *Phys. Rev.* B **43**, 1297 (1991).
3. J. Moodera, L. R. Kinder, T. M. Wong, and R. Meservey, *Phys. Rev. Lett.* **74**, 3273 (1995).
4. N. Garcia, M. Munoz, and Y.-W. Zhao, *Phys. Rev. Lett.* **82**, 2923 (1999).
5. H. D. Chopra and S.-Z. Hua, *Phys. Rev.* B **66**, 020403 (2002).
6. S.-Z. Hua and H. D. Chopra, *Phys. Rev.* B **67**, 060401 (2003).
7. For recent reviews of properties of doped manganites, see, for example, chapters by D. Khomskii and M. Viret in *Spin Electronics*, M. Ziese and M. J. Thornton (eds.) (Springer Verlag, Berlin, 2001).
8. S. A. Solin, T. Thio, D. R. Hines, and J. J. Heremans, *Science*, **289**, 1530 (2000).
9. T. Valet and A. Fert, *Phys. Rev.* B **48**, 7099 (1993).
10. H. A. M. van der Berg, W. Clemens, G. Gieres, G. Rupp, W. Schelter, and M. Vieth, *IEEE Trans. Magn.* **32**, 4624 (1996).
11. S. S. P. Parkin and D. Mauri, *Phys. Rev.* B **44**, 7131 (1991).
12. W. H. Butler, O. Heinonen, and X.-G. Zhang, in *The Physics of Ultra-High-Density Magnetic Recording*, M. Plumer, J. van Ek, and D. Weller (eds.) (Springer Verlag, Berlin, 2001).
13. S. S. P. Parkin, K. P. Roche, M. G. Sammant, P. M. Rice, R. B. Beyers, R. E. Scheurlein, E. J. O'Sullivan, S. L. Brown, J. Bucchiganno, D. W. Abraham, Y. Lu, M. Rooks, P. L. Trouiloud, R. A. Wanner, and W. J. Gallagher, *J. Appl. Phys.* **85**, 5828 (1999); R. C. Sousa, J. J. Sun, V. Soares, P. P. Freitas, A. Kling, M. F. da Silva, and J. C. Soares, *Appl. Phys. Lett.* **73**, 3288 (1998); S. Cardoso, V. Gehanno, R. Ferreira, and P. P. Freitas, *IEEE Trans. Magn.* **35**, 2952 (1999).
14. For a recent introductory review, see, for example, D. D. Awschalom and J. M. Kikkawa, *Physics Today*, June, 33 (1999).
15. For reviews of magnetic semiconductors, see, for example, H. Ohno, *Science*, **281**, 951 (1998); H. Ohno, *J. Magn. Magn. Mater.* **200**, 110 (1999); J. Furydna and J. Kossut, in *Diluted Magnetic Semiconductors*, vol. 25 of *Semiconductors and Semimetals* (Academic Press, New York, 1988).
16. D. J. Monsma, J. C. Lodder, Th. J. A. Popma, and B. Dieny, *Phys. Rev. Lett.* **74**, 5260 (1995).
17. S. Datta and B. Das, *Appl. Phys. Lett.* **56**, 665 (1990).

14

Elements of Magnetic Storage

Jordan A. Katine and Robert E. Fontana Jr.
IBM Almaden Research Center, San Jose, CA

14.1. INTRODUCTION TO MAGNETIC STORAGE

The magnetic hard drive is a technological wonder. An actuator arm positions a flying magnetic recording head over a disk rotating at up to 15,000 rpm with a physical spacing that approaches 10 nm. The tremendous efficiency of this technology lies in the fact that a single magnetic recording head with only two nanoscale features, a writer and a reader, can be used to respectively record and sense data stored magnetically at densities approaching 100 Gbit/in^2. This leverage makes it extremely unlikely that any competing technology will supplant the hard drive in the foreseeable future. The magnetic hard disk drive is the center of the multibillion dollar storage industry. The importance of this market has prompted the investment of tremendous resources which have sustained the evolution of this technology into the nanoscale regime.

The magnetic hard drive consists of two main components: the magnetic medium (usually a rotating disk platter), and the recording head. The magnetic platter is formed through sputter deposition of several metallic layers on top of a very smooth, flat substrate (often glass). The magnetic layers are covered by a protective overcoat (usually carbon-based), and by a lubricant layer. Since increasing the separation between the flying head and the magnetic media degrades the recording resolution, the overcoat and lubricant layers are kept very thin.

The recording head (also often referred to as the "slider") is used to record information on the disk, as well as to read the data that have been stored. While the slider is a three-dimensional object, it is fabricated using planar processing techniques similar to those used in semiconductor lines. The head is fabricated on a conductive ceramic substrate, usually AlTiC. This substrate facilitates the lapping process described below. Once the planar processing is completed, the wafer containing thousands of heads (approx. 0.5 mm^2 per head) is sawed into individual rows (Figure 14.1). In a process known as lapping, these

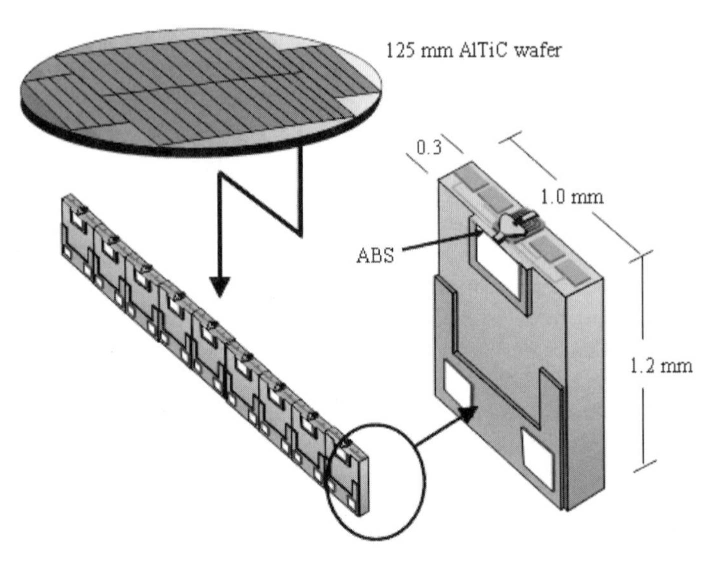

FIGURE 14.1. Following planar processing of the recording wafer, the lapping process dices the wafer into rows and then individual sliders. As a result, several critical dimensions are defined by the final position of the ABS.

rows are precisely polished, and a thin protective overcoat is applied to protect the air-bearing surface (ABS) that will fly over the disk. Interestingly, several critical dimensions are not defined lithographically, but rather by this lapping process. This is a concept likely unfamiliar to most readers, and an example will be given later in this chapter.

The process through which data is recorded is straightforward. A current pulse applied through the high-conductivity copper coil generates a magnetic flux. A magnetic yoke carrier this flux to the write head, which consists of a narrow gap defined between two poles. The fringing field emerging from the flux passing between these poles has a component parallel to the magnetic layers of the disk (Figure 14.2). This fringing field is strong enough over

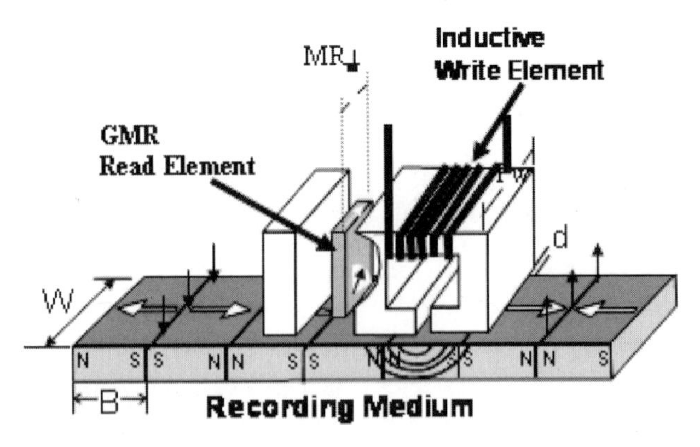

FIGURE 14.2. Schematic illustration of the magnetic recording process. A thin film head flies above the medium at a height d. A write element of width P_w orients the magnetization in the media. A read element of width MR_t detects the fringing field from the magnetic transitions in the media.

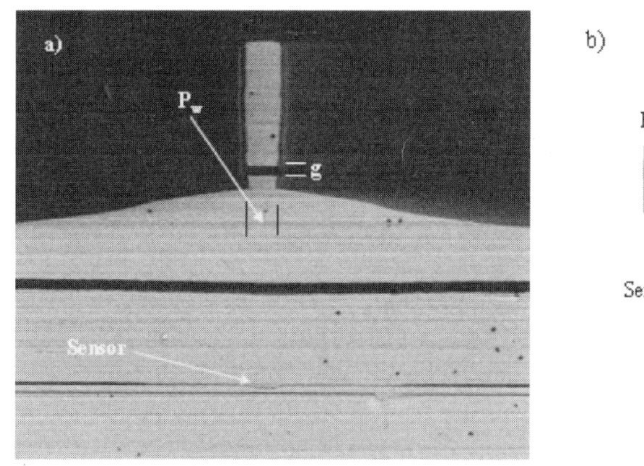

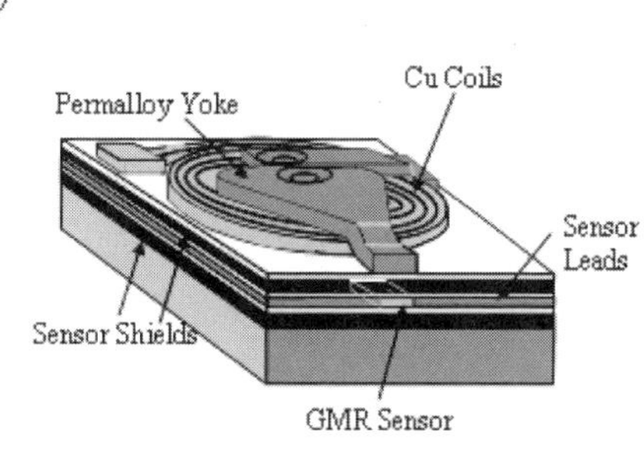

FIGURE 14.3. a) SEM image of a 40 Gbit/in^2 recording head viewed from the ABS. The sensor is located beneath the writer. The writer width, P_w, is 350 nm. The write gap, g, is 100 nm. b) Perspective view schematic of the recording head illustrating how flux is carried from the coils to the ABS.

some area (the bit) to orient the magnetization of the grains of the disk material. If the polarity of the current pulse to the coils is reversed, the bit is magnetized in the opposite direction. All commercial hard drive technologies today are longitudinal in nature, inasmuch as the medium is magnetized in-plane.

By applying current pulses of appropriate polarity as the writer flies over the disk, the recording head writes information in the form of magnetized bits. The string of bits recorded at a given radius onto the rotating disk is known as a *track*. The *linear density* of the bits recorded along the track is determined by the gaps in the writer and reader (Figs. 14.3 and 14.4). The width of the tracks (or *track density*) is determined by the lithographic width of the writer, P_w. The *areal storage density* is the product of linear density and track density.

Transitions between bits magnetized in opposite directions produce magnetic fields above the rotating disk. The recording head also contains a read element, which senses the

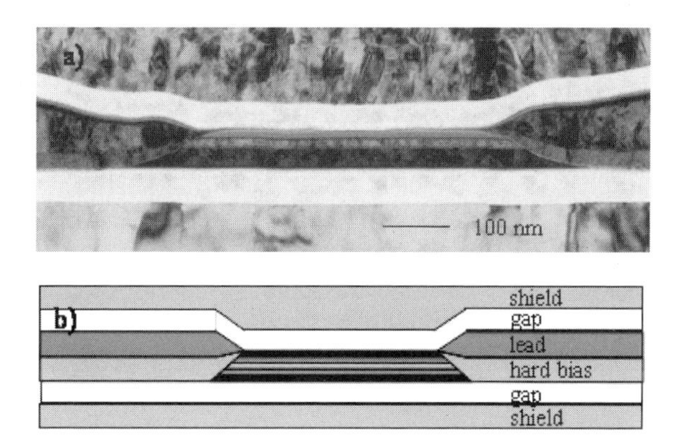

FIGURE 14.4. a) TEM image of a 40 Gbit/in^2 sensor viewed from the ABS. b) Schematic diagram of the sensor.

magnetic fields associated with these transitions, allowing the hard drive to read the data that have been stored on the disk. In early recording heads, inductive pick-up coils were used to sense these transitions. As will be discussed later, at high areal densities magnetoresistive sensors are more efficient than inductive sensors. In all commercial hard drives produced today, recording heads exploiting the giant magnetoresitive effect (GMR) are used to read the data stored in the media.

In section 2 of this chapter, we review the fundamentals of magnetism and their implications for the three critical components of the hard drive: the media, the writer, and the sensor. In section 3, we discuss fabrication technologies and scaling issues related to magnetic recording. In this section we will explore the challenges magnetic recording faces as it enters the truly nanoscale regime, and discuss technologies that may allow magnetic recording densities to climb towards a terabit per square inch.

14.2. FUNDAMENTALS OF MAGNETISM AND THEIR APPLICATION TO STORAGE

14.2.1. Fundamentals Pertaining to Media

A magnetic bit consists of a large number of grains magnetized in a same direction. For current 40 Gbit/in^2 products, typical grain diameters are 8 nm, and the typical bits have lengths of roughly 35 nm and widths of 250 nm, which corresponds to around 130 grains per bit. Statistical analysis shows that the signal-to-noise ratio in a given recording medium is actually proportional to the square root of the number of grains per bit. While a reduced bit size is required to achieve higher areal densities, the number of grains in each bit must be preserved in order to maintain an acceptable level of signal-to-noise. As a result, grain size must be decreased accordingly. Understanding the challenges involved in shrinking magnetic grains requires a concise overview of the concepts of coercivity and superparamagnetism.

The coercivity of a magnetic particle is a measure of the stability of its magnetization. The coercive field refers to the amplitude of the external field required to return the magnetization of the particle to zero. Materials that are magnetically soft, such as permalloy, have low coercive fields, while magnetically hard materials have much higher coercive fields. High coercivity is usually the result of a large crystalline anisotropy. All commercial media today is some variation of CoPtCr, often with a fourth or even fifth element present. Generally speaking, the Pt is there to increase the crystalline anisotropy and the Cr promotes grain segregation. The films are usually magnetron sputter deposited on elevated temperature substrates, with the choice of seedlayer heavily dependent of the disk material used.

From a signal-to-noise perspective, one would like to make the magnetic grains as small as possible. Unfortunately, as the volume of a magnetic particle becomes too small, thermal fluctuations may randomize the direction of its magnetization. Known as *superparamagnetism,* this phenomenon must be avoided to insure the stability of the information stored in the medium. The superparamagnetic relaxation time, τ, is given by:

$$\tau = \tau_o \exp(K_u V / kT) \tag{14.1}$$

where τ_o is a relaxation time constant, K_u is the anisotropy energy, V is the average grain volume, k is Boltzmann's constant, and T is temperature. Since we cannot operate the drives at cryogenic temperatures, the only way to prevent the superparamagnetic relaxation time from decreasing when reducing the grain volume is to increase the anisotropy of the media. However, this will also increase the coercivity of the media, requiring stronger fields to record data to the disk.

14.2.2. Fundamentals Pertaining to the Writer

The writer in a thin film recording head essentially consists of a miniature electromagnet (Figure 14.3b). A clockwise (counterclockwise) current in the coils will generate a downward (upward) magnetic field in the center of the coils. The magnetic field produced at the center of the coil is given by:

$$H = n\,\mu_0\,i/(2R) \tag{14.2}$$

where n is the number of turns, μ_0 is the permeability of vacuum ($4\pi \times 10^{-7}$ Tm/A), i is the coil current, and R is the radius of the coil. Since there is a permalloy ($Ni_{81}Fe_{19}$) yoke in the center of the coils, the total magnetic field, B, is the sum of the magnetic field H generated by the current, and the magnetization M of the yoke. The ratio of B to H is often expressed through the relationship:

$$B = H + M = \mu_r H \tag{14.3}$$

where the coefficient μ_r is the *permeability* of the material. For a soft magnetic material like permalloy, the value of μ_r at low fields is approximately 20,000. Note that the value of μ_r is indeed not constant. At large values of H, the magnetization of the yoke saturates and the permeability approaches unity.

The yoke must be designed to efficiently carry the magnetic flux produced by the coils to the pole tip at the air-bearing surface. Several materials properties will affect such efficiency. Magnetostriction is a phenomenon in which the magnetization of a ferromagnetic material is altered by stress. First, given that it is impossible to match the coefficients of thermal expansion of all of the materials used in the head, running large currents through the coils creates heating that stresses the yoke. In addition, the lapping processes used to produce the heads will also introduce stress in the material. Magnetic materials with very low magnetostriction coefficients are therefore desirable to insure reproducible magnetic behavior in the yoke and pole tip. Permalloy is virtually magnetostriction free and is therefore the material used in the yoke in all commercial thin film heads.

The saturation magnetization is the highest magnetization that can be attained by a material. Its origin can be understood at the atomic level. The magnetization of a ferromagnet is created by unpaired electron spins, each of which contributes a Bohr magnetron to the total magnetic moment. At saturation, all of the unpaired electron spins are aligned. The saturation magnetization can therefore be calculated from the number of unpaired electron spins per atom and the density of atoms (which is related to the lattice spacing and crystal structure).

As discussed earlier, increasing the areal density of the recording media means increasing its coercitivity, which in turn necessitates stronger recording fields. Unfortunaly, no matter how much current is fed to the coils, the write field is limited by the saturation magnetization of the pole tip. At saturation, the magnetic field in the media produced by the pole tip is approximately given by:[1]

$$B \sim 0.44\, B_s \tan^{-1}(g/2d) \tag{14.4}$$

where B_s is the saturation magnetization at the pole tip, g is the width of the gap between the poles of the writer, and d is the separation between the pole tip and the magnetic media. Of all the elements, iron offers the highest saturation magnetization ($B_s = 2.3$ T). In comparison, the values for Co and Ni are $B_s = 1.8$ and $B_s = 0.6$ T, respectively. In earlier generations of thin film recording heads, permalloy was used for the pole tip. Several years ago IBM switched its pole tip material from permalloy ($B_s = 1.2$ T) to $Ni_{45}Fe_{55}$ ($B_s = 1.7$ T). Pure iron cannot be used due to its very poor corrosion resistance and high magnetostriction. The $Ni_{45}Fe_{55}$ alloy therefore represents a compromise between the need for higher saturation magnetization, and the corrosion and magnetostriction constraints on pole tip performance. Today, efforts continue on finding materials with ever higher saturation magnetizations that are suitable for pole tips. Some alloys being considered have B_s values that would actually exceed that of pure Fe.

Perhaps the biggest challenge in fabricating pole tips comes from the geometry of the devices. The width of the pole tip at the air-bearing surface determines the trackwidth of the bit written to the magnetic media. With each generation of higher areal density drives, this trackwidth must decrease. For the 40 Gbit/in^2 products shipping in mid-2002, the pole tip trackwidth is approximately 350 nm. In addition, in order to efficiently transport flux to the ABS, the pole tip must be quite tall ($>2\ \mu$m in the 40 Gbit/in^2 designs). As will be discussed later, the best way to presently fabricate such high aspect ratio structures is through electrodeposition.

14.2.3. Fundamentals Pertaining to the Sensor

Earlier read/write heads used a single inductive coil both as the writing and reading device. As the slider flies over a written bit, the inductive voltage generated in the coil is given by Faraday's law:

$$E = -d\Phi/dt \tag{14.5}$$

where Φ is the magnetic flux passing through the coils, in Webers. As areal densities increased, the scaling of inductive read heads became problematic. With the bit dimension decreasing faster than the velocity of disk was increasing, the only way to maintain the electromotive force in the coil was to increase its number of turns, or even to stack coils on top of each other. Unfortunately, increasing the number of turns increases coil resistance and inductance, which adversely affects its performance during writing. Therefore, inductive sensors were replaced by magnetoresistive materials, which allowed the coil design to be optimized solely for writer performance.

As the name implies, a magnetoresistive material will exhibit a change in its resistivity in response to an applied magnetic field. The first magnetoresistive heads utilized an effect known as anisotropic magnetoresistance (AMR). The AMR effect originates from anisotropic scattering of electrons in the d-bands of magnetic transition elements. It manifests itself as a change in resistance which depends on the angle between the magnetization of the film and the direction of current flow.

Magnetic transitions in the recording medium induce out of plane magnetic fields. As the slider head flies over a written bit, these fields rotate the magnetization of the sensor film, which produces a resistance change detected by a current flowing in the sensor. The permalloy films used in AMR sensors have magnetoresistance coefficients of $\Delta R/R_{\text{min}} \sim 2\%$. Although some magnetic materials possess AMR coefficients as high as $\Delta R/R_{\text{min}} \sim 6\%$, they are not suitable as sensor materials due to their high magnetostriction coefficients and/or high crystalline anisotropy. Because high anisotropy materials have large coercive fields, the fields from the media (roughly 5 mT) are insufficient to rotate the magnetization of these high AMR coefficient films.

The giant magnetoresistive effect (GMR) was discovered in 1988.[2] GMR sensors function similar to AMR sensors, in that fields from the disk rotate the magnetization of a low coercivity sense layer, which produces a resistance change detected by a current flowing in the sensor. The physical origin of the GMR effect, though, is significantly different from that of the AMR effect, and a more complex magnetic multilayer is required to exploit the phenomenon. The complexity, however, is well worth the effort. Magnetoresistance coefficients in GMR sensors are approaching $\Delta R/R_{\text{min}} \sim 15\%$, and virtually all sensors used in hard drives manufactured today exploit the GMR effect.

Figure 14.4 shows a schematic of a GMR sensor. The resistance of the sensor is proportional to $\cos(\theta)$, where θ is the angle between the magnetization of the two ferromagnetic layers in the sensor. The ferromagnetic layers are typically CoFe or NiFe, less than 30 Å thick. The exact composition of the layers depends on the sensitivity, magnetostriction, and GMR amplitude requirements of the sensor. The two ferromagnetic layers are separated by a thin (~40 Å) Cu spacer layer. The GMR effect results from preferential transmission of electrons with parallel spins at the interfaces of ferromagnetic layers, and will be explained further in chapter IV-C. Copper is a good choice for the spacer, since electrons from the two ferromagnetic layers can travel long distances in the Cu layer without having their spin directions randomized. One of the layers of the sensor is pinned, which means that its direction of magnetization is not affected by the fields from the disk. This pinning is accomplished by exchange coupling that ferromagnetic layer to an antiferromagnet, such as IrMn or PtMn. The other ferromagnetic layer is free to rotate in response to the fields from the disk. The sensor stack is typically capped with a thin layer of Ta which protects the sensor from oxidation.

For predictable operation, it is important to stabilize the magnetization direction of the free layer. This is accomplished by abutting a hard magnetic layer (e.g., CoPtCr) against the edges of the sensor (Figure 14.4). The high-coercivity stabilization magnets produce a field that orients the free layer perpendicular to the pinned layer in the absence of an applied field. Since the resistance of the sensor is proportional to $\cos(\theta)$, the response of the sensor to excursions from the $\theta = 90^{\circ}$ bias point will be nearly linear.

To improve the resolution of the sensor, it is sandwiched between a top and bottom layer of soft magnetic material, typically permalloy or sendust. These soft layers, known as shields, isolate the sensor from the stray fields created by other nearby magnetic transitions. The sensor is separated from the magnetic shields by thin, insulating gaps—typically alumina. The spacing between the magnetic shields is known as the read gap, and determines the linear resolution of the sensor. Recall that the trackwidth of the bits is determined by the width of the pole tip. The trackwidth of the sensor, MR_w, is typically 70% of the pole tip width. By having a sensor narrower than the writer, the sensor will be less affected by the stray fields from adjacent tracks on the disk. For the 40 Gbit/in^2 areal density products, this corresponds to a MR_w of 250 nm.

14.3. FABRICATION TECHNOLOGIES AND SCALING

14.3.1. Moore's Law for Magnetic Recording

Progress in the magnetic recording industry is measured by the annual rate of increase in areal density of disk drives. At present, this increase is between 60% and 100% per year. For the sake of comparison, consider the integrated circuit industry where progress is measured by the annual increase in the number of transistors in a device. This increase is also between 60% and 100% per year. These measures of technological progress are geometric with time and are widely referred to as the Moore's Law of the particular technology.

Moore's Law requirements translate directly into minimum feature requirements for device structures. For magnetic recording applications, the areal density on the disk surface is related to the bit cell area which corresponds to the product of the read gap thickness and the magnetic trackwidth. These dimensions must decrease by 20% to 30% per year to sustain an areal density increase of 60% to 100% per year. For the integrated circuit industry, the transistor count in a device is related to the size of the device, the design of the transistor cell, and the minimum features of the transistor cell. Historically, each of these factors has contributed to the increase in transistor count. As a result, minimum features need be reduced only between 10% and 15% per year to achieve the 60% to 100% increase in number of transistors per device. Figure 14.5a shows the historical trend of minimum features of thin film heads and integrated circuits for the last 30 years using data through the year 2000. This figure predicts a convergence of film head and integrated circuit minimum features in the middle of this decade, a direct result of the different impact "Moore's Law" has on the minimum feature requirements of each technology. As seen in Figure 14.5a, around 1998 the rate of areal density increase for magnetic recording systems accelerated from 60% to 100% with the advent of the higher sensitivity GMR sensor. Figure 14.5b reflects this accelerated change with a more detailed view of the minimum feature requirements for thin film heads and of the convergence of integrated circuit and film head features. Note that in the middle of this decade, magnetic recording will require 50 nm features.

14.3.2. Media Scaling Issues

Advances in the media, the recording head, and the head/disk interface are all required if the areal density of magnetic storage is to continue increasing. Media scaling issues will be discussed extensively in chapter IVB, so here we only briefly highlight the challenges

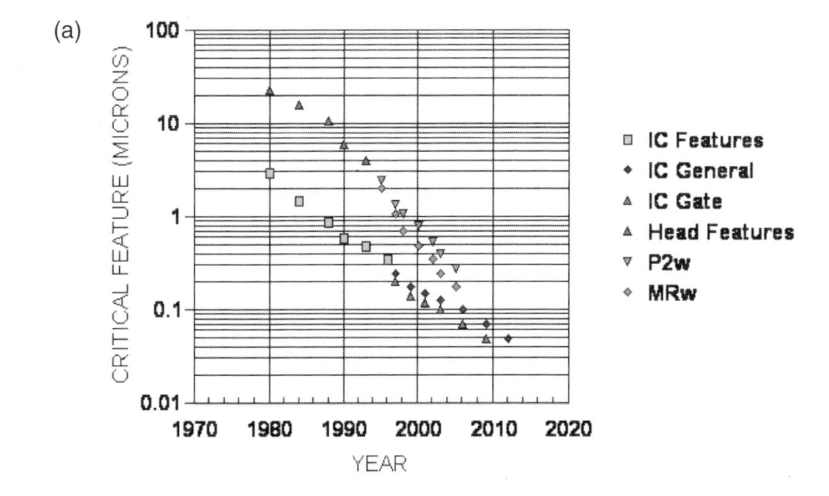

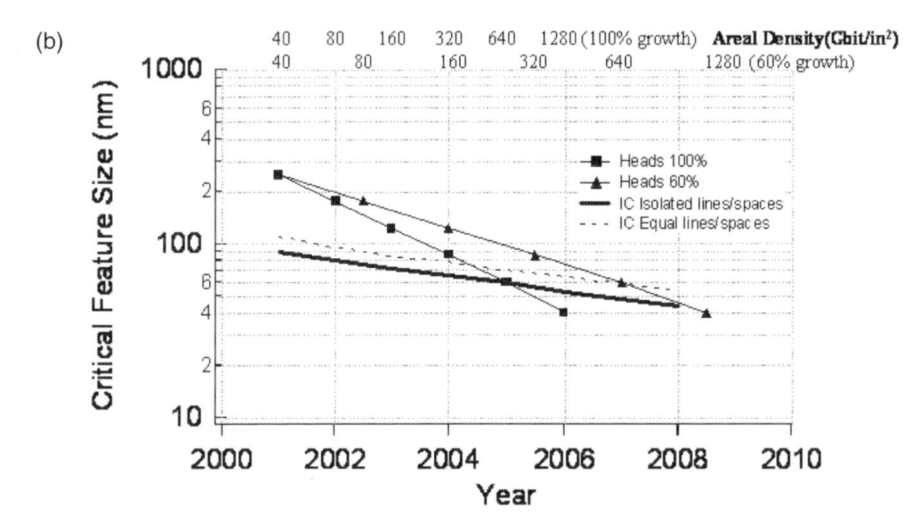

FIGURE 14.5. a) Historical trends in integrated circuit features and thin film head features (projection from 2000 based on 60% areal density growth and 50% decrease in integrated circuit features every 6 years). b) Scaling trends for this decade. The areal density corresponding to the feature size is shown above the graph.

involved in developing magnetic media suitable for areal storage densities greater than 100 Gbit/in^2.

The most significant recent breakthrough in media has been the rapid commercial introduction of antiferromagnetically coupled (AFC) media.[4] In AFC media, the active magnetic region is separated into two ferromagnetic layers separated by 6 Å of Ru. At that Ru thickness, the conduction electrons in the Ru mediate a very strong interaction that forces the ferromagnetic layers to align anti-parallel, i.e., antiferromagnetc coupling. Because of this, the effective magnetic thickness of the media can be quite small, independent of the actual layer thickness. This allows bits to utilize thicker grains with larger volumes, which, as shown in Eq. (1), improves thermal stability.

Another potential improvement in magnetic media could come from magnetic nanoparticles.[5] Traditional sputter deposition methods for media produce a distribution of grain sizes. It is possible to create an extremely tight size distribution of ferromagnetic nanoparticles, and then coat them in a uniform layer (or layers) over the disk. By tightening the volume distribution, one could go to a smaller average particle volume before reaching the onset of superparamagnetism.

An intriguing route to delay to onset of superparamagnetism is lithographically patterned media. Instead of the media being a continuous magnetic film, lithographic techniques are used to create isolated islands of ferromagnetic material. With the introduction of AFC media, it seems likely that conventional granular media can be extended to greater than 200 Gbit/in^2. Assuming square bits, the bit cell for 200 Gbit/in^2 media is slightly under 60 nm per side, which is below the resolution limits projected for optical lithography in the next decade. To be economically viable, any technique used to lithographically pattern media must cost no more than about one dollar per disk, so some form of high resolution imprint lithography or perhaps radially symmetric self-assembly must be developed before patterned media can be seriously considered.

No matter what form the media takes, the onset of superparamagnetism at high areal density can be delayed by increasing the coercivity of the media. Typically, the coercivity of the media is limited by the ability to write it, so improvements that allow the recording head to write high coercivity media will be instrumental in extending magnetic storage. Looking at Eq. (4), one obvious improvement would be to decrease the separation between the pole tip and the writer, and this is an area of ongoing research. Alternatively, nearly a factor of two improvements in the intensity of the writing field could be realized in perpendicular recording. In perpendicular recording, the write head has a single pole. The second pole is a soft magnetic underlayer on the disk. Therefore, the magnetic recording media lies directly in the write gap, where the field is much stronger than in the fringing fields used in longitudinal recording given by Eq. (4). The name perpendicular recording is taken from the fact that the write field is now perpendicular rather than parallel to the media, meaning the bits will be magnetized into or out of the plane of the disk. Another technique that may facilitate the writing of high coercivity media is thermally-assisted recording. When heated, the coercivity of many magnetic materials can be dramatically reduced. By integrating some localized heating mechanism into the write head, it is possible to momentarily reduce the coercivity of the media during the brief write pulse, after which the media temperature will quickly re-equilibrate, leaving the written data in a stable state.

14.3.3. Writer Scaling Issues

We have already addressed many of the techniques that will be used in scaling the writer: e.g., higher B_s pole tips and thermally-assisted recording. In this section we discuss the lithographic challenges associated with scaling the 2 μm tall pole tip to ever narrower dimensions. As mentioned earlier, the pole tip is created by electroplating a high B_s magnetic alloy into a resist stencil. The electrodeposition of these materials is itself an art. We note that some impressive early work in the fabrication of GMR nanowires was accomplished through electrodeposition,[6] and hope that future work in this area may benefit by incorporating some of the plating techniques discussed below.

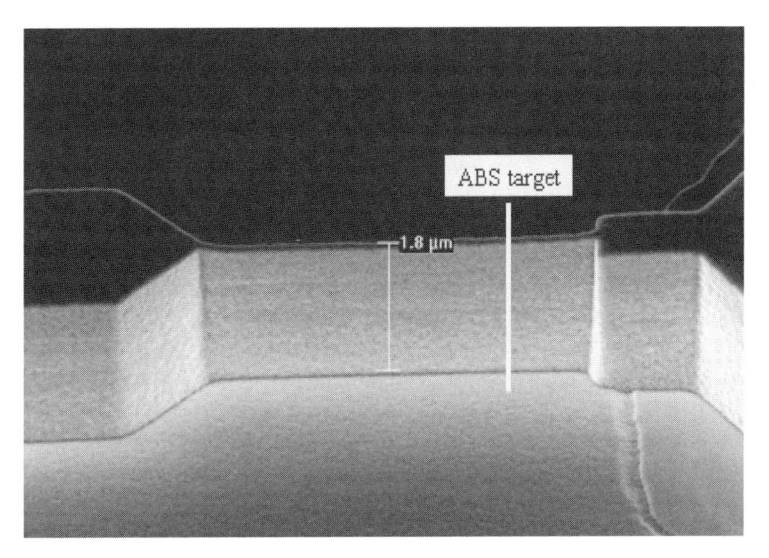

FIGURE 14.6. SEM image of plated pole tip feature. The resist stencil defined with 100 keV e-beam lithography was 2 μm tall and 200 nm wide. A line is drawn to show where lapping would define the ABS.

The choice of seed layer is critical in insuring proper film growth during plating, and the resist must also have excellent adhesion to prevent an "underplating" halo resulting from resist delamination. In plating the narrow pole tips, uniformity and control of the plating current is greatly enhanced by creating a large plating frame around the small pole tip feature. Without such a frame, nanoscale features are plated at high current densities where the deposition rates and uniformity are difficult to control. Of particular concern in plating alloys into high aspect ratio features is that mass transport limitations could create significant composition gradients during electrodeposition.

Of course, before one needs to worry about electroplating into a high aspect ratio resist stencil, the stencil itself must be created. With 100 keV direct write e-beam lithography, 2 μm tall tracks 200 nm in width have been created (Figure 14.6). Creating such high aspect ratio trenches in thick resist is impossible with optical lithography. One alternative could be to use a thin imaging resist and then transfer the pattern into a thick polymer underlayer through reactive ion etching. Alternatively, one could choose to plate a wider trackwidth pole tip, and then trim the pole tip width after electrodeposition. This technique eases the photolithography and plating constraints, but introduces the challenge of trimming the pole tip. Unfortunately, the high B_s alloys used in the pole tips are not susceptible to reactive ion etching, so only some form of physical sputtering can be used to trim the pole tip. One option commonly utilized in ultrahigh density magnetic recording demonstrations is trimming the pole tip with a focused ion beam (FIB). Although exquisitely narrow pole tips can be formed with FIB, the slow throughput makes FIB somewhat impractical for high volume manufacturing. In addition, as the trackwidth of the pole tips continues to decrease, the damage to the magnetic properties of the pole tip created by the implantation of Ga^+ during milling could become problematic.

Ion milling with argon ions is a long established means of etching magnetic materials. The accelerating voltages used in the mill are generally less than 1 kV, making implantation less of a concern. By milling at the proper angle, it is possible to etch the pole tip faster laterally than vertically, but there will inevitably be significant loss of the pole tip height during any trimming process with an ion mill. Therefore, the amount a pole tip can be trimmed with an ion mill is limited, and a high aspect ratio plated pole tip is required as a starting point.

Potential relief in the lithographic scaling of the writer trackwidth could come from moving towards perpendicular recording. Because the transport of flux is different in a single pole tip perpendicular recording head, the high aspect ratios currently required in longitudinal writers would not be necessary. In fact, electrodeposition may no longer be necessary, and, like the sensor described below, ion milling alone could be sufficient to define the writer.

14.3.4. Sensor Scaling Issues

As discussed earlier, the sensor is a multilayer stack of different materials. Thickness uniformity and interface quality are critical to the sensor performance. If one attempted to create a narrow trackwidth sensor by sputtering the layers into a resist stencil, and then lifting-off the resist, the quality of the multilayer stack would be inadequate. Instead, sensor processing begins with carefully controlled growth of the multilayer sensor stack, followed by subtractive processing to define the physical trackwidth.

Since most of the materials found in the sensor stack are not suitable for reactive ion etching, ion milling must be used in this subtractive processing step. The resist used in the sensor trackwidth definition step has two functions: it is the mask that protects the sensor during ion milling, and it is the lift-off stencil for the deposition of hard bias and lead material that occurs immediately after milling. Because the junction between the milled sensor edge and the hard bias is critical to sensor performance, multiple milling and deposition angles are often used to carefully craft the slope and cleanliness of this interface.

As shown in Figure 14.5, the minimum critical feature size in recording heads, i.e., the sensor trackwidth, has been historically larger than the minimum critical feature size on integrated circuits. This allowed the magnetic recording industry to borrow the lithographic processing technology that were already developed for IC manufacturing. For example, 248 nm DUV steppers were introduced in IC manufacturing in 1995, while such steppers were not used in the fabrication of recording heads sensors until 2001. Using phase shift technology and 193 nm steppers, it should be possible to lithographically define the *MR* sensor down to trackwidths approaching 50 nm. As shown in Figure 14.5, the sensor is shrinking faster than the gate width on IC chips, so the magnetic recording industry will likely reach the 50 nm node before the IC industry. The development of 157 nm or EUV technology to extend steppers beyond this node will be driven by the needs of the vastly larger IC industry, so in the not-too-distant future, lithographic tooling limitations could slow the areal density growth in magnetic recording. Direct-write electron beam lithography could be a path around these lithographic limitations. The density of high resolution features on recording head wafers is extremely sparse, and the daily wafer volumes are much lower than those in the IC industry, which may allow direct-write *e*-beam lithography to become a manufacturing solution for the storage industry.[7]

If, despite its throughput limitations, direct-write *e*-beam is a potential solution for defining MR sensor trackwidths, perhaps FIB should also be considered? On the surface, FIB would seem to be an attractive option—it has the required resolution, and would eliminate the ion milling step. Because edge effects will become increasingly important as trackwidth dimensions decrease, processing that minimizes edge degradation will be required. Hence, the large penetration and implantation depths inherent to FIB lithography make it an unlikely candidate for manufacturing ultra narrow trackwidth heads.

With all of the attention we have devoted to lithographic processing required to define the trackwidth, the observant reader may be wondering why this discussion has not also included the sensor stripeheight (the distance between the ABS and the back edge of the sensor), since the trackwidth and the stripeheight are comparable in size. The stripeheight is an example of one of the features we mentioned earlier, whose critical dimension is defined by the lapping process. In practice, a very wide stripeheight feature is defined, the front edge of which extends far beyond the targeted position of the ABS. The positioning of the back edge of this feature relative to the ABS is what actually determines the stripeheight of the sensor. Clearly controlling the stripeheight of the sensor in the sub-100 nm regime will require exquisite overlay accuracy for the lithographic positioning of the sensor back edge and extreme precision during the lapping process.

While reducing the lateral dimensions of the sensor improves the trackwidth resolution, the spacing between the shields above and below the sensor, i.e., the read gap, must be reduced to improve the linear resolution. In the present current-in-plane sensor geometry, there are only two ways to reduce the read gap: making the sensor stack thinner, or thinning the insulating gaps. Each ferromagnetic layer can perhaps be made 5 Å thinner, so reducing the thickness of the antiferromagnet (presently around 200 Å) is the only layer where a significant reduction may be realized. At narrow trackwidths, magnetic size effects may allow the pinning layer to be thinned or even eliminated, but the stability of the sensor could be compromised if the pinning strength is reduced too much. Presently, electrical shorts between the sensor and the shields prevent aggressive reduction of the insulating gaps (today each layer is about 150 Å). Perhaps superior gap materials and deposition quality will allow some gains to made in this area.

14.3.5. CPP Sensors

Ultimately, however, the most promising path for minimizing the read gap is switching to a current-perpendicular-to-the-plane (CPP) sensor geometry. As shown in Figure 14.7, the top and bottom magnetic shields are now the leads, so there are no insulating gaps above or below the sensor. If a conductive hard bias material is used to stabilize the sensor, a thin layer of insulation is required to prevent the hard bias from shunting current flow through the sensor.

There are several challenges that must be overcome before CPP sensors supplant the CIP geometry in commercial recording heads. The biggest challenge is finding a sensor material with the optimal combination of GMR amplitude and resistivity. Suppose we want a CPP sensor whose lateral dimensions are compatible with 200 Gbit/in^2 recording. The area of the sensor would be around 0.003 um^2. A CoFe/Cu/NiFe trilayer has a resistance-area product of only 1.0 mΩum^2, so the sensor resistance would be only 3 Ω, well below the 50 Ω target that gives best signal-to-noise performance in the recording head. Alternatively, a thin

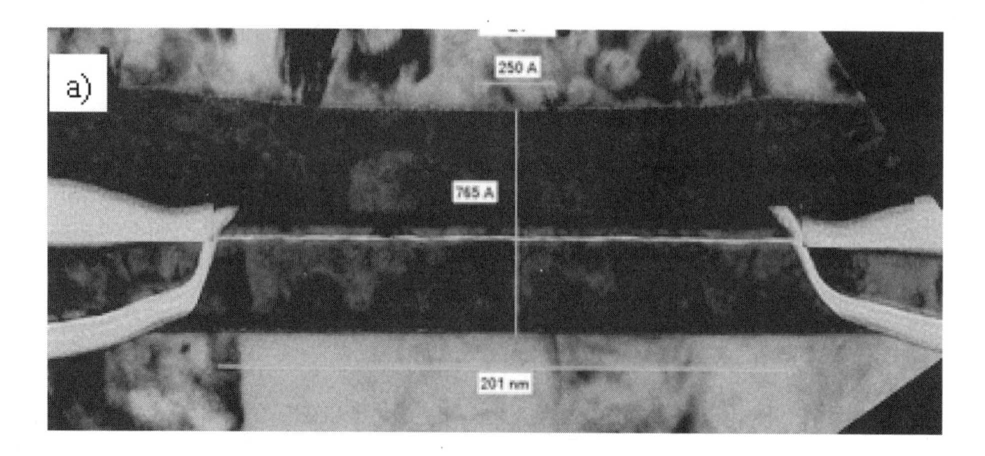

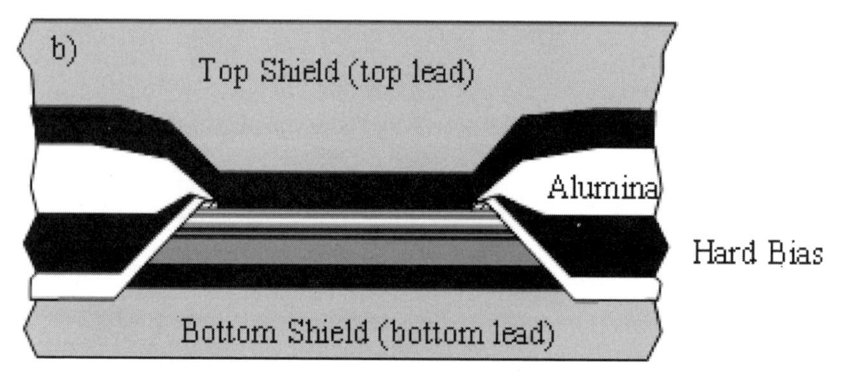

FIGURE 14.7. a) TEM image of a CPP sensor viewed from the ABS. b) Schematic diagram of the sensor. The hard bias is insulated from the sensor by a thin layer of alumina. The tunnel barrier in the sensor is the thin white line visible in the TEM micrograph.

tunneling barrier could replace the Cu spacer. Excellent MR ratios have been reported with Al_2O_3 barriers, but their resistance-area product is around 10 Ωum^2, which corresponds to a sensor resistance of 3000 Ω at the 200 Gbit/in^2 dimensions. Resistances this high are incompatible with high data rate recording.

Assuming the appropriate material is found for high areal density CPP sensors, the resistance of the top and bottom leads must be kept low to minimize parasitic resistance from current crowding into the narrow devices. If a tunnel barrier is used, to avoid shunting, no redeposition of highly conductive material can occur during the ion milling process that defines trackwidth. Also, conductive material cannot be smeared across the thin tunnel barrier during the lapping process that defines the air-bearing surface. The CPP geometry may also impose limitations on the amplitude of the currents used in the sensor. Unlike giant magnetoresistance in metals, the tunneling magnetoresistance amplitude drops sharply above a certain bias voltage threshold. In metallic CPP sensors with very narrow dimensions, the spin-transfer phenomenon could affect the stability of the free layer at high current densities.[8]

14.3.6. Coil Scaling Issues

The need to scale the writer and sensor dimensions for higher areal density is rather obvious. Important gains in recording head performance, however, can also be made by shrinking the pitch of the coils. The optimum number of coils turns, n, in modern recording heads is between four and eight. The inductance of the coil, which is proportional to n^2, would be reduced by decreasing the number of turns. Such reduction would be beneficial for very high data rate (>1 Gbit/sec) recording. Since the coil is no longer used as an inductive sensor, one would ideally like to move to a single-turn design. However, to produce the same field as an n-turn coil, the single turn coil would need to carry n-times as much current, which is more than standard IC drivers can deliver. Alternatively, reducing the coil radius improves writer performance on both fronts. As shown in Eq. (1), the average field strength within the coils grows as radius decreases. Also, the inductance per turn is proportional to the turn radius.

In state-of-the-art recording heads, the coils are made of electrodeposited Cu, and are roughly 0.5 μm wide, with a 1 μm pitch. Electrodeposition is required as the coils need to be quite thick (typically 3 μm) to carry currents as high as 25 mA without excessive heating. The processing technologies that have been developed for producing such narrow-pitch, high aspect ratio coils used in recording heads could have other applications. For example, similar coils might well be used as low resistance inductors in certain microwave electronics applications. In addition, Cu coils would be ideal for the nanofabricated electromagnets that are being experimented with in atom traps and mirrors[9] and for certain biological applications.

14.4. SUMMARY

Below we summarize some of the key points related to nanoscale processing in magnetic hard disk drives.

- The hard drive has several interrelated nanoscale compenents: the media, the sensor, the writer, and the coils; and all of these compenents must be scaled together in order to increase the areal density of the recording system.
- In scaling the media, one must balance the constraints of superparamagnetism against the ability of the recording head to write high coercivity media.
- Thermally-assisted recording and perpendicular recording may both offer paths to writing higher coercivity media.
- The sensor trackwidth is shrinking much fast than the smallest features found in ICs, which may place lithographic tooling constraints on the magnetic storage industry.

QUESTIONS

1a. From Eq. (14.1), for media to be thermally stable for over 10 years, $K_u V$ must be at least $40\ kT$. Assuming the media is composed of cylindrical grains 3 nm thick with $K_u = 1.5 \times 10^5$ J/m^3, what is the smallest grain diameter that would meet this stability criterion?

2a. Using Eq. (1), calculate the field, H, created by a 1-turn coil when a current pulse of 10 mA is passed through it; assume the coil has a diameter of 10 μm.
b. If the pole is fully saturated by a field of 10 mT, how many turns are required in this head?

3a. In an *MR* sensor with a resistance of 25 ohms, a sense current of 2 mA, and an *MR* coefficient of 15%, what signal voltage is generated by a transition that rotates the free layer between $\theta = 80^\circ$ and $\theta = 100^\circ$. What effects do you think limit the size of the sense current?
b. For the sake of simplicity, assume that a transition in the media creates a uniform flux across the area of an inductive head. If a 30-turn pick-up loop measures 100 nm per side, what is the peak signal voltage generated by a disk spinning such that the maximum field gradient is 10^8 T/s?
c. How large would the coil have to be before its signal was comparable to the GMR sensor in part a?
d. Note the head dimensions in Figure 14.5a corresponding to the introduction of the MR sensor in 1991. Does this make sense?
e. Magnetoresistive heads replaced inductive heads in tape drives prior to their introduction in hard drives. Why do you think this was the case?

REFERENCES

1. H. Neal Bertram, *Theory of Magnetic Recording*, Cambridge Univ. Press (1994).
2. M. N. Baibich, J. M. Broto, A. Fert, F. Nguyen Van Dan, and F. Petroff, *Phys. Rev. Lett.* **61**, 21 (1988).
3. R. E. Fontana, S. A. MacDonald, H. Santini, and C. Tsang, *IEEE Trans. Magn.* **35**, 806 (1999).
4. E. E. Fullerton, D. T. Marulies, M. E. Schabes, M. Carey, B. Gurney, A. Moser, M. Best, G. Zeltzer, K. Rubin, H. Rosen, and M. Doerner, *Appl. Phys. Lett.* **77**, 3806 (2000).
5. S. H. Sun, C. B. Murray, D. Weller, L. Folks, and A. Moser, *Science* **287**, 1989 (2000).
6. A. Blondel, J. P. Meier, B. Doudin, and J.-P. Ansermet, *Appl. Phys. Lett.*, **65** 3019 (1994).
7. R. E. Fontana, J. Katine, M. Rooks, R. Viswanathan, J. Lille, S. MacDonald, E. Kratschmer, C. Tsang, S. Nguyen, N. Robertson, and P. Kasiraj, *IEEE Trans. Magn.* **38**, 95 (2002).
8. J. A. Katine, F. J. Albert, R. A. Buhrman, E. B. Myers, and D. C. Ralph, *Phys. Rev. Lett.* **84**, 3149 (2000).
9. K. S. Johnson, M. Drndic, J. H. Thywissen, G. Zabow, R. M. Westervelt, and M. Prentiss, *Phys. Rev. Lett.*, **81**, 1137 (1998).

V

Nanotechnology in Integrative Systems

15

Introduction to Integrative Systems

Michael Gaitan
National Institute of Standards and Technology, Gaithersburg, MD

15.1. INTRODUCTION

These days it's hard to imagine life without the benefits of integration, computers, and the internet. Even simple tasks such as shopping for food have gone "high tech" with computerized checkout, laser scanners, digital scales, and electronic transactions. And during your drive home, you might use the onboard global positioning navigation system mounted in your dashboard and make a call using your mobile phone while feeling secure with your automatically controlled airbags in case of an accident. Such capabilities are what integrative systems are about; linking the functionality of information processing (computation) with gathering information (sensing) and acting on decisions (actuating).

Today's new integrative systems are referred to as MicroElectroMechanical Systems (MEMS) or MicroSystems Technology (MST). Generally speaking, MEMS and MST technologies are made up of systems of sensors and actuators that are monolithically integrated with digital and analog circuits on an integrated circuit (IC) chip. The term MEMS refers to a class of microsystems that are electrostatically actuated mechanical components. In contrast, MST is more inclusive and includes thermal, fluidic, chemical, biological, and optical components.

Two examples of MEMS devices, one a sensor and the other an actuator, are shown in Figure 15.1. The first example, shown in Figure 15.1a, is the Analog Devices accelerometer. This sensor is used to detect a sudden change in acceleration to control the release of airbags in automobiles in the event of a collision. Here, the sensor is a mechanical structure in the center of the chip. The mechanical structure can be visualized as two combs that are interleaved with each other and attached to springs. Surrounding the micromechanical element are circuits for sensing capacitance, signal conditioning, and output. The device has very high reliability; this is a critical requirement since nobody wants their airbags to explode unless they are in a collision.

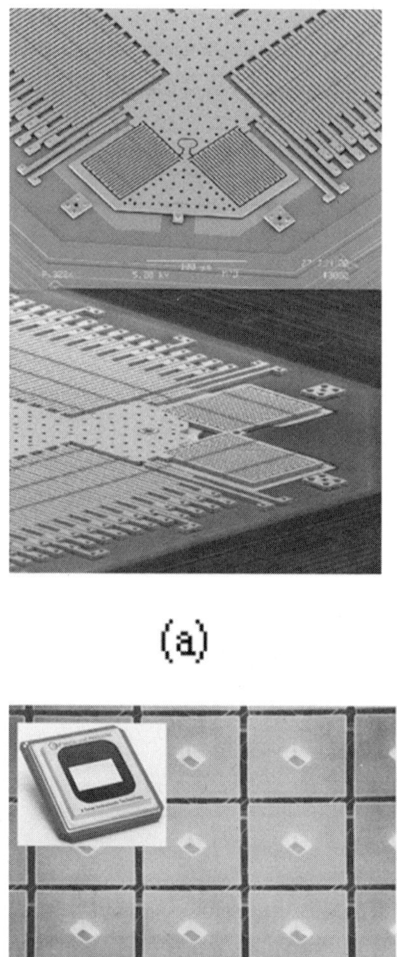

(a)

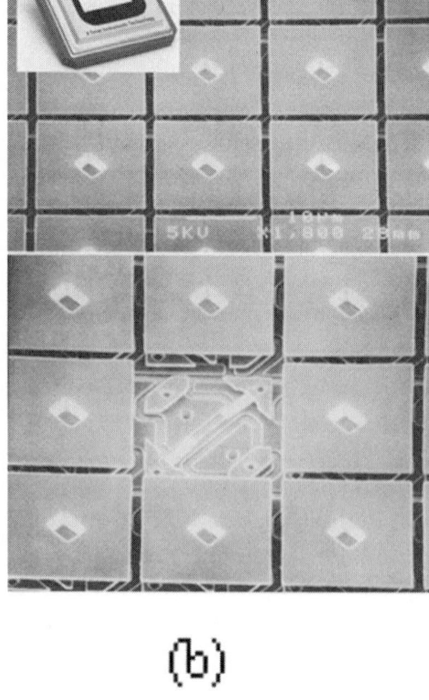

(b)

FIGURE 15.1. Two examples of MEMS/MST devices, the Analog Devices accelerometer (a), a sensor, and the Texas Instruments Digital Light Projector (DLP), an actuator.

The second example (Figure 15.1b), is the Texas Instruments Digital Light Projector (DLP) which is used to project video for presentations, is more recently being used for projection television, and may later be used in digital movie theaters. This device has a large array of tiny mirrors that are controlled by circuits integrated directly beneath each one. The size of the array is equal to the screen resolution which is typically 1024×780, corresponding to almost 1 million tiny mirrors. Each micro-mirror projects light and controls the brightness and color for a single pixel in the overall image. It is claimed that the DLP technology has a much higher dynamic range of brightness to produce more vivid pictures compared to liquid crystal display (LCD) technologies.

Both of these examples have one thing in common: they are mechanical systems that are monolithically integrated with microelectronics. This is a key feature of integrative systems and what MEMS and MST are about: *integration*. Over the past 20 years or more, many discrete microfabricated components have been developed and reported. Such microfabricated devices have been shown to have improved performance or enable totally new applications that traditional "macro-scale" devices cannot accomplish. However, it is the integrated systems such as the accelerometer and DLP that have had the greatest impact in commercial applications.

There are a few paradigms of the benefit of integrative systems that are important to note:

- Batch fabrication
- Monolithic integration with electronics
- Micro-scale devices (1 μm–100 μm)
- Large arrays that can work independently or together
- High reliability

Looking towards the new field of nanotechnology, the concept of integrative systems becomes increasingly important. Consider that as sensors become smaller (i.e., we can fit more in the same space), the number of sensors in a system would increase. For example, if we can currently fit 1 million tiny micro-scale mirrors on a single chip, as exemplified by the DLP, imagine how many nano-scale devices would fit in that same space. As we increase the density of device, the amount of information flowing rapidly increases. In order to exchange information efficiently between some central computer system and the vast amount of sensors that might be incorporated in such a system of nano-devices, one will require levels of hierarchy to control the volume and flow of information. Integrative systems are likely to be the key component to applications for nanotechnology.

The idea of hierarchy not only refers to signal processing or computation but also includes the manipulation side such as using micromechanical elements that bridge the macro world (the world of you and me) with the nano world (the world of nanotechnology) allowing us a means to "reach down" and access nano-scale processes. A perfect example of this concept is the atomic force microscope (AFM). The sharp tip of the AFM that allows one to sense force interactions at the atomic scale is a nano-measurement device. However, the AFM tip is supported by a micro-scale cantilever structure that is used to position the tip on the surface (manipulate) and transduce the measurement of force at the atomic scale. The micro-scale structures such by MEMS and MST may be the enabling technologies to bridge between the macro and nano world. In the following sections, we will have a review of MEMS and MST fabrication technologies, discuss methods for circuit integration, and finish this discussion with emerging commercial applications.

15.2. REVIEW OF MEMS AND MST FABRICATION TECHNOLOGIES

MEMS and MST are intimately linked to semiconductor electronics technology, which not only enables monolithic integration of circuits for on-chip information processing, but is also the basis of the fabrication approach used for miniaturization of MEMS or MST sensors and actuators. Semiconductor electronics fabrication processes are traditionally referred to as *microfabrication*. In addition to microfabrication, MEMS and MST commonly refer to an additional process called *micromachining*. This section is a summary of the two most common micromachining processes for silicon: bulk micromachining and surface micromachining.[1]

Integrated circuit fabrication and micromachining methods are based on three basic processes: deposition, lithography, and etching. Deposition refers to the process of depositing a thin film on a surface. Lithography (also called photolithography) refers to the process of patterning a photosensitive film (photo resist) using a photo mask to image the desired pattern on the resist. Lastly, etching simply refers to chemical processes to strip away a thin film.

The most common films that are deposited include silicon dioxide (SiO_2), silicon nitride (Si_3N_4), polysilicon, and various metals such as aluminum (Al), nickel (Ni), gold (Au), and chrome (Cr). The deposition techniques commonly used in microfabrication are thermal oxidation, chemical vapor deposition (CVD), evaporation, sputtering, and electroplating.

To illustrate the ideas of deposition, lithography, and etching we will step through a process to deposit and pattern a silicon dioxide film. Figure 15.2a shows an example of depositing (also referred to as growing) a silicon dioxide film (SiO_2) on silicon (Si). Many MEMS or MST researchers might choose to use a 75 mm (3 inch) diameter wafer for the

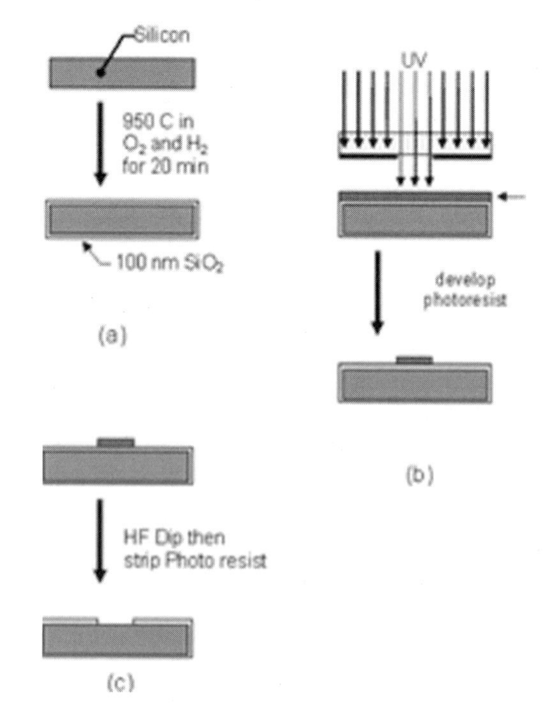

FIGURE 15.2. Example of deposition (a), lithography (b), etching (c).

substrate material in contrast to semiconductor manufacturers who might typically use 200 mm (8 inch) or larger diameter wafers. The wafer is placed in a furnace and elevated to a high temperature with a stream of oxygen (O_2) for dry oxidation or a mixture of oxygen and hydrogen (O_2 and H_2) for wet oxidation. The oxide film would grow at a rate that is related to the temperature of the furnace. A typical wet thermal oxidation might last for 20 min at 950°C and result in a 100 nm thick oxide film.[2]

The next step in patterning the SiO_2 film is the photolithographic process. Here, a photosensitive film (photo resist) is spin coated on the surface of the wafer. Figure 15.2b shows an example of using a negative photo resist for duplicating a pattern on a photo mask onto the photo resist. The photo resist might be spin coated at a few thousand rpm in order to produce a film thickness on the order of 1 μm. The desired pattern is transferred to the photo resist by using an aligner to project UV light through the photo mask onto the wafer surface and then developing the photo resist in a developer solution.

Once the photo resist has been patterned, it can be used as a mask to pattern the silicon dioxide film beneath it using hydrofluoric (HF) acid to etch the silicon dioxide. Figure 15.2c shows the process of using HF to etch the SiO_2 chemically, The photo resist masks the HF from etching the SiO_2 beneath it while the SiO_2 not covered by the photo resist will be removed. The last step in the process would be to remove the remaining photo resist by dissolving it in a solvent such as acetone. The procedures of deposition, lithography, and etching can be combined and repeated with different materials to produce many layers of patterned films. These types of procedures are used to form the complex network of interconnects in modern integrated circuits. As exemplified in Figure 15.3,

FIGURE 15.3. Example of interconnects in an integrated circuit generated by deposition, lithography, etch (Reproduced with kind permission of IBM, *http://www.chips.ibm.com/news/sa27.html*).

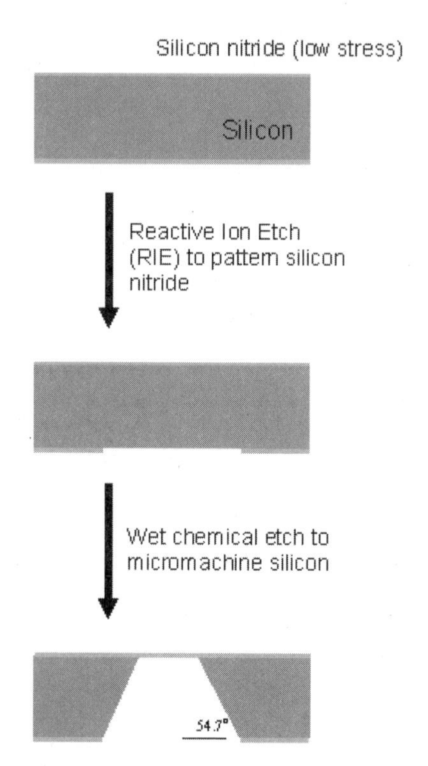

FIGURE 15.4. Example of bulk micromachine process for a pressure sensor.

modern integrated circuit technology uses photolithographic methods to create a complex network of metal lines that connect together the transistors in the integrated circuit. In contrast, a basic MEMS or MST process may not be as complex as a full CMOS process; however, in order to achieve the goal of an integrative system they should be compatible with it.

The "micromachining" processes for MEMS or MST devices can be thought of as a specialized set of etch processes. There are two predominant types that are used today in commercial applications: bulk and surface micromachining. Bulk micromachining is the process of chemically etching the bulk silicon material. An example of the bulk micromachining process to produce a thin membrane is shown in Figure 15.4. In this example, silicon nitride is used instead of silicon dioxide as described previously because of the desired mechanical property for the thin membrane: low residual stress. Low stress silicon nitride can be deposited by a process called low pressure chemical vapor deposition (LPCVD) by controlling the deposition temperature and the ratio of reactants, dichlorosilane ($SiCl_2H_2$) and ammonia (NH_4). This typically would be done at 850°C with a 5:1 ratio by volume of dichlorosilane to ammonia.

After the silicon nitride film is deposited, an opening is patterned on the back side of the wafer by a similar photolithographic process discussed for the SiO_2 film in the previous example. First, a photo resist is spin coated on the surface and then patterned. Instead of HF, the nitride film can be then etched using Freon (CF_4) plasma to result in an opening that exposes the silicon surface. Once the silicon nitride film is patterned, the silicon is etched through that opening. The bulk micromachining process involves using a wet anisotropic etchant such as potassium hydroxide (KOH)[3] or tetramethyl ammonium hydroxide (TMAH).[4] The silicon is etched completely through the wafer to the front side to create a thin membrane of low stress nitride. This basic bulk micromachining process can be used to create a differential pressure sensor.

The second predominate micromachining process in commercial use today is called surface micromachining.[5] Instead of etching the bulk silicon, a thin film "sacrificial material" is removed from underneath the mechanical structure in order to detach it from the substrate. Figure 15.5 shows an example of how the surface micromachining process works. This process typically starts by depositing a silicon nitride film to cover the surface of a silicon

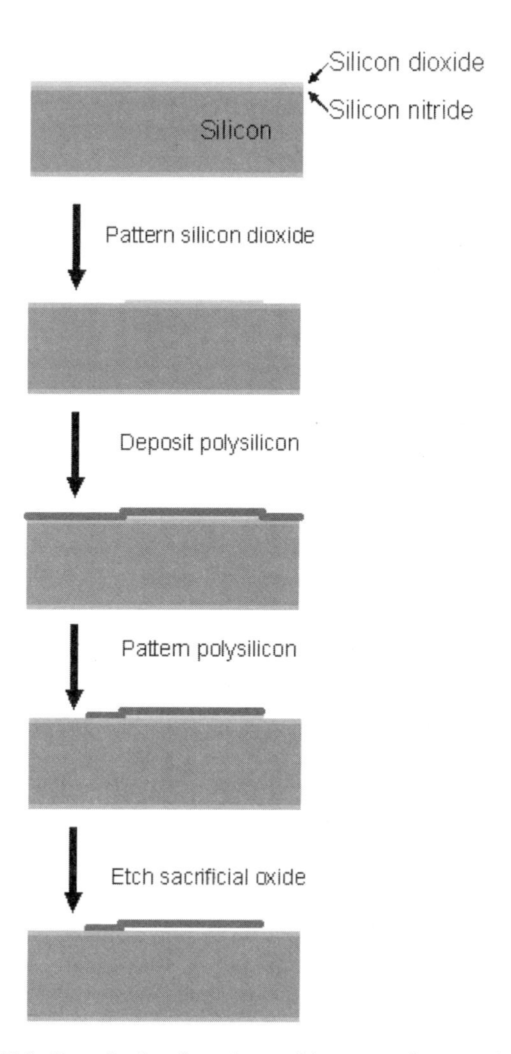

FIGURE 15.5. Example of surface micromachine process for an accelerometer.

wafer. Following this, a phosphosilicate glass (PSG) is deposited by LPCVD over the silicon nitride and then patterned. The PSG material is chosen as the sacrificial material that will be removed later to detach the mechanical structures from the substrate. The mechanical material, polysilicon, is deposited over the sacrificial material using an LPCVD process followed by an annealing process in order to relieve it of mechanical stress.[6] After film deposition, the polysilicon is patterned typically by using an RIE process. Once it is patterned, the regions where the polysilicon is in direct contact with the silicon nitride surface will serve as anchors for the mechanical structure. Finally, the PSG sacrificial material is etched away in hydrofluoric acid (HF). In this example, a cantilever structure has been patterned. The cantilever is attached to the substrate at the anchor and is released over the region where the sacrificial material was removed. This basic process could be used to fabricate the comb structures used in the accelerometer.

15.3. INTEGRATION OF MICROMACHINING WITH MICROELECTRONICS

A review of the two most common micromachining practices used in commercial MEMS or MST was presented in the previous section. These methods, called bulk micromachining and surface micromachining, were presented in a simplified form to illustrate how the basic concept works. In practice, the process for fabricating a micromachined device would be more complex in order to include electrical connections and additional layers for mechanical optimization, passivation, surface preparation, and sensing. In addition to the processing, the design and layout is optimized by modeling based on measurements of the thin film material properties to produce desired electronic and mechanical behavior.

The question to be addressed in this section concerns how one integrates micromachining with microelectronics. In pursuit of answering this, there are a few issues to consider. Consider some that are listed below:

- Compatibility with the IC process
- Fabrication approach
- Cost
- Time
- Packaging

One viewpoint concerning integrative systems is that the micromachining process must be compatible with the IC fabrication process. However, an attractive alternative is to fabricate the micromachined device on a separate chip from the IC and then package them together. Arguments can be made to support this approach based on the cost to bring a product to market. A good example to consider is the accelerometer which is a single micro sensor that is interfaced to a microelectronic circuit. In this case, it would be practical to fabricate a discrete accelerometer sensor on a custom micromachined chip and then package it together with an integrated circuit. Using this approach, the initial cost of making the device would be much lower than developing the fully integrated device because the discrete component (accelerometer) could be fabricated in a (relatively) simple micromachining process. The IC could then be designed and fabricated using a standard application specific integrated circuit (ASIC) process. In contrast, the approach used by Analog Devices was to embed the MEMS process in their commercial BiCMOS process. This approach would be expected to have a much higher startup cost because of the complexity of the process development. However, if the market were large enough, the economies of scale would yield the cheaper solution over the long term. In the end, the monolithically integrated device is the one that dominates the market for accelerometers, and integration was the key to success.

A second example to consider is the Texas Instruments Digital Light Projector (DLP) discussed in the previous section. Here, close to 1 million micro mirrors must be controlled independently to create the desired image. It would be impossible to control such a large number of micro devices on a single chip without monolithic integration of electronics. In this case, monolithic integration of electronics was the only method possible.

In looking beyond the answer that solves today's needs one might ask: "is there an optimal level of integration?" Historically speaking, as far as microelectronics is concerned, and over the shorter term MEMS and MST, an optimal level of integration has not been reached. The trend continues towards more and more integration. Gordon Moore, the co-founder of Intel, forecasted this trend back in 1965.[7] This idea is known as "Moore's Law," stating that there is a doubling of the transistor density every year. This trend of integration continues to hold true[8,9] as depicted in Figure 15.6 which plots the total number of transistors shipped by Intel by year from 1968 to 2002.

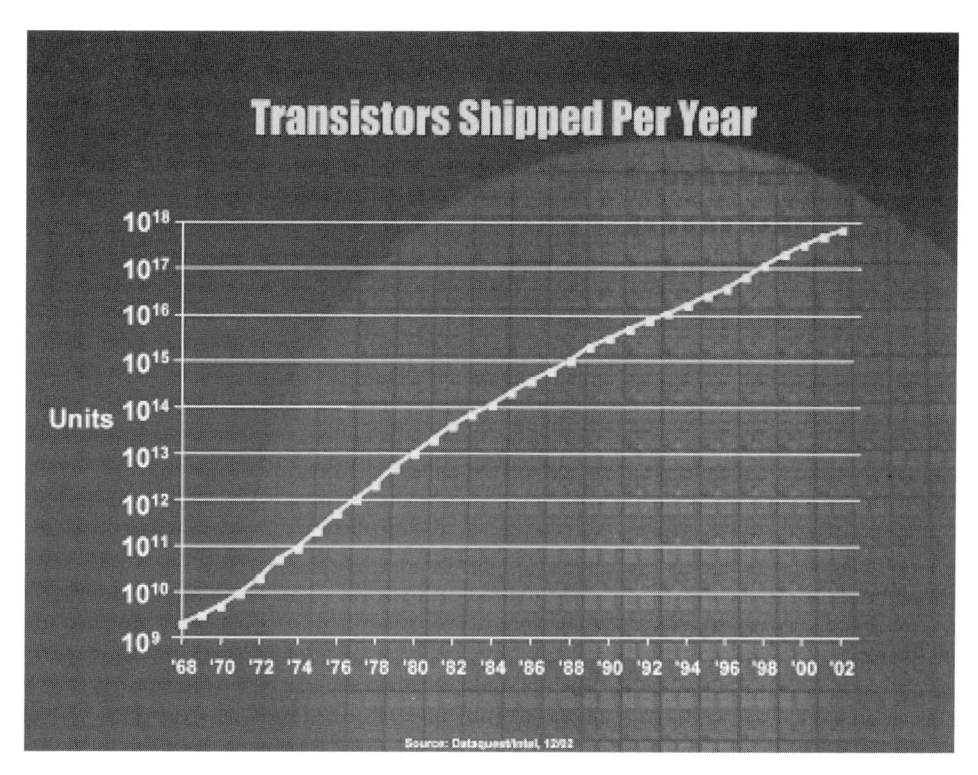

FIGURE 15.6. Plot of the total number of transistors shipped by Intel by year from 1968 to 2002. (Reproduced with kind permission of Intel, http://www.intel.com/research/silicon/mooreslaw.htm).

CMOS integrated circuit technologies are reaching their maturity as their critical dimensions approach 100 nm and will continue to be reduced. The cost of manufacturing CMOS ICs is increasing as the critical dimension is reduced. A time may come when quantum effects will limit any further reduction of the transistor as we now know it. The current projection is that the physical gate length of the transistor would approach 9 nm around the year 2017.[10] However, the integration trend is now also heading towards mixed technologies such and MEMS and MST. One point of view is that MEMS and MST are a fundamental next step in integration of the integrated circuit.

If you accept the premise that integrative systems are the key to commercial applications of MEMS, MST, and later for nanotechnology, then the important issue becomes what is the most cost effective method for integration. Presently, there are three main approaches to integration for MEMS technology:

- Embedded
- CMOS last
- CMOS first

The embedded approach is to interleave the micromachining fabrication process steps with the CMOS fabrication process steps. There are two big challenges here: on the technical side to develop the process and on the business side to convince a foundry to do it. A modern integrated circuit process is very complex and might involve 14 or more photo masks and hundreds of process steps. An integrated circuit foundry which offers a commercial CMOS process would have spent a great deal of time, effort, and money to optimize their process for the digital or analog performance of the circuits. Because of this, it is not an easy task to convince the foundry to make modifications to embed a MEMS device technology in their process unless (1) it could be fully dedicated to manufacturing that particular MEMS product at full capacity, (2) investment was available to develop the embedded MEMS process, and (3) the business plan to market the devices was convincing to the foundry and the investors. Still, if the market for the devices was large enough, then the cost per device in the end would be the least because of the economies of scale.

As pointed out earlier, the Analog Devices accelerometer is manufactured by what we call an embedded approach in their iMEMS[11] process. The primary application for the accelerometers is to detect the event of a collision to control the release of air bags in automobiles. The use of air bags in automobiles has resulted in a great advance in safety for automotive applications and is now required by law in every car that is sold in the US. It was reported that Analog Devices shipped its 100 millionth unit in 2002.[12] This is clearly an example of a high volume application where an embedded manufacturing approach is the best solution even though it might have a high initial cost.

In contrast, a low volume application or one where a prototype must be developed in order to demonstrate a concept would not be realizable by an embedded approach that required the development of a new process. The alternative is to leave the CMOS process completely intact and to add the MEMS process either before or after the full CMOS process.

In the CMOS last approach, a full MEMS process is run followed by a full CMOS process. This process was developed and demonstrated by Sandia National Labs.[13] Figure 15.7 is a drawing showing a cross section of a device that could be fabricated using this process. First, islands are etched into the silicon wafer where the MEMS devices will be fabricated. Then, a full surface micromachining MEMS process is carried out to fabricate the

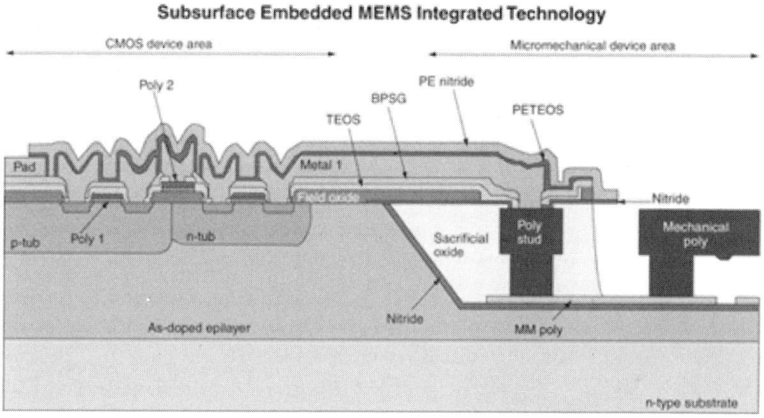

FIGURE 15.7. Drawing of the CMOS last approach developed by Sandia National Labs[13] (Reproduced with kind permission of Sandia National Laboratories, http://www.sandia.gov/mstc/technologies/micromachines/overview.html).

micromachined devices in these islands. After the MEMS process is completed, the islands are filled with a passivation oxide and then the entire wafer is planarized by a chemical mechanical process (CMP). From here, the full CMOS process is carried out to fabricate the circuits adjacent to the islands. Electronic connections are made between the CMOS circuitry and the MEMS devices. The last step is to release the MEMS devices by etching the sacrificial oxide in the islands. Figure 15.8 shows a system of gears that have been fabricated by Sandia's SUMMiT process.

Another interesting alternative is the CMOS first approach. There are a few implementations of this method but the two most widely used in the US may be the *cif*-MEMS (CMOS Integrated Circuit Foundry MEMS) through MOSIS[14] and the ASIMPS process.[15,16] These methods are also referred to a *post-processing* of CMOS ICs. The *cif*-MEMS approach is based on an opening[17] in the passivation film of the IC that is created by stacking all of the vias that are inherent in the standard process: the active area, the substrate or poly contact, the metal via, and the overglass cut for the bond pads. These mask levels are known in the Caltech Intermediate Form (CIF)[18] as CAA, CCA, CVA, and COG.

Using this technique, it is possible to fabricate openings with a minimum dimension of no less that 5 μm without any additional photolithographic steps. This method has been used to manufacture thermal-based structures such as thermal pixels, microhotplate gas sensors, convective accelerometers, flow sensors, vacuum sensors, and microchemical reactors. The method has also been used to demonstrate passive microwave elements such as spiral inductors, coplanar transmission lines, antennas, passive resonant filters, and power sensors. One commercial device that has a growing interest of late is the convective accelerometer shown in Figure 15.9.

One drawback of the *cif*-MEMS process is the limitation of the minimum feature size of the opening in the glass passivation. This is a major problem for manufacturing electrostatically actuated mechanical components since gaps between electrodes should be submicrometer in order to reach an acceptable range in voltage to actuate the device. A solution to this problem has been developed by the ASIMPS process from Carnegie Mellon

FIGURE 15.8. Photograph of Multiple Gear Speed Reduction Unit fabricated in Sandia's SUMMiT MEMS process (Reproduced with kind permission of Sandia National Laboratories, SUMMiT^{TM} Technologies, http://mems.sandia.gov/scripts/images.asp).

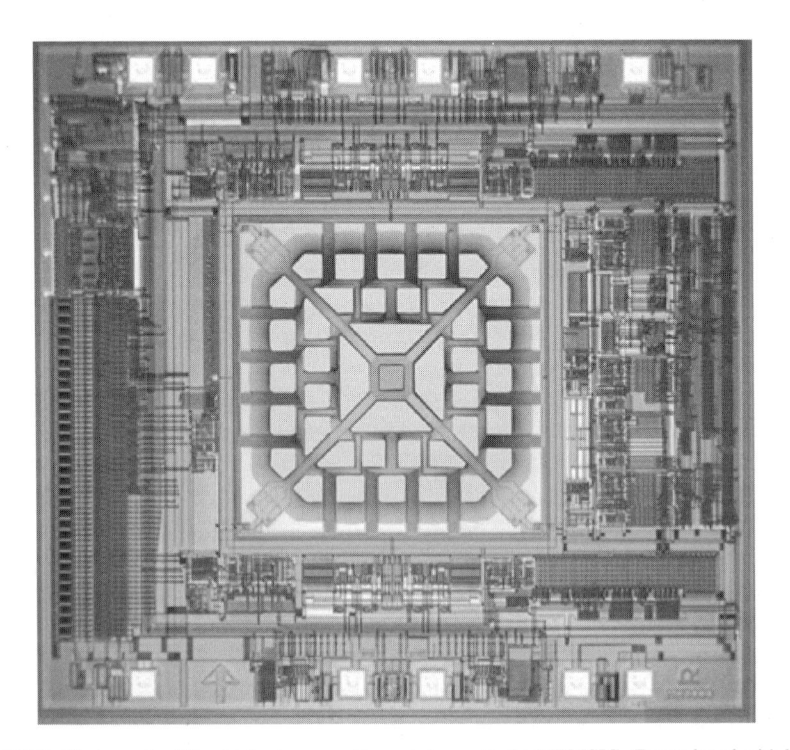

FIGURE 15.9. Photograph of the dual axis convective accelerometer by MEMSIC. (Reproduced with kind permission of MEMSIC, Inc., *http://www.memsic.com/memsic/*).

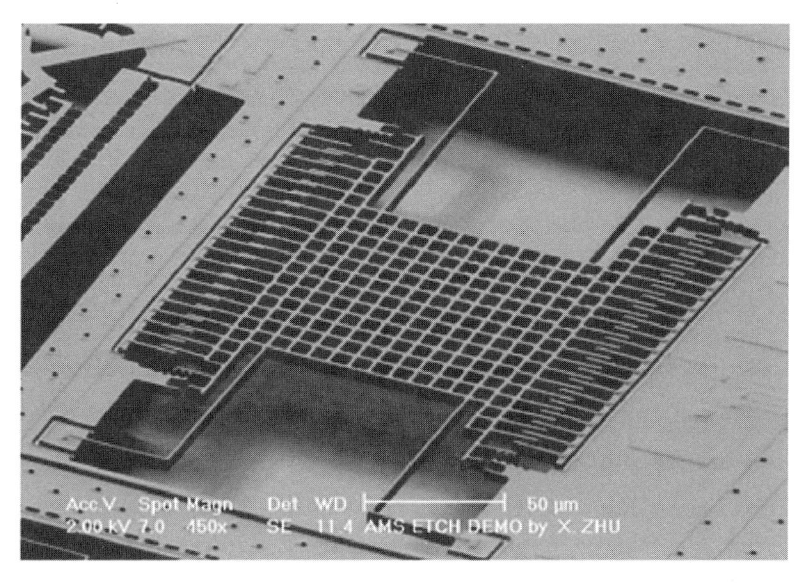

FIGURE 15.10. Scanning electron micrograph of electrostatically actuated resonator from the ASIMPS process. (Reproduced with kind permission of G. Fedder).

University. In this process, the top level of metalization (aluminum) is used as a mask for RIE of the oxide passivation layers below. In this case, the minimum feature size is determined by the top level metalization layer and can be patterned to a submicrometer resolution. After the RIE step is completed, an isotropic etch step using XeF_2 is used to undercut the structure. An example of an electrostatically actuated resonator fabricated using the ASIMPS process is shown in Figure 15.10.

15.4. OUTLOOK

There have been a number of market studies for MEMS and MST over the past several years that forecast rapid commercial growth for this technology. The recent SPC market study[19] reported an average growth rate of between 20%–30% per year. The SPC market study identified the following areas of growth in this industry, inertial measurement, microfluidics, optics, pressure measurement, and RF devices. Figure 15.11 shows a plot from the SPC market study per year between 1996 and 2003.

Another study by the European NEXUS group[20] reports a $14 billion world market in 1996, growing to $ 38 billion in 2002, and a potential of $ 68 billion by the year 2005. New products that MEMS and MST devices would grow were identified as:

- Read-Write Heads
- Ink jet Heads
- Heart Pacemeakers
- Biomedical diagnostics
- Hearing Aids
- Pressure Sensors

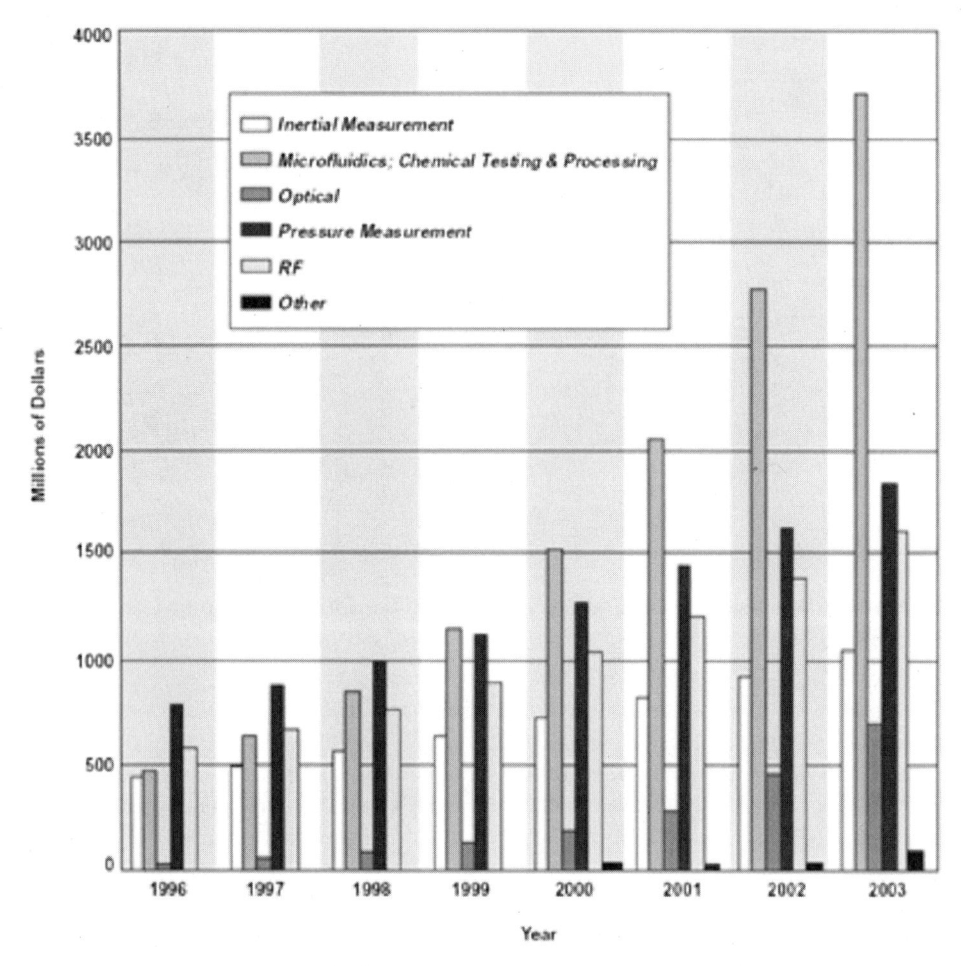

FIGURE 15.11. Areas of growth for MEMS and MST technologies. (From MEMS 1999 Emerging Applications and Markets by permission of System Planning Corporation.).

- Acceleration Sensors
- Gyroscopes
- Infrared Sensors, thermopiles, bolometers
- Flow Sensors
- Microdisplays
- Drug Delivery Systems
- MST devices for Chemical Analysis
- Optical Mouse
- Inclinometers
- Microspectrometers
- Optical MEMS
- RF MEMS
- Fingerprint biometric Sensors

- Micromotors
- Micro-optical Scanners
- Electronic Paper

Although there is some disagreement as to how great the growth rate will be, there is agreement in all of the market studies that there will be growth.

The applications for MEMS and MST are primarily for sensors. This could be due to the fact the "small" devices are not well suited to manipulate or actuate macroscale phenomena. The biggest potential market for actuators may be for optical applications; after all, it does not require much force to redirect light.

Perhaps this might change as we move towards commercial applications for nanotechnology. MEMS and MST would be well suited to actuate or position elements on a surface to interface to probe a nanoscale process analogous to the AFM. Other than by optical methods, MEMS or MST may be the only way to bridge macro to nano regimes.

QUESTIONS

1. What are the 4 mask levels required in a standard CMOS process to produce the stacked via passivation cut in the *cif*-MEMS process?
2. What is Moore's Law of integration?
3. What is the typical masking material and etch chemistry used in a CMOS compatible MEMS process to etch silicon?
4. What are the two predominate micromaching processes used in commercial MEMS manufacturing?

REFERENCES

1. For a more detailed review of microfabrication see for example: *Fundamentals of Microfabrication: The Science of Miniaturization*, Second Edition, Marc J. Madou, CRC Press, ISBN: 0849308267, 2002.
2. For a more detailed review on deposition methods see for example: *Physics and Technology of Semiconductor Devices,* Andrew S. Grove, John Wiley and Sons, ISBN: 0471329983, 1967.
3. H. Seidel, L. Csepregi, A. Heuberger, and H. Baumgartel, *Journal of the Electrochemical Society* **137** 3612 (1990).
4. O. Tabata, R. Asahi, H. Funabashi, K. Shimaoka, S. Sugiyama, *Sensors and Actuators* A, 34 (1) 51–57, July 1992.
5. R. Howe, J. Vac. Sci. Technol. B, 6 (6), 1809–1813, Nov 1988.
6. H. Guckel, J. J. Sniegowski, T. R. Christenson, S. Mohney, and T. F. Kelly, "Fabrication of Micromechanical Devices from Polysilicon Films with Smooth Surfaces," Sensors Actuators, 20 (1–2), 117–122, Nov 1989.
7. G. E. Moore, "Cramming More Components onto Integrated Circuits," Electronics 38, 114–117, April 19, 1965.
8. G. E. Moore, "Progress in Digital Integrated Electronics," Digest of the 1975 International Electron Devices Meeting, IEEE, New York, 1975, pp. 11–13.
9. G. Moore, "No Exponential is Forever ... But Forever can be Delayed," ISSCC 2003, February 9–13 2003, San Francisco, CA.
10. International Roadmap for Semiconductors 2001 Edition, Semiconductor Industry Association, http://public.itrs.net/

11. "Using iMEMS Accelerometers in Instrumentation Applications," Analog Devices tech. note.
12. Small Times, November 13, 2002.
13. J. Smith, S. Montague, J. Sniegowski, J. Murray, and P. McWhorter, "Embedded micromechanical devices for the monolithic integration of MEMS with COS," Proc. IEDM' 95, pp. 609–612, 1995.
14. J. C. Marshall, M. Parsmeswaran, M. E, Zaghloul, and M. Gaitan, "High-Level CAD Melds Micromachined Devices with Foundries," IEEE Circuits and Devices, 8 (6), 10–17, November 1992.
15. G. K. Fedder, S. Santhanam, M. L. Reed, S. C. Eagle, D. F. Guillou, M. Lu, L. Carley, "Laminated high-aspect-ratio microstructures in a conventional CMOS process," Sensors and Actuators A, 57 (2): 103–110 Nov 1996.
16. http://www.ece.cmu.edu/~mems/projects/asimps/index.shtml
17. M. Parameswaran, H. P. Baltes, L. Ristic, A. C. Dhadad, and A. M. Robinson, "A New Approach for the Fabrication of Micromechanical Structures," Sensors and Actuators, 19, 289–307, 1989.
18. C. Mead and L. Conway, "The Caltech Intermediate Form for LSI Layout Description," In Introduction to VLSI Systems, pages 115–127. Addison-Wesley, 1980.
19. MEMS 1999 Emerging Applications and Markets, © 1999 System Planning Corporation.
20. NEXUS Market Study, © 2000 Network of Excellence in Multifunctional Microsystems (NEXUS).

16

Nanoelectromechanical Systems

Stephane Evoy

Department of Electrical and Systems Engineering
University of Pennsylvania, Philadelphia, PA

Martin Duemling

Technische Universität Hamburg-Harburg, Germany

Tushar Jaruhar

Department of Electrical and Systems Engineering
University of Pennsylvania, Philadelphia, PA

16.1. OF MEMS AND NEMS

The integration of micromechanical structures, multifunctional materials, and micro-/opto-electronic circuits into single-chip *microsystems*[1] has opened new areas of application for silicon-based technology. These novel microsystems monolithically integrate transistor-based electronics with mechanical actuators, micro-pumps, valves, physical, biological and chemical sensors, potentially offering significant advantages with respect to size, power consumption, and to the lower cost inherent to batch processing. Such integration has allowed the development of single-chip systems that include the acquisition, processing, and communication of information. For instance, consider an airplane with a distributed network of embedded sensors, continually monitoring vibrations and stresses in critical components, and providing a stream of information on the operating conditions to a central processing core. Or imagine a bridge embedded during construction with numerous sensors that could alert engineers of developing weaknesses that could render the bridge unsafe in an earthquake or other disaster. Massively deployed networks of highly sensitive chemical/biological sensors could similarly provide early warning of accidents or attacks.

Integrative microsystems have also become firmly entrenched in the future of optical communications, as they offer the advantage of dynamically reconfigurable all-optical networks that can not be realized by any other technologies.[2] Integrated microsystems are also being aggressively pursued for single-chip wireless applications.[3,4,5,6] Indeed, the increasing demand for low-cost and high-efficiency wireless technologies would greatly benefit from such integration. Such needs have led to an explosion in the development of integrated circuit approaches in the RF/microwave area.

Nanomachining now routinely allows the fabrication of mechanical objects with lateral dimensions down to about 50 nm. Given their smaller volumes and higher surface-to-volume ratio, these *nanoelectromechanical systems* could enable a sensing technological platform offering enhanced sensitivities to outside forces or added mass. In addition, nanomachining also offers access to mechanical resonant frequencies now reaching the low GHz, a range of interest in single-chip RF signal processing systems. Finally, NEMS also allow access to length scales at which fundamental phenomena could uniquely be observed. In the millikelvin temperature range, the thermal energy of a GHz range NEMS would be smaller than the intrinsic quantum noise. Transducers that measure the position squared are only a factor of 100 away from the sensitivity that is needed to measure displacement in the quantum domain.[7,8]

This chapter presents an overview of the science and technology of nanoelectromechancial systems. We will first cover fabrication and detection technologies, with specific emphasis in the development of resonant devices reaching the RF range. We will then provide a primer of beam theories applicable to such devices, as well as a review of noise-inducing phenomena in NEMS. The chapter will then provide an overview of transduction and integration issues in NEMS, specifically concentrating on the interfacing between nanomechanical devices and quantum-based electronic devices. The chapter will close with an outlook towards molecular-scale and bottom-up approaches to NEMS design and integration.

16.2. SURFACE MACHINING AND CHARACTERIZATION OF NEMS

The fabrication of nanoelectrical and nanomechanical structures was first reported by Arney and McDonald[9] who developed a multi-level fabrication process involving the lateral oxidation of electron-beam patterned silicon structures. Initially targeting electrically-isolated nanometer-scale silicon conductors, the same group extended their process towards the fabrication of sub-micron mechanical actuators with integrated scanning tunneling probe tips.[10] Cleland and Roukes then reported a simplified version of this process that did not require the chemical vapor deposition of intermediate masking layers. This process also employed dry etching as final processing steps, greatly alleviating surface tension and stiction issues inherent to wet processing. The fabrication of 150 nm-wide mechanical resonators with a fundamental resonance frequency of 70.72 MHz were initially reported by this group.[11] Carr *et al.* followed suit with a further simplification of the nanomachining process that leveraged the already existing presence of sacrificial oxide in commercially available silicon-on-insulator (SOI) wafers.[12,13] In this process, a metal mask is first patterned on SOI using electron beam lithography, metal evaporation, and lift-off. The pattern is then transferred through the silicon layer using a plasma etching process that will stop at the buried oxide, thus defining the sidewalls of the released structure. The mechanical device is then released by underetching the oxide by immersion in HF. This simplified process allowed the fabrication of 50-nm structures, with resonant frequencies reaching 380 MHz (Figure16.1).[14] This "single buried sacrificial layer" approach was subsequently applied to a wide range of materials including GaAs,[15] Si_3N_4,[16] SiO_2,[17] poly-crystalline diamond[18] and SiC,[19] with GHz-range resonances recently reported in the latter.[20]

Two detection techniques have primarily been used for the analysis of NEMS resonant devices (Fig. 16.2a). In the interferometric approach,[12] a laser beam is illuminated onto the structures. Interferometric effects between the NEMS plane and the surface plane induce a displacement-sensitive modulation of the reflected signal that is picked up using an AC coupled photodetector. In this case, actual motion is induced either capacitively through application of AC signal between NEMS and substrate, or inertially by placing the chip onto a resonating piezoelectric stack. This technique offers reported sensitivities reaching the subnanometer range well into the 100s MHz range. In the magnetomotive approach,[11] the nanomechanical resonator is immersed in a intense magnetic field produced by a superconducting magnet. An AC signal is then fed to the device, creating a magnetomotive force that actuates the resonant motion (Fig 16.2b). Detection of motion is accomplished by either monitoring the reflected electrical signal, or through reverse effect by measuring any electromotive force generated by the motion across the terminals of another conductive path running across the device.

16.3. DYNAMICS OF NEMS

16.3.1. Euler-Bernoulli Model of Beams and Cantilevers

The theory described here assumes a homogeneous, straight and untwisted beam of constant cross-section (Fig 16.3). Assuming the beam thickness (d) and width (w) small compared to its length (l) reduces the system to a one-dimensional problem along the length of the beam. Furthermore, the normal stresses (σ_x and σ_y) in the lateral directions are considered negligible.[21] Finally, this model assumes a deflection smaller than the radius of gyration (K). If the maximum deflection approaches K, additional non-linear terms must be considered.[22] With these assumptions, the only remaining normal stress σ_z can be written as:

$$\sigma_z = kx \tag{16.1}$$

where k is a constant and $x = 0$ lies in the center of the beam. The total internal force has to be zero, and is given by:

$$F_{\text{int}} = \int_A \sigma_z \, dA = 0. \tag{16.2}$$

With no external momentum applied, the total bending moment is equal to the moment due to internal forces, which, from Eq. (16.1) are only non-vanishing in the y direction:

$$M = M_y = \int_A x\sigma_z dA = k \int_A x^2 dA. \tag{16.3}$$

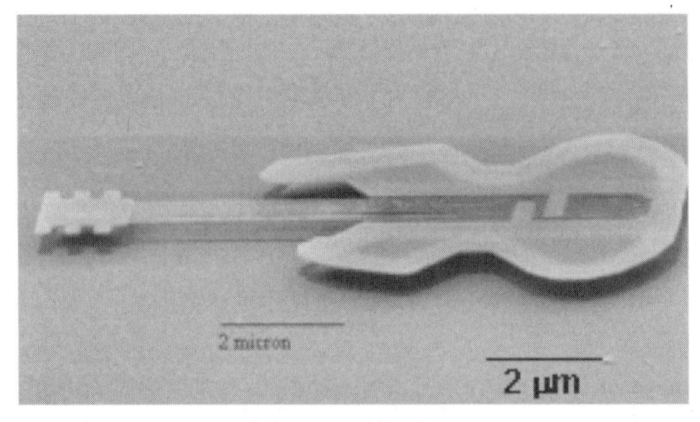

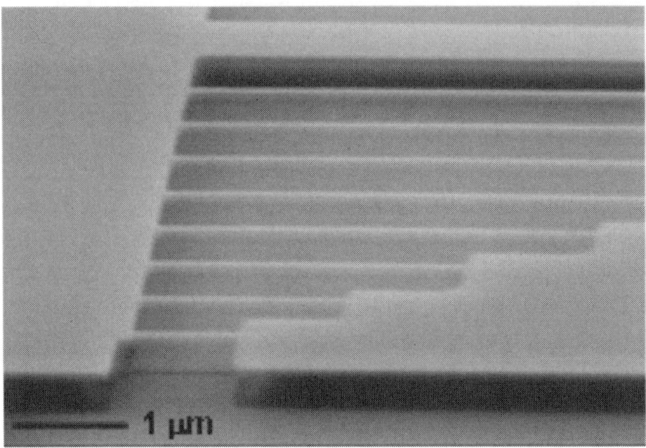

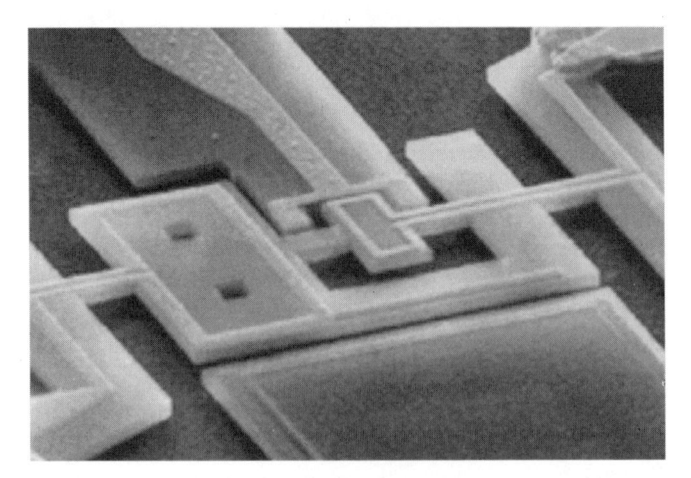

FIGURE 16.1. Si NEMS. Top: Silicon "nanoguitar" produced from machining of an SOI wafer. The strings consist of 50-nm wide suspended Si beams. (Reproduced with kind permission of D.W. Carr and H. G. Craighead.) Middle 100 nm wide Si beams produced from same process. (From Ref. 14 by permission of the American Institute o Physics.) Bottom: double-paddle electrometer produced from a similar SOI nanomachining process. (From A. N Cleland and M. L. Roukes, Nature **392**, 160 (1998) by permission of Nature Publishing Group.)

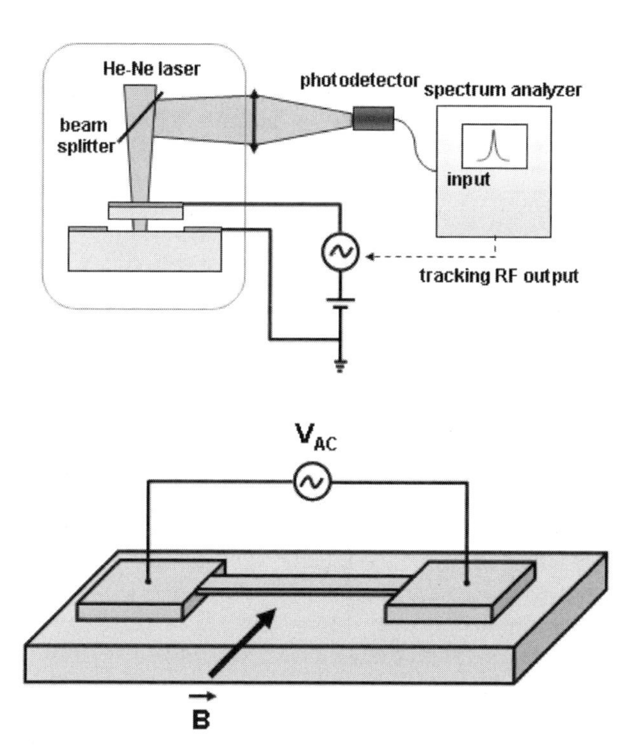

FIGURE 16.2. Actuation and detection of resonant NEMS. Top: Interferometric approach, Bottom: Magnetomotive approach.

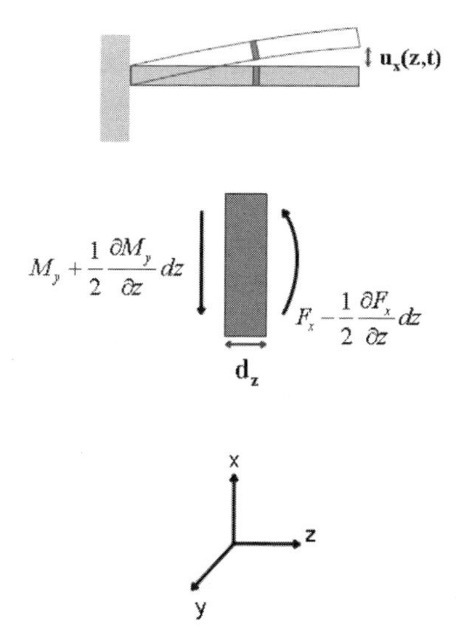

FIGURE 16.3. Flexural behavior of a straight beam and its stress distribution.

The moment of inertia is defined as:

$$I_y = \int_A x^2 dA. \tag{16.4}$$

By inspection of Eqs. (16.3)–(16.4):

$$k = \frac{M_y}{I_y} \tag{16.5}$$

and the stress of the cross section is given by:

$$\sigma_z = \frac{M_y x}{I_y}. \tag{16.6}$$

Using Hook's law, the strain is given by:

$$\varepsilon_z = \frac{\sigma_z}{E} = \frac{M_y x}{EI_y} \tag{16.7}$$

where E is Young's modulus.

If $u_x(z, t)$ is the displacement of the beam in x direction and the deflection is small ($du_x/dx \ll 1$), then the second derivative of the deflection is approximately the inverse of the radius of curvature r:

$$\frac{\partial^2 u_x(z, t)}{\partial z^2} \approx \frac{1}{r} \tag{16.8}$$

and the strain can be calculated as (Figure 16.4):

$$\varepsilon = \frac{dl - dl_0}{dl_0} = \frac{(r - x)d\theta - rd\theta}{r \sin d\theta} = \frac{-x}{r}. \tag{16.9}$$

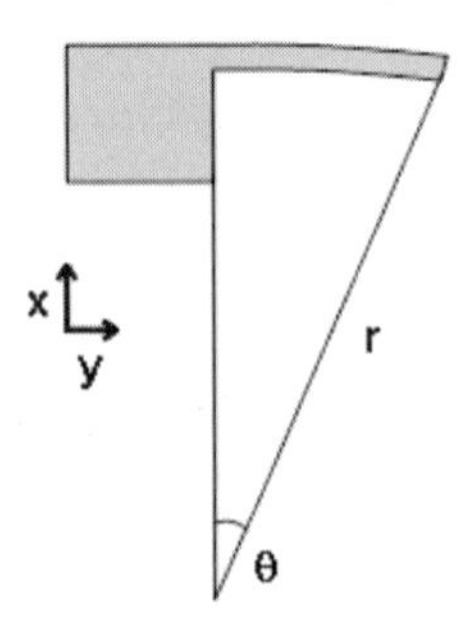

FIGURE 16.4. Strain in a cantilever.

Combining Eqs. (16.7)–(16.9) , we obtain the Euler-Bernoulli law of elementary beam theory:

$$M_y = -EI_y \frac{\partial^2 u_x(z,t)}{\partial z^2}. \tag{16.10}$$

If no external forces or bending moments are acting on the beam, the equation of motion becomes:

$$m \frac{d^2 u_x(z,t)}{dt^2} = \sum F_{\text{int}} \tag{16.11}$$

and the total momentum has to be zero

$$\sum M_{\text{int}} = 0. \tag{16.12}$$

The total force can be calculated from figure 3:

$$\sum F_{\text{int}} = \left(F_x + \frac{1}{2}\frac{\partial F_x}{\partial z}dz\right) - \left(F_x - \frac{1}{2}\frac{\partial F_x}{\partial z}dz\right) = \frac{\partial F_x}{\partial z}dz. \tag{16.13}$$

The sum of the bending moments is

$$\begin{aligned}\sum M_{\text{int}} = & \left(M_y + \frac{1}{2}\frac{\partial M_y}{\partial z}dz\right) - \left(M_y - \frac{1}{2}\frac{\partial M_y}{\partial z}dz\right) \\ & + \left(F_x + \frac{1}{2}\frac{\partial F_x}{\partial z}dz\right)\frac{dz}{2} - \left(F_x - \frac{1}{2}\frac{\partial F_x}{\partial z}dz\right)\left(-\frac{dz}{2}\right).\end{aligned} \tag{16.14}$$

Combining Eqs. (16.12)–(16.14) provides the relationship between the bending moment and the force:

$$F_x = -\frac{\partial M_y}{\partial z} \tag{16.15}$$

Using Eqs. (16.13)–(16.15), and with the mass of the beam given by $m = \rho A d_z$ (where ρ is the density of the beam, A is its cross section, and d_z its dimension along the z-direction), the equation of motion Eq. (16.11) becomes

$$\rho A \frac{d^2 u_x(z,t)}{dt^2} = -\frac{\partial^2 M_y}{\partial z^2}. \tag{16.16}$$

Now, substituting Eq. (16.10) leads to final equation of motion:

$$\rho A \frac{d^2 u_x(z,t)}{dt^2} + EI_y \frac{\partial^4 u_x(z,t)}{\partial z^4} = 0. \tag{16.17}$$

This harmonic linear 4th order differential equation can be solved using separation of variables.[21,23] Here, we are not concerned with the complete solution, but only the natural resonant frequency of the beam which is obtained using a Fourier transformation:

$$\rho A\,(i\omega)^2\, U_x(z,\omega) + EI_y \frac{\partial^4 U_x(z,\omega)}{\partial z^4} = 0, \tag{16.18}$$

which is rewritten as:

$$-\alpha^4 {\omega_x}^2 U_x(z,\omega) + \frac{\partial^4 U_x(z,\omega)}{\partial z^4} = 0 \tag{16.19}$$

with

$$\alpha = 4\sqrt{\frac{\rho A}{EI_y}}. \tag{16.20}$$

The solution of this differential equation is:

$$U_x(z,\omega) = B_1 \sin(\alpha z\sqrt{\omega}) + B_2 \cos(\alpha z\sqrt{\omega}) + B_3 \sinh(\alpha z\sqrt{\omega}) + B_4 \cosh(\alpha z\sqrt{\omega}). \tag{16.21}$$

In the case of a clamped-free cantilever, the boundary conditions at the clamped end are:

$$U_x(0,\omega) = 0 \qquad \frac{dU_x}{dz}(0,\omega) = 0, \tag{16.22}$$

while at the free end ($z = l$), no bending moments or shear forces act on the beam:

$$\frac{d^2 U_x(l,\omega)}{dz^2} = 0 \qquad \frac{dU_x^3(l,\omega)}{dz^3} = 0. \tag{16.23}$$

The first two boundary conditions forces $B_2 = B_4$ and $B_1 = -B_3$. In addition, application of the last two conditions reduces the solution to:

$$\frac{2 + 2\cos(\alpha \mathrm{l}\sqrt{\omega})\cosh(\alpha l\sqrt{\omega})}{\sin(\alpha \mathrm{l}\sqrt{\omega}) - \sinh(\alpha l\sqrt{\omega})} = 0. \tag{16.24}$$

A non-trivial solution is given if:

$$\cos(\alpha l\sqrt{\omega})\cosh(\alpha l\sqrt{\omega}) = -1. \tag{16.25}$$

This equation has no analytical solution, but can be solved numerically through the following substitution:

$$\beta = \alpha l\sqrt{\omega}. \tag{16.26}$$

The values for β_i can be found in Table 2.1. The natural resonant frequencies can be calculated from Eqs. (16.19)–(16.20):

$$\omega_i = \frac{\beta_i{}^2}{l^2}\sqrt{\frac{EI_y}{\rho A}} \tag{16.27}$$

The moment of inertia of a beam with circular cross section is given by:

$$I_y = \frac{\pi d^4}{64} \tag{16.28}$$

where d is the diameter of the beam. The inertia of a rectangular cross section is:

$$I_y = \frac{wt^3}{12}. \tag{16.29}$$

For a beam clamped on both sides the boundary conditions are:

$$U_x(0,\omega) = 0 \qquad \frac{dU_x}{dz}(0,\omega) = 0 \qquad U_x(L,\omega) = 0 \qquad \frac{dU_x}{dz}(L,\omega) = 0 \tag{16.30}$$

and for a beam free on both sides the boundary conditions are:

$$\frac{d^2U_x(0,\omega)}{dz^2} = 0 \qquad \frac{dU_x^3(0,\omega)}{dz^3} = 0 \qquad \frac{d^2U_x(l,\omega)}{dz^2} = 0 \qquad \frac{dU_x^3(l,\omega)}{dz^3} = 0. \tag{16.31}$$

The solution method is similar to the one for the cantilever. The final solution is the same for the clamped-clamped and the free-free beam and only differs from the cantilever in the factor β_i, (Table 16.1).

Coulomb damping (i.e., damping proportional to displacement) is then introduced in the Euler-Bernouilli derivation by defining a complex Young's modulus:

$$\hat{E} = E\left(1 + i\frac{1}{Q_{\text{Coulomb}}}\right). \tag{16.32}$$

TABLE 16.1. Solution of the equation of motion for a cantilever beam

i	β_i^2 clamped-free cantilever	β_i^2 clamped-clamped beam
1	3.516	22.373
2	22.034	61.678
3	61.701	120.903
4	120.912	199.860
5	199.855	298.526

The differential equation for this problem then becomes:

$$\rho \frac{\partial^2 u_x(z,t)}{\partial^2 t} + E\left(1 + i\frac{1}{Q_{\text{Coulomb}}}\right) I_y \frac{\partial^4 u_x(z,t)}{\partial z^4} = F(\omega, t). \tag{16.33}$$

If a time-dependent force is applied to the cantilever, the amplitude response (A_n) of the Coulomb-damped oscillator becomes:

$$A_n = \frac{|F_0 \alpha_n|}{|m|\sqrt{\left(\omega_n^2 - \omega^2\right)^2 + \left(\frac{\omega_n^2}{Q_{\text{Coulomb}}}\right)^2}} \tag{16.34}$$

where ω_n is the resonant frequency of the cantilever and α_n is a constant depending on the boundary condition (i.e., the form of the applied force). This equation has the same form as the solution of the simple harmonic oscillator.

16.3.2. Mechanical Resonators as Forced Harmonic Oscillators

Previous section showed that the equations of motion of a beam resonator are similar to those derived for the harmonic oscillator. In the ideal situation where no damping takes place:

$$M_{\text{eff}}\ddot{x} + k_{\text{eff}}x = F(\omega t) \tag{16.35}$$

where M_{eff} and k_{eff} are the effective mass and restoring stiffness of the system. For a simple beam, these quantities are equal to:

$$M_{\text{eff}} = \rho A l \tag{16.36}$$

$$k_{\text{eff}} = \frac{\beta_i^4 E I_y}{l^3} \tag{16.37}$$

where β_i depends on the mode of vibration (cf. Table 1).
The solution of this equation is:

$$x(t) = A_1 \sin\left(\sqrt{\frac{k_{\text{eff}}}{M_{\text{eff}}}}t\right) + A_2 \cos\left(\sqrt{\frac{k_{\text{eff}}}{M_{\text{eff}}}}t\right) \tag{16.38}$$

leading to a resonant frequency of:

$$\omega_0 = \sqrt{\frac{k_{\text{eff}}}{M_{\text{eff}}}}. \tag{16.39}$$

As discussed in previous section, coulomb damping (ie damping proportional to displacement) is introduced by defining a complex spring constant:[24]

$$\hat{k} = k\left(1 + i\frac{1}{Q_{\text{Coulomb}}}\right). \tag{16.40}$$

Viscous damping (i.e., damping proportional to velocity) is then introduced by adding a frictional force proportional to velocity:[25]

$$F_{\text{friction}} = \gamma \dot{x}(t) = \frac{m\omega_0}{Q_{\text{viscous}}}\dot{x}(t) \tag{16.41}$$

The equation of motion with a sinusoidal driving force becomes:

$$\ddot{x}(t) + \frac{\omega_0}{Q_{\text{viscous}}}\dot{x}(t) + \omega_0^2\left(1 + i\frac{1}{Q_{\text{Coulomb}}}\right)x(t) = \frac{F_0}{m}e^{i\omega t}. \tag{16.42}$$

The full solution of this equation is:

$$x(t) = \frac{F_0}{m}\left(\left(\frac{\omega_0\omega_{\text{f}}}{Q_{\text{viscous}}} - \frac{\omega_0^2}{Q_{\text{coulomb}}}\right)^2 + \left(\omega_f^2 - \omega_0^2\right)^2\right)^{-1/2} e^{i(\omega t + \psi)}. \tag{16.43}$$

The resulting mechanical resonance has an amplitude of:

$$A(\omega_{\text{f}}) = \frac{F_0}{m}\frac{1}{\sqrt{\left(\omega_0^2 - \omega_{\text{f}}^2\right)^2 + \left(\frac{\omega_0^2}{Q_{\text{coulomb}}} + \frac{\omega_0\omega_{\text{f}}}{Q_{\text{viscous}}}\right)^2}}. \tag{16.44}$$

and a phase shift of:

$$\psi = \tan^{-1}\left(\frac{\omega_0\omega_{\text{f}}}{Q_{\text{viscous}}\left(\omega_0^2 - \omega_{\text{f}}^2\right)} - \frac{\omega_0^2}{Q_{\text{coulomb}}\left(\omega_0^2 - \omega_{\text{f}}^2\right)}\right). \tag{16.45}$$

In the limit of a high quality factor (Q), the amplitude can be simplified as:

$$A_{\max}(\omega_f) = \frac{F_0}{k}\frac{1}{\sqrt{4\left(\frac{\omega_0 - \omega_{\text{f}}}{\omega_0}\right)^2 + \left(\frac{1}{Q_{\text{tot}}}\right)^2}} \tag{16.46}$$

with

$$\frac{1}{Q_{\text{tot}}} = \frac{1}{Q_{\text{coulomb}}} + \frac{1}{Q_{\text{viscous}}}. \tag{16.47}$$

Equation (16.46) has the form of a Lorentzian function of maximum amplitude

$$A_{\max} = \left| \frac{F_0}{k} \right| Q_{\text{tot}} \tag{16.48}$$

and of width (measured at half maximum amplitude) given by:

$$Q_{\text{tot}} = 1.73 \frac{f_0}{\Delta f_{\text{halfwidth}}} \tag{16.49}$$

$$Q_{\text{tot}} = \sqrt{3} \frac{\omega_0}{\omega_{\text{halfbandwith}}} = 1.73 \frac{f_0}{f_{\text{halfbandwith}}}. \tag{16.50}$$

16.3.3. Non-Linear Effects: Beam Stretching

Previous sections assumed small oscillations in which the restoring force is independent of displacement. This approximation no longer holds for large displacements where beam stretching can no longer be neglected. This nonlinear stretching is introduced using textbook approaches of suspended bridge cables.[26] When the beam is stretched down significantly from its rest position, its total length must change to accommodate the displacement. This model approximates that the beam remains straight while being stretched down by a end-displacement δ to an angle ϕ. The restoring force provided by this stretching is (Figure 16.5):

$$F_{\text{stretch}} = \varepsilon E a b \phi \tag{16.51}$$

where ε is the strain in the beam, ab is the cross-sectional area of the beam, and ϕ is its angular displacement. The strain is computed from geometrical arguments:

$$\varepsilon = \frac{\Delta L}{L} = \frac{(L + \Delta L) - L}{L} = \frac{L\left(\dfrac{1}{\cos\phi} - 1\right)}{L} \approx \frac{1}{\sqrt{1-\phi^2}} - 1 \approx \frac{\phi^2}{2} \tag{16.52}$$

which we replace in 16.51:

$$F_{\text{stretch}} = Eab\frac{\phi^3}{2}. \tag{16.53}$$

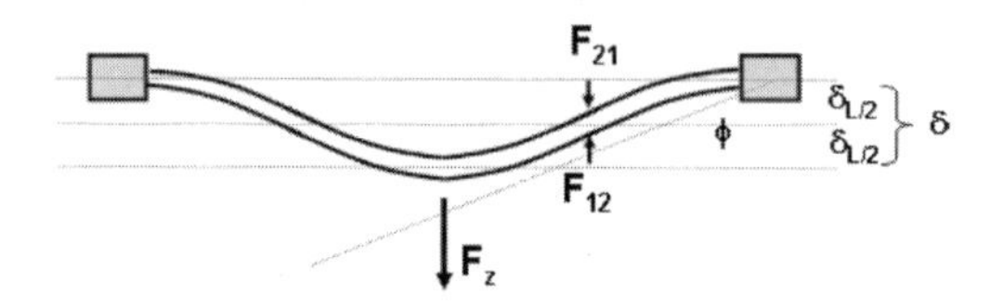

FIGURE 16.5. Restoring force applied to a stretched beam.

Given that $\delta \sim \phi L$, we obtain:

$$F_{\text{stretch}} = \frac{Eab}{2ML^3} \cdot \partial^3. \tag{16.54}$$

The equation of motion becomes:

$$M_{\text{eff}}\ddot{\partial} + k_1\partial + k_3\partial^3 + \frac{\omega_o\dot{\partial}}{Q_{\text{tot}}} = F(\omega t) \tag{16.55}$$

with $k_3 = Eab/2ML^3$. Such departure from the linear regime will in turn depart the frequency response from the symmetric Lorentzian response described by Eq. (16.46). Above a critical amplitude, the response actually becomes bi-stable in which three solutions may exist at a given frequency. This introduces a hysteretic shape of the frequency response. Such non-linear dynamics in nanomechanical oscillators were observed by Evoy *et al.*, in nanomechanical paddle oscillators of the type shown in Fig 16.6.[26] These devices were shown to have two fundamental modes of motion labeled "flexural" and "torsional", in which the micron-scale paddle would oscillate in an up-down or angular motion about the supporting nanoscale beam, respectively. In the case of the flexural motion, such devices can be made to oscillate at sufficient amplitude in order to observe non-linear beam stretching (Fig 16.7).

16.3.4. Non-Linear Effects: Tunability and Parametric Amplification

Sections 16.3.2 and 16.3.3 also assumed the actuating force $F(\omega t)$ to be a solely a function of time with no dependence on coordinate of motion. This situation does not hold

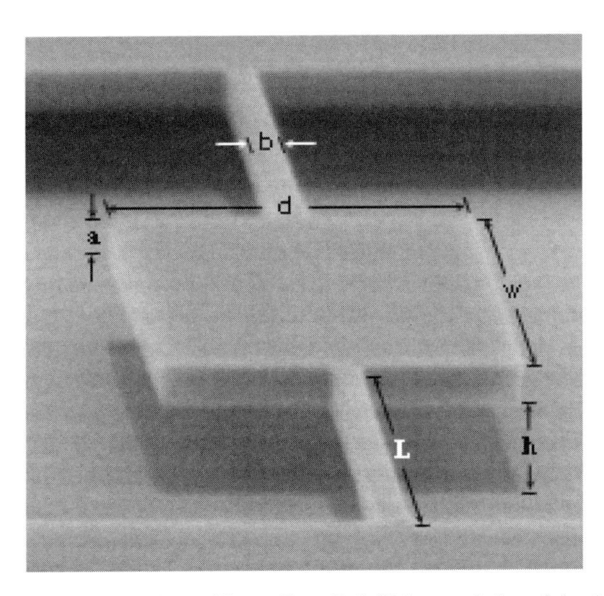

FIGURE 16.6. Nanomechanical paddle oscillator. (From Ref. 26 by permission of the American Institute of Physics.)

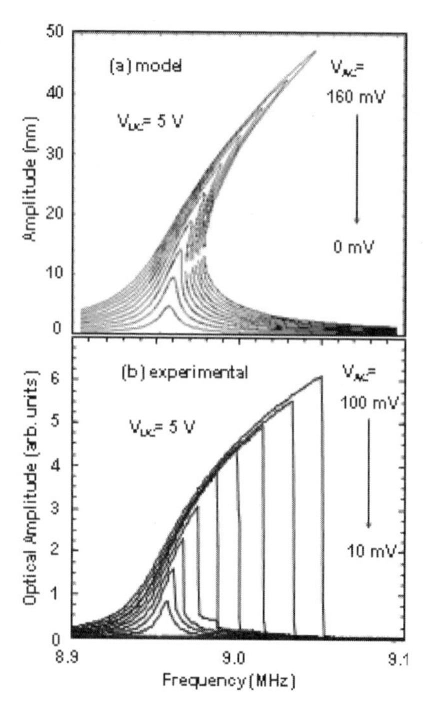

FIGURE 16.7. Modeling and experimental observation of onset of non-linear regime in flexural paddle resonators. (From Ref. 26 by permission of the American Institute of Physics.)

for all types of geometries and actuation methods. For instance, consider a torsional paddle capacitively actuated about its support beam through application of an AC signal between paddle and substrate (16.7). The general equation of motion of this torsional motion is written as:[26]

$$I_{\mathrm{eff}}\ddot{\theta} + \kappa_{\mathrm{mech}}\theta + \frac{\omega_o\dot{\theta}}{Q} = \tau(\omega t, \theta). \tag{16.56}$$

The right-hand external torque is computed by solving the Coulomb equation for the electrostatic force and by integrating over the area of the deflected paddle:

$$\tau\,(\omega t, \theta) = \frac{\varepsilon_0 V^2(\omega t) w}{2\theta^2}\left[\ln\left(\frac{h + d\sin\dfrac{\theta}{2}}{h - d\sin\dfrac{\theta}{2}}\right) + \frac{h}{h + d\sin\dfrac{\theta}{2}} - \frac{h}{h - d\sin\dfrac{\theta}{2}}\right]. \tag{16.57}$$

where h is the height of the paddle above surface, d is its length perpendicular to the support beams, and w its width along the beam axis. This expression can be closely approximated by a third order expansion:

$$\tau\,(\omega t,\theta)\approx\varepsilon_0 V^2(\omega t)w\left[\left(\frac{d^3}{12h^3}\right)\theta+\left(\frac{d^5}{40h^5}-\frac{d^3}{96h^3}\right)\theta^3\right]. \tag{16.58}$$

The external actuation is therefore asymmetric about origin, with its lower order term on the order of θ. By neglecting the contribution from the third order, the total equation of motion becomes:

$$I\ddot{\theta}+\kappa_{\mathrm{mech}}\theta+\frac{I}{\omega_0 Q}\dot{\theta}=\left[\varepsilon_0 w\frac{d^3}{12h^3}V^2(\omega t)\right]\cdot\theta \tag{16.59}$$

which simplifies to:

$$I\ddot{\theta}+(\kappa_{\mathrm{mech}}-\kappa_{\mathrm{ext}})\theta+\frac{I}{\omega_0 Q}\dot{\theta}=0 \tag{16.60}$$

with:

$$\kappa_{\mathrm{ext}}=\left[\varepsilon_0 w\frac{d^3}{12h^3}V^2(\omega t)\right]. \tag{16.61}$$

This form represents a conceptual departure from the usual derivation of forced oscillations inasmuch as the external drive is better understood as an external periodic modulation of the stiffness of the system rather than a time-dependant force. Such form opens possibility of inducing tunability and parametric amplification of the nanomechanical resonance. For instance, consider situation where a DC bias is added on top of the V_{AC} signal that actuates the motion. As shown by Eqs. (16.59)–(16.61), this added bias will result in a decrease of the overall stiffness of the system, resulting in a shift of resonant frequency:

$$\frac{f_0'}{f_0}=\sqrt{1-\frac{\alpha}{\kappa_{\mathrm{mech}}}V_{\mathrm{DC}}^2} \tag{16.62}$$

where α contains the geometry of the system. In the specific case of the paddle oscillators described in last section:

$$\alpha=\varepsilon_0 w\frac{d^3}{12h^3}. \tag{16.63}$$

Experimental observation of the effect is shown if Fig. 16.8.

These equations were derived for a paddle driven capacitively through application of signal between resonator and substrate. Consider the general case where a torsional resonator is driven at resonance, regardless of nature of actuating drive:

$$\tau(\omega t,\theta)=\tau_{\mathrm{o}}\cos(\omega_{\mathrm{o}}t+\theta) \tag{16.64}$$

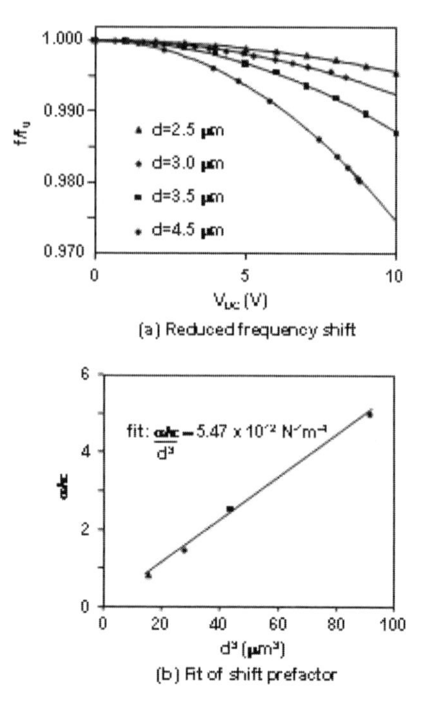

FIGURE 16.8. Experimental observation of DC tunability of torsional paddle nanoresonators. (From Ref. 26 by permission of the American Institute of Physics.)

and where the stiffness of the system is modulated at exactly twice the resonant frequency:

$$\kappa_{\text{ext}}(t) = \kappa_{\text{ext}}^{o} \cos(2\omega_o t) \tag{16.65}$$

where κ_{ext}^{o} is the amplitude of this external modulation, and θ is the phase angle between this external modulation and the independent actuating drive. Theory predicts[27,28] that the system will respond with an amplitude of oscillation A_0 given by

$$A_0 = \frac{\tau_0 Q}{k}\left[\frac{\cos^2\theta}{(1+Q\kappa_0'/2\kappa)^2} + \frac{\sin^2\theta}{(1-Q\kappa_0'/2\kappa)^2}\right]^{1/2}. \tag{16.66}$$

The first factor is the right hand side, $\tau_0 Q/k$, is the normal resonant response of a force oscillator in the absence of the parametric drive. The factor in parenthesis acts as a phase-dependent gain created by the modulation of the stiffness of the system. When the phase is $\theta = 0$, the system will be deamplified. When the phase is $\theta = \pi/2$, a positive gain exists, and may actually diverge to infinity when $\kappa_o' = 2\kappa/Q$.

Such parametric amplification was first reported in larger micromechanical resonators by Turner *et al.*[27] and was subsequently observed in capacitively-actuated nanomechanical paddle oscillators by Carr *et al.*[28] A similar effect was then induced in optically-pumped disk-shaped microresonators.[29] In this case, laser heating induced thermal stresses that modulated the effective spring constant, providing a mechanism for parametric amplification and self-excitation. Quality factors were observed to increase by an order of magnitude from $Q = 10\,000$ to $Q = 110\,000$.

16.4. DISSIPATIVE PROCESSES IN NEMS

16.4.1. Introduction

As previously mentioned, nanoelectromechanical systems offer a prospective platform for highly-sensitive mass- and force-sensors. In addition, access to RF-range mechanical resonances in a material compatible with transistor technologies would open possibilities for the deployment of NEMS as nanomechanical-based signal processors in single-chip wireless systems. Both these applications require low noise levels and high spectral purity.

The quality factor Q introduced in Section 16.3 is defined as:

$$\frac{1}{Q} = \frac{1}{2\pi} \frac{\text{Energy loss per cycle}}{\text{Total elastic energy}}. \tag{16.67}$$

A system with lower dissipation will have larger Q, thus sharper resonant response. While a noisy system showing high dissipative process will have lower Q, and broader resonant response. In general, different energy dissipation mechanisms contribute to the quality factor. The overall Q-factor of a system can then be found as the sum of the inverses of the individual Q-factors:

$$\frac{1}{Q_{\text{tot}}} = \sum \frac{1}{Q_i}. \tag{16.68}$$

We present here a review of these different dissipative processes that are known to affect nanomechanical resonators.

16.4.2. Atmospheric Damping

The impact of atmospheric damping is separated into two regions: molecular and viscous. At higher pressure, the gas acts as a viscous fluid. In general, the interaction of the beam with its surroundings is characterized by a drag force F_{drag}.[30] This force has the following form

$$F_{\text{drag}} = (\beta_1 + i\beta_2)\,\dot{u}_x = \beta_1 \dot{u}_x - \frac{\beta_2}{\omega}\ddot{u}_x = \gamma_1 L \dot{u}_x - \frac{\gamma_2 L}{\omega}\ddot{u}_x. \tag{16.69}$$

where $\dot{u}_x$ is the velocity and L the length of the beam. It can be shown that γ_1 is proportional to the quality factor and γ_2 is proportional to the frequency shift. Assuming that the air is incompressible and that the Reynolds number is small (no turbulences) the force on the surface can be calculated using the Navier-Stokes and the continuity equations. The beam will be approximated with a row of spheres vibrating independently from each other. The resulting force on the surface is then given as:

$$F_{\text{drag}} = \left[6\pi\mu r\left(1+\frac{r}{\delta}\right) - i\frac{2}{3}\pi r^3\rho_{\text{gas}}\left(1+\frac{9}{2}\frac{\delta}{r}\right)\omega\right]v \tag{16.70}$$

where μ the dynamic viscosity of the medium and ρ_0 the density of the gas.
For the ideal gas:

$$\rho_0 = \frac{M}{RT}p. \tag{16.71}$$

In addition, δ is the region around the beam where the motion of the gas is turbulent. It is approximated by:

$$\delta = \left(\frac{2\mu}{\rho_0\omega}\right)^{1/2}. \tag{16.72}$$

The Q factor related to this drag force then becomes:

$$Q = \frac{\rho_{\text{beam}}A}{\gamma_1}\omega_0 = \frac{\rho_{\text{beam}}wtl}{6\pi\mu r\left(1+\frac{r}{\gamma}\right)}\omega_0 \tag{16.73}$$

$$\frac{\Delta\omega}{\omega_0} = -\frac{1}{2}\frac{\gamma_2}{\rho_{\text{beam}}A} = -\frac{\pi r^3\rho_0}{3\rho_{\text{beam}}lwt}\left(1+\frac{9}{2}\frac{\delta}{r}\right) \tag{16.74}$$

where the radius r can be approximated with the width of the beam.

At low pressures, the collisions of the air molecules with the resonator surface are considered independent from each other. The force on the resonator is due to these collisions. The damping parameter γ_1 is proportional to the air pressure p and the width of the beam w

$$\gamma_1 = \left(\frac{32M}{9\pi RT}\right)^{1/2} wp \tag{16.75}$$

$$\gamma_2 = 0$$

where R is the gas constant and M the mass of the gas molecules ($M_{\text{air}} \approx 29$ g/mol). The quality factor becomes:

$$Q = \frac{\rho_{\text{beam}}A\omega_0}{\gamma} = \left(\frac{9\pi RT}{32M}\right)^{1/2}\frac{t\rho_{\text{beam}}\omega_0}{p} \tag{16.76}$$

where ρ is the density of the beam material. Blom *et al.*[30] found the transition from the viscous to the molecular regime around 1 torr and from molecular to the intrinsic regime between 10^{-2} and 10^{-3} torr for a cantilever with a cross section in the mm^2 range. While atmospheric damping therefore dominates NEMS performance in ambient condition, experimental work in this area usually involves operation in the 10^{-4} to 10^{-9} torr range, where such damping is considered to be small or negligible.

16.4.3. Clamping

Real mechanical structures never being perfectly rigid, energy can be dissipated from the resonator to the support structure where local deformations and microslip can occur. The energy loss per cycle due to this interaction is proportional to the inverse of the friction force at the contact area. While it is often not possible to increase the rigidity of the support, energy losses can be reduced with an optimized design. For instance, Olkhovets *et al.*[31] showed that a MHz range double beam with a flexible support improves the resonance quality by approximately 30% compared to a simple cantilever. Wang *et al.*[32] created a "free-free" beam with four torsional supports beam, each with a length of a quarter of the resonant frequency of the free-free beam. Since as a result of this design the torsional beams do not move at their support, no energy is dissipated into the support structure. Similarly, Roukes *et al.*[33] used a torsional double resonator to decouple the center resonator from its support.

16.4.4. Stress Relaxation

Stress relaxation results from a transition between two stable local structural configurations. This behavior can be illustrated by considering two states in a system with slightly different energy levels, separated by a potential barrier of height H. Before a stress is applied, the system is in its minimum energy state. If an external stress is applied, the energy levels change their position (Figure 16.9), and the other state is energetically more favorable. If the system can overcome the energy barrier H, a transition from state one to state two take place. The system is relaxed and the energy difference between the two states is lost.

Relaxation Through Motion of Point Defects Point defects in crystals produce a local non-uniform stress distribution.[34] It is possible to associate symmetry with each defect. If the defect symmetry is lower than the crystal symmetry, an elastic dipole originates.

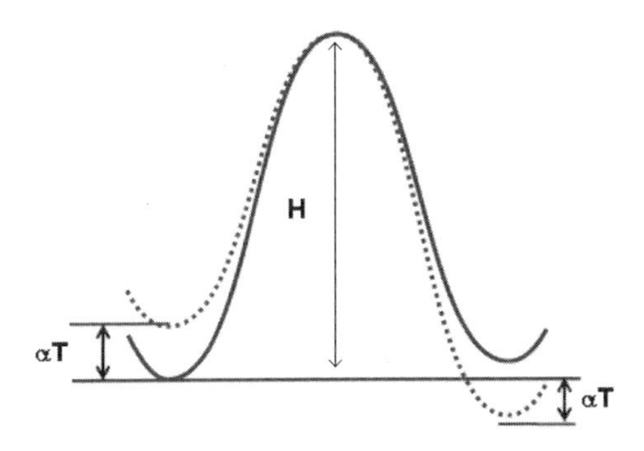

FIGURE 16.9. Influence of an applied stress to the energy levels of a system.

Consequently, there will be an interaction between the dipole and a homogeneous applied stress. If the related activation energy is surmounted as a result of this interaction, a reordering of the dipoles takes place. In bcc materials, the dipole is introduced by interstitial impurities (Snoek relaxation). The interstitial atoms occupy the octahedral sites and have a four-fold symmetry (tetragonal symmetry). When a dipole reorientates, the interstitials jump to the neighboring octahedral site. It can be found that the activation energy that controls the relaxation time is similar to the one for the migration of the interstitials.

Relaxation Through Motion of Dislocations The simplest model to describe dislocation relaxation is to consider the dislocation as a loop that is firmly pinned at its end, and executes vibration under an external periodical stress (like a vibrating string). A major limitation of this model is that stresses and strains outside the slip plane are neglected. The pinning points are assumed fixed, which is only true at low temperatures, where the diffusion is negligible. Dislocation damping can be nonlinear and shows strong amplitude dependence.

Relaxation Through Motion of Grains The grains of a polycrystalline material will slide back and forth under application of a periodical stress, resulting in a dissipation of energy.[35] The grain boundary consists of many disordered atom groups, separated by regions where the atoms fit (Figure 16.10). The disordered atoms can pass over each other by squeezing the atoms around them. This displacement has components parallel and perpendicular to the grain boundary and is therefore accomplished by a combination of sliding and migration. Migration is a volume diffusion process and sliding is a grain boundary diffusion process. Hence the activation energy for grain boundary relaxation is expected to be somewhere between the two activation energies.

The coefficient of viscosity of grain boundary sliding is similar to that of molten metal.[36] For larger grains, the relaxation peak shifts towards higher temperatures, but its height stays approximately the same.[37] Since a larger displacement can take place before the sliding is

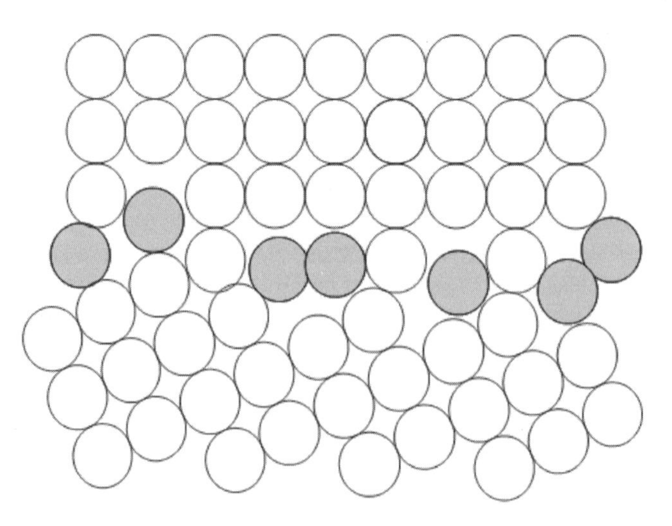

FIGURE 16.10. Schematic illustration of a disordered atom group model.

blocked by the grain edge for larger grains, the relaxation time increases. In a given volume the grain surface (S_{grain}) is proportional to the reciprocal of the grain size ($S_{grain} = \alpha/d_{grain}$).

The energy dissipation for each grain (E_{grain}) is proportional to the displacement at the grain boundary, and the maximum displacement is proportional to the grain size ($E_{grain} = \beta d_{grain}$). The total dissipated energy (ΔE) in a given volume is then

$$\Delta E = E_{grain} S_{grain} = \alpha\beta, \tag{16.77}$$

where α and β are constant. Equation (16.77) suggests that the total dissipated energy is independent of the grain size.

Experimental work in surface machined NEMS usually involves single crystal layers in which dissipation due to defect motion would be expected to be negligible. However, process-induced damage (through plasma processing, for instance) could potentially be a significant factor affecting NEMS performance. Such potential impact has yet to be thoroughly investigated.

16.4.5. Phonon-Phonon and Phonon-Electron Scattering

Phonon-phonon scattering takes place if the oscillatory wavelength is considerably larger than the mean free path of phonons. The phonons related to the oscillation (sound-waves) of the structure can interact with the phonons related to thermal vibration.[38] The values for ωQ are found to be in the order of 10^{14} to 10^{12} for aluminum at 4.2 K and room temperature respectively. In addition, the external periodical force will induce the positive ions to oscillate, resulting in an internal AC electric field. In metallic systems containing a large density of free electrons, this internal electric field will induce an oscillatory motion of the free electron gas. Given that an electron gas can be approximated as a viscous fluid, motion of this fluid will generate internal viscous losses of energy. For aluminum the ωQ is found to be 3×10^{11} at 273 K and 5×10^{13} at 4.2 K.

A semiconductor at non-zero temperature will also contain a finite density of electrons and holes due to thermal excitation of electrons from the valence band to the conduction band. The external strain will modulate the band structure in such a way that electrons may hop periodically from one band to another. Such transition results in energy dissipation. Intrinsic to the materials being used, these phenomena will limit the largest attainable Q-factor in a given system.

16.4.6. Surface-Related Phenomena

Carr *et al.*[14] observed an experimental correlation between damping in NEMS and surface-volume ratio, suggesting that a surface- or near-surface phenomena dominate the dissipative processes in single crystal Si NEMS. A few *possible* mechanisms are proposed here.

A thin amorphous layer of oxide is formed if a silicon device is exposed to air. The layer is approximately 2.5 nm thick.[39] Since SiO_2 needs twice the space that silicon does, additional stress is created at the interface, potentially resulting in additional thermoelastic losses.

Pohl *et al.*[40] developed a method to measure the quality factor of a thin overlayer deposited over macroscopic Si resonators. The relationship between the Q-factors is given as:

$$\frac{1}{Q_{\text{film}}} = \frac{G_{\text{Si}} t_{\text{Si}}}{3 G_{\text{Film}} t_{\text{Film}}} \left(\frac{1}{Q_{\text{Paddle and Film}}} - \frac{1}{Q_{\text{Paddle}}} \right) = \frac{G_{\text{Si}} t_{\text{Si}}}{3 G_{\text{Film}} t_{\text{Film}}} \frac{1}{\Delta Q} \tag{16.78}$$

where G_{Film} and G_{Si} are the rigidity modulus and t_{film} and t_{Si} the thickness of the deposited material and silicon respectively. They observed that the Q-factor of a SiO_2 film already differs significantly from the bulk SiO_2. The results suggest that for higher temperature, a more complex interaction between the two layers takes place and increases the losses. In addition, a 10 nm-thick water layer is known to form on a silicon surface with its native SiO_2 layer. Since water increases the internal friction of glass (SiO_2) by forming vibrating Si-OH groups,[41] it is also likely that it this overlayer increases the losses in silicon NEMS. Indeed, removing the SiO_2 and all surface contamination by heat treatment was shown to increase the Q-factor by almost a factor of 5.[42] Terminating the surface with hydrogen further increases the Q-factor.

16.4.7. Overview of Experimental Literature

Literature on micro- and nanomechanical resonators consistently points to a reduction of resonance quality as device dimension shrank from MEMS to NEMS. However, an overview of such data suggests that no single dissipative phenomenon dominates at all scales. At the micron scale, for instance, contour disk resonators have shown significantly higher resonance quality over end-clamped structures, suggesting the dominance of clamping issues at such scales.[43] At smaller scales, Yang *et al.* have recently reported a threefold decrease of energy dissipation, from $Q^{-1} = 3 \times 10^{-4}$ to $Q^{-1} = 1.24 \times 10^{-4}$ following the surface treatment of 60 nm thin, 5–7 μm wide, 10-80 μm long cantilevers.[44] These results suggest that surface effects account for 60% of the energy dissipation in these high surface-to-volume microcantilevers. Scaling down to NEMS dimensions, Evoy, Carr *et al.*, suggested that presence of oxide or other damaged layer dominate the energy dissipation of surface machined 200 nm × 200 nm × 1–8 μm megahertz range single-crystal Si NEMS.[26] While the exact nature of those surface losses has not been clearly established, the drastic impact of surface treatment on resonator quality does point towards the dominance of surface phenomena over all other issues. In addition, the relative importance of those loss mechanisms would be expected to increase for even smaller GHz range structures, making the topic a critical issue in nanomechanical resonators.

16.5. INTEGRATION OF NEMS WITH QUANTUM ELECTRONIC DEVICES

Another challenge in the deployment of NEMS in single-chip systems is the necessity of a sensitive, reliable, and integrative transduction mechanism. The high-frequency motion of a sub-micron mechanical structure generates a very weak capacitive signal that is easily lost in the background noise. Quantum phenomena such as single electron tunneling can offer better performance. As described in Chapter 11, the single electron transistor (SET)

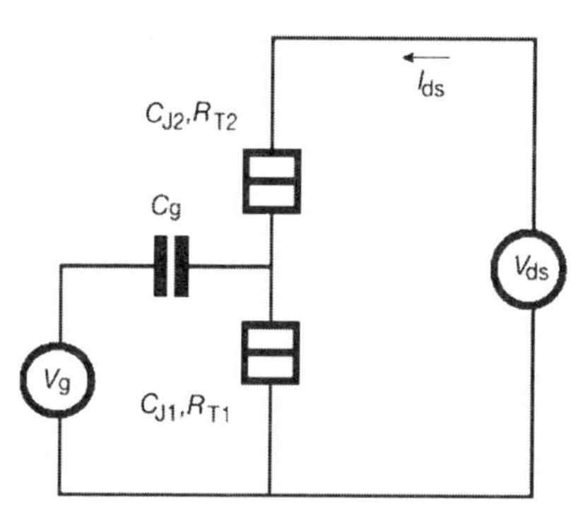

FIGURE 16.11. The circuit of a SET where Cg, CT1 and CT2 are the gate and junction capacitances respectively and RT1 and RT2 are the insulating barriers of the two tunnel junctions. (From Ref. 45 by permission of Nature Publishing Group.)

employs quantum tunneling for the transport of charges through a metal–insulator–metal junction. The electrical behavior of the tunnel junction is dependent on how effectively the barrier transmits electron waves and on the number of electron-wave modes that impinge on the barrier. A single-electron transistor exploits the fact that the transfer of charge through the barrier becomes quantized when the junction is made sufficiently resistive.[45,46,47] These devices have demonstrated transduction sensitivities as low as a few $10^{-5}e/\sqrt{Hz}$. Figure16.11 shows the circuit equivalent of such a device.[45] The integration of NEMS with quantum transistors offers a promising platform for the on-chip transduction of submicron range displacements. Tremendous advances have recently been made towards the integration of rf SET (a SET with a resonator tank) with mechanical resonators.

The schematic of a NEMS-rf SET displacement detector is shown in Figure 16.12.[48,49] One of the gate capacitor plates of the SET is placed on the resonator in such a way that for a fixed gate voltage bias, the mechanical displacement will be converted into a charge fluctuation. The lead stray capacitance C_s in contact with the SET, and an inductor L form a tank circuit with resonance frequency of $w_T = (LC_s)^{-0.5}$ and has a quality factor Q_T. This tank circuit is loaded by the SET A monochromatic carrier wave $v(t)$ is applied to the device, and at resonance (where the circuit impedance is small), the reflected power

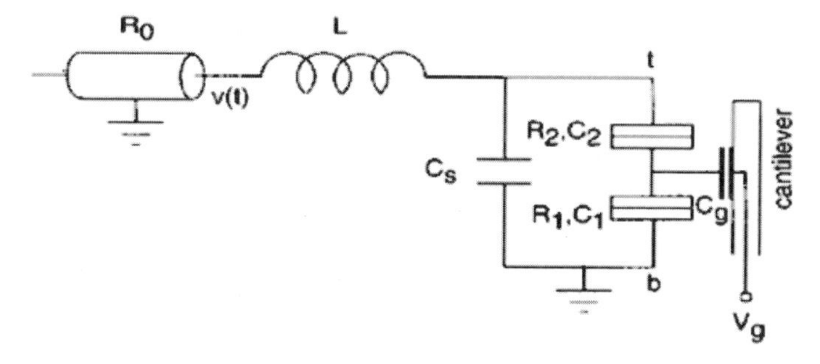

FIGURE 16.12. Schematic of a rf-SET displacement detector. (From Ref. 48 by permission of the American Institute of Physics.)

becomes a measure of the SET's differential resistance R_d. The biasing of the gate capacitor ensures that the mechanical motion of the resonator is converted into differential resistance changes, thereby modulating the reflected signal power.

The minimum detectable displacement at a mechanical signal frequency f_s, where $2\pi f_s < w_T/Q_T$, taking into account the shot noise due to SET tunneling current is given by:

$$\delta x = \sqrt{2e < |I_{\mathrm{SD}}(t)| \sin^2(wt) > \Delta f} / |< dI_{\mathrm{SD}}(t)/dx \sin wt >| \tag{16.79}$$

where I_{SD} is the source to drain current. The optimized displacement sensitivity improves with increasing gate voltages. With this method, a displacement sensitivity of $4 \times 10^{-16}\, m/\sqrt{Hz}$ is achievable.[50]

Aldridge *et al.*, recently reported a device consisting of a doubly clamped beam, coupled to the gate lead of the SET and driven at resonance using magnetomotive displacement (Figure 16.13).[51] The beam is biased with a dc voltage V_o, generating a coupling charge on the capacitor. The motion of the beam changes the charge coupled to the SET modulating the drain-source current in the transistor. The displacement of the resonator is transduced by the modulation of the coupling capacitance and the coupled charge. Sensitivity is determined by the charge sensitivity, coupling capacitance of the resonator displacement and by the magnitude of the SET back action on the mechanical resonator. This back action is caused by the force exerted on the resonator when the voltage on the center island fluctuates.[51] Displacement sensitivity is limited by the read out noise and the back action noise.[50]

Electron shuttling techniques can be employed to transduce resonator motion. Erbe *et al.*, report on the nanoelectromechanical resonator which they used as a mechanically flexible tunneling contact. This resonator is carved out of single crystal silicon on insulator substrate. Operating at 73 MHz at room temperature, it transfers electrons by mechanical

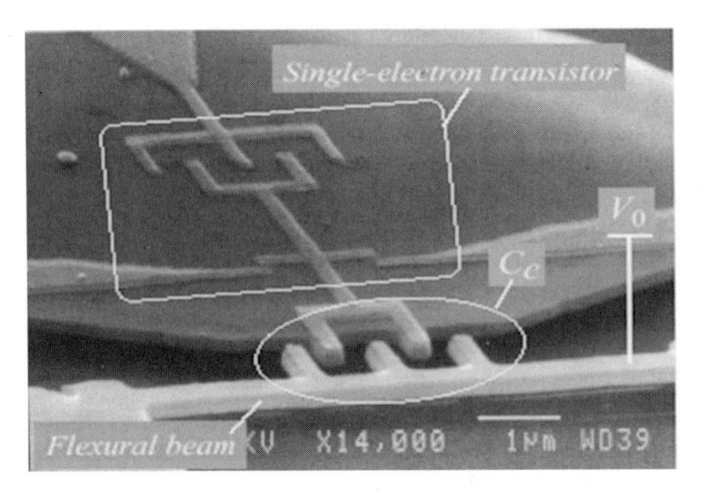

FIGURE 16.13. Micrograph of a doubly clamped beam coupled to a SET. The displacement x of the beam center point changes the coupled charge and modulates the current through the transistor. The resonating beam includes an interdigitated capacitor C(x) coupled to the gate of the SET. (From Ref. 51 by permission of the International Society for Optical Engineering.)

motion.[52] Datskos *et al.*, recently reported a device consisting of a beam resonator of length 4.5 μm and thickness of 0.5 μm positioned between two electrodes.[53] A dc bias is applied between the source and the drain electrodes, inducing the bending of the structure. No displacement current flows through the system since dc bias voltage is used. Upon excitation with an acoustic source, the resonator make contact with the source resulting in a transfer of the electronic charge from that electrode onto the resonator. When the cantilever makes contact with the drain, this charge is passed on to the second electrode. The group reported a current of 7 pA against a noise floor of 1.7 pA and a Q of 10 in air. If the resonance frequency can be measured with a resolution of 100 Hz, the minimum detection limit for adsorbed mass can be calculated as 10^{-19} g at resonance frequency of 5.2 KHz.[53]

16.6. "BOTTOM-UP" NEMS: CARBON NANOTUBE NANOMECHANICS

The pursuit of high-frequency and high-quality NEMS possesses similarities to the history of electronic and optoelectronic nanostructures. Early fabrication of quantum confined optoelectronic devices was indeed based on surface microfabrication processes that had been scaled down to nanoscale dimensions. However, plasma-induced damage prevented the production devices of sufficient *quality* for the target applications. While post-processing annealing, regrowth, and passivation were briefly explored, dry etching of quantum structures was eventually abandoned in favor of *self-assembled* epitaxial methods. In parallel to the history of *high-quality* quantum devices, surface machining *may* therefore prove inadequate for the production of *viable* GHz-range NEMS in applications demanding high spectral purity. Interestingly, surface and near-surface phenomena are once again at the center of the issue. Surface machining is currently limited to electronic materials whose natural oxides, surface strain fields, and processing-induced roughness are already known to severely affect the durability and performance of *both* MEMS and NEMS.

Recent advances in "bottom-up" nanoscience offer increasingly interesting alternatives to surface machining by expanding the range of materials and functions that can be designed into a nanostructure through careful control of the synthesis technique. For instance, carbon nanotubes offer very high Young's modulus, low density and good conductive properties. Theoretical[54,55] and experimental[56,57,58,59] studies have shown that such nanostructure offer a Young's modulus in the 1.5 to 5.0 TPa range. In addition, such nanostructures potentially offer perfect surface lattices that could therefore yield high-quality resonances at RF frequencies.

Cumings and Zettl reported on low friction and low wear bearing created from a multiwall nanotube (MWNT). This group studied the shell-to-shell interaction forces by attaching a nanomanipulator to the core layers, and pulling them out from the outer shell. The group observed a weak friction force between shells. Hence, the core will rapidly retract within the outer shell due to van der Waals interactions.[60] Zheng *et al.*[61] have calculated that the core indeed possesses a maximum and minimum potential energy at the fully extruded and fully inserted positions, respectively. When this shell is released it passes through a point of minimum potential energy and then extrudes to the other end. This causes the core to oscillate in the low GHz range with respect to the position of minimum.

Quin *et al.*, have calculated that a 1% change of resonant frequency in SWNT would correspond to an added mass of 800 amu, or approximately the mass of a single C_{60}

molecule.[62] Poncharal *et al.*, have used such a resonating nanotube cantilever to estimate the mass of an attached femtogram nanoparticle.[63] Nonlinear force-deflection curve of nanotubes under certain conditions can result in bistable response. Yu *et al.*, observed four parametric resonances of the fundamental mode of boron nanowires.[64] It is possible that by exploiting parametric resonances, highly sensitive molecular sensor could be produced.

On-chip transduction technology and viable integration techniques must be developed to allow the deployment of such devices in single-chip systems. However, advances in these directions are being made, which could eventually allow their deployment as *integrative nanosystems*.

QUESTIONS

Note: materials parameters are to be found in appropriate literature

1. What is the lowest resonant frequency of a Si beam of length of 1 μm, width 100 nm, and thickness of 50 nm? What does the frequency becomes if the beam is machined out of GaA? SiC? Diamond?
2. The resonator of Question #1 has a Q of 1000, and is driven at resonance using an oscillatory force amplitude of $F_o = 10$ nN amplitude. What is resulting amplitude of motion? What does this amplitude of motion becomes in GaAs? Diamond? Discuss.
3. Consider the Si beam of Question #1. At what amplitude of vibration will non-linear stretching effects appear? Assume an onset of non-linear behavior to occur when the non-linear stretching force equals the linear restoring force.
4. Consider the Si beam of Question #1. What maximum value of Q will ever be possible to attain given fundamental limits of phonon-phonon, and phonon-electron scattering?
5. Propose three enabling technologies that could enable the placement and/or controlled growth of carbon nanotube resonators on pre-established site of a silicon circuit. Enumerate advantages and disadvantages for each.
6. Are there other "bottom-up" nanosystems other than carbon nanotube materials that could offer an equally promising enabling platform for the integration of high-frequency high-quality nanomechanical resonators? Discuss their advantages and disadvantages over surface machined Si resonators.

REFERENCES

1. S. T. Picraux and P. J. McWhorter, *IEEE Spectrum* **25**, 24 (1998).
2. J. A. Walker, *Journal of Micromechanics and Microengineering* **10**, R1 (2000).
3. L. E. Larson, IEEE J. of Solid-State Circuits **33**, 387 (1998).
4. J.-F. Luy and G. E. Ponchak, *IEEE Trans. on Microwave Theory and Techniques* **46**, 569 (1998).
5. C. T.-C. Nguyen, Proceedings, 1998 Sensors Expo, San Jose, CA, May 19–21, 1998, pp. 447–455.
6. W. B. Kuhn, N. K. Yanduru, and A. S. Wyszynski, *IEEE Trans. on Microwave Theory and Techniques* **46**, 2577 (1998).
7. R. H. Blick, A. Erbe, A. Tilke, and A. Wixforth, *Phys Bl* **1**, 31 (2000).
8. M. L. Roukes, *Phys. World* **14**, 25 (2001).
9. S. C. Arney and N. C. McDonald, *J. Vac. Sci. Technol. B* **6**, 341 (1988).

10. J. J. Yao, and S. C. Arney, *IEEE J. of Microelectromechanical Systems* **1**, 14 (1992).
11. A. N. Cleland and M. L. Roukes, *Appl. Phys. Lett.* **69**, 2653 (1996).
12. D. W. Carr and H. G. Craighead, 41st Electron, Ion and Photon Beam Technology and Nanofabrication Conference (Dana Point, Calif, 1997).
13. D. W. Carr and H. G. Craighead, *J. Vac. Sci. Technol. B* **15**, 2760 (1997).
14. D. W. Carr, S. Evoy, L. Sekaric, J. M. Parpia, and H. G. Craighead, *Appl. Phys. Lett.* **75,** 920 (1999).
15. T. S. Tighe, J. M. Worlock, and M. L. Roukes, *Appl. Phys. Lett.* **70**, 2687 (1997).
16. L. Sekaric, D. W. Carr, S. Evoy, J. M. Parpia, and H. G. Craighead, Sens. Act. A **101**, 215 (2002).
17. D. M. Tanenbaum, A. Olkhovets, and L. Sekaric, *J. Vac. Sci. Tech. B* **19**, 2829 (2001).
18. L. Sekaric, J. M. Parpia, H. G. Craighead, T. Feygelson, B. H. Houston, and J. E. Butler, *Appl. Phys. Lett.* **81**, 4455 (2002).
19. Y. T. Yang, K. L. Ekinci, X. M. H. Huang, L. M. Schiavone, C. A. Zorman and M. Mehregany, and M. L. Roukes, *Appl. Phys. Lett.* **78**, 162 (2001).
20. X. M. H. Huang, C.A. Zorman, M. Mehregany, and M. L. Roukes, *Nature* **421**, 496 (2003).
21. G. Genta, Vibration of Structure and Machines, Springer 1999.
22. H. A. C. Tilmans, M. Elwenspoek, and J. H. J. Fluitman, Sens. and Actuators A **30**, 35 (1992).
23. A. A. Shabana, "Vibration of Discrete and Continuous Systems" Springer-Verlag 1997.
24. D. W. Carr, Nanoelectomechanical Resonators, Ph.D. thesis Cornell University 2000.
25. A. A. Shabana, "Theory of Vibration Volume I" Springer-Verlag 1990.
26. S. Evoy, D. W. Carr, L. Sekaric, A. Olkhovets, J. M. Parpia, and H. G. Craighead, *J. Appl. Phys.* **86**, 6072 (1999).
27. K. L. Turner, S. A Miller, P. G. Hartwell, N. C. MacDonald, S. H. Strogartz, and S. G. Adams, *Nature* **396**, 149 (1998).
28. D. W. Carr, S. Evoy, L. Sekaric, A. Olkhovets, J. M. Parpia, and H. G. Craighead, *Appl. Phys. Lett.* **77**, 1545 (2000).
29. M. Zalalutdinov, A. Zehnder, A. Olkhovets, S. Turner, L. Sekaric, B. Ilic, D. Czaplewski, J. M. Parpia, and H. G. Craighead, *Appl. Phys. Lett.* **79**, 695 (2001).
30. F. R. Blom, S. Bouwstra, M. Elwenspoek, and J. H. J. Fluitman, *J. Vac. Sci. Technol. B* **10**, 19 (1992).
31. A. Olkhovets, S. Evoy, D. W. Carr, J. M. Parpia, and H. G. Craighead, *J. Vac. Sci. Technol B* **18**, 3549 (2000).
32. K. Wang, A. C. Wong, and C. T. C. Nguyen, *J. Microelectromech. Syst.* **9**, 347 (2000).
33. D. A. Harrington, P. Mohanty, and M. L. Roukes, *Phys. B* **248-288**, 2145 (2000).
34. R. De Batist, Internal friction of structural defects in crystalline solids, North-Holland Publishing Company, 1972.
35. T. S. Ke, Metallurgical and Materials Transaction A **30**, 2267 (1949).
36. T. S. Ke, Phys. Rev. **71**, 553 (1947).
37. T. S. Ke, Phys. Rev. **72**, 41 (1947).
38. V. B. Braginskey, V. P. Mitrofanov, and V. I. Panov, Systems with small dissipation, The University of Chicago Press 1985.
39. D. W. Carr, Nanoelectomechanical Resonators, Ph.D. thesis Cornell University 2000.
40. B. E. White, and R. O. Pohl, *Phys. Rev. Let.* **75**, 4437 (1995).
41. W. A. Zdaniewski, G. E. Rindone, and D. E. Day, *J. Mat. Sci.* **14**, 763 (1979).
42. J. Yang, T. Ono, and M. Esashi, *J. Vac. Sci. Technol. B* **19**, 551 (2001).
43. J. R. Clark, W.-T. Hsu, and C. T.-C. Nguyen, 2000 Intl. Electron Devices Meeting Technical Digest, pp. 493–496.
44. J. Yanga, T. Ono, and Masayoshi Esashi, *J. Vac. Sci. Technol. B* **19**, 551 (2001).
45. M. H. Devore, and R. J. Schoelkopf, *Nature* **406**, 1039 (2000).
46. H. Grabert *Phys. Rev.* **50**, 17364 (1994).
47. H. Schoeller and G. Schoen, *Phys. Rev.* B**50**, 18436 (1994).
48. B. Starkman, T. Henning, T. Claeson, P. Delsing and A. N. Korotkov, *J. Appl Phys.* **86**, 2132 (1999).
49. A. N. Korotkov and M. A. Paalanen, *Appl. Phys. Lett.* **74**, 4052 (1999).
50. R. Knobel and A. N. Cleland, *Appl. Phys. Lett.* **81**, 2258 (2002).
51. J. S. Aldridge, R. S. Knobel, D. R. Schmidt, C. S. Yung, and A. N. Cleland, *SPIE Proceedings* **4591**, 11 (2001).
52. A. Erbe, R. H. Blick, A. Tilke, A. Kriele and J. P. Kotthaus, *Appl. Phys. Lett* **73**, 3751 (1998).
53. P. G. Datskos and T. Thundat, *J. of Nanoscience and Nanotechnology* **2**, 369 (2002).

54. M. J. Tracey, T. W. Ebbessen, and J. M. Gibson *Nature* **381**, 678 (1992).
55. G. Overney, W. Zhong, and D. Tomanek, *Z. Physic D* **27**, 93 (1993).
56. G. H. Gao, T. Cagin, and W. A. Goddard, Nanotechnology **9,** 184 (1998).
57. B. I. Yakobson, C. J. Brabec, and J. Bernholc, *Phys. Rev. Lett.* **76,** 2511 (1996).
58. E. Hernandez, C. Goze, P. Bernier, and A. Rubio, *Phys. Rev. Lett.* **80,** 4502 (1998).
59. X. Zhou, J. J. Zhou, and Z. C. Ou-Yang, *Phys. Rev. B* **62**, 13692 (2000).
60. J. Cummings and A. Zettl, Science **289**, 602 (2000).
61. Q. Zheng and Q. Jiang, *Phys. Rev. Lett.* **88,** 045503 (2002).
62. D. Qian, G. J. Wagner, W. K. Liu, M.-F. Yu, and R. S. Ruoff, *Appl. Mech. Rev.* **55**, 495 (2002).
63. P. Poncharal, Z. L. Wang, D. Ugarte, and W. A. de Heer, *Science* **283,** 1513 (1999).
64. M. F. Yu, G. J. Wagner, R. S. Ruoff, and M. J. Dyer, *Phys. Rev. B* **66**, 073406 (2002).

17

Micromechanical Sensors

P. G. Datskos, N. V. Lavrik, and M. J. Sepaniak
Oak Ridge National Laboratory and
University of Tennessee, Knoxville, TN

17.1. INTRODUCTION

During the last two decades, advances in microlectromechanical (MEMS) and nanoelectromechanical (NEMS) systems have facilitated development of sensors based on new transduction principles that involve mechanical energy and rely heavily on mechanical phenomena. Functionality of NEMS and MEMS devices is based on mechanical motions of their components, such as single-clamped suspended beams (cantilevers), double-clamped suspended beams ("bridges") or suspended diaphragms. Cantilever structures similar to probes used in atomic force microscopy (AFM) are some of the simplest MEMS. They can also be considered as basic building blocks for the whole variety of more complex MEMS and NEMS devices. While MEMS and NEMS transducers span a great variety of designs, devices with very simple cantilever-type configurations appeared to be especially suitable as transducers of physical, chemical and biological stimuli into readily measured signals. Since the advent of scanning probe microscopy, refinement of approaches to fabrication and characterization of microscopic cantilevers useful as AFM probes has been a subject of extensive research efforts. Broader interest in MEMS transducers can be explained by their potential for applications in optical imaging, telecommunications, and data storage. As a result of recent advances in several converging areas of science and technology, not only a variety of sophisticated probes became available for scanning probe microscopy but also an innovative family of physical, chemical and biological sensors based on cantilever technology was shaped out.[1–10]

A general idea behind all MEMS sensors is that a certain parameter of the ambient environment (which can represent a physical, chemical or biological entity) can affect mechanical characteristics of the micro- or nano-mechanical transducer in such a manner that this change can be measured using electronic, optical or other means. In particular,

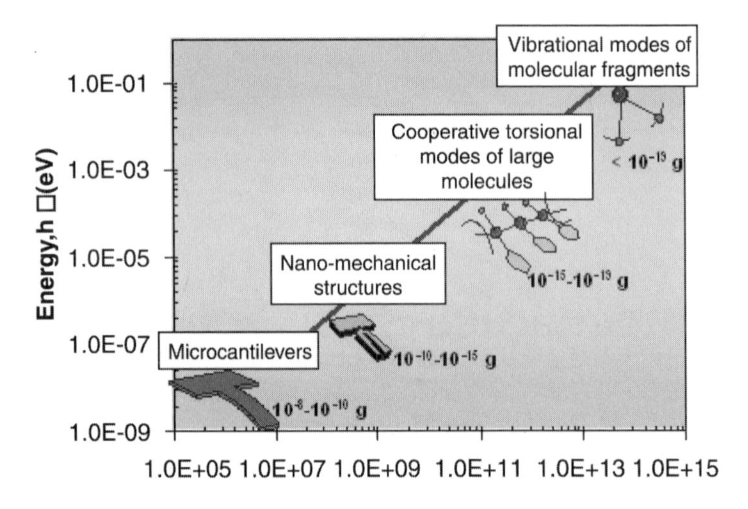

FIGURE 17.1. Evolution from MEMS to NEMS to molecular structures. Nanostructures may have a total mass of only a few femtograms. In the nanomechanical regime, it is possible to attain extremely high fundamental frequencies approaching those of vibrational molecular modes.

microfabricated cantilevers together with read-out means that are capable of measuring 10^{-12} to 10^{-6} m displacements can operate as detectors of surface stresses,[10–17] extremely small mechanical forces,[18–21] charges,[22,23] heat fluxes,[24–25] and IR photons.[27–31] More importantly, as devices approach the nanoscale, their mechanical behavior starts resembling vibrational modes of molecules and atoms (Figure 17.1). At a certain level of microminiaturization, nanomechanical resonators similar to microcantilevers can be envisioned as very large molecules that interact controllably with both their molecular environment and readout components.

This chapter focuses primarily on MEMS sensors with transducers in a form of cantilevers or analogous structures with more complex shapes and one or several anchoring points. We will use the terms "cantilever" and "bridge" throughout the text of this section to denote devices analogous to, respectively, single-clamped and double-clamped suspended beams of various sizes and shapes. For simplicity, we will mainly use the term "MEMS", although derived terms, such as NEMS, micro-opto-electromechanical systems (MOEMS), and biological microlectromechanical (Bio-MEMS) could also be justified in this content to emphasize specific features of certain sensors based on micromechanical transducers. The section is structured largely around the four aspects of MEMS sensors: (i) operation principles and models, (ii) figures of merit, (iii) fabrication, and (iv) applications. Since many MEMS are truly multi-faceted devices that may integrate several transduction modes a significant part of the review is devoted to principles of their operation.

17.2. MECHANICAL MODELS

Operation of MEMS sensors may involve measurements of deflections, resonance frequencies and, in some cases, damping characteristics of suspended structural components

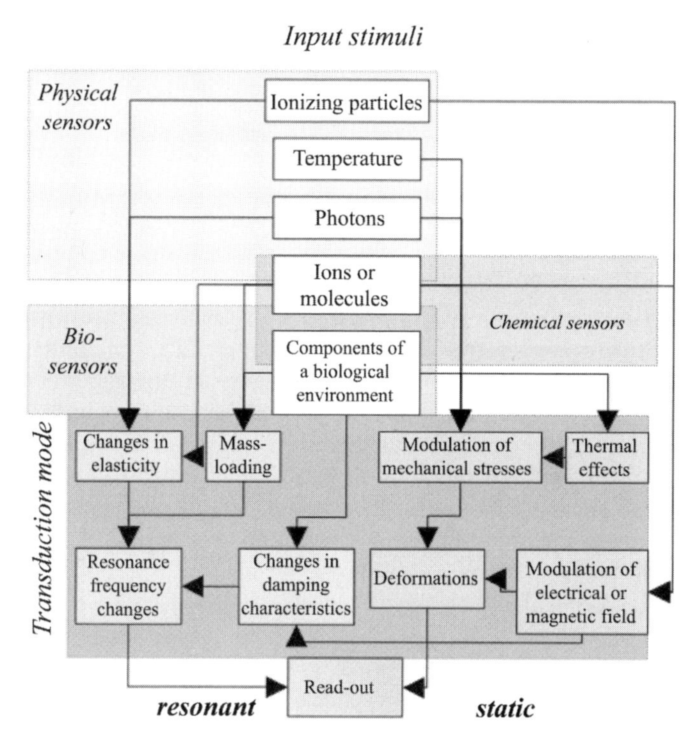

FIGURE 17.2. Different transduction mechanisms for a cantilever that can provide conversion of input stimuli into output signals. Depending on the measured parameter-structural deformations or resonance frequency changes—the mode of sensor operation can be refereed to as either *static* or *resonant*. Each of these modes, in turn, can be associated with different input stimuli and transduction scenarios.

The variety of transduction mechanisms that provide conversion of possible input stimuli into output signals is depicted in Figure 17.2. Depending on the measured parameter—structural deformations or resonance frequency changes-the mode of sensor operation can be refereed to as either *static* or *resonant*. The term "static" emphasizes that the transducer does not exhibit functional movements unless there is a change in the input stimulus. Each of these modes, in turn, can be associated with different input stimuli and transduction scenarios (Figure 17.2). Deformations in MEMS can be caused by either external forces exerted on the cantilever (as in AFM probes) or intrinsic stresses generated on the cantilever surface or within the cantilever. Cantilever sensors operating in the resonant mode are essentially mechanical oscillators, resonance characteristics of which depend upon the attached mass as well as viscoelastic properties of the medium. For instance, adsorption of analyte molecules on a resonating cantilever results in lowering of its resonance frequency due to the increased suspended mass of the resonator. Depending on the nature of the input stimuli, MEMS sensors can be referred to as *physical, chemical* or *biological* sensors (Figure 17.2). The variety of transduction modes stems from the fact that a stimulus of each type may affect mechanical state of the transducer directly or may undergo one or several transformations before the measured mechanical parameter of the transducer is affected. For instance, IR photons can be detected by measuring mechanical stresses produced in a MEMS detector as a direct consequence the photon absorption process. More commonly, however, detection of IR photons by MEMS detectors consists in detecting the temperature increases associated with absorption of IR photon. Molecular and biochemical interactions can also be detected by MEMS sensors due to the heat effects of exothermic reactions or molecular adsorption processes. Alternatively, chemical and biochemical species can induce direct responses due to surface stress changes or mass loading effects.

17.2.1. Static Deformations

In the absence of external gravitational, magnetic and electrostatic forces, cantilever deformation is unambiguously related to a gradient of mechanical stress generated in the device. For instance, cantilevers made of two layered materials with different coefficients of thermal expansion undergo deformation upon changes in temperature. This deformation resulting from a strain gradient due to unequal thermal expansion of each layer has been used extensively as an operation principle of thermostats. For instance, theoretical evaluation of bimetal thermostats performed by Timoshenko[32] provides an analytical expression for the radius of curvature of a bimaterial plate as a function of a temperature change. More recently, various modifications of this model have been used to predict thermally induced deflections of microscopic bimaterial cantilevers.[24,33]

As applied to chemical and biological sensors, cantilever based calorimetry enables two transduction scenarios (Figure 17.2). First, the presence of analyte species can be detected due to the heat associated with their adsorption on the transducer. Second, the heat produced in the course of a subsequent chemical reaction on the cantilever surface can be characteristic of the analyte presence. However, molecular adsorption processes and interfacial chemical reactions may also affect mechanical stresses in thin plates more directly and independently of the thermal effects. It has been known since the 1960s that molecular and atomic adsorbates on atomically pure faces of single crystals tend to induce significant surface stress changes. Long before the first micromachined cantilevers were created, changes in surface stresses in these systems had been studied by measuring minute deformations of relatively thin (up to 1 mm) plates. Using this method, often referred to as the beam-bending technique,[12,34,35] Kosch *et al.*, studied[36,37] surface stress changes induced by adsorption of atoms on atomically pure surfaces in vacuum. Another class of materials in which adsorbate and chemically induced interfacial stresses have been traditionally studied is colloidal systems. Important examples of colloidal phenomena associated with the surface stress changes include swelling of hydrogels upon hydration or formation of surfactant monolayers at the air-water interface.[38] Fundamental studies of adsorption and absorption induced mechanical phenomena, however, had limited implications for chemical sensors until mass produced AFM probes became widely available. As compared to their macroscopic predecessors, microcantilevers coupled with the optical lever readout greatly simplified real-time measurements of surface stress changes in the low mN m^{-1} range.

A cantilever intended for chemical sensing is normally modified so that one of its sides i relatively passive while the other exhibits high affinity to the targeted analyte. Consequently changes in differential surface stress can be governed primarily by changes in Gibbs fre energy associated with adsorption (surface interaction) or absorption (bulk phase interactio with thin films) processes on the active side. Apart from fundamental interest in the dire conversion of chemical energy into mechanical energy, this mechanism means that MEM transducers are compatible with many responsive phases and can function in both gas an liquid environments. In order to understand how different coatings provide responses c

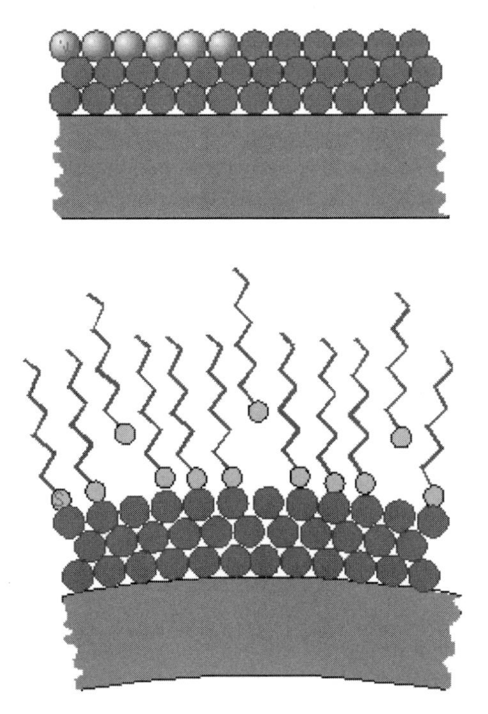

FIGURE 17.3. Schematic depiction of chemisorption of straight-chain thiol molecules on a gold coated cantilever. Spontaneous adsorption processes are driven by an excess of the interfacial free energy, and accompanied by reduction of the interfacial stress.

cantilever sensors working in the static bending mode, it is useful to consider the three distinctive models.

The first model is most adequate when interactions between the cantilever and its environment are predominantly surface phenomena. An example of this situation is given in Figure 17.3, where chemisorption of straight-chain thiol molecules on a gold coated cantilever is schematically depicted. Since spontaneous adsorption processes are driven by an excess of the interfacial free energy, they are typically accompanied by reduction of the interfacial stress. In other words, surfaces usually tend to expand (see Figure 17.3) as a result of adsorptive processes. This type of surface stress change is defined as compressive, referring to a possibility of return of the surface into the original compressed state. The larger the initial surface free energy of the substrate, the greater the possible change in surface stress results from spontaneous adsorption processes. In many cases, adsorbate-induced deformations of thin plates can be accurately predicted using a modification of the relationships originally derived by Stoney:[39,40]

$$\frac{1}{R} = \frac{6(1-\upsilon)}{Et^2}\delta\sigma \tag{17.1}$$

where R is the cantilever's radius of curvature, υ and E are Poisson's ratio and Young's modulus for the substrate, respectively, t is the thickness of the cantilever, and $\delta\sigma$ is the differential surface stress. When adsorbate-induced stresses are generated on ideal smooth

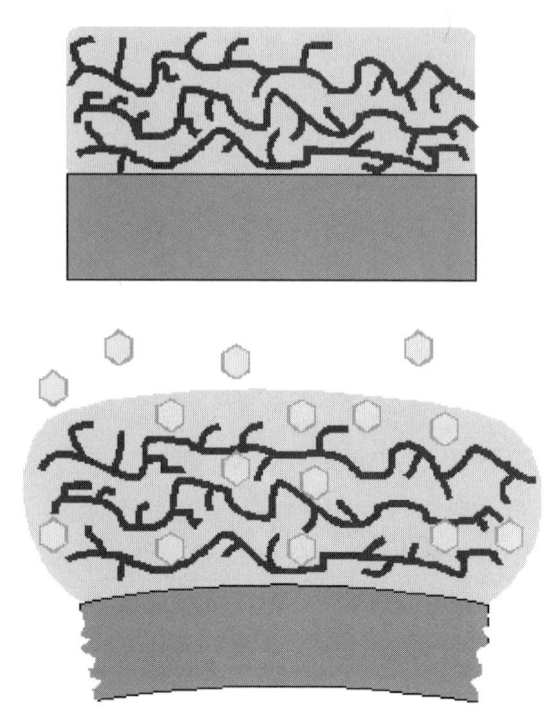

FIGURE 17.4. Schematic depiction of the case for analyte-induced stresses when the cantilever surface is modified with a much than a monolayer analyte-permeable coating. Interactions of the analyte molecules with the bulk of the responsive phase lead to coating swelling and can be quantified using approaches employed in colloidal and polymer science.

surfaces or within coatings that are very thin in comparison to the cantilever, the analysis according to Equation 17.1 is rather straightforward. Using Eq. 17.1, the predictions for the cantilever bending can be based on the expected surface stress change. Alternatively, responses of cantilever sensors converted into surface stress changes can be analyzed as the measure of the coating efficiency independently of the transducer geometry.

When a cantilever is modified with a much thicker than a monolayer analyte-permeable coating,[41,42] the second model of analyte-induced stresses (Figure 17.4) may appear to be more useful. Taking into account interactions of the analyte molecules with the bulk of the responsive phase, a predominant mechanism of cantilever deflection can be described as deformation due to analyte-induced swelling of the coating (Figure 17.4). Such swelling processes can be quantified using approaches developed in colloidal and polymer science, i.e., by evaluating molecular forces acting in the coating and between the coating and the analyte species. In general, dispersion, electrostatic, steric, osmotic and solvation forces,[38] acting within the coating can be altered by absorbed analytes. Depending on whether it is more appropriate to describe the responsive phase as solid or gel-like, these altered forces can be put into accordance with, respectively, stress or pressure changes inside the coating.

The third model (Figure 17.5) is most relevant to structured (heterogeneous) interfaces and coatings that have been recently recognized as a very promising class of chemically

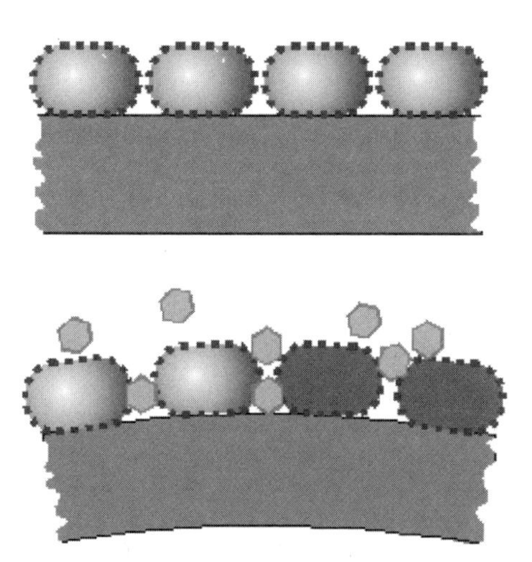

FIGURE 17.5. Schematic depiction of the case for structured phases (molecular sponges). Analyte-induced deflections of cantilevers with structured phases combine mechanisms of bulk, surface, and inter-surface interactions.

responsive phases for MEMS sensors.[6] Many of these structured phases exhibit behaviors of *molecular sponges*. Analyte-induced deflections of cantilevers with these phases (Figure 17.5) combine mechanisms of bulk, surface, and inter-surface interactions.[38] A combination of these mechanisms facilitates efficient conversion of the energy of receptor-analyte interactions into mechanical energy of cantilever bending. Recent studies demonstrated that up to two orders of magnitude increases in cantilever responses can be obtained when receptor molecules are immobilized on nanostructured instead of smooth gold surfaces.[43–45] Furthermore, nanostructured responsive phases offer an approach to substantially increase the number of binding sites per cantilever without compromising their accessibility for the analyte. Although deflections of cantilevers with nanostructured coatings or thicker hydrogel layers cannot be accurately predicted using the models mentioned above, estimates for the upper limit of the mechanical energy produced by any cantilever transducer can always be based on simple energy conservation. This upper limit in available energy is given by the product of the energy associated with the binding site-analyte interaction and the number of such interactions on the cantilever surface.

17.2.2. Externally Driven and Thermal Oscillations

Cantilever transducers in the air or vacuum can be treated as weakly damped mechanical oscillators. Their resonant behavior can be observed using excitation in alternated electric, electromagnetic or acoustic fields. Furthermore, minute sizes and mass of microfabricated cantilevers makes them susceptible to thermally induced noise, which has the same origin as Brownian motion of small particles in liquids. Therefore, the cantilever sensors may operate in the resonant mode either with or without external excitation.

As a rule, Eq. 17.39 gives a fair approximation of the resonance frequency, f_0, of a weakly damped mechanical resonator, such as a microscopic cantilever in air. The effective suspended mass of a cantilever can be related to the total mass of the suspended portion of the beam, m_b, through the relation: $m_0 = nm_b$, where n is a geometric parameter. For a rectangular cantilever, n has a typical value of 0.24 and the spring constant is given by:[46]

$$k = \frac{Ewt^3}{4L^3} \tag{17.2}$$

where E is the modulus of elasticity for the material composing the cantilever and w, t, L are the width, thickness and length of the cantilever, respectively. Equation 17.2 shows that longer cantilevers tend to have significantly smaller spring constants. As a result, longer cantilevers are more sensitive for measuring both externally exerted forces and interfacial stress changes. However, cantilevers with low stiffness are also more susceptible to all types of noise, including thermally induced noise. Any cantilever in equilibrium with its thermal environment has a "built-in" source of white thermal noise $\Psi_{\text{th}}(f)$. The amplitude of the resulting thermally induced oscillation of a cantilever beam is proportional to the square root of the thermal energy:

$$\delta_n = \sqrt{\frac{2k_B TB}{\pi k f_o Q}} \tag{17.3}$$

Here, k_B is the Boltzmann constant (1.38×10^{-23} J/K), T is the absolute temperature (300 K at room temperature), B is the bandwidth of measurement, f_o is the resonance frequency of the cantilever, and Q is the quality factor (which has been discussed in previous chapter). As it follows from Eq. 17.3, lower cantilever stiffness corresponds to higher amplitudes of thermal noise. As a result of the dynamic exchange between cantilever mechanical energy and the ambient thermal energy, the actual frequency, f, of thermally induced cantilever oscillations at any given moment can noticeably deviate from the resonance frequency, f_0. The amplitude of such frequency fluctuations, δf_o, due to the exchange between mechanical and thermal energy is:[47]

$$\delta f_0 = \frac{1}{A}\sqrt{\frac{2\pi\, f_0\, k_B\, TB}{kQ}} \tag{17.4}$$

where A is the amplitude of the cantilever oscillations. Equation 17.4 predicts increased absolute fluctuations of the resonance frequency, Δf_0, as the resonance frequency f_0 increases. However, relative frequency instability, $\Delta f_0/f_0$, decreases in the case of higher frequency oscillators

$$\frac{\Delta f_0}{f_0} = \frac{1}{A}\sqrt{\frac{2\pi\, k_B\, TB}{kQ\, f_0}} \tag{17.5}$$

Although Eqs. 17.4–17.5 are valid for thermally excited cantilevers, they can also be used to evaluate effects of thermal noise on the frequency instability of the externally driven cantilevers.[47] As applied to cantilever sensors operating in the resonance mode, an important implication of Eqs. 17.4–17.5 is that frequency instability due to effects of thermal noise can be minimized by driving the transducer with the highest possible amplitude.

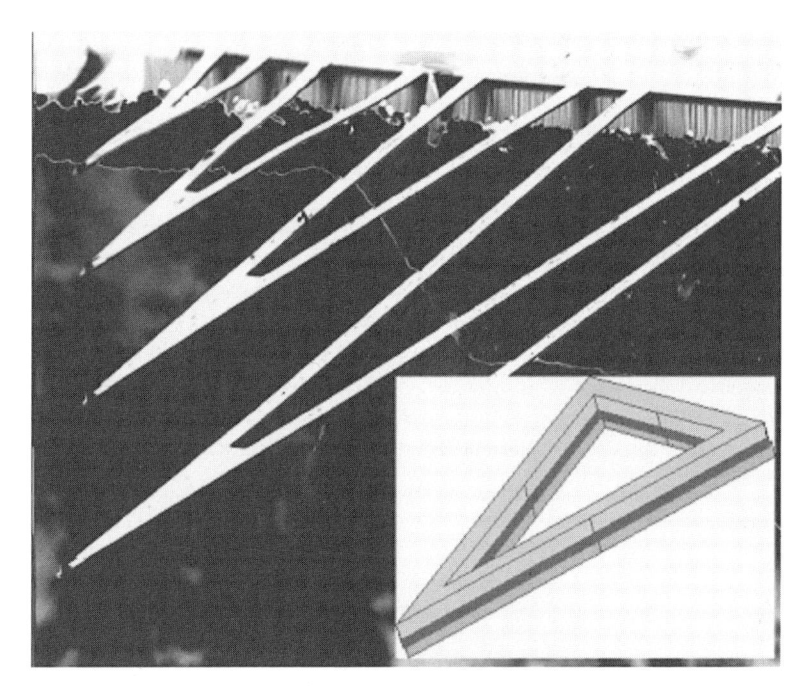

FIGURE 17.6. Examples of typical triangular cantilevers used as standard AFM probes.

17.3. FABRICATION AND READOUT

Commercially available AFM probes made of silicon or silicon nitride have been used extensively in research on cantilever based sensors.[1–9] In fact, main structural and geometrical requirement to cantilever transducers for sensor applications are similar to those applicable in AFM. In analogy to AFM cantilevers (Figure 17.6), cantilever transducers for MEMS sensors are usually fabricated from silicon or silicon nitride and have typical thicknesses in the range of 0.5 to 5 μm. The typical lengths of cantilevers for both AFM and sensor applications are in the range of 10 to 500 μm and approximately correspond to spring constants of, respectively, 1 to 0.01 N m^{-1}. Fabrication of AFM probes is based on well established and cost efficient process flows that provide low cost, high yield and good reproducibility of the resulting devices. However, AFM cantilevers are designed and fabricated to satisfy a number of the application specific requirements, which become partly redundant in the case of cantilever transducers for sensor applications. The most notable of such redundant features are the presence of a sharp tip on the cantilever end and the accessibility of the tip for a sample surface.

Generally, fabrication of a suspended microstructure, such as a cantilever transducer, consists of deposition, patterning, and etching steps that define, respectively, thickness, lateral sizes, and the surrounding of the cantilever. In particular, fabrication of suspended structures such as cantilevers and bridges often rely on bulk micromachining of single crystal silicon. One of the frequently used process flows begins with deposition of a structural silicon nitride layer on a single-crystal silicon wafer using a low-pressure chemical vapor deposition (LPCVD) process. By varying the conditions of the LPCVD process, the stress and stress gradient in the film can be minimized so that suspended structures do not exhibit significant deformation when released. The cantilever shapes can be defined by patterning the silicon nitride film on the top surface using photolithography followed by reactive ion etch (RIE). Photolithographic patterning of the silicon nitride on the bottom surface is used to define mask for anisotropic bulk etch of Si. The silicon substrate is then etched away to produce free-standing cantilevers. Single crystal silicon cantilevers can be produced using a similar sequence of processes except that doping of silicon or epitaxy of a doped silicon layer substitutes deposition of a silicon nitride layer; the p-doped silicon plays a role of an etch stop layer.[48,49]

In order to avoid any bulk machining, such as through-etch of silicon in KOH, various cantilever fabrication processes based on the use of a sacrificial layer were developed. These processes frequently rely on silicon oxide as a material for the sacrificial layer.[50,51] The use of a sacrificial layer for fabrication of silicon nitride membranes, bridges and cantilevers is illustrated in Fig. 17.7. While the use of sacrificial layer introduces additional restrictions

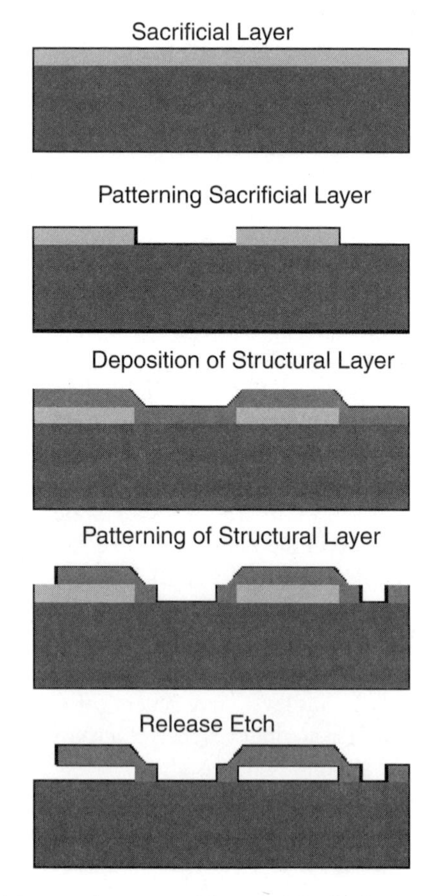

FIGURE 17.7. Illustration of the steps in a process flow used for fabrication of silicon nitride membranes bridges and cantilevers. The process begins with deposition of a structural silicon nitride layer on a single-crystal silicon wafer. The cantilever shapes can be defined by patterning the silicon nitride film on the top surface using photolithography followed by reactive ion etching (RIE).

on the material choice, it enables process flows that are fully compatible with standard complementary-metal-oxide-semiconductor (CMOS) chip technology.

In the case of chemical and biological sensors, noble metal coatings provide surfaces that can be selectively modified with synthetic or biological receptors using thiol-gold reaction schemes.[43,44] It was found that palladium and gold coatings can be used in MEMS sensors in order to achieve chemical specificity to hydrogen gas and mercury vapor, respectively.[52,53] Polymeric and macrocyclic compounds in a form of 5 nm to 5 μm thick films were shown to provide sensitivity to various organic compounds in a vapors phase[41,44,54,55] as well as organic compounds[45] and ionic species in water.[56–58]

Using optical, piezoresistive, piezoelectric, capacitance, or electron tunneling methods, deformations and resonance frequency shifts of a cantilever transducer can be measured with high precision. All these signal transduction methods are compatible with array format. One of the unique advantages of the MEMS sensors is that deformations and resonance frequency shifts measured simultaneously provide complementary information about the interactions between the transducers and the environment (see Figure 17.2).

17.3.1. Optical Methods

It is noteworthy that cantilever sensors inherited not only the unique advantages of a microfabricated AFM probe, but also the elegant "optical lever" read-out scheme commonly used in modern AFM instruments. Optical methods most extensively used for measurements of cantilever deflections in AFM include optical beam deflection (also referred to as the "optical lever" method) and optical interferometry. In optical beam deflection technique, a laser diode is focused at the free end of the cantilever. This optical detection scheme (Figure 17.8) can discern extremely small changes in the cantilever bending; measurements of 10^{-14}m displacements were reported. Absence of electrical connections to the cantilever, linear response, simplicity and reliability are important advantages of the optical lever method. As this method has been used in vast majority of the work on cantilever sensors, its limitations are well recognized. For instance, changes in the optical properties of the medium surrounding the cantilever may interfere with the output signal. The effect of the refractive index change as well as other interfering factors can be further suppressed by

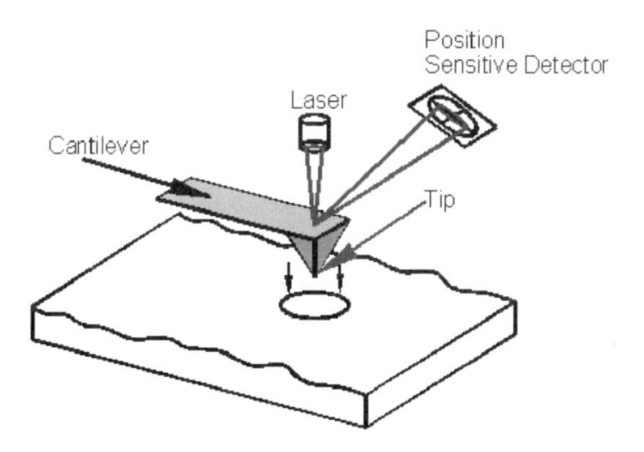

FIGURE 17.8. Optical lever readout commonly used to measure deflections of microfabricated cantilever probes in AFM.

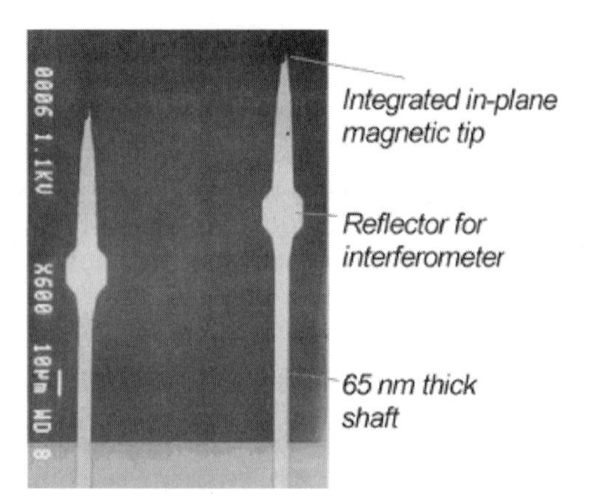

FIGURE 17.9. Ultrasensitive cantilevers for magnetic resonance microscopy. The cantilevers are about 200 microns long and 65 nm thick. Interferometric readout was used to measure their deflections. At liquid helium temperatures, these cantilevers permitted detection of forces as small as 3 attonewtons (3×10^{-18} N) in a 1 Hz bandwidth. (Reproduced with kind permission of T. Kenny and T. Stowe).

using differential pairs or arrays of cantilevers. However, applications of cantilever sensors with the optical lever readout are limited to analysis of low opacity, low turbidity media. Another limitation of the optical lever method is related to the bandwidth of PSDs, which typically on the order of several hundred KHz.

As the requirements of the high bandwidth become more critical in the case of smaller and stiffer cantilevers that operate in the resonant mode, alternatives to the optical lever readout were explored. For instance, motion of a microscopic structure, such as a cantilever, illuminated with a tightly focused laser beam, produces change in the spatial distribution of the reflected and/or scattered light. A simple spot photodetector alone or in combination with a knife edge obstacle can be used to monitor these intensity fluctuations.[59] The readout bandwidth of this method can be extended into the GHz range by using a small area, high-speed avalanche photodiode. Approaches based on a single photodetector and light scattering, however, suffer from the interference with ambient light, nonlinear response, and a poorly controllable optical gain. More accurate high-band width optical measurements of cantilever deflections can be carried out using interferometric schemes. Notably, interferometry was the first optical technique used for measurements of cantilever deflections in AFM. Rugar *et al.* used interefrometry to measure subnanometer deflections of the ultrasensitive cantilevers (see Figure 17.9) designed for ultrasensitive force measurements that could ultimately permit single-spin magnetic resonance microscopy.[18–20]

17.3.2. Piezoresistive and Piezoelectric Methods

Piezoresistivity is the phenomenon of changes in the bulk resistivity with applied stress. One of the most common materials that exhibit a strong piezoresistive effects is doped silicon. When a silicon cantilever with an appropriately shaped doped region is deformed, the change in the resistance of the doped region reflects the degree of the deformation. This variation in the resistance is typically measured by including the cantilever into a dc-biased

Wheatstone bridge. Piezoelectric readout technique requires deposition of piezoelectric material, such as ZnO, on the cantilever. Due to a piezoelectric effect, transient charges are induced in the piezoelectric layer when a cantilever is deformed. The main disadvantage of the piezoelectric as well as piezoresistance readout is that they require electrical connections to the cantilever. An additional disadvantage of the piezoelectric technique, is that obtaining appreciable output signals may require thickness of the piezoelectric film to be well above the values that correspond to optimal mechanical characteristics. Furthermore, the piezoelectric readout is inefficient when slowly changing cantilever deflections need to be measured. Because of the aforementioned disadvantages, application of piezoelectric readout to MEMS sensors is somewhat limited.

17.3.3. Capacitive Method

Capacitance readout is based on measuring the capacitance between a conductor on the cantilever and another fixed conductor on the substrate that is separated from the cantilever by a small gap. Changes in the gap due to cantilever deformation result in changes in the capacitance between the conductors. Since the capacitance of a flat condenser is inversely proportional to the separation distance, sensitivity of this method relies on a very small gap between the cantilever and the substrate. Capacitance readout suffers from interference with variations in the dielectric constant of the medium. While differential schemes may eliminate this interference, electrically conductive media, such as electrolytes, make capacitance readout very challenging if possible at all. One of the main advantages of capacitance readout is that it can be used in integrated MEMS devices that are fully compliant with standard CMOS technology. An interesting variation of the capacitance methods is the "electron shuttling" regime that is especially promising for NEMS. For instance, Blick et al. reported on the "quantum bell"[60] that consists of five metal-coated cantilever structures and operates in the radio frequency range (Figure 17.10).

17.4. PERFORMANCE OF MICROMECHANICAL SENSORS

An important advantage of microcantilever sensors is that they can operate in vacuum, gases, and liquids. A compelling feature of the cantilever-based sensors operating in the resonant mode is that four response parameters (resonance frequency, phase, amplitude, and Q-factor, measured simultaneously) may provide complementary information about the interactions between the sensor and the environment. The damping effects of a liquid medium, however, reduces the resonance response of a resonating MEMS. In most common liquids, such as water, the amplitude of the observed resonance response is approximately an order of magnitude smaller as compared to the same resonating cantilever operating in air. On the other hand, operation in the static mode is unaffected by viscous properties of the medium. Tehrefore, microcantilever sensors operating in the static mode are especially attractive as a platform for nanomechanical biochemical assays and other biomedical applications. It is anticipated that microfabricated cantilevers can provide a versatile platform for real-time, in-situ measurements of physical, chemical, and biochemical properties of physiological fluids.

FIGURE 17.10. The "quantum bell" fabricated and studied by Blick *et al.* This NEMS operates at about 30 MHz and provides signal transduction in the electron shuttle regime. (From Ref. 60 by permission of Elsevier Science).

Cantilever sensors offer improved dynamic response, greatly reduced size, high precision, and increased reliability compared with conventional sensors. They are some of the simplest micromechanical systems that can be mass-produced using conventional micromachining techniques. The advantages of cantilever sensors can be fully realized by arranging individual cantilever transducers into large multi-sensor arrays integrated with on-chip electronic circuitry. Because typical thermal masses of microfabricated cantilevers are extremely small, they can be heated and cooled with a thermal time constant of less than a millisecond. This is advantageous for rapid reversal of molecular absorption processes and regeneration purposes. In general, the MEMS platform offers an unparalleled capability for the development and mass production of extremely sensitive, low-cost sensors suitable for rapid analysis of many chemical and biological species. Cantilever sensors with progressively increased mass sensitivity can be fabricated by simply reducing the transducer dimensions. While transition from MEMS to NEMS is expected to lead to better energy efficiency, shorter response times, and increased sensitivity, the challenges of NEMS fabrication and readout should not be underestimated.

Very important aspects of any sensor include sensitivity, selectivity, and the ability for regeneration. Similarly to other more conventional types of sensors developed previously, chemical selectivity of MEMS sensors relies on the use of selectively responsive phases, such as certain polymers, self-assembled monolayers, or biological receptors. As pointed out earlier, MEMS sensors can operate in the two distinctive modes, resonant and static, based, respectively, on measurements of resonance frequency variations and adsorption-induced deformations. Transduction efficiency of the static mode increases when the stiffness of the cantilever is reduced. Therefore, longer cantilevers with very small spring constants are preferable for the use in the static mode. On the other hand, the sensitivity of the resonant mode increases progressively with the operation frequency.

17.4.1. Sensitivity of the Resonance Frequency-Based Approach

Although adsorption-induced stresses were extensively explored as a transduction principle in many cantilever sensors, the advantage of the resonant operation is that it can potentially provide mass detection at the single molecule level. The resonance frequency of a cantilever beam depends on its geometry as well as the elastic modulus and density of its material. By changing cantilever dimensions, its resonance frequency can be varied from hundreds of Hz to a few GHz (see Figure 17.1). For a given cantilever mass, higher spring constants correspond to higher resonance frequencies. For a given cantilever thickness, shorter cantilevers have higher spring constants. Depending on the cantilever material, GHz resonance frequencies can be achieved, when the cantilever length is less than a few microns. Very short cantilevers with high resonance frequency are, therefore, promising in extending the detection limit down to a few molecules.

The dependence of the fundamental frequency on the cantilever parameters can be expressed as:

$$f_0 = \frac{t}{2\pi(0.98)L^2}\sqrt{\frac{E}{\rho}} \tag{17.6}$$

where ρ is the density of the cantilever material. The mass of the adsorbed material can be determined from the initial and final resonance frequency and the initial mass of the cantilever as:

$$\frac{f_0^2 - f_1^2}{f_0^2} \approx \frac{\delta m}{m} \tag{17.7}$$

where f_0 and f_1 are the initial and final frequency respectively and δm and m are adsorbed mass and initial mass of the cantilever, respectively. If the adsorption is confined to the free end of the cantilever, Eq. 17.7 should be modified in order to take into account the effective mass of the cantilever.

The mass sensitivity of a cantilever sensor can be defined as:

$$S_m = \lim_{\Delta m \to 0} \frac{1}{f_0}\frac{\Delta f}{\Delta \Gamma} = \frac{1}{f_0}\frac{df}{dm} \tag{17.8}$$

where $\Delta\Gamma$ and dm are normalized to the active sensor area of the device ($\Delta\Gamma = \delta m/A$, where A is the area of the cantilever). As can be seen from Eq. 17.8, the sensitivity is the fractional change of the resonant frequency of the structure with addition of mass to the sensor. When applying this definition to the case of the cantilever sensor, the sensitivity is defined as:

$$S_m = \frac{1}{\rho_a t_a} \times \frac{\Delta f}{f_o} \tag{17.9}$$

where ρ_a and t_a are, respectively, the density and the thickness of the adsorbate. Note that the resonance frequency, f_o, of a cantilever sensor depends on its material density, thickness, length, and elasticity.

Another figure of merit of a cantilever sensor is the smallest detectable mass. By rewriting Eqs. 17.8–17.9, the lowest detectable surface density of the adsorbate can be defined as:

$$\Delta\Gamma_{\min} = \frac{1}{S_m}\frac{\Delta f_{\min}}{f} \tag{17.10}$$

where $\Delta\Gamma_{\min}$, $\Delta f_{\min}$ are the minimum detectable surface density and minimum detectable frequency change, respectively.

By changing the physical dimension of a cantilever one can increase its detection limits by many orders of magnitude. For a given cantilever design, the smallest (thermal noise limited) detectable change in the surface density is given by:

$$\Delta\Gamma = 8\sqrt{\frac{2\pi^5\, k\, k_B\, TB}{f_0^5\, Q}} \tag{17.11}$$

17.4.2. Sensitivity of the Static Deflection Approach

As mentioned earlier, adsorption-induced cantilever bending is a preferable mode of cantilever operation in liquids. Using thin microfabricated cantilevers a sum of weak intermolecular forces involved in molecular adsorption processes on surfaces can be converted into readily measured displacements. Adsorbate-induced stresses and associated deformations should be distinguished from the bulk effects, such as changes in volume of thicker polymer films, which also lead to cantilever deformations. A relationship between the cantilever tip displacement and the differential surface stress is given by a modified Stoneys' equation:

$$\Delta z = \frac{3l^2(1-\upsilon)}{Et^2}\Delta\sigma \tag{17.12}$$

According to Eq. 17.12, there is a linear relationship between the cantilever bending and the differential surface stress.

Using the Shuttleworth equation,[61] surface stress, σ, and surface free energy, γ, can be interrelated

$$\sigma = \gamma + \left(\frac{\partial\gamma}{\partial\varepsilon}\right) \tag{17.13}$$

where σ is the surface stress. The surface strain $d\varepsilon$ is defined as ratio of change in surface area, $\partial\varepsilon = dA/A$. In many cases, the contribution from the surface strain term can be neglected and the free energy change approximately equals the change in surface stress.

17.5. APPLICATIONS OF CANTILEVERS SENSORS

17.5.1. Gas Phase Analytes

Detection of mercury vapors reported by Thundat *et al.*[52] was one of the first gas sensor applications of microscopic cantilevers. It was found that both resonance frequencies and static deflections of the gold coated cantilevers underwent changes in presence of mercury vapor (30 μg m^{-3}) added to a nitrogen carrier gas. When one side of the cantilevers was completely coated with gold, the resonance frequency of the cantilevers increased as a result of exposure to mercury vapors. This rather unexpected result was explained by competing effects of the absorbed mercury on the cantilever force constant and on the cantilever suspended mass. It was concluded that interaction of mercury with the gold coating led to an increase in the cantilever resonance frequency as a result of a relatively small increase in the cantilever effective mass and a more significant increase in the cantilever force constant.

Dual (static/dynamic) mode responses of gold coated were reported for several other gaseous phase analytes, in particular, 2-mercaptoethanol.[62] In the case of 2-mercaptoethanol, analyte-induced deflections rather than changes in the resonance frequency of gold-coated AFM cantilevers was found to be a preferable mode of sensor operation. Measurements of cantilever deflections permitted detection of mercaptoethanol vapors at the concentrations down to 50 part per billion (ppb). The calibration curve obtained in the static deflection mode had a slope of 0.432 nm per ppb in the concentration range of 0-400 ppb.

Fairly high sensitivity and selectivity demonstrated in the early works on cantilever sensors relied on properties of some metals used as active coatings. For instance, gold is a very chemically inert metal that, nevertheless has very high reactivity towards mercaptans (or thiols), i.e., substances with one or more sulfohydryl (–SH) groups. High solubility of hydrogen in palladium and palladium based alloys is one of the few other mechanisms that lead to selective interaction of metal coatings with gas-phase analytes. Good sensitivity of Au and Pd coated cantilevers to, respectively, mercury and hydrogen was subsequently used to implement a palm-sized, self-contained sensor module with spread-spectrum-telemetry reporting.[63]

As inorganic coatings alone can not provide the selectivity patterns sought in many applications, modification of cantilevers with chemically selective organic layers has been a subject of more recent studied. One of the first cantilever sensors with organic coatings was a humidity sensor described by Thundat *et al.*[64] In these studies, silicon nitride AFM cantilevers were coated with gelatine, by contacting one side of the cantilever with a 0.1% gelatine solution in distilled water. When thus prepared cantilever was exposed to atmosphere of gradually increased humidity, both cantilever deflections and increases in the resonance frequency were observed. Another design of a cantilever humidity sensor employed cantilevers with integrated piezoresistive read-out.[54] The design included both humidity sensitive and reference cantilevers as a part of a Wheatstone bridge. The layered silicon/silicon oxide cantilevers were 200 μm long, 50 μm wide 1.5 μm thick with deflection sensitivity, $\Delta R/R$, of approximately 10^{-6} nm^{-1}. Using a glass capillary and a micromanipulator, the active (humidity sensitive) cantilever was additionally coated with 10 μm photoresist. Swelling of the photoresist layer in presence of water vapor provided sensor responses that were nearly proportional to RH% in the range of 2 to 60%. The reference cantilever provided temperature compensation and could also be used for temperature measurements.[54]

In analogy to chemical sensors based on SAW transducers,[65] cantilevers coated with various commercially available polymers were proposed for distinguishing between different volatile organic compounds (VOCs) in air. Lang *et al.*[1,55] reported on a multi-cantilever sensor, in which signals were collected in a quasi-simultaneous (time-multiplexing) manner from eight individual cantilever transducers, each modified with a different coating.[66] This design allowed the researchers to transfer the concept of a "chemical nose" from more conventional transduction principles[67] to innovative nanomechanical devices. Polymethylmethacrylate (PMMA) as well as Pt metal coatings were used in some of these studies in order to demonstrate versatility of the cantilever arrays. Using a cantilever sensor with a PMMA coating, responses to a series of alcohols were obtained in both resonance and static deflection mode. Based on the differences in the shapes of response curves (either static deflection or resonance frequency change plotted as a function of time), the presence of different alcohols could be differentiated. In this case, the observed selectivity was primarily related to the fact that alcohols with different molecular weight and/or molecular structure have different diffusion rates in the PMMA coating. Therefore, the use of a multi-cantilever array with different polymeric coatings was the next logical step in developing a "chemical nose" based upon the cantilever platform. It was shown that cantilevers coated with several readily available "generic" polymers, such as polymethylmethacrylate, polystyrene, polyurethane and their blends or co-polymers, respond differently to various VOCs.[66] By applying principal component and artificial neuron network analysis to response patterns from arrays of such polymer-modified cantilevers, the concept of an artificial nose was successfully implemented. In order to create cantilever sensors with even more distinctive selectivity patterns with regard to different classes of VOCs, sol-gel coatings as well as covalently attached or evaporated films of synthetic receptors were found to be useful.[5,43,44] Thin films of sol-gels were formed on one side of 600 nm thick silicon cantilevers using aqueous solutions of organosilane precursors and spin coating procedures.[42] The cantilevers with sol-gel coatings exhibited strong bending in response to vapors of polar VOCs, in particular ethanol, while normalized sensitivity to less polar compounds was relatively low.

In addition to sensors that utilize static or dynamic (resonance) cantilever responses due to adsorption (or absorption) of analyte molecules, bimaterial cantilevers can detect local temperature changes associated with a chemical reaction that involves analyte molecules and is catalyzed by a catalyst on the cantilever surface. One of the first implementations of this detection scheme was reported by Barnes *et al.*[24] For a standard AFM bi-material cantilever and AFM optical cantilever read-out, the limits of detection were estimated to be 1 pJ of thermal energy and 10^{-5} K of local temperature differences.[24] Even higher sensitivity of this method can be achieved using modified silicon or silicon nitride cantilevers with increased thermal isolation between their active regions (catalytic areas) and supporting bases ("heat sinks"). For instance, we have shown in our recent studies that analytes present on cantilevers in a form of thin coatings (ca. 100 nm average thickness) can be detected in a calorimetric spectroscopy mode.[68] While such cantilever based spectroscopic instruments may not satisfy more rigorous definitions of chemical sensors they offer excellent portability combined with inherent to vibrational spectroscopy differentiating power.

17.5.2. Liquid Phase Analytes

Early works on cantilever based chemical detection in liquids involved standard AFM cantilevers and AFM heads for their read-out. For instance, alkylthiols terminated with different chemical groups were most extensively used as modifying agents for gold-coated cantilevers. The other modification procedures involved silane-oxide chemistry and spontaneous oxidation of evaporated aluminum films. When pH responses of cantilevers modified with carboxylic acid, hydroxyl and amino groups were analyzed,[57] reasonable correlation between the experimental calibration plots and expected protonation-deprotonation behavior of the cantilever surfaces was found. The reported pH responsivities varied from 15 to 50 nm/pH, depending on the surface treatment, cantilever type, and pH range.

Some of the most impressive figures of merit demonstrated with cantilever sensors are related detection of heavy metal ions. In particular, Ji *et al.*[69] reported on highly sensitive and selective detection of Cs^+ ions using a cantilever sensor with a self-assembled responsive layer of the molecular recognition type. The responsive layer for this sensor was formed using a newly synthesized receptor compound which combined calixarene and crown-ether macrocycles and had a reactive –SH group that provided its covalent attachment to gold surfaces. Using a cantilever transducer with this responsive layer, Cs^+ ions could be detected in the concentration range of 10^{-11} to 10^{-7} M. By modifying gold coated cantilever transducers with another self-assembled responsive monolayer, triethyl-12-mercaptododecylammonium bromide, a detector of trace amounts of CrO_4^{2-} was implemented.[56] As an extension of this concept, detection of trace levels of Ca^{2+} was also achieved using cantilever transducers modified with Ca^{2+} selective self-assembled responsive layers.[70] Self-assembled monolayers of yet another long-chain thiol compound were found to improve selectivity of gold coated cantilevers towards Hg^{2+} cations. Hg^{2+} at concentrations as low as 10^{-11} could be detected using this approach, while other cations, such as K^+, Na^+, Pb^{2+}, Zn^{2+}, Ni^{2+}, Cd^{2+}, Cu^{2+}, and Ca^{2+} had little or no effect on the cantilever deflections.

17.5.3. Biosensors

One of the first attempts to combine the biosensor concept and a cantilever transducer took advantage of the ultrahigh calorimetric sensitivity of a bimaterial microcantilever.[5] By immobilizing glucose oxidize on the surface of 200-μm long, gold coated silicon nitride, Thundat *et al.* created a glucose sensor that responded to presence of glucose in the aqueous medium due to the enzyme-induced exothermic processes.[71] Another indirect method of detecting biological species using micromachined cantilevers was proposed by Colton *et al.*[72] The proposed force amplified biological sensor (FABS) utilized a micromachined cantilever placed in a strong magnetic field. Similarly to many conventional biological assays, such as the enzyme-linked immunosorbent assay (ELISE), the FABS method relied on labeled biological material, however, magnetic beads rather than enzymes or fluorophores were used as a label. It was shown that an important advantage of the FABS method over existing bioassays is its capability to detect trace amounts of extremely dilute biological samples.

More recently, significant attention has been drawn to direct conversion of various biological receptor-ligand interactions into mechanical responses using cantilever. Butt *et al.* explored high sensitivity of cantilever transducers to interfacial stress changes [73] in their work on a biosensor for a herbicide.

Despite the excellent sensitivity of cantilever transducers and their capability to detect receptor-ligand interactions directly, the static deflection mode is not free from long-term drifts and instabilities inherent to other types of biosensors. In addition to temperature-induced drifts, it was also established that both specific binding and nonspecific adsorption of proteins on various surfaces are accompanied with very slow surface stress changes.[74] Moulin *et al.*[74] used microfabricated cantilevers to measure surface stress changes associated with nonspecific adsorption of immunoglobulin G (IgG) and bovine serum albumin (BSA) on gold surfaces. Compressive and tensile surface stress changes were observed upon adsorption of, respectively, IgG and BSA. This difference was attributed to different packing and deformation of each protein on the gold surface. Taking into account extremely high sensitivity of cantilever bending to interfacial biomolecular binding events, Moulin *et al.*[4] proposed a clinically relevant cantilever biosensor for differentiation of low density lipoproteins (LPL) and their oxidized form (oxLDL).

A significant milestone in developing cantilever based biosensors was demonstration of their applicability to DNA analysis. Fritz *et al.* reported on sensitive and specific monitoring of oligonucleotide hybridization using arrays of functionalized cantilevers[75] and optical read-out of their deflections. Arrays of 1 μm thick, 500 μm long rectangular silicon cantilevers, custom designed and microfabricated at IBM Zurich Research Laboratory were used in these studies. A thin layer of gold on one side of the cantilevers permitted controllable immobilization of thio-modified oligonucleotides. When 12-mer oligo-nucleotides with different degree of complementarity were used in the hybridization assay, a single base pair mismatch was clearly detectable. The use of a differential pair of cantilever transducers, i.e., functionalized and "blank", and analysis of the differential deflections was an important refinement that minimized interfering effects of temperature, mechanical vibrations and fluid flow in the cell and, therefore, provided more reliable differentiation of the responses that accompanied specific biomolecular interactions.

Under carefully controlled experimental conditions (temperature, pH, ionic strength, etc.), even a single cantilever transducer provides a sensitive means for detection of various biomolecular interactions. For instance, Thundat and coworkers succeeded in differentiating a single-nucleotide mismatch using a cantilever transducer placed in a thermally stabilized flow cell.[76] The same group of researchers also reported[76] on detection of ultra-low concentrations (0.2 ng/ml) of prostate specific antigen (PSA) using a similar thermoelectrically stabilized cell housing a single cantilever transducer.

Biotin-streptavidin is yet another example of high-affinity biomolecular interactions that was successfully monitored using cantilever transducers. Raiteri *et al.* used biotin functionalized silicon nitride cantilevers and measured their deflection responses in presence of 100 nM streptavidin.[77] These responses that reached approximately 50 nm in magnitude within 10 min were largely reversible. In the case of high-affinity streptavidin-biotin interactions, the reversible nature of the responses is especially unexpected and apparently indicates a nontrivial relationship between the surface coverage of streptavidin molecules on the cantilever surface and the associated surface stress change.

17.6. SUMMARY

Adsorption-induced cantilever deflection and resonance frequency variation can form the basis of a universal platform for real-time, in-situ measurement of physical, chemical, and biochemical properties. A plethora of physical, chemical, and biological sensors based on the micromachined cantilever platform have already been demonstrated. Because cantilever bending and resonance frequency can be measured simultaneously, sensors can be based on adsorption-induced resonance frequency shifts and/or cantilever bending. The resonance frequency shifts and bending of a cantilever can be measured with very high precision using different readout techniques[78] such as optical beam deflection, variations in piezoresistivity, capacitance, and piezoelectric properties.

As the technology to fabricate nanosize mechanical structures further develops we envision a plethora of new application where these systems can play an important role. As the frequency of these devices approaches or even exceeds GHz, nanomechanical devices will be in the same time and frequency domain reserved now only for electronic devices.

One very important issue that needs to be addressed is the efficient readout of nanocantilevers not only because of the small size but also their high density in the arrays. Conventional techniques of measuring the resonance frequency, such as optical beam deflection, fall short when applied to micromachined nanocantilevers. For example, in optical beam deflection the cantilever motion is measured by reflecting a laser diode off the free end of a cantilever into a position sensitive detector. The shortcomings of optical techniques are simply due to lack of sufficient reflected (or scattered) optical signal from the cantilever beam. Optical beam deflection is extremely sensitive when used for cantilevers that are 50 μm to a few hundreds of micrometers long while the electron transfer signal transduction is extremely sensitive for cantilevers that are a few hundred nanometers to a few microns in length.

The applicability of electron transfer signal transduction for aqueous environments is extremely challenging. Presence of electro active ions in the water can cause a large Faraday leakage current that can overwhelm electron transfer signal. The leakage current, however, can be significantly reduced using proper insulation, reduced bias voltage, and reduced number of charge carriers in the solution. However, this technique may be applicable for biosensors that can be operated in humid atmosphere.

QUESTIONS

1. Estimate frequency shifts of two gravimetric micromechanical sensors resonating at 10 kHz and 1 MHz due to adsorption of adsorbate with the effective mass equal 1% of the initial suspended mass.

2. A gold coated silicon cantilever is used as a temperature sensor. Estimate the displacement of the cantilever tip per 1 K temperature change if the cantilever thickness and length are 1 mm and 500 mm, respectively, and the thickness of the gold coating is 100 nm. Coefficients of thermal expansion for gold and silicon are $1.4 \times 10^{-5}K^{-1}$ and $2.6 \times 10^{-6}K^{-1}$, respectively. Young's moduli of gold and silicon are 77 and 156 GPa, respectively.

3. Four different rectangular silicon cantilevers are 100 mm long and 0.5 mm thick and are coated with 500 nm of, respectively, gold, silicon, silicon nitride, and aluminum. Describe

what will happen to the cantilever resonance frequency in each case (increase, decrease or no change).

4. Estimate the amplitude of spontaneous cantilever oscillation at 300 K within 1 Hz band away from its resonance by taking into account the following cantilever parameters:
   — cantilever spring constant $k = 0.01$ N m^{-1}
   — $Q = 100$
   — Resonance frequency $f_0 = 1$ MHz

REFERENCES

1. R. Berger, C. Gerber, H. P. Lang, and J. K. Gimzewski, *Microelectronic Engin.* **35**, 373 (1997).
2. E. A. Wachter and T. Thundat, *Rev. Sci. Instr.* **66**, 3662 (1995).
3. H. L. Tuller and R. Mlcak, Current *Opinion in Solid State & Materials Science* **3**, 501 (1998).
4. A. M. Moulin, S. J. O'Shea, and M. E. Welland, *Ultramicroscopy* **82**, 23 (2000).
5. P. G. Datskos, M. J. Sepaniak, C. A. Tipple, and N. Lavrik, *Sens. Act. B-Chemical* **76**, 393 (2001).
6. M. Sepaniak, P. Datskos, N. Lavrik, and C. Tipple, *Analytical Chemistry* **74**, 568A (2002).
7. T. Thundat, P. I. Oden, and R. J. Warmack, *Microscale Thermo. Engin.* **1**, 185 (1997).
8. H. G. Craighead, *Science* **290**, 1532 (2000).
9. R. D. Pereira, *Biochemical Pharmacology* **62**, 975 (2001).
10. H. J. Butt, *J. of Colloid and Interface Science* **180**, 251 (1996).
11. J. Samuel, C. J. Brinker, L. J. D. Frink, and F. van Swol, *Langmuir* **14**, 2602 (1998).
12. L. J. D. Frink and F. van Swol, *Colloids & Surfaces A Physicochemical & Engineering Aspects* **162**, 25 (2000).
13. R. Berger, E. Delamarche, H. P. Lang, C. Gerber, J. K. Gimzewski, E. Meyer, and H. J. Guntherodt, *Appl. Phys. A* (Materials Science Processing) **66**, S 55 (1998).
14. J. E. Sader, *J. Appl. Phys.* **91**, 9354 (2002).
15. J. E. Sader, *J. Appl. Phys.* **89**, 2911 (2001).
16. M. Godin, V. Tabard-Cossa, P. Grutter, and P. Williams, *Appl. Phys. Lett.* **79**, 551 (2001).
17. R. Raiteri, H. J. Butt, and M. Grattarola, Electrochimica Acta 46, 157 (2000).
18. K. J. Bruland, J. L. Garbini, W. M. Dougherty, and J. A. Sidles, *J. of App. Phys.* **83**, 3972 (1998).
19. D. Rugar, C. S. Yannoni, and J. A. Sidles, *Nature* **360**, 563 (1992).
20. H. J. Mamin and D. Rugar, *Appl. Phys. Lett.* **79**, 3358 (2001).
21. P. Streckeisen, S. Rast, C. Wattinger, E. Meyer, P. Vettiger, C. Gerber, and H. J. Guntherodt, *Appl. Phys. A-Materials Science & Processing* **66**, S341 (1998).
22. A. N. Cleland and M. L. Roukes, *Nature* **392**, 160 (1998).
23. A. C. Stephan, T. Gaulden, A. D. Brown, M. Smith, L. F. Miller, and T. Thundat, *Rev. Sci. Instrum.* **73**, 36 (2002).
24. J. R. Barnes, R. J. Stephenson, M. E. Welland, C. Gerber, and J. K. Gimzewski, *Nature* **372**, 79 (1994).
25. R. Berger, C. Gerber, J. K. Gimzewski, E. Meyer, and H. J. Guntherodt, *Appl. Phys. Lett.* **69**, 40 (1996).
26. R. Berger, H. P. Lang, C. Gerber, J. K. Gimzewski, J. H. Fabian, L. Scandella, E. Meyer, and H. J. Guntherodt, *Chem. Phys. Lett.* **294**, 363 (1998).
27. P. G. Datskos, S. Rajic, M. J. Sepaniak, N. Lavrik, C. A. Tipple, L. R. Senesac, and I. Datskou, *J. Vac. Sci. Technol.* B **19**, 1173 (2001).
28. P. I. Oden, P. G. Datskos, T. Thundat, and R. J. Warmack, *Appl. Phys. Lett.* **69**, 3277 (1996).
29. E. A. Wachter, T. Thundat, P. I. Oden, R. J. Warmack, P. G. Datskos, and S. L. Sharp, *Rev. Sci. Instrum.* **67**, 3434 (1996).
30. P. G. Datskos, S. Rajic, and I. Datskou, *Ultramicroscopy* **82**, 49 (2000).
31. T. Perazzo, M. Mao, O. Kwon, A. Majumdar, J. B. Varesi, and P. Norton, *Appl. Phys. Lett.* **74**, 3567 (1999).
32. S. P. Timoshenko, Analysis of Bi-metal Thermostats, *J. Opt. Soc. Am.* **11**, 233 (1925).
33. A. M. Moulin, R. J. Stephenson, and M. E. Welland, *J. Vac. Sci. Technol.* B **15**, 590 (1997).
34. R. E. Martinez, W. M. Augustyniak, and J. A. Golovchenko, *Phys. Rev. Lett.* **64**, 1035 (1990).
35. H. Ibach, *J. Vac. Sci. Technol. A-Vacuum Surfaces and Films* **12**, 2240 (1994).

36. R. Koch, *Journal of Physics-Condensed Matter* **6**, 9519 (1994).
37. R. Koch, *Applied Physics a-Materials Science & Processing* **69**, 529 (1999).
38. J. Israelachvili, *Intermolecular and Surface Forces* (Academic Press, San Diego, 1991).
39. G. G. Stoney, *Proceedings of the Royal Society of London A* **82**, 172 (1909).
40. F. J. von Preissig, *J. Appl. Phys.* **66**, 4262 (1989).
41. T. A. Betts, C. A. Tipple, M. J. Sepaniak, and P. G. Datskos, *Analytica Chimica Acta* **422**, 89 (2000).
42. B. C. Fagan, C. A. Tipple, Z. L. Xue, M. J. Sepaniak, and P. G. Datskos, *Talanta* **53**, 599 (2000).
43. N. V. Lavrik, C. A. Tipple, M. J. Sepaniak, and P. G. Datskos, *Biomedical Microdevices* **3**, 33 (2001).
44. N. V. Lavrik, C. A. Tipple, M. J. Sepaniak, and P. G. Datskos, *Chem. Phys. Lett.* **336**, 371 (2001).
45. C. A. Tipple, N. V. Lavrik, M. Culha, J. Headrick, P. Datskos, and M. J. Sepaniak, *Anal. Chem.* **74**, 3118 (2002).
46. D. Sarid, *Scanning Force Microscopy* (Oxford University Press, New York, 1991).
47. T. R. Albrecht, P. Grutter, D. Horne, and D. Rugar, *J. Appl. Phys.* **69**, 668 (1991).
48. A. J. Steckl, H. C. Mogul, and S. Mogren, *Appl. Phys. Lett.* **60**, 1833 (1992).
49. J. Brugger, G. Beljakovic, M. Despont, N. F. deRooij, and P. Vettiger, *Microelectronic Engineering* **35**, 401 (1997).
50. J. Buhler, F. P. Steiner, and H. Baltes, *J. of Micromechanics and Microengineering* **7**, R1 (1997).
51. Y. Zhao, M. Y. Mao, R. Horowitz, A. Majumdar, J. Varesi, P. Norton, and J. Kitching, *J. of Microelectromechanical Systems* **11**, 136 (2002).
52. T. Thundat, E. A. Wachter, S. L. Sharp, and R. J. Warmack, *Appl. Phys. Lett.* **66**, 1695 (1995).
53. Z. Y. Hu, T. Thundat, and R. J. Warmack, *J. Appl. Phys.* 90**,** 427 (2001).
54. A. Boisen, J. Thaysen, H. Jensenius, and O. Hansen, *Ultramicroscopy* **82**, 11 (2000).
55. M. K. Baller, H. P. Lang, J. Fritz, C. Gerber, J. K. Gimzewski, U. Drechsler, H. Rothuizen, M. Despont, P. Vettiger, F. M. Battiston, J. P. Ramseyer, P. Fornaro, E. Meyer, and H. J. Guntherodt, *Ultramicroscopy* **82**, 1 (2000).
56. H. F. Ji, T. Thundat, R. Dabestani, G. M. Brown, P. F. Britt, and P. V. Bonnesen, *Anal. Chem.* **73**, 1572 (2001).
57. H. F. Ji, K. M. Hansen, Z. Hu, and T. Thundat, *Sens. Act. B-Chemical* **72**, 233 (2001).
58. X. H. Xu, T. G. Thundat, G. M. Brown, and H. F. Ji, *Anal. Chem.* **74**, 3611 (2002).
59. D. W. Carr, S. Evoy, L. Sekaric, H. G. Craighead, and J. M. Parpia, *Appl. Phys. Lett.* **75**, 920 (1999).
60. A. Erbe and R. H. Blick, *Physica B* **272**, 575 (1999).
61. R. Shuttleworth, *Proc. Phys. Soc. London* **63A**, 444 (1950).
62. P. G. Datskos and I. Sauers, *Sens. Act. B Chemical* **61**, 75 (1999).
63. C. L. Britton, R. L. Jones, P. I. Oden, Z. Hu, R. J. Warmack, S. F. Smith, W. L. Bryan, and J. M. Rochelle, *Ultramicroscopy* **82**, 17 (2000).
64. T. Thundat, G. Y. Chen, R. J. Warmack, D. P. Allison, and E. A. Wachter, *Anal. Chem.* **67**, 519 (1995).
65. J. W. Grate, 1995.
66. H. P. Lang, R. Berger, F. Battiston, J. P. Ramseyer, E. Meyer, C. Andreoli, J. Brugger, P. Vettiger, M. Despont, T. Mezzacasa, L. Scandella, H. J. Guntherodt, C. Gerber, and J. K. Gimzewski, *Appl. Phys. A* (Materials Science Processing) **66**, S 61 (1998).
67. J. Janata, M. Josowicz, and D. M. Devaney, Chemical Sensors, *Anal. Chem* **66**, R207 (1994).
68. E. T. Arakawa, N. V. Lavrik, and P. G. Datskos, *Applied Optics* (in press).
69. H. F. Ji, E. Finot, R. Dabestani, T. Thundat, G. M. Brown, and P. F. Britt, *Chem. Comm.* **6**, 457 (2000).
70. S. Cherian, A. Mehta, and T. Thundat, *Langmuir* **18**, 6935 (2002).
71. A. Subramanian, P. I. Oden, S. J. Kennel, K. B. Jacobson, R. J. Warmack, T. Thundat, and M. J. Doktycz, *Appl. Phys. Lett.* **81**, 385 (2002).
72. D. R. Baselt, G. U. Lee, K. M. Hansen, L. A. Chrisey, and R. J. Colton, *Proc. IEEE 85*, **672** (1997).
73. R. Raiteri, G. Nelles, H. J. Butt, W. Knoll, and P. Skladal, *Sens. Act. B-Chemical* **61**, 213 (1999).
74. A. M. Moulin, S. J. O'Shea, R. A. Badley, P. Doyle, and M. E. Welland, *Langmuir* **15**, 8776 (1999).
75. J. Fritz, M. K. Baller, H. P. Lang, H. Rothuizen, P. Vettiger, E. Meyer, H. J. Guntherodt, C. Gerber, and J. K. Gimzewski, *Science* **288**, 316 (2000).
76. G. H. Wu, R. H. Datar, K. M. Hansen, T. Thundat, R. J. Cote, and A. Majumdar, *Nature Biotechnology* **19**, 856 (2001).
77. R. Raiteri, M. Grattarola, H. J. Butt, and P. Skladal, *Sens. Act. B-Chemical* **79**, 115 (2001).

VI

Nanoscale Optoelectronics

18

Quantum-Confined Optoelectronic Systems

Simon Fafard*

Cyrium Technologies Inc., Ottawa, Ontario, Canada

This chapter discusses quantum-confined semiconductor structures and their role in optoelectronic systems, with an emphasis on quantum dot nanostructures. We first start with an overview of quantum optoelectronic devices in the more traditional epitaxial systems, opening with a brief discussion of semiconductor lasers and other devices, and describe the impact and advantages of quantum confinement for these devices. The chapter will then focus on the science and technology of quantum dots and related devices, in particular Stranski-Kranstanow growth, self-assembling quantum dots, and the engineering of the quantum dots. We conclude with the challenges and the outlook of quantum dot optoelectronic devices.

18.1. INTRODUCTION

Dimensional control, smoothness, and uniformity of epitaxial films have been the cornerstones in the development of two-dimensional quantum structures. They have enabled the engineering of many quantum structures and devices including quantum well heterostructures, high mobility two dimensional electron gas, and resonant tunneling devices. These devices have changed our every-day life because of their common use in our telecommunication systems and in the optoelectronic technologies we use regularly (for example, CD and DVD players/recorders).

*Also at Department of Physics, University of Ottawa, Ottawa, Ontario, Canada K1N 6N5.

18.1.a. Two-Dimensional Carrier Confinement

The use of optical fibers to carry information has revolutionized the world's telecommunications. In particular, the past 10 years have generated numerous technological advancements. Optical bits are now transported at rates of 10 Gbit/s and higher between buildings (Enterprise Networks, Storage Area Networks), within a city (Metropolitan Networks), between cities and between continents (Long Haul and Ultra Long Haul Networks). The guiding of light signals into optical fibers is based on confinement of photons into glass materials with a different index of refraction between the core and the cladding. That is, photon confinement in passive materials. Similarly, the devices which are used for generating (laser diode transmitters) and detecting (photodetector receivers) the light signals are based on confinement of electrons and holes into semiconductor materials having different bandgaps. That is, electron confinement in active materials. The dispersion of the silica in the fibers imposes some constraints on the spectral linewidth of the semiconductor lasers which can be used to propagate the bits over long distances. The dispersion tolerances are particularly stringent as the data rate gets higher. Also, the choice of direct modulation or external modulation influences the chirp of the laser diodes and consequently the distance the information can travel before regeneration of the signals is necessary. Also, the temperature stability of material becomes essential with the thrust to dense wavelength division multiplexing (WDM), especially with channel spacing of 50 GHz or 25 GHz in the C-band (1529 nm–1565 nm) and/or L-band (1565 nm–1610 nm). In concert with the optical confinement in the semiconductor diodes, carrier confinement was key in achieving the necessary performance for various transmitters such as Fabry-Perot lasers, DFB lasers, distributed Bragg reflector (DBR) lasers, vertical cavity surface emitting lasers (VCSELs), and even LEDs. Similarly, for the case of semiconductor optical amplifiers (SOA), quantum confinement leads to favorable gain properties and places SOA as an interesting approach to compete with Erbium Doped Fiber Amplifier (EDFA) for the amplification of the optical signals. This is especially important in cost-driven applications such as in the metropolitan segment of telecom. For non-WDM applications, it can be suitable as an in-line booster or as a pre-amp on the detection side. For WDM systems, the SOA can be used as a pre-amp placed before the detection using semiconductor *p-i-n* or avalanche photodiode (APD) devices.

Carrier confinement is also crucial to obtain efficient laser diodes to pump EDFA with hundreds of milliWatts at 980 nm and/or around 1480 nm. The arrival of further integration using various material platforms for Mux/DeMux functions and for active components will also benefit from material flexibility. For example, III–V detectors or lasers are now hybrid-integrated with silica arrayed waveguide gratings (AWG) and with variable optical attenuators (VOA) to obtain higher level functionalities while using a reduced footprint. Reduced size is also important to minimize the packaging challenges and thus keep the cost of the modules down. Innovative designs also need flexibility to accommodate wavelength conversion and routing needs while reducing the number of optical-electrical-optical (OEO) conversions.

18.1.b. Three-Dimensional Carrier Confinement

In the past decades, much interesting research progress has been made using quantum dot semiconductor nanostructures which confine the carriers in all 3 dimensions (thus described as zero-dimensional structures). Ideal zero-dimensional quantum systems have a deep confining potential yielding several observable discrete levels. The observation of a number of well-resolved excited states is therefore the principal figure of merit for semiconductor heterostructures based on quantum dots. Studies of self-assembled quantum dots of high quality have revealed a number of resolved electronic shells in state-filling spectroscopy.[1–9] Recently, it has been demonstrated that self-assembled growth can be controlled to systematically and reproducibly fabricate quantum dot structures which have well-defined excited state transitions, resembling an artificial atom, and to manipulate their energy levels to tailor the number of confined states and their intersublevel energy-spacing.[10–11] Moreover, the structural quality of the self-assembled quantum dots and their compatibility with conventional III–V technology has permitted the demonstration of high performance prototype devices such as semiconductor quantum dot lasers.[12–27] These developments are very promising for quantum dot devices which can yield novel nanostructures with unique properties due to their atomic-like shell configurations. For example, quantum dot infrared detectors are sensitive in normal incidence detection due to the geometry of the quantum dots. Also, optical memories based on coupling of quantum dots can have large storage density.

For some applications, it is desirable to have multiple layers of quantum dots. Uniformity in the quantum dot ensemble can sometimes be a problem, especially when multiple stacked-layers[25] are used for the purpose of trying to increase the gain in the active region, or to increase the photoresponse. In the following sections, various aspects of semiconductor self-assembled quantum dots will be reviewed, as well as how size and shape engineered single layers and stacks[13,28] of uniform quantum dots can be used to obtain optoelectronic devices, such as quantum dot lasers, with sharp adjustable zero dimensional shells.

18.2. SIZE AND SHAPE ENGINEERING OF QUANTUM DOTS

As discussed in Chapter 7, nanoscale optoelectronic devices can be fabricated using self-assembled growth with most epitaxy machines. Molecular beam epitaxy (MBE), chemical beam epitaxy (CBE), and metal-organic vapor phase epitaxy are commonly used. For example, the III–V layers are often grown in an MBE system using a nominal As_2 or As_4 molecular flux.[28] The self-assembled quantum dots are most commonly obtained using the spontaneous island formation in the initial stages of the Stranski-Krastanow growth mode during the epitaxy of highly strained InAs on (Al, Ga)As layers on (100) GaAs substrates.[29–31] The GaAs-based laser structures typically consist of a thick ($\sim$2 μm) n^+ $Al_xGa_{1-x}As$ contact layer, followed by an $Al_yGa_{1-y}As$ bottom cladding layer with a lower doping and with $y < x$ to form a 2-step separate confinement cladding region, followed by the active region. The active region is usually made of thin undoped GaAs on each side of a stack of self-aligned InAs quantum dots. This section will give examples of results for lasers obtained with stacks of 7 or 14 layers with GaAs spacers of $\sim$10 nm, for $x = 0.7$ and $y = 0.35$. The active region is followed by the symmetric step-graded cladding and p-doped contact layers, and terminated with a p^+ GaAs cap. For samples designed specifically for optical and structural studies, a GaAs buffer layer (for example 0.8 μm thick) is grown below the quantum dots and they are covered with a thin GaAs cap, typically between 30 nm and 100 nm.

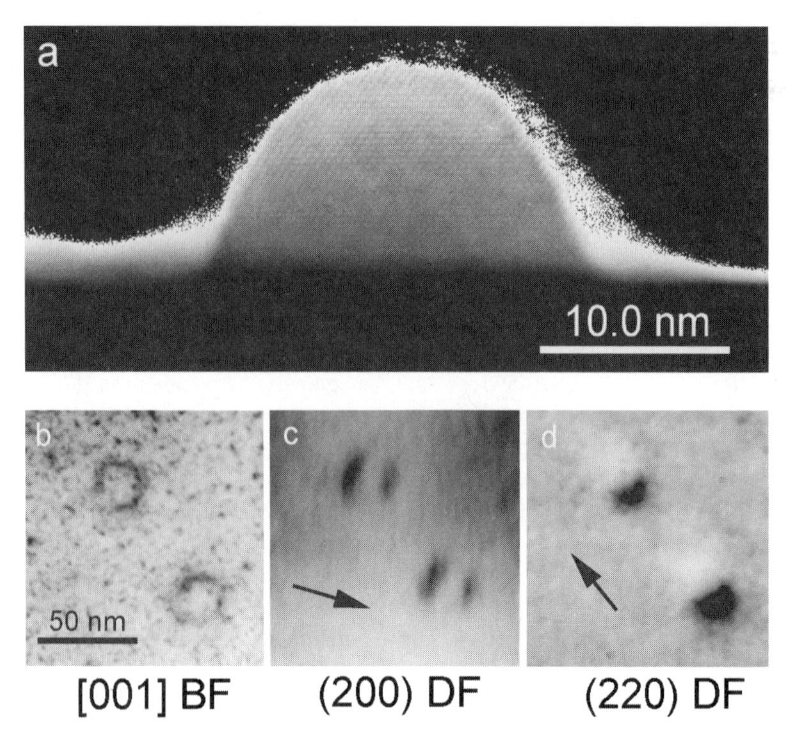

FIGURE 18.1. Transmission electron microscope (TEM) images of InAs/GaAs self-assembled quantum dots. (a) Cross-section [011] view of an uncapped quantum dot and (b–d) plan view of 2 capped quantum dots from different bright field (BF) and dark field (DF) diffraction conditions. (From Ref. 32 by permission of American Institute of Physics and Ref. 44 by permission of WILEY-VCH Verlag GmbH & Co.)

Transmission Electron Microscopy (TEM) is a useful tool to assess the structural aspect of the quantum dots. For example, Fig. 18.1a shows a cross-section view of an InAs/GaAs quantum dot which was grown without a cap layer.[32] The sample was grown on (100) GaAs, at a substrate temperature of ~515°C, with a deposition of the equivalent of 0.54 nm of InAs at a growth rate of 0.02 nm/s, and followed with a growth interruption of 60 s. For the uncapped quantum dots this can be followed by a rapid cool down. TEM gives a nice illustration of the island growth and it can also be useful for evaluating the diameter of the quantum dots, but is not representative of the height of the quantum dots since the latter can be adjusted by doing some size and shape engineering during the growth of the cap layer.[13,28,33–35] To properly evaluate the shape and the size of the quantum dots, plane view TEM can be used on capped layers. Such imaging is complicated by the fact that contrast mainly originates from the strain profile and it changes with the diffraction condition used for the imaging as seen in Fig. 18.1b–d. To interpret the apparent shape, accurate modeling is required.[36–38] It was found that the (220) DF condition gives very little differentiation in the shape, but the [001] BF condition can give useful information about the quantum dot perimeter.[32]

Further information about the shape and composition can be obtained using scanning tunneling microscopy (STM).[39–43] For example, Fig. 18.2 shows a cross-sectional STM image.[39] Such studies provide information with atomic resolution about the shape and

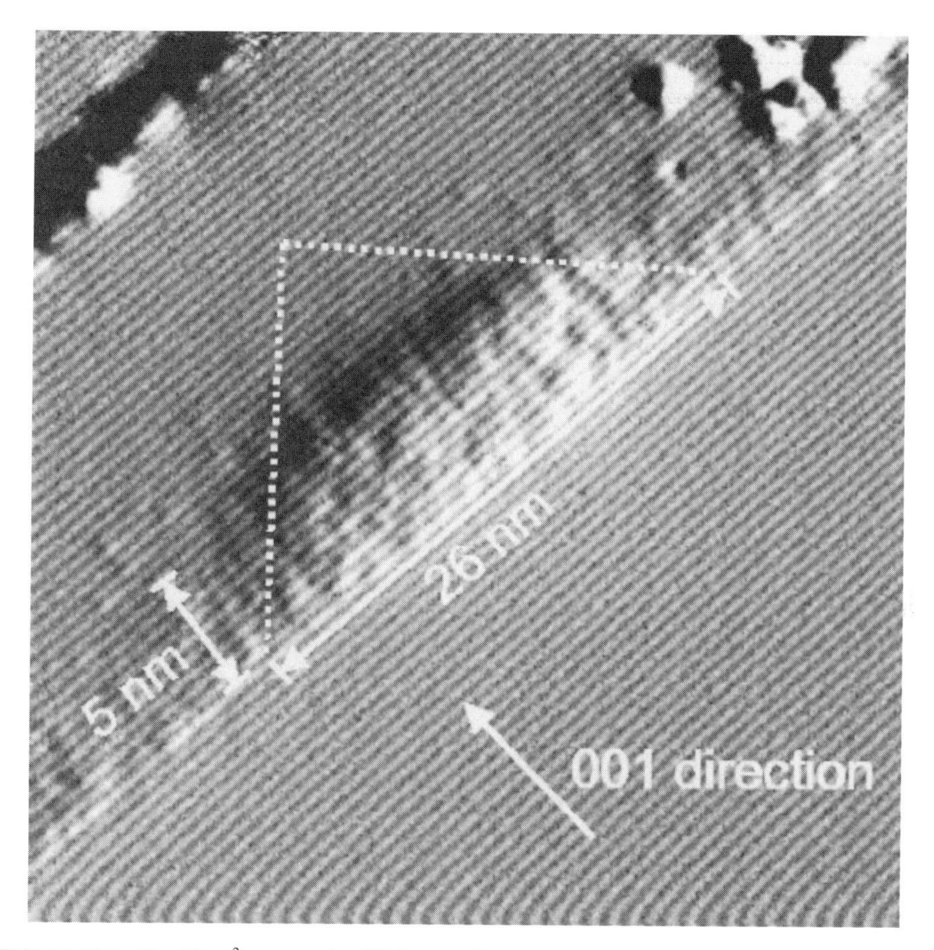

FIGURE 18.2. 40×40 nm^2 cross-section STM current image of a cleaved InAs quantum dot and the wetting layer. (From Ref. 39 by permission of American Institute of Physics.)

the composition of the nanostructures, giving details about the height and the diameter of the strained material. Measurements of the lattice constant also give details concerning the strain and the local indium concentration.

Mutlilayer stacks of quantum dots can also be grown with good uniformity. For example, Fig. 18.3 shows the active region of a quantum dot laser grown with 14 layers where the quantum dots are quite uniform from one layer to the next.[44] This is useful for improving the gain saturation of quantum dot lasers, but it is also of great interest for coupled quantum dot structures[45] as will be discussed in Section 18.6f. Indeed, a prerequisite for observing the level splitting caused by quantum mechanical coupling between overlapping states is that the levels of the uncoupled quantum dots should be close in energy. Size, composition, and strain changes in the stacks can often cause energy level variations in the adjacent quantum dots. This leads to larger inhomogeneous broadening (increased absorption and emission linewidths due to a broad distribution of different energy levels) in ensembles of uncoupled

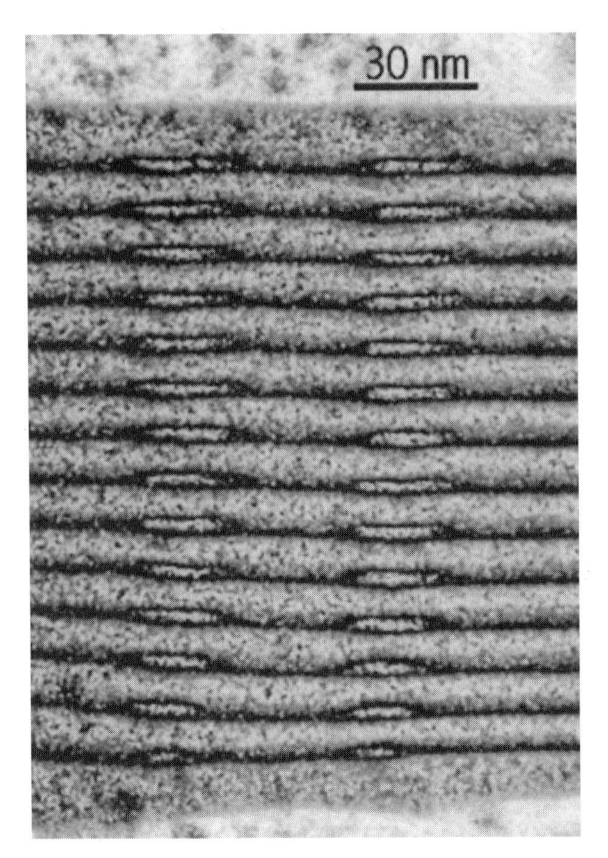

FIGURE 18.3. Laser diode with shape-engineered self-aligned quantum dots. Cross-section TEM of the active region of the laser diode containing a stack of 14 layers of InAs quantum dots indium-flushed at 5.0 nm and separated by 10.0 nm GaAs spacers. (From Ref. 44 by permission of WILEY-VCH Verlag GmbH & Co.)

quantum dot stacks, and to charge transfer between asymmetric quantum dots instead of level splitting for the closely stacked quantum dots.

18.3. OPTICAL PROPERTIES OF SELF-ASSEMBLED QUANTUM DOTS

Structural information can be useful to model the energy levels of the quantum dots For example, Fig. 18.4 depicts a quantum dot having the shape of an hemispherical cap o top of the wetting layer (WL).[46] Such a quantum dot would give a round perimeter in plan-view TEM under the [001] bright field condition of Fig. 18.1b. For example, using quantum dot potential as in Fig. 18.4, the wavefunctions and the quantum dot energy level can be calculated.[46–47] To exemplify, Fig. 18.5 depicts the electron orbitals of a lens-shape quantum dot.[48] In similarity to real atoms, the orbitals can be classified into s, p, d, f, etc, shells according to their quantum numbers $l = m + n$, where n and m would be th radial and angular quantum numbers respectively. For example, for an effective harmoni oscillator potential, there is a single-degenerate $l = 0$ for the s-shell, double-degenera $l = 1$ for p-shell, triple-degenerate $l = 2$ for d-shell, etc. A schematic of the energy leve

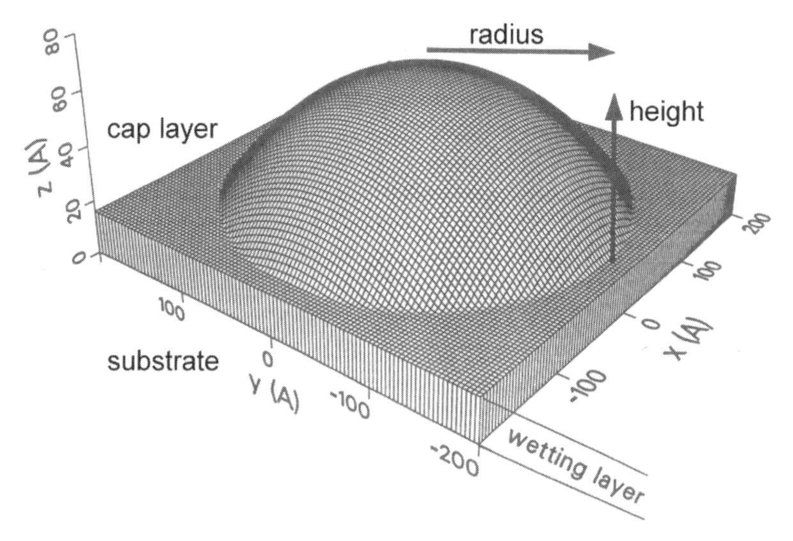

FIGURE 18.4. Modeling of the self-assembled quantum dot potential using a hemispherical cap of InAs on top of an InAs wetting layer embedded in a GaAs substrate and cap layer. (From Ref. 46 by permission of American Physical Society.)

of an InAs/GaAs self-assembled quantum dot having 5 electron and hole shells ($v = s$, p, d, f, g) is illustrated in Fig. 18.6. There is a nearly constant energy separation between the shells. This is reminiscent of the effective parabolic-potential, and it leads to a degeneracy $g_v = 2(l + 1)$ where $l = 0$ for s, $l = 1$ for p, etc, and the factor 2 is for the spin. The resulting interband transition will be mainly between excitons formed from electrons and holes in shells with the same quantum numbers, yielding S, P, D, etc transitions in photoluminescence experiments.[8] Fig. 18.6 also shows that the dynamics of the state-filling involves the interband recombination time for the various shells (τ_v), the intersublevel relaxation rates (γ_{ISL}) which will depend on the filling of the lower shells because of the Pauli exclusion principle, and also the carrier diffusion (τ_D) and capture (τ_c) times.[8] This is a simplified single particle picture which neglects the effects of multiexciton complexes as the number of excitons in the quantum dot is changed. The latter can lead to variation in the emission energy over tens of meV.[47,49] Nevertheless, condensation of the electrons and holes into coherent many-exciton states due to hidden symmetries yields emission spectra with electronic shells which can be associated with s, p, d, etc, electronic shells similar to the simple picture presented above for the single particle energy levels.[47,4] These electronic shells can be observed for quantum dot ensembles for which the inhomogenous broadening leads to nergy fluctuation smaller than the lateral quantization energy: i.e., an emission linewidth ith full-width at half-maximum (FWHM) smaller than the intersublevel energy spacing.

It has been found that there are five main parameters to control during the growth to obin good quality quantum dots with well-resolved excited-state emission: (1) the substrate mperature is selected to control the size of the quantum dots and thus to obtain the desired tersublevel energy-spacing;[10,11,50] (2) the amount of strained material deposited (here As) has to be precisely determined for the chosen growth temperature to control the density quantum dots and thus obtain well-resolved excited-states.[10,51,52] For an optimized density, in addition to a number of quantum dot states, the wetting layer photoluminescence (PL)

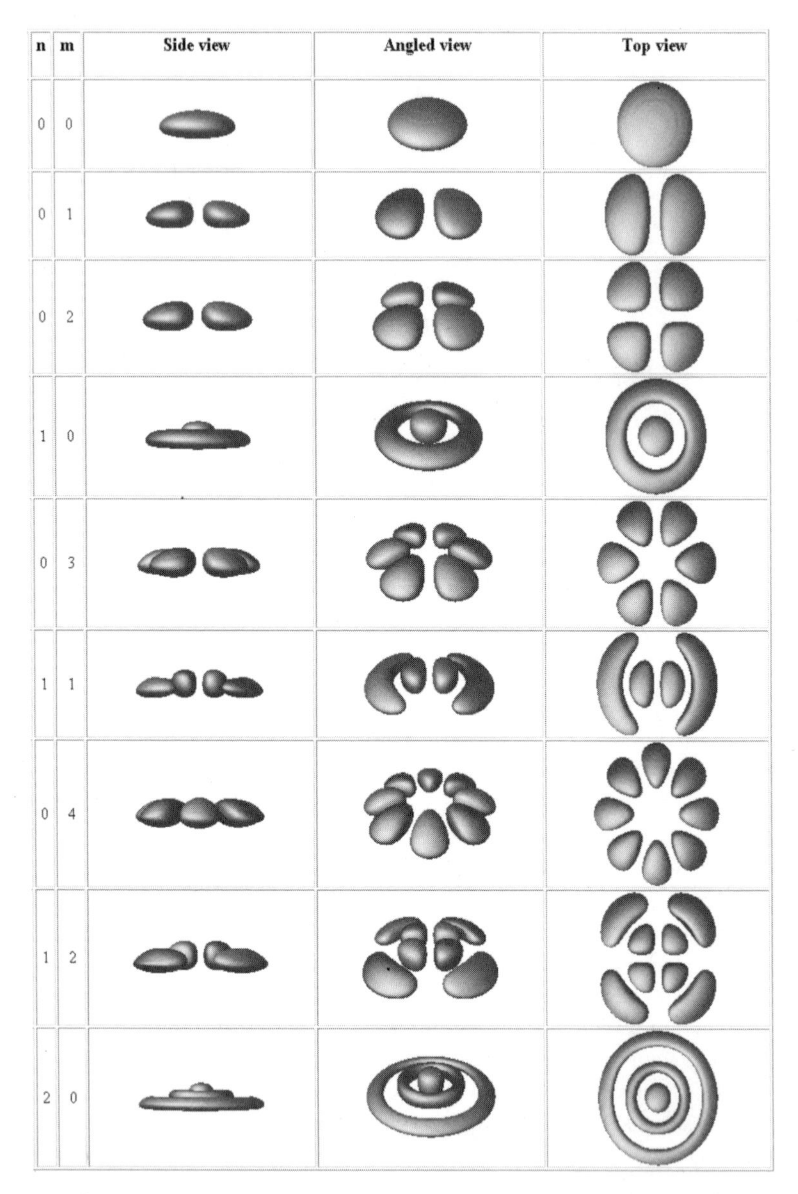

FIGURE 18.5. Orbitals in a lens-shaped quantum dot. The probability density of the electron wavefunction are displayed for the various radial (n) and angular (m) quantum numbers as viewed from different perspectives. In similarity to real atoms, the orbitals can be classified into s, p, d, f, g, etc, shells according to their quantum numbers $l = m + n$. For example, single-degenerate $l = 0$ for the s-shell, double-degenerate $l = 1$ for p-shell, triple-degenerate $l = 2$ for d-shell, etc. Each level is double-degenerate for the spin. (From Ref. 48 by permission of SPIE.)

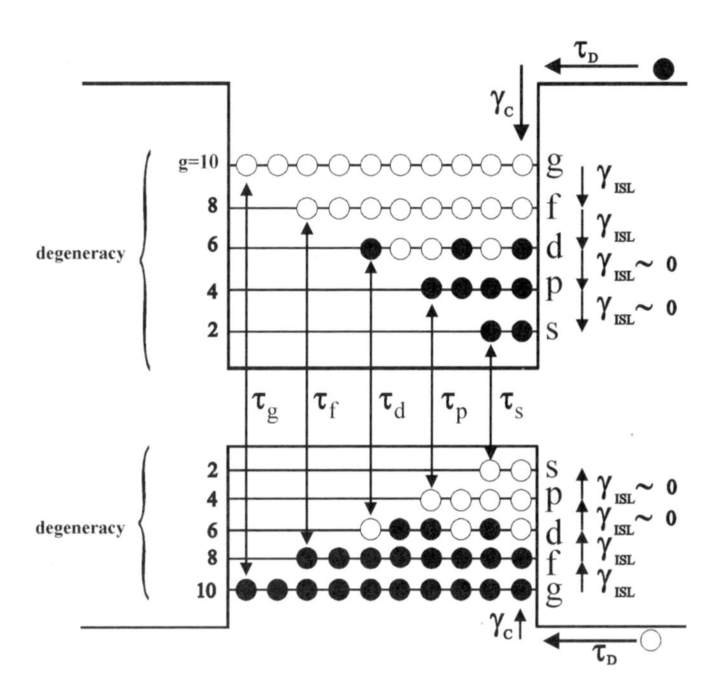

FIGURE 18.6. Schematic of the energy levels in an InAs/GaAs self-assembled quantum dot having 5 electron and hole shells ($v = s, p, d, f, g$) with a quasi-constant energy separation between the shells reminiscent of an effective parabolic-potential, and with a degeneracy $g_v = 2(l+1)$ where $l = 0$ for s, $l = 1$ for p, etc. The dynamics of the state-filling involve the interband recombination time for the various shells τ_v, the intersublevel relaxation rates γ_{ISL}, and the carrier diffusion (τ_D) and capture (τ_c) times. (From Ref. 8 by permission of American Physical Society.)

is typically observed in state-filling spectroscopy; (3) depending on the growth rate chosen for the strain material, a growth interrupt (anneal) must be used to let the quantum dots evolve to their equilibrium shape and obtain good ensemble uniformity;[10] (4) the group V pressure must be controlled to obtain the proper growth dynamics;[53] (5) an *indium-flush* technique can be used during the capping of the quantum dots for additional size and shape engineering.[10,13,28] Such shape-engineered stacks of self-aligned quantum dots with improved uniformity are useful to increase the gain in the active region of quantum dot lasers having well-defined excited-states, as was shown above in Fig. 18.3.

Fig. 18.7 shows an example of state-filling spectroscopy. Photoluminescence spectra were obtained at different excitation intensities. For low excitation, only a low energy peak is observed, indicating a fast carrier relaxation toward the lowest states of the quantum dots (*s*-shell). Indeed, the intersublevel relaxation rates are very fast compared to the interband radiative lifetime for the exciton recombination for the various shells. The carrier energy relaxation in quasi-zero-dimensional structures has been widely studied in recent years[54–57] due to its important physical implications in the improvement of the quantum-dot laser performance. The discrete atomic-like energy spectrum of the quantum dots imposes constraints on the allowed inelastic relaxation mechanisms. In particular, single longitudinal-optical (LO) phonon emission is forbidden unless the energy levels of the dots are separated by exactly the fixed LO-phonon energy. The blocking of this normally important relaxation

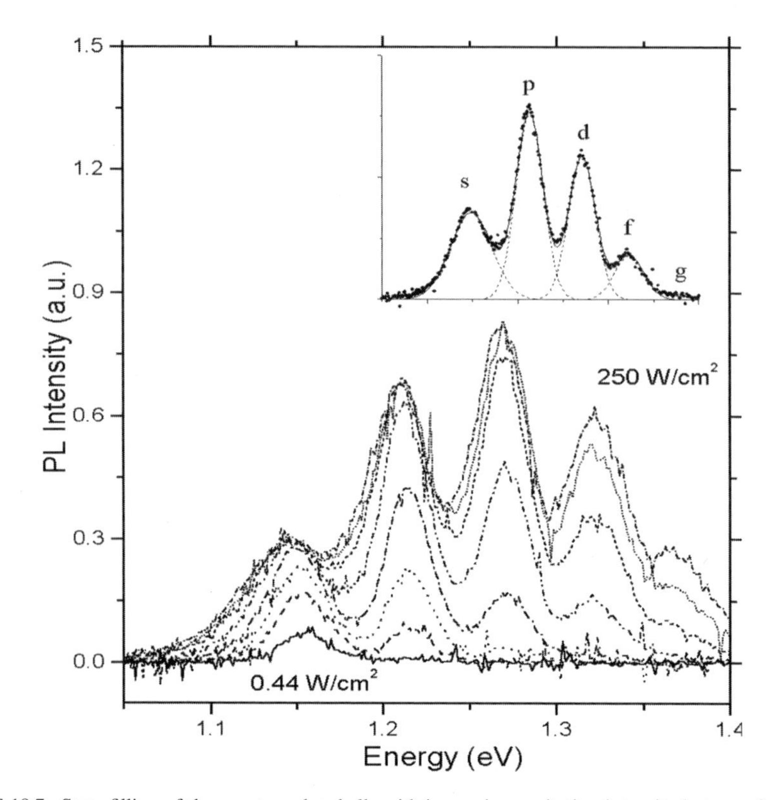

FIGURE 18.7. State-filling of the quantum dot shells with increasing excitation intensity between 0.44 W/cm^2 and 250 W/cm^2 in low temperature photoluminescence (PL) spectroscopy. Absolute saturation of the lower shells is observed when the excitation beam has a constant intensity profile. The inset shows a Gaussian fit used to deconvolute the contributions from the various shells. (From Ref. 8 by permission of American Physical Society.)

channel was expected to cause a strong reduction in the carrier relaxation scattering rate.[58] The observed good carrier relaxation efficiency can be explained by mechanisms such as Auger-type processes[59–61] and multiphonon processes.[1,55,62]

The excited state radiative lifetimes can be deduced from the state-filling spectroscopy of Fig. 18.7.[61] As the excitation intensity is increased, the lower shells are filled as depicted in Fig. 18.6, and excited-state emission starts to appear, and may eventually show stronger emission than the ground state depending on the shell degeneracy and the radiative lifetime (τ_ν) of the upper shells. However, for accurate state-filling spectroscopy, the excitation profile of the probe beam must be uniform. For example, for a typical Gaussian profile of excitation, a continuous range of excitation intensities are effectively used because the quantum dots situated in the center of the focused spot will be highly excited but the quantum dots in the wings of the Gaussian beam will receive a lower excitation intensity. For non-uniform excitation profiles, the *s*-shell emission is favored over that of the upper shells because a large area is probed with low intensities, and also the *s*-shell emission will not saturate because the number of quantum dots effectively probed increases as the low intensity wings of the Gaussian beam are expanding. In Fig. 18.7, a probe beam with uniform intensity was used and an absolute saturation of the *s*, *p*, and *d* shells is observed for the intensities tested. For such steady-state uniform excitation in the high-excitation limit, the ratio of the radiative lifetimes of any two shells is proportional to the ratio of their saturation intensities and inversely proportional to the ratio of the degeneracy of these shells.

Depending on the quantum dot density and on the photocarrier diffusion length, the wetting layer emission can also be observed. For example, time-resolved studies of the wetting layer PL can be combined with state-filling spectroscopy of the quantum dot and of the wetting layer emission to obtain carrier transfer rates from the wetting layer to the quantum dots. This provides a method to measure the capture rates and to experimentally determine the Auger capture coefficient in self-assembled quantum dots.[61] The results show that the capture efficiency increases with the carrier concentration in the wetting layer, indicating the important role of the Auger processes in the capture dynamics. Such studies combining state-filling spectroscopy of the quantum dots and time-resolved PL of the wetting layer provide a unique method for studying the carrier dynamics with a regime of constant carrier concentration instead of with probe pulses in which case the number of carriers is varying with time. It allows quantitative comparison of the relative importance of multi-phonon and Auger processes.

Probing the quantum dots with a magnetic field can further provide information about the symmetry of the quantum dots and also about the carrier capture. For example, Fig. 18.8 shows the state-filling spectroscopy at 4 K, for fields between 0 T and 18 T, on a quantum dot ensemble displaying sharp electronic shells. Only the s, p, and part of d shells are observed in Fig. 18.8, due to the somewhat lower excitation density used here, because a fiber with a large diameter was used to deliver the excitation in this magneto-PL experiment. Nevertheless, the spectra clearly show a splitting into 2 of the p-shell and suggest a splitting

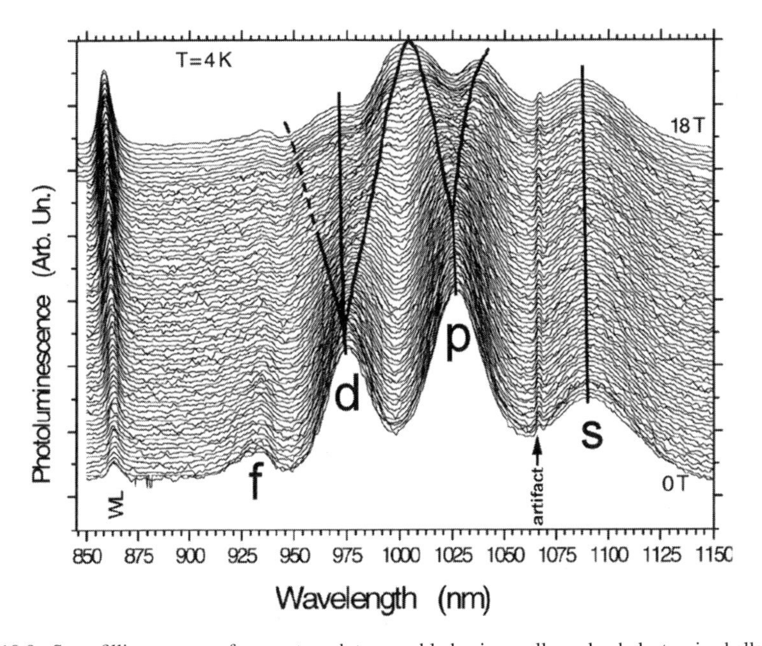

FIGURE 18.8. State-filling spectra of a quantum dot ensemble having well-resolved electronic shells in a magnetic field between 0 T and 18 T. The magnetic field lifts the degeneracy of the upper shells and leads to shell reconstruction. (From Ref. 48 by permission of SPIE.)

into 3 of the *d*-shell with a shell reconstruction at higher magnetic field according to the various angular momenta. The recombination can be understood in terms of a gas of weakly interacting excitons. Such spectra of exciton droplets in zero-dimensional systems can be compared with modeling of coherent many-exciton states and the destruction of hidden symmetry by the magnetic field.[4,46]

It is also clear in the magneto-PL spectra that the diffusion and the capture of the photocarriers transferred from the wetting layer to the quantum dots is affected by their cyclotron orbit in the magnetic field. All the spectra in Fig. 18.8 are obtained at a constant excitation intensity and displayed on the same scale (with an offset for clarity). Clearly the amplitude of the wetting layer peak (WL) increases as the field increases. At higher fields, the photocarriers are forced to orbit with a diameter that becomes smaller relative to the mean quantum dot separation and diffusion length.

18.4. ENERGY LEVEL ENGINEERING IN QUANTUM DOTS

Strain-induced vertical self-alignment has been observed for several years in the Stranski-Krastanow epitaxy. As mentioned above, for optoelectronic devices it can be desirable to have multiple layers of quantum dots, and also the electronic coupling between nanostructures is of great interest. However, it is only recently that it has been possible to observe well-resolved electronic shells in an ensemble of quantum dots having a number of correlated layers by using an indium-flush technique.[13] For example, this is demonstrated in Fig. 18.9 for correlated stacks of 7 layers separated with 10 nm spacers of GaAs, grown with and without the indium-flush. In Fig. 18.9d for no indium-flush, no sharp electronic shell

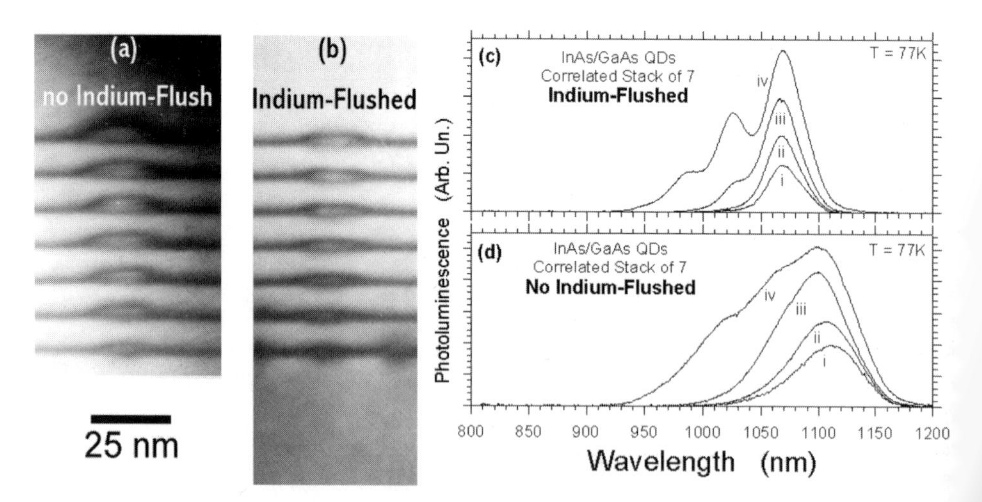

FIGURE 18.9. Using the Indium-Flush technique to make InAs/GaAs quantum dot stacks more uniform: (a) and (b) are cross-section TEM images of shape-engineered InAs/GaAs quantum dots grown at 515°C, and (c) and (d) are the corresponding state-filling spectroscopy. Stacks Indium-Flushed at 5.0 nm show better uniformity than stacks with no Indium-Flush. The GaAs spacers are 10 nm. The PL is excited with various intensities up to a few kW/cm^2. (From Ref. 13 by permission of American Institute of Physics.)

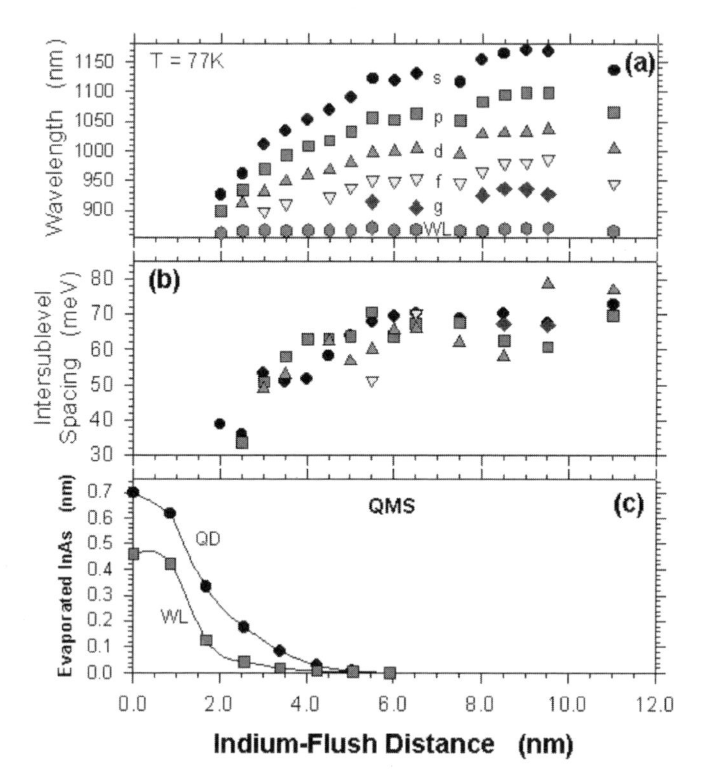

FIGURE 18.10. Tuning (a) of the energy levels and (b) of the intersublevel energy-spacing with the position of the Indium-Flush during the capping of the InAs/GaAs quantum dots. Single layer of quantum dots, 1.9 monolayers of InAs grown in 27 s at 515°C, growth interruption of 60 s, and the total GaAs cap thickness is 100 nm. (c) The InAs removal can be monitored with in-situ QMS measurements for quantum dot layers (circles) or wetting layers (square). The zero dimensional shells are labeled with the atomic notation: *s*, *p*, *d*, etc. (From Ref. 13 by permission of American Institute of Physics.)

can be observed because of the change in the quantum dot size from one layer to the next, as observed from the TEM images (Fig. 18.9a). The uniformity problem with correlated stacks can be circumvented with the indium-flush, as illustrated in Fig. 18.9b and 9c which was for a sample grown under the same conditions but with an indium-flush executed at 5.0 nm in the middle of the GaAs barrier. The number of stacked layers can be further increased while preserving the well-resolved electronic shell structure as was seen in Fig. 18.3. The indium-flush technique can also be used to do some size and shape engineering in the case of single layer samples and to tailor the quantum dot shell structure. This is demonstrated in Fig. 18.10 which shows that for an indium-flush executed after between 2.5 nm and ~5.5 nm, the quantum dots can be continuously adjusted from having a more disk-like shape to the more standard lens-shape. This is illustrated in Fig. 18.11 which shows the state-filling spectra for various values of indium-flush. The total number of bound zero dimensional states changes from three for the sample indium-flushed at 2.5 nm to five for the sample indium-flushed at more than ~5.5 nm. For this growth temperature, the intersublevel energy-spacing was tuned from ~35 to ~65 meV (see Fig. 18.10b). For the

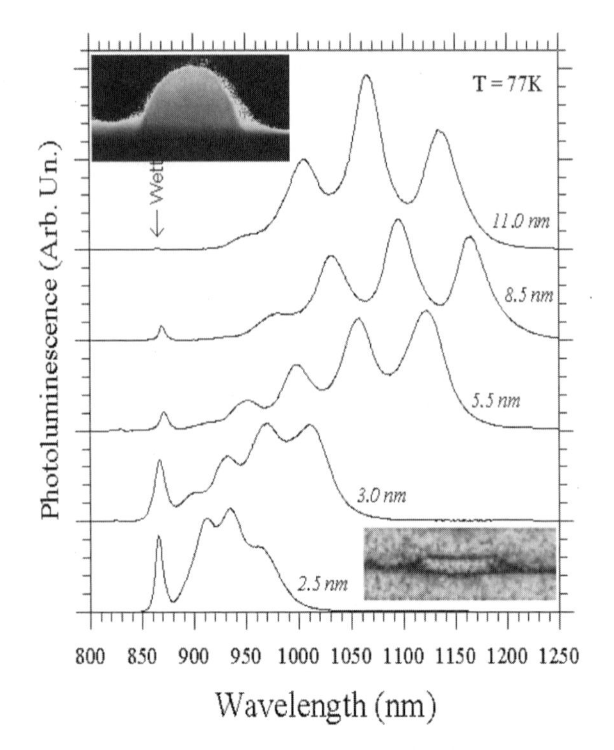

FIGURE 18.11. Size and shape engineering of InAs/GaAs quantum dots, and tuning of the intersublevel energy spacing: the state-filling spectroscopy shows the evolution with an indium flush executed after the deposition of thin GaAs caps between 2.5 and 11.0 nm. Otherwise, the InAs/GaAs quantum dots were grown with optimized growth conditions with 1.9 ML of InAs grown in 27 s at 515°C, with an anneal of 60 s. During the indium flush, the growth is interrupted and the substrate temperature is raised to ~610°C, after which the GaAs cap was completed to a thickness of 100 nm in all cases. The photoluminescence was excited above the barrier energy with a few kW/cm^2. (From Ref. 10 by permission of American Physical Society.)

samples indium-flushed after a GaAs thickness larger than ~5.5 nm, little InAs is removed and the intersublevel spacing remains about the same, but the energy level positions shift slightly due to changes in the confinement potential caused by the combined effects of quantum dot intermixing and indium segregation in the barrier. The InAs removal can be monitored with in-situ quadrupole mass spectrometer (QMS) measurements, as shown in Fig. 18.10c. It demonstrates that below ~2 nm, most of the indium is evaporated (flushed), and that more InAs is removed in the case of the quantum dots compared to the wetting layer

As mentioned above, the arsenic pressure can also have a significant influence on the optical properties and the uniformity of the quantum dot ensemble. For example, Fig. 18.12 shows results which demonstrate that the electronic shell structure of the quantum dots can be adjusted by controlling the arsenic pressure.[53] The reduced wetting layer luminescence relative to the quantum dot intensity in the samples at $P_{As} > 2.0\ P_0$ is attributed to a substantially increased quantum dot density, and was verified by scanning electron microscopy (SEM). As P_{As} is raised from 0.59 P_0 to 1.3 P_0, the quantum dot emission lines redshift by as much as 40 meV. And as P_{As} is further raised to 5.2 P_0 the shift in PL reverses, and the emission lines blueshift by more than 180 meV. This trend is observed for all the resolved

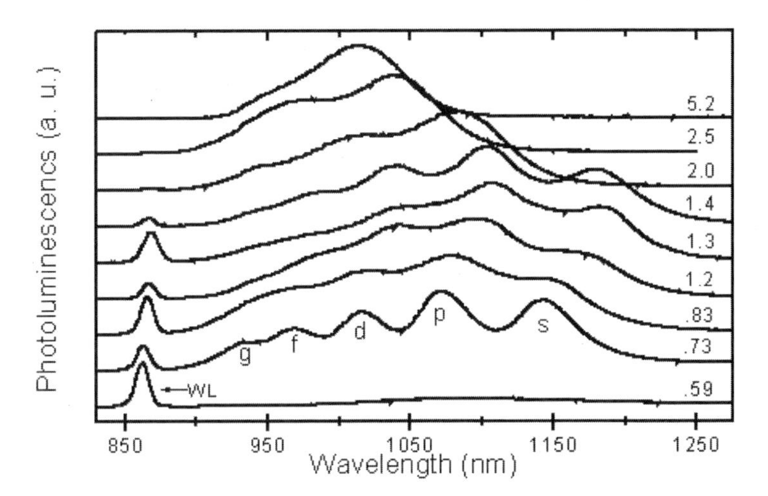

FIGURE 18.12. High excitation state-filling photoluminescence spectra at low temperature for nine quantum dot samples grown with a uniform InAs coverage. The arsenic pressure at which the InAs was deposited and annealed was different for the various samples (as indicated relative to P_0, the pressure at which the unreconstructed (1×1) GaAs surfaces is observed with no sign of (4×2) reconstruction). (From Ref. 53 by permission of Elsevier B.V.)

electronic shells. Note that for low P_{As}, five confined shells are observed, but for high P_{As}, as the emission lines are blueshifted, the total number of electronic shells is reduced from five to three. As P_{As} is raised, the full width at half maximum almost doubles, going from ~35 meV to ~70 meV, while the intersublevel spacing increases only slightly resulting in poorly resolved shells for the samples grown at high P_{As}. In this example, the other relevant growth parameters such as the amount of strained material deposited, the growth temperature, and the anneal time were identical and adjusted for optimized quantum dot growth. It is therefore clear that simultaneous control of all the parameters, including the arsenic pressure, is necessary to engineer quantum dot ensembles. To produce quantum dot ensembles with well resolved electronic shells, a low arsenic pressure was found to be necessary.

Another parameter which must be controlled, and which can be used to manipulate the energy levels of the quantum dots, is the substrate temperature which influences the equilibrium of the growth dynamics. The main effect of the substrate temperature is to change the size of the quantum dots and consequently their inter-sublevel energy spacing. This is illustrated in Fig. 18.13 which shows the state-filling spectroscopy for four different InAs/GaAs quantum dot samples and illustrates how the electronic shell structure can be tuned during the formation of the quantum dots. Larger quantum dots with smaller intersublevel energy spacings are obtained at higher growth temperatures.[50] The photoluminescence spectra, excited at various intensities, are displayed for (a) $T_{growth} = 535°C$, (b) $T_{growth} = 515°C$, (c) $T_{growth} = 500°C$, and (d) $T_{growth} \sim 480°C$. The measured intersublevel energy spacings were 57 meV, 65 meV, 75 meV and, 90 meV respectively. This demonstrates that the size of the quantum dots can be used to control the intersublevel energy spacing in a simple step during growth. It should be noted however that the size is not the only parameter which is modified when changing the substrate temperature. For example, the larger intersublevel energy spacings (tight lateral confinement) are associated with the quantum dots emitting at lower energies and are obtained by growing at lower substrate temperatures, and therefore

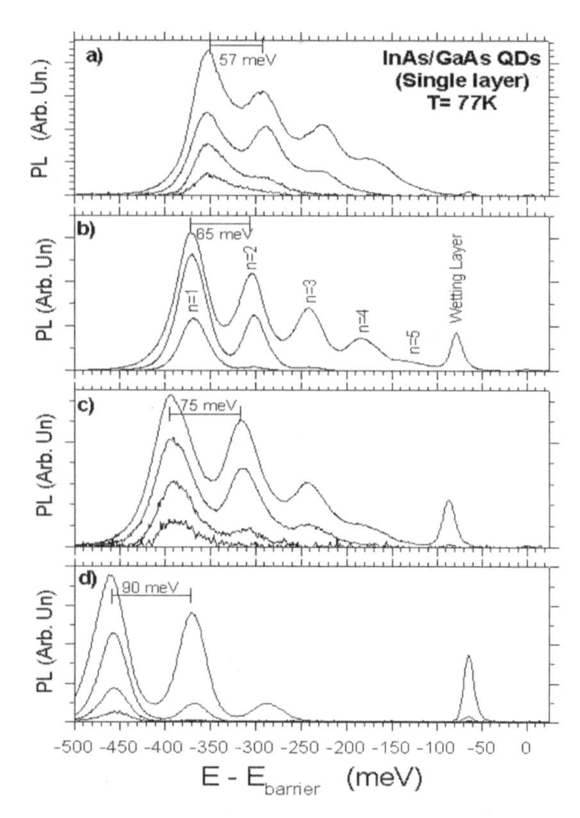

FIGURE 18.13. Tuning of the intersublevel energy spacing with the substrate temperature during the growth of InAs/GaAs QDs. Larger QDs with smaller intersublevel energy spacings are obtained at higher temperatures: (a) $T_{\text{growth}} = 535°\text{C}$, (b) $T_{\text{growth}} = 515°\text{C}$, (c) $T_{\text{growth}} = 500°\text{C}$, and (d) $T_{\text{growth}} \sim 480°\text{C}$, giving an intersublevel energy spacing adjustable between 57 meV and 90 meV. The state-filling spectroscopy is obtained with the photoluminescence at 77 K, with the highest excitation of a few kW/cm^2 above the barrier energy. (From Ref. 10 by permission of American Physical Society.)

are less affected by the possibility of intermixing during growth as will be discussed in section 18.6.a.

18.5. SINGLE QUANTUM DOT SPECTROSCOPY

The optical properties discussed so far are the properties of a large ensemble of quantum dots. Even though the inhomogenous broadening can be narrow enough to display well-resolved excited states, this inhomogenous broadening hides details about the extremely narrow homogeneous linewidths of single dots and the energies of the multi-exciton emission. The homogenous linewidth of the quantum dot emission is extremely narrow, about 0.1 meV and can only be observed by reducing the size of the ensemble of quantum dots probed.[1,64] In recent years, much research has been devoted to single quantum dot spectroscopy.[49,65–68] A wealth of interesting properties are revealed by probing single quantum dots. As discussed

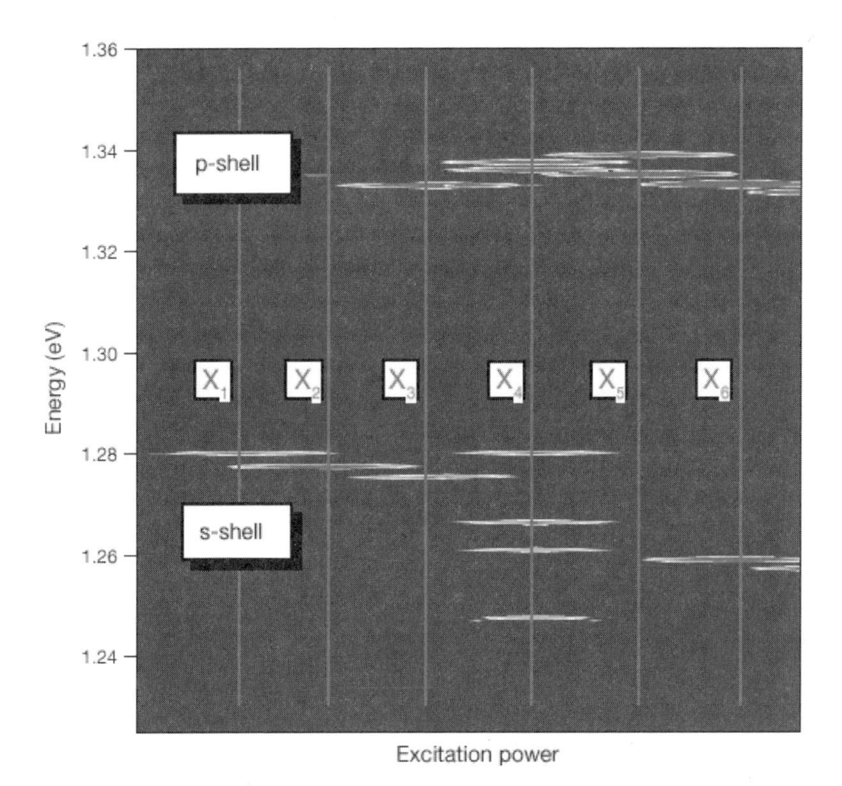

FIGURE 18.14. Multi-exciton spectroscopy for the *s*-shell and the *p*-shell of a single quantum dot are obtained with a contour plot measuring the variation of the emission of an $In_{0.60}Ga_{0.40}As$ single quantum dot as a function of excitation power and with energy. Bright regions indicate strong emission intensities, blue regions low intensities. To reduce multiple phonon emission and heating, optical excitation energy close to the bottom of the wetting layer was used. The excitation power was varied between 50 nW and 5 mW to excite between 1 and 6 excitons. (From Ref. 49 by permission of Macmillan Magazines Ltd.)

in previous sections, the *s*-shell is two-fold spin-degenerate, and can be occupied by not more than two excitons. Increasing the excitation intensity results in complete *s*-shell filling and saturates its emission. Simultaneously, emission from the *p*-shell appears. However, in the ensemble spectra the effects of Coulomb interactions are blurred by inhomogeneous broadening. Single quantum dot spectroscopy has been used to suppress inhomogeneous broadening and to resolve correlation effects. This is exemplified in Fig. 18.14 which shows the emission intensity from the *s* and *p*-shells of a single dot as a function of excitation power and energy. The low-excitation spectra exhibit a single sharp emission line, X1 which can be attributed to recombination of a single electron–hole pair in the *s*-shell. With increasing illumination, biexciton emission X2 becomes dominant at the low-energy side of X1 while the exciton line fades away. Then the biexciton emission decreases, while a new *s*-shell emission line emerges at slightly lower energies. Simultaneously, intense emission from the *p*-shell appears at an energy 50 meV higher. In the further evolution of the spectra with excitation power, two characteristic features are to be noted. (1) In the *p*-shell emission, only a very restricted number of spectral lines are observed. First, a single emission line appears (X3). This emission line is replaced by two lines (X4), followed by two other lines with very similar energies (X5). Finally, these two lines are replaced by a single line (X6). The energies of all these lines are approximately the same, and the splittings between them are small compared to the splitting between the quantum-dot shells. In contrast, there are dramatic changes in the number of lines of the s-shell emission and in their energies. At low p-shell population a single s-shell emission line (X3) is observed. This line is replaced by a broad band (X4) consisting of four strong emission lines, when the emission from the p-shell splits into two. On increasing the illumination, it comes to a situation (X5) in which the s-shell emission almost completely vanishes. Increasing the excitation further leads to a single emission line (X6) from the s-shell when there is also a single line from the p-shell. Calculations have been performed to help interpret such results and revealed the fundamental principle of hidden symmetries as a rule determining the electronic structure of electron–hole complexes in quantum dots. In this sense, the hidden symmetries are analogous to Hund's rules in atomic physics.[49,46,47,67]

18.6. QUANTUM DOT DEVICES

In this section, we discuss various quantum dot devices and the potential of such nanostructures for applications in optoelectronics. Indeed, several unique properties associated with quantum dots could benefit optoelectronic applications: broadband gain/tunability spectra, low chirp, low temperature sensitivity, narrow homogeneous linewidth, and a gain medium robust to material defects.

18.6.a. Near Infrared Quantum Dot Lasers (900 nm-1.3 μm)

The indium-flush technique discussed in Section 18.4 can be used to engineer quantum dot lasers and detectors with improved uniformity similar to that seen by TEM in Fig. 18.3. Such quantum dots can be grown in the active region of a semiconductor p-i-n laser diode, as illustrated in Fig. 18.15. The bottom layers of the laser are made of higher bandgap materials which are doped n-type. The active region is typically undoped, with a lower bandgap. The active region is sandwiched between the bottom n-type layers and the top p-type layers as illustrated in the band diagram at the bottom of Fig. 18.15. The electrons of the n-layers and the holes of the p-layers are both pushed toward the central active region by applying a forward bias between a narrow strip contact on top and the back contact. The electrons and holes relax in the quantized states and recombine emitting photons. The lasing occurs from stimulated emission resulting from the optical feedback in the cavity formed by the strip contact laterally, the front and back facets longitudinally, and from the higher index of refraction of the active region in the growth direction. Using an indium-flush study as in Fig. 18.10, a quantum dot laser diode can be designed with an indium-flush performed at 3.0 nm to provide a ground state emission at $\lambda \sim 980$ nm, with uniform self-aligned quantum dot stacks having 7 layers separated by 10.0 nm, and which yields sharp excited-state emission. Such results are shown in Fig. 18.16 for 77 K, and in Fig. 18.17 for 300 K. The zero dimensional shells involved in the stimulated emission can unambiguously be assigned when the injection current is increased above threshold. The lasing is observed in

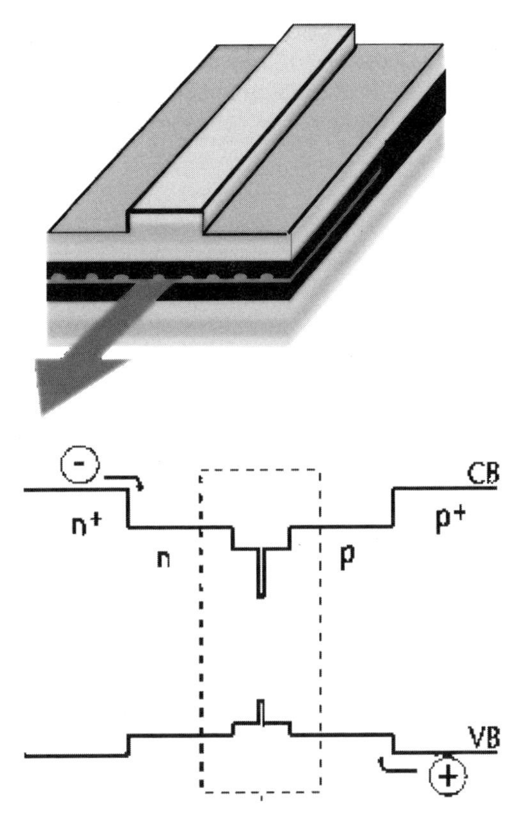

FIGURE 18.15. Schematic of a semiconductor *p-i-n* laser diode having quantum dots in its active region. The active region is represented as the lighter region between the dark regions in the middle of the top diagram. The corresponding band diagram is show at the bottom for the conduction band (CB) and valence band (VB).

the upper quantum dot shells for short laser cavities, and progresses towards the quantum dot ground states when the cavity length is increased (larger gain medium). Very low current density thresholds (J_{th}) can be obtained. For example at 77 K, $J_{\text{th}} = 15$ A/cm^2 for a 2 mm cavity lasing in the first excited state (p-shell) and $J_{\text{th}} = 125$ A/cm^2 for a 1 mm cavity lasing in the d-shell. At 290 K for a 5 mm cavity, the threshold in this example was increased to 430 A/cm^2 with lasing in the d-shell, whereas a shorter cavity of 1 mm has $J_{\text{th}} = 490$ A/cm^2 in the f-shell.[13]

It is necessary to grow the top cladding layers of the quantum dot laser structure at lower temperatures (~540°C) to avoid additional *in situ* quantum dot intermixing. Indeed, it is also important to control the alloy intermixing between the quantum dot and barrier materials during the growth, particularly for laser diodes which typically require long growth times after the deposition of the quantum dots. For laser structures grown with the upper cladding at a temperature around 540°C, the spontaneous emission of the diode matches well the targeted values deduced from the indium-flush study as shown in Fig. 18.10. However, it has been demonstrated that quantum dot intermixing can have a pronounced effect on

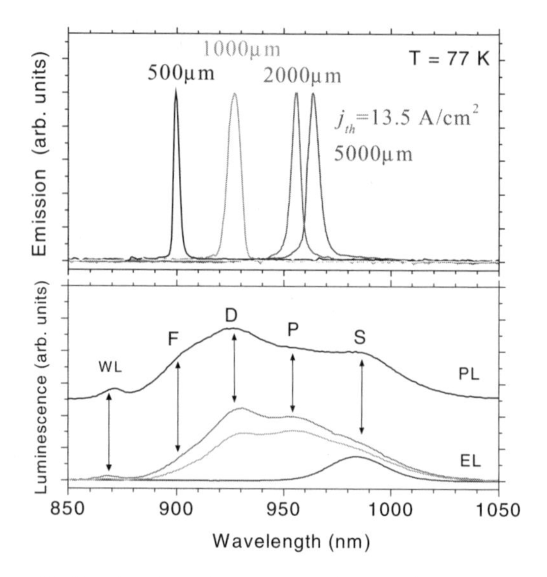

FIGURE 18.16. Lasing at 77 K in broad area lasers having shape-engineered quantum dot stacks with well-resolved excited-states: The electroluminescence (EL) and PL are shown (bottom) to identify which of the electronic shells (s to f and wetting layer) are lasing for the various cavity lengths (top). The devices consist of stacks of 7 InAs/GaAs quantum dots with indium-flush at 2.8 nm and 10.4 nm GaAs spacers. (From Ref. 13 by permission of American Institute of Physics.)

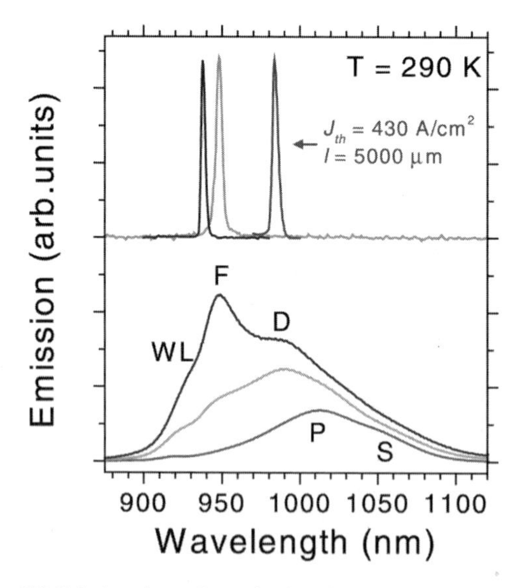

FIGURE 18.17. Lasing at 290 K in broad area lasers having shape-engineered quantum dot stacks with well-resolved excited-states: The EL spectra below threshold are shown (bottom) to identify which electronic shell is lasing (top). Same sample as Fig. 18.16. Cavity lengths are 0.5 mm, 1 mm, and 5 mm from left to right. (From Ref. 13 by permission of American Institute of Physics.)

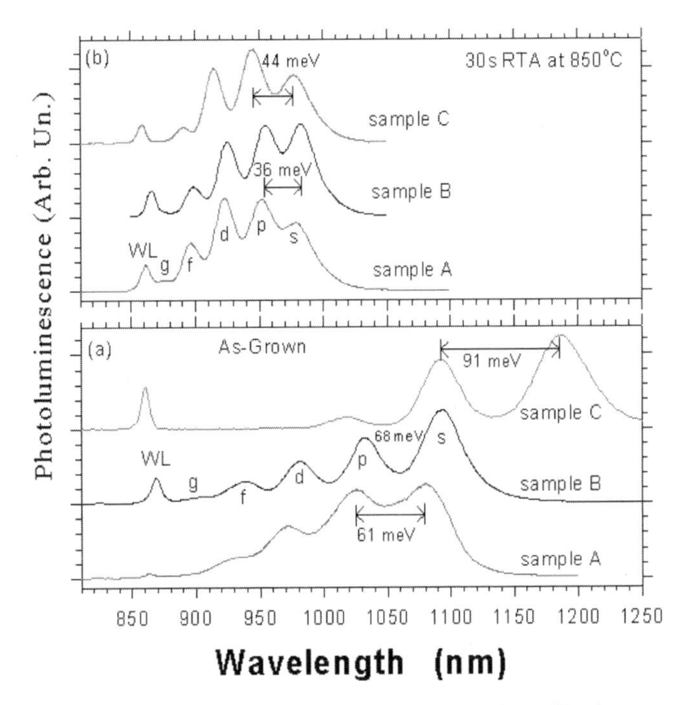

FIGURE 18.18. Thermal intermixing of quantum dots having different intersublevel energy spacings. (a) PL showing the state-filling spectra of the as-grown samples A, B, and C obtained at substrate temperatures of 530°C, 515°C, and ~480°C respectively. (b) Same samples intermixed with a RTA of 30 s at 850°C. (From Ref. 68 by permission of American Institute of Physics.)

the optical properties of nanostructures. For example, a test quantum dot structure grown with an *in situ* anneal at a substrate temperature of 620°C for 30 minutes has been used to simulate the growth of cladding layers at temperatures often used for GaAs growth by MBE. The quantum dot structure grown with the *in situ* anneal had its *s*-shell shifted by 94 meV compared to a control structure grown without the anneal (see Ref. 13). Also the intersublevel energy spacing shifted from 68 meV to 40 meV.

More detailed studies on quantum dot intermixing have revealed that there are 3 main effects for the intermixing of the InAs/GaAs self-assembled quantum dots.[68,3] Using rapid thermal annealing (RTA) to induce alloy intermixing on as-grown samples: (1) their intersublevel can be tuned between ~90 meV and 25 meV, (2) strong blueshifts (up to ~200 meV) can be obtained, and (3) a pronounced narrowing of the inhomogeneously broadened emission (down to ~12 meV) is observed. For example, Fig. 18.18a presents results for different as-grown InAs/GaAs quantum dot samples having intersublevel spacings of 61 meV (sample A), 68 meV (sample B), and 91 meV (sample C). The various intersublevel energy spacings of the as-grown samples were obtained by adjusting the substrate temperature during the quantum dot formation to change their equilibrium size.[10,11,50] Fig. 18.18b shows the effects of quantum dot intermixing using a 30 s RTA at 850°C to interdiffuse the InAs quantum dots with the GaAs barriers. It is clear that the energy levels of the intermixed quantum dots from the different samples are more alike after intermixing.

For example the energy difference between ground states (s-shell) of the various samples went from over 100 meV for the as-grown samples to just a few meV after intermixing. Nevertheless, in this example, sample C originally with the largest intersublevel energy spacing is still the one with the largest lateral quantization energy after intermixing. The intense and sharp shell structures observed in the photoluminescence of the intermixed samples indicate that the quantum dots retained their zero-dimensional density of states after the diffusion of the potential. No pronounced changes were observed in the intensity of the PL for the intermixed quantum dots. It has also been reported that quantum well like behaviors were observed for the higher intermixing conditions and/or a reduction in the PL were observed and attributed to the formation of dislocations. Such detrimental effects are apparently not present in the case of high quality nanostructures with low in-plane density in the initial quantum dot ensemble.[10,51,52]

Quantum dot lasers can also be grown for ~1.3 micron using the InAs/GaAs material system. Indeed, the 1.3 micron quantum dot lasers have received a lot of attention because of the optoelectronic applications for telecom. In fact, the 1.3 micron quantum dot lasers are just a different flavor of the near infrared InAs/GaAs quantum dots using slight variations during the growth of the quantum dots or cladding materials. Room temperature emission at 1300 nm can be obtained using a distant indium-flush (see Section 18.4, it leads to more indium segregation) and/or equivalently by using some low concentration InGaAs alloys near the quantum dots.[69–76] Alternatively, some have experimented using InGaNAs alloys for the quantum dot formation.[77–78] Generally, for the InAs/GaAs quantum dot lasers, the best performances have been obtained for the longer wavelength emitters. This is in part because the low energy emission wavelengths combined with high barriers leads to large intersublevel energy spacings and a good carrier confinement for separate confinement heterostructures, and consequently a low thermionic emission and a low temperature quenching of the radiative emission. For the quantum dot lasers emitting around 1.3 micron, record current density thresholds have therefore been observed, together with good stability of the characteristic temperatures (T_o).[79–81] The fact that 1.3 micron quantum dot lasers outperformed their quantum well counterparts is exciting, but on the other hand the quantum well technology is very mature and well anchored for the 1.3 micron emitters and unlikely to be replaced in the near future.

18.6.b. Red-Emitting Quantum Dot Lasers

III–V quantum dot alloys can be varied to obtain emission at a wide range of wavelengths. For example, visible stimulated emission in semiconductor quantum dot lasers has been demonstrated.[12] The red-emitting self-assembled quantum dots are obtained by using highly strained InAlAs[82,64] in AlGaAs barriers on a GaAs substrate. For example Fig. 18.19 shows that the carriers injected electrically from the doped regions of a separate confinement heterostructure can be thermalized efficiently into the zero-dimensional quantum dot states, and stimulated emission around 700 nm is observed with a threshold current density in the kW/cm^2 range for the first red-emitting quantum dot lasers[12] which had an external efficiency of ~10% at low temperature and a quasi-CW (continuous wave output, as opposed to pulsed) peak power greater than 200 mW. Further development helped improve the thresholds and the temperature dependence of the threshold.[24,66,83] This depends on the barrier height for visible dots, which have relatively shallow confinement, according to

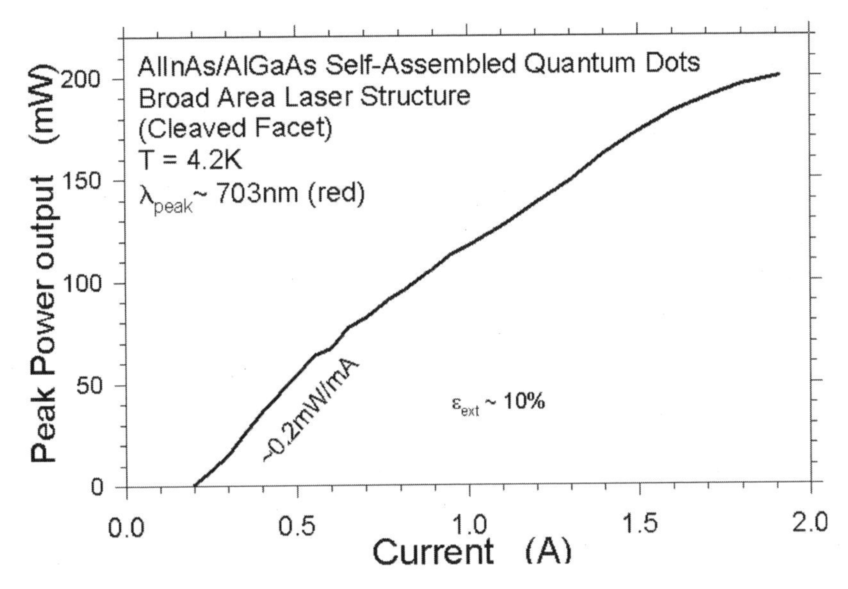

FIGURE 18.19. Lasing properties of the first red quantum dot laser emitting up to 200 mW in quasi-CW at low temperature. (From Ref. 12 by permission of American Association for the Advancement of Science.)

the activation energy (E_b) and the inhomogeneous distribution [Ref. 1e]

$$\mathrm{PL}(E,T) = \frac{Ce^{-((E-E_o)/\Gamma)^2}}{1+\alpha e^{-(E_b-E)/kT}} \tag{18.1}$$

where the intensity of the PL depends on the temperature (T) and the energy (E) relative to the center of the distribution (E_o) having a broadening (Γ). Further studies also helped improve the uniformity of quantum dots formed with ternary alloys, where random alloy fluctuations can cause additional inhomogeneous broadening.[84]

18.6.c. Quantum Dot Lasers for ~1.5 Micron Emission

As discussed in Section 18.1.a, the most important optoelectronics applications are for telecom components, predominantly in the C-band (1520 nm to 1565 nm) and also in the L-band (1565 nm to 1610 nm). For 3-dimensional confinement, this important telecom range can be addressed by using the InAs/InP material system.[85] The substrate is InP while the quantum dots are still made of strained InAs. The barrier can be made of quaternary InGaAsP[86–88] or of InGaAlAs[26] (of course, barriers made of the lower bandgap InGaAs will result in long wavelength emission and have been found to emit at around 1.9 micron[89]). For example, Fig. 18.20 shows a plot of the power output as a function of current density for an InAs/InGaAsP/InP quantum dot laser grown by CBE, emitting at ~1640 nm. A cross-section TEM picture of the active region is also shown where the quantum dots are evident and fairly uniform. There is still progress required to obtain very low threshold values as was the case for GaAs based quantum dot lasers,[79–81] but it is promising given that the InAs/InP material system has not been studied as much as the InAs/GaAs system.

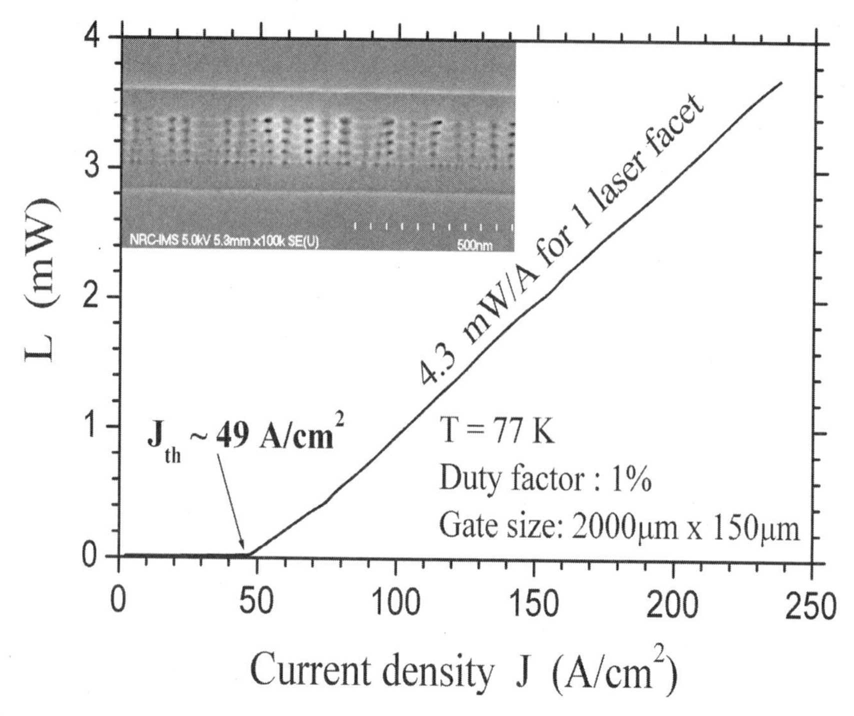

FIGURE 18.20. The telecom wavelength range is reached with InAs/InGaAsP/InP quantum dot lasers. A plot of the power output as a function of current density of a QD laser grown by CBE emitting at ~1640 nm. Cross-section SEM picture of the active region in insert. (From Ref. 86 by permission of American Institute of Physics.)

A common problem for applications using directly modulated quantum well laser diodes is the chirp. The chirp causes the output wavelength spectrum to be broadened and the laser linewidth increases for high frequency on and off cycles. Narrow linewidths are required to increase the tolerance on chromatic dispersion in the fiber and increase the propagation distance of telecom transmitters. Therefore, another interesting property unique to quantum dot lasers is the observation of a smaller chirp for modulated lasers[90] or of a small linewidth enhancement factor, $\alpha = -4\pi/\lambda(dn/dN)(dg/dN)^{-1}$, where g is the gain. Large values of α can result in antiguiding in narrow stripe lasers, self-focusing and filamentation in broad-area lasers, and chirp under modulation.[91] It was found that in contrast to the case of quantum wells which have their highest gain at a low current density and gradually decreasing with increasing density, for the quantum dot lasers the gain increased superlinearly up to the point of gain saturation. The small α values are a consequence of the symmetric gain spectrum of the quantum dots.

Also of interest is the fact that quantum dot lasers are found to be more robust to radiation and more resilient to defect propagation because the gain is localized in space.[92,93] Irradiated quantum dot laser diodes demonstrated lasing action over 2 orders of magnitude beyond the maximum radiation dose sustainable by quantum well devices. The improved damage response of quantum dot based structures results from efficient collection and localization of electrons and holes by quantum dots in the active region, which limit carrier transfer to nonradiative centers. It suggests the suitability of quantum dot device architectures for use in radiation environments and in high power applications, wherever nonradiative processes promote the degradation or failure of traditional quantum well devices.

Finally, the most interesting property, which is unique to quantum dots and atomic-like systems, is the wide tuning range which will be discussed in the following section.

18.6.d. Widely Tunable Quantum Dot Lasers

Quantum confinement leads to one of the most important and unique properties of the quantum dot lasers: their widely tunable spectrum. The broad quantum dot gain spectrum is applicable to tunable laser sources and to semiconductor optical amplifiers (SOA). Indeed, quantum dot lasers owe their wide gain spectrum to their resemblance to artificial atoms and to the intrinsic property that their density of states can be saturated over a broad range of energy. The simplest demonstration of the quantum dot broad gain is obtained when pumping a Fabry-Perot laser at high current densities, in which case broadband lasing over tens of nanometers has been observed.[24] The lasing at various wavelengths can also be obtained by using different laser cavity lengths as was shown for the near infrared lasers in Figs. 18.16 and 18.17 above. For example, Fig. 18.21 illustrates tuning over 85 nm, obtained by changing the cavity length of broad area Fabry-Perot InAs/GaAs quantum dot lasers. To control the lasing wavelength, a configuration using an external cavity can be used. The output facet of the laser diode needs to be anti-reflection coated and the resonant wavelength is fedback from a dispersive element which can be tuned to select the laser frequency. In nanostructures with 3-dimensional confinement such as self-assembled quantum dots, it is possible to saturate all the zero-dimensional states from the ground state (s-shell) up to the wetting layer states. This filling of all the quantum dot states has been shown to yield tuning over a wavelength range as wide as 200 nm or even in excess of 300 nm depending on the quantum dot energy level engineering.

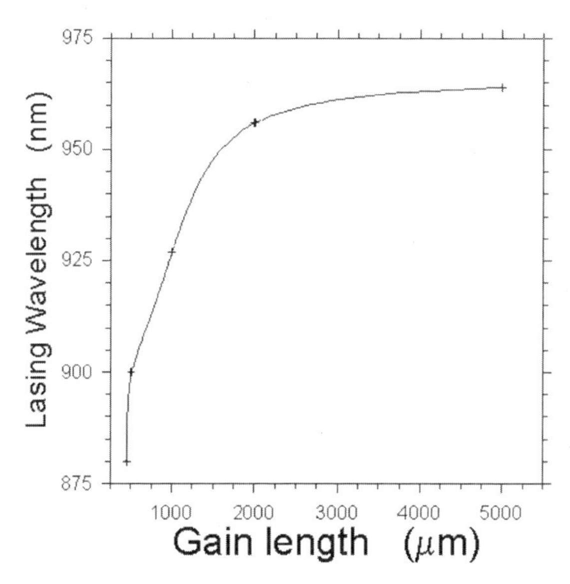

FIGURE 18.21. Wide tuning of quantum dot lasers (over 85 nm) by changing the cavity lenghts.

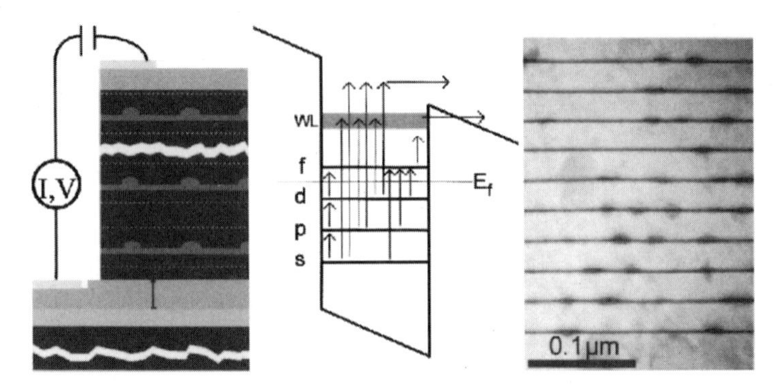

FIGURE 18.22. Schematic of a quantum dot infrared photdetector (IP) structure (left), of its energy diagram (center), and cross-section TEM view of the bottom 10 layers of a 50-layer InAs/GaAs quantum dot IP device (sample A). The self-assembled quantum dots are separated with barriers which are modulation-doped to populate the lower shells of the quantum dots. Transitions from the occupied quantum dot states to the WL or to the continuum result in infrared detection with normal incidence response. (Adapted from Ref. 94, by permission of American Institute of Physics).

18.6.e. Quantum Dot Photodetectors

The schematic of a quantum dot infrared photodetector (QDIP) structure (based on intraband transitions, or equivalently, inter-sublevel transitions) is shown in Fig. 18.22, on the left. Similar to a quantum well infrared photodetector (QWIP), the quantum dot infrared photodetector device comprises multiple layers between a collector and an emitter.[94–96] The emitter and collector are doped to act as a reservoir of charged carriers and to conduct the current during detection and for operation under an applied bias. The charged carriers are introduced in the quantum dot layers by either doping the barriers and/or the dot layers, and after redistribution of the charged carriers and adjustment of the Fermi level (E_f). The level of doping in the various regions of the device and the scheme of doping are adjusted to set the Fermi level such that the desired number of quantum dot states are occupied with charged carriers to achieve the targeted range of detection wavelengths. For example, the middle of Fig. 18.22 depicts the energy band diagram for a quantum dot with four bound shells below the 2-dimensional wetting layer (WL) states. Results are shown below for InAs self-assembled quantum dots separated with Ga(Al)As barriers which are modulation-doped to populate the lower shells of the quantum dots. As discussed in the previous sections, the size and the number of quantum dots per unit area can be adjusted from the growth parameters in conjunction with the doping to achieve the desired detection range while optimizing the detection efficiency for the wavelengths of interest. For example, the growth temperature can be used to adjust the size/shape (see Fig. 18.13) while the amount of strained material deposited can be used to control the density of quantum dots.[10,51,52,97] Transitions from the occupied quantum dot states to the wetting layer or to the continuum states will result in infrared detection as depicted in Fig. 18.22. Similarly, the choice of the barrier material, height, and thickness is adjusted in conjunction with the quantum dot size to set the detection range. For example, the right of Fig. 18.22 shows a cross-section TEM for "sample A" with InAs quantum dots separated with 30 nm GaAs barriers grown on a (100) GaAs substrate. Molecular beam epitaxy was used to deposit the equivalent of 0.54 nm of InAs to form the

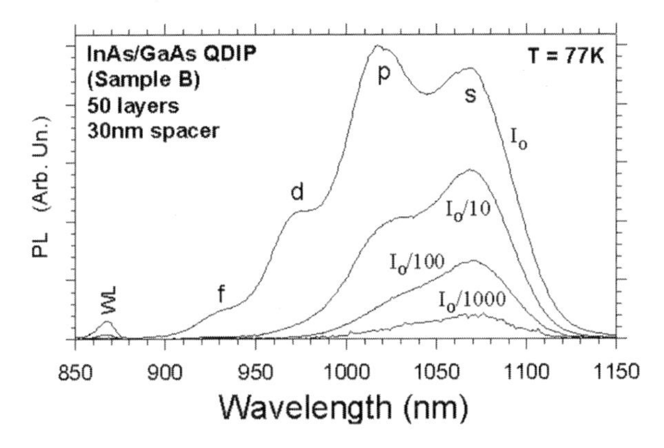

FIGURE 18.23. State-filling spectroscopy displaying the zero-dimensional states of an InAs/GaAs quantum dot infrared photodetector. The photoluminescence is obtained at 77 K excited with $\lambda_{ex} = 532$ nm and $I_0 \sim$ few kW/cm^2. (From Ref. 94 by permission of American Institute of Physics.)

quantum dots at $T \sim 520°$C, followed with a growth interruption of 60 s.[10,28] These quantum dot infrared photodetector devices were fabricated with 50 layers of quantum dots. Mesa devices were made using standard GaAs micro-fabrication techniques. The mesas were defined by wet chemical etching. The top and bottom contacts were made by depositing Ni/Ge/Au and then annealing. For optical measurements, the devices were mounted in a variable temperature cryostat. The spectra were taken by a Fourier transform spectrometer with a globar infrared source. The TEM of Fig. 18.22 shows the bottom 10 layers, and is representative of other TEM images obtained elsewhere in the upper layers. The quantum dots have a diameter of 20 ± 3 nm and are not correlated (as expected for a spacer of 30 nm): there is essentially no vertical alignment observed between the various layers.

Fig. 18.23 shows the results of low temperature state-filling spectroscopy for a similar quantum dot infrared photodetector structure also made with 50 InAs quantum dot layers with GaAs spacers 30 nm thick ("sample B"). By increasing the excitation intensity from a few W/cm^2 to a few kW/cm^2, the lowest quantum dot energy levels are progressively filled and one observes well-resolved atomic-like s, p, d, f shells in PL with an excitonic intersublevel energy spacing of about 58 meV, as well as the wetting layer emission, as discussed in Section 18.3. The observation of the well-resolved excited-states clearly demonstrates that the quantum dots are quite uniform from one layer to the next, even for a stack of 50 layers with a total thickness of $\sim$1.5 μm.

Fig. 18.24a displays the highest excitation intensity PL spectra of these samples A and B with an energy axis plotted relative to the wetting layer transition and can be compared with the measured infrared quantum dot infrared photodetector response which is plotted in Fig. 18.24b. For the InAs/GaAs quantum dots of sample A and B, the highest energy difference between the s-shell excitons and the wetting layer is smaller than $\sim$300 meV ($\lambda \sim 4.1$ μm). The transition energy of electrons excited from the s-shell to the wetting layer is actually smaller than that since the PL probes the total of the electron and hole quantization energies, whereas the infrared detection only involves the electron states. The quantization energies of the holes are expected to be no more than about half of that of the electrons.[47] The quantum dot infrared photodetector response in Fig. 18.24b

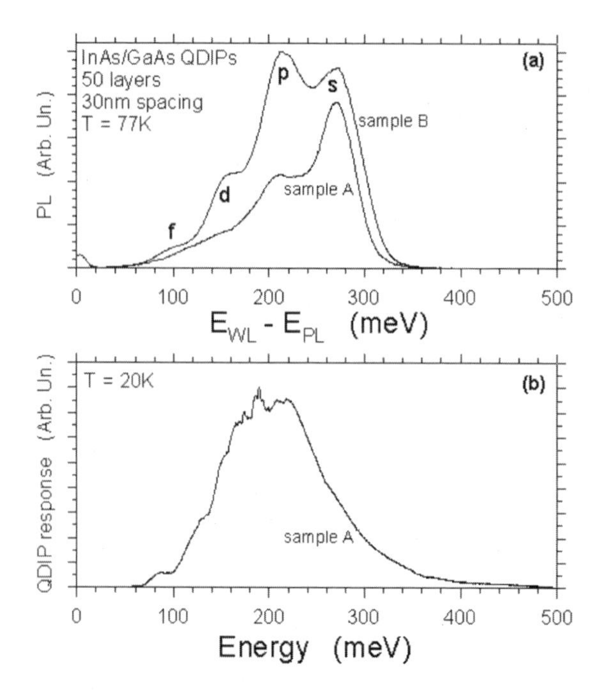

FIGURE 18.24. Normal incidence detection for InAs/GaAs quantum dot infrared photodetectors. (a) The position of the electronic shells relative to the WL energy. (b) The quantum dot IP response measured with sample A. (From Ref. 94 by permission of American Institute of Physics.)

therefore peaks at an energy smaller than the s-shell distribution plotted in Fig. 18.24a with a maximum between the energy range of ~140 meV to ~280 meV. This suggests that the maximum quantum dot infrared photodetector response corresponds mainly to transitions of electrons occupying the s and p shells excited to the wetting layer. The noisy features in the quantum dot infrared photodetector response in the 160–220 meV region are attributed to residual atmospheric absorption. The high energy tail which extends up to ~500 meV would correspond to transitions from the quantum dot states directly into continuum states above the GaAs barriers as illustrated in the energy diagram of Fig. 18.22. The low energy cutoff is determined by the level of doping which raises the Fermi level closer to the wetting layer energy. For example, the small shoulder at ~90 meV in Fig. 18.24b is probably originating from a partially populated f-shell. It takes more than 6 electrons per quantum dot to start to populate the f-shell, and the nominal doping concentration was 250 electrons/μm^2. From these two values, the quantum dot density was roughly estimated to be lower than ~60 quantum dots/μm^2 for one layer. This is consistent with the density typically observed with plan-view TEM for such growth conditions.[10]

An important aspect of QDIPs compared to quantum well-based structures is that QDIPs can detect normal incidence light. Since the direction of confinement in QWs is perpendicular to the substrate, light that is at normal incidence to the substrate must have an electric field that is in a direction parallel to the substrate and thus cannot excite the quantized levels of the QW. QWIPs can only detect light that has a component of the polarization perpendicular to the substrate.

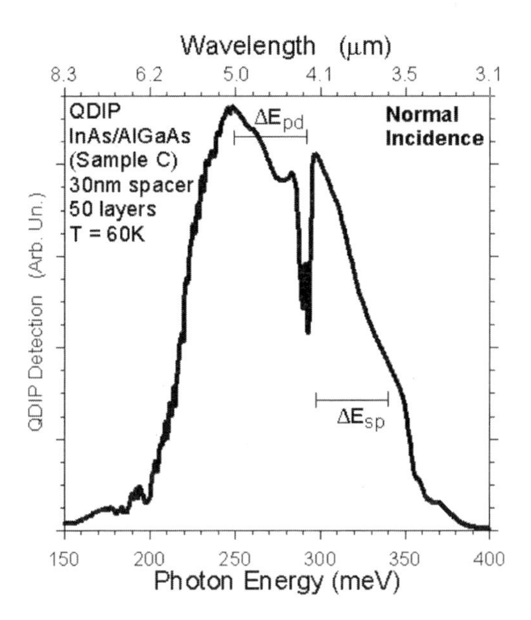

FIGURE 18.25. Normal incidence detection in a InAs/AlGaAs quantum dot infrared photodetector at 60 K. (From Ref. 94 by permission of American Institute of Physics.)

As mentioned above, the barrier material can be chosen to change the energy range of the detection in the quantum dot infrared photodetector. Fig. 18.25 shows another example of quantum dot infrared photodetector ("sample C") with normal-incidence detection. In this case, the InAs quantum dots were embedded in $Al_{0.33}Ga_{0.67}As$ barriers with a modulation doping of $n \sim 1 \times 10^{11} cm^{-2}$. The spectral response is shifted to higher energies, peaking at ~250 meV (5.0 μm). The spectrum also contains a dip due to atmospheric absorption, but also appears to have 3 distinct broad features. Assuming the 3 features correspond to transitions from the *s*, *p*, and *d* shells to the same upper level (e.g., the wetting layer), then one can extract an intersublevel energy spacing for the electrons $\Delta E_{sp} \sim 42$ meV and $\Delta E_{pd} \sim 50$ meV. The responsivity was measured at 80 K, and values above 0.1 A/W were obtained for voltages of ~ 3 V at $\lambda_{detection} = 5$ μm.

The quantum dot infrared photodetector photodetectors have a broader response range in the infrared compared to typical quantum well infrared photodetector designs. Indeed, the self-assembled quantum dots naturally grow (or can be grown purposely) with an inhomogeneous broadening due to the size, composition, or strain. More importantly, in addition to the inhomogeneous broadening, the quantum dot infrared photodetector provides a larger response range because the reduced density of states yields a variety of possible transitions which, depending on the occupancy of the quantum dot states, can participate in the detection and which effectively leads to a larger homogeneous broadening. For comparison, the spectral width of a typical quantum well infrared photodetector is about $\Delta\lambda/\lambda \sim 20\%$ at half-point coverage, whereas $\Delta\lambda/\lambda \sim 40\%$ for the quantum dot infrared photodetector of Fig. 18.25. For many applications, the overall responsivity would actually be proportional to the wavelength-integrated response, which would be higher by a factor directly proportional to the bandwidth of the response. Therefore, because quantum dot infrared photodetectors

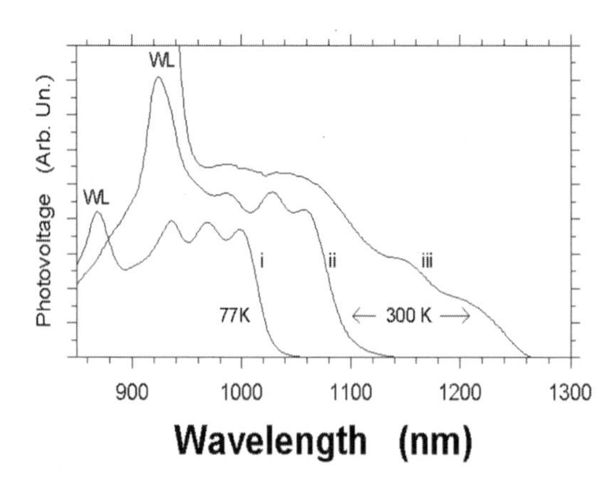

FIGURE 18.26. Interband zero-dimensional transitions measured with open circuit photovoltage across quantum dot laser diodes: (i) 77 K for a stack of 14 indium-flushed at 5.0 nm (the laser shown in TEM in Fig. 18.3), (ii) the same laser at 300 K, and iii) a laser diode with a single layer indium-flushed at 8.5 nm with a quantum dot ground state at close to 1.3 μm at 300 K. (From Ref. 48 by permission of SPIE.)

are sensitive at normal incidence and have broad response ranges, they are promising for applications and are becoming a field of interest for several research groups.[98–100] For detection in the visible and in the near infrared, quantum dots can be grown into a suitable *p-i-n* or *n-i-n* structure which can exploit the excitonic interband absorption in zero-dimensional nanostructures.[101]

Fig. 18.26 shows the low temperature (curve i) and room temperature (curve ii) open circuit photovoltage measured on the quantum dot laser diode shown by TEM in Fig. 18.3. Well-resolved shell structures are observed and can be compared with the results obtained in PL or in electroluminescence (EL). The photovoltage gives a measure of the absorption when the zero-dimensional density of state is empty of carriers. The photovoltage therefore gives results about the shell structure when it is empty. This should be closer to the single particle picture (as discussed in Figs. 18.5 and 18.6), and the photovoltage spectra can be contrasted with the state filling spectroscopy and the lasing spectra to better understand multi-exciton interactions. Photovoltage is a useful tool to characterize the zero-dimensional states in the laser diodes. For example, curve iii of Fig. 18.26 shows a laser diode with a single layer indium-flushed at 8.5 nm (e.g., see Fig. 18.10) with a quantum dot ground state at close to 1.3 μm at 300 K, and quantum dot excited-states below the wetting layer (WL) which peaks at $\lambda \sim 920$ nm.

18.6.f. Coupled Quantum Dots and Artificial Excitonic Molecules

The self-aligned uniform quantum dots obtained with the indium-flush technique seen in Fig. 18.3 opened up an opportunity for producing coupled quantum dots by reducing the spacer thickness further to grow in close proximity 2 or more layers of quantum dots which are as nearly identical as possible.[45] It has been demonstrated that artificial molecules can be obtained using such coupled quantum dot ensembles with well-defined electronic shells. The coupling strength between the zero-dimensional states can be varied by changing the distance between 2 layers of stacked self-assembled InAs/GaAs quantum dots. For strongly

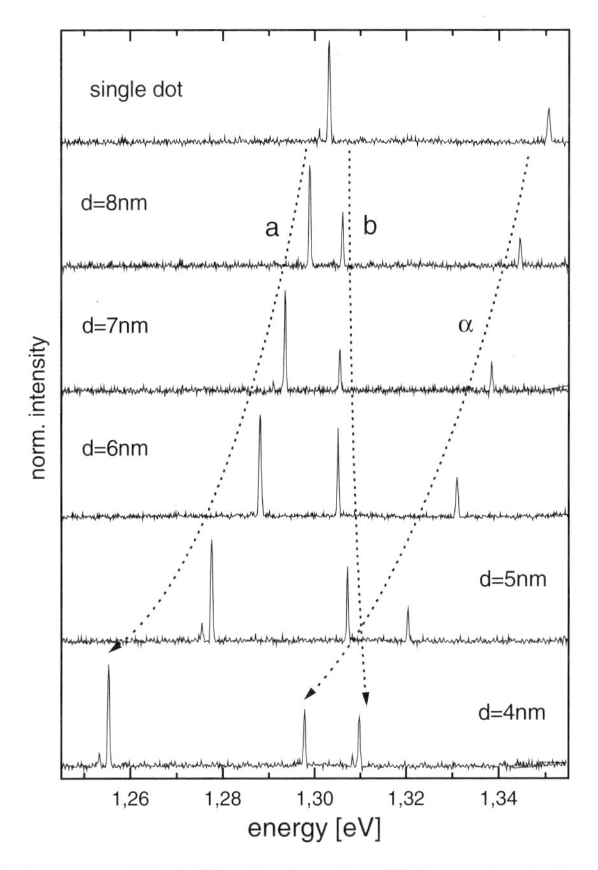

FIGURE 18.27. Emission spectra of single QD molecules at $T = 60$ K for varying dot layer separation d from 8 down to 4 nm. As reference, also a single QD was studied (top trace). The coupling of the states is evident with an energy splitting of 30 meV for the closest separation of 4 nm. (From Ref. 102 by permission of American Association for the Advancement of Science.)

coupled quantum dots grown with a 4 nm spacer, state-filling spectroscopy reveals a shift of the quantum dot symmetric state to lower energies by ~23 meV. The wetting layer states are also strongly coupled because of the shallow confinement, resulting in a red-shift of its symmetric state by ~26 meV.[45] Single quantum dot spectroscopy[49] was also performed on double coupled quantum dots with different spacer thickness to further study the entangling of the low-dimensional quantum states in the vertically self-aligned quantum dots. The entanglement was followed as a function of the coupling strength, and the coupling was measured as a splitting of the emission line from a single pair of entangled quantum dots. Such results are shown in Fig. 18.27, where an energy splitting exceeding 30 meV was observed for quantum dot layers separated by 4 nm.[102] Further studies can be done and devices could be made to tune the coupling with an applied electric field. There is much interest in quantum information processing using such coupled quantum dots, where the key building block of a quantum processor is a quantum gate, which is used to entangle the states of two quantum bits.[103]

18.6.g. Electric Field Effects and "Optical Memories"

Semiconductor self-assembled quantum dots efficiently capture photo-carriers into spatially localized zero-dimensional states. Most semiconductor surfaces have states localized to their surface. However, the carriers trapped at the surface typically recombine non-radiatively, wasting photo-carriers generated near the surface in the absence of passivation. The surface is also effectively a high potential which can perturb the energy levels of a near-surface quantum heterostructure.[104] For these reasons, near-surface quantum wells have been studied to observe shifts of the quantized energy levels and study the properties of devices having finite cap thicknesses.

It is also interesting to study how the surface influences the optical properties of quantum dots[105–106] for which case the in-plane motion of the carriers is eliminated. For that purpose samples with strained InAs/GaAs quantum dots exhibiting sharp shell structures have been grown with thin GaAs cap layers. Furthermore, such studies also provide useful information for the related case when small mesas are used to delineate the number of self-assembled quantum dots probed in single quantum dot studies.[1,49] The quantum dot states are found to initially red-shift when approaching closer proximity to the surface, because the thin cap allows partial relaxation of the quantum dot strain. Competition between the capture of the photo-carriers in the quantum dots and the surface states leads to a pronounced decrease in the quantum dot emission for caps thinner than ~10 nm. Also, the transfer of carriers between the quantum dots and the surface gives rise to interesting optical memory effects lasting over time-scales of several minutes.

The energy levels of the near-surface quantum dots are revealed by observing the electronic shells of the quantum dot ensemble using state-filling spectroscopy.[1–9,47] For example, Fig. 18.28 shows the low temperature photoluminescence (PL) spectra, obtained at an excitation intensity of a few kW/cm^2, for quantum dots indium-flushed at 5.0 nm with a cap thickness between 100 nm and 5.0 nm. The spectra reveal 5 well-resolved shells (s, p, d, f, g), as can be typically observed for a quantum dot ensemble having a good homogeneity, together with the emission from the wetting layer (WL). The self-assembled quantum dots with the thick cap have an harmonic oscillator potential yielding an intersublevel

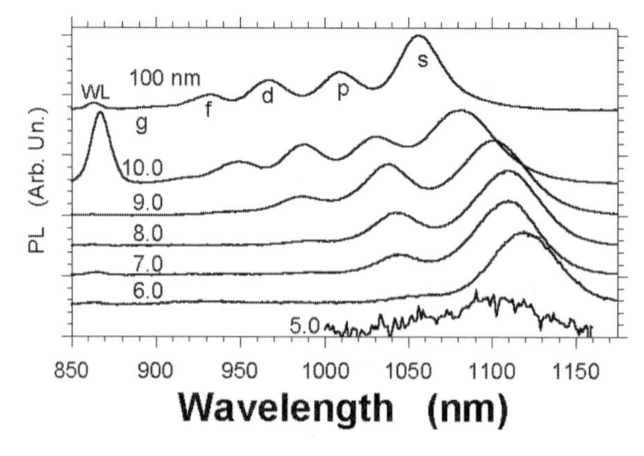

FIGURE 18.28. State-filling spectroscopy of near-surface InAs/GaAs quantum dots covered with the indicated cap thickness. The 77 K PL spectra are excited with a few kW/cm^2 and vertical offsets have been used for clarity. (From Ref. 106 by permission of American Institute of Physics.)

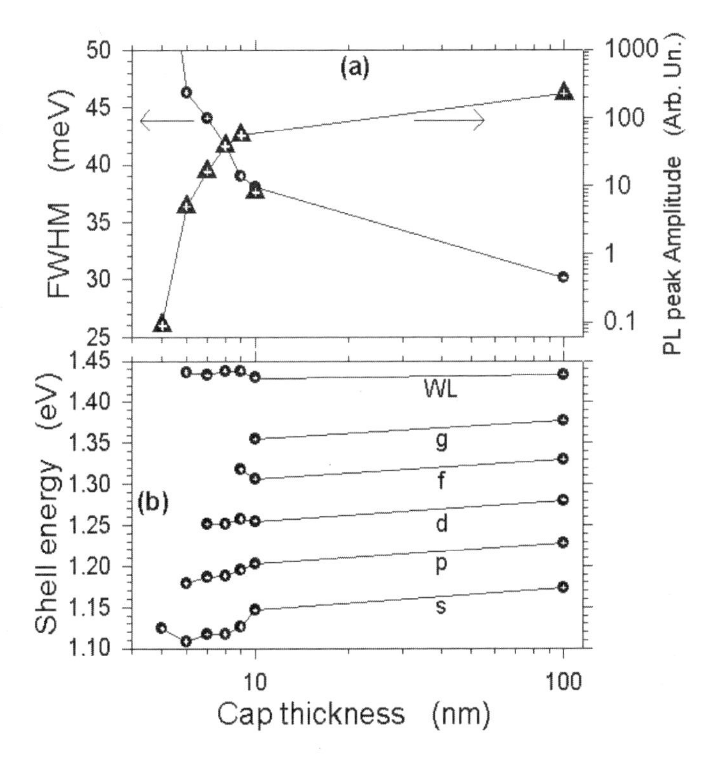

FIGURE 18.29. Evolution with the cap thickness of the FWHM (full-with-half-maximum) and the PL amplitude in a), and of the position of the s, p, d, f, g shells, and the WL in b). (From Ref. 106 by permission of American Institute of Physics.)

energy-spacing of ~55 meV. As the cap thickness is reduced from 10.0 nm to 6.0 nm, a clear change in the optical properties of the near-surface quantum dots is the red-shift of all the zero-dimensional states, as seen in Fig. 18.28. Similar red-shifts have also been observed in InAs quantum dots embedded with thin InGaAs caps.[105] The shift to lower energies is explained if the thinner cap allows a partial relaxation of the InAs strain. However, in the limit where the cap is thin enough to allow the high surface potential to perturb the quantum dot wavefunctions, a blue-shift of the quantum dot energy levels would be expected. For the thinner cap, the two effects (a red-shift from strain-relaxation and a blue-shift from high surface potential) are competing, however, for the 5.0 nm cap, the intensity of the PL is very weak.

The evolution of the optical properties with the cap thickness is summarized in Fig. 18.29. Fig. 18.29a shows the changes in the inhomogeneous broadening of the ensemble (left axis), and the changes in the PL intensity (right axis). There is a progressive increase in the inhomogeneous broadening as the cap thickness is reduced, increasing the PL linewidth of the s-shell from 30 meV to above 46 meV. The larger inhomogeneous broadening indicates more variations in the electronic environment for the quantum dots which are closer to the surface. It is unlikely that the larger inhomogenous broadening can be caused by an increase in the homogeneous linewidth of the individual quantum dots which is typically smaller by more than 2 orders of magnitude. The partial strain relaxation in the thin cap layer, which causes the red-shifts, is most likely the origin of the increased inhomogeneous broadening. For example, the energy levels of the near-surface quantum dots might be more sensitive to the neighboring quantum dot distribution. It has been observed that the random in-plane distribution of the quantum dots contributes to the inhomogeneous broadening of the ensemble.[52,97]

The right axis of Fig. 18.29a shows the continuous reduction in the PL intensity over 3 orders of magnitude as the quantum dots are grown closer to the surface. The PL quenching for the near-surface quantum dots is similar to the PL quenching observed with near-surface quantum wells. Indeed, the radiative lifetimes and the capture times are comparable for quantum dots or for quantum wells, relative to the non-radiative surface recombination rates. Fig. 18.29b gives more details of the position of the quantum dot shells. In addition to the red-shift of the energy levels discussed above, a subtle increase in the intersublevel energy-spacing (from 55 meV to more than ~65 meV) can be observed. The increase in the intersublevel energy-spacing, as well as the slight blue-shift for the thinnest cap, are likely the consequence of the high potential influencing the energy levels of the quantum dot layers grown close to the surface.

Fig. 18.30 shows the evolution with time of the PL amplitude of the *s*-shell when excited with a constant intensity of a few kW/cm^2. For a cap thickness of 10.0 nm or larger (100 nm), the PL intensity does not vary with time. In contrast, for the samples with a cap 9.0 nm or smaller, a remarkable decrease of the steady-state PL with time is observed, with optical memory effects lasting over several minutes. The quenching of the photo-generated carriers is faster for smaller cap thicknesses and for higher temperatures. The memory effect remains for several minutes (e.g., "off" in Fig. 18.30), but the intensity of the PL gradually recovers completely when the quantum dots are not photo-excited for tens of minutes. Very low excitations can also be used to monitor or to "read" the recovery of a "written" quantum dot region. This effect is the opposite of a photo-carrier screening.[107] Here, for the near-surface quantum dots, the photo-carriers contribute to the formation of a built-in electric field. The field arises from the charge separation after the carriers transfer between the quantum dot and surface states.

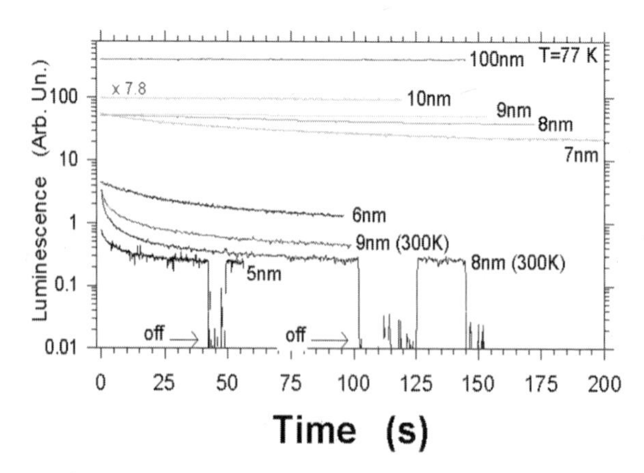

FIGURE 18.30. Optical memory effects lasting over time-scales of several minutes in near-surface quantum dots at 77 K and 300 K. The PL is excited with a constant intensity of a few kW/cm^2, and decreases with time for cap thicknesses 9.0 nm and smaller. The photo-carrier transfer is faster for smaller cap thicknesses or for higher temperatures. (From Ref. 106 by permission of American Institute of Physics.)

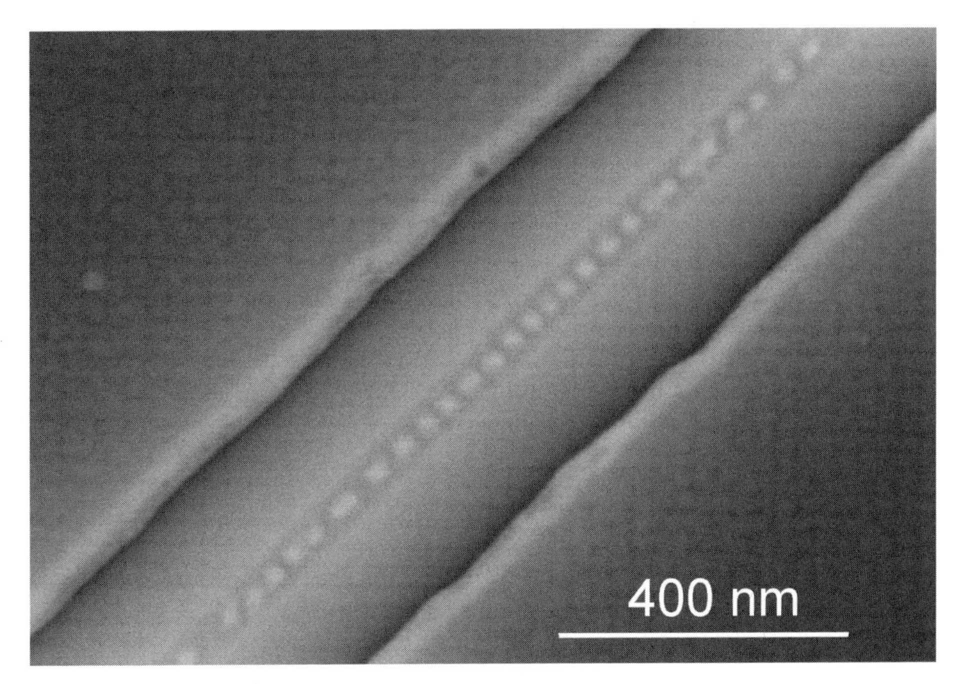

FIGURE 18.31. Nucleation site engineering of InAs/GaAs self-assembled quantum dots using a ridge structure. (From Ref. 108 by permission of American Vacuum Society.)

18.7. SITE ENGINEERING OF QUANTUM DOT NANOSTRUCTURES

For some applications, it is desirable to be able to position the quantum dots in space in a determined order or pattern. Nucleation site engineering can be used to achieve the spatial ordering of self-assembled quantum dots. For example, selective area epitaxy and the anisotropic indium mobility have been used to create narrow ridges and promote the formation of quantum dots positioned in a single row on such ridges.[108–110] Fig. 18.31 gives a nice demonstration of such a line of quantum dots. Similarly, an organized array of self-assembled quantum dots has been fabricated in Petroff's group, where an average of 3 to 4 quantum dots regrouped in order on microfabricated mesas as demonstrated in Fig. 18.32.[111]

18.8. SUMMARY

Semiconductor quantum dot nanostructures are an important application of nanotechnology to optoelectronics. Self-assembled quantum dots have demonstrated unique properties related to their similarity to artificial atoms. They are compatible with standard III–V semiconductor technology and have no additional defects. These nanostructures have small size which yields large confinement energies (larger than the thermal energy at room temperature). The challenge for optoelectronic applications is not simply to exceed the performance of two-dimensional quantum well devices. For example, it would be challenging to displace quantum well optoelectronic components for telecom applications for WDM lasers at 1.5 micron or for lasers at 1.3 micron. These traditional devices are well established in

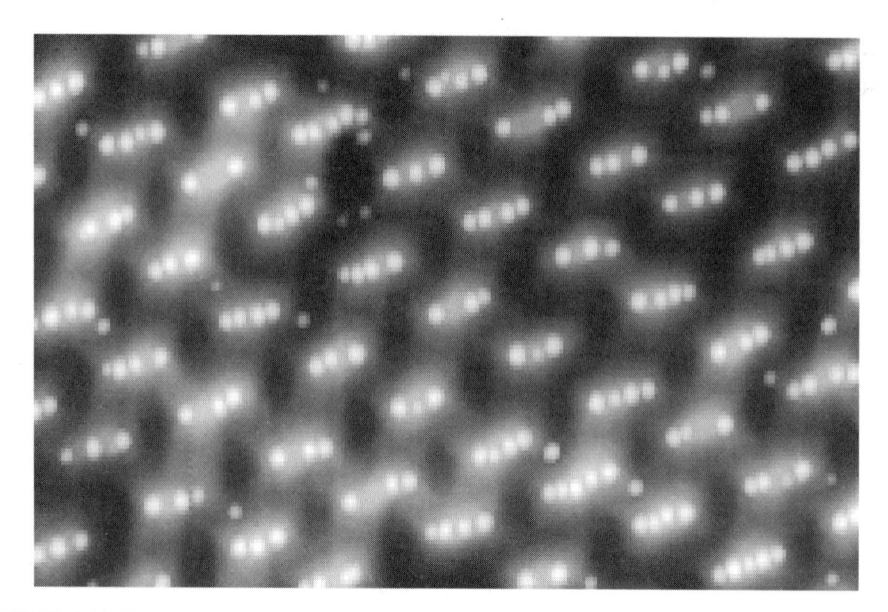

FIGURE 18.32. Nucleation site engineering of InAs/GaAs self-assembled quantum dots using an array of mesa structures. (From Ref. 111 by permission of American Institute of Physics.)

a mature market and the cost of qualifying new nanoscale devices represents a significant barrier. The opportunities are rather to look for unique properties such as broadband gain/tunability spectra, low chirp, low temperature sensitivity, narrow homogeneous linewidth, and a gain medium robust to material defects.

Such nanostructures based on self-assembled quantum dots devices with genuine zero-dimensional properties, with good uniformity, and with strong carrier confinement can be realized by exercising good control of the growth parameters during the epitaxy of self-assembled nanostructures. The observation of a number of well-resolved excited states is the main figure of merit for semiconductor heterostructures based on quantum dots. It is clear that one can develop quantum dot lasers having a sharp electronic shell structure and with low lasing thresholds in forward bias. Similarly, potential applications include the possibility for fast detection in reverse bias, quantum dot infrared photodetectors with broad normal incidence response, and nanostructures with interesting memory effects using near-surface quantum dots or coupled quantum dot structures. Furthermore, alloy intermixing can be used to produce extensive changes in the electronic properties of quantum dots, including a pronounced blue-shift and a narrowing of the inhomogeneous broadening and of the intersublevel energy spacing for the InAs/GaAs material system.

QUESTIONS

1. Density of States and State Filling

The Volmer-Weber epitaxial growth mode permits the direct formation of quantum dot structures (without any 2-dimensional growth beneath the quantum dot). Imagine such an ensemble of perfect quantum dots can be grown (with all the dots having the same size and same density of states), with a density of 2×10^{10} cm^{-2}, in the middle of a quantum well structure which is situated at 100 nm below the sample surface. Assume that the resulting position of the energy levels are as depicted schematically below:

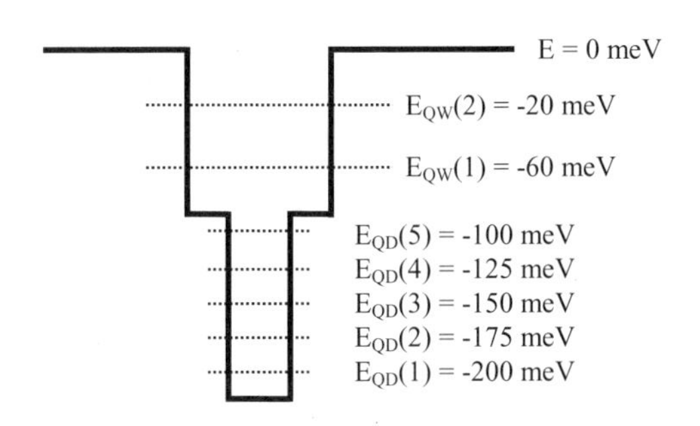

a. Assume that the material properties of the heterostructure are similar to GaAs and that the quantum dots have a harmonic oscillator energy spectrum. Calculate the electron density of states of the structure and the number of available states before saturation in the 3-dimensional states.

b. The green line of an Argon laser is used to do some state filling spectroscopy. Assuming an efficient carrier relaxation to the lowest available states and a radiative interband recombination rate of 1.0 ns with an infinite reservoir of holes, estimate the excitation intensity required to (i) fill all the 0-dimensional states and (ii) fill up to 10 meV above the 2nd quantum well state [i.e., $E_{QW}(2)$].

c. Briefly discuss some issues and possible mechanisms related to the efficient relaxation of the carriers to the lowest available states in such a structure.

2. *Thermionic emission in a quantum dot ensemble:*

a. Modify eqn. 18.1 of section 18.6.b to simulate the emission of a quantum dot ensemble having 4 zero-dimensional states (S, P, D, F).

b. Plot the photoluminescence emission of such a quantum dot ensemble at various temperatures assuming an equi-distant inter-sublevel energy spacing of 50 meV, with the highest quantum dot shell 50 meV below the wetting layer, and assuming thermionic loss of the carriers via the wetting layer.

3. *Intermixing:*

The alloy intermixing as discussed in Fig. 18.18 is often modeled using a Fickian diffusion model. The diffused alloy composition along the growth axis with z-coordinate can be written as:

$$X(z) = X_o + 1/2 \sum_i X_i \left\{ \text{erf}\left[\left(L_z^i + 2\left(z - z_o^i\right)\right)/4L_d\right] + \text{erf}\left[\left(L_z^i - 2\left(z - z_o^i\right)\right)/4L_d\right]\right\}$$

where X_o is the as-grown alloy concentration in the barrier, L_d is the diffusion length, X_i is the as-grown alloy concentration in the i-th potential segment of width L_z^i centered at $z = z_o^i$, and erf[] denotes the error function. The bandgap profile and the position-dependent effective masses are then deduced from the composition profile $X(z)$. Plot the alloy concentration for various diffusion length L_d for a 3.0 nm InAs/GaAs quantum well.

REFERENCES

1. a) S. Fafard, R. Leon, D. Leonard, J. L. Merz, and P. M. Petroff, *Phys. Rev.* B **52**, 5752 (1995); b) *Ibidem, Phys. Rev.* B **50**, 8086 (1994); c) *Ibidem*, Superlattices and Microstructures, **16**, 303 (1994); d) S. Fafard, D. Leonard, J. M. Merz, and P. M. Petroff, *Appl. Phys. Lett.* **65**, 1388 (1994); e) S. Fafard, S. Raymond, G. Wang, R. Leon, D. Leonard, S. Charbonneau, J. L. Merz, P. M. Petroff, and J. E. Bowers, *Surface Science* 361, 778, (1996).
2. K. Mukai, N. Ohtsuka, H. Shoji, M. Sugawara, *Appl. Phys. Lett.* **68**, 3013 (1996).
3. R. Leon, S. Fafard, P. G. Piva, S. Ruvimov, Z. Liliental-Weber, *Phys. Rev.* B **58**, R4262 (1998).
4. S. Raymond, P. Hawrylak, C. Gould, S. Fafard, A. Sachrajda, M. Potemski, A. Wojs, S. Charbonneau, D. Leonard, P. M. Petroff and J. L. Merz, *Solid State Communications* **101**, 883 (1997).
5. G. Park, O. B. Shchekin, D. L. Huffaker, D. G. Deppe, *Appl. Phys. Lett.* **73**, 3351 (1998).
6. Y. Sugiyama, Y. Nakata, T. Futatsugi, M. Sugawara, Y. Awano, N. Yokoyama, *Jpn. J. Appl. Phys.* **36**, L158 (1997).
7. M. Grundmann, N. N. Ledentsov, O. Stier, D. Bimberg, V. M. Ustinov, P. S. Kopev, and Zh. I. Alferov, *Appl. Phys. Lett.* **68**, 979 (1996).
8. S. Raymond, X. Guo, J. L. Merz, and S. Fafard, *Phys. Rev.* B **59**, 7624 (1999).
9. H. Lipsanen, M. Sopanen, J. Ahopelto, *Phys. Rev.* B **51**, 13868 (1995).
10. S. Fafard, Z. R. Wasilewski, C. Nì. Allen, D. Picard, M. Spanner, J. P. McCaffrey, and P. G. Piva, *Phys. Rev.* B **59**, 15368 (1999).
11. S. Fafard, Z. R. Wasilewski, C. Nì Allen, D. Picard, P. G. Piva, and J. P. McCaffrey, *Superlat. and Microst.* **25**, 87 (1999).
12. S. Fafard, K. Hinzer, S. Raymond, M. Dion, J. McCaffrey, Y. Feng, and S. Charbonneau, *Science* **274**, 1350 (1996)
13. S. Fafard, Z. R. Wasilewski, C. Nì. Allen, K. Hinzer, J. P. McCaffrey, and Y. Feng, *Appl. Phys. Lett.* **75**, 986 (1999).
14. S. Fafard, Photonics Spectra **31**, 160 (1997).
15. R. Mirin, A. Gossard, and J. Bowers, *Electronics Lett.* **32**, 1732 (1996).
16. Q. Xie, A. Kalburge, P. Chen, and A. Madhukar, *IEEE Photonics Technology Letters* **8**, 965 (1996).
17. H. Shoji, K. Mukai, N. Ohtsuka, M. Sugawara et al., *IEEE Photonics Technology Letters* **12**, 1385 (1995).
18. H. Shoji, Y. Nakata, K. Mukai, Y. Sugiyama, M. Sugawara, N. Yokoyama, and H. Ishikawa, *Jap. J. Appl. Phys. II, Lett.* **35**, L903 (1996); Ibidem, *Electonics Lett.* **32**, 2023 (1996).
19. H. Saito, K. Nishi, I. Ogura, S. Sugou, and Y. Sugimoto, *Appl. Phys. Lett.* **69**, 3140 (1996).
20. K. Kamath, P. Bhattacharya, T. Sosnowski, T. Norris, and J. Phillips, *Electronics Lett.* **32**, 1374 (1996).
21. D. G. Deppe and H. Huang, *Appl. Phys. Lett.* **75**, 3455 (1999).
22. G. Park, O. B. Shchekin, S. Csutak, D. Huffaker, and D. G. Deppe, *Appl. Phys. Lett.* **75**, 3267 (1999).
23. N. N. Ledentsov, V. M. Ustinov, V. A. Shchukin, P. S. Kop'ev, Zh. I. Alferov, and D. Bimberg, Semicond **32**, 343 (1998).
24. K. Hinzer, J. Lapointe, Y. Feng, A. Delage, and S. Fafard, A. J. SpringThorpe and E. M. Griswold, *J. Appl Phys.* **87**, 1496 (2000).
25. S. Fafard, K. Hinzer, A. J. SpringThorpe, Y. Feng, J. McCaffrey, S. Charbonneau, and E. M. Griswold Material *Science and Engineering* **51**, 114 (1998).
26. S. Fafard, J. McCaffrey, Y. Feng, C. Ni Allen, H. Marchand, L. Isnard, P. Desjardins, S. Guillon, and R. A Masut, Proc. SPIE 3491, 272 (1998).
27. M. Sugawara, K. Mukai, Y. Nakata, H. Ishikawa, and A. Sakamoto, *Phys. Rev.* B **61**, 7595 (2000).
28. Z. R. Wasilewski, S. Fafard, J. P. McCaffrey, and J. Crystal Gr. 201, 1131 (1999).

29. D. Leonard, M. Krishnamurthy, C. M. Reaves, S. P. Denbars, and P. M. Petroff, *Appl. Phys. Lett.* **63**, 3203 (1993).
30. D. Leonard, S. Fafard, K. Pond, Y. H. Zhang, J. M. Merz, and P. M. Petroff, *J. Vac. Sci. Technol.* B **12**, 2516 (1994).
31. D. Leonard, M. Krishnamurthy, S. Fafard, J. M. Merz, and P. M. Petroff, *J. Vac. Sci. Technol.* B **12**, 1063 (1994).
32. J. P. McCaffrey, M. D. Robertson, Z. R. Wasilewski, E. M. Griswold, L. D. Madsen, and S. Fafard, Determination of the size, shape, and composition of indium-flushed self-assembled quantum dots by transmission electron microscopy, *J. Appl. Phys.* **88**, 2272 (2000).
33. J. M. Garcia, T. Mankad, P. O. Holtz, P. J. Wellman, and P. M. Petroff, *Appl. Phys. Lett.* **72**, 3172 (1998).
34. G. D. Lian, J. Yuan, L. M. Brown, G. H. Kim, and D. A. Ritchie, *Appl. Phys. Lett.* **73**, 49 (1998).
35. H. Saito, K. Nishi, and S. Sugou, *Appl. Phys. Lett.* **73**, 2742 (1998); ibidem, vol **74**, 1224 (1999).
36. J. P. McCaffrey, M. D. Robertson, P. J. Poole, B. J. Riel, and S. Fafard, "Interpretation and Modelling of Buried InAs Quantum Dots on GaAs and InP Substrates," *J. Appl. Phys.* **90**, 1784 (2001).
37. J. P. McCaffrey, M. D. Robertson, Z. R. Wasilewski, S. Fafard, and L. D. Madsen. *Inst. Phys. Conf. Ser.* **164**, 107 (1999).
38. X. Z. Liao, J. Zou, D. J. H. Cockayne, R. Leon, and C. Lobo, *Phys. Rev. Lett.* **82**, 5148 (1999).
39. D. M. Bruls, J. W. A. M. Vugs, P. M. Koenraad, H. W. M. Salemink, J. H. Wolter, M. Hopkinson, M. S. Skolnick, and G. S. P. A. Fei-Long, Determination of the shape and indium distribution of low-growth-rate InAs quantum dots by cross-sectional scanning tunneling microscopy, *Appl. Phys. Lett.* **81**, 1708 (2002).
40. U. Hakanson, M. K. Johansson, J. Persson, J. Johansson, M. E. Pistol, L. Montelius, and L. Samuelson, Single InP/GaInP quantum dots studied by scanning tunneling microscopy and scanning tunneling microscopy induced luminescence, *Appl. Phys. Lett.* **80**, 494 (2002).
41. T. Yamauchi, Y. Ohyama, Y. Matsuba, M. Tabuchi, and A. Nakamura, Observation of quantum size and alloying effects of single InGaAs quantum dots on GaAs(001) by scanning tunneling spectroscopy, *Appl. Phys. Lett.* **79**, 2465 (2001).
42. P. Ballet, J. B. Smathers, H. Yang, C. L. Workman, and G. J. Salamo, Scanning tunneling microscopy investigation of truncated InP/$GaInP_2$ self-assembled islands, *Appl. Phys. Lett.* **77**, 3406 (2000).
43. T. K. Johal, R. Rinaldi, A. Passaseo, R. Cingolani, A. Vasanelli, R. Ferreira, and G. Bastard, Imaging of the electronic states of self-assembled $In_xGa_{1-x}As$ quantum dots by scanning tunneling spectroscopy, *Phys. Rev.* B **66**, 075336 (2002).
44. K. Hinzer, M. Bayer, J. P. McCaffrey, P. Hawrylak, M. Korkusinski, O. Stern, Z. R. Wasilewski, S. Fafard, and A. Forchel, "Optical Spectroscopy of Electronic States in a Single Pair of Vertically Coupled Self-Assembled Quantum Dots", *Physica Status Solidi* B **224**, 385 (2001).
45. S. Fafard, M. Spanner, J. P. McCaffrey, and Z. R. Wasilewski, *Appl. Phys. Lett.* **76**, 2268 (2000).
46. A. Wojs, P. Hawrylak, S. Fafard, and L. Jacak, *Phys. Rev.* B **54**, 5604, (1996); A. Wojs, P. Hawrylak, Solid State Comm. 100, 487 (1996).
47. P. Hawrylak, *Phys. Rev.* B **60**, 5597 (1999); A. Wojs, P. Hawrylak, S. Fafard, L. Jacak, Physica E 2, 603 (1998).
48. S. Fafard, H. C. Liu, Z. R. Wasilewski, J. McCaffrey, M. Spanner, S. Raymond, C. Nì. Allen, K. Hinzer, J. Lapointe, C. Struby, M. Gao, P. Hawrylak, C. Gould, A. Sachrajda, and P. Zawadzki, Quantum dots devices, SPIE 4078, 100 (2000).
49. M. Bayer, O. Stern, P. Hawrylak, S. Fafard, and A. Forchel, Hidden symmetries in the energy levels of excitonic artificial atoms, *Nature* **405**, 923 (2000).
50. R. Leon, C. Lobo, A. Clark, R. Bozek, A. Wysmolek, A. Kurpiewski, and M. Kaminska, *J. Appl. Phys.* **84**, 248 (1998).
51. R. Leon and S. Fafard, *Phys. Rev.* B **58**, R1726, (1998).
52. S. Fafard, Z. R. Wasilewski, and M. Spanner, *Appl. Phys. Lett.* **75**, 1866 (1999).
53. B. J. Riel, K.Hinzer, S. Moisa, J. Fraser, P. Finnie, P. Piercy, S. Fafard, Z. R. Wasilewski, and J. Cryst. Growth. **236**, 145 (2002).
54. G. Wang, S. Fafard, D. Leonard, J. E. Bowers, J. M. Merz, and P. M. Petroff, *Appl. Phys. Lett.* **64**, 2815 (1994).
55. S. Raymond, S. Fafard, S. Charbonneau, R. Leon, D. Leonard, P. M. Petroff, and J. L. Merz, *Phys. Rev.* B **52**, 17238 (1995); J. Arlett, F. Yang, K. Hinzer, S. Fafard, Y. Feng, S. Charbonneau, R. Leon, *J. Vac. Sc. Technol.* B **16**, 578 (1998).

56. S. Raymond, S. Fafard, A. Wojs, P. Hawrylak, S. Charbonneau, D. Leonard, R. Leon, P. M. Petroff, and J. L. Merz, *Phys. Rev.* B **54**, 11548 (1996).
57. D. Morris, N. Perret, and S. Fafard, *Appl. Phys. Lett.* **75**, 3593 (1999).
58. H. Benisty, C. M. Sotomayor-Torres, and C. Weisbuch, *Phys. Rev.* B **44**, 10945 (1991).
59. A. V. Uskov, J. McInerney, F. Adler, H. Schweizer, and M. H. Pikuhn, *Appl. Phys. Lett.* **72**, 58 (1998).
60. Al. L. Efros, V. A. Kharchenko, and M. Rosen, *Solid State Commun.* **93**, 281 (1995).
61. S. Raymond, K. Hinzer, S. Fafard, and J. L. Merz, *Phys. Rev.* B **61**, 15 Jun. (2000).
62. T. Inoshita and H. Sakaki, *Phys. Rev.* B **46**, 7260 (1992)
63. D. Morris, N. Perret, D. Riabinina, J. Beerens, V. Aimez, J. Beauvais, and S. Fafard, Dynamics of Photo-Excited Carriers in Self-Assembled Quantum Dots, SPIE proceedings Photonics North 2002.
64. R. Leon, P. M. Petroff, D. Leonard, and S. Fafard, Spatially Resolved Visible Luminescence of Self-Assembled Quantum Dots, *Science* **267**, 1966 (1995).
65. M. Bayer, A. Forchel, P. Hawrylak, S. Fafard, and G. Narvaez, Excitonic states in In(Ga)As self-assembled quantum dots, Physica Stat. Sol. B **224**, 331 (2001).
66. K. Hinzer, P. Hawrylak, M. Korkusinski, S. Fafard, M. Bayer, O. Stern, A. Gorbunov, A. Forchel, "Optical spectroscopy of a single AlInAs/AlGaAs quantum dot", *Phys. Rev.* B **63**, 75314 (2001).
67. M. Bayer, G. Ortner, O. Stern, A. Kuther, A. A. Gorbunov, A. Forchel, P. Hawrylak, S. Fafard, K. Hinzer, T. L. Reinecke, S. N. Walck, J. P. Reithmaier, K. Klopf, and F. Schäfer, Fine structure of neutral and charged excitons in self-assembled In(Ga)As/(Al)GaAs quantum dots, *Phys. Rev.* B **65**, 195315-1 (2002); For further references on single quantum dot spectroscopy, see also Ref. 1 to Ref. 50 of this reference.
68. S. Fafard and C. Nì. Allen, *Appl. Phys. Lett.* **75**, 2374 (1999).
69. Y. Q. Wei, S. M. Wang, F. Ferdos, J. Vukusic, A. Larsson, Q. X. Zhao, and M. Sadeghi, Large ground-to-first-excited-state transition energy separation for InAs quantum dots emitting at 1.3 micron, *Appl. Phys. Lett.* **81**, 1621, (2002).
70. M. Kuntz, N. N. Ledentsov, D. Bimberg, A. R. Kovsh, V. M. Ustinov, A. E. Zhukov, and Yu. M. Shernyakov, Spectrotemporal response of 1.3 micron quantum-dot lasers, *Appl. Phys. Lett.* **81**, 3846, (2002).
71. O. B. Shchekin, D. G. Deppe, 1.3 micron InAs quantum dot laser with To = 161K from 0 to 80 degrees C, *Appl. Phys. Lett.* **80**, 3277 (2002).
72. A. Markus, A. Fiore, J. D. Ganiere, U. Oesterle, J. X. Chen, B. Deveaud, M. Ilegems, and H. Riechert, Comparison of radiative properties of InAs quantum dots and GaInNAs quantum wells emitting around 1.3micron, *Appl. Phys. Lett.* **80**, 911 (2002).
73. Y. Qiu, P. Gogna, S. Forouhar, A. Stintz, and L. F. Lester, High-performance InAs quantum-dot lasers near 1.3 micron, *Appl. Phys. Lett.* **79**, 3570 (2001).
74. K. Mukai, Y. Nakata, K. Otsubo, M. Sugawara, N. Yokoyama, and H. Ishikawa, High characteristic temperature of near-1.3- mu m InGaAs/GaAs quantum-dot lasers at room temperature, *Appl. Phys. Lett.* **76**, 3349, (2000).
75. P. B. Joyce, T. J. Krzyzewski, G. R. Bell, T. S. Jones, E. C. Le-Ru, and R. Murray, Optimizing the growth of 1.3 mu m InAs/GaAs quantum dots, *Phys. Rev.* B **64**, 235317 (2001).
76. J. Tatebayashi, M. Nishioka, and Y. Arakawa, Over 1.5 micron light emission from InAs quantum dots embedded in InGaAs strain-reducing layer grown by metalorganic chemical vapor phase deposition, *Appl. Phys. Lett.* **78**, 3469, (2000).
77. M. Sopanen, H. P. Xin, and C. W. Tu , Self-assembled GaInNAs quantum dots for 1.3 and 1.55 μm emission on GaAs, *Appl. Phys. Lett.* **75**, 994, (2000).
78. US patent Office, patent No. 6, **177**, 684, Y. Sugiyama (2001).
79. G. Park, O. B. Shchekin, D. L. Huffaker, and D. G. Deppe, IEEE Photo-nics Technol. Lett. 12, 230 (2000); O. B. Shchekin, G. Park, D. L. Huffaker, and D. G. Deppe, Discrete energy level separation and the threshold temperature dependence of quantum dot lasers, *Appl. Phys. Lett.* **76**, 466 (2000).
80. X. Huang, A. Stintz, C. P. Hains, G. T. Liu, J. Cheng, and K. J. Malloy, *Electron. Lett.* **36**, 41(2000).
81. A. Stintz, G. T. Liu, H. Li, L. F. Lester, and K. J. Malloy, "Low-threshold current density 1.3 μm InAs quantum-dot lasers with the dots-in-a-well (DWELL) structure," *IEEE Photon. Technol. Lett.*, **12**, 591 (2000).
82. R. Leon, S. Fafard, D. Leonard, J. L. Merz, and P. M. Petroff, Visible Luminescence from large Semiconductor Quantum Dot Ensembles, *Appl. Phys. Lett.* **67**, 521 (1995).
83. H. Y. Liu, I. R. Sellers, R. J. Airey, M. J. Steer, P. A. Houston, D. J. Mowbray, J. Cockburn, M. S. Skolnick, B. Xu, Z. G. Wang, Room-temperature, ground-state lasing for red-emitting vertically aligned InAlAs/AlGaAs quantum dots grown on a GaAs(100) substrate, *Appl. Phys. Lett.* **80**, 3769 (2002).

84. C. Ni. Allen, P. Finnie, S. Raymond, Z. R. Wasilewski, and S. Fafard, Inhomogeneous broadening in quantum dots with ternary aluminum alloys, *Appl. Phys. Lett.* **79**, 2701 (2001).
85. S. Fafard, Z. Wasilewski, J. McCaffrey, S. Raymond, and S. Charbonneau, InAs Self-Assembled Quantum Dots grown by Molecular Beam Epitaxy on InP substrate, *Appl. Phys. Lett.* **68**, 991 (1996).
86. C. Nì. Allen, P. J. Poole, P. Marshall, J. Fraser, S. Raymond, and S. Fafard, *Appl. Phys. Lett.* **80**, 3629 (2002).
87. C. Paranthoen, N. Bertu, O. Dehaese, A. Le Corre, S. Loualiche, B. Lambert, and G. Patriarche, Height dispersion control of InAs/InP quantum dots emitting at 1.55 μm, *Appl. Phys. Lett.* **78**, 1751 (2001).
88. Wang RH, Stintz A, Varangis PM, Newell T. C., Li H, Malloy K. J., and Lester L. F., *IEEE Photonics Technology Letters* **13**, 767 (2001).
89. V. M. Ustinov, A. E. Zhukov, A. Yu. Egorov, A. R. Kovsh, S. V. Zaitsev, N. Yu. Gordeev, V. I. Kopchatov, N. N. Ledentsov, A. F. Tsatsul'nikov, B. V. Volovik, P. S. Kop'ev, Z. I. Alferov, S. S. Ruvimov, Z. Liliental-Weber, and D. Bimberg, *Electron. Lett.* **34**, 670 (1998).
90. H. Saito, K. Nishi, A. Kamei, and S. Sugou, *IEEE Photonics Technology Letters* **12**, 1298 (2000).
91. T. C. Newell, D. J. Bossert, A. Stintz, B. Fuchs, K. J. Malloy, and L. F. Lester, *IEEE Photonics Technology Letters* **11**, 1527 (1999).
92. P. G. Piva, R. D. Goldberg, I. V. Mitchell, D. Labrie, R. Leon, S. Charbonneau, Z. R. Wasilewski, and S. Fafard, Enhanced Degradation Resistance of Quantum Dot Lasers to Radiation Damage, *Appl. Phys. Lett.* **77**, 624 (2000).
93. R. Leon, G. M. Swift, B. Magness, W. A. Taylor, Y. S. Tang, K. L. Wang, P. Dowd, and Y. H. Zhang, *Appl. Phys. Lett.* **76**, 2074 (2000).
94. H. C. Liu, M. Gao, J. McCaffrey, Z. R. Wasilewski, S. Fafard, Quantum Dot Infrared Photodetectors, *Appl. Phys. Lett.* **78**, 79 (2001).
95. A. D. Stiff-Roberts, S. Chakrabarti, S. Pradhan, B. Kochman, and P. Bhattacharya, Raster-scan imaging with normal-incidence, midinfrared InAs/GaAs quantum dot infrared photodetectors, *Appl. Phys. Lett.* **80,** 3265 (2002).
96. L. Chu, A. Zrenner, M. Bichler, and G. Abstreiter, Quantum-dot infrared photodetector with lateral carrier transport, *Appl. Phys. Lett.* **79**, 2249 (2001).
97. R. Leon, S. Marcinkevicius, X. Z. Liao, J. Zou, D. J. H. Cockayne, and S. Fafard, *Phys. Rev.* B **60**, R8517 (1999).
98. S. Sauvage, P. Boucaud, F. H. Julien, J. M. Gérard, and V. Thierry-Mieg, *Appl. Phys. Lett.* **71**, 2785 (1997).
99. D. Pan, E. Towe, and S. Kennerly, *Appl. Phys. Lett.* **73**, 1937 (1998).
00. L. Chu, A. Zrenner, G. Bohm, and G. Abstreiter, *Appl. Phys. Lett.* **75**, 3599 (1999).
01. F. Yang, K. Hinzer, C. Ni. Allen, S. Fafard, G. C. Aers, Yan Feng, J. McCaffrey, and S. Charbonneau, Superlattices and Microstructures **25**, 419 (1999).
02. M. Bayer, P. Hawrylak, K. Hinzer, S. Fafard, M. Korkusinski, Z. R. Wasilewski, O. Stern, and A. Forchel, Coupling and Entangling of Quantum States in Quantum Dot Molecules, SCIENCE **291**, 451 (2001).
03. P. Hawrylak, S, Fafard, and Z. Wasilewski, Engineering Quantum States in Self-Assembled Quantum Dots for Quantum Information Processing, Condensed Matter News **7**, 16 (1999).
04. S. Fafard, Phys. Rev. B 50, 1961 (1994); S. Fafard, *Phys. Rev.* B **46**, 4659 (1992).
05. T.-E. Nee, N.-T. Yeh, J.-I. Chyi, C.-T. Lee, *Solid State Electronics* **42**, 1331 (1998).
06. S. Fafard, *Appl. Phys. Lett.* **76**, 2707 (2000).
07. S. Fafard, E. Fortin, and J. L. Merz, *Phys. Rev.* B **48**, 11062 (1993).
08. J. Lefebvre, P. J. Poole, G. C. Aers, D. Chithrani, and R. L. Williams, Tunable emission from InAs quantum dots on InP nano-templates, *J. Vac. Sci. Technol.* B, **20**, 2173 (2002).
09. J. Lefebvre, P. J. Poole, J. Fraser, G. C. Aers, D. Chithrani, and R. L. Williams, Self-assembled InAs quantum dots on InP nano-templates, J. Cryst. Growth **234**, 391 (2002).
10. R. L. Williams, G. C. Aers, P. J. Poole, J. Lefebvre, D. Chithrani, and B. Lamontagne, Controlling the self-assembly of InAs/InP Quantum Dots, J. Cryst. Growth **223**, 321–331 (2001).
11. H. Lee, J. A. Johnson, M. Y. He, J. S. Speck, and P. M. Petroff, Strain-engineered self-assembled semiconductor quantum dot lattices, *Appl. Phys. Lett.* **78**, 105 (2001).

19

Organic Optoelectronic Nanostructures

J. R. Heflin

Department of Physics
Virginia Tech, Blacksburg, VA

19.1. INTRODUCTION

Synthetic polymers are ubiquitous features of everyday life in forms such as plastic containers and packages as well as fibers in clothing. Their excellent mechanical properties together with their easy processability, light weight, and low cost are primarily responsible for their widespread use in such passive applications. The discovery in the late 1970's of conjugated polymers with conductivities as large as metals established the possibility of using polymers as active electronic, optical, and optoelectronic materials. Extensive research and development efforts have been required to convert that promise into reality, but organic and polymeric optoelectronic materials have now begun to move out of the laboratory and into commercial products. Light-emitting diodes (LEDs) made from small, conjugated organic molecules are now used as display elements in products such as car stereos, cell phones, and digital cameras, for example. It is likely that flexible, plastic electronics and organic solar cells will soon follow.

A molecule is referred to as conjugated when it possess alternating single and double bonds. This occurs, for example, in a series of carbon atoms in which each carbon is bound to three other atoms. Of the four valence electrons of carbon, three of them form sp^2 hybrid orbitals known as σ orbitals that bond with the atom's three neighbors. The remaining valence electrons from each carbon form π-electron orbitals, which are delocalized over the molecule. The delocalized π-electrons result in strong optical absorption, large polarizability, and large conductivity. A polymer is a long molecule in which a basic unit is repeated hundreds or thousands of times. The structures of a number of important conjugated

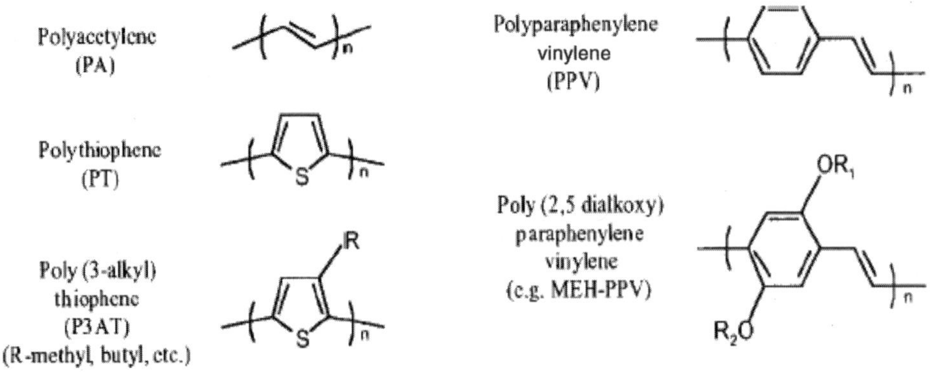

FIGURE 19.1. Molecular structures of several conjugated polymers. (From Ref. 1 by permission of American Physical Society.)

polymers are shown in Figure 19.1. The conductivities of conjugated polymers span the range from 10^{-10} to 10^{7} S/m, approaching those of conventional metals. The conducting and semiconducting properties of conjugated polymers together with their strong optical absorption bands and large polarizabilities make them of great interest for optoelectronic applications.[1]

While the size of a typical small conjugated molecule or the repeat unit of a polymer chain is on the order of a nanometer, they are (perhaps unfairly) generally not considered nanomaterials in and of themselves but rather as the standard subjects of organic chemistry. However, in many cases nanoscale control of the assembly and organization of these materials can result in dramatic improvement of their optoelectronic properties. This chapter will focus primarily on examples where nanoscale self-assembly and self-organization as well as combination with nanomaterials such as fullerenes or inorganic nanoparticles have been utilized to provide enhanced performance in organic and polymeric light-emitting diodes, solar cells, and nonlinear optical materials. While there has been extensive work done in each area, we are limited here to discussing a few representative examples.

19.2. ORGANIC AND POLYMERIC LIGHT-EMITTING DIODES

Many conjugated molecules and polymers are highly photoluminescent. The quantum efficiency (ratio of photons emitted to photons absorbed) can be nearly 100%. Generally the energy of the emitted photon is smaller (and, correspondingly, the wavelength is longer than that of the absorbed photon as some of the energy is retained by the molecule, usually in the form of vibrational modes. The difference in the two photon energies is referred to a the Stokes shift. Absorption in the visible part of the electromagnetic spectrum correspond to excitation from the ground state to one of the lower-lying electronic excited state The visible wavelength range of 400 to 700 nm corresponds to 1.7 to 3.0 eV. Becaus various organic molecules and polymers have excitation energies that span this entire range there is much interest in developing full color (RGB) organic light-emitting diode (OLED displays.

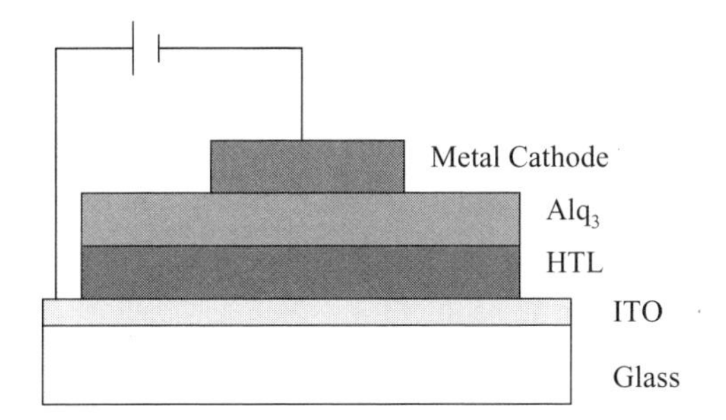

FIGURE 19.2. Device structure of an organic light-emitting diode.

In an OLED, the electrons injected from the cathode (negative electrode) recombine with holes injected from the anode (positive electrode) at a single molecular site. If there is a large quantum efficiency for radiative decay, then the applied current results in emitted light. However, electrons and holes that pass from one electrode to the other without undergoing recombination result in higher currents and lower light emission and, therefore, lower device efficiency. Tang and VanSlyke demonstrated the first efficient OLED by sandwiching two different organic materials between the anode and cathode.[2] Tris(8-hydroxyquinoline) aluminum (Alq_3) was used as the active light-emitting material. In order to prevent transport of electrons across the Alq_3 layer without undergoing recombination, an aromatic diamine layer was deposited between the anode and Alq_3. The diamine preferentially transports holes but not electrons, so the interface between the diamine and Alq_3 layers becomes the primary region for electron-hole recombination. The general device structure is shown in Figure 19.2. In order to allow the emitted light to escape the device, a transparent conductor such as indium tin oxide (ITO) is usually used as the anode. The ITO is coated on a glass substrate, and the emitted light is viewed through the glass. The cathode is usually a common metal such as aluminum, gold, magnesium, or calcium. The organic layers and the cathode are typically deposited by vacuum evaporation. Tang and VanSlyke used a diamine hole transport layer (HTL) 75 nm thick and an Alq_3 layer 60 nm thick. Luminances greater than 1000 cd/m^2 were obtained at less than 10 V applied bias. (For comparison, the luminance of a computer monitor is $\sim$100 cd/m^2.) The external quantum efficiency (number of photons emitted per number of electrons injected) of the device was about 1%. Since this initial demonstration, several approaches have been demonstrated for improving OLED efficiency including incorporation of electron injection[3] and hole-blocking[4] layers between the emissive layer and cathode, insertion of a layer between the anode and HTL to suppress hole-injection,[5] and doping with fluorescent[6] or phosphorescent[7] dyes. The injection of holes in OLEDs is generally more efficient than injection of electrons. It is believed that the insertion of a layer between the ITO and HTL that suppresses the hole injection leads to better electron-hole balance and thus improves the device quantum and power efficiencies.

Various methods have been employed to deposit an interlayer between the ITO anode and the HTL including chemical treatment, oxide overlayer physical vapor deposition, and vacuum deposition of oganic layers. An approach that exercises nanoscale control of the interlayer is layer-by-layer, self-limiting chemisorptive siloxane self-assembly as depicted in Figure 19.3.[8] Layers of octachlorotrisiloxane are chemisorbed from solution, hydrolyzed,

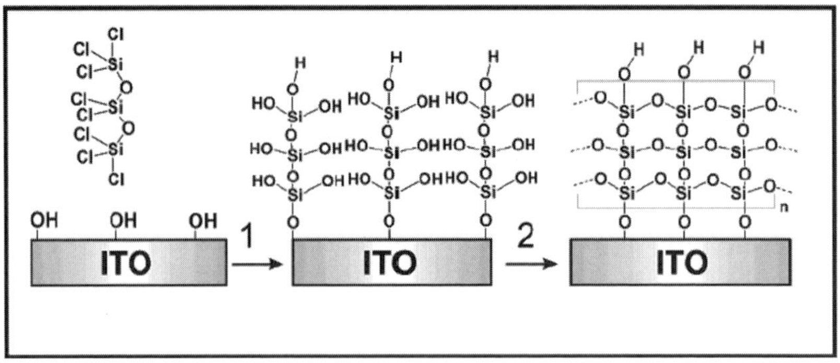

FIGURE 19.3. Layer-by-layer chemisorptive self-assembly of siloxane dielectric layers on an ITO anode. (From Ref. 8 by permission of American Chemical Society.)

and thermally cured/cross-linked. Each iteration results in deposition of a layer 0.83 nm thick. By varying the number of layers deposited prior to vacuum deposition of the HTL, Alq_3, and metal anode layers, a detailed examination of the role of the interlayer was made possible. It was found that the deposition of one through four layers resulted in a monotonic decrease in the hole injection efficiency with the largest decreases occuring after the first and second layers. With an aluminum cathode, the addition of one layer increased the turn-on voltage at which light emission if first observed from 9 V to 15 V. Addition of the second and third layers resulted in successive decreases in the turn-on voltage followed by an increase after deposition of the fourth layer. The decreased turn-on voltage for the second and third layers is ascribed to a built-in field from the interlayer that increases the electron-injection efficiency at the cathode. The external quantum efficiency and luminous efficiency both increased with addition of the self-assembled layers, as illustrated in Figure 19.4. Because the hole injection efficiency is much larger than the electron injection efficiency without incorporation of the dielectric layers, the inclusion of the layers improves the electron-hole balance and, correspondingly, the device efficiency.

The first demonstration of a relatively efficient polymer LED (PLED) was made by Friend and co-workers in 1990.[9,10] The device structure is similar to that of small-molecule OLEDs; the semiconducting polymer poly(p-pheneylene vinylene) (PPV, see Figure 19.1) was sandwiched between an ITO anode and an aluminum cathode. Two potential advantages of PLEDs over OLEDs are that the materials are easily processed by spin-casting from solutions and that the potential exists for flexible devices since the polymers are largely amorphous. While the quantum efficiency of the initial demonstration devices was ~0.01% dramatic improvements have been achieved to values comparable to those of OLEDs (several percent) by using lower work function cathodes such as calcium and magnesium and by employing polymers with higher luminescence efficiencies such as poly(2-methoxy-5-(2'-ethylhexyloxy)-1,4-phenylenevinylene) (MEH-PPV). As in the case of OLEDs, judicious choice of interlayers between the emissive layer and electrodes can enhance device efficiency, though the physical mechanisms are somewhat different. A schematic energy level diagram for a PPV LED is shown in Figure 19.5. The injection barrier for electrons (ΔE_e) is decreased with lower work function cathodes. On the anode side, it is common practice to spin-cast a layer of the conducting polymer poly(3,4-ethylenedioxythiophene

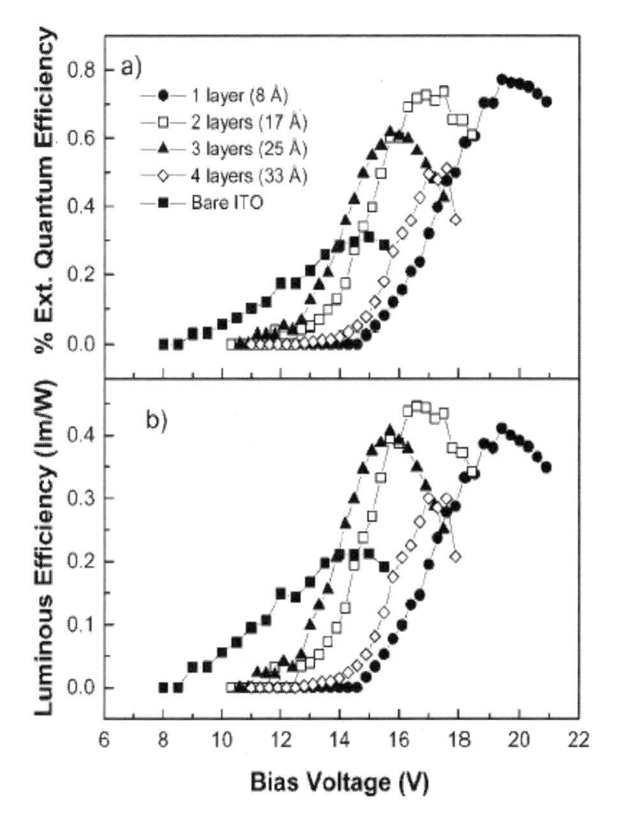

FIGURE 19.4. a) External quantum efficiency and b) luminous efficiency as a function of the number of self-assembled siloxane layers. (From Ref. 8 by permission of American Chemical Society.)

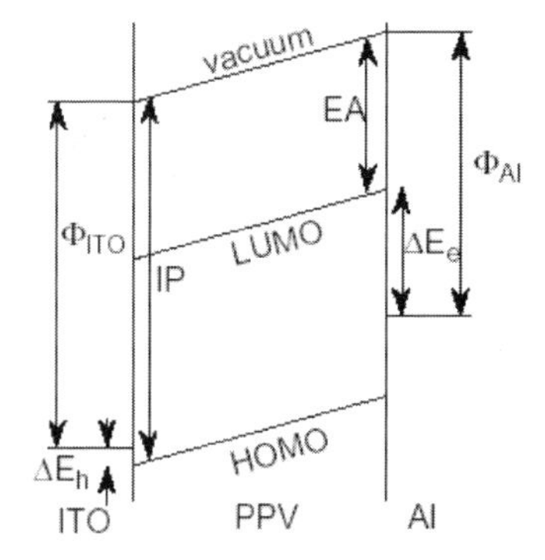

FIGURE 19.5. Schematic energy level diagram of an ITO/PPV/Al LED indicating the electron affinity (EA), ionization potential (IP), workfunction (Φ), and injection barriers ΔE. (From Ref. 10 by permission of Macmillan Magazines Ltd.)

(PEDOT) doped with poly(4-styrene sulfonate) (PSS) onto the ITO. In addition to lowering the hole injection barrier, it is also believed that the PEDOT improves electrical contact to the emissive polymer by smoothening the relatively rough ITO surface.

While derivatives such as MEH-PPV are soluble in many organic solvents, pristine PPV is not and must be processed in the form of a precursor. A nonconjugated, cationic tetrahydrothiophenium precursor is often used. After spin-casting of a film of the precursor polymer, conversion to the conjugated PPV is achieved through a thermal elimination step at >200°C for >10 hours in vacuum. Rubner and coworkers have used the cationic precursor to fabricate polymer LEDs from self-assembled multilayer films using the layer-by-layer electrostatic deposition process (Section 2.4.1).[11,12] The polyanions used were poly(methacrylic acid) (PMA) and poly(styrene sulfonic acid) (PSS). While the polyanion is used primarily as a "glue" for build up of the multilayer structures, it was found that LEDs fabricated using the two different polyanions exhibited dramatically different performance. Devices were fabricated using ten through fifty bilayers in each case. For PPV/PMA films, a thickness of 1.6 nm per bilayer was obtained while PPV/PSS films had 0.8 nm bilayer thickness. The PPV/PMA devices exhibited a rectification ratio of up to 10^5 as expected for diode behavior and luminances of 10–20 cd/m^2. In contrast, devices comprising PPV/PSS multilayers showed fairly symmetric I–V curves, higher current densities, and one order of magnitude lower luminances. The differences were ascribed to doping of the PPV by the sulfonic acid groups of PSS, which leads to increased hole injection efficiency under both positive and negative bias. The increased hole injection efficiency of PPV/PSS was utilized in a self-assembled heterostructure to further increase the device efficiency. Five bilayers of PPV/PSS were deposited between the ITO and 15 bilayers of PPV/PMA in order to increase the efficiency of hole injection into the PPV/PMA emissive layer. The 4 nm thick PPV/PSS multilayer increased the luminance by nearly an order of magnitude (100–150 cd/m^2) relative to PPV/PMA devices fabricated directly on ITO.

Friend and coworkers have demonstrated a similar approach in which layer-by-layer electrostatic self-assembly is used to deposit a hole injection layer on the ITO substrate followed by spin-casting of the thicker emissive polymer layer.[13] The PEDOT:PSS complex was used as the polyanion and the tetrahydrothiophenium PPV precursor was the polycation. An additional degree of nanoscale engineering was exercised that is not possible with spin-casting. PEDOT:PSS was partially de-doped at three different levels with hydrazine hydrate. In this manner, a graded interface was fabricated in which PEDOT:PSS layers with successive degrees of de-doping were deposited on the ITO, as illustrated in Figure 19.6. Holes injected from the ITO anode thus undergo a more continuous transition into the valence band of the emissive polymer, reducing the barrier to hole injection indicated as ΔE_h in Figure 19.5. A second feature incorporated into the device design was an electron blocking layer using partially converted PPV from the cationic precursor. The lower electron affinity of this layer relative to the emissive polymer layer results in confinement of electron to the emissive polymer, thus increasing efficiency of radiative recombination of the elec trons and holes. These two features are particularly useful for blue-emitting polymers wit large ionization potentials such that the hole-injection barrier ΔE_h is especially large. Wit poly(9,9-dioctylfluorene) as the emissive polymer, devices with self-assembled, graded in terlayers exhibited external quantum efficiencies of 6.5% and luminances of 1000 cd/ m 40 times larger than for devices with spin-cast PEDOT:PSS on ITO. Self-assembly offe

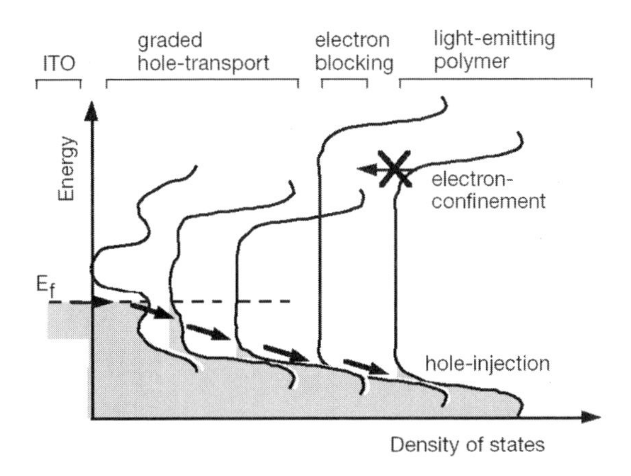

FIGURE 19.6. Schematic of the electronic density of states across a graded interlayer fabricated using electrostatic layer-by-layer assembly. (From Ref. 13 by permission of Macmillan Magazines Ltd.)

an attractive route for nanoscale engineering of the interface between the electrodes and emissive material in organic and polymeric LEDs.

19.3. PHOTOVOLTAIC POLYMERS

While OLEDs and PLEDs rely on the large luminescence quantum efficiencies of conjugated organic materials, the propensity towards fluorescence is certainly not beneficial for photovoltaic applications, in which it is desired to convert incident optical energy into electrical energy. An important advance in polymer photovoltaics was made with the observation by Heeger and coworkers of photogenerated charge separation in PPV-C_{60} composites.[14] Upon photoexcitation, rapid electron transfer occurs from the polymer to the high electron affinity C_{60}. This charge-transfer process prevents radiative electron-hole recombination. The fluorescence is quenched and the separated electron and hole can be collected at the electrodes as a photocurrent. However, photoexcited electron-hole pairs at distances larger than ~10 nm from the fullerene acceptor recombine before charge separation occurs.[15] Early polymer-fullerene photovoltaic devices consisted of a ~100 nm spin-cast polymer layer followed by a ~100 nm C_{60} layer deposited by vacuum sublimation. In this configuration, only photoexcitations within 10 nm of the interface undergo efficient charge transfer. Nanoscale compositional control of the electron donor and acceptor species is clearly important to optimizing the performance of polymeric photovoltaics.

The photovoltaic polymer device geometry is similar to that of polymer LEDs. The active organic layer is sandwiched between a metal cathode and a transparent ITO anode on a glass substrate. The light illluminates the polymer through the glass substrate and ITO and the photocurrent or photovoltage is measured between the ITO and the metal cathode, which is often aluminum, magnesium, or calcium. Figure 19.7 illustrates a typical curve of the current versus an applied voltage for a photovoltaic device under illumination and identifies several key quantities. The short circuit current (I_{SC}) is the current measured with zero resistance (e.g., an ammeter) between the anode and cathode. The open circuit voltage

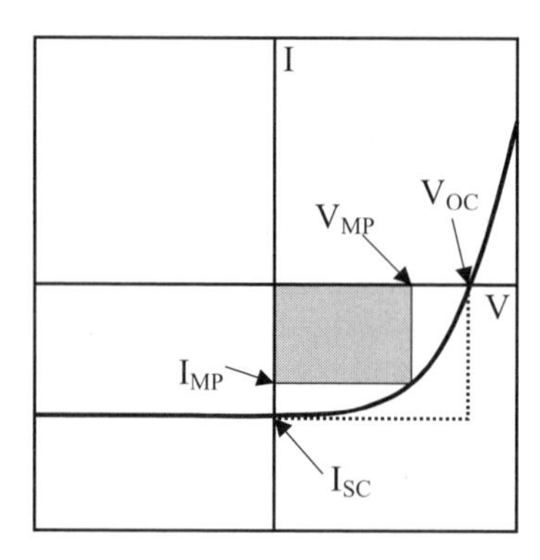

FIGURE 19.7. Typical current-voltage characteristics of a solar cell under illumination. The identified quantities are discussed in the text.

(V_{OC}) is the value of the applied external bias at which the current is zero. This is because the applied voltage is then exactly opposite the voltage generated by the device under illumination. This is thus a measure of the voltage generated by the device with infinite resistance (e.g., a voltmeter) between the anode and cathode. Ideally, the device would exhibit a rectangular I–V curve between these two points, corresponding to maximum power efficiency for any external load (resistance). In reality, devices exhibit a curve such as that shown by the solid line. Since the power output is the product of the current and the voltage ($P = IV$), there will be a maximum power output for some values of the current and voltage that are less than I_{SC} and V_{OC}, respectively. One thus defines a quantity known as the fill factor (FF) which is the ratio of the actual maximum power output to that for the ideal case ($I_{SC}V_{OC}$). Thus, $FF = (I_{MP}V_{MP})/(I_{SC}V_{OC})$ where $I_{MP}$and V_{MP} are the current and voltage at the maximum power point. Clearly, one wants to maximize I_{SC}, V_{OC}, and FF in order to make the most efficient solar cell.

One approach to ensure proximity between the donor and acceptor species in order to provide efficient charge-transfer is to spin-cast a blended film from a solution containing both components. However, C_{60} is only weakly soluble in the organic solvents from which semiconducting polymers are spin-cast and is also prone to large-scale aggregation in such blends. This has been addressed by using a second semiconducting polymer or a derivatized fullerene as the acceptor, both of which are highly soluble. The ionization potential and electron affinity of a cyano derivative of PPV (CN-PPV) are increased by $\sim$0.5 eV relative to MEH-PPV. Films spin-cast from mixtures of MEH-PPV and CN-PPV show phase separation on a length scale of 10–100 nm.[16] The blend film thus forms an interpenetrating polymer network in which holes can efficiently propagate through the electron donor MEH-PPV phase to the ITO anode and electrons can efficiently propagate through the electron acceptor CN-PPV phase to the metal cathode. The blend devices exhibit I_{SC} values more than two orders of magnitude larger than those of devices containing either polymer alone.

Derivatization of C_{60} can dramatically increase its solubility, as in the case of the derivatives denoted as PCBM is Figure 19.8, which also shows the typical configuration of a polymer photovoltaic device. Blends of MEH-PPV and PCBM also serve as bulk heterojunctions in which the interfacial area between the donor and acceptor components is dramatically increased.[17] Figure 19.9 shows semilog plots of the photocurrent density

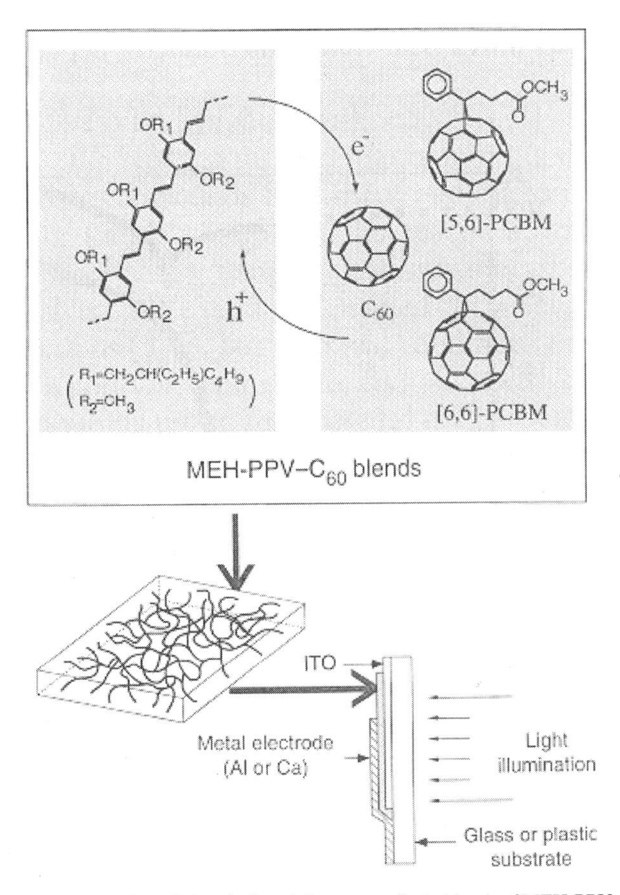

FIGURE 19.8. Schematic illustration of photoinduced charge transfer in blends of MEH-PPV and C_{60} derivatives. Also illustrated are the phase separation into a bicontinuous network and the general configuration of polymer photovoltaic device. (From Ref. 17 by permission of American Association for the Advancement of Science.)

versus the applied external bias for MEH-PPV devices with and without PCBM. The short circuit current is increased by two orders of magnitude in the blend device, demonstrating the efficient charge separation in this case. The decrease in the V_{OC} from 1.6 V to 0.8 V is not desirable, but is a consequence of the electron transfer to the lowest unoccupied molecular orbital (LUMO) of PCBM. The monochromatic power conversion efficiency (ratio of electrical power out to optical power in) is ~3% for illumination at 430 nm. For evaluation as a solar cell that converts sunlight into electrical power, one must also take into consideration the spectrum of solar radiation at the earth's surface rather than evaluate the device at the wavelength of its peak efficiency. Because of the relatively narrow absorption bands of conjugated polymers, the total energy conversion efficiency under illumination is typically about an order of magnitude smaller than the peak mononchromatic value. Continued advances in the materials and their morphology has led to total energy conversion efficiciencies under solar illumination conditions as high as 2.5% for blends of MDMO-PPV and PCBM.[18] For comparison, the power conversion efficiencies of commercial silicon solar cells are typically 10–15%.

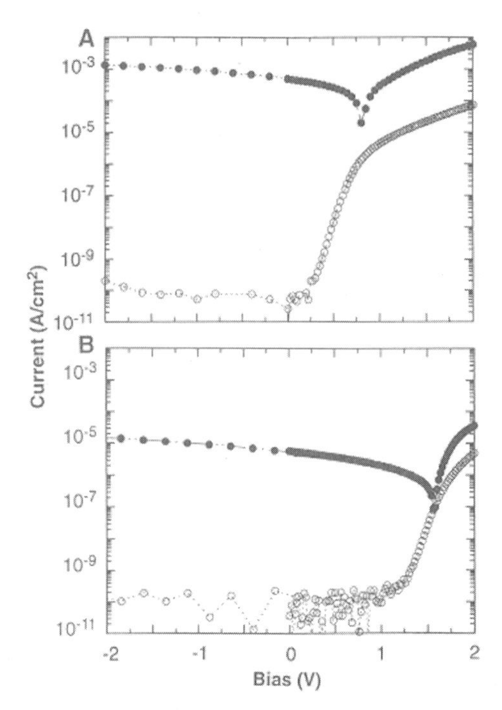

FIGURE 19.9. I–V characteristics of a (A) Ca/MEH-PPV:[6,6]PCBM/ITO device and (B) Ca/MEH-PPV/ITO device in the dark (open circles) and under 20 mW/cm^2 illumination at 430 nm (solid circles). (From Ref. 17 by permission of American Association for the Advancement of Science.)

While the interpenetrating network approach maximizes the charge-transfer efficiency, it might be expected that charge transport and collection at the electrodes would be improved by a graded structure that is donor-rich at the anode and acceptor-rich at the cathode. Several approaches have been developed to provide such a graded donor-acceptor concentration profile. One method involves spin-casting first the donor layer and then spin-casting the acceptor layer from a solvent in which the donor is partially soluble. Poly[2-methoxy-5-(3′,7′-dimethyloctyloxy-1,4-phenylene vinylene] (MDMO-PPV) is highly soluble in xylene at 90°C but only slightly so at room temperature. Devices were fabricated by spin-casting MDMO-PPV from hot solution and then spin-casting PCBM from a room temperature xylene solution.[19] The solar spectrum power conversion efficiency was 0.5%, five times larger than that of blend devices from the same two materials.

A second approach is a lamination technique in which the acceptor is deposited on the cathode on one substrate and the donor is deposited on the anode on a second substrate. The two are pressed together while one of the films is at elevated temperature in order to provide some interpenetration of the two layers.[20] Laminated devices made with a cyano PPV derivative on a Ca cathode and a polythiophene derivative on the ITO anode exhibited total energy conversion efficiencies of 1.9%, an order of magnitude larger than that of blend devices of the two polymers.

A third method is to start with two discrete layers and utilize thermally-enhanced interdiffusion to create a gradient, bulk heterojunction.[21] A C_{60} layer was sublimed on a

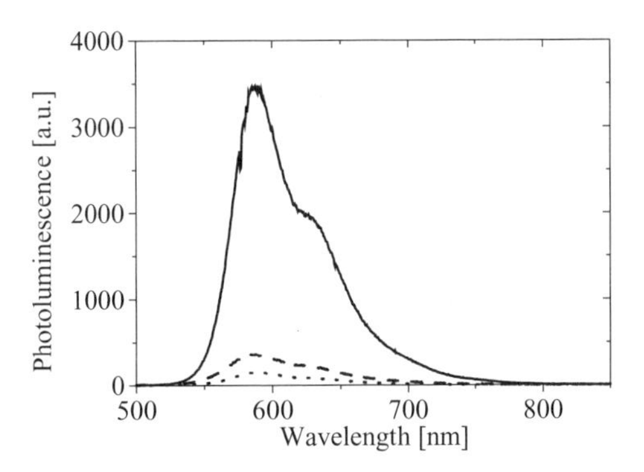

FIGURE 19.10. Photoluminescence of an unheated MEH-PPV/C_{60} bilayer device (solid) and two heated MEH-PPV/C_{60} devices (5 minutes at 150°C (dotted) and 5 minutes at 250°C (dashed)). (From Ref. 21 by permission of American Institute of Physics.)

spin-cast MEH-PPV layer, each ~100 nm thick. Interdiffusion was induced by heating in the vicinity of the MEH-PPV glass transition temperature (230°C) for five minutes. Figure 19.10 shows the decrease in the photoluminescence intensity by an order of magnitude for devices heated at 150°C and 250°C compared to unheated devices, demonstrating the improved proximity of the donor and acceptor species throughout the device. While the devices heated at 150°C do not show a significant increase in the peak photoresponsivity (short circuit current density divided by optical intensity), the devices heated at 250°C exhibit an order of magnitude increase in the photocurrent throughout most of the visible spectrum, as illustrated in Figure 19.11. This indicates that while charge transfer is efficient for the devices heated at 150°C, charge transport out of the device only becomes efficient at the higher level of interdiffusion induced at 250°C.

In addition to C_{60}, nanomaterials such as carbon nanotubes and semiconductor nanorods have also been used in polymer photovoltaics for enhanced charge transfer and charge transport. Single-walled carbon nanotubes have been incorporated in blends with

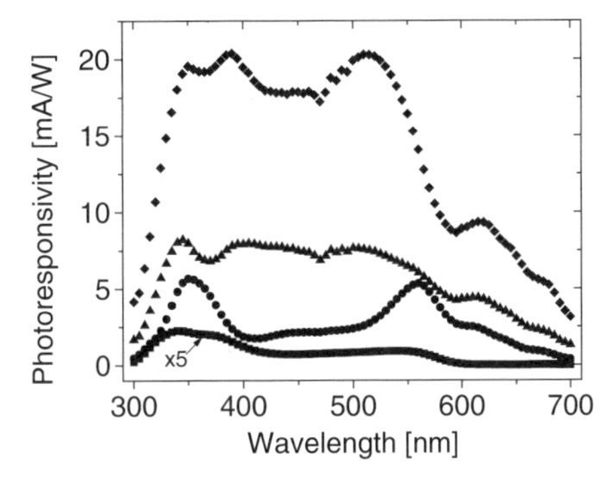

FIGURE 19.11. Photoresponsivity of a MEH-PPV device (squares magnified by 5×), an unheated MEH-PPV/C_{60} bilayer device (circles), a MEH-PPV/C_{60} bilayer device that was heated at 150°C for 5 minutes (triangles) and a MEH-PPV/C_{60} bilayer device that was heated at 250°C for 5 minutes (diamonds). (From Ref. 21 by permission of American Institute of Physics.)

poly(3-octylthiophene).[22] The polymer and SWNTs were co-dissolved in chloroform and spin-cast as 60 nm films onto ITO-coated glass substrates and covered with an Al cathode. Devices with the SWNTs exhibited short circuit currents more than two orders of magnitude larger than those without SWNTs and an increase in V_{OC} from 0.35 V to 0.75 V. These results indicate that nanotubes may serve as both efficient electron acceptors and charge transport agents in polymer photovoltaics.

Inorganic semiconductor nanoparticles and nanorods combine tunablility of the band gap by particle size with larger electron mobilities than those typically found in organic materials. CdSe nanorods were co-dissolved with poly-3(hexylthiophene) (P3HT) in a mixture of pyridine and chloroform.[23] Films were 200 nm thick and contained 90% by weight CdSe nanorods. Variation of the aspect ratio of the nanorods from 7 nm diameter by 7 nm length to 7 nm diameter by 60 nm length allows tuning of both the optical absorption edge and the distance over which charge can transport before hopping is required from one nanorod to the next. Figure 19.12 shows TEM images of 20% by weight 7 nm by 7 nm and 7 nm by 60 nm nanorods in the polymer as well as cross-sectional images of nanorod-polymer films. The best photovoltaic performance was obtained in 200 nm thick films of 90 wt % 7 nm by 60 nm CdSe nanorods in P3HT. The high aspect ratio nanorods exhibited optical absorption to longer wavelength and improved charge transport due to their larger length.

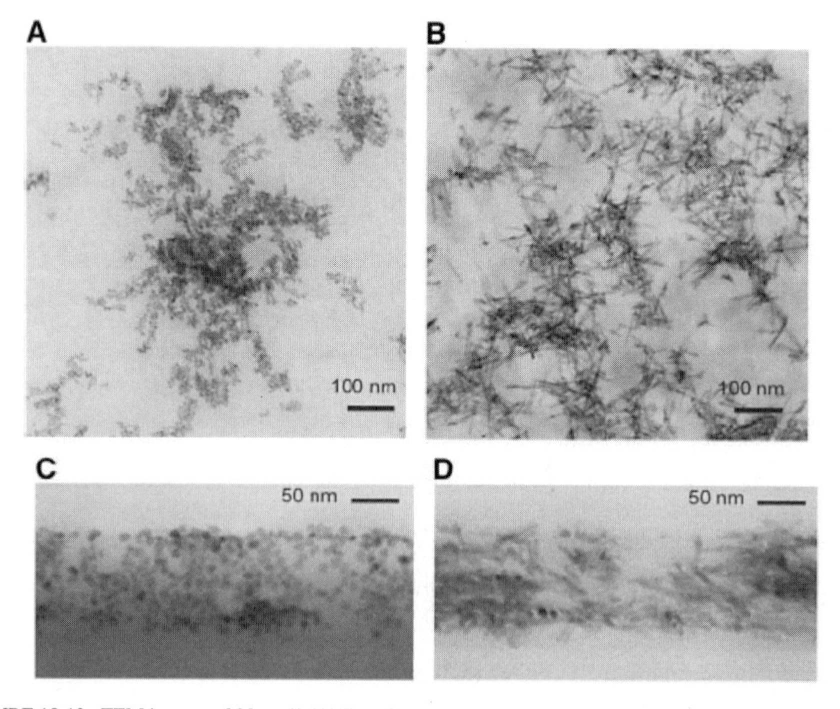

FIGURE 19.12. TEM images of 20 wt % (A) 7 nm by 7 nm and (B) 7 nm by 60 nm CdSe nanocrystals in P3HT and cross-section TEMs of (C) a 110 nm thick film of 60 wt % 10 nm by 10 nm nanocrystals and (d) a 100 nm thick film of 40 wt % 7 nm by 60 nm nanorods in P3HT. (From Ref. 23 by permission of American Association for the Advancement of Science.)

A power conversion efficiency of 1.7% for simulated solar spectrum and V_{OC} of 0.7 V were obtained.

19.4. SELF-ASSEMBLED ORGANIC NONLINEAR OPTICAL MATERIALS

Nonlinear optics (NLO) refers to phenomena in which the polarization of the medium has a nonlinear dependence on the incident electric and/or magnetic fields. As a result, the optical properties of a material such as refractive index or absorption can be dependent on the incident intensity or an applied voltage. This allows for the fabrication of optically- or electrically-controlled optical switches and modulators. In order to possess nonzero even-order nonlinear optical susceptibilities, a material must lack a center of inversion at the macroscopic level.[24] Of particular interest is the macroscopic second order (quadratic) susceptibility $\chi^{(2)}$, which governs the nonlinear polarization of the medium at frequency ω_3 in response to (optical) electric fields at frequencies ω_1 and ω_2 through

$$P_i^{\omega_3} = \chi_{ijk}^{(2)}(-\omega_3;\ \omega_1,\ \omega_2) E_j^{\omega_1} E_k^{\omega_2}, \tag{19.1}$$

where the subscripts refer to the directions of polarization of the fields. An example of an important application of a $\chi^{(2)}$ material is the Mach-Zehnder waveguide modulator illustrated in Figure 19.13. Light is coupled from an optical fiber into the waveguide from the left, and the waveguide branch separates the signal into two beams of equal intensity. Through the $\chi^{(2)}$ process known as the electro-optic effect, the refractive index of the material is changed by the application of a dc or low-frequency voltage across one arm of the device which, correspondingly, changes the optical path length of that arm and modulates the optical signal at the output end. The voltage required to vary the output from maximum value to zero is called the halfwave voltage (V_π) and is related to $\chi^{(2)}$ by

$$V_\pi = (nd\lambda)/\left(2L\chi^{(2)}\right) \tag{19.2}$$

where d is the waveguide thickness, λ is the wavelength, L is the length of the waveguide arms, and n is the refractive index.

Since second order, $\chi^{(2)}$, nonlinear optical effects, such as harmonic generation, optical parametric oscillation, and the electro-optic effect, are quadratic in the applied fields, symmetry considerations require that a material must be noncentrosymmetric in order to possess

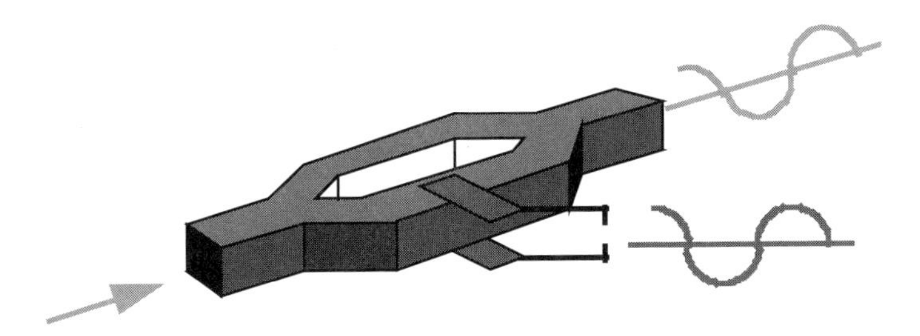

FIGURE 19.13. Mach-Zehnder waveguide modulator. Through $\chi^{(2)}$, the electrical input modulates the optical output.

a nonzero $\chi^{(2)}$. The most common $\chi^{(2)}$ materials are ferroelectric, inorganic crystals such as lithium niobate, potassium dihydrogen phosphate (KDP), and beta-barium borate (BBO). Growth of high-quality crystals, however, is difficult, time-consuming, and expensive.

Organic materials have many potential advantages over inorganic crystals for $\chi^{(2)}$ applications: (1) large $\chi^{(2)}$ values; (2) low dielectric constant, which allows electro-optic modulation at higher frequencies; (3) high transparency; (4) ultrafast response times ($<10^{-15}$ sec) as a result of nonresonant mechanisms; (5) easy, low-cost fabrication; (6) simple waveguide patterning by photobleaching. As a result of the multitude of potential frequency conversion, optical modulation, and optical switching applications that stem from the $\chi^{(2)}$ second order susceptibility, several novel methods for creating noncentrosymmetric materials incorporating organic molecules with large β molecular susceptibilities have been developed. These include electric field poled polymers, Langmuir-Blodgett films, and self-assembled monolayer structures.

There has been extensive research and development effort for the past two decades in the area of poled polymers. These consist of a glassy polymer matrix that contains an NLO chromophore either as a guest dopant or as a covalently-bonded substituent.[25] A typical organic NLO chromophore consists of a long, conjugated molecule with electron-donating and electron-accepting moieties located on opposite ends, yielding a highly polarizable, polar structure. In the isotropic amorphous state, there is no $\chi^{(2)}$ response because the material is macroscopically centrosymmetric, despite the requisite noncentrosymmetry at the molecular level. To obtain a nonzero $\chi^{(2)}$, the polymer is heated above its glass transition temperature, T_g, and an electric field is applied to achieve polar ordering of the NLO chromophores. The polymer is then cooled to room temperature while the field remains applied in order to freeze in the noncentrosymmetric ordering. The exceptional potential of organic electro-optic materials is illustrated in the demonstrations in poled polymers of full optical modulation at <1.0 V and at >100 GHz.[26] While the eventual randomization of the orientation back to the isotropic state has proven a challenging problem, advances continue to be made through use of higher T_g hosts, covalent attachment of the chromophore to the polymer, cross-linked polymers, and dendrimer structures. For example, recent materials have shown stability for >1000 hours at 85°C.[27]

Because of the challenges in obtaining temporal and thermal stability in poled polymers, alternative approaches for fabricating noncentrosymmetric organic materials are being developed that do not require trying to "freeze in" an alignment that is not thermodynamically-stable. Of particular interest here are self-assembly methods for obtaining polar order in organic NLO materials. One approach involves using layer-by-layer covalent deposition such that each reaction step self-terminates after deposition of a monolayer. Methods have been developed employing zirconium phosphate-phosphonate[28] and siloxane[29] chemistry. In the former case, illustrated in Figure 19.14, an amine-terminated or hydroxyl-terminated surface on a glass substrate is phosphorylated and then zirconated by immersion in two appropriate solutions for 5 min each. Immersion for 10 min in a solution containing the organic NLO chromophore 4-{4-[N,N-bis(2-hydroxyethyl)amino]phenylazo}phenylphoshonic acid results in the covalent attachment of a dye monolayer. The process is then repeated to build up a multilayer structure. Films were fabricated up to 30 layers. The linear growth of the ellipsometrically-measured thickness and the absorbance with the number of layers indicates the uniform and regular growth of successive layers. The second order nonlinear optical response of the films was measured via second harmonic generation (SHG), also

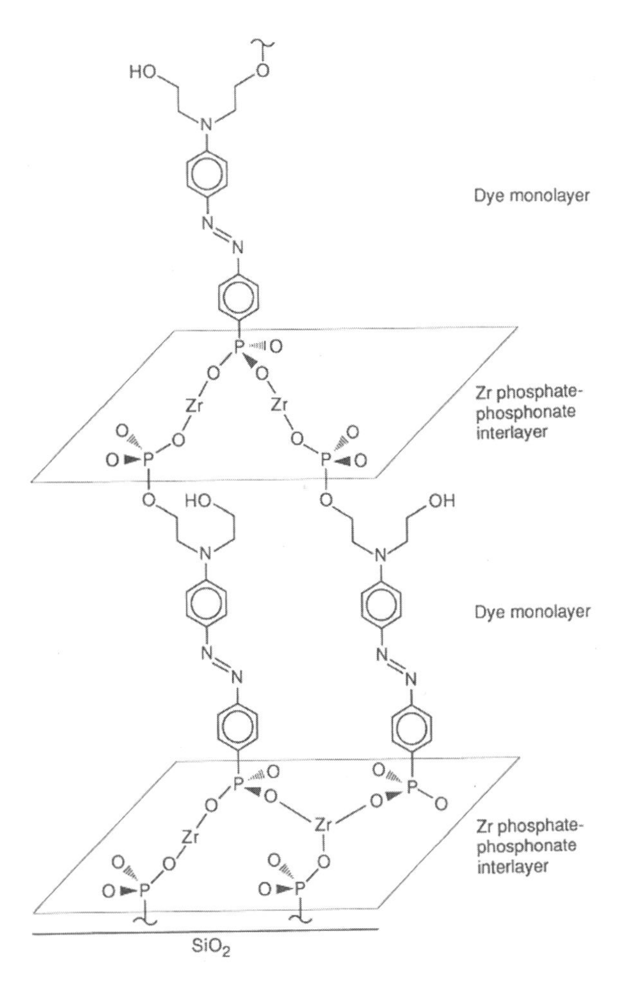

FIGURE 19.14. Schematic illustration of polar, self-assembled multilayers grown with Zr phosphate-phosphonate interlayers. (From Ref. 28 by permission of American Association for the Advancement of Science.)

known as frequency doubling. In this $\chi^{(2)}$ process, the material generates light at frequency 2ω in response to incident light at frequency ω. For films thinner than the coherence length l_c of the material ($l_c = \lambda/[4(n^{2\omega} - n^{\omega})]$, where λ is the wavelength and n is the refractive index at each frequency), the SHG intensity is expected to grow quadratically with the thickness of the film. Since l_c is typically on the order of 10 μm and the layers are 1.6 nm thick, that condition is satisfied in the present case. The films did exhibit a quadratic dependence on the number of layers, demonstrating that each successive layer has the same degree of polar order. Comparison of the SHG intensity from the self-assembled multilayers to that from a material with a known $\chi^{(2)}$ value, such as quartz, allows determination of $\chi^{(2)}$ of the films. The measured $\chi^{(2)}$ value was 50×10^{-9} esu. For comparison, the $\chi^{(2)}$ value of the common inorganic NLO crystal lithium niobate is 200×10^{-9} esu.

Several variations have been developed for growth of polar, sef-assembled multilayers using siloxane chemistry. In an early approach,[30] the substrate was immersed in a solution

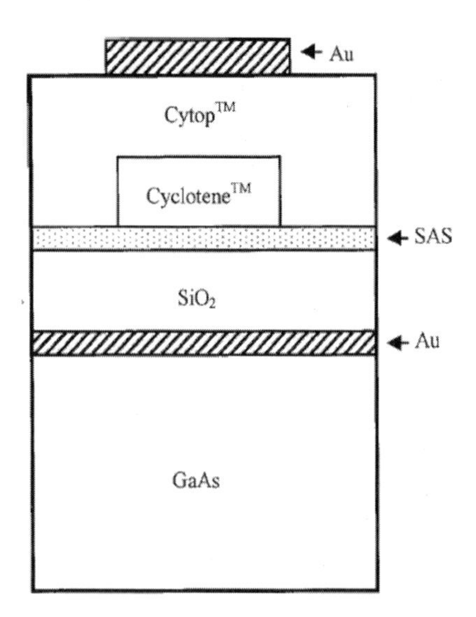

FIGURE 19.15. Schematic cross-section of an electro-optic modulator waveguide with an active organic self-assembled superlattice (SAS). (From Ref. 32 by permission of American Institute of Physics.)

containing a silanizing reagent for 18–64 h, cured at 115°C for 15 min, and then immersed in a solution containing a (dialkylamino)stilbazole NLO chromophore for 20 h. A $\chi^{(2)}$ value of 300×10^{-9} ese was obtained by SHG measurements. A more rapid deposition procedure has been developed that yields similarly large NLO responses. A monolayer of a $SiCl_2I$ derivative of an azobenzene chromophore is assembled onto the substrate by immersion in solution for 15 min. Immersion in octachlorotrisiloxane solution for 25 min then removes a protecting group from the free end of the molecule and caps it with a polysiloxane layer.[31] The procedure is repeated to build up multilayers. Quadratic growth of the SHG intensity with the number of layers was observed, and a $\chi^{(2)}$ value of 430×10^{-9} esu was obtained. Covalent self-assembled multilayers of this sort have been incorporated into an electro-optic modulator device using the waveguide structure illustrated in Figure 19.15.[32] Because the refractive index is lower in the surrounding media, the light is confined in the Cyclotene™ and self-assembled superlattice (SAS) by total internal reflection. The SAS consists of 40 layers with a total thickness of 150 nm. This is too thin to serve as an effective waveguide, so the Cylcotene™ layer is also used. However, that means that most of the light is not interacting with the electro-optic medium. As a result, the voltage required for modulation is much higher than would be required if the waveguide were formed exclusively from electro-optic material. Nonetheless, full optical modulation was observed with application of 340 V, demonstrating the principle of electro-optic modulation from self-assembled organic NLO materials.

Polar organic NLO films have also been fabricated using the layer-by-layer electrostatic deposition process (Section 2.4.1).[33,34,35,36,37] Several variations have been employed, but the most common one involves use of a polyelectrolyte that contains the conjugated NLO choromophore as a side group with a terminal ionic moiety. Usually, the oppositely-charged polyelectrolyte does not have an NLO chromophore and serves primarily as a "glue" for

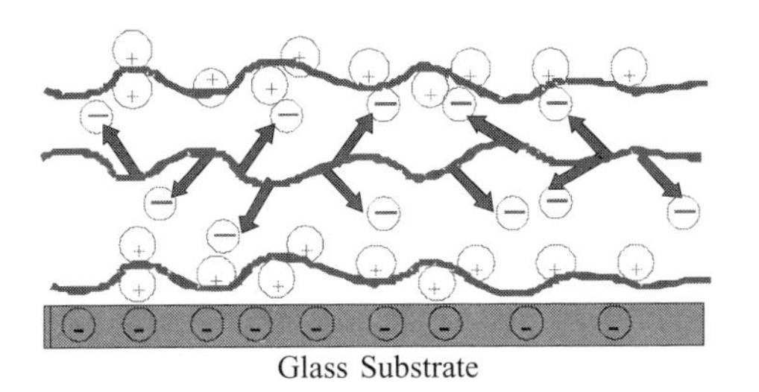

FIGURE 19.16. Schematic illustration of polar order in a film fabricated by the layer-by-layer electrostatic deposition process. The NLO chromophores are represented by the arrows.

growth of the multilayers. The electrostatic bonding between the chromophore and the underlying oppositely-charged layer provides directional orientation of the chromophore preferentially towards the substrate. The structure is schematically illustrated in Figure 19.16. The figure also illustrates that the adsorption of the next oppositely-charged layer requires some fraction of the chromophores to be oriented away from the substrate. It is expected that this competition of chromophores oriented in opposite directions leads to a reduction of the net polar order and, correspondingly, $\chi^{(2)}$ value of the film. SHG measurements demonstrate, however, that the cancellation is not complete and net polar order does exist in self-assembled films fabricated by this method. For example, Figure 19.17 shows the linear growth of the absorbance and quadratic growth of the SHG intensity with the film thickness up to 100 bilayers for films made from a polyelectrolyte with an azobenzene side group.[38] The $\chi^{(2)}$ value of these films is a relatively modest 1.3×10^{-9} esu, presumably due to the partial cancellation of chromophores oriented in opposite directions. The temporal and thermal stability of the films is excellent, though. The $\chi^{(2)}$ values exhibited no loss at room temperature over several years nor 20 hours at 150°C.[39] This self-assembly approach also provides rapid adsorption of layers, typically in the range of one to five minutes. Methods

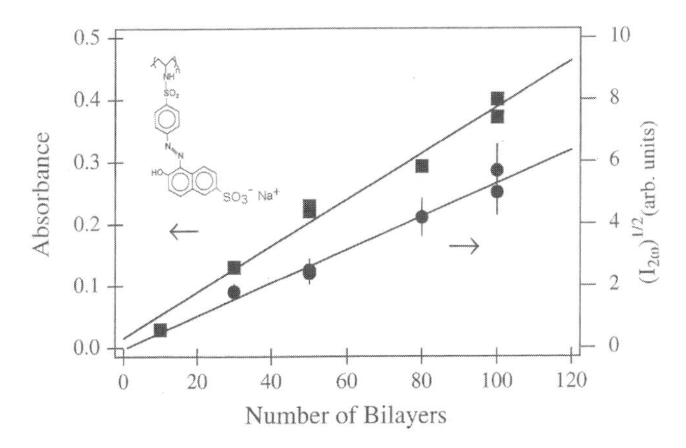

FIGURE 19.17. Absorbance at 500 nm and square root of the SHG intensity as a function of the number of bilayers for films made by layer-by-layer polyelectrolyte deposition. (From Ref. 38 by permission of American Institute of Physics.)

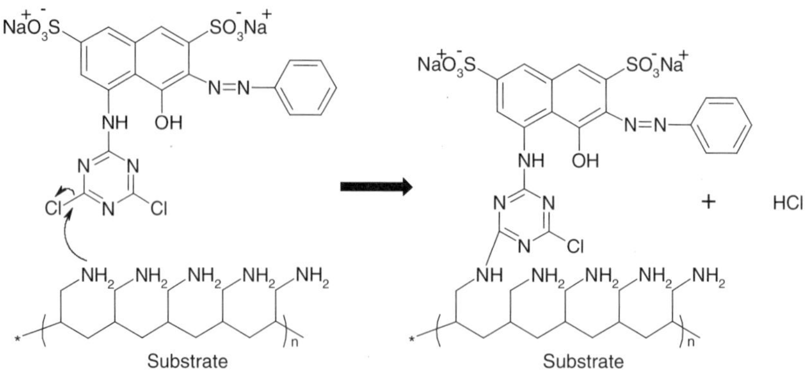

FIGURE 19.18. Schematic illustration of a hybrid covalent/electrostaic deposition process for fabricating polar self-assembled multilayers. (From Ref. 40 by permission of WILEY-VCH Verlag GmbH & Co.)

are therefore being developed to eliminate the competition in chromophore orientation. One technique involves using monomeric, rather than polymeric, chromophores and two different adsorption mechanisms to selectivey orient the chromophore. Figure 19.18 shows one scheme in which the Procion Red MX-5B (PR) chromophore is covalently attached to poly(allylamine hydrochloride) (PAH) by immersion in a solution for three minutes at a pH of 10.5 (at which PAH is unprotonated).[40] A layer of PAH is then electrostatically bound to the sulfonates of PR by immersion for three minutes in a solution at pH 7 such that PAH is protonated. A $\chi^{(2)}$ value of 11×10^{-9} esu was obtained for PR/PAH films fabricated by this hybrid covalent/electrostatic deposition process. Substantially higher values should be possible with the use of chromophores with larger NLO susceptibility.

19.5. SUMMARY

- Conjugated organic and polymeric materials have large conductivities, absorption strengths, and polarizabilities that make them interesting optoelectronic materials.
- Self-assembled monolayers provide nanoscale control of electron and hole injection in organic light-emitting diodes as a method to enhance device performance.
- Photovoltaic polymers rely on efficient charge transfer between electron donors and electron acceptors, which must be within 10 nm of one another.
- Interpenetrating networks of two semiconducting polymers or a semiconducting polymer and a fullerene provide the required donor-acceptor proximity for photovoltaic efficiency.
- Gradient donor-acceptor concentration profiles in photovoltaic polymers can be produced by a number of processing methods.
- Second order nonlinear optical responses, useful in electro-optic modulators and switches, require noncentrosymmetric materials.
- Covalent and electrostatic self-assembly methods can produce polar organic materials with large nonlinear optical responses and excellent temporal and thermal stability.

QUESTIONS

1. The data in Figure 19.9A show a short circuit current density $\sim 5 \times 10^{-4}$ A/cm^2. Using the information given in the caption, calculate the external quantum efficiency (fraction of electrons collected as current per incident photon.)
2. While the ideal I-V behavior of a solar cell under illumination is illustrated by the dashed curve in Figure 19.7, devices typically have a fill factor significantly less than 1. As a first approximation, one might expect that the worst case observed would correspond to a linear (Ohmic-like) response between the short circuit current and open circuit voltage. What is the value of the fill factor for this case? (In reality, it is even possible, though highly undesirable, to have a lower value for the fill factor because the I-V curve may have a negative second derivative in this region.)
3. Consider an electromagnetic wave with frequency ω incident on a material with a nonzero $\chi^{(2)}$, such that the electric field at a particular point in space can be described by Et = $E_\omega \cos(\omega t)$.
   a) As indicated in Eqn. 19.1, the second order polarization is proportional to E^2. By rewriting it in terms of only linear trigonometric functions, determine the frequency components of this polarization.
   b) Now consider that, in addition to the wave at frequency ω, there is also a dc field E_0 applied to the material. As in part a), determine the frequency components of the second order polarization resulting from the total electric field.
   c) The part of the second order polarization at frequency ω in part b) results in a modification of the material's refractive index by the applied dc field. Considering both the linear polarization ($P = \varepsilon_0 \chi E$) and the second order polarization at frequency ω, determine the modified refrative index as a function of E_0, and show that it depends linearly on E_0 when the second order polarization is much smaller than the first order polarization.

REFERENCES

1. See, for example, A. J. Heeger, *Rev. Mod. Phys.* **73**, 681 (2000).
2. C. W. Tang and S. A. VanSlyke, *Appl. Phys. Lett.* **51**, 913 (1987).
3. G. E. Jabbour, Y. Kawabe, S. E. Shaheen, J. F. Wang, M. M. Morrell, B. Kippelen, and N. Peyghambarian, *Appl. Phys. Lett.* **71**, 1762 (1997).
4. Z. Y. Xie, L. S. Hung, and S. T. Lee, *Appl. Phys. Lett.* **79**, 1048 (2001).
5. E. W. Forsythe, M. A. Abkowitz, and Y. Gao, *J. Phys. Chem. B* **104**, 3948 (2000).
6. C. W. Tang, S. A. VanSlyke, and C. H. Chen, *J. Appl. Phys.* **65**, 3610 (1989).
7. M. A. Baldo, D. F. O'Brien, Y. You, A. Shoustikov, S. Sibley, M. E. Thompson, and S. R. Forrest, *Nature* **395**, 151 (1998).
8. J. E. Malinsky, J. G. C. Veinot, G. E. Jabbour, S. E. Shaheen, J. D. Anderson, P. Lee, A. G. Richter, A. L. Burin, M. A. Ratner, T. J. Marks, N. R. Armstrong, B. Kippelen, P. Dutta, and N. Peyghambarian, *Chem. Mater.* **14**, 3054 (2002).
9. J. H. Burroughes, D. D. C. Bradley, A. R. Brown, R. N. Marks, K. Mackay, R. H. Friend, P. L. Burns, and A. B. Holmes, *Nature* **347**, 539 (1990).
10. For a review, see R. H. Friend, R. W. Gymer, A. B. Holmes, J. H. Burroughes, R. N. Marks, C. Taliani, D. D. C. Bradley, D. A. Dos Santos, J. L. Bredas, M. Loglund, and W. R. Salaneck, *Nature* **397**, 121 (1999).
11. A. C. Fou, O. Onitsuka, M. Fereira, M. F. Rubner, and B. R. Hsieh, *J. Appl. Phys.* **79**, 7501 (1996).
12. O. Onitsuka, A. C. Fou, M. Fereira, B. R. Hsieh, and M. F. Rubner, *J. Appl. Phys.* **80**, 4067 (1996).
13. P. K. H. Ho, J. S. Kim, J. H. Burroughes, H. Becker, S. F. Y. Li, T. M. Brown, F. Cacialli, and R. H. Friend, *Nature* **404**, 481 (2000).

14. N. S. Sariciftci, L. Smilowitz, A. J. Heeger, and F. Wudl, *Science* **258**, 1474 (1992).
15. J. J. M. Halls, K. Pichler, R. H. Friend, S. C. Moratti, and A. B. Holmes, *Appl. Phys. Lett.* **68**, 3120 (1996).
16. J. J. M. Halls, C. A. Walsh, N. C. Grenham, E. A. Marseglia, R. Friend, S. C. Moratti, and A. B. Holmes, *Nature* **376**, 498 (1995).
17. G. Yu, J. Gao, J. C. Hummelen, F. Wudl, and A. J. Heeger, *Science* **270**, 1789 (1995).
18. S. E. Shaheen, C. J. Brabec, N. S. Saricftci, F. Padinger, T. Fromherz, and J. C. Hummelen, *Appl. Phys. Lett.* **78**, 841 (2001).
19. L. Chen, D. Godovsky, O. Inganäs, J. C. Hummelen, R. A. J. Janssens, M. Svensson, and M. R. Andersson, *Adv. Mater.* **12**, 1367 (2000).
20. M. Granstrom, K. Petrisch, A. C. Arias, A. Lux, M. R. Andersson, and R. H. Friend, *Nature* **395**, 257 (1998).
21. M. Drees, K. Premaratne, W. Graupner, J. R. Heflin, R. M. Davis, D. Marciu, and M. Miller, *Appl. Phys. Lett.* **81**, 4607–4609 (2002).
22. E. Kymais and G. A. J. Amaratunga, *Appl. Phys. Lett.* **80**, 112 (2002).
23. W. U. Huynh, J. J. Dittmer, and A. P. Alivisatos, *Science* **295**, 2425 (2002).
24. See, for example, R. W. Boyd, *Nonlinear Optics*, Academic Press, Boston, 1992.
25. K. D. Singer, J. E. Sohn, and S. J. Lalama, *Appl. Phys. Lett.* **49**, 248 (1986).
26. Y. Shi, C. Zhang, H. Zhang, J. H. Bechtel, L. R. Dalton, B. H. Robinson, and W. H. Steier, *Science* **288**, 119 (2000).
27. H. Ma, S. Liu, J. Luo, S. Suresh, L. Liu, S. H. Kang, M. Haller, T. Sassa, L. R Dalton, and A. K. Y. Jen, *Adv. Funct. Mater.* **12**, 565 (2002).
28. H. E. Katz, G. Scheller, T. M. Putvinski, M. L. Schilling, W. L. Wilson, and C. E. D. Chidsey, *Science* **254**, 1485 (1991).
29. D. Li, M. A. Ratner, T. J. Marks, C. Zhang, J. Yang, and G. K. Wong, *J. Am. Chem. Soc.* **112**, 7389 (1990).
30. S. Yitzchaik, S. B. Roscoe, A. K. Kakkar, D. S. Allan, T. J. Marks, Z. Xu, T. Zhang, W. Lin, and G. K. Wong, *J. Phys. Chem.* **97**, 6958 (1993).
31. P. Zhu, M. E. van der Boom, H. Kang, G. Evmenenko, P. Dutta, and T. J. Marks, *Chem. Mater.* **14**, 4982 (2002).
32. Y. G. Zhhao, A. Wu, H. L. Lu, S. Chang, W. K. Lu, S. T. Ho, M. E. van der Boom, and T. J. Marks, *Appl. Phys. Lett.* **79**, 587 (2001).
33. Y. Lvov, S. Yamada, and T. Kunitake, *Thin Solid Films* **300**, 107 (1997).
34. X. Wang, S. Balasubramanian, L. Li, X. Jiang, D. Sandman, M. F. Rubner, J. Kumar, and S. K. Tripathy, *Macromol. Rapid Commun.* **18**, 451 (1997).
35. A. Laschewsky, B. Mayer, E. Wisheroff, X. Arys, A. Jonas, M. Kauranen, and A. Persoons, *Angew. Chem. Int.* Ed. **36**, 2788 (1997).
36. J. R. Heflin, C. Figura, D. Marciu, Y. Liu, and R. O. Claus, *SPIE Proc.* **3147**,10 (1997).
37. M. J. Roberts, G. A. Lindsay, W. N. Herman, and K. J. Wynne, *J. Am. Chem. Soc.* **120**, 11202 (1998).
38. J. R. Heflin, C. Figura, D. Marciu, Y. Liu, and R. O. Claus, *Appl. Phys. Lett.* **74**, 495 (1999).
39. C. Figura, P. J. Neyman, D. Marciu, C. Brands, M. A. Murray, S. Hair, R. M. Davis, M. B. Miller, and J. R. Heflin, *SPIE Proc.* Vol. **3939**, pp. 214–222 (2000).
40. K. Van Cott, M. Guzy, P. Neyman, C. Brands, J. R. Heflin, H. W. Gibson, and R. M. Davis, *Angew. Chem. Int.* Ed. **41**, 3236 (2002).

20

Photonic Crystals

Younan Xia, Kaori Kamata, and Yu Lu
Department of Chemistry, University of Washington, Seattle, WA 98195

20.1. INTRODUCTION

Photonic (band gap) crystals refer to spatially periodic structures fabricated from dielectric materials having different refractive indices.[1] The concept of this relatively new class of materials was first proposed independently by Yablonovich[2] and John[3] in 1987. The major feature of a photonic crystal is the existence of a band gap in its photonic structure that is able to influence the propagation of electromagnetic waves in a similar way as the electronic band gap of a semiconductor does for electrons. In the years since, the concept of photonic band gap has been exploited as a convenient and powerful tool to confine, control, and manipulate photons in all three dimensions of space. A variety of applications have been demonstrated or envisioned for this class of structured materials.[4] A photonic crystal can, for example, serve as a filter to block the propagation of photons (irrespective of their polarization and/or direction of motion) within a specific range of frequencies. It can also be exploited as an effective means to localize photons to a specific region at restricted frequencies; to inhibit the spontaneous emission of an excited chromophore; to modulate a stimulated emission process; and to serve as a waveguide to direct the flow of photons along a specific path with essentially no energy loss. All these properties are technologically important because they can be further explored to fabricate light-emitting-diodes (LEDs) displaying coherent outputs, to generate semiconductor diode lasers with no threshold, and to significantly enhance the performance of many other types of optical, electro-optical, and quantum electronic devices.[5–7]

Research on photonic crystals has been extended to cover all three dimensions (Figure 20.1), and the spectral range from ultraviolet to radio frequencies. The major driving force in this area has been the quest for a complete photonic band gap around 1.55 μm—the wavelength used as the standard in optical fiber communications.[8] Through an interplay between computational and experimental studies, this goal has now been successfully

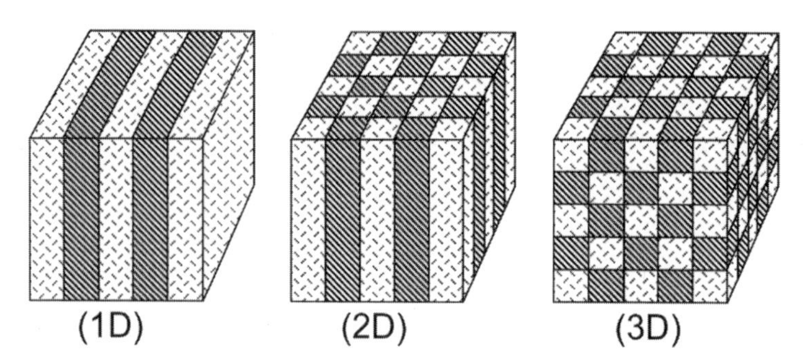

FIGURE 20.1. Schematic illustrations of 1D, 2D, and 3D photonic crystals patterned from two different types of dielectric materials.

accomplished with the fabrication of photonic crystals suitable for operation in the vicinity of 1.55 μm.[9] Perhaps different from other areas of materials science, computational studies have played the most important role in defining and guiding the research directions of this area. For example, all crystal structures (including the symmetry and volume fraction of their lattice points, and the minimum contrast in refractive index between the high and low dielectric regions) for achieving complete band gaps were discovered through computational simulations, and then realized experimentally.[7,10] The fundamental challenge of this area is still one of materials science: that is, the fabrication of a proposed crystal structure in a controllable fashion, and at reasonable expense. In this regard, fabrication methodology development has been a major thrust of research efforts in this area. Thanks to the continuous efforts of many groups, photonic crystals with a range of structures can now be readily fabricated using either top-down (microlithography) or bottom-up (self-assembly) approaches. This chapter presents an overview of the fabrication and characterization of three-dimensional (3D) photonic crystals, as well as the intriguing applications of this new class of functional materials.

The chapter is organized into five sections: The first section provides a brief introduction on photonic band structures and band gaps, using one-dimensional (1D) systems to illustrate the concept. The second section discusses the use of microlithographic techniques in generating 3D photonic crystals, with examples including both layer-by-layer microfabrication and holographic patterning. The third section fully illustrates the potential of self-assembly as a generic route to long-range ordered lattices that are suitable for use as photonic crystals. We will concentrate on self-assembly of monodispersed colloidal particles, and the concept will be illustrated using three specific examples: opals made of spherical colloids, inverse opals consisting of interconnected air balls; and crystalline lattices composed of nonspherical colloids. The fourth section highlights a number of unique applications associated with photonic crystals, with an emphasis on those systems displaying tunable properties. The last section concludes with some speculations on the directions towards which future research on this new class of materials might be directed.

20.2. PHOTONIC BAND STRUCTURES AND BAND GAPS

The photonic band structure of a crystalline lattice can be readily computed by solving Maxwell's equations where the dielectric constant is expressed as a spatially periodic

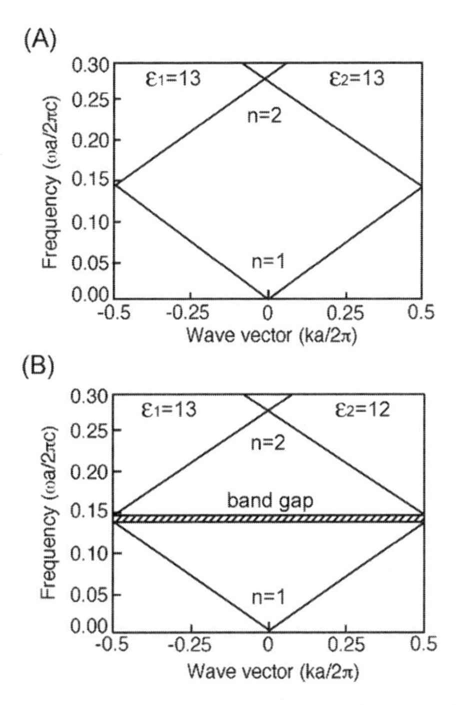

FIGURE 20.2. Photonic band structures (normal incidence) calculated for a typical 1D photonic crystal made of alternating thin layers of two dielectric materials. The periodicity of the lattice is a, and the dielectric constants are (A) $\varepsilon_1 = 13$, $\varepsilon_2 = 13$; and (B) $\varepsilon_1 = 13$, $\varepsilon_2 = 12$. The hatched region represents a photonic band gap. Any electromagnetic wave whose frequency falls anywhere in this gap will not be able to propagate through this crystal.

function.[1] Figure 20.2 shows the photonic band structures (at normal incidence to the surface) calculated for a 1D system, that can be fabricated by alternately stacking thin films of two different materials in a layer-by-layer fashion. Because Maxwell's equations enjoy scale invariance, we can shift the bands and gaps theoretically to any frequency range simply by scaling the feature size of a periodic structure. As a result, the frequency is usually expressed in terms of the periodicity (a) of this multilayered structure to reflect this salient feature. The band structure shown in Figure 20.2A was obtained by setting the dielectric constants of both layers to 13, and in this case, the periodic lattice is reduced to a homogeneous thin film made of GaAs. For this particular sample, there is no band gap in the photonic band structure, and the frequency of light is simply related to wave vector through the dispersion equation: $\omega(k) = ck/\varepsilon^{1/2}$. Since the wave vector is supposed to repeat itself outside the Brillouin zone, the dispersion lines have to fold back into the zone when they reach the edges. Figure 20.2B shows the band structure calculated for a multilayered film where the two adjacent layers have a dielectric constant of 13 (for GaAs) and 12 (for AlGaAs), respectively. In this case, there exists a frequency gap between the lower ($n = 1$) and upper ($n = 2$) bands. Within this gap, no mode (regardless of k value) can propagate through the multilayered film in the direction normal to the surface. As the contrast in dielectric constant is further increased from 13/12 to 13/1 (e.g., when the sample is made of GaAs and air), this band gap will be significantly widened. Such a discontinuity in the frequency domain is referred to as the photonic band gap of this periodic lattice. The appearance of a photonic band gap can also be understood in terms of modes that correspond to standing waves having a wavelength of $2a$, or twice the lattice constant of the 1D structure. When electromagnetic waves with frequencies located in the band gap are shined onto such a photonic crystal, they will be bounced back as reflected waves with an efficiency as high as 100%. In this regard, 1D photonic crystals can act as perfect dielectric mirrors, and have found widespread use in fabricating the optical cavities of semiconductor diode lasers.[11]

For a 1D system, there always exist an infinite number of gaps (between adjacent sets of bands) in the band structure as long as there is a dielectric contrast. For 2D and 3D systems, we have to distinguish between complete and incomplete band gaps. A complete (or full, true) gap is defined as one that extends throughout the entire Brillouin zone in the photonic band structure. An incomplete one is often referred to as a pseudo gap (or the so-called stop band) because it only appears in the transmission spectrum along a certain propagation direction. A 3D complete gap can also be considered as the combination of a set of stop bands that overlap in frequency when the propagation direction is varied across all three dimensions of space. In this case, the evolution of a true band gap is strongly dependent on the crystallographic structure, as well as the dielectric contrast (i.e., the strength of the potential well). The presence of a periodic modulation in refractive index may not necessarily lead to the formation of complete photonic gaps in the band structures.

In principle, the photonic band structure of a 3D crystal can also be obtained by solving Maxwell's equations, with the dielectric constant expressed as a three-dimensionally periodic function. In his pioneering work on photonic crystals,[2] Yablonovich used a scalar model to argue that such a face-centered cubic lattice should be favorable for the formation of a complete band gap because the Brillouin zone of its reciprocal lattice has the least deviation from a spherical shape. In following studies by other research groups, it was demonstrated that the scalar model was no longer adequate in describing the photonic band structures of a 3D system, and rigorous models had to be introduced to account for the vectorial nature of both electrical and magnetic components of electromagnetic waves.[12] Among various models, the plane wave expansion method (PWEM) has become the most commonly used tool for 3D systems whose dielectric constants do not contain imaginary parts (due to intrinsic absorption).[13] For systems (e.g., metals) with significant absorptions, one has to use an alternative method that can be considered as the photonic analogue of the Korringa-Kohn-Rostoker (KKR) approach developed for electronic systems.[14] For simple cubic and hexagonal lattices, a detailed description of the computational results was recently summarized by Haus *et al.*[15] and John *et al.*,[16] and the computational codes can also be downloaded from the web sites of several research groups. It is generally accepted that for a face-centered cubic lattice consisting of spherical building blocks, there only exists a pseudo gap in the photonic band structure, no matter how high the dielectric contrast is. The existence of a complete gap in this simple cubic system is prohibited by a symmetry-induced degeneracy at the W-point. When the symmetry of such a lattice is reduced to a diamond structure, a complete gap will evolve between the 2nd and 3rd bands in the photonic band structure, with a maximum gap to mid-gap ratio approaching 15.7% obtained at a volume fraction of ~37% and a dielectric contrast of ~13 (or ~3.6, in terms of refractive index).[17,18]

20.3. PHOTONIC CRYSTALS BY MICROFABRICATION

Fabrication of 1D or 2D photonic crystals is relatively simple and straightforward: for example, 1D systems can be generated through programmed deposition of alternating layers of different dielectric materials;[11] while 2D systems can be routinely produced via selective etching of the underlying substrates through masks.[19] As a result, 1D (and 2D) photonic crystals have already found widespread use as active components in fabricating a rich variety of commercial devices that include optical notch filters,[20] dielectric mirrors or optical resonance cavities,[21] and Bragg gratings (on optical fibers).[22] At the current stage of development, it still remains a grand challenge in applying the conventional patterning techniques to the generation of pre-designed, 3D periodic lattices that meet all the criteria required for creating complete photonic band gaps. Since the position of a photonic band gap is scaled by half to one third of the periodicity of a 3D lattice, the fabrication task is expected to become even more challenging as the mid gap positions move towards the visible or ultraviolet regions. In these cases, one will need to pattern the dielectric materials into microstructures of 100–300 nm in size along all three dimensions, and to accomplish registration among the structures in different layers with an accuracy better than tens of nanometers. Nevertheless, a number of relatively simple 3D lattices amenable to layer-by-layer fabrication or holographic patterning have recently been produced in the prototype form by several research groups and some of them were shown to exhibit the signature of complete band gaps down to the near-infrared region. The following two sections provide a very brief coverage on these dedicated fabrication methods.

20.3.1. Layer-by-Layer Fabrication

In 1994,[23] Ho and coworkers proposed a simple cubic lattice (Figure 20.3A) that could be fabricated by stacking dielectric rods (with rectangular, circular, or ellipsoidal cross-sections) into the so-called woodpile structure in a layer-by-layer fashion. Depending on the filling ratio, such a lattice could possess a complete band gap (between the 2nd and 3rd band, see Figure 20.3B) at a refractive index contrast as low as 1.9. In a related study, the photonic band structures of woodpile lattices having a range of different symmetries were computed, and the signatures of complete band gaps were found in a number of systems.[24] These promising results stimulated a number of research groups to fabricate the proposed structures, in an effort to experimentally probe their photonic band structures. To this end, 3D photonic crystals were successfully fabricated with complete band gaps in the millimeter-wave region by stacking together silicon wafers (or metal sheets) that had been patterned with parallel arrays of micromachined rods.[25] An average attenuation of 50 dB was measured in the band gap, whose position could be easily tuned in the spectral region ranging from 30 GHz to 3 THz. This approach was later extended to structures with much smaller feature sizes,[26] and photonic crystals with complete band gaps at optical communication wavelengths were fabricated with single crystalline GaAs and InP as the materials for the building blocks.[27] A band gap attenuation of >40 dB (corresponding to a reflection of 99.99%) was successfully achieved with an eight-layer crystalline lattice.

In a related approach, cubic woodpile structures were fabricated from polycrystalline silicon (poly-Si) using the conventional microlithographic technique.[28] Figure 20.4 shows

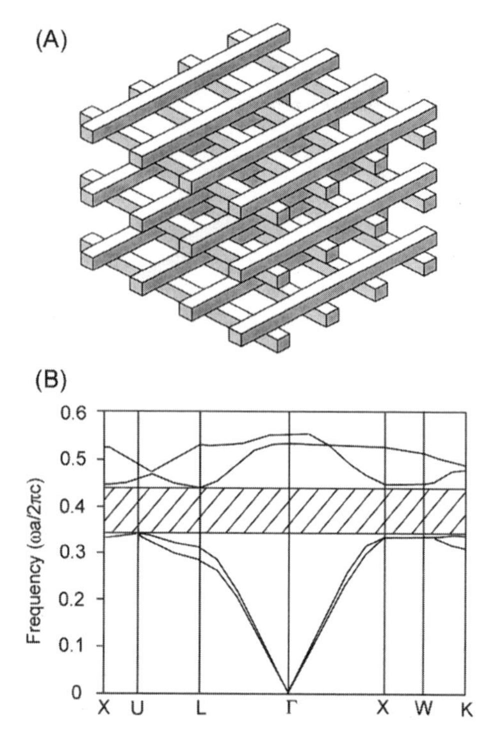

FIGURE 20.3. (A) An illustration of the 3D woodpile lattice and (B) its photonic band structure calculated using the PWEM method. The filling fraction of the dielectric rods is 26.6%, and the contrast in refractive index is set to be 3.6/1.0. (From Ref. 23 by permission of Elsevier B.V.)

a schematic illustration of the procedure that began with the patterning of a thin film of poly-Si (deposited on a Si wafer) with standard photolithography and reactive ion etching. After filling the recessed regions (between poly-Si lines) with silica (SiO_2) and planarizing the surface (via chemical mechanical polishing, or CMP), a second layer of poly-Si was deposited and then patterned (with the long axis of Si lines oriented perpendicular to those

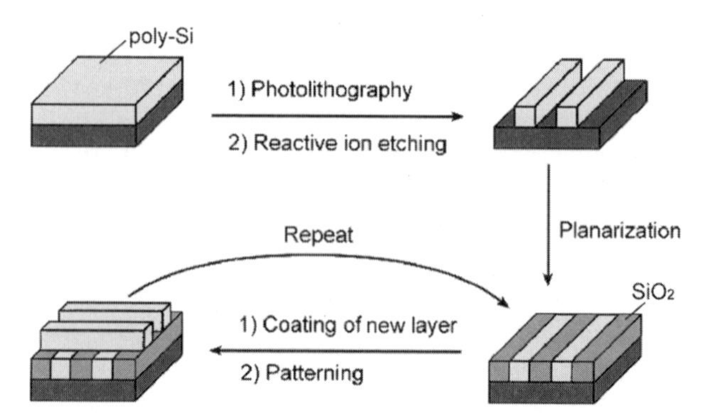

FIGURE 20.4. A schematic illustration showing the microfabrication of a 3D woodpile lattice of polycrystalline silicon in a layer-by-layer fashion.

in the first layer). This step of filling/planarization/deposition/patterning was then repeated to build up the 3D structure layer-by-layer. Finally, a woodpile lattice made of poly-Si was obtained by selectively removing the silica regions through wet etching in HF. Crystals of up to 6 layers in thickness were fabricated, and a strong attention of light (between 10 and 14.5 μm) was observed with strength of ~12 dB per unit cell. By introducing a vacancy defect into such a photonic crystal, the localization of light was also demonstrated to less than a cubic wavelength in all there dimensions.[29] In a most recent demonstration, this fabrication method was extended to generate 3D photonic crystals made of metals such as tungsten.[30] A very strong attenuation (30 dB per unit cell at 12 μm) was measured in the photonic band gap that extended from 8 to 20 μm. Such metallic photonic crystals are expected to find use in various applications related to black body radiation (e.g., thermophotovoltaic).

20.3.2. Other Microfabrication Methods

Although the layer-by-layer fabrication method discussed in the above section provides an effective route to woodpile structures with complete band gaps in the infrared region, it is still a grand challenge to extend it to the visible region where the critical dimension of building blocks has to be controlled on the scale of 100–300 nm. In addition, it seems to be infeasible to apply this time-consuming method to the fabrication of 3D photonic crystals containing more than 10 layers, or to high-volume production. For these reasons, alternative structures and techniques have been constantly sought to solve these problems. In 1991,[31] Yablonovitch proposed a new face-centered-cubic lattice (made of nonspherical building blocks) that could lift the degeneracy at the *W*-point of the Brillouin zone to generate a complete band gap. More importantly, such a periodic lattice (now commonly referred to as Yablonovite) could be readily generated in the surface of a solid substrate by chemically-assisted ion beam etching (CAIBE) through a mask containing a hexagonal array of holes.[32] Photonic crystals up to four lattice periods in depth have been fabricated, and a forbidden gap lying in the near infrared (1.1 to 1.5 μm) was observed in the transmission spectra. The mid gap attenuation was found to be highly sensitive to structural errors (mismatches) that might be introduced into the lattice during the fabrication step.[33] In addition to ion beams, X-ray lithography was also demonstrated to generate the Yablonovite lattice in a poly(methyl methacrylate) film.[34] In this case, the film was consecutively exposed to an X-ray beam three times through a triangular lattice of holes. Up to 6 crystal periods were obtained in a polymer film 6.2 μm thick. When the lattice constant was controlled at 1.3 μm, a stop band around 2.4 μm was observed in the reflectance/transmission spectra measured at normal incidence.

In another set of experiments, it was demonstrated that the Yablonovite lattice could also be generated in a photosensitive film using holographic patterning. When two mutually coherent laser beams interfere, the resulting pattern in field intensity can be faithfully recorded in a thin film of photoresist.[35] After development, an array of parallel lines will be created in the resist film, with the spacing between the lines being $\lambda/2n \sin(\theta/2)$, where λ is the wavelength of the laser, n is the refractive index of the photoresist, and θ is the angle of intersection of the beams. This technique is widely used in industry to manufacture holographic diffraction gratings and anti-reflection coatings. By exposing the same photoresist film to the interference pattern twice (orthogonal to each other), a 2D array of posts will be generated. More complex patterns can be formed by intersecting more than two laser beams

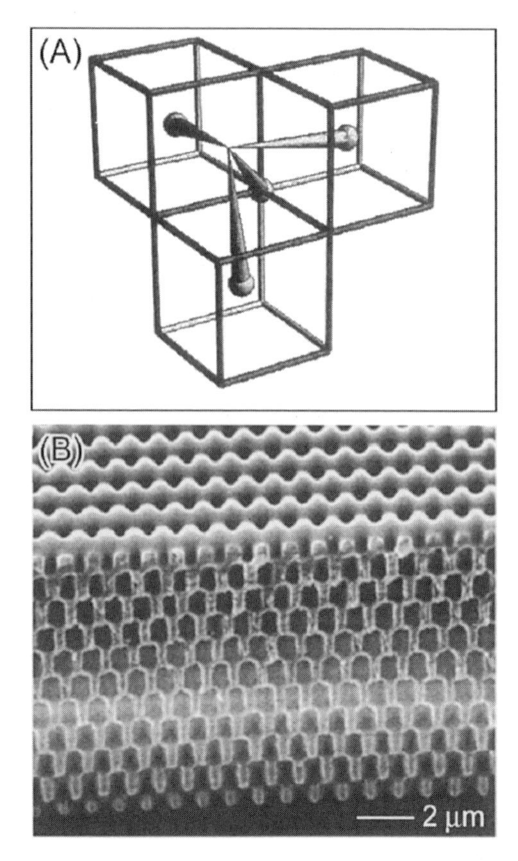

FIGURE 20.5. (A) A schematic illustration of the experimental setup for holographic patterning that involves the use of four coherent laser beams. (From Ref. 36 by permission of Nature Publishing Group.) (B) The scanning electron microscopy image of a 3D periodic lattice generated in a photoresist through photo-induced polymerization. (From Ref. 37 by permission of American Chemical Society.)

or by using multiple sequential exposures. For example, when four laser beams are focused onto the same spot (Figure 20.5A), a 3D structure (Figure 20.5B) similar to the Yablonovite lattice will be generated after the exposed regions have been selectively removed through wet etching.[36,37] The major advantage of this method is that it is a relatively fast process, and the exposure and etching steps could be completed in a few minutes. In this regard, this method is well-suited for high throughput production. As a major drawback associated with this technique, the materials that can be directly patterned have to be photosensitive polymers that often have low refractive indices, and it is necessary to backfill with inorganic semiconductors in order to create photonic crystals with complete band gaps.

There are also a number of other alternative structures and fabrication techniques that have been explored for generating photonic crystals. For example, John *et al.* proposed a new type of structure with spiral chains as the building blocks that is expected to exhibit a complete band gap between the 2nd and 3rd bands.[38] Such a chiral lattice could be potentially fabricated using the glancing angle deposition (GLAD) technique.[39] In a set of recent demonstrations, two-photon lithography has been demonstrated by several groups to generate 3D lattices with arbitrary structures.[40] Although this method may lack the speed for high-volume production, it provides a potentially useful method for writing waveguiding structures (or other types of defect structures) inside the 3D crystalline lattice of a photonic crystal.[41]

20.4. PHOTONIC CRYSTALS BY SELF-ASSEMBLY

The technical challenges and barriers to extending conventional microlithography from 2D to 3D patterning have made it possible to consider alternative approaches to the fabrication of long-range ordered lattices to be used as photonic crystals. In addition to the holographic and two-photon techniques discussed in Section 20.3, self-assembly represents another alternative route that has been extensively explored to generate 3D periodic lattices with well-defined structures. The key components of self-assembly (see Chapter 2) are building blocks—that is, pre-designed molecules or objects on various scales—that spontaneously organize into a relatively stable (often long-range ordered) structure through non-covalent interactions.[42] The final structure is usually determined by the characteristics of the building blocks since the information that defines the self-assembly process has already been coded in the building blocks in the form of topology, shape, or surface functionality. Because the final structure is often at (or sufficiently close to) a thermodynamic equilibrium state, such a process has the tendency to reject defects and thus lead to the formation of structures having greater order than could be reached in non-self-assembling systems. More importantly, the inherently parallel nature of self-assembly makes it a promising candidate for large-scale production where low-cost and high throughput represent two major requirements.

Self-assembly is especially evident in the biological world.[43] Much of the inspiration for studies on this subject comes directly from the admiration of biological structures that include, for example, the lipid bilayer membranes of cells, the DNA helical structures, and the tertiary or quaternary structures of proteins. In the past several decades, the concept of self-assembly has been extended to many non-biological systems, in an effort to generate highly ordered structures with their feature sizes ranging from molecular to mesoscopic and macroscopic scales.[44] Self-assembly methods that have been demonstrated for use with the fabrication of photonic crystals include the following: i) phase separation of organic block-co-polymers,[45] ii) crystallization of colloidal particles (both spherical and nonspherical in shape) that are monodispersed in size;[46] and iii) template-directed synthesis against closely packed lattices of colloidal particles.[47] Here we only concentrate on the last two approaches, with opals and inverse opals as examples to demonstrate the concept and potential of this approach.

20.4.1. *Opals: Crystalline Lattices of Spherical Colloids*

Colloids are usually referred to as small particles with at least one characteristic dimension in the range of a few nm to one μm.[48] Thanks to many years of continuous efforts, a wealth of colloids can now be synthesized as monodispersed samples in which the size and shape of the particles, and the charges chemically fixed on their surfaces, are all identical to within 1–2%.[49–51] Spherical particles, in particular, has long been established as a class of monodispersed colloids. Over the past several decades, many chemical procedures have been developed for generating spherical colloids from various materials that include organic polymers and inorganic ceramics. One of the most attractive features associated with spherical colloids is that they can be readily organized into long-range ordered, crystalline lattices (or the so-called opals).[52] The availability of such crystalline lattices has enabled the observation of interesting functionality not only from the constituent material of the spherical colloids, but also from the long-range order intrinsic to the lattices. As a matter of fact, self-assembly of spherical colloids into opals seems to represent the most convenient route to the fabrication of photonic crystals.[46,53]

Many different methods have been demonstrated for crystallizing monodispersed spherical colloids into opaline lattices.[52–56] Several of them are, in particular, fruitful in producing opals with relatively large domain sizes and well-defined structures. For example, sedimentation under the influence of gravitational field represents a simple, convenient method for crystallizing spherical colloids with diameters >500 nm.[57] Repulsive electrostatic interactions between highly charged spherical colloids have been widely exploited to organize these colloids into body-centered cubic or face-centered cubic crystals with thicknesses up to hundreds of layers.[58] Attractive capillary forces (as caused by solvent evaporation), on the other hand, have been used in a layer-by-layer fashion to generate opals with well-controlled thicknesses.[59] The procedure we commonly use involves the use of a microfluidic cell (Figure 20.6) that was specifically designed to combine physical confinement and shear flow to crystallize spherical colloids into 3D opaline lattices with thicknesses up to several hundred layers.[60–65]

In a typical procedure, the microfluidic cell was fabricated by sandwiching a square frame (patterned in a thin film of photoresist or cut from the commercial Mylar film) between two glass substrates. Small channels between the gasket and the glass substrates could be easily fabricated using a number of microlithographic methods: for example, by

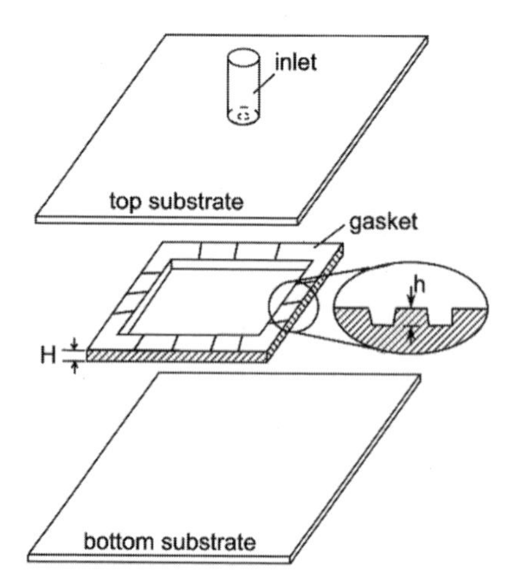

FIGURE 20.6. A schematic illustration of the microfluidic cell used to fabricate opaline lattices from monodispersed spherical colloids (polystyrene beads or silica spheres).

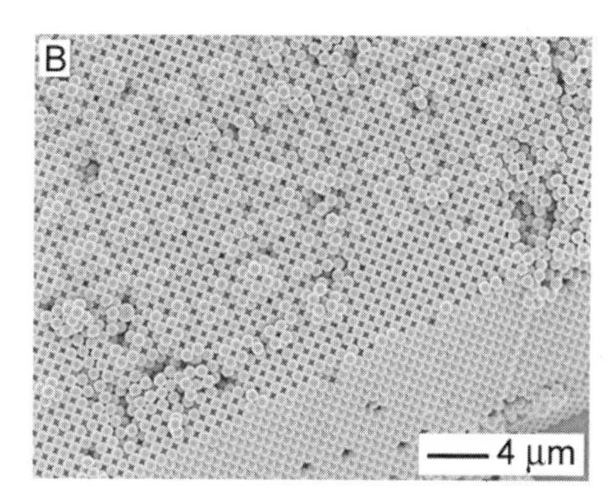

FIGURE 20.7. Scanning electron microscopy images of two typical examples of 3D opaline lattices that were crystallized from polystyrene beads, with their (111) and (100) crystallographic planes oriented parallel to the surfaces of the supporting substrates, respectively. (From Ref. 63 by permission of WILEY-VCH Verlag GmbH & Co. and from Ref. 66 by permission of American Chemical Society.)

using two-step photolithography; by scraping both sides of the Mylar film with a piece of soft paper to generate channels via abrasion; and by coating both surfaces of the Mylar film with colloids smaller than those to be crystallized in the cell.[65] The aqueous dispersion of spherical colloids could be injected into the cell through a glass tube glued to the small hole (~3 mm in diameter) drilled in the top glass substrate. Colloids with a diameter larger than the depth of channels in the gasket surfaces could be concentrated at the bottom of the microfluidic cell and crystallized into a 3D opaline lattice.

Using this procedure, monodispersed spherical colloids (both polymer latexes and silica beads) were routinely assembled into opaline lattices over areas of several square centimeters that were characterized by a homogeneous structure, well-controlled thickness, and long-ranged ordering. Both scanning electron microscopy and optical diffraction studies indicate that the opaline lattices fabricated using this approach exhibited a cubic-close-packed structure, with the (111) crystallographic planes oriented parallel to the surfaces of supporting substrates. Figure 20.7A shows the scanning electron microscopy (SEM) image of a typical opal sample that was crystallized from 220-nm diameter polystyrene beads. The cross-sections of this opaline lattice also confirmed that it had an "ABC" stacking along the direction perpendicular to the surface of the substrate. Most recently, we and other research groups also demonstrated the fabrication of opaline lattices with the (100) crystallographic planes oriented parallel to the surfaces of substrates by templating against various relief structures.[66–69] Figure 20.7B shows the SEM image of such an example that was fabricated by templating polystyrene beads against a 2D array of pyramidal pits anisotropically etched on the surface of a Si (100) wafer. The key to the success of this approach was to precisely control the lateral dimensions of and separations between the square pyramidal pits to match the diameter of the spherical colloids.

The photonic band structures of opals have been extensively studied (both theoretically and experimentally). It is generally accepted that the traditional scalar-wave approximation is not adequate in fully describing many of the important features of a 3D crystalline lattice. In this case, the plane wave expansion method (PWEM) has to be used, although it will not

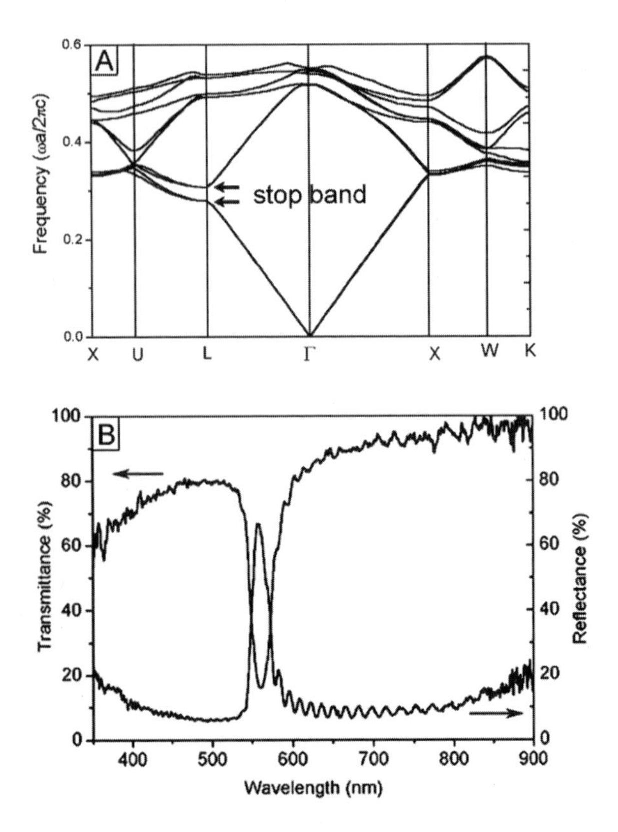

FIGURE 20.8. (A) The photonic band structure calculated for a 3D opaline lattice consisting of spherical colloids with a refractive index of 3.0. This system only exhibits a stop band between the 2nd and 3rd bands. (From Ref. 70 by permission of WILEY-VCH Verlag GmbH & Co.) (B) The transmission and reflectance spectra recorded from an opaline lattice made of polystyrene beads. The fine features on the spectra were caused by interference of light bounced back from the front and back surfaces of the sample.

be valid for 3D systems whose dielectric constants contain large imaginary parts (due to absorption). Figure 20.8A shows the photonic band structure calculated (using the PWEM method) for a face-centered cubic lattice of spherical colloids, indicating that this simple system does not possess a full bandgap. There are only a number of pseudo bandgaps (or stop bands), as marked by two arrows on the graph.[70] In this case, the existence of complete bandgap is inhibited by the degeneracy at W- or U-point which is, in turn, induced by the spherical symmetry of the lattice points. It is impossible to completely eliminate by increasing the refractive index contrast, or by changing the filling fraction of spheres from 74% (closely packed) to any other values. In this regard, the spherical symmetry is not good for generating full gaps in the photonic band structures, although spherical colloids represent the simplest building blocks to synthesize as monodispersed samples and/or to crystallize into 3D opaline lattices. The stop band of a face-centered cubic lattice made of spherical colloids can be easily measured from its transmission or reflectance spectrum.

Figure 20.8B shows two such spectra that were taken from a 3D colloidal crystal assembled from polystyrene beads. The peaks observed in transmission and reflectance spectra matched well in position and intensity, indicating the high quality (low density of defects) of this sample. The position of the peak (or stop band) could be calculated from the lattice constants and mean refractive index by using Bragg's diffraction equation.[64] Although 3D opaline lattices of spherical colloids do not have full band gaps, they offer a simple and easily prepared model system to experimentally probe the photonic band diagrams of certain types of 3D lattices.

20.4.2. Inverse Opals: Crystalline Lattices of Interconnected Air Balls

Computational studies have suggested that a porous material consisting of an opaline lattice of interconnected air balls (embedded in an interconnected matrix with a higher refractive index) might lead to the formation of a complete gap in the photonic band structure.[71] To optimize the photonic effect, the volume fraction of the solid matrix has to be controlled within the range of 20–30%. As shown in Section 20.3, it was possible to build up such porous structures layer-by-layer through conventional microlithographic techniques, albeit it has been a challenge to achieve this goal as the critical feature size becomes comparable to the wavelength of visible or UV light. Processing difficulties have also limited the formation of such porous structures with thicknesses more than 10 layers, or from materials other than those allowed by conventional microfabrication techniques. In comparison, templating against opaline arrays of spherical colloids provides a promising route to the fabrication of such porous structures from essentially all solid materials and with well-controlled feature sizes down to ~30 nm.[72]

Figure 20.9 shows a schematic procedure commonly used for this templating method. After a cubic-close-packed lattice of spherical colloids has been dried, there is 26% (by volume) of void spaces within the lattice that can be infiltrated (partially or completely) with another material (in the form of a gaseous or liquid precursor, as well as a colloidal suspension) to form a solid matrix around the spherical colloids. Selective removal of the template (through selective dissolution or calcination) will leave behind a 3D porous material containing a highly ordered architecture of uniform air balls interconnected to each other through small "windows". Because this porous material has a structure complementary to that of an opal, it is also often referred to as an inverse opal.[73] In this template-directed synthesis, the opaline lattice simply serves as a scaffold around which many other kinds of

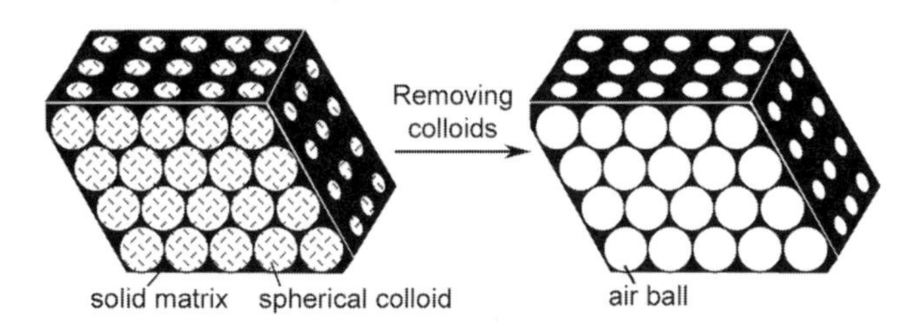

GURE 20.9. A schematic illustration of the experimental procedure that generates 3D inverse opals by templating against an opaline lattice of spherical colloids, followed by selective removal of the colloidal templates. rom Ref. 46 by permission of WILEY-VCH Verlag GmbH & Co.)

materials are formed. This method is, in particular, attractive for the fabrication of photonic crystals because the periods of resultant structures can be conveniently tuned from tens of nanometers to several micrometers, and a variety of materials with relatively high refractive indices can also be easily incorporated into the procedure.

A number of procedures have been demonstrated to fill the voids within an opaline lattice with the desired functional material.[74–80] Liquid prepolymers that can be UV- or thermally cured are particularly attractive for such applications because of their low shrinkage in volume (usually $< 1\%$).[81] When a few drops of such a prepolymer is applied along the edges of an opaline lattice, the voids within the opaline lattice will be spontaneously filled (as driven by capillary action[82]). Similar to organic liquid prepolymers, a sol-gel precursor solution can also be adopted for such a process, and a ceramic material with the desired composition will be obtained upon hydrolysis and condensation of the precursor.[83] The major difference is that that a sol-gel solution is often very dilute in terms of solid concentration. Even though the voids among spherical colloids are completely filled with the precursor solution, the amount of ceramics produced in a sol-gel process might only be enough to form thin coatings on the surfaces of spherical colloids after the solvent has evaporated. In this regard, multiple insertions of the precursor solution are necessary in order to completely fill the voids in the 3D crystalline lattice. Otherwise, ceramic shells rather than an inverse opal will be likely obtained as the final product.[84] In addition to homogeneous solutions, the void spaces within an opaline lattice could be directly filled with a suspension of nanoparticles with dimensions much smaller than the spherical colloids.[85–87] After the dispersing medium has been evaporated, a compact solid matrix made of the nanoparticles will be formed. Essentially, all materials (e.g., Au, Ag, CdSe, SiO_2, TiO_2, and Fe_2O_3) that could be processed as nanoscale particles could be included in this processing to obtain inverse opals made of materials exhibiting various functionalities. Figure 20.10 shows the scanning electron microcopy images of several typical examples of inverse opals. Note that the distinctive feature of these structures is a highly ordered array of air balls interconnected through small windows.

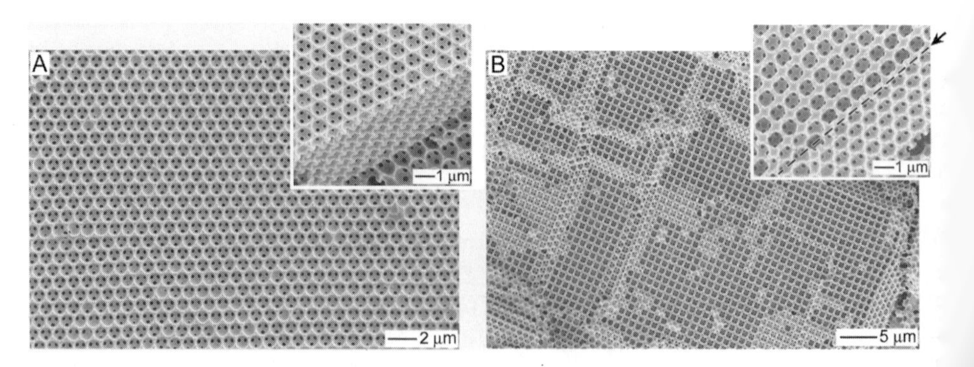

FIGURE 20.10. Scanning electron microscopic images of two typical examples of inverse opals that were fabr cated by templating the UV-curable liquid organic prepolymer to poly(acrylate methacrylate) against an opalir lattices of polystyrene beads. The polystyrene beads had been selectively removed by etching with toluene. Tl (111) and (100) plane of these two samples were oriented parallel to the supporting substrate, respectively. (Fro Ref. 66 by permission of American Chemical Society.)

Electrochemical methods have also been explored to deposit metals and semiconductors within the voids of an opaline lattice. For example, Colvin *et al.* have demonstrated the use of electroless deposition to generate macroporous materials of nickel, copper, silver, gold, and platinum.[88] In a typical procedure, they coated the surfaces of silica spheres with an activator, 3-mercaptopropyltrimethoxysilane (3-MPTMS), before they were assembled into opaline lattices. Gold nanoparticles (of a few nm in diameter) were then attached to the surfaces of these silica colloids through coupling reactions with the thiol groups. Finally, the template was immersed in an electroless deposition bath to generate a metal within the voids of this 3D opaline lattice. One of the major advantages associated with this new approach is that there exist standard recipes for the electroless deposition of essentially all metals and alloys. In another demonstration, Braun and Wiltzius have fabricated inverse opals from CdS and CdSe by electrochemically depositing these materials directly into the voids of opaline lattices.[89,90]

Recent studies indicate that the voids within an opaline lattice can also be readily filled with a solid material decomposed from a vapor-phase precursor. For example, CdSe inverse opals have been synthesized using methods based on chemical vapor deposition (CVD).[91] These demonstrations opened the door to the fabrication of inverse opals from a rich variety of elemental or compound semiconductors with relatively high refractive indices. In a follow-up study, Silicon inverse opals were synthesized and the signature of a complete bandgap was observed in this system.[92] It has also been demonstrated that it is possible to further pattern these inverse opals into well-defined microstructures by reactive ion etching through physical masks.[93] More recent studies by scanning electron microscopy revealed that the semiconductor was deposited around each spherical colloid in a layer-by-layer fashion until the voids were completely filled.[94] In general, this is a method suitable for use with many vapor-phase precursors that have already been developed for microelectronics.[95] One of the potential problems associated with this method is that the material deposited on the exterior surface of an opaline sample may block the entrance of precursor molecules in the subsequent steps and eventually lead to the formation of an inhomogeneous film (in particular, when the sample is relatively thick).

Figure 20.11A shows the photonic band structure of an inverse opal made of a material whose refractive index is around 3.0.[70] This diagram indicates the existence of a complete band gap between the 8th and 9th bands (the hatched region) that extends over the entire Brillouin zone. Computational simulations further suggest that the minimum contrast in the refractive index at which a complete bandgap will be observed in an inverse opal is around 2.8. It will be possible to accomplish this requirement using a number of inorganic materials such as group IV or II–VI semiconductors (Si, Ge, or CdSe), rutile (TiO_2), and iron oxides (α-Fe_2O_3). Figure 20.11B shows the transmission and reflectance spectra taken from an inverse opal made of magnetite.[96] For this sample, there only existed a stop band (although its bandwidth is broader than that of an opal based on silica spheres or polystyrene beads). Part of the reason lies in the fact that the filling of void spaces among spherical colloids was not complete and the resultant materials might not be sufficiently dense to acquire a refractive index close to that of the bulk solid. For example, Vos *et al.* have estimated the refractive index of their anatase porous structure to be 1.18–1.29, a value well below that of single crystalline anatase (~2.6).[97] It seems to be that the definitive signature of a complete photonic band gap could only be observed in inverse opals that were fabricated from semiconductors with high refractive indices (e.g., Si or Ge) using the CVD method.

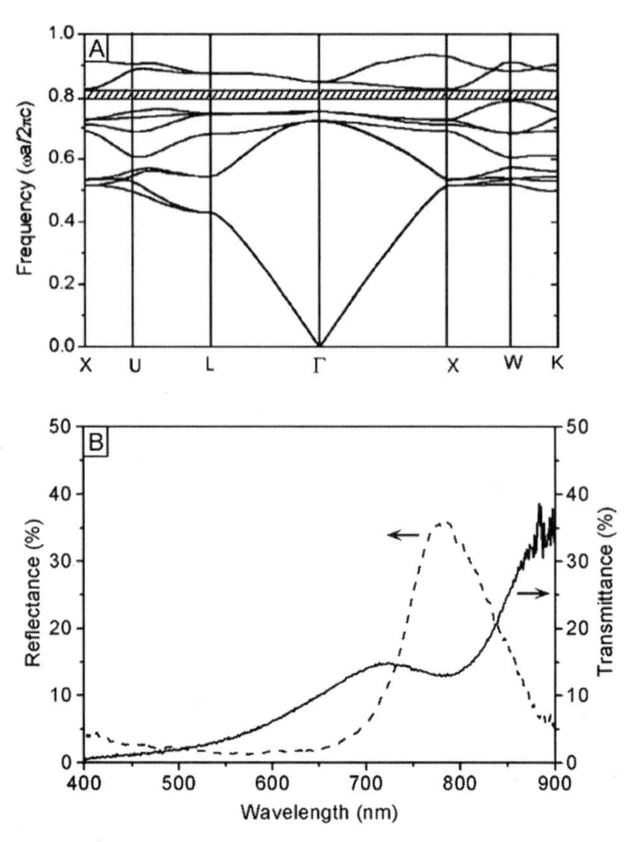

FIGURE 20.11. (A) The photonic band structure calculated for an inverse opal with a refractive index contrast of 3.0/1.0. Note that there exists a full band gap between the 8th and 9th bands when the contrast in refractive index is higher than 2.8. (From Ref. 70 by permission of WILEY-VCH Verlag GmbH & Co.) (B) The transmission and reflectance spectra recorded from an inverse opal that was fabricated by templating magnetite nanoparticles against an opaline lattice of polystyrene beads of 480 nm in diameter. (From Ref. 96 by permission of WILEY-VCH Verlag GmbH & Co.)

20.4.3. Crystalline Lattices of Nonspherical Colloidal Particles

The other approach that can potentially lead to the formation of a complete photoni band gap is to reduce the symmetry of an opaline lattice from the face-centered cubi to the diamond structure. In this case, a complete gap was shown to develop betwee the 2nd and 3rd bands in the photonic band structure, with a maximum gap to mid-ga ratio around 15.7%, corresponding to a filling ratio of 37% and a contrast in refractiv index of 3.60.[98] Unfortunately, it has been extremely difficult (in fact, never observed) obtain the diamond structure when spherical colloids are crystallized into 3D lattices. Late computational studies by several groups found that the symmetry-induced degeneracy the W- or U-point could also be lifted by employing nonspherical objects as the buildi blocks to construct a cubic lattice via self-assembly.[18] Their results indicated that a comple band gap could develop between the 2nd and 3rd bands in the photonic structure as t spherical lattice points are replaced with dimeric units consisting of two interconnect

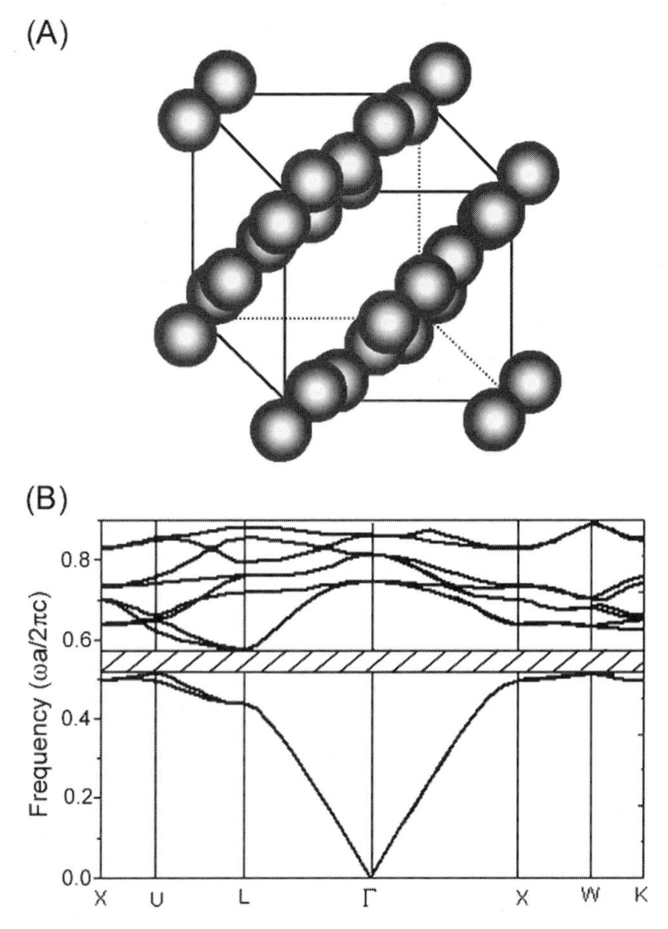

:IGURE 20.12. (A) A schematic illustration of the diamond-like lattice that could be produced via self-assembly)f dimers consisting two monodispersed spherical colloids. (B) The photonic band structure of this lattice, with :he filling fraction of the dimeric building blocks being 34%, and their longitudinal axes being directed towards he ⟨111⟩ direction. The contrast in refractive index is set to be 3.01/1.00. (From Ref. 70 by permission of VILEY-VCH Verlag GmbH & Co.)

iielectric spheres (Figure 20.12A). Such a band gap could have a gap to mid-gap ratio as nuch as 11.2%, when the filling ratio and the contrast in refractive index approached 30% nd 3.60/1.00, respectively. Figure 20.12B shows the photonic band structure calculated or a face-centered cubic lattice constructed from peanut-shaped colloids of ferric oxide, a ommon inorganic substance with a refractive index around 3.01.[70] As indicated by a shaded one on the photonic band structure, this lattice does exhibit a complete gap that can be xtended throughout the entire Brillouin zone. The computational studies also revealed that e minimum contrast in refractive index that is required for generating a complete bandgap such a 3D system is only ~2.40, a relatively low value that could be easily achieved using a rge number of inorganic materials (such as titania, II–VI semiconductors, and selenium) as ng as they can be prepared as nonspherical colloids with monodispersed sizes and shapes.

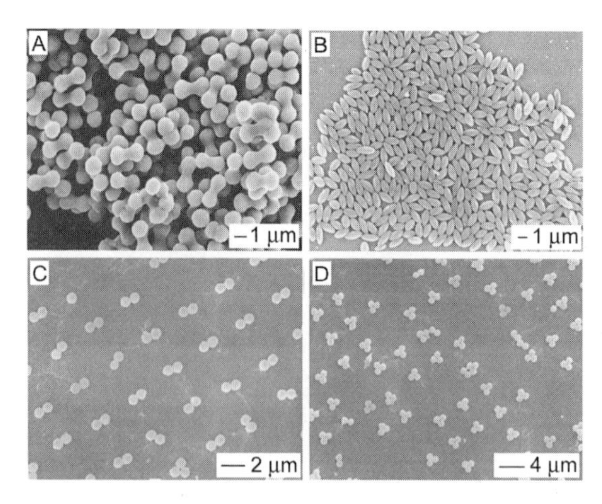

FIGURE 20.13. Scanning electron microscopy images of several typical examples of colloids with various non-spherical shapes: (A) peanut-shaped iron oxide colloids synthesized using a wet chemical method; (From Ref. 108 by permission of WILEY-VCH Verlag GmbH & Co.) (B) ellipsoidal polystyrene beads fabricated by stretching a polymer thin film containing spherical polymer beads; (From Ref. 107 by permission of American Chemical Society.) (C, D) dimeric and trimeric units assembled from spherical polymer beads through the confinement of physical templates. (From Ref. 13 by permission of American Physical Society.)

Preparation of colloidal particles with nonspherical shapes and as monodispersed samples remains to be a challenging subject in colloidal science. There are only a limited number of methods (direct and indirect) available for producing nonspherical colloids with well-controlled sizes and shapes. The direct methods generate nonspherical colloids from smaller subunits such as molecular species by controlling the nucleation and growth processes. These methods were largely developed with iron oxides as typical examples,[99–101] and have recently been extended to a number of other metal oxides and metals.[102] The indirect methods form nonspherical colloids through the use of spherical colloids as the precursors.[103–106] The success of these methods relies on the availability of copious quantities of polymer latexes and silica spheres as monodispersed samples. Figure 20.13 shows the scanning electron microscopy images of several typical examples, illustrating the concep and potential of each method.

Although the molecular world is replete with examples where building blocks as com plex as proteins or other kinds of macromolecules can be self-assembled into highly ordere crystalline lattices with very few defects, very little is known about the crystallization pro cess that involves nonspherical building blocks on the scale of 0.2 to 1 µm that is mos important to photonic crystals. Nevertheless, it should be possible to assemble nonspherica colloids (such as polystyrene spheroids or iron oxide peanuts) into 3D crystalline lattice exhibiting both positional and orientational order.[107] For example, one can derivatize th surfaces of the building blocks with positively (or negatively) charged functional group and then make use of the repulsive, electrostatic interactions between these charged objec to organize them into 3D lattices with positional ordering (that is, the ellipsoidal colloi will be separated by a distance comparable to their dimensions, with the particle at eac lattice point randomly orientated in the space). In the second step, directional ordering ca be achieved by applying an external (magnetic, electric, and/or hydrodynamic) field.[1

Once the assembly is completed, the dispersion medium can be slowly removed (while the external field is still on) to condense this crystalline lattice into a closely packed structure, or be cross-linked into a jelly network. The external field can be turned off afterwards, while the orientational order can be preserved in this lattice due to a lack of space for the relaxation of these building blocks.

20.5. PHOTONIC CRYSTALS WITH TUNABLE PROPERTIES

The wavelength of light diffracted from the surface of a 3D crystalline lattice of spherical colloids (or air balls) can also be expressed in terms of the Bragg equation:[64]

$$m \cdot \lambda = 2 \cdot n \cdot d_{\mathrm{hkl}} \cdot \sin\theta$$

where m is the order of the diffraction, λ is the wavelength of diffracted light (or the so-called stop band), n is the mean refractive index of the 3-D array, d_{hkl} is the interplanar spacing along the [hkl] direction, and θ is the angle between the incident light and the normal to the (hkl) planes. This equation indicates that the position of the stop band is directly proportional to the interplanar spacing and the mean refractive index. Any variation in these two parameters should lead to an observable shift in the diffraction peak, and thus be used to control and fine-tune the photonic properties of a photonic crystal. In addition, a crystalline lattice of spherical colloids may serve as the platform to fabricate optical sensors that will be able to measure and display environmental changes through color changes.

20.5.1. Tuning of Lattice Constants

There are a number of ways to tune the lattice constants of a crystalline lattice of spherical colloids. For example, a gel has been used as a transducer to change the lattice constants of a crystal through the application of an external stimulus. When spherical colloids are surrounded by a gel made of poly(*N*-isopropyl acrylamide), the reversible shrinkage of this polymer gel between 10–35°C changes the lattice spacing and subsequently results in a shift for the Bragg diffraction peak.[109] When a hydrogel responsive to Pb^{2+}, Ba^{2+}, and K^{+} ions was fabricated by incorporating crown-ether (4-acryloyl aminobenzo-18-crown-6) groups into the gel, the crown-ethers selectively trapped these cations, drawing in counterions and increasing the Donnan osmotic pressure of the hydrogel and swelling the matrix.[110] The magnitude of this reversible swelling depended on the number of charge groups covalently attached to the gel. A ~150 nm red-shift was observed when the concentration of Pb^{2+} was increased from ~20 ppb to ~2,000 ppm. An enzymatic based glucose sensor was also demonstrated with reversible swelling in response to the presence of 10^{-12} M (in the absence of O_2) to 0.5 mM glucose. In this case, the avidine-functionalized glucose oxidase was attached to a biotinylated composite of crystalline colloidal spheres embedded in a polyacrylamide gel. Conversion of glucose to gluconic acid by the enzyme swells the gel through the formation of anions, increasing the lattice constant of the crystalline framework.[111]

In related demonstrations, opals have been fabricated from spherical colloids made of hydrogels, allowing fine-tuning of the color of diffracted light by changing temperature or by applying an electric field.[112,113] Colloidal crystals embedded in appropriate polymer

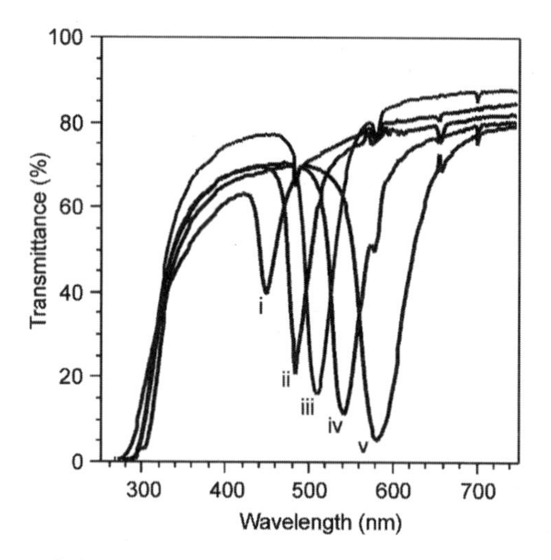

FIGURE 20.14. UV-Vis transmission spectra taken from a 3D opaline lattice (made of 175-nm polystyrene beads, and embedded in an elastomeric matrix) before (curve *i*) and after (curves *ii–v*) it had been swollen with silicone liquids of different molecular weights (and viscosities): *ii*) T12 (20 cSt), *iii*) T11 (10 cSt), *iv*) T05 (5 cSt), and *v*) T00 (0.65 cSt). (From Ref. 117 by permission of WILEY-VCH Verlag GmbH & Co.)

films have been shown to serve as mechanical sensors for measuring strains applied through uniaxial stretching or compressing.[114,115,116] Based on a similar color tuning mechanism, a photonic paper/ink system has been fabricated that allowed for color writing on and with colorless materials only.[117] In this case, the "paper" was a 3D crystal of polymer beads embedded in an elastomer matrix made of poly(dimethylsiloxane), and the "ink" was a liquid (e.g., a silicone fluid or an organic solvent such as octane) capable of swelling the elastomer matrix. As the elastomer matrix was swollen by the ink, the lattice constant (and thus the wavelength of diffracted light) was increased. For a crystal made of 175-nm polystyrene beads, the wavelength of light diffracted could be tuned to cover the spectral region from 450 to 580 nm by swelling it with silicone fluids of various molecular weights (Figure 20.14). Colored patterns could be formed on the surface of a thin film of colloidal crystal by writing with a Pilot pen, or by contact printing with an elastomer stamp.

20.5.2. Tuning of Refractive Indices

The photonic band gap is also sensitive to the change in mean refractive index for a 3D opaline lattice. Computational studies have shown that the band gap of an inverse opal could be readily tuned if the surface of this porous structure was coated with a few layers of an optically birefringent material such as a nematic liquid crystal.[118] In this case, the band gap could be either opened or closed by applying an external electric field that was able to rotate the liquid crystal molecules with respect to the normal to the surface of the inverse opal. Such dependence could also be employed to fabricate optical sensors. For example, a ceramic inverse opal has been used to detect organic solvents through the change in refractive index.[119] The reversible color tuning of a colloidal crystal could be potentially adopted for biosensing.[120,121] In another study, it was shown that tuning of the stop band could also be achieved through the application of an electrical potential to an opaline film infiltrated with redox-active polymer.

20.6. SUMMARY

This chapter provided a brief overview of a variety of methods that have been developed for fabricating 3D photonic crystals. We broadly divided these methods into two major categories, in which i) microlithography is used to generate the periodic structures in a top-down fashion; and ii) self-assembly is used to build the periodic lattices in a bottom-up manner. Each approach has its specific merits and inevitable weaknesses. For example, the layer-by-layer method based on conventional microfabrication is able to generate 3D photonic crystals (as woodpiles) having extremely uniform structures and feature sizes. This method, however, only works well for a limited set of materials that can be patterned using photolithography and reactive ion etching. The high cost associated with this method also makes it accessible only to a few research groups and it still remains a grand challenge to extend this method to features smaller than 500 nm, and to periodic lattices other than the woodpile structure. These problems could be partially solved by using the holographic method, but this alternative approach can only be directly applied to polymers whose refractive indices are often too low to generate complete photonic band gaps. In comparison, methods based on self-assembly of spherical colloids provide a simple and versatile route to 3D crystals with lattice constants controllable in the range of tens of nanometers to a few micrometers. The template-directed methods, in particular, have enabled the fabrication of 3D photonic crystals from essentially all classes of materials, including those that exhibit complete band gaps in the optical regime. However, removal of the template through a post-fabrication process may cause damage to the periodic lattice. In general, the types of periodic structures that could be generated using the self-assembly methods are limited to the face-centered (or body-centered) cubic lattices which are not necessarily the best candidates for creating full band gaps. The opaline and inverse opaline structures may also contain poorly defined defects because the self-assembly process is seldom performed under an equilibrium condition.

This chapter also provided a brief discussion on some interesting properties associated with 3D photonic crystals, in the context of various intriguing applications. It is obvious that the greatest (but not necessarily the most realizable) opportunity for 3D photonic crystals is in the area of optoelectronics, where a tight control and good manipulation of photons means higher efficiencies and more compact designs for light emitting and guiding devices. It is also worth pointing out that there are a range of other applications for 3D photonic crystals in areas outside optoelectronics. Examples may include, for instance, diffractive optical sensors, tunable filters, switches, and display components. Recent demonstrations of photonic papers and inks represent two good examples of such applications, and they have the potential to become the first commercial products based on photonic crystals. These applications, together with a fundamental interest in nanoscale photonics will continuously provide compelling motivation for research into techniques for generating 3D crystals with better controlled structures.

Photonic crystals and most of the applications derived from these materials are still in an early stage of technical development. There are a number of issues that remain to be addressed before these materials can reach their potential in core industrial applications. First of all, there still exists the need to develop a method capable of generating 3D periodic lattices with tightly controlled structures rapidly and at low cost. Preliminary demonstrations from many research groups suggest that the ultimate use of photonic crystals strongly depends on the ability to precisely control and fine tune their crystallographic structures, lattice constants, and densities/types of defects. Judged against these metrics, all the methods described in this chapter need to be greatly improved before they will find widespread use in commercial applications. As the utilization of photonic crystals grows in importance in a broad range of areas, the demand for new (and more effective) fabrication methods will certainly increase. The second challenge is to design and fabricate a photonic crystal whose band gap can be selectively closed or opened by applying an external stimulus. Only such a capability will allow for the confinement or releasing of photons on demand. The third challenge is to build waveguiding structures inside photonic crystals and to integrate them with other photonic components such as optical sources and detectors. Only such integration will enable the fabrication of truly compact and highly efficient photonic devices. In this regard, it might be the most important priority associated with this research area to achieve a tight control over the types and densities of defects within a photonic crystal. Last but not least, it will be of paramount importance to demonstrate radically new applications for 3D photonic crystals in an effort to greatly expand the scope of areas that these materials will impact. As we have seen from many other kinds of functional materials (e.g., quantum dots and conducting polymers), the first commercial product is not necessarily related to the one that was envisioned when these materials were invented. This might also be the case for photonic crystals, and only future developments will tell.

QUESTIONS

1. What is a photonic crystal? Briefly describe one approach that has been demonstrated for fabricating a 1D, 2D, or 3D photonic crystal.
2. What are the major advantages associated with photonic crystals when compared to conventional optical components such as mirrors and waveguides?
3. What are some of the requirements one has to fulfill in order to obtain complete band gaps in a 3D sysstem?

REFERENCES

1. J. D. Joannoupoulos, R. D. Meade, and J. N. Winn, *Photonic Crystals* (Princeton University Press, USA, 1995).
2. E. Yablonovitch, *Phys. Rev. Lett.* **58**, 2059–2062 (1987).
3. S. John, *Phys. Rev. Lett.* **58**, 2486–2489 (1987).
4. C. M. Soukoulis, *Photonic Band Gap Materials* (Kluwer Academic Publishers, USA, 1996).
5. S. John, *Physics Today,* May, 32–40 (1991).
6. E. Yablonovitch, *J. Opt. Soc. Am.* B **10**, 283–295 (1993).
7. J. D. Joannopoulos, P. R. Villeneuve, and S. Fan, *Nature* **386**, 143–149 (1997).
8. O. Krauss, *DWDM and Optical Network* (Publicis Corporate Publishing: Erlangen, USA, 2002).
9. B. G. Levi, *Phys. Today,* January, 17-19-1516 (1999).

10. E. Yablonovitch, *Sci. Am.* December, 47–55 (2001).
11. J. D. Rancourt, *Optical Thin Films: User Handbook* (SPIE Optical Engineering Press: Bellingham, WA, USA, 1996).
12. K. M. Ho, C. T. Chan, and C. M. Soukoulis, *Phys. Rev. Lett.* **65**, 3152–3155 (1990).
13. Z.-Y. Li and Y. Xia, *Phys. Rev.* A **63**, 043817 (2001).
14. A. Moroz and C. Sommers, *J. Phys.: Condens. Matter* **11**, 997–1008 (1999).
15. J. W. Haus, *J. Mod. Opt.* **41**, 195–207 (1994).
16. K. Busch and S. John, *Phys. Rev.* **58**, 3896–3908 (1998).
17. Z.-Y. Li, J. Wang, and B.-Y. Gu, *Phys. Rev.* B **58**, 3721–3729 (1998).
18. Z.-Y. Li, J. Wang, and B.-Y. Gu, *J. Phys. Soc. Jpn.* **67**, 3288–3291 (1998).
19. T. D. Happ, M. Kamp, F. Klopf, J. P. Reithmaier, and A. Forchel, *Semicond. Sci. Tech.* **16**, 227–232 (2001).
20. B. A. Usievich, A. M. Prokhorov, and V. A. Sychugov, *Laser Phys.* **12**, 898–902 (2002).
21. M. Fujita and T. Baba, *Appl. Phys. Lett.* **80**, 2051–2053 (2002).
22. A. M. Vengsarkar, *Laser Focus World* June, 243–248 (1996).
23. K. M. Ho, C. T. Chan, C. M. Soukoulis, R. Biswas, and M. Sigalas, *Solid State Commun.* **89**, 413–416 (1994).
24. H. S. Sözüer and J. P. Dowling, *J. Modern Opt.* **41**, 231–239 (1994).
25. E. Özbay, E. Michel, G. Tuttle, R. Biswas, M. Sigalas, and K. M. Ho, *Appl. Phys. Lett.* **64**, 2059–2061 (1994).
26. S. Noda, N. Yamamoto, and A. Sasaki, *Jpn. J. Appl. Phys.* **35**, 909–912 (1996).
27. S. Noda, K. Yomoda, N. Yamamoto, and A. Chutinan, *Science* **289**, 604–606 (2000).
28. S.-Y. Lin, J. G. Fleming, D. L. Hetherington, B. K. Smith, R. Biswas, K. M. Ho, M. M. Sigalas, W. Zubrzycki, S. R. Kurtz, and J. Bur, *Nature* **394**, 251–253 (1998).
29. S.-Y. Lin, J. G. Fleming, M. M. Sigalas, R. Biswas, and K. M. Ho, *Phys. Rev.* B **59**, 579–582 (1999).
30. J. G. Fleming, S.-Y. Lin, I. El-Kady, R. Biswas, and K. M. Ho, *Nature* **417**, 52–55 (2002).
31. E. Yablonovitch, T. J. Gmitter, and K. M. Leung, *Phys. Rev. Lett.* **67**, 2295–2298 (1991).
32. C. C. Cheng, V. ArbetEngels, A. Scherer, and E. Yablonovitch, *Phys. Scripta* **T68**, 17–20 (1996).
33. A. Chelnokov, K. Wang, S. Rowson, P. Garoche, and J. M. Lourtioz, *Appl. Phys. Lett.* **77**, 2943–2945 (2000).
34. C. Cuisin, A. Chelnokov, J. M. Lourtioz, D. Decanini, and Y. Chen, *Appl. Phys. Lett.* **77**, 770–772 (2000).
35. J. Nole, Laser Focus World, May, 209 (1997).
36. M. Campbell, D. N. Sharp, M. Harrison, R. G. Denning, and A. J. Turberfield, *Nature* **404**, 53 (2000).
37. S. Yang, M. Megans, J. Aizenberg, P. Wiltzius, P. M. Chaikin, and W. B. Russel, *Chem. Mater.* **14**, 2831 (2002).
38. O. Toader and S. John, *Science* **292**, 1133–1135 (2001).
39. K. Robbie, and M. J. Brett, *J. Vac. Sci. Ttechnol.* **B15**, 1460–1465 (1997).
40. B. H. Cumpston, S. P. Ananthavel, S. Barlow, D. L. Dyer, J. E. Ehrlich, L. L. Erskine, A. A. Heikal, S. M. Kuebler, I.-Y. S. Lee, D. McCord-Maughon, J. Qin, H. Röckel, M. Rumi, X.-L. Wu, S. R. Marder, and J. W. Perry, *Nature* **398**, 51–54 (1999).
41. W. Lee, S. A. Pruzinsky, and P. V. Braun, *Adv. Mater.* **14**, 271–274 (2002).
42. P. Ball, *The Self-Made Tapestry: Pattern Formation in Nature* (Oxford University Press, New York, 1999).
43. A. Klug, *Angew. Chem. Int. Ed. Engl.* **22**, 565–582 (1983).
44. L. Isaacs, D. N. Chin, N. Bowen, Y. Xia, and G. M. Whitesides, in *Supermolecular Technology*, edited by D. N. Reinhoudt (John Wiley and Sons, New York, 1999), pp. 1–46.
45. Y. Fink, J. N. Winn, S. Fan, C. Chen, J. Michl, J. D. Joannopoulos, and E. L. Thomas, *Science* **282**, 1679–1682 (1998).
46. Y. Xia, B. Gates, Y. Yin, and Y. Lu, *Adv. Mater.* **12**, 693–713 (2000).
47. D. V. Orlin and M. L. Abraham, *Current Opin. Colloid Interf. Sci.* **5**, 56–63 (2000).
48. W. B. Russel, *The Dynamics of Colloidal Systems* (The University of Wisconsin Press, Madison, Wisconsin, 1987).
49. E. Matijevic, *Chem. Mater.* **5**, 412–426 (1993).
50. *Emulsion Polymerization,* edited by I. Piirma (Academic Press, New York, 1982).
51. R. K. Iler, *The Chemistry of Silica* (Wiley-Interscience, New York, 1979).
52. C. A. Murray and D. G. Grier, *Am. Sci.* **83**, 238–245 (1995).
53. Photonic Crystals, a special issue in *Adv. Mater.* **13**, 361–452 (2001).
54. P. Pieranski, *Contemp. Phys.* **24**, 25–73 (1983).
55. From Dynamics to Devices: Directed Self-Assembly of Colloidal Materials, edited by D. G. Grier, a special issue in *MRS Bull.* **23**, 21 (1998).

56. A. D. Dinsmore, J. C. Crocker, and A. G. Yodel, *Curr. Opin. Colloid Interface* **3**, 5–11 (1998).
57. H. Míguez, C. López, F. Meseguer, A. Blanco, L. Vázquez, R. Mayoral, M. Ocaña, V. Fornés, and A. Mifsud, *Appl. Phys. Lett.* **71**, 1148–1150 (1997).
58. R. J. Carlson and S. A. Asher, *Appl. Spectrosc.* **38**, 297–304 (1984).
59. P. Jiang, J. F. Bertone, K. S. Hwang, and V. L. Colvin, *Chem. Mater.* **11**, 2132–2140 (1999).
60. S. H. Park, D. Qin, and Y. Xia, *Adv. Mater.* **10**, 1028–1032 (1998).
61. S. H. Park and Y. Xia, *Langmuir* **15**, 266–273 (1999).
62. B. Gates, D. Qin, and Y. Xia, *Adv. Mater.* **11**, 466–466 (1999).
63. S. H. Park, B. Gates, and Y. Xia, *Adv. Mater.* **11**, 462–469 (1999).
64. Y. Xia, B. Gates, and S. H. Park, *IEEE J. Lightwave Technol.* **17**, 1956–1962 (1999).
65. Y. Lu, Y. Yin, B. Gates, and Y. Xia, *Langmuir* **17**, 6344–6350 (2001).
66. Y. Yin, Z.-Y. Li, and Y. Xia, *Langmuir* **19**, 622–631 (2003).
67. A. van Blaaderen, R. Ruel, and P. Wiltzius, *Nature* **385**, 321–324 (1997).
68. K. H. Lin, J. C. Crocker, V. Prasad, A. Schofield, D. A. Weitz, T. C. Lubensky, and A. G. Yodh, *Phys. Rev. Lett.* **85**, 1770–1773 (2000).
69. S. M. Yang and G. A. Ozin, *Chem. Commun.* **2000**, 2507–2508.
70. Y. Xia, B. Gates, and Z.-Y. Li, *Adv. Mater.* **13**, 409–413 (2001).
71. R. Biswas, M. M. Sigalas, G. Subramania, C. M. Soukoulis, and K. M. Ho, *Phys. Rev.* **61**, 4549–4553 (2000).
72. S. A. Johnson, P. J. Ollivier, and T. E. Mallouk, *Science* **283**, 963–965 (1999).
73. J. V. Sanders, *Nature* **204**, 1151–1152 (1964).
74. O. D. Velev, T. A. Jede, R. F. Lobo, and A. M. Lenhoff, *Nature* **389**, 447–448 (1997).
75. B. T. Holland, C. F. Blanford, and A. Stein, *Science* **281,** 538–540 (1998).
76. P. Yang, T. Deng, D. Zhao, P. Feng, D. Pine, B. F. Chmelka, G. M. Whitesides, and G. D. Sturcky, *Science* **282**, 2244–2246 (1998).
77. K. M. Kulinowski, P. Jiang, H. Vaswani, and V. L. Colvin, *Adv. Mater.* **12**, 833–838 (2000).
78. A. Imhf and D. J. Pine, *Nature* **389**, 948–951 (1997).
79. P. V. Braun, R. W. Zehner, C. A. White, M. K. Weldon, C. Kloc, S. S. Patel, and P. Wiltzius, *Adv. Mater.* **13**, 721–724 (2001).
80. Y. Chen, W. T. Ford, N. F. Materer, and D. Teeters, *J. Am. Chem. Soc.* **122**, 10472–10473 (2000).
81. Y. Xia, J. J. McClelland, R. Gupta, D. Qin, X.-M. Zhao, L. S. Sohn, R. J. Celotta, and G. M. Whitesides, *Adv. Mater.* **9**, 147–150 (1997).
82. E. Kim, Y. Xia, and G. M. Whitesides, *Nature* **376**, 581–584 (1995).
83. L. L. Hench and J. K. West, *Chem. Rev.* **90**, 33–72 (1990).
84. Z. Zhong, Y. Yin, B. Gates, and Y. Xia, *Adv. Mater.* **12**, 206–209 (2000).
85. O. V. Velev, P. M. Tessier, A. M. Lenhoff, and E. W. Kaler, *Nature* **401**, 548–548 (1999).
86. P. M. Tessier, O. D. Velev, A. T. Kalambur, J. F. Rabolt, A. M. Lenhoff, and E. W. Kaler, *J. Am. Chem. Soc.* **122**, 9554–9555 (2000).
87. Y. A.Vlasov, N. Yao, and D. J. Norris, *Adv. Mater.* **11**, 165–169 (1999).
88. P. Jiang, J. Cizeron, J. F. Bertone, and V. L. Colvin, *J. Am. Chem. Soc.* **121**, 7957–7958 (1999).
89. P. V. Braun and P. Wiltzius, *Nature* **402**, 603–604 (1999).
90. P. V. Braun and P. Wiltzius, *Adv. Mater.* **13**, 482–485 (2001).
91. V. N. Astratov, Y. A. Vlasov, O. Z. Karimov, A. A. Kaplyanskii, Y. G. Musikhin, N. A. Bert, V. N. Bogomolov, and A. V. Prokofiev, *Phys. Lett.* A **222**, 349–353 (1996).
92. A. Blanco, E. Chomski, S. Grabtchak, M. Ibisate, S. John, S. W. Leonard, C. Lopez, F. Meseguer, H. Miguez, J. P. Mondia, G. A. Ozin, O. Toader, and H. M. van Driel, *Nature* **405**, 437–440 (2000).
93. Y. A. Vlasov, X.-Z. Bo, J. C. Sturm, and D. J. Norris, *Nature* **414**, 289–293 (2001).
94. H. Miguez, E. Chomski, F. Garcia-Santamaria, M. Ibisate, S. John, C. Lopez, F. Meseguer, J. P. Mondia, G. A. Ozin, O. Toader, and H. M. Van Driel, *Adv. Mater.* **13**, 1634–1637 (2001).
95. F. Meseguer, A. Blanco, H. Miguez, F. Garcia-Santamaria, M. Ibisate, and C. Lopez, *Colloids Surf.* A **202**, 281–290 (2002).
96. B. Gates and Y. Xia, *Adv. Mater.* **13**, 1605–1608 (2001).
97. J. E. G. J. Wijnhoven and W. L. Vos, *Science* **281**, 802–804 (1998).
98. F. Garcia-Santamaria, C. Lopez, F. Meseguer, F. Lopez-Tejeira, J. Sanchez-Dehesa, and H. T. Miyazaki, *Appl. Phys. Lett.* **79**, 2309–2311 (2001).
99. N. Sasaki, Y. Murakami, D. Shindo, and T. Sugimoto, *J. Colloid Interface Sci.* **213**, 121–125 (1999).

100. T. Sugimoto, M. M. Khan, A. Muramatsu, and H. Itoh, *Colloids Surf.* A **79**, 233–247 (1993).
101. E. Snoeks, A. van Blaaderen, T. van Dillen, C. M. van Kats, M. L. Brongersma, and A. Polman, *Adv. Mater.* **12**, 1511–1514 (2000).
102. Y.-G. Sun and Y. Xia, *Science* **298**, 2176–2179 (2002).
103. Y. Yin and Y. Xia, *Adv. Mater.* **13**, 267–271 (2001).
104. Y. Yin, Y. Lu, B. Gates, and Y. Xia, *J. Am. Chem. Soc.* **123**, 8718–8729 (2001).
105. Y. Yin, Y. Lu, and Y. Xia, *J. Am. Chem. Soc.* **123**, 771–772 (2000).
106. Y. Yin, Y. Lu, and Y. Xia, *J. Mater. Chem.* **11**, 987–989 (2001).
107. Y. Lu, Y. Yin, Z.-Y. Li, and Y. Xia, *Langmuir* **18**, 7722–7727 (2002).
108. Y. Lu, Y. Yin, and Y. Xia, *Adv. Mater.* **13**, 415–420 (2001).
109. M. Weissman, H. B. Sunkara, A. S. Tse, and S. A. Asher, *Science* **274**, 959–960 (1996).
110. J. H. Holtz and S. A. Asher, *Nature* **389**, 829–832 (1997).
111. J. H. Holtz, J. S. W. Holtz, C. H. Munro, and S. A. Asher, *Anal. Chem.* **70**, 780–791 (1998).
112. Z. B. Hu, X. H. Lu, and J. Gao, *Adv. Mater.* **13**, 1708–1712 (2001).
113. J. D. Debord, S. Eustis, S. B. Debord, M. T. Lofye, and L. A. Lyon, *Adv. Mater.* **14**, 658–662 (2002).
114. J. M. Jethmalani, W. T. Ford, and G. Beaucage, *Langmuir* **13**, 3338–3344 (1997).
115. S. H. Foulger, P. Jiang, A. C. Lattam, D. W. Smith, and J. Ballato, *Langmuir* **17**, 6023–6026 (2001).
116. K. Sumioka, H. Kayashima, and T. Tsutsumi, *Adv. Mater.* **14**, 1284–1286 (2002).
117. H. Fudouzi and Y. Xia, *Adv. Mater.* **11**, 892 (2003).
118. E. Yablonovitch, *Nature* **401**, 539–541 (1999).
119. C. F. Blanford, R. C. Schroden, M. Al-Daous, and A. Stein, *Adv. Mater.* **13**, 26–29 (2001).
120. W. Qian, Z.-Z. Gu, A. Fujishima, and O. Sato, *Langmuir* **18**, 4526–4529 (2002).
121. T. Cassagneau, and F. Caruso, *Adv. Mater.* **14**, 1629–1633 (2002).

VII

Nanobiotechnology

21

Biomimetic Nanostructures

Dennis E. Discher
Department of Chemical and Biomolecular Engineering
University of Pennsylvania, Philadelphia, PA

21.1. INTRODUCTION: WATER, CELL INSPIRATIONS, AND COPOLYMERS

H_2O is the predominant compound of most biological systems—typically ~70% by mass, and so it is not surprising that a major driving force for molecular assembly in biology exploits water. The hydrophobic effect that makes oil coalesce in water drives proteins to fold into functional forms and also bind one another. Such binding generates complex machines such as protein-synthesizing ribosomes, reinforcing scaffolds like filamentous cytoskeletons, and many other structures. The membranes that define and delimit both cells and cellular sub-compartments like the nucleus and mitochondria also form largely due to the hydrophobic effect. All of these many structures form because the constituent molecules, lipids, are generally polymers (if small) composed of both oily, hydrophobic monomers and water-soluble, polar monomers. For example, of the nineteen natural amino acids that make up polypeptides or proteins, nine are hydrophobic and thus more soluble in carbon-based oils and solvents than in H_2O. Likewise, the lipids that make up cell membranes have a dual hydrophobic-polar (or *H-P*) nature, but this is generally arranged in a more segmented or block fashion (e.g., *HHH–PP*) than found for the amino acids in proteins. The sequential or block arrangement is generally referred to as amphiphilic. What emerges from such simplistic overviews of hydrophobicity in molecular biology are naïve templates for creating biomimetic nanostructures through aqueous self-assembly.

Whether one is conjuring synthetic mimics of cell components or theorizing about how cells work, a host of physicochemical properties ranging from energetics to stability and fluidity all have potentially important roles. Depending on one's viewpoint, such properties are largely reflected in the evolved chemistry of the system or else they dictate it. The crawling of a cell on a substrate examplifies some of these issues. Figure 21.1 shows the roughened outline of a cell which is—as always—delimited by a confining membrane that displaces

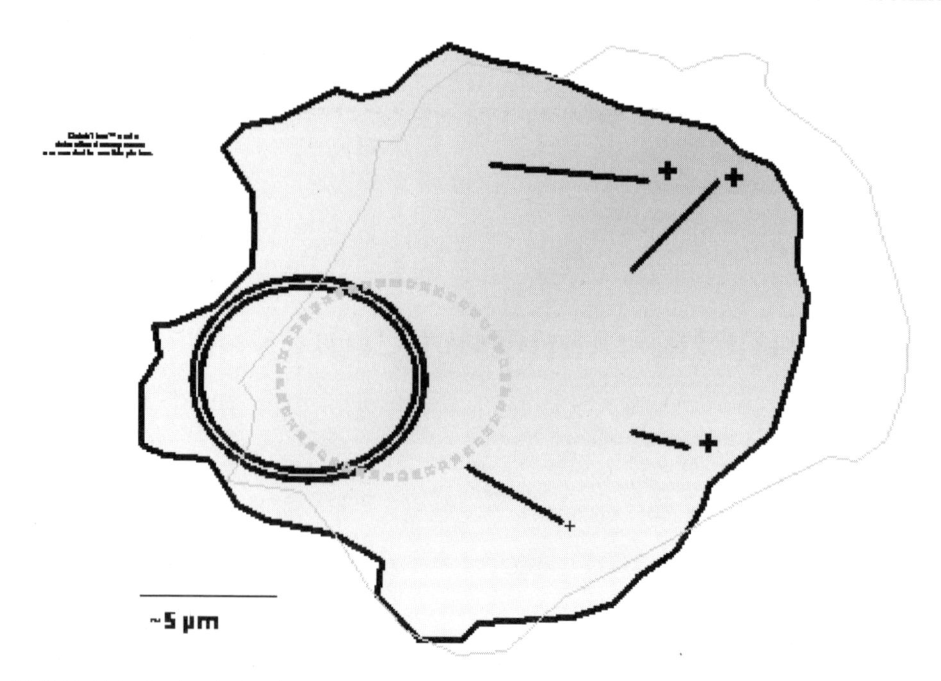

FIGURE 21.1. One inspiration for synthetic polymer membranes and rods: cell crawling. Outlines of a cell traced from images of an actin-labeled cell crawling on an adhesive surface (inset). The cell's plasma membrane and inner nuclear (double) membrane are both displaced from left to right over a period of ~1 min. Forward movement occurs primarily by polymerization of stiff actin filaments which are selectively shown as black rods or as dense gray regions. Polymerization is faster at the biased, (+)-ends of filaments. Anchoring of the opposite ends of the filaments extends through the membrane in cell adhesion to the substrate. Because the filaments are stiff, filament polymerization drives the membrane forward. Because the cell membrane is robust, it can be torn away from the substrate at the rear of the cell. (From Ref. 14 by permission of Elsevier Science.)

during crawling. The filamentous assemblies of proteins make up the cytoskeleton and are non-covalent assemblies that undergo constant polymerization and de-polymerization. These filaments clearly have central roles in driving the cell forward. The prevailing hypothesis explaining such crawling suggests that a preferential polymerization at the membrane proximal end of a filament locally rectifies the nano-scale Brownian fluctuations of the soft cell membrane.[1] Except for such living polymerization processes, filamentous and membrane cell structures with tunable cell-like hydrodynamic responses are now within synthetic reach, as illustrated by the worm micelle and polymer vesicle structures described here.

Since Bangham first isolated lipid and reconstituted it into vesicles or liposomes in the 1960's,[2] tremendous progress has been made in terms of both integrating biomolecules into the membranes[3] or encapsulating biomolecules that function within vesicles.[4] Both cited references are only representative, but the latter is particularly notable as it demonstrates entrapment of the polymerizable cytoskeleton protein tubulin in lipid vesicles, and vividly shows that work on the membrane can be exerted by relatively stiff microtubules. Like cell crawling, which the system models in primitive form, the study again illustrates a synergistic complementarity between filaments and membranes. These systems are not particularly robust, however. Purified proteins often unfold and agglomerate over time, and many lipid vesicles become leaky. While cells have metabolic pathways of remedying such problems and effectively healing themselves, fully synthetic polymer systems would offer advantages of robustness as well as simplicity. Beyond a fascination for structural mimicry, such goals are also motivated by the prospects of engineering synthetic cells that move in response to stimuli. The results here with polymer membranes and worms begin to raise questions of physical principles requisite for a soft, synthetic 'living' system. Indeed, the results serve to highlight questions about the minimal physical basis for mimicking cell functions such as crawling.

Segmented *A-B* polymers, or block copolymers,[5] have been synthesized since the 1970's with one motivation being to use them to molecularly bond a pure sample of *A* polymer to an otherwise distinct and immiscible sample of pure *B* polymer. Typical applications pursued thus mostly involve adhesives and solubilizing agents. Even in bulk, however, pure block copolymers are known to self-organize into periodic microphases such as arrays of rods and stacks of lamellar sheets. Liquid crystals share some of the same phase behavior. Whether microphases form with block copolymers depends principally on both the *A*-to-*B* ratio of segment lengths, and the energy differences between *A* interacting with *A* versus *A* interacting with *B*. On a relative scale of energy per mass, hydrophobic-polar (*H–P*) interactions that drive structure formation in water are strong, though non-covalent. They are certainly strong enough to drive structure formation even with small molecular weight molecules of a few hundred Daltons or less—as often exploited in biology. Returning to effects of molecular weight later, the history of block copolymers—where molecular weights are thousands of Daltons or more—has been such that research on solvent-diluted copolymer systems began in earnest only in the 1990's. Among the important discoveries was the formation of cell-mimetic sacs or vesicles in aqueous solution using various amphiphilic block copolymers.[6] Rod-like aggregates referred to as worm micelles have also been described and resemble some of the linear protein assemblies found both inside cells (e.g., cytoskeleton filaments) and outside cells (e.g., collagen fibers). This chapter will cover these two biomimetic morphologies (Figure 21.2), their underlying principles, and the applications that might be developed around them.

21.2. WORM MICELLES AND VESICLES FROM BLOCK COPOLYMERS

For a simple amphiphile in aqueous solution, a time-average molecular shape (Figure 21.1) in the respective form of a cylinder, wedge, or cone dictates whether a membrane, rod-like, or spherical morphology will form.[7] This average molecular shape is most simply a reflection of the average polar fraction f, although solvent and other variables can also have an influence. In other words, block copolymer amphiphiles that assemble into worm micelles and polymer membranes are part of a morphological landscape in f. Figure 21.3 illustrates the basic architecture of a two segment, *H–P* diblock copolymer where one block is oxygen-rich, polar PEO (polyethyleneoxide, which is equivalent to PEG, polyethyleneglycol) and the second is a hydrocarbon-based hydrophobic block. It should be pointed out that PEO is non-ionic and is a widely favored polymer for surface or interface treaments of intended biomaterials as it tends to make them more biocompatible. As discussed later in section 8, however, PEO does not make materials bio-inert and truly compatible; it only delays biological reactions that typify the so-called foreign body response of organisms to synthetic materials as well as micro-organisms.[8]

By making the fraction of the PEO block $f \approx 25–42\%$, bulk copolymer added to water will generally lead to the spontaneous formation of vesicles or polymersomes. As currently

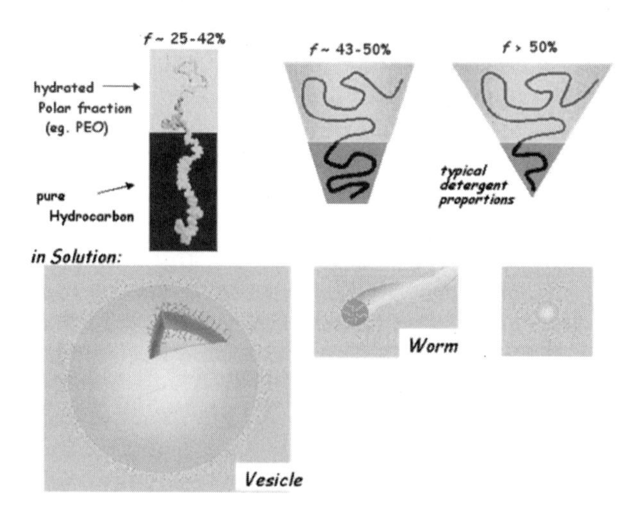

FIGURE 21.2. Schematics of molecular shapes that arise with hydration of the polar chains of amphiphiles. Such shapes underlie the formation of various morphological phases: vesicles, worms, and spheres. Vesicles predominate when the polar fraction, f, of most simple amphiphiles is in the indicated range. Polyethyleneoxide (PEO) is a typical, non-ionized polar chain with a high oxygen content (red atoms) that facilitates hydration. The hydrocarbon segments of the amphiphiles interact with each other, driving aggregation and excluding water to create a hydrophobic core. A vesicle is sketched with a small section removed to reveal the core's nano-scale thickness d. Rod-like worm micelles form when the polar segments are made slightly longer, while spheres form for even longer polar segments. (From Ref. 14 by permission of Elsevier Science.)

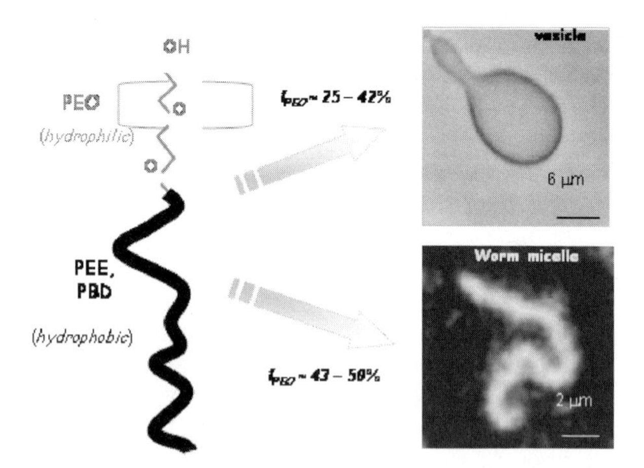

FIGURE 21.3. Synthetic polymer membranes and worms made by self-assembly of PEO-based diblocks in water. The weight or volume fraction of the hydrophilic PEO (polyethylene oxide) block is specified by f. The hydrophobic, hydrocarbon block of the copolymers thus far studied consists of either PEE (polyethylethylene) or its crosslinkable analog PBD (polybutadiene). Note that an increased f by just a few percent leads to worm micelles instead of vesicles. Cryo-TEM has already shown that the worm micelles, made of block copolymers of molecular weight $MW \sim 4$ kDa, have hydrophobic cores of $\sim$10 nm.

understood, hydration of the PEO (e.g., hydrogen bonding of water's hydrogens with PEO's oxygens) draws in enough water to generate sufficient osmotic force when aggregated with other diblocks to balance the hydrophobic volume of the remainder of the chain. However, with a relatively small increase in the PEO fraction to $f \approx 42$–50%, hydration of the PEO overcompensates, expands like a swollen sponge, and induces a curvature in the aggregate, generating rod-like worm micelles. Aggregate formation is strongly driven by the relatively high molecular weight of the hydrocarbon segment in either case. This creates an interfacial tension, γ, which separates the core from PEO as well as the bulk aqueous phase. Consistent with the strong segregation mechanism of aggregate formation, it is already clear from micromanipulation measurements on individual vesicles that the interfacial tension, γ, is independent of molecular weight, *MW*, as well as relatively small changes in hydrocarbon chemistry.[9]

In phase contrast microscopy, giant polymersomes generally appear as diverse in shape as lipid vesicles.[10] Starfish, tube, pear (Figure 21.3A), and string-of-pearl shapes[11] are all prominent and in part reflect a relatively well understood vesicle-scale imbalance between the number of molecules in one leaflet of a membrane bilayer versus the number in the other leaflet. In other words, the membranes of these polymer vesicles appear to be bilayers, although the exact partitioning of amphiphiles between leaflets and the location of the midplane is not yet well controlled or understood. Nonetheless, images from cryo-transmission electron microscopy (cryo-TEM) also prove consistent with a bilayer in that they show a membrane thickness, d, which not only tends to be significantly greater than the 3–4 nm thickness of lipid bilayers but is also too thick for a single, convoluted copolymer to fully span.[9]

Although the average shapes of polymersomes appear similar to liposomes, it should be made clear that the temperature range over which amphiphilic structures can be expected to be stable (i.e., 0°C–100°C, or 273 K–373 K) is generally high enough to excite thermal bending modes or flexural Brownian motion. While these membrane undulations have been clearly documented in lipid vesicles[12] and simple blood cells,[13] they are minimal with the polymer vesicle membranes. As described further below, the membrane bending resistance is expected to increase with polymer *MW* through membrane thickness, d. Additional properties, especially vesicle stability and in-plane hydrodynamic properties will be shown below to depend strongly and non-linearly on copolymer *MW*.

Worm micelles likewise assembled by hydration and dilution of these non-ionic copolymers (again $MW \sim 4$ kg/mol) have also been visualized: fluorescent labeling of the core with a hydrophobic fluorophore and confinement to $\sim$1 μm high chambers shows worms up to 30 μm long. The worms appear highly flexible and dynamically contorted (see snapshot in Figure 21.3B). A diameter of about 10 nm for these worms is far less than optical resolution given by the fluorescence wavelengths of $\lambda \sim 500$ nm. However, the contour is almost always well visualized because the worms are stiff on length scales smaller than their characteristic persistence length, p, and here $p \sim \lambda$.[14] This is unlike DNA[15] which is stiff on length scales of $\sim$50 nm and smaller, that is: $p_{\mathrm{DNA}} \ll \lambda$. DNA is also much more complex in terms of its large negative charge, its helical structure, and many other features.

Despite having a similar diameter as the polymersome membrane thickness, d, the Brownian dynamics of worm micelles appear highly pronounced in contrast to the membranes. Autocorrelation analyses of the end-to-end distance for the microns-long worms yields relaxation times of seconds. These are clearly fluid assemblies, as more fully elaborated below for the polymersomes, but the stability of the worms is equally clear and appears fully consistent with the high γ that drives membrane formation and underlies the stability of polymersomes. Non-fluid worm and membrane assemblies are discussed later.

21.3. SOLVENT, SIZE, ENERGETICS, AND FLUIDITY

In either aqueous solutions or organic (i.e., oily or carbon) solvents, aggregate morphologies as well as aggregate fluidity and stability are all governed by chain *MW*, interfacial tensions, and selective interactions with one preferred fraction of the amphiphile.[16] Lipids segregate and form vesicles in many aqueous solutions, for example, but they fail to do so in solvents that are blind to their amphiphilicity (e.g., chloroform: $CHCl_3$). Solubility also depends in general on chain *MW*, suggesting that novel worms and vesicles can also be made by aggregation of weakly hydrophobic polymers.

Lipids and small amphiphiles can differ considerably in their polar 'head' group, but they almost always contain one or two strongly hydrophobic chains composed of multiple ethylene $(-CH_2-CH_2-)_n$ units; for most common lipids, $n = 5$–10, but there are some naturally occuring chains found at low concentrations in the membranes of most species that are considerably longer.[17] A simple measure of vesicle stability is provided by the critical micelle concentration $C_{CMC} \sim \exp(-n\varepsilon_h/k_BT)$. In this, k_BT is the thermal energy, and ε_h is the monomer's effective interaction energy with the bulk solution. Only at surfactant concentrations above the C_{CMC} do aggregates such as vesicles form. For ethylene groups at biological $T_{biol}(\sim 300\text{ K})$, $\varepsilon_h \approx 1 - 2k_BT_{biol}(\sim 4\text{–}8\text{ pN nm})$ so that the C_{CMC} for lipids and related amphiphiles in aqueous solutions ranges from micromolar to picomolar, i.e., aggregates are stable in high dilution. Amphiphile exchange rates between aggregates are also generally proportional to C_{CMC}, with characteristic exchange times for phospholipids estimated in hours. Polymer amphiphiles with hydrophobic MW_h an order of magnitude or more larger than those typical of lipids result in kinetic quenching and metastable aggregates. Such effects have motivated some to investigate co-solvents which fluidize and activate the polymer systems (e.g., by reducing ε_h).[18]

21.3.1. Interfacial Energy Underlies Self-Assembly

Many responses of the assembled aggregates are predicated not on n or *MW* but on the most interfacial hydrophobic monomers, as these see both head group moieties and bulk solution. As a result, interfacial monomers typically have screened interface energies $\varepsilon_I \leq \varepsilon_h$. Thus, on the scale of the monomer area a ($\sim 0.5\text{ nm}^2$ for lipid in water), the interfacial tension that is created has a magnitude $\gamma \approx \varepsilon_I a^{-1}$ which is invariably much less than the $\sim$70 mN/m between water and air. This tension is balanced by fluid-like packing[19] and suggests worm and vesicle formation processes as well as vesicle shapes—like the varied shapes of cells[10]—are generally influenced by hydrodynamic-type forces such as mechanical shear. Entangled and glassy polymers can be expected, however, to slow down lateral or in-plane rearrangements of molecules (i.e., convection and diffusion) and thereby impede morphological responses of polymer vesicles.

Despite the potential for non-equilibrium flow effects, a molecular-scale force balance is also expected to underlie the basic stability of at least a curved, vesicular aggregate just as domed buildings must balance their internal stresses in defying collapse under gravity. In place of gravity, however, the osmotic pressure set up by a vesicle's semi-permeable membrane largely dictates vesicle volume up to the point of actually tensing the vesicle's membrane.[20] Assuming no osmotically induced tension and a simple spherical vesicle of radius R, the local curvature nonetheless leads to a slight molecular splay and a change in exposed area Δa which is dependent on membrane thickness d. The associated cost $\gamma \Delta a$ generates a local curvature energy classically viewed as proportional to $1/R^2$.[21] Integrated over a vesicle area $4\pi R^2$, however, the total curvature energy, $F_c \sim \gamma d^2 4\pi R^2 \times 1/R^2$, is independent of R. From this perspective, vesicle size in a given solution thus depends less on thermodynamics and more on non-equilibrium aspects of the formation process.

In addition to the classical curvature energy above, vesicle bilayers as well as multi-lamellar vesicles embedded one within the other like an onion can also possess a strain energy analogous to that in a bimetallic strip held flat while heated. The lateral strain that arises from a mismatched number of molecules in one leaflet or layer relative to the other(s) leads to an area difference elasticity.[22] 'Flip–flop' of amphiphiles from one monolayer leaflet to another relaxes the strain. In cells the process is catalysed by specific proteins called 'flipases',[23] which underscore the cost of transferring the hydrophilic part of an amphiphile through a membrane's hydrophobic core. However, synthetic polymers are polydisperse in *MW*, so that a small fraction of less hydrophilic chains will be more prone to flip-flop.[24]

21.4. POLYMERSOMES FROM BLOCK COPOLYMERS IN AQUEOUS SOLUTION

Building on the many years of experience in the manufacture and characterization of lipid vesicles or liposomes, diblock copolymer vesicles have been made and studied in strictly aqueous solutions. Maintaining the folding stability of proteins both encapsulated inside the vesicles and attached to the surface for cell-targeting purposes are significant practical motivations. The types of polymer vesicles reviewed below have involved use of copolymers with average *MW*'s up to ~100-fold greater than those of lipids as well as a far broader distribution of *MW*.

The peptide containing amphiphile PS_{40}–poly(isocyano-L-alanine-L-alanine)$_m$ is an early example of a diblock that assembles directly under solvent-free aqueous conditions.[25] It is semi-synthetic in the sense that it contains naturally occurring peptide moieties (L-alanine). Although the peptide-polymer design is expanded upon later, it is important to point out here that biomedical application of such copolymers is likely to be limited by the fact that polypeptides tend to be far more immunogenic than purely synthetic polymers. Nonetheless, under mildly acidic conditions and for $m = 10$ (but not $m = 20$ or 30), collapsed vesicular shells 10's–100's of nm in diameter were observed coexisting with rod-like filaments as well as chiral superhelices.

Using fully synthetic diblock copolymers of PEO-polybutadiene PEO_m–PBD_n (Figure 21.3A) and the hydrogenated homolog PEO-polyethylethylene (PEO_m–PEE_n), more mono-morphic, unilamellar vesicles referred to as 'polymersomes' have been made under a variety of aqueous conditions.[26] Figure 21.4A shows how addition of water to a lamellar

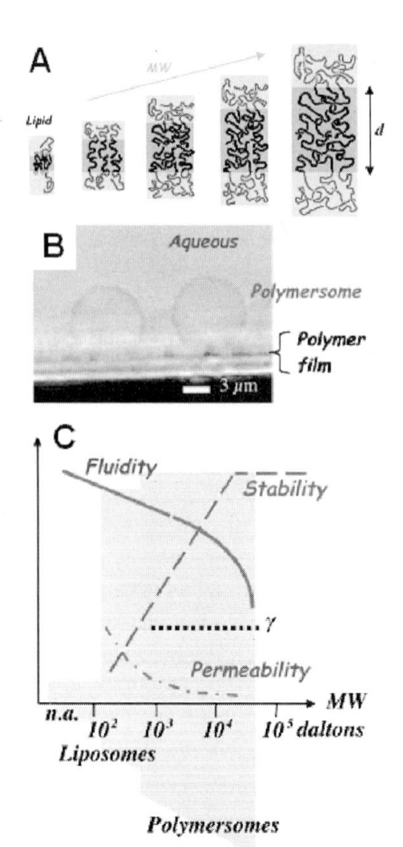

FIGURE 21.4. Polymersome membrane structure scheme, vesicle formation, and intrinsic membrane property trends. (A) Schematic of a polymersome membrane series (plus lipid) made from increasing molecular weight copolymers. The hydrophobic core thickness is d. (B) Analogous to liposomes, diblock copolymer vesicles can be formed by hydration of a thin, dried film of copolymer. Aqueous solutions can vary from phospate-buffered saline to 1 M sucrose or distilled water. (C) Schematic of membrane properties versus amphiphile molecular weight based on single vesicle measurements and ranging from liposomes to polymersomes, with n.a. denoting non-aggregating systems. Liposomes and related vesicles are generally made with amphipiles with *MW* less than 1 kDa; polymersomes can be made in aqueous solution with larger amphiphiles. Difficulties at high *MW* might be attributed to a rapid decrease in fluidity associated with entangling chains. At least for strongly segregating diblocks, the membrane thickness increases with *MW*, and makes polymersomes decreasingly permeable. Stability by a number of measures also increases, but only up to a limit set by the interfacial tension that arises with membrane formation.

film, microns-thick, generates polymersomes spontaneously. Injection of chloroform solutions of copolymer into water also generate vesicles, and the diluted chloroform can be thoroughly removed by dialysis. Initial quantitative studies of small solute and protein encapsulation indicate efficiencies comparable to liposomes and suggest the potential for novel, polymer-based controlled release vesicle systems.

Based on the examples above and more below, we should return to the unifying rule for polymersomes in water which is a phospholipid-like ratio of hydrophilic to total mass $f_{\mathrm{hydrophilic}} \approx 33\% \pm 10\%$.[9,11] A cylindrically shaped molecule which is asymmetric with

$f_{hydrophilic} < 50\%$ presumably reflects the ability of hydration to balance a disproportionately large hydrophobic fraction. Molecules with $f_{hydrophilic} > 45$–50% can be expected to form micelles, whereas molecules with $f_{hydrophilic} < 20$–25% can be expected to form inverted microstructures. Sensitivities of these rules to chain chemistry and MW have not yet been fully probed. Moreover, cryo-TEM of 100–200 nm vesicles shows membrane cores increase with MW from $d \approx 8$ to 21 nm.[9] Lipid membranes have a far more limited range of $d \approx$ 3 to 5 nm (schematically compared in Figure 21.4A), which is clearly compatible with the many integral membrane proteins in cells. Polymersome membranes thus present a novel opportunity to study membrane properties as a function of d (or MW).

21.4.1. MW-dependent Properties of Polymersomes

Retention of encapsulants (from dextrans to sucrose to physiological saline) over periods of months has been observed with both ~100 nm polymersomes prepared by liposome-type extrusion techniques[26] as well as ~10 μm giant vesicles (Figure 21.4B). The latter allow detailed characterization by single vesicle micromanipulation methods as summarized in Figure 21.4C. Lateral diffusivity as well as apparent membrane viscosity,[11, 27, 28] indicate that membrane fluidity (distinct from flip-flop) decreases with increasing MW. Moreover, the decreases are most dramatic when the chains are long enough to entangle. Area elasticity measurements of γ (≈ 25 mN/m) appear independent of MW and indicate that opposition to interface dilation is dictated by polymer chemistry and solvent alone. Electromechanical stability increases with membrane thickness up to a limit reflective of γ and reaching almost 10 volts.[11] Phospholipid membranes rupture well below such a limit simply because their small d makes them more susceptible to fluctuations and defects. To exceed this interfacial limit of any self-assembly, one must introduce additional interactions such as covalent crosslinking within the membranes[29]—an idea long-recognized[30] but only rarely realized with even small liposomes.[31] Permeation of water through the polymersome membranes has been be measured and shows, in comparison to phospholipid membranes, a significantly reduced transport rate, which is consistent with early measurements on liposomes by Bangham on a relatively narrow MW-series of lipids.[2] What appears most suggestive from the studies to date is that biomembranes are optimized less for stability than for fluidity. In this context, cholesterol is an intriguing component of cell membranes since it both toughens and fluidizes.[10]

21.4.2. Vesicles from Multiblock Copolymers or Bio-degradable Blocks

Several triblock copolymers also appear to fall within the class of polymersome-formers, and differences in membrane properties offer important insights. At least one commercial triblock known as a 'pluronic' with a relatively large poly(propyleneoxide) midblock (PEO_5-PPO_{68}-PEO_5) yields small vesicles in water with relatively thin membranes of $d = 3$–5 nm and stability of only hours.[32] Such results likely reflect the weakly hydrophobic mid-block PPO and the interfacial localization of the mid-block's oxygens. Indeed, factoring the oxygen into an effective hydrophilic fraction, shows this vesicle-forming super-amphiphile has an $f_{hydrophilic}$ in the same range as lipids. Another vesicle-forming triblock with a lipid-like $f_{hydrophilic}$ ($\approx 40\%$) consists of a hydrophobic midblock of poly(dimethylsiloxane) (PDMS) and two water-soluble blocks of poly(2-methyloxazoline)

(PMOXA) terminating in crosslinkable methacrylate groups.[33] Polydispersity in *MW* is intrinsic to all synthetic polymers and that of this copolymer is notably broad, implying that polydispersity is not a severe limitation to vesicle formation. Lastly, a first example of a vesicle-forming pentablock composed of PEO and a polysilane[34] is also intriguing although the electron microscopy is thus far suggestive of incomplete, collapsed shells.

Increasing the functionality of self-assembling block copolymers is clearly in the designs of emerging polymersome systems. These include suitably proportioned copolymers of PEO-poly(propylenesulfide) (PEO-PPS) whose PPS blocks are susceptible, in principle, to oxidative degradation.[35] Kilodalton-size large block copolymers of PEO-polylactide (PEO-PLA) are known to make micelles[36] and are of great interest because PLA is susceptible to hydrolytic biodegradation—as widely exploited in the field of controlled drug release.[37] Hence, biodegradable vesicles for controlled release of encapsulants seem achievable, and preliminary results indicate that this is indeed feasible if $f_{hydrophilic}$ is lipid-like.[6]

21.5. STIFFNESS AND STABILITY TUNING OF WORMS AND MEMBRANES

Worm micelles in a flow field rather than under quiescent conditions show that they respond in a way roughly in agreement with present theory for polymers under flow.[38] Stuck occasionally at a single point to a coverslip, the worms increasingly straighten under high flow and their thermally accessed conformational space is always narrowest near the point of attachment (Figure 21.5A). Simple observation as well as more careful analyses of fluctuation modes[14] under quiescent conditions indicate a sub-micron persistence length, *p*, in agreement with estimations from neutron scattering.[39] The fact that the worms themselves are fluid and the imposed flow should slip rather than stick as a boundary condition does not appear obvious in these responses. It is clear, however, that the worms can withstand

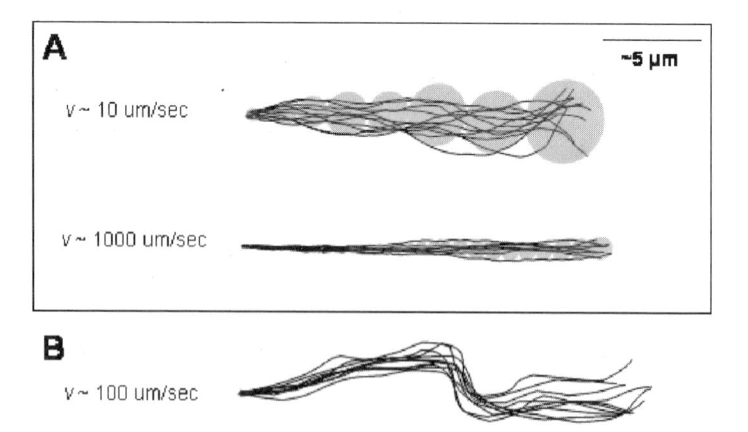

FIGURE 21.5. Point-attached worm micelles under flow. (A) Trumpet envelopes of conformations exhibited b non-crosslinked worms. The mean flow velocity is indicated; higher velocities narrow the trumpets. Lack of wor fragmentation under all such flows is consistent with strongly associated systems. (B) Configuration envelope a crosslinked worm under flow. Note the persistent kink which resists straightening under flow. (From Ref. 14 b permission of Elsevier Science.)

very high flow fields which can be estimated to impose tensions <1 mN/m. This places a lower bound on the γ holding the worms together.

Double bonds in the hydrophobic block of PBD allow crosslinking to be introduced by solution free radical polymerization into the worm cores.[40] Worms can thus be made even more stable and solid, emulating a classic covalent polymer chain but at a more mesoscopic scale. Such crosslinked worms often appear kinked (Figure 21.5B). This occurs simply because the polymerization within the core is done while the worm is flexing in solution. When stuck at a point and examined under flow, such fully crosslinked worms generally just wobble and pivot about the attachment, with any kink remaining locked in and nearly unperturbed by the flow-imposed stresses.

The two types of worms, fluid or crosslinked, are two extremes at either end of a continuous stiffness scale that can be experimentally realized by blending saturated PEE copolymer with the crosslinkable PBD copolymer. The two types of copolymer were shown with membranes to be fully miscible, and the PBD was also successfully crosslinked to give an equally wide range of stabilities and stiffnesses.[41] For both worms and membranes made with similar *MW* copolymers, an interesting percolation of the rigidity (e.g., p) appears at relatively low mole fractions of PBD ($\sim$20%). For worms, this represents a quasi-one-dimensional fluid-to-solid transition with potentially interesting dynamics of worms blended at or near the transition point. For the copolymer membranes, stability below percolation proved to be lower than even uncrosslinked membranes, which explains why crosslinking of polymerizable lipids has met with very limited success.[31] The fully crosslinked giant vesicles proved stable in chloroform and could even be dehydrated and re-hydrated like microballons without rendering the $\sim$9 nm thick membrane core; the results imply defect-free membranes many microns-squared in area. At the very least, the various crosslinking results provide general insights into covalent crosslinking within self-assembled nanostructures.

21.6. VESICLES IN INDUSTRY

Vesicle development has been most prominent, perhaps, in the cosmetics and pharmaceutical industries with progress in the latter predicated on years of phased testing and governmental approval. Nonetheless, both stealth and conventional liposomes loaded with anti-cancer agents have overcome cost, scale-up, sterility, and stability hurdles, and have also demonstrated clinical efficacy.[42] For topical or cosmetic application, novasomes are perhaps the latest multi-lamellar incarnation of the earlier synthetic niosomes and are reported to have the processability of many simple emulsifications.[43] The polymer processing and materials revolution of the twentieth century might likewise extend to vesicle encapsulators of the future.

21.7. ADDITIONAL POLYMER INTERACTIONS AND OTHER HOLLOW SHELLS

Most liposomes are composed of charged, typically zwitterionic or doubly-charged amphiphiles such as phosphatidylcholines which probably add a counter-charge pairing interaction to the interfacial tension, γ. Vesicles made from charge-compensated surfactants such as pH-tuned fatty acids[44] or cationic plus anionic surfactants[45] appear exceptional in depending very heavily on electrostatics. However, the vesicle phase for these systems may not be such a large region in a suitable phase diagram with both ionic strength and pH axes added as independent variables. Diblocks of poly(ethylethylene)-poly(styrene-sulfonic acid) (PEE-PSSH), are prime candidates to study polyelectrolyte affects,[46] but many additional charged block copolymers and mixtures are conceivable. Templated shells of layered polyelectrolytes[47] might, for example, inspire new vesicle designs.

The effects of chain flexibility in vesicle membranes are also not yet clear. In lamellar phases of pure block copolymers, chains tend to be laterally squeezed and effectively stretched by the interfacial tension. Polymers also have an instantaneous major axis (like any liquid crystal molecule) and packing tends to align adjacent axes. Either effect gives fractal scaling for $d \sim MW^b$, with $b = 0.5$–0.66;[9] rod-like polymers have $b = 1$. One diblock of note[48] used rigid ring monomers of phenylquinoline (PQ) to make a rod-coil copolymer with PS (specfically PS_{300}–PPQ_{50}) which gives collapsed monolayer shells (rather than bilayer or interdigitated membranes) when precipitated out of solvent. One feature of interest with this system is a novel photoluminescent coupling in the PPQ that arises in an aggregate specific manner. More generally, however, this diblock suggests the vesicle-forming potential of stiff, rod-like chains, although one difficulty that might be anticipated with long stiff hydrophobic blocks is the inability of a membrane sheet to bend and seal its edges for all but the largest, giant vesicles.

Finally, mimicking the complex filament-in-cell structures of Figure 21.2 with synthetic worms contained within copolymer membranes is not as difficult as it might seem. Rapidly emerging work shows that mixtures of copolymers illustrated in Figure 21.1 with suitably broad f can phase separate into filament-in-bag structures that have a rudimentary resemblance to snapshots in Figure 21.2. Many future opportunities in synthetic chemistry lie in maximizing a phase separation of worms from vesicles by making, for example, the hydrophobic blocks highly incompatible with very different γ: hydrocarbon-based worms could segregate but internalize within fluorocarbon-based membranes. Far more difficult challenges lie in the direction of creating living copolymer worms which assemble preferentially at one and thus might drive a membrane forward. In the context of section 3, the need is to create a worm with a C_{CMC} at one end which is different from C'_{CMC} at the other end. This is, however, the essence of a key biopolymer principle that underlies cell crawling of Figure 21.2. With cargo and targeting mechanisms discussed in the next section, one can imagine (if not yet realize) a degradable vesicle of PEO-PLA that crawls in the body through worm polymerization in order to release a drug cargo at a desired site such as a tumor or an atherscelrotic lesion. Environmental applications in bio-toxin clean-up and many other applications would be equally feasible.

21.8. INTERFACING BIOLOGICAL STRUCTURES AND FUNCTIONS

In a somewhat surprising result, integral membrane proteins 3–5 nm high have been compatibly inserted into PDMS-PMOXA membranes ($d \approx 10$ nm).[49] At least two features of polymer membranes facilitate protein miscibility:[50] (i) polymer chains can be significantly compressed, and (ii) polydispersity allows small chains to segregate around a membrane protein. Inserted channel proteins can also effectively dock with viruses and facilitate transfer-loading of viral DNA into the polymer vesicle (Figure 21.6A).[51] DNA delivery would, of course, seem important to explore provided release mechanisms can b

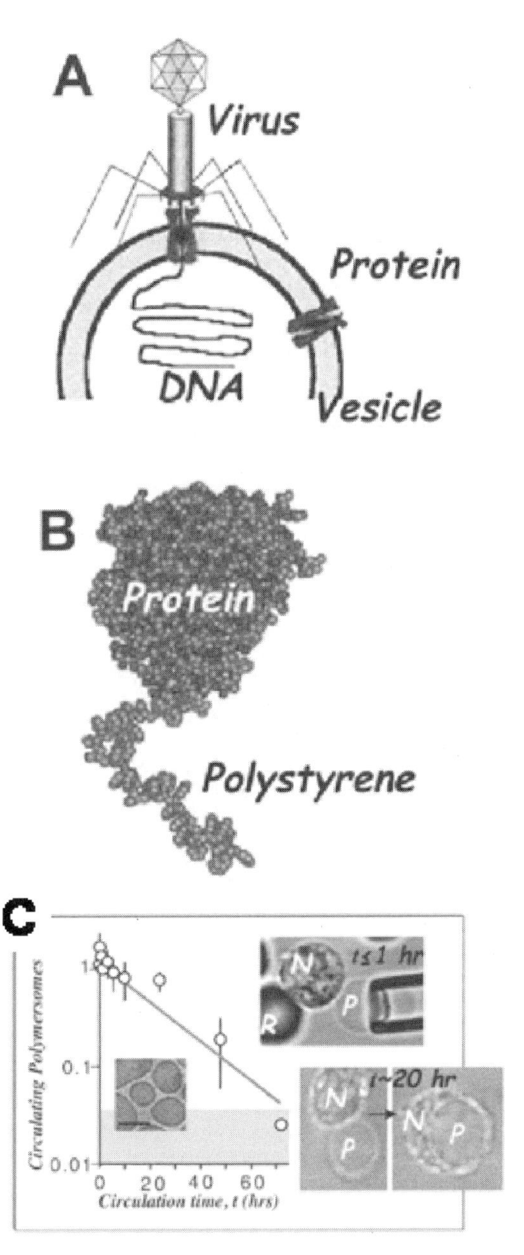

IGURE 21.6. Polymer amphiphile systems at the interface with biology. (A) Virus-assisted loading of a triblock opolymer vesicle containing a natural channel protein in its membrane; for liposomes, see Lambert *et al.*[55] 3) Protein-polymer hybrid amphiphile which assembles in a THF/Water solution into rod-like micelles. (C) iocompatibility studies of PEO based polymersomes.[6,54] The exponential decay in circulating polymersomes a rat model indicates a circulation half-life very comparable to times found for stealth liposomes.[42] The inset low the curve is a cryo-TEM image of ~100 nm (scale bar) polymersomes. The two optical micrographs at right ow giant polymersomes, *P*, placed in contact with a phagocytic neutrophil, *N* by a micropipette (of diameter μm). *R* denotes a red blood cell. The vesicles have been pre-incubated in plasma for the indicated times and ly after tens of hours are the polymersomes engulfed by the phagocytic neutrophils. (From Ref. 6 by permission the American Association for the Advancement of Science.)

incorporated. From a fundamental perspective as well, the broadened osmotic and rupture characteristics of polymer vesicles might be used to confirm emerging estimates of internal pressures exerted by encapsulated DNA.

Extending their work on dipeptide-containing block copolymers above, Nolte and coworkers have covalently attached a PS_{40} chain to an enzymatically active, water soluble protein of $MW = 40$ kDa ($f_{\text{hydrophilic}} \sim 50\%$) (Figure 21.6B). In solution, the conjugate generated worm-like micelles.[52] To obtain vesicles or sheets, longer chains of PS might be required. Nonetheless, the general idea is clearly to have multi-enzyme scaffolds in which enzyme A passes its product to adjacent enzyme B and so on, cooperating in a localized metabolic chain. Photosynthesis and polypeptide extension by ribosomes are two among many reactions that exploit enzymatic proximity. A synthetic cell that senses and enzymatically acts on its environment, like a biomembrane's glycocalyx does,[10] might also be developed around these bio-hybrid super-amphiphiles.

In more complex biological milieu, polymersomes are also showing initial promise. Injection of ~100 nm PEO-PEE polymersomes[53] into the blood circulation of rats yields results (Figure 21.6C) very similar to those for stealth liposomes which circulate for ~15–20 hours and end up being engulfed by 'phagocytic' cells of the liver and spleen.[42] Parallel test-tube studies involving timed pre-incubations in cell-free blood plasma (Figure 21.4C - inset) help elucidate the delayed, liposome-like clearance mechanism whereby plasma protein slowly accumulates on the polymersome membrane,[54] and eventually mediates cell attachment. The PEO brush provides an effective delay to this process and thus acts somewhat like a biomembrane glycocalyx. The results highlight the biomedical promise of polymersomes, and raise the question—what new fields await?

21.9. SUMMARY

With the increasing exploration and development of polymer aggregate systems, one underlying goal is to expand upon and clarify the properties of natural cell systems that have evolved over eons. It is already clear, for example, that many biological membrane processes can be faithfully mimicked by synthetic polymer vesicles (e.g., protein integration, fusion, DNA encapsulation, compatibility). It is equally clear that biomembranes are not optimized for either stability or equilibrium. However, they do not need to be since they encapsulate sufficient machinery to maintain a dynamic, animated state. Perhaps the mos useful concept to mimic and extend with block copolymers is the amphiphilicity templat provided by lipids and related. In light of the material versatility of polymers in terms o their molecular weight, polydispersity, reactivity, and synthetic diversity, a strictly polyme approach to the self-assembly of biomimetic nanostructures would seem to offer significan possibilities.

QUESTIONS

1. Consider two large glasses of water. Into one glass you inject with a syringe a 1 ml droplet of vegetable oil (assume neutral buoyancy); into the second glass you inject two 0.5 ml drople of the same oil.

a. Assuming a typical water-oil interfacial tension $\gamma = 40$ mN/m, calculate the total interfacial energy of the two systems as described.
b. What happens to each system given enough time? Hint: one system changes, the other does not. Explain.
c. Given that oil is a hydrocarbon lacking any oxygen and water, H—O—H, interacts strongly with itself through non-covalent interactions between each oxygen's lone-pair electrons and positively-polarized hydrogens of adjacent waters, try to explain or sketch the basis for γ.

2. It was demonstrated that the total bending or curvature energy, F_c, of a spherical vesicle is independent of vesicle size R because $F_c \sim \gamma d^2 4\pi R^2 \times 1/R^2$. Whereas $4\pi R^2$ is obviously a vesicle's area, the product $f_c \sim \gamma d^2 \times 1/R^2$ is the local curvature energy which is clearly dependent quadratically on both the curvature $(1/R)$ and membrane thickness, d. To understand whether a membrane bends locally, one needs to compare $k_B T (\sim 4 \times 10^{-21}$ J) to the energy scale, γd^2, for bending a fluid membrane and exposing more interface.
a. Calculate the ratio $\gamma d^2 / k_B T$ for a typical lipid membrane of $d = 4$ nm and the thickest polymersome membrane $d = 20$ nm; use $\gamma = 25$ mN/m as measured for polymersomes, independent of *MW*.
b. Which membrane system is more susceptible to thermal or Brownian undulations?

3. Because of the thermal fluctuations of polymer chains, polymer systems generally exhibit non-trivial or 'fractal' length scales. For example, with the membrane-forming copolymer systems, one finds that $d \sim MW^{0.5}$. The simplest model for this can be illustrated by sketching on a piece of paper a connected sequence of 1 cm line segments, randomly directed and without concern for overlaps. After drawing N such segments and doing this 100 times or more, we want an estimate for the root-mean-square average of the distance, L, between the ends. Denote each segment i by the vector a_I so that $L = a_1 + a_2 + \cdots + a_N$. Then determine $< L * L >$ by using the fact that $< a_1 * a_2 >$ and related terms vanish although each self-product is the same, i.e., $< a_1 * a_1 > = < a_2 * a_2 > = a^2$.

REFERENCES

1. A. Mogilner and G. Oster, *Biophys. J.* **71**, 3030 (1996).
2. A. D. Bangham, *Chem. Phys. of Lipids* **64**, 275 (1993).
3. B. Cornet, E. Decroly, D. Thines-Sempoux, J. M. Ruysschaert, and M. Vandenbranden, *AIDS Res. & Hum. Retroviruses*. **8**, 1823 (1992).
4. D. K. Fygenson, J. F. Marko, and A. Libchaber, *Phys. Rev. Lett.* **79,** 4497 (1997).
5. F. S. Bates and G. H. Fredrickson, *Phys. Today* **52**, 32 (1999).
6. D. E. Discher and A. Eisenberg. *Science* **297**, 967 (2002).

J. Israelachvili, *Intermolecular and Surface Forces, Ed. 2* (Academic Press, New York, 1991).

J. M. Anderson, *Annu. Rev. Mater. Res.* **31,** 81 (2001).

H. Bermudez, A. K. Brannan, D. A. Hammer, F. S. Bates, and D. E. Discher, *Macromolecules* **35**, 8203 (2002).

R. Lipowsky and E. Sackmann, Eds., *Structure and Dynamics of Membranes–From Cells to Vesicles.* (Elsevier Science, Amsterdam, 1995).

H. Aranda-Espinoza, H. Bermudez, F. S. Bates, and D. E. Discher, *Phys. Rev. Lett.* **87**, 208301 (2001).

R. Hirn, T. M. Bayer, J. O. Radler, and E. Sackmann, *Faraday Discussions* **111**, 17 (1998).

F. Brochard and J. F. Lennon, *J. Phys.-Paris* **36** , 1035 (1975).

P. Dalhaimer, F. S. Bates, H. Aranda-Espinoza, and D. E. Discher, *Comptes Rendus de d'Academie des Sciences - Serie IV Physique* **4,** 251 (2003).

J. F. Marko, *Europhys. Lett.* **38**, 183 (1997).

F. K. Bedu-Addo, P. Tang, Y. Xu, and L. Huang, *Pharmaceutical Res.* **13**, 710 (1996).

17. I. Eggens, T. Chojnacki, L. Kenne, and G. Dallner, *Biochim. Biophys. Acta.* **751**, 355 (1983).
18. L. Zhang and A. Eisenberg, *Science* **268**, 727 (1995).
19. S. C. Semple, A. Chonn, and P. G. Cullis, *Adv. Drug Delivery Rev.* **32**, 3 (1998).
20. E. Evans and R. Skalak, *Mechanics and thermodynamics of biomembranes*. CRC press, Boca Raton, Fla. (1980).
21. M. Bloom, E. Evans, and O. G. Mouritsen, *Rev. Biophys.* **24**, 293 (1991).
22. H. G. Dobereiner, E. Evans, M. Kraus, U. Seifert, and M. Wortis, *Phys. Rev. E* **55**, 4458 (1997).
23. C. F. Higgins, *Cell* **79**, 393 (1994).
24. L. Luo and A. Eisenberg, *Ang. Chem. Int. Ed.* **41**, 1001 (2002).
25. J. J. L. M. Cornelissen, M. Fischer, N. A. Sommerdijk, and R. J. M. Nolte, *Science* **280**, 1427 (1998).
26. J. C.-M. Lee, *et al., Biotech. and Bioeng.* **43**, 135 (2001).
27. J. C.-M. Lee, M. Santore, F. S. Bates, and D. E. Discher, *Macromolecules* **35**, 323–326 (2002).
28. R. Dimova , U. Seifert, B. Pouligny, S. Forster, and H.-G. Döbereiner, *Eur. Phys. J. E* **7**, 241 (2002).
29. B. M. Discher *et al. J. Phys. Chem.* B **106**, 2848 (2002).
30. H. Ringsdorf, B. Schlarb, and J. Venzmer, *Angew. Chem Int. Edit.* **27**, 113 (1988).
31. S. C. Liu and D. F. O'Brien, *J. Am. Chem. Soc.* **124**, 6037 (2002).
32. K. Schillen, K. Bryskhe, and Y. S. Mel'nikova, *Macromolecules* **32**, 6885 (1999).
33. C. Nardin , T. Hirt, J. Leukel, and W. Meier, *Langmuir* **16**, 1035 (2000).
34. N. A. Sommerdijk, J. M., S. J. Holder, R. C. Hiorns, R. G. Jones, and R. J. M. Nolte. *Macromolecules* **33**, 8289 (2000).
35. A. Napoli, N. Tirelli, G. Kilcher, and J. A. Hubbell, *Macromolecules* **34**, 8913 (2001).
36. S. A. Hagan *et al., Langmuir* **12**, 2153 (1996).
37. R. P. Batycky, J. Hanes, R. Langer, and D. A. Edwards, *J. Pharm. Sci.* **86**, 1464 (1997).
38. F. Brochard-Wyard, *Europhys. Lett.* **23**, 105 (1993).
39. Y. Y. Won, K. Paso, H. T. Davis, and F. S. Bates, *J. Phys. Chem. B* **105,** 8302 (2001).
40. Y. Y. Won, H. T. Davis, and F. S. Bates, *Science* **283**, 960 (1999).
41. B. M. Discher, H. Bermudez, D. A. Hammer, D. E. Discher, Y.-Y. Won, F. S. Bates, *J. Phys. Chem. B* **106**, 2848 (2002).
42. D. D. Lasic, D. Papahadjopoulos, Eds., *Medical applications of liposomes*. (Elsevier Science, Amsterdam 1998).
43. R. Mathur and P. Capasso, *ACS Sym. Ser.* **610,** 219 (1995).
44. J. M. Gebicki and M. Hicks, *Nature* **243**, 232 (1973).
45. H. T. Jung, B. Coldren, J. A. Zasadzinski, D. J. Iampietro, and E. W. Kaler, *P. Natl. Acad. Sci. USA* **98**, 135[illegible] (2001).
46. M. Regenbrecht, S. Akari, S. Forster, and H. Mohwald, J. Phys. Chem. B **103**, 6669 (1999).
47. F. Caruso, D. Trau, H. Mohwald, and R. Renneberg, *Langmuir* **16**, 1485 (2000).
48. Jenekhe S. A., X. L. Chen, Science **279**, 1903 (1998).
49. R. W. Meier, C. Nardin, and M. Winterhalter, *Angew. Chem. Int. Ed.* **39**, 4599 (2000).
50. N. Dan and S. A. Safran, *Macromolecules* **27**, 5766 (1994).
51. A. Graff , M. Sauer, P. V. Gelder, and W. Meier, *P. Natl. Acad. Sci. USA* **99**, 5064 (2002).
52. K. Velonia, A. E. Rowan, and R. J. M. Nolte, *J. Am. Chem. Soc.* **124**, 4224 (2002)
53. J. C.-M. Lee, M. Santore, F. S. Bates, and D. E. Discher, *Macromolecules* **35**, 323–326 (2002).
54. P. Photos, B. M. Discher, L. Bacakova, F. S. Bates, and D. E. Discher, unpublished.
55. O. Lambert, L. Letellier, W. M. Gelbart, and J.-L. Rigaud, *P. Natl. Acad. Sci. USA* **97**, 7248 (2000).

22

Biomolecular Motors

Jacob Schmidt and Carlo Montemagno
Department of Bioengineering
University of California at Los Angeles, CA

22.1. INTRODUCTION

Biomolecular motor are proteins that, central to their natural function, produce or consume mechanical energy. Working in concert, motor proteins can exert forces exceeding kilonewtons, enabling massive animals such as whales and elephants to move in their native habitats. They can also act individually; piconewton forces from single molecules ferry cargo between different points in a cell, transport ions across membranes, or generate biochemical fuel necessary for other cellular activities. Molecular motors are necessary and vital for cell functions and life processes, transporting essential solutes and organelles when diffusion is insufficient. The ability of biological systems to structure and organize materials hierarchically enables forces ranging over orders of magnitude to arise from the same components.

Although motor proteins have existed in nature for millions of years, much of what we know about them has only been learned in the last decade. Various motor proteins can act as rotary motors, linear steppers, and screws. Individual motors range in size from five to hundreds of nm and can operate over a large range of speeds: different subtypes of Myosin move from 0.06 to 60 μm/sec![1] Nature has indeed created some highly efficient (some close to 100% efficient)[2] compact, and often non-intuitive designs for her molecular motors. Due to their unique sizes, speeds, and functions, the function and mechanisms of motor proteins at the single molecule level have been the subject of much recent scientific and technological interest. Our knowledge is poised to increase rapidly as a result of the Human Genome Project and other whole-organism sequencing efforts.

Production of motor proteins in large quantities is straightforward, and assembly of protein-based engineered devices can take place in a test tube or Petri dish, rendering most aspects of device production feasible and economical. Their sizes, speeds, functions, and manufacture cannot be matched by conventional micro- and nano-electromechanical systems (MEMS and NEMS) devices. However, they also have limitations in certain aspects of operation and manufacturing in which MEMS systems are superior. The best of both worlds may result from hybrid combinations of biological and inorganic components: devices with biologically active elements at fundamental size limits in direct communication with and controllable by the macroscopic world.

Some exciting experiments involving the direct manipulation of individual biomolecular machines have recently provided a glimpse of the future: fabrication of hybrid organic/inorganic nanomachines, thousands of which could fit inside a single living cell. Further refinement and application of the techniques used to create these systems may lead to revolutions in medicine and technology. This chapter discusses the functional characteristics of selected biomotors, their manufacturing and engineering processes, recent work integrating them into fabricated devices, and some of the improvements which must be made to these devices if useful applications of biomotors are to be realized. Finally, we conclude with an outlook of future directions; speculation on what the future may bring to the field and what the field may bring to the future.

22.2. OF MEMS AND BIOMOLECULAR MOTORS

From an engineering perspective, motor proteins are an attractive alternative to current MEMS devices because of their small sizes (length scales hundreds of times smaller than those of equivalent MEMS devices), high speeds, high efficiencies, and ease of mass-production. The possibility of incorporating biological components into hybrid devices offers functions and flexibilities in design superior to purely inorganic devices. Further, the sizes and efficiencies of motor proteins allow performance densities otherwise unattainable. The three-dimensional geometry and diverse energy requirements of the class of motor proteins facilitate assembly of compact structures. Finally, while the motor proteins themselves are not self-replicating, the biological machinery that makes them is self-replicating, which leads to low economic and time costs. However, the precise placement and attachment of biological molecules in complex devices is problematic. In addition, external interfacing and control of such devices is difficult and not currently achievable with complex biological processes. These shortcomings do not exist for MEMS, which draws on the tremendous body of knowledge and experience in materials science and semiconductor lithography developed for and by the electronics industry. These points are discussed further in the following sections:

22.2.1. Size, Materials, Geometry, and Efficiency

For the manufacture of some nanodevices, molecular motors may be the only possibl component choice due to size considerations. They are some of the smallest controllabl moving structures known. For example, the rotary motor F_1-ATPase is 12 nm in diamete with a 3 nm rotor which is capable of rotating over 130 times per second while providin a constant torque greater than 40 pNnm at close to 100% efficiency.[3] A MEMS analog t this motor does not exist.

Conventional MEMS devices are made using techniques of semiconductor lithography, which expose and develop selected planar regions of resist on a surface. Inorganic materials such as metals, alloys, semiconductors, and oxides can be deposited or etched using a variety of techniques. As electronic feature sizes decrease into the nanoscale, MEMS devices made using the same "top-down" fabrication techniques will be hard pressed to follow. The bulk additive and subtractive material modification methods necessary to produce these parts will require increasingly exacting tolerances to ensure that structures are made as desired. Further, because of the inherent planar nature of exposure-based lithography, the geometry of possible MEMS devices is limited.

In contrast, proteins are nanometer-scale, atomically precise structures produced through universal biological manufacturing processes. The composition of proteins and affiliated biological components is limited to the 20 amino acids and some biomolecules, which are able to interface with ions, metals, and inorganic materials. Although the primary structure of proteins is a linear polymer, these polymers can fold into a variety of three-dimensional shapes. The folded shape of the protein determines its mechanical properties as well as its catalytic specificity and function. In addition, whole proteins are often composed of separate distinct molecular subunits that self-assemble in solution, circumventing many limitations on protein geometry imposed by linear peptide structure.

Furthermore, the surface forces present in MEMS devices increase in relative magnitude as their size decreases: machines at the micro- and nanoscale do not work the same as their macroscale counterparts. Forces previously negligible on the macroscale (such as friction and capillarity) can dominate the performance of MEMS and NEMS devices. However, biological motors efficiently operate for millions of cycles at high speeds without losing a single atom! As a result of these intriguing properties, molecular motors are not only active research subjects at present, but also highly promising candidates for use in mechanical nanodevices in the future.

2.2.2. *Energy Sources*

Biological systems adapt themselves to their environment; if they do not, they will not survive. Because of the diverse terrestrial environments, different organisms have adapted their energy requirements to exploit their surroundings best. As a result, there are proteins and systems of proteins known that can convert between optical, electro-osmotic, electronic, chemical, and mechanical forms of energy. The flexibility of energy sources is a particular advantage in device design, as the system geometry and components often dictate the forms of energy produced or consumed. For example, mobile devices may be best powered by optical or chemical energy, while devices immobilized on a substrate may be best powered by electrical energy. Using the same components for each kind of device would be problematic if not for the existence of the energy conversion pathways listed above. This ability to transform energy is an advantageous characteristic of biological systems compared to wholly inorganic MEMS components, which are presently able to be powered using electricity only.

2.3. *Manufacturing Equipment and Economics*

Most of the equipment necessary for protein fabrication is relatively inexpensive and can fit on a laboratory benchtop. As the complicated machinery for making nanometer-sized proteins is contained in a single cell, the primary tasks center on giving these cells instructions to manufacture protein and extracting these products. Due to the large quantity of protein made using bacterial expression systems (this process is described below), the unit cost of biomolecular motors can be extremely small. Further, little time is required: a standard 3.5 liter production run of F_1-ATPase requires approximately four days from initial transfection of the bacteria to use in experiments. This process typically produces about 20 g wet weight of cells, resulting in 6–10 mg of protein, $\sim 1.5 \times 10^{16}$ motors (over 10^{14} motors made per hour!). The production process can be scaled up with relative ease, requiring only the inexpensive growth of more bacteria. Customized proteins for which simple modifications are made using site-directed mutagenesis (described later), can be mass-produced from start to finish in as little as a week.

In contrast, typical MEMS fabrication technologies require completely dust-free clean rooms containing many hundreds of millions of dollars of complex lithography equipment, etchers, evaporators, and ovens which fabricate devices entirely on-chip. The production capacity of this equipment is limited by comparison. For example, Intel recently announced the shipping of its 100 millionth boxed processor.[4] If all of these were Pentium IV microprocessors (each of which contains 28 million transistors), 2.8×10^{15} transistors have been produced. Whether the processor or the transistor is considered as the production unit, a small biology laboratory is able to cheaply out-produce a billion dollar production facility in a significantly shorter time, making the economics of biological manufacturing very attractive indeed.

22.2.4. *Conclusions*

We have shown that nanodevices made by biology and MEMS fabrication have many complementary disparities in size, performance, materials, energy, manufacturing, and economics. Certain applications may favor the use of biological components, while others may favor MEMS. A combination of the ability of biological systems to produce and self assemble large numbers of relatively complex components with the capability of MEMS fabrication to precisely position device elements and substrate features could therefore resul is a very powerful and versatile manufacturing technology.

22.3. OPERATION AND FUNCTION OF MOTOR PROTEINS

Many proteins act as enzymes and therefore have affinities for the biochemicals involve in the reactions they catalyze. This affinity manifests itself as a strong and exclusive bindir of the protein to a biochemical substrate, which is accompanied by a change in shape of th protein. In some instances, this change of shape is incidental to the operation of the protei in others, like motor proteins, it is essential to its function, as it causes motion to occur a force to be exerted.

The coupling of the motor proteins' shape change to substrate binding and relea and their enzymatic function catalyzing biomolecular synthesis and hydrolysis highlig the importance of energy and entropy. A thermodynamic analysis of motor proteins il minates much of their operation. Below we will describe the three most widely studi motor proteins, Kinesin, Myosin, and F_1-ATPase as well as briefly introduce some le well understood motors. The functions performed by these motor proteins are as varied as the structures and mechanisms of the proteins themselves. Taken together, they form an excellent example of the many ways Nature solves a particular problem.

22.3.1. Thermodynamics of Motor Proteins

Most motor proteins bind to nucleotide triphosphates, resulting in these molecules' production or consumption. The most common of these biochemical fuel molecules is adenosine triphosphate (ATP), which is hydrolyzed in the reaction: ATP $\leftrightarrows$ ADP + P_i, where the high-energy phosphate bond is broken, leaving adenosine diphosphate (ADP) and inorganic phosphate (P_i). There are also other chemical states involving complexes with Mg^{2+}, water, and other molecules, but the spirit of the concepts is demonstrated in a simplified analysis.

The free energy change in the synthesis or hydrolysis of one ATP is given by

$$\Delta G = \Delta G_0 - k_B T \ln \frac{[\mathrm{ATP}]}{[\mathrm{ADP}][\mathrm{P}_i]} \tag{22.1}$$

where ΔG_0 is the standard free energy (the free energy at 1 M concentration of all reactants) and $k_B T$ is the thermal energy. ΔG_0 for ATP is -51×10^{-21} J[5] and at room temperature $k_B T$ is 4.14×10^{-21} J. Both of these quantities are more intuitively accessible and useful as -51 pNnm and 4.14 pNnm, respectively. In a typical intracellular environment, the concentrations of ATP, ADP, and P_i are 2 mM, 10 μM, and 2 mM, respectively, resulting in a free energy of ≈ -100 pNnm[5], about 25 times thermal energy at room temperature. We can also see that about half of the free energy ATP hydrolysis provides results from the high energy bonds of ATP and half results from the high cellular concentrations of ATP, forcing the reaction strongly in the forward direction.

How can motion be produced merely by the binding of ATP to a motor protein, without its hydrolysis? Furthermore, since binding and unbinding of the ATP to the motor is a reversible event, how can motion in one direction result? Powerstrokes (force-producing movements) of motor proteins can be grouped into the following categories: motion on ATP binding, motion on hydrolysis of ATP, and motion on ADP or P_i unbinding. Whether this motion occurs immediately following binding changes or is caused by thermal motion proceeding over an energy barrier lowered by binding changes, the overall direction of the reaction determines the net direction of motion. The cyclical reaction sequence of a hypothetical motor may be idealized as:

$$\text{Motor} + \text{ATP} \leftrightarrows \text{Motor ATP} \leftrightarrows \text{Motor}\,(\text{ADP} + \text{P}_i) \leftrightarrows \text{Motor ADP} \leftrightarrows \text{Motor.}$$

The motor and ATP begin free in solution, the motor binds to ATP and catalyzes its hydrolysis to ADP and P_i, the P_i is released, and the ADP is released bringing the cycle to a close. The force or motion generating step can be linked to any of these events.

We can now see why the motor will only walk in one direction and that, although all of the chemical reactions are reversible, the motion isn't. Since the cell keeps the concentrations of ATP so far out of equilibrium, the overall reaction and, therefore, the motion is biased in one direction. By measuring the work performed by the motor over a catalytic cycle and comparing this to the free energy of ATP hydrolysis, the efficiency of a particular motor protein can be determined.

22.3.2. *Myosin, Kinesin, and F_0F_1-ATP Synthase*

Myosin, kinesin, and F_0F_1-ATP synthase are compact, highly efficient motor proteins which have received a significant amount of attention recently in efforts to understand their motion and explore their engineering potential at the single molecule level. All three of these motors are powered by ATP, although the powerstrokes and details of motility are different for each. In contrast to the rotary motor F_0F_1-ATP synthase, myosin and kinesin are linear motors which generate motion through propagation down a complementary track. Because the binding to the track is asymmetric, and the ATP hydrolysis is effectively irreversible, the motors travel in only one direction along their complementary tracks. This makes the motion highly reliable and enables motor propagation to be specified merely by orienting the track.

Some linear motors are "walkers" and some are "runners". This is an expression of the *processivity* of the motor, which defines whether some part of the motor remains in contact with its track at all times. Walkers maintain constant contact with the track (processive), whereas runners do not (non-processive). Different motor subtypes can vary in their processivity: myosin II is non-processive, whereas myosin V is processive. Obviously, non-processive motors must work together in teams to produce motion or transduce force for any significant length of time. Similarly, processive motors must operate in small numbers or work alone to accomplish their functions. The class of non-processive motors even includes single headed motors.

22.3.2.1. Myosin Myosin motor proteins are characterized by binding affinity and motility associated with F-actin, a double-stranded helical filament formed by the polymerization of G-actin protein monomers. These filaments are asymmetric, with "plus" anc "minus" ends. A depiction of the structure of myosin and its interaction with F-actin is shown in Fig. 22.1. ATP binding to myosin affects its actin binding affinity and vice versa It is this modulation of binding affinities that allows ATP to serve as an energy source fo myosin's motion. The binding of myosin to actin filaments is also asymmetric so that, a myosin hydrolyzes ATP, it advances toward the plus end of the actin filament (left side o image).

The powerstroke of myosin II, found in skeletal muscle, occurs following the unbindin of P_i from the actomyosin complex.[1] A complete mechanical cycle of myosin II is a follows: the head is bound to actin following the previous powerstroke cycle. Upon bindin of ATP to the actomyosin complex, the head unbinds from the actin which stimulates th hydrolysis of ATP. This causes a conformational change in the head in preparation for th next powerstroke. The hydrolysis increases the binding affinity of the head for the actin, an the head docks with the actin. The P_i then unbinds from the head, triggering the powerstrok and advancing the head by one step size relative to the actin. The ADP unbinds from th head, allowing ATP to bind to the head, which decreases the affinity for actin and retur the cycle to the beginning.

The step size of myosin varies with the subtype, depending on the neck linker leng (red and yellow in the figure). Each motor individually provides 1–10 pN of force.[1] Althou

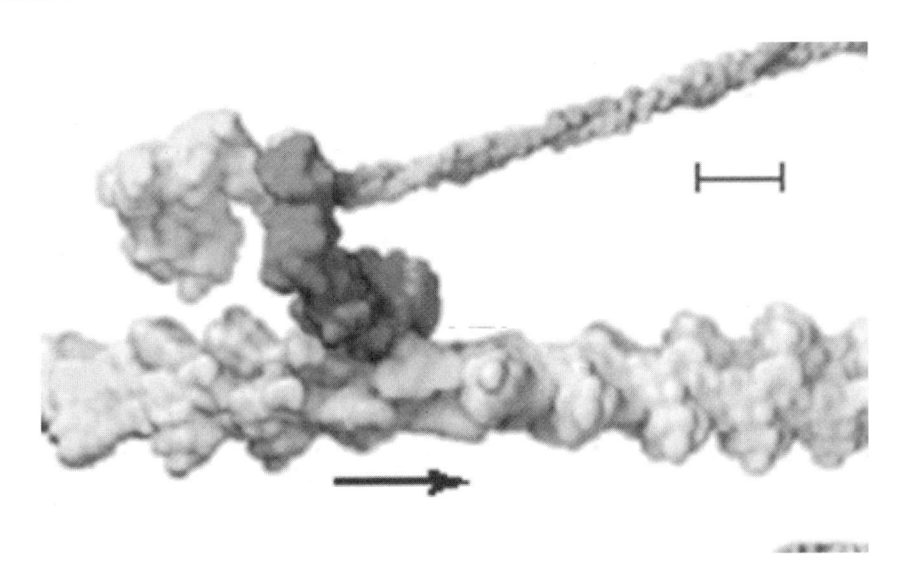

FIGURE 22.1. During its power stroke, Myosin exerts a force on an actin filament, propelling the motor to the left. Scale bar = 6 nm. (From R. D. Vale and R. A. Milligan, *Science* **288**, 88 (2000) by permission of the American Association for the Advancement of Science.)

this is only enough to lift a weight of 5×10^{-13} kg, it is almost one billion times the motor's weight! Macroscopic amounts of work can be realized by millions of these motors acting in concert in one muscle cell, and millions of muscle cells in concert provide the forces necessary to move our bodies and interact with our surroundings.

22.3.2.2. Kinesin The kinesin family of proteins is characterized by binding affinity and motion associated with microtubules. Microtubules (also with plus and minus ends) are asymmetric multi-filamentous protein polymers formed by the polymerization of α- and β-tubulin dimers (shown as grey and white in Figure 22.2). ATP and its hydrolysis products also modulate the binding of kinesin to microtubules, allowing unidirectional motion to occur. There are a large number of similarities between kinesin and myosin, including structural details, and it is thought that they are diverged evolutionarily millions of years ago from a common ancestor.[6] A common function of kinesin is the transport of cellular organelles. This is a particularly important function, as large cells, like the sciatic nerve, cannot transport material from one end of the cell to another via diffusion. Active transport, such as is provided by kinesin, is required to effect this.

Kinesin is a processive motor. To ensure this behavior, the heads of kinesin are cooperative: the chemical or mechanical state of one head can influence the state of the other. A complete mechanical cycle is as follows: with the leading head empty and bound to the microtubule and the trailing head bound to ADP, ATP binding to the empty head throws the trailing ADP head forward. The binding of this head to the microtubule triggers ADP release followed by hydrolysis of ATP in the now trailing head. Upon the release of P_i, the trailing head detaches and is ready to be thrown forward when another ATP is bound. "Conventional" kinesin has a step size of 8 nm and stall forces measured between 4–8 pN, resulting in 30–60% maximum efficiency.[7–9]

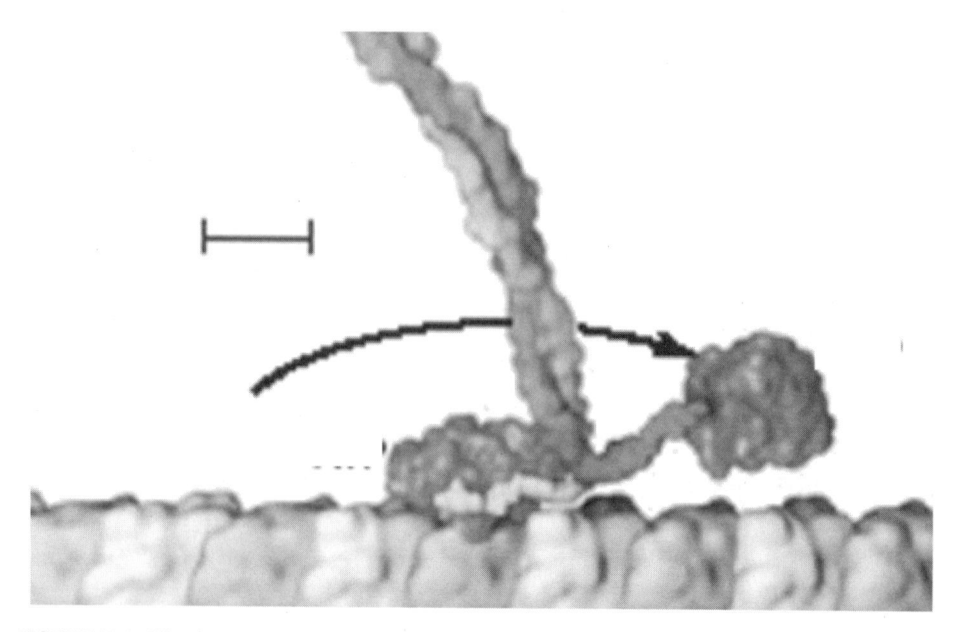

FIGURE 22.2. Kinesin steps forward, transporting its cargo toward the plus end of a microtubule. Scale bar = 4 nm. Figure from Vale and Milligan (2000). (From R. D. Vale and R. A. Milligan, *Science* **288**, 88 (2000) by permission of the American Association for the Advancement of Science.)

22.3.2.3. F_0F_1-ATP Synthase F_0F_1-ATP Synthase is a 12 nm diameter membrane protein complex which produces ATP from ADP and P_i in the presence of a trans-membrane proton gradient, depicted in Figure 22.3. Other cellular membrane proteins transport protons, creating a large electroosmotic potential caused by the large proton concentration on the lower side of the membrane. The membrane-bound F_0 component (pink) relieves this electroosmotic pressure by transporting the protons to the other side of the membrane, and using their energy to turn the rotor (tan). The rotor is coupled to the F_1 component and the

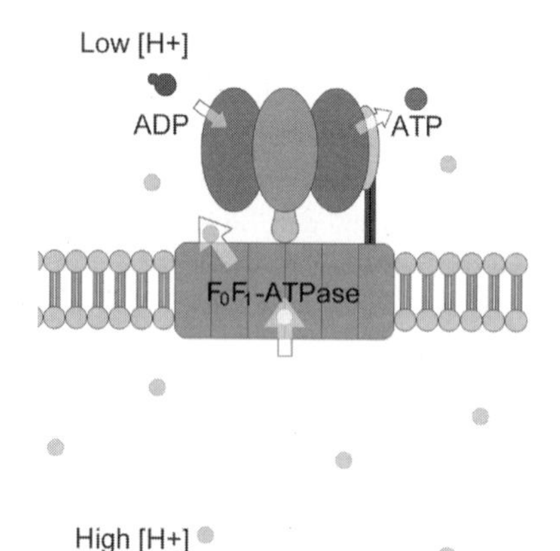

FIGURE 22.3. A depiction of F0F1-ATP Synthase. The free energy of high proton concentrations is used to generate ATP from ADP by mechanically coupling proton movement to conformational changes in the protein.

rotor's motion induces conformational changes of the α and β subunits (brown and light brown). ADP and P_i are bound to the F_1 component and these induced conformational changes result in the synthesis of ATP from ADP. Interestingly, this protein does not have a mechanical function; the rotation of the rotor is a mechanical linkage between the F_0 and F_1 components that converts energy from electroosmotic to mechanical to chemical.

The F_1 component of F_0F_1-ATP synthase is capable of hydrolyzing ATP and functioning as a motor independently. The model of ATP synthesis and hydrolysis is the binding change model proposed by Boyer,[10] supported by recent direct experimental evidence by Yoshida and coworkers.[3] F_1-ATPase has three catalytic sites available for substrate binding, and this binding exhibits cooperative behavior relative to the advancement of the rotor. A complete mechanical cycle is as follows: two of the binding sites are empty and the rotor is associated with the third site bound to ATP. The site counterclockwise (when viewed from the F_0 side) from the ATP site then binds ATP, causing the rotor to advance 90°. The ATP at the first site hydrolyzes, and P_i is released. When ADP is released from the first site, the rotor advances the remaining 30° and the motor is at the initial state, but 120° counterclockwise. F_1-ATPase is capable of exerting 40 pNnm of torque while rotating, resulting in a maximum efficiency of at least 80%.

22.3.2.4. "Unconventional" Motors The bacterial flagellar motor, a huge protein complex found in bacterial membranes, is an excellent example of an "unconventional" motor protein.[11] Instead of utilizing ATP as its fuel source, it uses a proton gradient, which drives the rotation of a biomolecular turbine as the protons are transported across the membrane. Flagella, long helical whip-like protein assemblies, are attached to the rotor and create thrust when rotated. These motors can move at over 1000 rotations per second and supply over 500 pNnm of torque. In a single bacterial cell, there are ten to fifteen flagellar motors in operation simultaneously. When the motion of the motors is directing fluid away from the cell, the flagella intertwine and rotate in unison, driving the cell in one direction. After a period of time, the motors reverse direction, the filaments untangle, and the cell "tumbles". After the motors restart in the forward direction, the cell proceeds in a newly randomized direction. In this way, the bacterium can explore its local environment, sampling a variety of chemical conditions.

Another excellent example of an unconventional motor has only been studied recently. The bacteriophage ϕ29 portal DNA packaging motor takes viral DNA and packs it into its protein container at nearly crystalline density.[12] The motor cyclically rotates its point of contact with the DNA strand. As the DNA is helical, this circular motion is translated into linear motion and the DNA packs into the viral capsule. The motor must work against a large pressure emerging from the huge decrease in entropy and large electrostatic energies involved with such high-density packing of negatively charged DNA. Single molecule studies performed recently show that the motor has a stall force of 55 pN, the greatest measured to date.

22.4. BIOTECHNOLOGY OF MOTOR PROTEINS

The natural protein manufacturing process occurs through the transcription of DNA to RNA and the translation of RNA to protein. Once the protein is made, it folds into a functional form, ready to work in support of the cell and whole organism. To study or engineer a selected motor protein, it can be purified from its native organism (a long process often with low yield) or the gene encoding it can be copied into a protein expression system. Protein expression systems are capable of producing the protein at high efficiency, in large quantities, and with less stringent conditions.

Once the gene isolating the protein has been identified and sequenced, the protein can be genetically engineered and the amino acid sequence of the protein altered. These changes can alter functional groups, change the folded state of the protein, or aid in purifying the protein. After production, the proteins must be extracted and purified, separating them from the thousands of other proteins present in their host cells. Biochemical techniques may then be used to modify the finished proteins through addition, modification, or removal of chemical groups in the protein.

22.4.1. Biology of Protein Production: From DNA to Protein

The primary, linear structure of a protein is encoded in the base pair sequence of DNA encoding the gene for the protein. There are four different bases, adenine (A), cytosine (C), guanine (G), and thymine (T), which each form complementary hydrogen bonds with T, G, C, and A, respectively, in the double-stranded form of DNA. As genetic information, the sequence is grouped into three-base units called codons, which encode for a specific amino acid. As there are only 20 amino acids, some have more than one codon. The start of the gene is denoted by AUG, and the end by UAA, UAG, or UGA.

Genetic expression begins when RNA polymerase binds at a promoter site upstream of the start codon and makes a single-stranded copy of the DNA in RNA (called messenger RNA, or mRNA), proceeding until the stop codon is reached. This mRNA is fed into the ribosome, where complementary RNA codons (called transfer RNA, or tRNA) attached to their corresponding amino acids hybridize with the mRNA. The ribosome covalently attaches the new amino acid to the end of the growing peptide chain and the process continues.

The ribosome produces a linear chain of peptides which are denatured (unfolded) protein. From the 20 common natural amino acids comes the rich variety of protein structures: each amino acid has a different side chain functional group and interacts with neighboring peptides and the external environment of lipids, ions, and water differently. As a result of these interactions, the protein folds into an equilibrium shape that minimizes its free energy. The resulting shape of the protein dictates its function.

22.4.2. A Short Overview of Recombinant DNA Technology

Bacteria, such as *Escherichia coli*, have a much smaller genome and simpler cell structure than eukaryotes. The biology and genetics of *E. coli* have been long studied and is well understood. When growing in optimal conditions, *E. coli* cells can double in number every five minutes. These factors make it a very powerful and versatile system for genetic engineering and protein expression.

The process of genetic engineering begins with the isolation and extraction of the gene of interest from the host DNA. The host DNA is first digested with restriction enzymes, proteins that cleave double stranded DNA at specific base pair sequences, making "sticky ends" which readily hybridize with complementary sequences generated with the same technique (Figure 22.4). If the DNA containing the gene of interest is combined with other

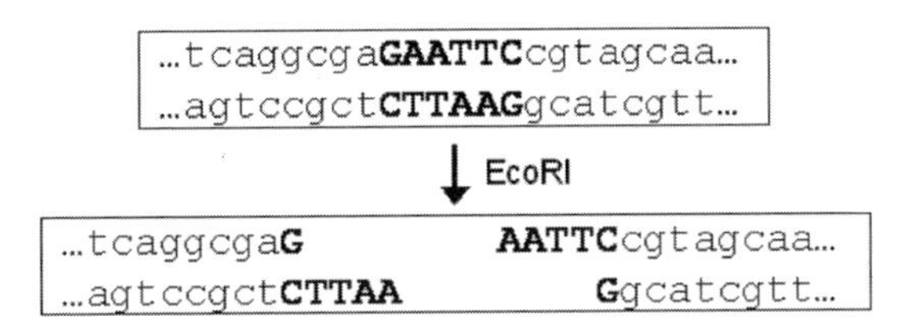

FIGURE 22.4. A sequence of double stranded DNA containing a six nucleotide sequence (shown in boldface) recognized by the restriction enzyme EcoRI. The action of EcoRI is to cut the strands as indicated, making "sticky ends" which have a high affinity for their complements. Mixture of two different strands cut by the same restriction enzyme can result in chimeras: sequences containing DNA from two different sources.

DNA modified with the same restriction enzyme, the gene can hybridize with that DNA. This new chimeric DNA can be mass-produced using Polymerase Chain Reaction (PCR), a standard laboratory technique capable of replicating trace amounts of DNA over 10^{12}-fold.

These recombinant techniques can be used to insert the gene of interest into a plasmid (Figure 22.5). The direct insertion of a stretch of genetic material into another sequence using

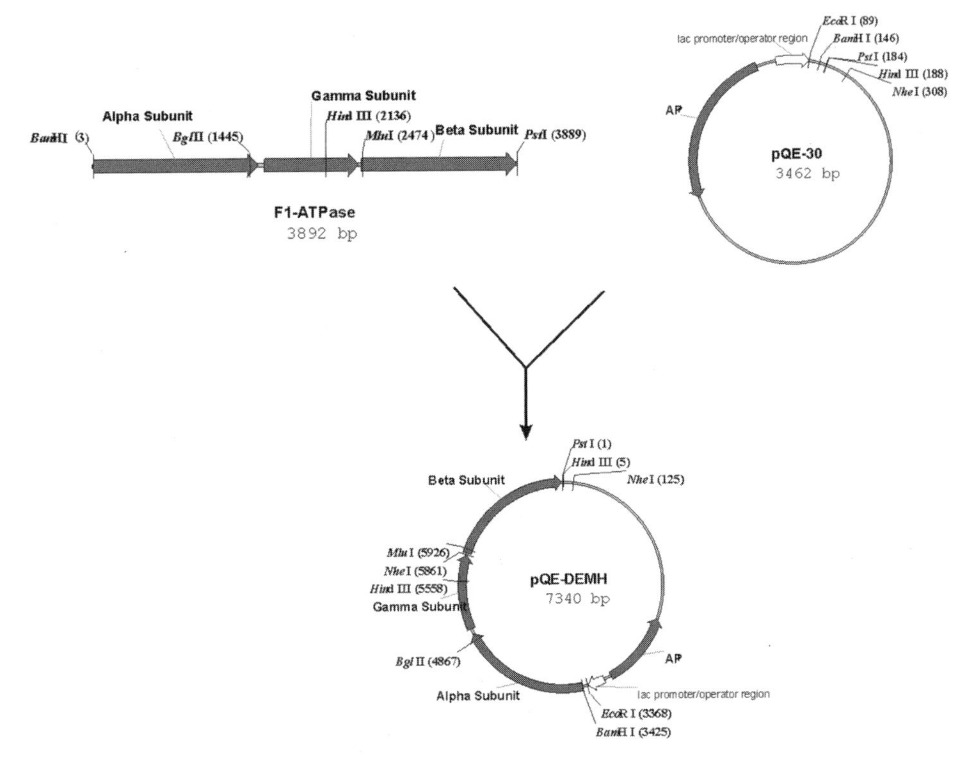

FIGURE 22.5. Insertion of the F1-ATPase gene into a plasmid. The gene encoding the motor is flanked by two restriction enzyme sites, BamHI and PstI. The plasmid pQE-30 contains a number of restriction sites, including BamHI and PstI and a gene encoding for ampicillin resistance (Ampicillin is a potent anti-bacterial agent). Incubation of the gene and the plasmid with the restriction enzymes results in some new plasmids containing the ATPase gene. Insertion of these plasmids into cells and growth of these cells in media containing ampicillin selects only cells containing the new genes. Addition of IPTG to the cell culture induces expression of the genes after the lac promoter, in this case, the genes encoding the α, β, and γ subunits of F1-ATPase.

this technique is known as cassette mutagenesis. When this engineered plasmid is complete, it can be inserted into bacteria which can express the genetic information contained on it, producing the designated proteins. If certain desired amino acids are not present or occur at inconvenient locations in the protein, the nucleotide sequence may be changed to add, subtract, or substitute various peptides in the primary sequence of the protein using site-directed mutagenesis,[5] an easy and useful technique for changes involving a small number of peptides.

Because bacteria typically reproduce in minutes, large quantities of bacteria containing the foreign gene can be grown in a short time. By using another gene that promotes protein expression, the *lac* operon, large quantities of the desired protein can be made by the addition of isopropyl-beta-D-thiogalactopyranoside (IPTG), which binds to a molecule repressing this gene. After the addition of IPTG, each cell can over-express the genes contained next to *lac* in the inserted plasmid, in quantities up to 50% of the total protein produced by the cell.

22.4.3. Biochemistry: Post-Production Modification

Once a motor protein and its gene are obtained in quantities sufficient for experimentation, the protein may be modified to enable easier manipulation and precise attachment to another device component or substrate surface. One avenue is direct biochemical modification,[13] using naturally occurring reactive groups on the protein. However, proteins offer limited options for site-specific chemical reactions. Reaction with the terminal amino and carboxyl groups are complicated by the presence of the amino acids such as lysine and glutamate, which can interfere with reagents intended for the terminal groups. Indiscriminate chemical attachment to these groups is often used to tether proteins to surfaces.[14] However, denaturation or impairment of functional activity can result as the protein may be immobilized or cross-linked during the modification.[13] The aliphatic amino acids are relatively chemically inert and often buried inaccessibly due to their hydrophobicity. Cysteine, with its unique sulfhydryl group, offers opportunities for site-specific chemical modification. It is often used to attach fluorescent dyes, biological molecules, and other proteins using sulfur-specific moieties such as malemides and vinylsulfones. Proteins containing small numbers of structurally insignificant cysteines are easily engineered: naturally occurring but inconveniently located cysteine residues can be changed to chemically inert peptides and inert peptides changed to cysteines at desired locations on the protein.

Site-specific links to proteins can be also be made through purely biological means. Streptavidin is a protein with four binding pockets that bind strongly and specifically to the biomolecule biotin.[15] The strong binding force results from the high degree of surface matching, which allows large numbers of weak bonds to cooperate. In addition, the four binding sites allow biotin-streptavidin-biotin linkages. As a result, this powerful system is very popular and functionalized biotin is commercially available with a variety of linking chemistries.

Another versatile and powerful method of protein binding utilizes the antigen/antibody interaction. Antibodies, produced by the immune system, are molecules that have regions with high affinity for a specific antigen and no other molecules. Isolation and modification of these antibodies can be used to link the antigen to an external surface. There are also proteins that interface with specific biomolecules in the course of normal healthy activity.

For example, actin-capping proteins, proteins covering the ends of actin filaments, nucleate polymerization and prevent depolymerization.[16] Isolation and modification of these capping proteins may be useful for attaching specific ends of actin filaments to substrate surfaces in a similar way that the α-tubulin antibody has been used to tether and orient microtubules on substrate surfaces.[17]

22.5. SCIENCE AND ENGINEERING OF MOLECULAR MOTORS

As the engineering work with biomotors is quite recent, there has been much more progress in the science of the motors: bulk and single molecule studies done by biochemists and biophysicists have illuminated the mechanisms by which some of these motors achieve motion. From these experiments, we can determine stalling force, speed versus load, and other parameters necessary for the design and synthesis of a successful device. The initial engineering efforts have consisted of proof-of-concept demonstrations; no useful devices using molecular motors have been made to date. The work is incredibly exciting because of the rapid progress and incredible promise represented. Linear and rotary biomotor systems are being engineered and controlled. Progress has been also been made toward the manufacture of self-sustaining power sources for these systems. This section touches on some exciting recent science and engineering with molecular motors.

22.5.1. Single-Molecule Biophysics

Advances in the science and technology of molecular motors are primarily due to the development and application of experimental tools capable of studying the motors at the single molecule level. Measurements studying the dynamic behavior or mechanical properties of individual motor proteins center on time-resolved measurements of force and distance.[18] To function properly, these tools require precise coupling and commensurate compliances with the molecules of interest, which are also essential for successful use of these proteins as mechanical elements in engineered devices. These experimental tools can also be used to verify the successful assembly and operation of an individual device or component.

22.5.1.1. Laser Tweezers and Cantilever-Based Systems Force measurement methods fix the protein of interest to a compliant surface and measure the movement of this surface as the protein moves or is mechanically manipulated. Laser tweezers utilize a well-focused laser beam where, at the focal point, matter can become stably trapped due to scattering forces from the beam.[19] The magnitude of the trapping force is dependent on a number of factors such as the beam power, indices of refraction, numerical aperture, and so on. The restoring force is spring-like, allowing forces to be determined by measuring the particle distance to the trap center. Displacements of less than 1 nm can be measured at kHz rates using split photodiodes or video averaging techniques. The spring constants of laser tweezers are generally .01–10 pN/nm, which allow force measurements in the range of .01–200 pN.

Typically, commercially obtained polystyrene or silica spheres μm in size are used. They can be chemically functionalized, for example, presenting carboxylate or amino

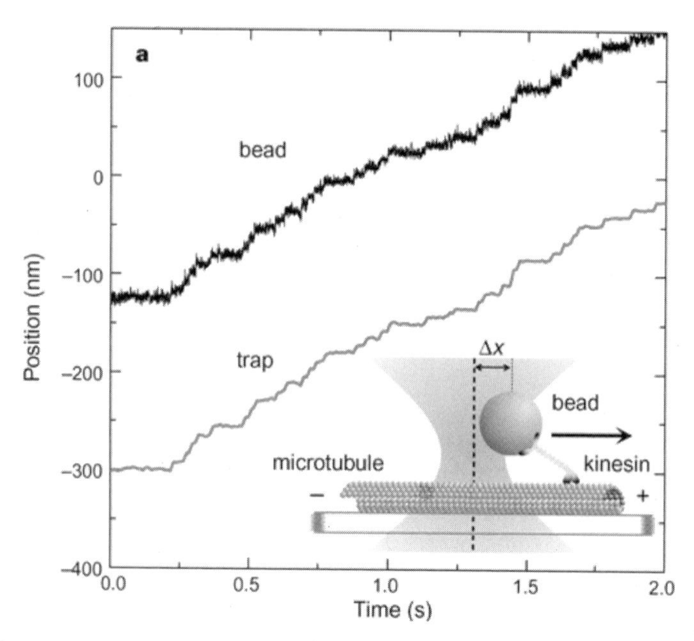

FIGURE 22.6. Laser tweezer position measurements of kinesin stepping along a microtubule. Constant force feedback moves the optical trap as the motor moves, maintaining a constant displacement of the bead from the trap center 8 nm steps can be seen. (From K. Visscher, M. J. Schnitzer, and S. M. Block, *Nature* **400**, 6740 (1999) by permission of Nature Publishing Group.)

endgroups. Attaching molecular motors or crosslinkers to these endgroups (as discussed in Section 2.3) is straightforward and allows the motor proteins to be strongly linked to the optical trap.

These instruments have been widely used to measure dynamic properties of motor proteins as well as for stretching studies of DNA and RNA. Shown in Figure 22.6 are measurements of the stepping behavior of kinesin on microtubules.[8] The motor took 8 nm steps against a 6.5 pN force, performing mechanical work of 52 pN-nm for each step. The ATP and ADP concentrations were similar to cellular concentrations, resulting in a measured efficiency of approximately 50%. The force necessary to stall the kinesin was 8 pN.

Force and distance measurements using cantilever-based systems operate similarly to optical tweezers, except that the spring-like restoring force results from mechanical stress in a deflected cantilever. Commercially obtained MEMS cantilevers or custom-made glass microneedles are typically used, with spring constants in the range of 10^{-3} to 10^5 pN/nm. Deflection measurements of cantilevers utilize the same detection techniques as optical tweezers: photodiodes measure the movement of a laser spot reflected from the back of the cantilever and video averaging can measure the position of the end of a microneedle. Glass microfibers have been used to measure single molecule steps of myosin on actin, measuring a step size of 5.5 nm.[20]

22.5.1.2. Optical and Fluorescence Microscopy The resolution of optical microscopy is diffraction-limited; however, this applies only to resolving the separation of two point-like

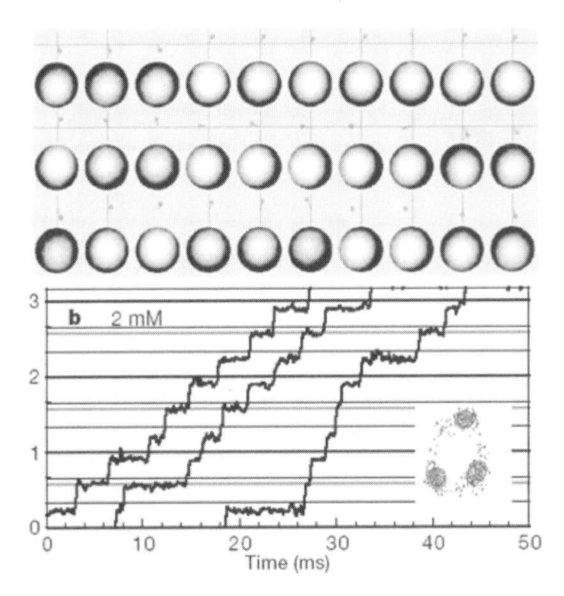

FIGURE 22.7. (a) Darkfield images of gold beads attached to the rotor of F1-ATPase. Centroid positions are shown above the images at 3× magnification. The interval between images is 0.5 ms. (b) Rotation versus time at 2 mM ATP. (From Ref. 3 by permission of Nature Publishing Group.)

objects. The movement of individual objects can be detected and measured to nm precision with video averaging and computer analysis. An example of dark-field optical microscopy measurements on a single molecular motor is shown in Figure 22.7. In these measurements, 20 nm gold spheres were attached to the rotor of F_1-ATPase and the rotation of the motor on a sub-ms timescale was measured by recording the movement of the gold spheres.[3] Images of the spheres were captured at 8 kHz using a fast CCD camera, digitzed, and analyzed. The high temporal resolution was able to identify substeps in the rotation; each 120° step comprises a 90° and a 30° substep.

Other forms of optical microscopy, such as Differential Interference Contrast (DIC) microscopy and fluorescent microscopy, are commonly used in measurements of microtubule and actin filament gliding assays. In these experiments, lawns of kinesin or myosin are deposited on the surface of a microscope slide in a flow cell. Microtubules or actin filaments, respectively, are introduced into the flow cell along with ATP, and the resulting movement of the filaments is visualized and recorded. The gliding velocity of the filaments can be studied as a function of motor concentration, fuel concentration, or specific mutations in the motor designed to probe protein function.

22.5.1.3. Fluorescence Resonant Energy Transfer (FRET) A third technique capable of measuring nanometer distances is Fluorescence Resonant Energy Transfer (FRET),[21] which utilizes two different fluorophores on the molecule of interest. The excitation and emission spectra of the fluorophores are chosen such that, when the molecules are sufficiently near, the energy resulting from the excitation of the higher frequency fluorophore will be transferred to the lower frequency fluorophore. A judicious choice of the fluorophores ensures that the excitation of the lower frequency molecule takes place only in these conditions.

If these molecules are situated on a mechanical protein, the movement of this protein can bring them closer together or further apart, thereby modulating strength of the resonant transfer. By exciting the first molecule and measuring the emission of the second, a high signal-to-noise ratio may be achieved. The intensity of the emitted light is a direct measurement of distance between the two fluorophores. This technique has been used to measure the mechanical movement of individual myosin motor proteins stepping along an actin filament.[22] The proximity measurements of different parts of the motor to each other during different stages of the catalytic and mechanical cycles is important in elucidating the mechanisms of motor movement.

22.5.2. *Engineered Devices*

In addition to experiments elucidating the mechanisms of motor proteins, there is a growing interest in investigating the possibilities of using them as force generating elements in engineered devices. For this to occur, these proteins must be interfaced with other mechanical components and positioned at desired locations on a substrate. The size of these components and the force constants associated with their linkages must be compatible with the forces generated by the motor, and the linkages themselves must be strong enough to enable the device to operate for a sufficiently long time.

Combining semiconductor fabrication techniques with modern biotechnology, a number of groups have begun to develop hybrid devices incorporating motor proteins. Using myosin, kinesin, and F_1-ATPase, they have shown that it is possible to control the positioning and activity of biomotors while combining them with inorganic components in an entirely non-biological environment.

22.5.2.1. Manufacturing of Inorganic Components The inorganic components are typically patterned using conventional optical or electron beam lithography (described in Chapter 1). Variations on the synthetic and lithographic processes are possible through alternative fabrication technologies, such as imprint lithography,[23] micro-contact printing,[24] and ink-jet lithography,[25] among others. The sizes and shapes of these inorganic components are chosen to be compatible with the function and capabilities of the designed device. By modifying the surface properties of these components through materials deposition or chemistry, proteins can be deposited precisely, with the spatial precision of attachment varying with the size of the defined area, as small as tens of nm.

When these inorganic surfaces are chemically modified, they become more or less amenable to protein conjugation. Both specific and non-specific interactions can be used to attach proteins where desired and to prevent them from attaching in undesired locations. Each has its own unique advantages and disadvantages. Specific binding implies an exact conjugation of a functional group on the protein to a functional group on the surface. The most common surfaces used to effect these attachments are oxides and gold. The hydroxyl group displayed by silicon oxide is relatively unreactive with proteins and biomolecules. Change of the surface group is easily done through the formation of self-assembled monolayers (SAMs) of linear bifunctional molecules,[26] such as silanes, which have one end reactive toward the surface –OH groups and the opposite end having a variable functionality. In this way, the hydroxyl surface may be transformed into an amine surface, a carboxylate surface, a hydrophobic surface, and so on. Gold binds specifically and strongly to sulfhydryl groups and modification of the surface functionality of gold surfaces can be likewise accomplished using bifunctional thiols.

Non-specific binding arises from generic interactions, which can occur when there are van der Waals, hydrophobic, or electrostatic forces between the protein and the surface sufficiently large to immobilize the protein. Non-specific interactions are advantageous in that they are easy ways to attach biomolecules to inorganic surfaces. However, this ease can also pose a problem, as other proteins in solution will also experience the same attraction to the surface as the target protein. This can also be used to prevent a protein from binding. Bovine serum albumin (BSA) has an extremely large non-specific binding, and may competitively interfere with the non-specific adsorption of the second protein. Incubation of BSA conjugated with biotin is also an easy way to produce biotinylated surfaces.[27] Non-specific adhesion properties can be also be affected by changing the surface chemistry. For example, conjugation of the silane diethylenetriaminopropyltrimethoxysilane (DETA) to silicon oxide surfaces attaches six amino groups to each silanated surface hydroxyl group. At physiological pH, this can present a rather large positive charge and thus attract or repel charged proteins.[28] Alternatively, covering the surface with SAMs of poly(ethylene glycol) (PEG) derivatives will also decrease the tendency to non-specifically bind.[29] PEG layers render surfaces resistant to protein adsorption, because they prevent the proteins from reaching the surface. This is due in part to steric repulsion by the long polymers and also the hydrophilicity of the polymers, creating stable interfacial water layers which further frustrate the solute proteins from any surface interactions.[30]

Rather than chemically engineering the surface, the protein itself can be modified by introducing or removing chemically active amino acids in the protein, as discussed in section 2.2. Addition of repeating histidine residues is an established technique in biochemistry for protein purification, since these repeats form strong bonds with metal ion complexes[31] and do not occur naturally. Recent work has also demonstrated the affinity of histidine tags for various solid metal substrates.[32] Short peptide sequences have been discovered which adhere specifically to gold and chromium,[33] iron oxide,[34] and various semiconductors (Si(100), InP(100), GaAs(100), GaAs(111)A, and GaAs(111)B).[35] This technique of tethering proteins to solid inorganic substrates has only begun to be exploited and has already been used to assemble microscopic particles of different materials.[36]

22.5.2.2. Devices Designing fabricated components with the desired interfacial chemistries and assembly of these components with engineered motor proteins are the first steps toward the synthesis of useful hybrid nanodevices. The techniques described above have been applied in initial efforts to integrate molecular motors into engineered devices:

22.5.2.2a. Linear Stepping Motors The natural translational systems utilizing kinesin and microtubules are very versatile, Such systems govern biological regulation of the translocation of different cargoes, active assembly and disassembly of microtubules, and complex processes like mitosis and meiosis. Engineering simplified versions of these systems is a first step toward replicating and adapting natural processes for artificial ends.

Initial efforts have centered on deposition of the motors and tracks on patterned substrates. Using a hydrophobic interaction between proteins and a patterned substrate, kinesin and myosin were deposited along predefined directions through shear deposition of poly(tetrafluoroethylene) (PTFE) on a glass surface.[37,38] Incubation of the motors followed by the addition of the complementary filaments and ATP resulted in translation along the PTFE alignment direction. The hydrophobic interaction was also used to deposit myosin in 10 μm lines of e-beam resist.[39] Fluorescently labeled actin was observed to move along the lines after addition of ATP. Other strategies for the directed deposition of microtubules have used electrostatic interactions between the negatively charged microtubules and positively charged SAMs of DETA.[17,28] Recent work investigating engineered material transport has utilized kinesin and microtubules. A demonstration system was constructed in which chips of silicon were micromachined using optical lithography and coated with kinesin.[40] Microtubules were incubated in a fluid cell with DETA-coated surfaces. Introduction of the kinesin chips into the flow cell followed by the addition of ATP resulted in translation and rotation of the chips.

Since kinesin translates unidirectionally, no cargo transport system will work very well without proper alignment and orientation of the microtubules. To address this issue, Hess and coworkers aligned microtubules using motility-induced collisions with patterned surfaces.[41] They adsorbed kinesin onto patterned microfabricated channels of polyurethane 2 μm wide and 1 μm deep. When the microtubules hit the channel walls, they were deflected along the channel direction. This is a possible method for the preparation of defined patterns of aligned microtubules. In this system, any microtubules that do not start with the patterned alignment acquire it as a consequence of their motion.

Although these patterned substrates could align microtubules, they were not oriented. By making the micromachined patterns asymmetric, it is possible to align and orient microtubules. Shown in Figure 22.8 is an optical micrograph of patterns in 1 μm thick photoresist.[42] By creating asymmetric patterns, the collisions of microtubules with sidewalls of the photoresist can rectify their motion and eventually force all of the microtubules to travel in the same direction. This results in the alignment and orientation of the microtubules. Subsequent fixing of the microtubules to the surface, will force the motion of any cargo-carrying kinesin to proceed unidirectionally along the direction of the pattern.

Directed orientation has also been successfully achieved using fluid flow. Stracke and coworkers aligned and oriented microtubules gliding on kinesin in a flow cell.[43] A more directed approach utilized antibodies specific for the plus end of the microtubule.[17] In these experiments, surfaces were coated with a surfactant containing Ni-NTA ligands, to which

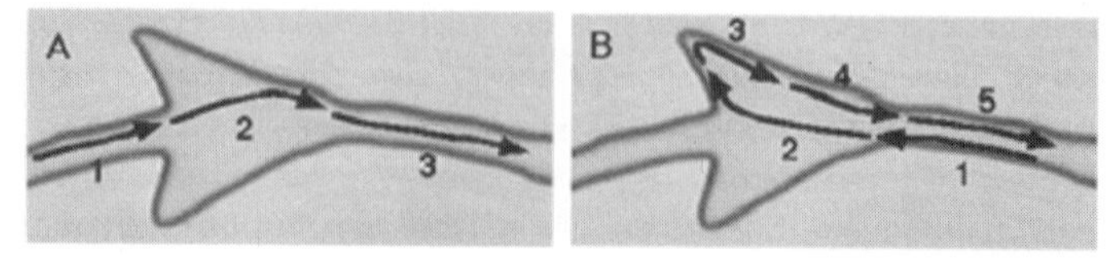

FIGURE 22.8. Patterned "ratchets" sort microtubules. Due to asymmetrically patterned photoresist, microtubules entering the triangular area exit to the right irrespective of their side of entry. This results in aligned and oriented microtubules. (From Ref. 42 by permission of the Biophysical Society.)

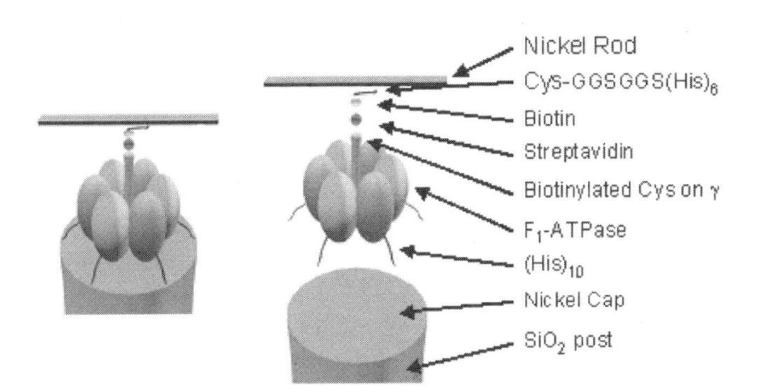

FIGURE 22.9. Left: Depiction of an assembled hybrid biomolecular nanodevice based on the rotary motor F1-ATPase (not to scale). Right: Exploded view showing all structural and linking components. The device self-assembles in multiple steps.

was bound a Histidine-tagged α-tubulin antibody. Introduction of microtubules into this chamber bound the α-tubulin-displaying plus end of the microtubules to the surface at one point, allowing the fluid flow to align and orient the microtubules.

Further work engineering the kinesin/microtubule system has involved optical control through the use of caged ATP.[41] Caged ATP is a non-hydrolyzable form of ATP which can become active upon illumination of UV light.[44] The light breaks the bonds forming the cage and frees the ATP for hydrolysis. After exposing the caged ATP to a controlled pulse of UV light, kinesin was activated for a limited time (a 30 second pulse liberated 20% of the caged ATP in the system), with the kinesin velocity falling as the ATP concentration decreased following the cessation of each pulse.

22.5.2.2b. F_1-ATPase-Based Hybrid Nanodevices There have also been recent successes centered on the creation of hybrid devices using F_1-ATPase. The devices were created by attaching the motor proteins to a patterned substrate and subsequently attaching nanofabricated nickel rods to the rotor of the motor. Utilizing many of the techniques outlined previously, all parts of the device were engineered: the gene encoding the protein, the manufactured protein, and the inorganic components of the device. Using video microscopy, rods were observed to rotate after the addition of ATP and functioned for several hours before failure.[45]

The devices (Figure 22.9) were designed to be placed precisely at specific locations on the substrate surface, arranged by patterning 50 nm circles of nickel on a glass coverslip using electron-beam lithography. The small size of the circles ensured a high probability of single motor attachment to each area. Concerns over surface interactions of the rotating rod during device rotation prompted a 100 nm-deep etch of the patterned glass surface. Since the nickel is resistant to this etching, posts are formed having nickel caps. To enable selective attachment of the motors to the nickel regions, $(\text{Histidine})_{10}$ tags were genetically engineered onto the "bottom" of the bases of the alpha and beta subunits of the protein.

Attachment of the nickel rod to the gamma subunit is made possible using sulfur chemistry to a non-naturally occurring cysteine in the gamma subunit. The wild type F_1-ATPase contains an undesirably located cysteine on each alpha subunit, which were replaced by serine using site-directed mutagenesis. Another serine on the "top" of the rotor was replaced with cysteine. Following these changes, the modified plasmid (see Figure 22.5) containing the F_1-ATPase gene was inserted into a bacterial expression system and the protein was manufactured in large quantities. Chemical modification of the individual self-assembled protein proceeded by attachment of biotin maleimide to the gamma cysteine, and subsequent ligation with streptavidin. The modified motors were then incubated with the nickel posts.

Nickel rods, 75 nm in diameter and 750 nm long, were fabricated using electron beam lithography and incubated with short polypeptides CysGlyGlySerGlyGlySer$(His)_6$. The histidine tags adhered to the nickel rod and the sulfur chemistry described above was used to attach biotin maleimide to the cysteine. The biotinylated rods, when incubated with the motor/post assemblies, linked to the streptavidin present on the gamma subunit. The device was then fully assembled, awaiting ATP for activation. The dimensions of the rods attached to the motor were chosen so that they would be large enough to observe with optical microscopy. Components much smaller than this are possible, but their observation is problematic.

Further work with F_1-ATPase has resulted in an engineered chemical switch, enabling it to be turned on and off as desired.[46] Using metal binding protein techniques,[47,48] three zinc ion binding sites were engineered into the motors at locations separate from the ATP binding sites. During normal function, the alpha and beta subunits move relative to each other; the new binding sites were engineered at the cleft of the alpha-beta dimer at a location of large relative displacement. When zinc ions are present, they are bound to the site and, since the alpha and beta subunits are both bound to the ion, they are effectively bound to each other. These motors can be reversibly and repeatedly switched off and on through the addition of a zinc salt and a chelator, which removes the zinc from the protein.

22.6. ENABLING MOLECULAR MOTORS IN TECHNOLOGICAL APPLICATIONS

The early attempts to incorporate molecular motors into devices discussed above have suffered from low yields, crude control mechanisms, short lifetimes, dependence on chemical fuel, and high sensitivity to environmental conditions. Finding solutions to these problems will enable the creation of entire classes of devices, and will allow the field to move forward rapidly.

22.6.1. *Directed Device Assembly*

Biomolecular hybrid devices produced by self-assembly alone rely on diffusion of the component parts, a process that works exceptionally well for biological and chemical systems. However, when certain component parts are relatively few in number, have small diffusion constants, or are contained in a large solution volume, the time necessary to construct the devices can become prohibitively great. Diffusion-based self-assembly will not suffice for devices with parts having poor diffusion characteristics or requiring subsequent manipulation. It is not enough that some proteins collide with their intended targets—they must be oriented and bonded specifically to them and no other.

Although collision rates can be improved through increasing the number of particles or decreasing the assembly chamber size, this will not suffice for rare parts or unusual device geometries. One possible approach is to use electrophoresis, which has been used to immobilize DNA[49] and microscopic fabricated components[50] at specific locations of fabricated substrates. For uncharged proteins and fabricated components in ionic buffers, AC dielectrophoresis (DEP) can be successfully substituted for DC electrophoresis. DEP has been used to trap protein structures like the 250 nm-diameter herpes simplex virus,[51] as well as latex spheres down to 14 nm in diameter.[52] In addition, traveling-wave dielectrophoresis (TWDEP) can be used to move trapped particles between phased electrodes, enabling controllable collection and manipulation of particles.[53] Enhanced docking rates may be thus be achieved by immobilizing the proteins in specific locations using DEP and controllably introducing other components to the proteins with TWDEP, repeating the process until the devices are fully assembled. Assembled devices can be removed using TWDEP and the process repeated.

22.6.2. Control

Control of a device increases its functionality and utility. Some motors, like kinesin and myosin, have natural mechanisms that enable them to hydrolyze ATP up to 5000 and 200 times faster if they are in contact with microtubules and actin, respectively.[1] This prevents the motors from functioning and wasting fuel when the motors are not in contact with their substrates. Engineering controls and switches into any hybrid device would be useful for the same reason and also for a variety of others: turning the motors on and off may be essential for a specific application, to regulate the occurrence of certain activities, to direct motion, or control speed.

Substrate-bound devices can be electrically addressable, which facilitates connection with the outside world and rapidly conveys signals to the device. Externally applied electric fields can cause proteins to deform sufficiently that their operation is affected. This can happen because of charge or polar groups in the protein, which cause parts of the protein to move in response to an applied field. This shape change can alter the catalytic activity of the protein, slowing it down or stopping it altogether. There are a large number of membrane channel and pore proteins which gate their conductances and transport properties in response to applied voltages. These proteins themselves can be incorporated into devices, or their motifs for electrical sensitivity could be used to modify existing proteins. The response times of some of these proteins can be very short: a newly discovered auditory protein, prestin, shows activity at kilohertz frequencies.[54]

Mobile devices may be better served by chemical or optical switches. Protein modification such as the zinc switch discussed in section 3.2.2b could result in devices which respond presence of ions or other analytes in their environment. Optical controls can alter the fuel supply such as the caged-ATP system discussed in Section 3.2.2a, or they can directly interact with the proteins. Although direct engineering of these controls has not been realized yet, there are a large number of proteins which are optically active, such as bacteriorhodopsin, rhodoposin, and so on, which have light-induced activities or conformational changes upon illumination by light of certain wavelengths.

Mechanical regulation of motor activity can be achieved through alteration of the load through application of magnetic or electric fields. Any moving part attached to the motor, if susceptible to an external field, may be used to increase the drag force experienced by the motor. If the motor operates at constant efficiency, any change in the load will be met by an opposing change in speed.

22.6.3. *Energy Sources*

Continuously supplying ATP to bionanomachines is a major concern, and the possibility of renewable energy sources of ATP is very attractive. Recent work has resulted in light-powered ATP production in lipid vesicles using F_0F_1-ATP synthase with bacteriorhodopsin[55–57] or artificial photosynthetic complexes.[58] Bacteriorhodopsin (BR) is a natural optically active proton pump that transports protons across the bacterial membrane and out of the cell upon the absorption of a photon of green light. As protons are pumped out of the cell, a charge and pH gradient forms across the cell membrane, forming an electro-osmotic potential. This potential provides the energy to power ATP synthase, which, in transferring the protons back across the bacterial membrane, uses their electrochemical energy to produce ATP. Because these fuel-generating systems are self-contained, they can be mixed with already existing devices to supply a virtually unlimited quantity of fuel.

22.6.4. *Device Fabrication Materials*

Construction of machines made from biological and inorganic components combine two classes of materials with very different processing and handling requirements. To bridge this compatibility gap, special fabrication techniques and unconventional materials must be developed. Besides developing inorganic fabrication processes and molecular biology protocols for engineering of the biomolecular motors, interfacial protocols must be developed that will allow seamless merging of nanofabrication and biology. In general, the fabrication environment is not amenable to biological materials and molecular biology procedures and materials are detrimental to semiconductor fabrication equipment.

Membrane-bound proteins incorporated into engineered devices may be incorporated into biocompatible polymers. Use of polymer membranes is desirable for the following reasons: they have a longer lifetime than lipid membranes, they are more rugged, and properties such as electronic and ionic conductivity and permeability can be tailored to suit each application. A large number of materials have demonstrated compatibility and can incorporate proteins while maintaining their functionality, such as gelatin, hydrated polymers, sol-gels, poly(vinyl alcohol), and poly(acrylamide).[59] For devices in which the protein environment must replicate the two dimensional nature of natural membranes, vesicle and membrane forming amphiphilic polymers can be used.[60]

22.6.5. *Engineering Issues*

Before a successfully functioning device can be made, its design must acknowledge the limitations imposed by the performance and composition of each of its individual parts. Motor proteins cannot be loaded beyond the stall force, devices cannot be exposed to forces which would cause their components to dissociate or rupture, and they cannot be exposed to environments in which the protein will denature. The environmental conditions the hybrid devices are exposed to can be characterized by temperature, pH, ionic composition and concentration, solvent, and well as extra-solute factors such as external light (UV, IR etc), sound, and so on.

An additional concern is internal structural integrity of a given protein. A particularly well-studied example of this is the protein titin, whose tertiary structure consists of a series of small globular domains, which "unzip" as tension is applied to this linear protein[61]. For hybrid motor devices in which the motor is attached to one object while moving another, it is useful to bear in mind the forces and stresses each bond individually is likely to experience.

22.7. CONCLUSION

The number of motor proteins known today is quite large (for example, 145 myosins, 268 kinesins, and counting) and will expand further in the future as more whole genomes are sequenced. Many organisms have evolved to fill a natural niche, and their protein constituents will have adapted to support it there. The number of organisms and environments result in a large number of different motors, as each motor has different properties and activities as a function of pH, temperature, and other environmental conditions. As increasing numbers of motor structures are known, general principles regarding motor function will be elucidated and refined. Purification of these proteins enables crystallization and structure identification, which in turn, guides any choices for engineering and modification. As knowledge of these principles progresses, the possibility of designing motor proteins *de novo* becomes more realistic. A purely artificial motor may be designed for a specific purpose and, as such, may be easier to manipulate, fabricate, handle, mount, and operate. The science of motor proteins will inform the design and synthesis of non-protein motors that will fully take advantage of the full range of organic and inorganic chemistry.

When constructing engineered hybrid biomotor devices, the ultimate aim is to construct systems of nanomechanical machines that are able to carry out tasks more efficiently or rapidly than their macro- or microscale counterparts. Most engineering applications (cargo transport, actuation, pumping, sensing, etc.) are required for the normal functioning of living systems and their constituent cells. Because of this, these interdependent assemblies of cooperative components- on a scale of complexity beyond our current ability to design- have existed naturally for millions of years. If Nature's evolved designs are not the best solution to their given task, they can at least inform us as to a vastly different way to think about how things move and interact on the nanometer scale- greatly improving any artificial designs that we may devise.

The outlook for this field is very bright. Natural nanomachines can self-assemble and self-replicate, making structures as complex and powerful as the human brain, composite lightweight materials like wood, and information storage systems complex and compact enough to fit the genome encoding an entire organism into a cellular organelle. It seems that the power to understand, duplicate, and improve on these systems would result in revolutionary capabilities to construct and change the world around us. To date, we have isolated only a few bionanomachines and used them to perform a small number of simple tasks. To expand our capability to make different kinds of devices for diverse purposes, a larger toolbox of linking molecules is needed to assemble the various parts in each device. As more motors are discovered, motors that have unique properties, operating conditions, speeds, and functionalities will enable devices not previously possible to be constructed. With advances in these areas and more experience in designing and constructing hybrid bionanomachines, ever more complex devices will be realized.

FURTHER READING

A great introduction to the biology and physics of motor proteins: *Mechanics of Motor Proteins and the Cytoskeleton*, Jonathon Howard, (Sinauer Associates Inc., Sunderland, 2001).

A comprehensive source on cytoskeletal and motor proteins: *Cytoskeletal and Motor Proteins*, T. Kreis and R. Vale, (Oxford University Press, New York, 1999).

Biochemistry: *Biochemistry*, L. Stryer (W. H. Freeman and Company, New York, 1995).

Mechanisms and operation motor proteins in the cell: *The Molecular Biology of the Cell*, B. Alberts, D. Bray, J. Lewis, M. Raff, K. Roberts, J. D. Watson (Garland Publishing, New York, 1994).

QUESTIONS

1. What are the differences between biological and MEMS mechanical devices?
2. Compare and contrast the catalytic cycles of kinesin, myosin, and F_1-ATPase.
3. Design the construction and operation of a cargo transportation system using kinesin and microtubules which transports and combines two different cargos, bringing them together in a designated reaction zone.
4. Name some shortcomings in hybrid devices made to date and some proposed improvements.

REFERENCES

1. J. Howard, *Mechanics of Motor Proteins and the Cytoskeleton* (Sinauer Associates, Inc., Sunderland, 2001).
2. K. Kinosita, *Faseb Journal* **14**, 1567 (2000).
3. R. Yasuda, H. Noji, M. Yoshida, K. Kinosita, and H. Itoh, *Nature* **410**, 898 (2001).
4. Intel Corporate Press Release, 2002.
5. L. Stryer, *Biochemistry* (W. H. Freeman and Company, New York, 1995).
6. R. D Vale and R. J. Fletterick, *Annual Review of Cell and Developmental Biology* **13**, 745 (1997).
7. K. Svoboda and S. M. Block, *Cell* **77**, 773 (1994).
8. K. Visscher, M. J. Schnitzer, and S. M. Block, *Nature* **400**, 184 (1999).
9. F. Gittes, E. Meyhofer, S. Baek, and J. Howard, *J. Biophysical Journal* **70**, 418 (1996).
10. P. D. Boyer and W. E. Kohlbrenner, in *Energy Coupling in Photosynthesis* (eds. Selman, B. R. & Selman-Reimer, S.) 231–240 (Elsevier, Amsterdam, 1981).
11. D. J. DeRozier, *Cell* **93**, 17–20 (1998).
12. D. E. Smith, S. J. Tans, S. B. Smith, S. Grimes, D. L. Anderson, and C. Bustamante, *Nature* **413**, 748 (2001)
13. G. Hermanson, (Academic Press, San Diego, 1996).
14. S. S. Wong, *Chemistry of protein conjugation and cross-linking* (CRC Press, Boca Raton, 1991).
15. R. Merkel, P. Nassoy, A. Leung, K. Ritchie, and E. Evans, *Nature* **397**, 50 (1999).

16. J. A. Cooper, M. Hart, T. Karpova, and D. Schafer, in *Guidebook to the Cytoskeletal and Motor Proteins* (eds. Kreis, T., & Vale, R.) (Oxford University Press, New York, 1999).
17. L. Limberis, J. Magda, and R. J. Stewart, *Nanoletters* **1**, 277 (2001).
18. C. Bustamante, J. C. Macosko, and G. J. L. White, *Nature Reviews Molecular Cell Biology* **1**, 130 (2000).
19. M. P. Sheetz, *Laser Tweezers in Cell Biology* (Academic Press, San Diego, 1998).
20. K. Kitamura, A. Ishijima, M. Tokunaga, and T. Yanagida, *JSAP International* **4**, 4 (2001).
21. S. Weiss, *Nature Structural Biology* **7**, 724 (2000).
22. W. M. Shih, Z. Gryczynski, J. R. Lakowicz, and J. A. Spudich, *Cell* **102**, 683 (2000).
23. S. Y. Chou, P. R. Krauss, and P. J. Renstrom, *Journal of Vacuum Science & Technology* B **14**, 4129 (1996).
24. R. Kane, S. Takayama, E. Ostuni, D. Ingber, and G. Whitesides, *Biomaterials* **20**, 2363 (1999).
25. S. Hong and C. A. Mirkin, *Science* **288**, 1808 (2000).
26. D. L. Allara in *Nanofabrication and biosystems: integrating materials science, engineering, and biology* (eds. Hoch, H. C., Jelinski, L. W., & Craighead, H. G.) (Cambridge University Press, New York, 1996).
27. V. T. Moy, E.-L. Florin, and H. E. Gaub, *Science* **266**, 257 (1994).
28. C. M. Kacher, I. M. Weiss, R. J. Stewart, C. F. Schmidt, P. K. Hansma, M. Radmacher, and M. Fritz, *European Biophysics Journal* **28**, 611 (2000).
29. F. E. Bailey, Jr., and J. Y. Koleske, *Poly(Ethylene Oxide)* (Academic Press, New York, 1976).
30. P. Harder, M. Grunze, R. Dahint, G. M. Whitesides, and P. E. Laibinis, *Journal of Physical Chemistry* B **102**, 426 (1998).
31. F. Kienberger, G. Kada, H. Gruber, V. Pastushenko, C. Riener, M. Trieb, H.-G. Knaus, H. Schindler, and P. Hinterdorfer, *Single Molecules* **1**, 59 (2000).
32. J. J. Schmidt, X. Jiang, and C. D. Montemagno, *Nanoletters* **2**, 1229 (2002).
33. S. Brown, *Nature Biotechnology* **15**, 269 (1997).
34. S. Brown, *Proceedings of the National Academy of Sciences* **89**, 8651 (1992).
35. S. Whaley, D. English, E. Hu, P. Barbara, and A. Belcher, *Nature* **405**, 665 (2000).
36. S. Brown, *Nanoletters* **1**, 391 (2001).
37. H. Suzuki, K. Oiwa, A. Yamada, H. Sakakibara, H. Nakayama, and S. Mashiko, *Japanese Journal of Applied Physics* **34**, 3937 (1995).
38. J. R. Dennis, J. Howard, and V. Vogel, *Nanotechnology* **10**, 232 (1999).
39. D. V. Nicolau, H. Suzuki, S. Mashiko, T. Taguchi, and S. Yoshikawa, *Biophysical Journal* **77**, 1126 (1999).
40. L. Limberis, and R. J. Stewart, *Nanotechnology* **11**, 47 (2000).
41. H. Hess, J. Clemmens, D. Qin, J. Howard, and V. Vogel, *Nanoletters* **1**, 235 (2001).
42. Y. Hiratsuka, T. Tada, K. Oiwa, T. Kanayama, T. Q. P Uyeda, *Biophysical Journal* **81**, 1555 (2001).
43. R. Stracke, K. J. Bohm, J. Burgold, H.-J. Schacht, and E. Unger, *Nanotechnology* **11**, 52 (2000).
44. J. A. Dantzig, H. Higuchi, and Y. E. Goldman, *Methods in Enzymology, Caged Compounds* **291**, 307 (1998).
45. R. K. Soong, G. D. Bachand, H. P Neves, A. G. Olkhovets, H. G. Craighead, and C. D. Montemagno, *Science* **290**, 1555 (2000).
46. H. Liu, J. J. Schmidt, G. D. Bachand, S. S. Rizk, L. L. Looger, H. W. Hellinga, and C. D. Montemagno, *Nature Materials* **1**, 173 (2002).
47. L. Regan, *Annual Reviews of Biophysics and Biomolecular Structure* **22**, 257 (1993).
48. H. W. Hellinga and F. M. Richards, *Journal of Molecular Biology* **222**, 763 (1991).
49. C. F. Edman, D. E. Raymond, D. J. Wu, E. Tu, R. G. Sosnowski, W. F. Butler, M. Nerenberg, and M. J. Heller, *Nucleic Acid Research* **25**, 4907 (1997).
50. C. F. Edman, R. B. Swint, C. Gurtner, R. E. Formosa, S. D. Roh, K. E. Lee, P. D. Swanson, D. E. Ackley, J. J. Coleman, and M. J. Heller, *IEEE Photonics Technology Letters* **12**, 1198–1200 (2000).
51. H. Morgan, M. P. Hughes, and N. G. Green, *Biophysical Journal* **77**, 516 (1999).
52. T. Muller, A. Gerardino, T. Schnelle, S. G. Shirley, F. Bordoni, G. De Gasperis, R. Leoni, and G. Fuhr, *Journal of Physics D: Applied Physics* **29**, 340 (1996).
53. M. P. Hughes, R. Pethig, and X.-B. Wang, *Journal of Physics D: Applied Physics* **29**, 474 (1996).
54. M. C. Liberman, J. Gao, D. Z. Z. He, X. Wu, S. Jia, and J. Zuo, *Nature* **419**, 300 (2002).
55. B. Pitard, P. Richard, M. Dunach, and J.-L. Rigaud, *European Journal of Biochemistry* **235**, 779 (1996).
56. P. Richard and P. Graber, *European Journal of Biochemistry* **210**, 287 (1992).
57. P. Richard, B. Pitard, and J.-L. Rigaud, *Journal of Biological Chemistry* **270**, 21571 (1995).

58. G. Steinberg-Yfrach, J.-L. Rigaud, E. N. Durantini, A. L. Moore, D. Gust, and T. A. Moore, *Nature* **392**, 479 (1998).
59. R. Birge, N. Gillespie, E. Izaguirre, A. Kusnetzow, A. Lawrence, S. Singh, W. Song, E. Schmidt, J. Stuart, S. Seetharaman, and K. Wise, *Journal of Physical Chemistry* B **103**, 10746 (1999).
60. C. Nardin, J. Widmer, M. Winterhalter, and W. Meier, *Eur. Phys. J.* E **4**, 403 (2001).
61. M. Rief, M. Gautel, F. Oesterhelt, J. M. Fernandez, and H. E. Gaub, *Science* **276**, 1109 (1997).

23

Nanofluidics

Jongyoon Han
Department of Electrical Engineering and Computer Science
Division of Biological Engineering
Massachusetts Institute of Technology, Cambridge, MA

23.1. INTRODUCTION

The study of fluidic motion at small size scales has always been a part of several engineering areas. For example, water or oil transport through porous material (e.g., rocks and soil) has been important in petroleum engineering for a long while. The introduction of the 'lab-on-a-chip' or the 'micro total analysis system (μTAS)' concept has generated a renewed interest in micro/nanofluidic technologies among engineers and biologists over the last decade. Most chemical and biological analyses are currently done in 'wet' laboratories, and require bulky and expensive analysis tools as well as skilled labor. In addition, these analysis methods require large amount of sample because of the sample loss during the complicated sample preparation/purification steps they entail. These problems would be greatly alleviated if a molecular analysis system could be scaled down to very small dimensions. Also the scaling laws of molecular separation processes generally favor miniaturization in terms of speed and separation resolution.[1] Such concept was first introduced by Terry *et al.*[2,3] when they devised a gas chromatography system micromachined on a silicon wafer. However, it was only in 1990 when the advantages of micromachined chemical analysis system were fully appreciated by Manz *et al.*[1] Micro total analysis systems have many advantages, such as lower power consumption, smaller reagent consumption, smaller overall size, lower cost, portability and disposability. Such prospects became of specifically high interest in the 1990s with the advent of large biological undertakings such as the Human Genome Project.[4,5]

So far, microfluidic chemical and biological analysis has been successfully applied to DNA sequencing, protein separation, single cell analysis, sample preparation/preconcentration, biomolecular sorting, and even single molecule detection. Apart from the

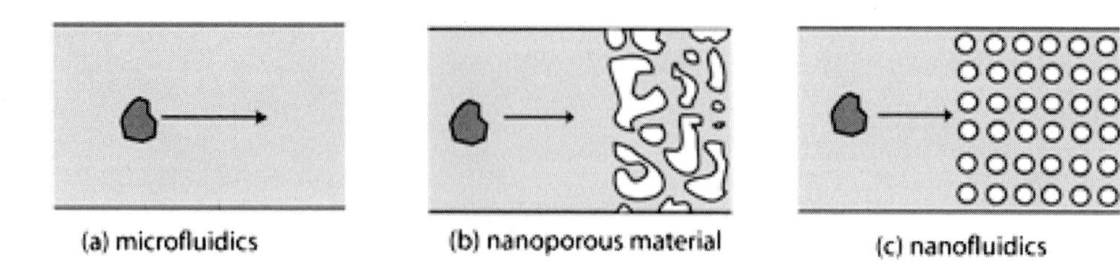

FIGURE 23.1. Comparison between microfluidic and nanofluidic biomolecule separation. (a) In microfluidic device, friction between liquid and the molecule determines the molecular mobility. (b) In nanoporous material such as gel, molecules are filtered or sieved by random nanostructures. (c) In nanofluidic devices, the molecular sieve structure is well defined and regular.

development of individual analysis components, efforts are also under way to integrate these components into a complete chemical/biological analysis system. Reviews on the general μTAS research can be found in references,[6,7] as well as in conference proceedings.[8,9]

While μTAS research decreased the size of analysis system drastically, most microfluidic systems demonstrated so far have dimensions of 10 to 500 μm, which is still substantially larger than the dimension of various biological particles or organelles, not to mention biomolecules like protein and DNA. In such an analytical system, separation is achieved mainly by the interaction between the solvent and the molecules to be analyzed (Fig. 23.1a). In many cases, the basic mechanism of operation is the same as the conventional analysis techniques, thus retaining their inherent limitations. Traditionally, the molecular separation processes used a random porous material (a random polymer matrix, as in Fig. 23.1b) for sieving/filtering nanometer-size biomolecules. Polymers gels (polyacrylamide or agarose gel) with various pore sizes have been used for biomolecule separation. However, the microscopic structure of these systems is inherently random, and critical dimensional parameters (pore sizes) are not easy to measure or control. This poses a problem for theoretical and experimental studies aimed at improving or optimizing the technique for better separation.

In contrast, nanofluidic systems fabricated by micro/nanofabrication techniques provide unique capability in biomolecule analysis and control. It is now possible to fabricate regular nanostructures or nanoconstrictions of dimensions ranging between 10 and 1000 nm (Figure 23.1c). Fluid at this size scale will have distinct properties that cannot be found at the macro- or microfluidic size scale. These distinct properties can be exploited for new engineering concepts, such as in electrokinetic pumping. Molecules within the fluid with such a small dimension are expected to behave differently, although there is a general lack of theoretical and experimental researches in this area. The size of many organelles or even larger components in cells falls in this range, such as lysosome (200∼500 nm), mitochondria (∼500 nm), secretory vesicles (50∼200 nm), ribosome (∼30 nm), virus particles (∼50 nm), and even DNA wrapped around histones (11 nm). In such a nanofluidic system, the interaction between the fluidic structure (or inner surface of the fluidic device) and molecules plays a dominant role over the interaction between the molecules and surrounding solvent. Since the fluidic structure is relatively easy to control and change (as opposed to the liquid-molecule interaction), this provides opportunities to control these biomolecules or bioparticles by carefully designed nanofluidic devices or nanostructures.

The possibility of advanced molecular control, as well as the novel nanofluidic properties that can be used in various applications, are the two key drivers for current nanofluidic research. Regular, micromachined nanofluidic structures have a clear advantage over traditional random nanoporous materials in the fact that one can carefully design and control the fluidic motion and molecular motion within the structure.

23.2. FLUIDS AT THE MICRO- AND NANOMETER SCALE

23.2.1. Low Reynolds Number Fluidics

Almost all physical phenomena will become qualitatively different when the length scale of the system is changed. Both airplanes and flies can fly, but their flying mechanisms are quite different. Within micro- or nanoscale fluidic channels, fluid behaves quite differently than in larger channels or pipes, mainly because relevant forces scale differently with size. In general, the inertia of the fluid becomes negligible while the (viscous) friction force and surface tension become dominant forces at the small length scale. In fluid mechanics, the Reynolds number is the dimensionless parameter that determines the relative importance of inertia compared to frictional force due to viscosity. It is written as:

$$N_R = \frac{\text{inertial force}}{\text{friction (viscous) force}} = \frac{LV\rho}{\mu} \tag{23.1}$$

where L is a relevant length of the fluidic system, V is the relevant speed, ρ and μ are the density and viscosity of the liquid, respectively. The Reynolds number for a typical microfluidic system (assuming the physical dimension in the order of 10 μm, viscosity and density of water, fluid velocity in the order of 1~100 μm/s) is between $10^{-3}\sim10^{-5}$. A system is generally considered to be in the laminar flow regime (or low-Reynolds number regime) when N_R is less than $\sim10^2$, therefore it is the case in microfluidics and nanofluidics (Figure 23.2). A good analogy for a low Reynolds number fluidics is the dynamics of a highly viscous fluid, such as molasses.

The Navier-Stokes Equation for the low-Reynolds number regime, given that one can ignore the inertial term in the low Reynolds regime, is reduced to the following simple, linear equation.

$$\eta\nabla^2\vec{v} = \nabla P \tag{23.2}$$

where η is viscosity of the fluid, and P is the pressure. This equation has been solved for several geometries relevant to microfluidics.[10] It also possesses several unique characteristics.

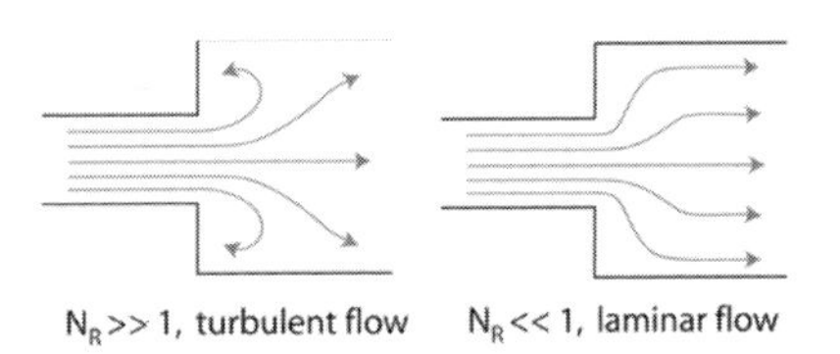

IGURE 23.2. High and Low Reynolds number fluidics. When the Reynolds number is low, viscous interaction etween the wall and the fluid is strong, and there is no turbulences or vortices.

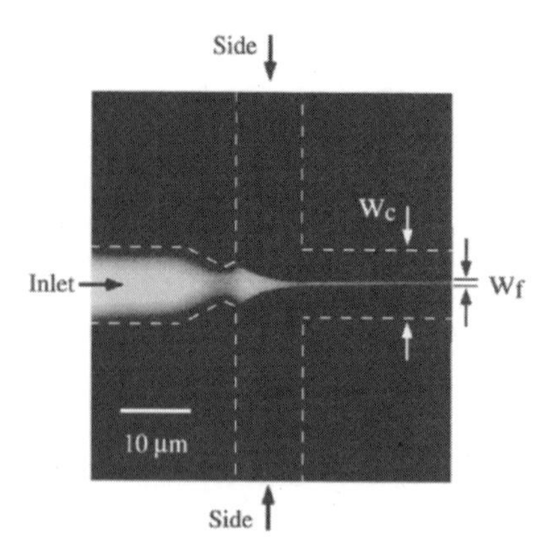

FIGURE 23.3. Hydrodynamic focusing of a liquid stream in a microfluidic channel. Fluid containing fluorescent molecules is driven from the inlet to meet with two other non-fluorescent liquid streams from the side channels. The width of the liquid stream can be controlled by changing the pressures applied to side and inlet channels. In the microchannel, only diffusional mixing can occur, and by narrowing down inlet stream width (wf) one can achieve fast diffusional mixing. (Adapted from Ref. 11 by permission of the American Physical Society.)

First, this equation is time-independent, which stems from the absence of inertia terms. The fluid therefore only reacts to the external forces (such as electric forces or hydraulic pressures), and when this external force is reversed the fluid motion is also reversed without any mixing or irreversible change. This turbulence-free fluidic environment is ideal for studying and manipulating fragile polymeric biomolecules such as chromosomal DNA. On the other hand, it is not trivial endeavor to design a fluid mixer in microfluidic systems. The lack of turbulence also means that no other mechanism except diffusion is present to mix two different fluids.

Indeed, transport by diffusion can be a very effective means of mixing in the low Reynolds number regime. Approximate diffusion time of an analyte or fluid molecule is given as $\sim(\text{distance})^2/(\text{diffusion constant})$. The squared dependence on the length of diffusion means that diffusion would be very efficient in a short length scale, while very slow in large length scale. For example, Knight *et al.*, developed a microfluidic system for fast and controlled mixing of liquids by diffusion[11] (Figure 23.3). Controlled diffusional transport was also used for biomolecule size separation in micro/nanofluidic environment.[12,13]

23.2.2. *The Effect of Surface Charge and Debye Layer*

The high surface to volume ratio of micro- and nanofluidic channels means the dominance of surface effects on fluidic motion. Most surfaces have some surface charges when they are in contact with an electrolyte solution, because the surface chemical group can be protonated or unprotonated depending on the pH of the electrolyte. For example, the surface of silicon dioxide or glass is mostly terminated by silanol groups (—Si—OH), and they can be unprotonated at pH higher than ~2. Therefore, at most pH conditions, the surface of glass or silicon dioxide contains negative surface charges, and the positive ions in the solution develop a charge-screening layer called Debye layer (Figure 23.4). Within the Debye layer, electroneutrality rule is not satisfied and the electric potential is not zero. According

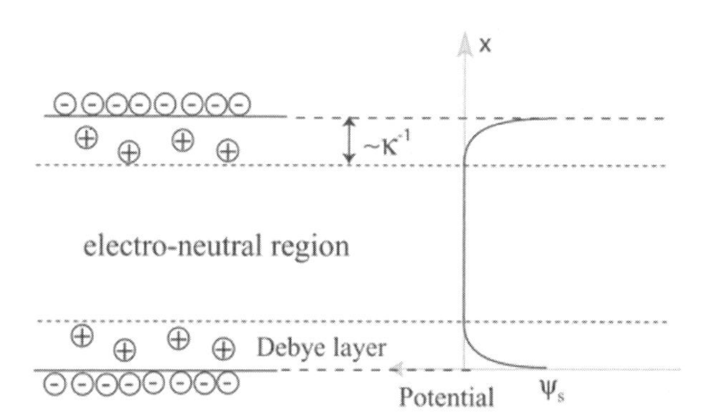

FIGURE 23.4. Charge screening and Debye length. The surface charges are screened by the counterions in the aqueous solution, and the screened charge layer contains mobile, net charges. These mobile charges can drift under the influence of external electric field.

to the Gouy–Chapman model, this screening charge density decays exponentially from the surface. The characteristic length of this is called as the Debye length, and is given as,[14]

$$\kappa^{-1} = \frac{3.04}{z\sqrt{M}} \times 10^{-10}(m) \tag{23.3}$$

where M is the molarity of a (z:z) electrolyte in the buffer. As the ionic strength (or the number of charged ions) of the electrolyte increases, Debye length decreases.

For a univalent electrolyte ($z = 1$) and $M = 10^{-4}$, the Debye length is calculated to be about 30 nm. Therefore, at low buffer concentrations and with nanofluidic channels, Debye layer thickness can be comparable to the channel dimension. This could change fluid or biomolecule motion qualitatively in the nanofluidic channel.[15] The phenomenon continues to be the subject of ongoing research.

23.2.3. *Electroosmosis and Streaming Potential*

Electroosmosis consists in the motion of liquids in a porous structure under an applied lectric field, driven by the motion of Debye layer charges. While the bulk fluid has no net harge, this electroneutrality rule breaks down near the walls of the porous structure or mirochannel. However, given that the screening charges within this Debye layer are mobile, ey can be driven by applying an electric field. For instance, in the case of negative surface harges (as is the case for SiO_2 or glass microfluidic channels), the fluid within the Debye yer will be driven to the cathode under an electric field. The entire fluid column can then e dragged by this layer due to viscous interaction, resulting in a net fluid flow through the annel. Streaming potential is the potential generated when a fluid is forced to flow through channel or porous material, a reversal of electroosmosis. The fluid flow drags and accumutes the mobile layer charges, resulting in a potential difference. Electroosmotic flow can sily be generated by applying electric potential across a micro- or nanofluidic channel, d is widely used as a flow generation mechanism in these devices. However, the use of

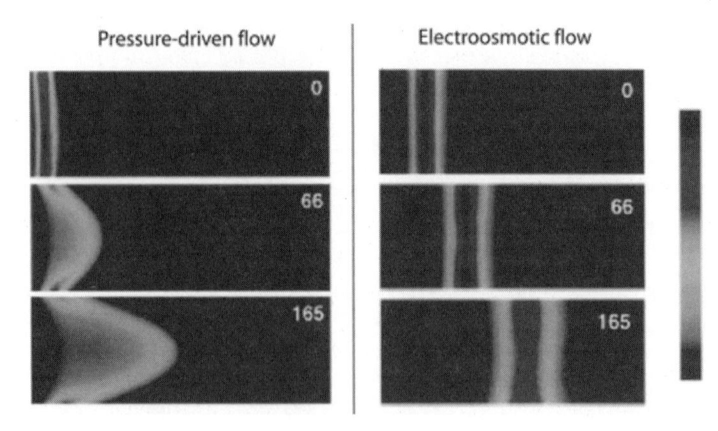

FIGURE 23.5. Pressure-driven flow vs. Electroosmotic flow. These images were taken by caged dye techniques. At $t = 0$, an initial flat fluorescent line is generated in microchannel by pulse-exposing and breaking the caged dye, rendering them fluorescent. These dyes were transported by the fluid flow generated by pressure-driven flow (left column) or electroosmotic flow (right column), showing the flow profile. (Adapted from P. H. Paul, M. G. Garguilo, and D. J. Rakestraw, *Anal. Chem.* **70**, 2459 (1998) by permission of the American Chemical Society.)

streaming potential is limited by the high pressure necessary to force liquids through nanoporous structures.

Electroosmotic flow velocity depends on the applied electric field, the viscosity of the fluid, and the surface potential due to the surface charge density. The electroosmotic flow speed is given as:

$$v = -\frac{\varepsilon \psi_s E}{\eta} \tag{23.4}$$

where ε is the permittivity, η is the viscosity, and ψ_s is the surface (zeta) potential. Interestingly, flow velocity does not depend on the size or shape of the channel. Because of this, electroosmotic flow is the only feasible mechanism to induce fluid flow through very thin nanofluidic channels. In addition, electroosmotic flow possesses a preferred flat-shaped 'plug flow' profile, while pressure-driven flow has a parabolic (Poiseiulle) flow profile. A flat profile is preferred for molecular separation since it does not feature the dispersion of the pressure-driven (or Poiseiulle) flow profile (Figure 23.5).

Measuring electroosmotic flow is also a simple way to measure the surface potential (or surface charge density) of micro- and nanofluidic channels. This is important in practical point of view, since the surface charge density of microchannels and microfabricated surfaces can vary a lot depending on the exact processing conditions and methods. I microfluidic or nanofluidic MEMS, characterizing the surface charge density or the electroosmotic flow is essential to insure reliable and reproducible results. Electroosmotic flow in micro- and nanochannels can also be affected by other factors, such as surface roughness pH, and the ionic strength of the liquid (Debye layer thickness). In general, the increase buffer concentration or ionic strength will decrease the electroosmotic flow velocity, sinc it will decrease the thickness of Debye layer and the number of mobile screening charge that is responsible for electroosmosis.

In some applications, it is highly desirable to suppress or change the direction of electroosmotic flow. This can be done by a number of ways, including simple methods such as lowering pH or increasing buffer concentration, and more elaborate methods such as coating the surface with polymers, or dynamic coating methods.[16,17] Through an appropriate coating method, it is even possible to change the polarity of the surface charge, therefore reversing the electroosmotic flow direction. In addition to the coating techniques, it is possible to alter the surface potential by applying external electric potential to the channel surface.[18]

23.2.4. Biomolecule Sieving, Ogston and Reptation Models

There is a great need in modern biology for a technology that can analyze and separate biomolecules quickly, inexpensively, and accurately. For instance, the human gemome consists of approximately 3 billion basepairs of nucleotides (molecular code of *A, G, T* and *C*), and contains roughly 30,000 different genes. A typical human liver cell contains about 10,000 different proteins expressed at any given time. These numbers demonstrate enormous analytical challenges for understanding biological systems.

Biomolecules in aqueous solution usually carry charges, or can be coated with charged molecules such as sodium dodecyl sulfate (SDS). While applied electric fields will induce their drift in the solution, their interaction with either surrounding fluid or nanoporous structures will result in different drift velocities for different molecules. Such motion of charged molecules in an electric field (known as electrophoresis) is of great interest due to its widespread use in molecular analysis. Under physiological conditions, the charges on the molecule are generally shielded by counterions (ions with opposite sign) in the aqueous solution. Under an external electric field, the counterions and the molecule will move in opposite directions, and resulting friction between the molecule and its counterions will determine the mobility of the molecule. For example, for uniformly charged polymeric molecules (e.g., DNA) the mobility becomes independent of length (sometimes called as 'free-draining')[19] since the number of charges and counterions on the molecule are both proportional to the length of the molecule. This means that it is generally difficult to separate uniformly charged polymeric biomolecules in a free solution using electrophoresis. Electrophoresis through constrictions such as those provided by a nanoporous gel (polyacrylamide or agarose) is therefore required to separate these molecules efficiently.

Ogston[20] has suggested a model to describe the retardation of molecules by the randomly distributed fibers of a gel (Figure 23.6). This model calculates the pore size

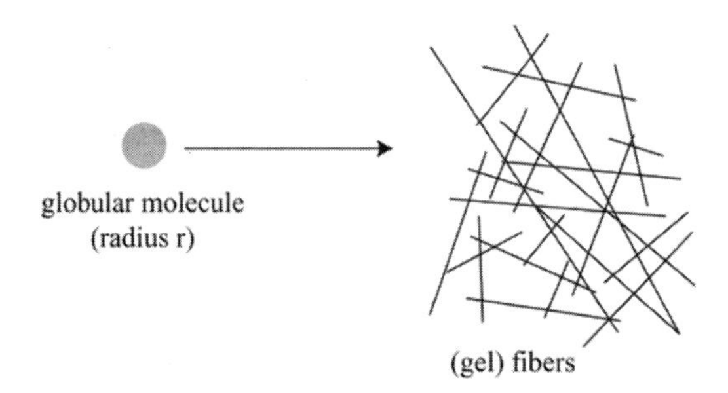

FIGURE 23.6. Ogston Sieving model. Gel is modeled as a random distribution of fibers in space.

distribution as a function of gel concentration. According to this calculation, the probability of gel pores to be larger than r (so that any molecule with effective radius r can pass through) is given by

$$P(r) \sim \exp\left(-2\pi CLr^2 - \frac{4\pi}{3}Cr^3\right) \tag{23.5}$$

where C is the concentration of fiber molecule, and L is the length of the fiber. In the structure like polyacrylamide gel, L is much larger than r (typical size of the pore), therefore the second term in the exponential factor is negligible. Therefore, the probability for a molecule with effective radius r to pass through gel pores is given as the above equation, and this would be proportional to the mobility of the molecule in the gel. Therefore,

$$\log\left(\frac{\mu(r)}{\mu_0}\right) = -CK_R \tag{23.6}$$

where μ_0 is the mobility when there is no sieving effect ($r = 0$ limit), and K_R is a retardation coefficient for a particular gel. Ogston model predicts $K_R \sim r^2$, and it is well established experimentally.

In fact, many biomolecules are macromolecules, and it is uncertain what would be the effective radius of the molecule r should be for a given DNA or protein molecules. Short, single stranded DNA (or RNA) molecules generally treated as rod-like molecules in a physiological solution because electrical charges in their backbone make them stretched out. In such cases, the effective radius should be the length of the molecule. For longer DNA molecules (longer than its persistence length) or globular protein molecules, however, the effective radius of the molecule would be smaller than the stretched length of the molecule. The Ogston sieving model is known to work well when the average pore size is comparable to or larger than the molecular size.

When the molecular dimension is much larger than the pore dimension (or nanofluidic channel dimension), the behavior of molecules within nanoporous structures or nanofluidic channels will change drastically. One example is a long polymer, where the width of the polymer chain is smaller than the pore size while the extended length of the chain is much longer than the pore size. For the case of a polymer molecule passing through a nanoporous material with much smaller pores than its length, a 'reptation model' was developed.[21,22] A long polymer molecule drifting through a maze of obstacles, such as gel, acts like a snake moving through a bush (Figure 23.7). Since the polymer backbone chain is confined by the existence of the obstacles (gel fibers), polymer molecules can move only in the direction of its contour. According to the reptation model, the relaxation time and the diffusion constant is given as

$$\tau \sim \frac{N^3}{k_B T}, \quad D \sim \frac{k_B T}{N^2} \tag{23.7}$$

where N is the parameter representing the length of the polymer molecule. Note the stronger N dependence of the relaxation time and the diffusion constant. Due to the existence of gel matrix, longer DNA will have a harder time in relaxation, and will be slower in diffusion

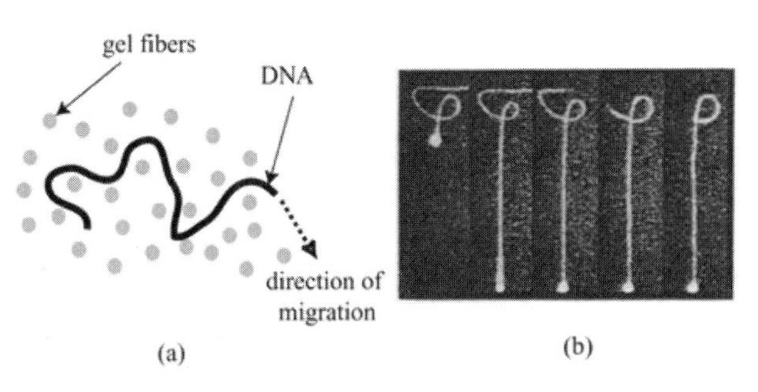

FIGURE 23.7. Reptation of long polymer (DNA) in gel. (a) Two dimensional schematics of reptation dynamics. Due to the constriction imposed by the gel fibers, the polymer (DNA in this case) can only move along the line of its backbone. (b) Experimental demonstration of reptation-like motion of DNA by Perkins *et al.* (From Ref. 25 by permission of the American Association for the Advancement of Science.)

From the above result and the Einstein relation ($D = u_e kT/ze$, u_e: electrophonetic mobility), the electrophoretic mobility would be inversely proportional to N, therefore one could separate DNA molecules using the gel electrophoresis.[23,24] Perkins *et al.* visualized the reptation-like motion of DNA molecules in an entangled polymer solution by a fluorescence microscopy technique (Figure 23.7).[25]

Reptation model assumes that the conformation of a DNA, even when embedded in a random gel, is that of a random coil. This is only true for relatively short DNA molecules at a moderate field. It was shown experimentally and theoretically that for high electric field, reptation dynamics do not work and the drift velocity becomes independent of the molecular length again.[26–28] When under a high electric field, polymer conformations are no more that of a random coil, and DNA molecules are more likely to align in the direction of the field. This will make the DNA mobility length-independent in gel electrophoresis, especially for long DNA molecules. Therefore, separation of long DNA is currently done with pulsed field gel electrophoresis,[29] where the electric field is repeatedly switched on and off during the run to affect the conformation of molecules.

23.2.5. *Macromolecules Moving Through Nanopores or Nanofluidic Channels*

While random network of fibers (nanoporous gels) are most frequently used for biomolecule separation, most nanoporous membrane materials are modeled as an array of straight nanopores in membrane theory. The pore size could be comparable to the molecular size, although it is still larger than the molecular diameter, as seen in Figure 23.8a. Such a pore system can be a useful theoretical model to describe molecular filtering in membranes.

Despite of its apparent simplicity, it is not trivial to study such systems. Experimentally, it has not been easy to fabricate a pore system with a well-defined structure, so one can correlate experimental data with theoretical modeling only by phenomenological parameters. Theoretical studies or modeling have often been complicated by the existence of near-surface liquid boundary layers (such as Debye layer), which could affect the molecular motion in a small, nanometer-sized pores.

The existence of pores could modify molecular motion or drift through various mechanisms. Simply put, molecules are hindered because the available space is limited by the

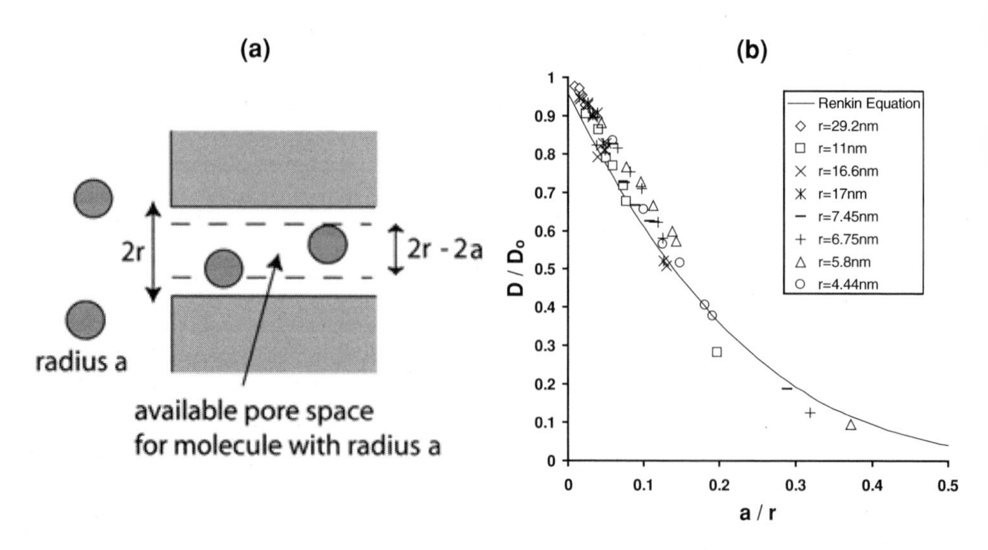

FIGURE 23.8. Steric hindrance effect of globular molecules in nanopores. (a) Schematic diagram demonstrating the steric hindrance effect. The available pore space for the molecule with a will be limited due to the steric hindrance effect. (b) Experimental result by Beck and Schultz. They measured modified (hindered) permeability of molecules with various sizes (0.2 to 2 nm) through nanopores with various pore sizes (10~60 nm). The ordinate is the hindered diffusion coefficient through the nanopore divided by the free diffusion coefficients in bulk solution. The abscissa is the ratio between molecular and pore radius ($2\frac{a}{r}$). The solid line represents the Renkin's calculation. (From Ref. 32 by permission of the American Association for the Advancement of Science.) Plotted from data by Beck and Schultz (*Biochim. Biophys. Acta* **255**, 273–303 (1972)).

existence of pore walls. This steric hindrance effect was studied by Renkin,[30] who calculated the approximate diffusion coefficient of a solute molecule with radius r through a nanopore with radius a.

$$\frac{D}{D_0} = \left(1 - \frac{a}{r}\right)^2 \left[1 - 2.104\left(\frac{a}{r}\right) + 2.\left(\frac{a}{r}\right)^3 - 0.95\left(\frac{a}{r}\right)^5\right] \tag{23.8}$$

Here, D_0 is the diffusion coefficient of a solute in a free solution. The first factor o the right hand side represents the steric hindrance effect caused by the pore.[31] This resul was experimentally confirmed by Beck and Schultz.[32,33] (Figure 23.8b) Steric hindrance fo other geometries were also calculated theoretically.[34] From this experiment, it is clear tha the existence of nanopore structure or equivalent nanostructures can affect the molecula motion even though the pore size is larger than the actual size of the molecule.

The problem becomes more complicated for polymer molecules because of their in ternal degree of freedom (conformation). Extensive theoretical works have been publishe on the topic due to its widespread relevance. The statics[35] and dynamics[36] of a polyme molecules within restricted environment (1D or 2D confinement) has been studied the oretically, as well as the motion of polymer molecules through a small, molecular-siz pores.[37] More recently, the entropic trapping and hindrance of polymers has been studied. Although extensive theoretical background has been developed, only a few experiment studies exist. This was partly due to the difficulty of producing a regular, well-define nanofluidic structure to study these problems. Olgica *et al.*[39] showed slowed dynamics of polymers when they are confined in a two dimensional space. Han *et al.*[40–42] used entropic trapping to separate long DNA molecules. Turner *et al.*[43] studied entropic recoil forces of a long polymer at the boundary of restricted and unrestricted space for polymer molecules.

23.3. FABRICATION OF NANOPOROUS AND NANOFLUIDIC DEVICES

23.3.1. Random Nanoporous Materials

Gel is a solid-like colloid material with typically less than ~20% in solid content. Such high porosity is useful for molecular sieving purposes. Currently, various types of gel are used widely in biomolecular separation. Electron micrograph of these gels shows a random network of crosslinked polymers, with many pores in the structure with diverse pore sizes.[44,45] They are typically made from a monomer solution by initiating a polymerization process chemically (using initiators) or physically (by heating, for example). Polyacrylamide gel is typically used for the separation of single stranded DNA or proteins. The pore dimensions in a given gel must be closely related to the size and nature of molecules to be separated. The average pore size can be controlled by changing the concentration of monomer solution. The pore sizes of polyacrylamide gel are expected to be around 1~3 nm depending on the concentration of gel solution. For the separation of larger double-stranded DNA molecules, agarose gels with larger pores (100~300 nm) are typically used.

It is generally hard to characterize or control the pore structure of gels, which is critical in molecular sieving applications. However, gel polymerization can be spatially controlled using light-sensitive initiator molecules. This provides a way to incorporate a pattern of gel within microsystem. Patterned hydrogel was used to control the fluid flow in microfluidic channels[46] also as a molecular sieve for DNA and protein separation.[47,48] Also, by changing the composition of gel solution, the chemical properties of hydrogels can be modified to create an actuator with different sensitivity to environmental pH.[49]

Controlled polymerization process offers great potential for the fabrication of random nanoporous materials with engineered chemical and physical properties. By carefully controlling the sol-gel transition process during polymerization, one can produce a random nanoporous material with desired chemical and physical properties (Figure 23.9). In such processes, chemical properties of nanostructure can be controlled by the choice of monomers with known chemical group, while pH and other conditions of sol-gel solution determine the pore size.[50–52] These polymer monoliths have been used for sieving materials in various bioseparation processes.[53,54] For some other applications, it is also feasible to use commercially available nanopore filter membranes fabricated by nuclear-track etching methods.[32,33] These pores have relatively uniform, cylindrical shapes, and the pore size can be controlled within ~15%.

23.3.2. Regular, Engineered Nanofluidic Devices

23.3.2.1. Fabrication of Micro/Nanofluidic Devices Micromachined nanofluidic structures offer several advantages over random porous material discussed previously. The

FIGURE 23.9. Scanning electron micrograph of a polymer nanoporous material. The scale bar below is 10 μm. (From Ref. 53 by permission of Wiley Periodicals, Inc.)

nanofluidic element is regular and periodic, and its structural parameters can be precisely controlled. It is therefore much easier to establish a theoretical model based on micro/nanofabricated devices. In addition, the fabrication is generally compatible with standard microfabrication processes.

Turner *et al.* used electron beam lithography to pattern ~50 nm pillar array as a DNA sieve structure.[55] While this is the most accurate way to define a complex nanofluidic structure, this fabrication technique requires expensive and slow e-beam lithography. Park *et al.* used colloid block copolymer to define regular hexagonal nanostructure.[56] A regular stack of microbead was used as a template to fabricate regular porous structure with submicrometer pore sizes.[57] Alternatively, nanoscale features can be produced *vertically* without the need for high-resolution lithography. For instance, Han *et al.* used standard photolithography to define a very thin channel (~75 nm) as an entropic trap for DNA.[40]

Another important issue in nanofluidic fabrication is the sealing of the channel (Figure 23.10). Anodic bonding technique is one of the most commonly used technique for such sealing.[58] The bonding is assisted by the electric force between carrier ions at the bonding interface which is maintained at high temperature (~ 400°C). However, this technique is generally limited to glass materials (such as borosilicate or PYREX) with good thermal matching to Si and decent electrical conductivity at elevated temperatures. Other techniques

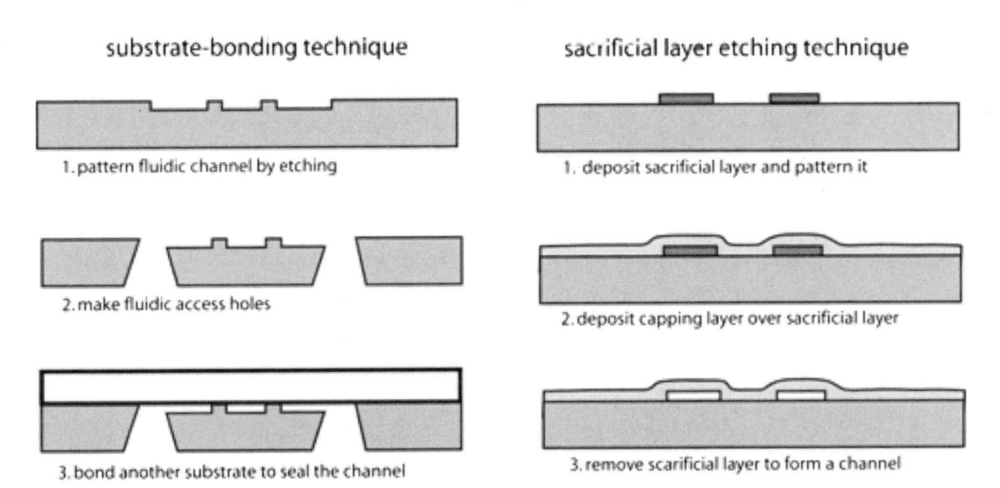

FIGURE 23.10. Two most common method for fabrication of micro/nanofluidic devices. (a) substrate-bonding method (b) sacrificial layer etching method.

are also available for bonding glass-glass or glass-Si substrates. Oxide surface of glass or Si wafer can be chemically activated by treating them with wet chemicals or oxygen plasma.[59] These activated surfaces can then be sealed at room temperature by the weak hydrogen bonds between surface atoms. This non-permanent bonding can then be strengthened by annealing at 500∼1000°C. Unlike the anodic technique, the flatness and cleanliness of the wafer surface are critical to the quality of bonding. Several manufacturers can provide glass wafers suitable for this thermal bonding. Also, relatively high annealing temperature for the thermal bonding techniques generally makes glass materials soft, which could generate problems such as sagging or warping of the bonding wafers. This method is routinely used for the production of microfluidic electrophoresis chips for biomolecule separation.

Most bonding processes are not compatible with standard VLSI processing due to their need of high-temperature annealing of glass, which contains various metal ions. They also do not allow the fabrication of more complex multi-layer fluidic structures. As alternative, monolithic approaches for microfluidic fabrication have been developed.[55,60] In one technique, a first sacrificial layer is deposited and patterned. A second capping layer is then deposited over the sacrificial layer. The sacrificial layer material is then removed by high-selectivity wet etching techniques to form empty and sealed fluidic channels. Turner *et al.*[55] used polysilicon layer as the sacrificial layer, which was later removed by tetramethyl ammonium hydroxide (TMAH) etching. A similar process, using plastic polymer material (parylene C) as the capping material and photoresist as a sacrificial layer, was also developed by Webster *et al.*[60]

23.3.2.2. Fluidic Devices with Plastic Substrate Non-traditional MEMS materials such as plastics are actively pursued in microfluidics due to their low cost and amenability to large-volume manufacturing. Polydimethylsiloxane (PDMS), polymethylmathacrylate (PMMA), polycarbonate (PC) and cycloolefin copolymer (COC) are amongst the most frequently used substrate materials.[61]

Plastic substrates could be patterned by photolithography and etching techniques, but they are better suited for more standard manufacturing techniques such as injection molding. Most polymer-based plastics have relatively low glass transition temperatures (100 ~ 250°C). Near or above this transition temperature, the plastic is easily molded to a master to form the microstructures. Alternatively, in hot embossing, the plastic material is heated to near the glass transition temperature, and another master substrate (typically made of metal) is pressed onto the substrate to imprint the pattern. After the patterning, another flat substrate is bonded to form a sealed microchannel. Bonding of plastic substrates can be achieved by heating the two substrates near the glass transition temperature under high pressure.

These polymer materials are widely used as a substrate for microfluidic device for various purposes. However, the softness of plastic materials as well as thermal bonding process can be problematic for the fabrication of nanofluidic systems with submicrometer size features and depths, since the dimension of ultra-small features can vary a lot during the thermal bonding process due to the distortion caused by surface melting.

PDMS (polydimethylsiloxane, silicone) is an elastomer widely used in microfluidics. It is best suited for prototyping complex microfluidic system inexpensively.[62] PDMS structure can be casted at room temperature on a solid master (usually made by photolithography), and it can be bonded to another PDMS, Si or glass substrate reversibly or irreversibly, at room temperature. The irreversible bonding is facilitated by a well-documented oxygen plasma treatment.[63] Since the bonding of PDMS requires neither high temperature nor high pressure, PDMS microfluidics potentially could be extended to very thin and narrow fluidic features (~100 nm or less). These plastic materials are generally softer (lower Young's modulus) than glass or Si, and can be used to fabricate moveable components. For example, a thin flexible PDMS membrane was used to fabricate a microfluidic valve.[64,65] It is also permeable to many gases, which allows the easy removal of trapped air bubbles in the microfluidic channel. PDMS microfluidic devices, however, have several drawbacks. The surface condition of plasma-treated PDMS can be uncontrollable and change over time, which could produce unpredictable electroosmotic flows. The irreversible PDMS bonding is not as strong as to support some high-pressure fluidics applications. Also, plastic materials including PDMS are generally less chemically inert than glass to various solvents (acetone or acetonitrile) that are frequently used in molecular separation.

23.4. APPLICATIONS OF NANOFLUIDICS

23.4.1. Nanofluidic Biomolecular Sieving and Sorting

Molecular sieve structures or filters with nanometer-sized pores are repeatedly used in biomolecular separation processes. Nanofluidic devices, with regular nanostructure filter or sieve, could be used instead of these random nanoporous materials. However, several issues impede such development. First, it is still not easy to fabricate nanostructures with comparable pore sizes (1–10 nm) for molecular sieving. Second, the interaction between nanostructures and biomolecules is still not fully understood. For instance, very little data exist on the stochastic motion of biomolecules within nanostructures or nanoconstrictions. The situation becomes more complex when one considers polymeric macromolecules, where molecular conformation should also be taken into consideration.

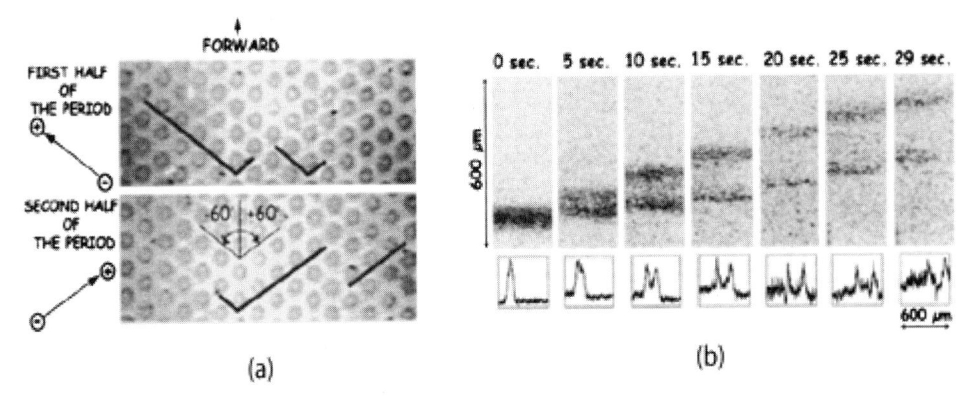

FIGURE 23.11. Pulsed field electrophoresis of long DNA in artificial system. (a) Principles of operation, presented over the optical micrograph of the actual device used. Electric field is switched in direction by 120 degree as shown in the figure. While shorter DNA strand can progress forward, longer DNA strand gets stuck for longer amount of time. After many repetition of this cycle, DNA molecules with different lengths are separated. (b) Separation of two DNA (166 kbp and 48 kbp) within the system. (From Ref. 69 by permission of the American Chemical Society.)

Regular nanofluidic channels can still be useful in many applications, especially for large molecules or particles. Separation of large DNA molecules is a good example where nanofluidic devices provide interesting possibilities. Separation of long double stranded DNA molecules (20 kbp$\sim$10 Mbp) is important in DNA fingerprinting and genome mapping, but current technology (pulsed field gel electrophoresis) is very inefficient and time consuming.[66] The size of these DNA molecules is relatively large. A 40 kbp double stranded DNA molecule has a radius of gyration (approximate size in a aqueous solution) of $\sim$1 μm, and the extended length of $\sim$16 μm. It is therefore feasible to fabricate molecular sieve structures for these large DNA molecules with standard MEMS technology.

Volkmuth and Austin first suggested the concept of an "artificial gel" as an efficient way to separate long DNA molecules.[67] This group fabricated an array of pillars within microfluidic channels as obstacles for DNA motion. When DNA is driven through the channel, the molecular chains are hooked and retarded by the nanostructures. By applying appropriate set of AC electric field to the DNA, these authors developed a very efficient AC-DNA separation system, which can separate megabasepair DNA molecules within a minute.[68–70] (Figure 23.11)

Han and Craighead designed, fabricated and tested a nanofluidic channel with submicron constrictions to separate double-stranded DNA molecules.[40,41,71] This channel consisted of alternating shallow and deep regions, which were fabricated by silicon based lithography and etching techniques (Figure 23.12). Because the radius of gyration of DNA used was much larger than the shallow region gap, DNA molecules were trapped when they moved from a deep to a shallow region. This entropic trapping determined the mobility of DNA in the system. Interestingly, longer DNA molecules were found to escape entropic traps faster than shorter ones, contrary to conventional wisdom.[41] This phenomenon can be explained by appreciating the importance of polymer conformation in the dynamics of DNA molecules. This size-dependent trapping process creates electrophoretic mobility differences between long and short DNA molecules, thus enabling efficient separation without

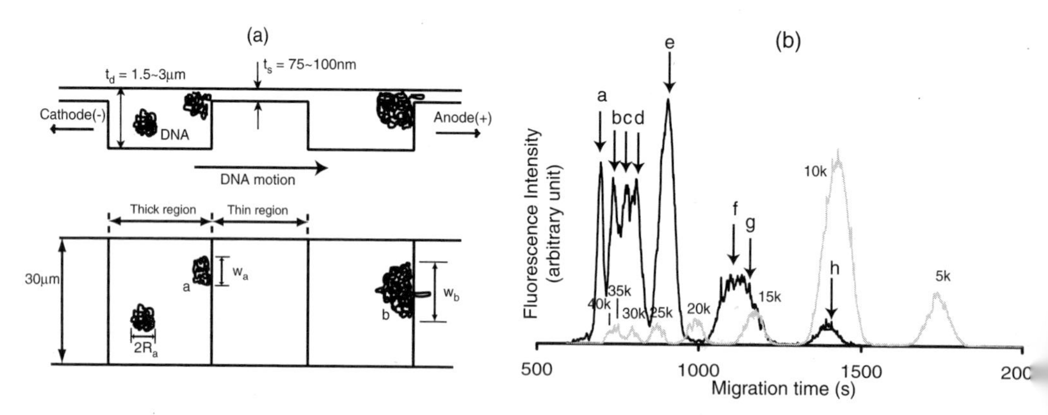

FIGURE 23.12. Entropic trap DNA separation. (a) Schematic diagram of entropic trap DNA separation system. (b) Separation result of long DNA molecules. (From Ref. 40 by permission of the American Association for the Advancement of Science.)

using gel or pulsed electric fields. Samples of long DNA molecules (5 to 166 kbp) were efficiently separated into bands in 15-millimeter-long channels, typically within 30 minutes.[40]

In addition to DNA separation, nanofluidic devices are a good model system for studying complex polymer dynamics problems, since its pore shapes can be designed and characterized precisely. Turner *et al.* studied entropic recoil force of long DNA molecule within nanofluidic channel.[43] Nykypanchuk *et al.* used a regular submicron porous structure made from bead array to study the hindered Brownian dynamics of a long DNA molecule.[57] Also, dynamics of DNA molecule in a thin nanofluidic channel was studied by Olgica *et al.*[39]

23.4.2. Nanopore Molecular Scanner/Detectors

Nanoscale pores or nanofluidic channels form a very narrow path for the passage of biomolecules or biological particles. At moderate concentration, transport of molecules through these channels can be discrete, allowing the detection of individual molecules. This provides a unique opportunity for a new kind of biosensors in which molecules are detected individually as they sieve through nanopores.

Saleh and Sohn[72,73] recently developed a nanofluidic particle coulter counter using submicrometer deep fluidic channels (Figure 23.13). Since the size of the particle is comparable to the pore dimension, its passage will disrupt the electrical property of the channel as it passes through the pore. By measuring the current between the two reservoirs, the authors were able to detect single crossing events of submicron particles, and estimate the size of the particle crossed. The same group used the similar principle to make a cell capacitive cytometer using bigger pores.[74] Nanofluidic channels can be also used as a mean to control the conformation of polymer molecules in the fluid. Foquet *et al.*[75] and Tegenfeldt *et al.*[76] used a thin (~200 nm) and narrow (~1 μm) channel to stretch out long DNA molecules, enabling the measurement of the DNA length as well as other properties.

This idea can be pushed to the molecular level in an attempt to scan and sequence single stranded DNA molecules. Kasianowicz *et al.*[77] used a membrane pore protein that has 2.6 nm pore as the only channel where single stranded DNA molecules can go through. The passage decreased the ionic current transiently, opening possibilities for the real-time monitoring of the molecular signature for different bases on the DNA molecules (Figure 23.14). The

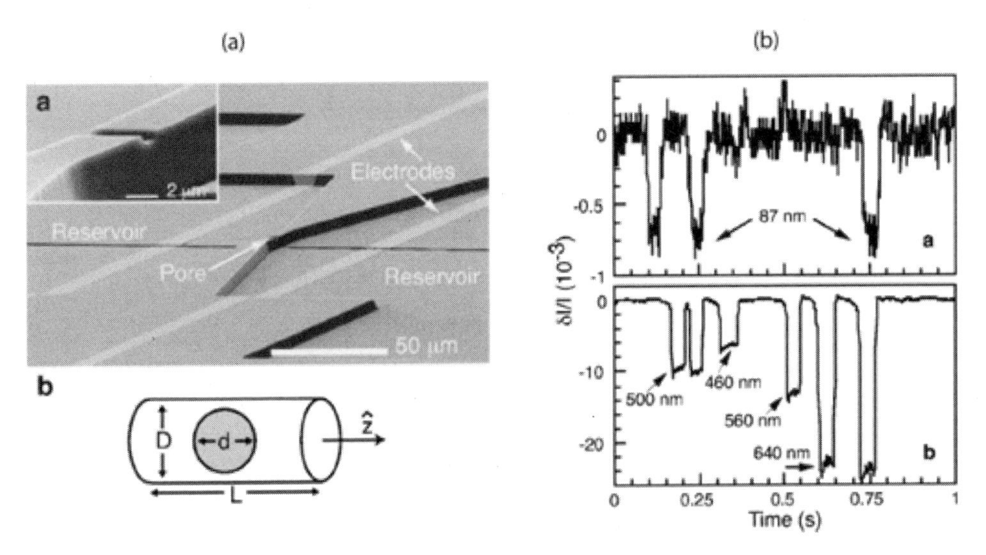

FIGURE 23.13. Nanofluidic Coulter counter for submicrometer particles. (a) Scanning Electron Micrograph of the nanofluidic channel. (b) Detected signal during the passage of several submicrometer particles (bead). (From Ref. 72 by permission of the American Institute of Physics.)

molecular crossing event occurs within less than a millisecond, therefore requiring fast detection abilities. The same group then employed this technique to achieve discrimination between polynucleotides (such as poly$(dA)_{100}$ vs. poly$(dC)_{100}$)[78,79] but have yet to achieve true, real-time sequencing of single stranded DNA. The fabrication of more sophisticated nanofluidic pores with defined geometries might improve the detection sensitivity. Still, it is regarded as one of the most promising candidate for a future, ultra-fast DNA sequencing technique.

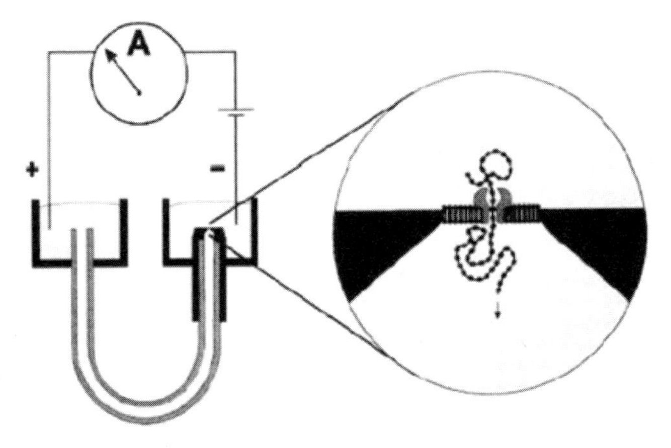

FIGURE 23.14. Nanopore DNA sequencing. DNA molecules are forced to pass through the nanopore membrane protein, and the current between the two reservoir is monitored. (From Ref. 79 by permission of the Biophysical Society.)

23.4.3. *Single Molecule Detection*

Optical fluorescence detection is a very sensitive technique to visualize molecules in a liquid, and now it is possible to detect fluorescence signal from a single fluorophore. Recently, various single molecule detection techniques based on optical fluorescence detection have been developed.[80] These emerging techniques demonstrate that it is now possible to detect and identify individual molecules, an ability that would enable a wide number of applications.[81–83] Experiments involving detection of individual molecules, as opposed to macroscopic measurements, allow the investigation of molecular properties without the need of averaging over a large number of molecules.

Single molecule detection of fluorophores in liquid is typically achieved by localizing the excitation light into a small confocal volume, so only a few fluorophores can be found in the volume at a given time. The fluorescence signal from such a small detection volume would represent the characteristic of molecules within that volume. This can be achieved by various methods, such as confocal optics,[84] two-photon excitation,[85,86] surface evanescent waves excitation,[87] and near-field scanning optical microscopy.[88–90]

The most widely used technique is confocal (or two-photon) fluorescence correlation spectroscopy (FCS), where a laser light is focused into a diffraction-limited volume and the fluorescence signal coming out of that volume is detected. If only a small number of molecules are in that volume, the relative fluctuation of fluorescence signal compared to the total fluorescence becomes significant. This fluctuation contains information about molecular stochastic motion, such as the diffusion coefficient. By measuring the correlation function of the fluorescence signal from the detection volume, one can measure the diffusion coefficient of the molecules in the liquid.

The typical confocal volume using high N.A. optics is about 1 fl or less, and the average number of molecules in such a volume reaches 1 at a concentration of $\sim$1 nM. Higher concentration degrades the signal to noise ratio, because the number of molecules in the volume increases and the relative fluctuation becomes very small. Various nanostructures can be used to decrease the detection volume further, either by confining the liquid volume or by limiting the illuminated volume of the sample. Levene *et al.*[91] used nanometer-sized metallic structures to limit the confocal detection volume of fluorescence correlation spectroscopy further by orders of magnitudes, achieving higher signal to noise ratio. Due to the increased signal to noise ratio and higher permitted sample concentration, this technique could be used for many interesting applications of molecular sensing and analysis, such as a novel DNA sequencing technique.[92,93]

23.4.4. *Electrokinetic Fluid Control*

Electroosmosis has a unique property that the flow speed does not depend on the size of the channel. While it takes a huge hydraulic pressure to drive liquid through nanoporous materials or nanofluidic channels, electroosmotic flow speed only depends on the surface properties and applied potential. This means that electroosmosis can be a source for a high-pressure generation device.

Paul *et al.*[94] used a glass capillary packed with micron-size silica bead to generate electroosmotic flow, which in turn generated a very high pressure (up to 8000 psi). Alarie *et al.*, used a 100 nm deep channel as a salt bridge junction to make an electroosmotically induced

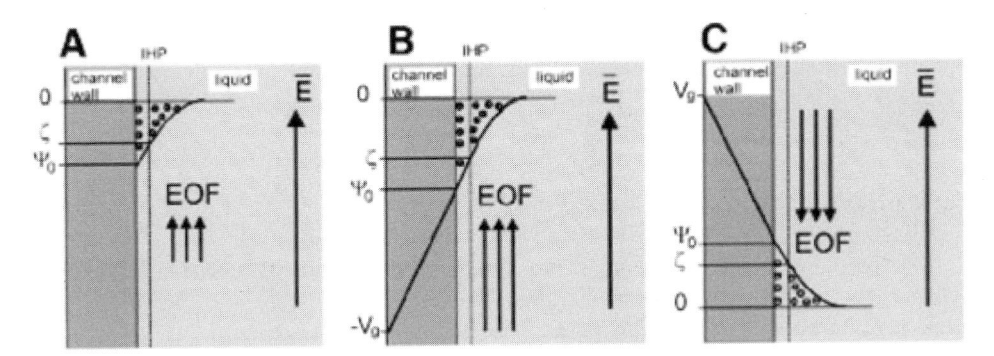

FIGURE 23.15. Control of electroosmotic flow by external electric field (gate voltage). By applying external potential to the outside surface of a fluidic channel, one can control the electroosmotic flow just as one can control the charge inversion region conductances by controlling gate voltage in MOSFET. (a) normal state. (b) applying negative potential to increase the electroosmotic flow. (c) applying positive potential to decrease or reverse the electroosmotic flow. (From Ref. 18 by permission of the American Association for the Advancement of Science.)

hydraulic pump for microchip separation.[95] Electroosmotic pump has many potential use in microfluidic devices, such as in HPLC system,[96] on-chip pressure pump for mechanical actuation,[97] nanotitration and mixing,[98] and electrospray mass spectrometry.[99,100]

Electroosmotic flow depends on the surface potential of the microfluidic and nanofluidic channels. This surface potential of a fluidic channel can be controlled by applying external field. Schasfoort *et al.,* used an external potential to control the surface potential of a microfluidic channel.[18] (Figure 23.15). This is conceptually very similar to the field effect transistors in microelectronics (so they called this as 'flowFET', Figure 23.16), where gate voltage controls the conductance of the surface inversion layer of the field effect transistor. By applying external potential to offset or reverse inherent surface potential, they were able to change the sign and magnitude of the electroosmotic flow. It is expected that the

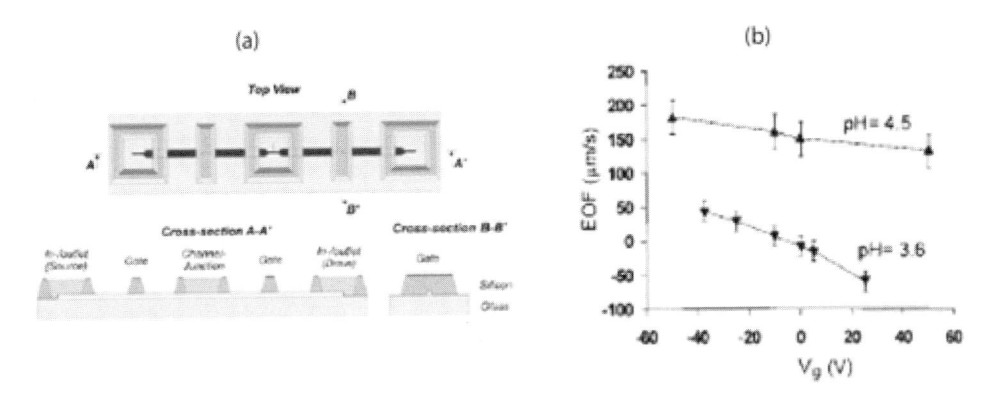

FIGURE 23.16. FlowFET device fabricated by Schasfoort *et al.* (a) Schematic diagram of the FlowFET device. The channel wall was 390 nm silicon nitride membrane (equivalent to the gate oxide in MOSFET), and external potentials were applied to the gate. (b) Change of electroosmotic flow versus the gate voltage. (From Ref. 18 by permission of the American Association for the Advancement of Science.)

extension of this concept could lead to the development of more complex flow controlling circuit, which then can control a complex chemical or biochemical reaction in the microchannel.

23.5. SUMMARY

Nanofluidics hold a promising future for developing novel molecular manipulation systems such as biomolecule sieves, which are essential in the context of the micro total analysis system. Unlike conventional nanoporous molecular filters or sieves, nanofluidic filters can be carefully designed to have a regular structure that can be used to manipulate biomolecules and other particles in a well-defined way. Implications of such devices are significant, and could lead to a new concept for biomolecule analysis and separation. Fluidic properties and molecular diffusion at the nanoscale is quite different from that of micro or macroscale, and nanofluidic devices can be designed to exploit these novel properties to provide solutions to various engineering problems. However, further basic and applied research should be done to investigate the science of fluid and molecules in fluid at the nanoscale.

QUESTIONS

1. Consider metal particles with spatial dimension in the order of 100 μm, 1 μm and 10 nm, respectively. Using the physical laws as guidance, give a rough estimate on the ratio between gravitational force and viscous force acting on the particle. Do the same thing for surface tension and gravitational force. What can you say from these estimations?
2. Consider micro/nanochannels with a rectangular cross section with 30 μm width and 10 μm, 1 μm, and 100 nm height, respectively. Calculate the pressure required to push water through these channels to get a flow speed of 1 μm/s.
3. Microorganisms are always living in the world of microfluidics. Consider *Escherichia coli*, a microorganism with the length of roughly 2 μm. They swim through water at the speed of ~30 μm/s. At the 1 nM nutrient concentration, calculate the average time for an *E. coli* to swim to get a single nutrient molecule ($D = 10^{-7}$ cm^2/s). Next, assume that this particular *E. coli* cannot swim and has to wait until nutrient molecules diffuse to itself. Estimate how long the *E. coli* should wait to get a single nutrient molecule in this case. What can you say from these estimations? (For an entertaining discussion regarding microfluidic environment and biology, read Purcell's article.[101])
4. Both ends of a glass microfluidic channel of length 1 cm, width 30 μm and height 10 μm are connected to the reservoirs with cathode and anode electrode, respectively. The fluid near cathode contained in a closed vessel, meaning the microchannel is the only fluidic path leading to that reservoir. The anodic reservoir is open to the atmospheric pressure. When a voltage (100 V) is applied between both electrodes, the electroosmotic flow transports fluid to the cathode reservoir, increasing the pressure. Assuming water as the fluid, calculate the maximum pressure that the cathode reservoir can reach. Assume the zeta potential of the channel as 25 mV.

REFERENCES

1. A. Manz, N. Graber, and H. M. Widmer, *Sensors and Actuators* **B1**, 244 (1990).
2. S. C. Terry (Stanford ICL, Technical Report #4603-1, 1975).
3. S. C. Terry, J. H. Jerman, and J. B. Angell, *IEEE Trans. Electron Devices* **ED-26**, 1880 (1979).
4. J. C. Venter, *et al.*, *Science*, **291**, 1304 (2001).
5. International Human Genome Sequencing Consortium, *Nature* **409**, 860 (2001).
6. D. R. Reyes, D. Iossifidis, P.-A. Auroux, and A. Manz, *Anal. Chem.* **74**, 2623 (2002).
7. P.-A. Auroux, D. Iossifidis, D. R. Reyes, and A. Manz, *Anal. Chem.* **74**, 2637 (2002).
8. *Proceedings of the Micro Total Analysis Systems 2001* (Kluwer Academic Publishers, Dordrecht, The Netherlands, 2001).
9. *Micro Total Analysis Systems 2000, Proceedings of the μTAS 2000 Symposium* (Dordrecht, The Netherlands, 2000).
10. J. P. Brody, P. Yager, R. E. Goldstein, and R. H. Austin, *Biophys. J.* **71**, 3430 (1996).
11. J. B. Knight, A. Vishwanath, J. P. Brody, and R. H. Austin, *Phys. Rev. Lett.* **80**, 3863 (1998).
12. C.-F. Chou, *et al.*, *Proc. Natl. Acad. Sci. U.S.A.* **96**, 13762 (1999).
13. A. v. Oudenaarden and S. G. Boxer, *Science* **285**, 1046 (1999).
14. T. F. Weiss, *Cellular Biophysics. Volume 1: Transport* (MIT Press, 1995).
15. S. C. Jacobson, J. P. Alarie, and J. M. Ramsey, in *Proceedings of the microTAS 2001 Symposium*, edited by J. M. Ramsey and A. v. d. Berg (Kluwer Academic Publishers, Monterey, CA, 2001), p. 57–59.
16. J. Horvath and V. Dolnik, *Electrophoresis* **22**, 644 (2001).
17. P. G. Righetti, C. Gelfi, B. Verzola, and L. Castelletti, *Electrophoresis* **22**, 603 (2001).
18. R. B. M. Schasfoort, S. Schlautmann, J. Hendrikse, and A. v. d. Berg, *Science* (Washington, D.C., 1883–) **286**, 942 (1999).
19. D. Long, J.-L. Viovy, and A. Ajdari, *Phys. Rev. Lett.* **76**, 3858 (1996).
20. A. G. Ogston, *Trans. Faraday Soc.* **54**, 1754 (1958).
21. P. G. DeGennes, *J. Chem. Phys.* **55**, 572 (1971).
22. P. G. DeGennes, *Scaling Concepts in Polymer Physics* (Cornell University Press, Ithaca, NY, 1979).
23. L. S. Lerman and H. L. Frisch, *Biopolymers* **21**, 995 (1982).
24. O. J. Lumpkin and B. H. Zimm, *Biopolymers* **21**, 2315 (1982).
25. T. T. Perkins, D. E. Smith, and S. Chu, *Science* **264**, 819 (1994).
26. G. W. Slater and J. Noolandi, *Phys. Rev. Lett.* **55**, 1579 (1985).
27. T. A. J. Duke, A. N. Semenov, and J. L. Viovy, *Phys. Rev. Lett.* **69**, 3260 (1992).
28. T. A. J. Duke, J. L. Viovy, and A. N. Semenov, *Biopolymers* **34**, 239 (1994).
29. G. F. Carle, M. Frank, and M. V. Olson, *Science* **232**, 65 (1986).
30. E. M. Renkin, *Journal of General Physiology* **38**, 225 (1954).
31. J. C. Giddings, *Unified Separation Science* (John Wiley & Sons, Inc, New York, NY, 1991).
32. R. E. Beck and J. S. Schultz, *Science*, **170**, 1302 (1970).
33. R. E. Beck and J. S. Schultz, *Biochimica et Biophysica Acta* **255**, 273 (1972).
34. W. M. Deen, *AIChE J.* **33**, 1409 (1987).
35. M. Daoud and P. G. DeGennes, *J. Physique* **38**, 85 (1977).
36. F. Brochard, *J. Physique*, **38**, 1285 (1977).
37. S. Daoudi and F. Brochard, *Macromolecules* **11**, 751 (1978).
38. M. Muthukumar, *J. Non-Cryst. Solids* **131–133**, 654 (1991).
39. O. B. Bakajin, *et al.*, *Phys. Rev. Lett.* **80**, 2737 (1998).
40. J. Han and H. G. Craighead, *Science* **288**, 1026 (2000).
41. J. Han, S. W. Turner, and H. G. Craighead, *Phys. Rev. Lett.* **83**, 1688 (1999).
42. J. Han and H. G. Craighead, *Anal. Chem.* **74**, 394 (2002).
43. S. W. P. Turner, M. Cabodi, and H. G. Craighead, *Phys. Rev. Lett.* **88**, 128013 (2002).
44. S. Waki, J. D. Harvey, and A. R. Bellamy, *Biopolymers* **21**, 1909 (1982).
45. T. K. Attwood, B. J. Nelmes, and D. B. Sellen, *Boipolymers* **27**, 201 (1988).
46. D. J. Beebe, *et al.*, *Nature (London)* **404**, 588 (2000).
47. S. N. Brahmasandra, *et al.*, *Electrophoresis* **22**, 300 (2001).

48. J. Han and A. K. Singh, in *Proceedings of the microTAS 2002 Symposium*, edited by A. van den Berg (Kleuwer Academic Publishers, Nara, Japan, 2002), Vol. 1, p. 596.
49. Q. Yu, J. M. Bauer, J. S. Moore, and D. J. Beebe, *Applied Physics Letters* **78**, 2589 (2001).
50. C. J. Brinker, *et al.*, *J. Membr. Sci.* **94**, 85 (1994).
51. C. J. Brinker, *Curr. Opin. Solid St. M.* **1**, 798 (1996).
52. N. K. Raman, M. T. Anderson, and C. J. Brinker, *Chem. Mater.* **8**, 1682 (1996).
53. C. Yu, M. C. Xu, F. Svec, and J. M. J. Frechet, *J. Polym. Sci. Pol. Chem.* **40**, 755 (2002).
54. D. J. Throckmorton, T. J. Shepodd, and A. K. Singh, *Anal. Chem.* **74**, 784 (2002).
55. S. W. Turner, A. M. Perez, A. Lopez, and H. G. Craighead, *J. Vac. Sci. Technol.*, B **16**, 3835 (1998).
56. M. Park, *et al.*, *Science*, **276**, 1401 (1997).
57. D. Nykypanchuk, H. H. Strey, and D. A. Hoagland, *Science* **297**, 987 (2002).
58. G. Wallis and D. I. Pomerantz, *J. Appl. Phys.* **40**, 3946 (1969).
59. P. Barth, *Sensors and Actuators*, **A21–A23**, 919 (1990).
60. J. R. Webster, M. A. Burns, D. T. Burke, and C. H. Mastrangelo, *Anal. Chem.* **73**, 1622 (2001).
61. H. Becker and C. Gärtner, *Electrophoresis* **21**, 12 (2000).
62. D. C. Duffy, J. C. McDonald, O. J. A. Schueller, and G. M. Whitesides, *Anal. Chem.* **70**, 4974 (1998).
63. B.-H. Jo, L. M. V. Lerberghe, K. M. Motsegood, and D. J. Beebe, *Journal of Microelectromechanical Systems*, **9**, 76 (2000).
64. B.-H. Jo, J. Moorthy, and D. J. Beebe, in *Proceedings of Micro Total Analysis Systems 2000* (Kluwer Academic Publishers, Dordrecht, The Netherlands, 2000), p. 335–338.
65. M. A. Unger, *et al.*, *Science* **288**, 113–116 (2000).
66. F. H. Kirkpatrick, edited by E. Lai and B. Birren (Cold Spring Harbor Laboratory Press, Cold Spring Harbor, NY, 1990).
67. W. D. Volkmuth and R. H. Austin, *Nature (London)* **358**, 600 (1992).
68. T. A. J. Duke, R. H. Austin, E. C. Cox, and S. S. Chan, *Electrophoresis* **17**, 1075 (1996).
69. O. Bakajin, *et al.*, *Anal. Chem.* **73**, 6053 (2001).
70. L. R. Huang, *et al.*, *Nat. Biotechnol.* **20**, 1048 (2002).
71. J. Han and H. G. Craighead, *J. Vac. Sci. Technol.*, A **17**, 2142 (1999).
72. O. A. Saleh and L. L. Sohn, *Rev. Sci. Instrum.* **72**, 4449 (2001).
73. O. A. Saleh and L. L. Sohn, in *Proceedings of the Micro Total Analysis Systems 2001*, edited by J. M. Ramsey and A. v. d. Berg (Kluwer Academic Publishers, Dordrecht, The Netherlands, 2001), p. 54–56.
74. L. L. Sohn, *et al.*, *Proc. Natl. Acad. Sci. U.S.A.* **97**, 10687 (2000).
75. M. Foquet, *et al.*, *Anal. Chem.* **74**, 1415 (2002).
76. J. O. Tegenfeldt, *et al.*, *Phys. Rev. Lett.* **86**, 1378 (2001).
77. J. J. Kasianowicz, E. Brandin, D. Branton, and D. W. Deamer, *Proc. Natl. Acad. Sci. U.S.A.* **93**, 13770 (1996).
78. A. Meller, *et al.*, *Proc. Natl. Acad. Sci. U.S.A.* **97**, 1079 (2000).
79. M. Akeson, *et al.*, *Biophys. J.* **77**, 3227 (1999).
80. S. Nie and R. N. Zare, *Annu. Rev. Biophys. Biomol. Struct.* **26**, 567 (1997).
81. T. Basché, S. Nie, and J. M. Fernandez, *Proc. Natl. Acad. Sci. U.S.A.* **98**, 10527 (2001).
82. M. Eigen and R. Rigler, *Proc. Natl. Acad. Sci. U.S.A.* **91**, 5740 (1994).
83. S. Weiss, *Science* (Washington, D.C., 1883–) **283**, 1676 (1999).
84. H. Qian and E. L. Elson, *Applied Optics* **30**, 1185 (1991).
85. K. M. Berland, P. T. C. So, and E. Gratton, *Biophysical Journal* **68**, 694 (1995).
86. P. T. C. So, C. Dong, K. M. Berland, and E. Gratton, *Biophysical Journal* **64**, 218a (1993).
87. T. Hirschfeld, *Appl. Opt.* **15**, 2965 (1976).
88. E. Betzig and J. K. Trautman, *Science* **257**, 189 (1992).
89. E. Betzig and R. J. Chichester, *Science* **262**, 1422 (1993).
90. E. Betzig, R. J. Chichester, F. Lanni, and D. Taylor, *BioImaging*, **1**, 129 (1993).
91. M. J. Levene, *et al.*, *Science*, **299**, 682 (2003).
92. J. Korlach, *et al.*, *Biophys. J.* **82**, 507a (2002).
93. J. Korlach, *et al.*, *Biophys. J.* **80**, 147a (2001).
94. P. H. Paul, D. W. Arnold, and D. J. Rakestraw, in *Proceedings of Micro Total Analysis Systems 1998* (Kluwer Academic Publishers, Dordrecht, The Netherlands, 1998), p. 49–52.
95. J. P. Alarie, *et al.*, in *Proceedings of the microTAS 2001 Symposium*, edited by A. v. d. Berg (Kluwer Academic Publishers, Monterey, CA, 2001), p. 131–132.

96. P. H. Paul, D. W. Arnold, D. W. Neyer, and K. B. Smith, in *Proceedings of the Micro Total Analysis Systems 2000* (Kluwer Academic Publishers, 2000), p. 583–590.
97. Y. Takamura, *et al.*, in *Proceedings of the microTAS 2001 Symposium*, edited by A. v. d. Berg (Kluwer Academic Publishers, Monterey, CA, 2001), p. 230–232.
98. O. T. Guenat, D. Ghiglione, W. E. Morf, and N. F. d. Rooij, *Sensors and Actuators, B: Chemical Sensors and Materials* **72**, 273 (2001).
99. I. M. Lazar, *et al.*, *Journal of Chromatography*, A **892**, 195 (2000).
100. I. M. Lazar and B. L. Karger, in *Proceedings of the microTAS 2001 Symposium*, edited by A. v. d. Berg (Kluwer Academic Publishers, Monterey, CA, 2001), p. 219.
101. E. M. Purcell, *Am. J. Phys.* **45**, 3 (1977).

Index